Morse Theory of Gradient Flows, Concavity and Complexity on Manifolds with Boundary

Morse Theory of Gradient Flows, Concavity and Complexity on Manifolds with Boundary

Gabriel Katz

Massachusetts Institute of Technology, USA

World Scientific

NEW JERSEY • LONDON • SINGAPORE • BEIJING • SHANGHAI • HONG KONG • TAIPEI • CHENNAI • TOKYO

Published by

World Scientific Publishing Co. Pte. Ltd.
5 Toh Tuck Link, Singapore 596224
USA office: 27 Warren Street, Suite 401-402, Hackensack, NJ 07601
UK office: 57 Shelton Street, Covent Garden, London WC2H 9HE

British Library Cataloguing-in-Publication Data
A catalogue record for this book is available from the British Library.

MORSE THEORY OF GRADIENT FLOWS, CONCAVITY AND COMPLEXITY ON MANIFOLDS WITH BOUNDARY

ISBN 978-981-4368-75-9

For any available supplementary material, please visit
https://www.worldscientific.com/worldscibooks/10.1142/8282#t=suppl

To Sarah, Ami, and their young tribes

Preface

As I exit the space
Into the abandoned garden of quantities,
I tear apart
The seeming constancy and self-awareness of causes.

In the well of solitude, Infinity,
I read your manual, —
A leafless and primeval medicine book,
A problem set of gigantic roots.

Osip Mandelstam (translation by G. K.)

This monograph is an account of the author's thoughts about gradient vector flows on compact manifolds with boundary. It is not a depiction of a recognized mathematical field or even a subfield, although many structures and constructions we propose fit comfortably in the framework of Morse Theory and, more generally, of the Singularity Theory of smooth maps, being applied to boundary value problems for ordinary differential equations (ODEs).

Recall that the classical Morse Theory links the singularities of Morse functions with the topology of closed differentiable manifolds. Specifically, the singularities of Morse functions $f : X \to \mathbb{R}$ on a manifold X cause *discontinuities* of the f-gradient flow, and the homology or even the topological type of X can be expressed in terms of such discontinuities [C]. These terms include descending and ascending manifolds of the f-critical loci and the spaces of flow trajectories that connect them.

On manifolds with boundary, an additional source of flow discontinuity occurs: it comes from a particular *concave* geometry of the boundary ∂X with respect to the flow. In fact, on manifolds with boundary, one can *trade* f-singularities in the interior of X for these boundary effects. So, in our approach, the boundary effects take central stage, while the singularities remain in the background.

Originally, the Morse theory of smooth functions and their gradient flows on manifolds *with boundary* has been studied (under general boundary conditions) in the papers of Morse [Mo] and of Morse and Van Schaack [MS]. Much later, Goresky and MacPherson have developed the Morse Theory on quite general stratified spaces ([GM], [GM1], and [GM2]). In a sense, Morse Theory on manifolds with boundary is a very special case of the stratified Morse Theory, however, it is a special case with a lot of unique flavor. The paper [BNR] gives a vivid account of fundamental properties such theory. We would to acknowledge also the influence on our worldview of Kronheimer's and Mrowka's work on Morse complexes on manifolds with boundary ([KM], section 2.4).

It turns out that the geometric and combinatorial structures, arising from the interactions of the vector flow with the boundary, are surprisingly rich. This geometric situation leads organically to many encounters with Singularity Theory, Combinatorics, Differential Topology, Differential Geometry, Dynamical Systems, and especially to an encounter with boundary value problems for ordinary differential equations. It is this diversity of connections that motivated us to write this monograph in the first place.

At certain times, our enterprise may sound as a one-man-orchestra; the author does not claim the mastery of each instrument. On many occasions, we do not hesitate to share with the reader the questions that animate our research and the conjectures that mark its borders. Perhaps, the experts in the relevant fields may find our arguments open to purifications and simplifications, and conjectures easy to validate. However, we try to keep the *interplay* between the boundary value problems for families of ODEs and the topology of manifolds as the unifying theme throughout the book.

Let X be a smooth compact connected manifold with boundary. In the book, we are preoccupied with the following basic question. Suppose we restrict a priori the combinatorial patterns that characterize the way in which the trajectories of a smooth non-vanishing gradient vector flow interact with the boundary ∂X. *What are the constraints, imposed on the topology of X by such restrictions?*

Counting gradient flow trajectories of given combinatorial types leads to new invariants of the smooth structure on X. We think of these invariants as measuring the *complexity* of flows on X. Thus the word "Complexity" in the title of the book. *Concavity* of the boundary ∂X with respect to flows is the geometric incarnation of their complexity.

We are also interested in the flow-generated structures on the boundary ∂X that make it possible to reach conclusions about the topological nature of X and the flow. Such structures are reminiscent of holographic pictures and thus deserve the name "holographic". In fact, holographic structures reflect *the topological rigidity* of boundary value problems for a vast class of ODEs. In this monograph, we will encounter just few appearances of the holographic phenomenon. Its full investigation belongs to a different book [K13].

Throughout the text, we presume that the reader is familiar with the basic notions and classical tools of the Differential and Algebraic Topology and Differential Geometry.

Few words about the organization of the book. It is divided into two parts: the first one deals with flows in dimensions two and three. By its nature, this part is more "pictorial". The second part is devoted to the flows in any dimension. It employs more analytical tools from the field of Thom-Bordman Singularity Theory [Bo].

The book consists of nine chapters. We regard Chapter 1 as a detailed introduction to the rest of the book. Each chapter starts with a description of its purpose and main results. This organization provides our reader with *independent entrances* into different chapters. As a result, the book has certain redundancy in its design: occasionally, some constructions and notions that first appear in the 3-dimensional setting, reappear in their full generality later.

The chart below indicates several alternative pathways through the chapters.

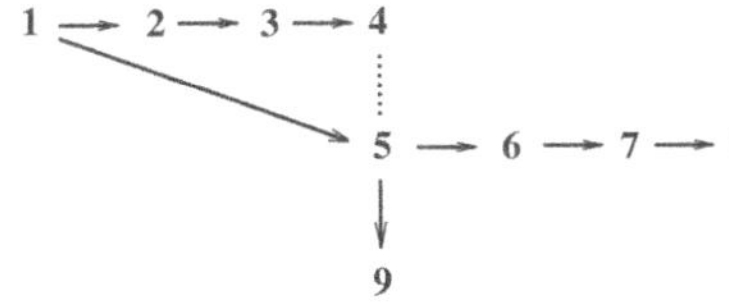

Fig. 0.1

Acknowledgments

During my work on this book, I enjoyed the hospitality of Massachusetts Institute of Technology, Stanford University, Hebrew University of Jerusalem, and Weizmann Institute. It is my pleasure to acknowledge their support.

Over the span of my professional life, many mathematicians helped me to overcome the challenges of academic carrier and provided inspiration. My deep gratitude goes to Sylvain Cappell, Julius Shaneson, Shmuel Weinberger, William Browder, Jerome Levine, Victor Kač, Shlomo Sternberg, and Yakov Eliashberg.

I would also like to thank Rajesh Babu and Tan Rok Ting for their able help with the technical support and editing of this book.

Finally and most importantly, I would not be able to complete this enterprise without the steady support of my wife Adela Katz. It is my real pleasure to thank Adela for her sacrifice, patience, and wisdom.

Contents

Preface vii

Acknowledgments xi

Flows in $2D$ and $3D$ **1**

1. A Prelude in *2D*: Flows in Flatland 3

1.1 In this chapter ... 3
1.2 On Morse Theory on surfaces with boundary and beyond ... 3
1.3 Vector fields and Morse stratifications on surfaces ... 9
1.4 Concavity and complexity of vector flows in *2D* ... 13
1.5 On spaces of vector fields ... 16
1.6 On the graph-theoretical approach to the concavity of traversing fields in *2D* ... 17
1.7 Combinatorics of tangency for traversing flows in *2D* ... 25

2. Vector Fields, Morse Stratifications, and Gradient Spines of 3-folds 27

2.1 In this chapter ... 27
2.2 Morse stratifications on 3-folds with boundary ... 28
2.3 Surgery on Morse stratifications of 3-folds ... 34
2.4 Morse strata, convexity, and concavity ... 37
2.5 Cascades, $2D$-spines, and concavity ... 46
2.6 Abstract gradient spines ... 70

3. Concavity and Complexity of Traversing Flows on 3-folds 85

3.1 In this chapter ... 85
3.2 Combinatorial and gradient complexities of 3-folds 86
3.3 Which spines are gradient? 111
3.4 The 3-folds of gradient complexity 0 and 1 120

4. Deformations of Traversing Flows in *3D* and Modifications of Gradient Spines 145

4.1 In this chapter... 145
4.2 On the combinatorics of $3D$ and $4D$ traversing vector flows 146
4.3 One-parameter families of traversing flows on 3-folds 148
4.4 Local algebraic models of traversally generic $4D$-flows and their $3D$-slices 151
4.5 Algebraic models of elementary surgery for truly generic deformations of traversing $3D$-flows 154
4.6 How truly generic deformations of traversing vector fields affect the gradient spines 161
4.7 Philips' Theorem and the spaces of vector fields in $3D$ 182
4.8 Eliminating cusps by surgery on traversing vector fields, while preserving their complexity 192
4.9 On the spaces of traversing vector fields in $3D$ with no cusps 199

High-dimensional Flows 211

5. Stratified Convexity and Concavity of Flows on Manifolds with Boundary 213

5.1 In this chapter... 213
5.2 Revisiting the Morse stratification $\{\partial_j^+ X(v)\}$ 214
5.3 Deforming the Morse stratification 226
5.4 Boundary convexity and concavity of vector fields 235
5.5 Morse stratifications of the boundary 3-convex and 3-concave vector fields 250
5.6 On the concave traversing flows whose trajectories are of the combinatorial types (11) and (121) only 260

6. Traversally Generic and Versal Flows: Semi-algebraic Models of Tangency 267

6.1 In this chapter ... 267
6.2 Morin's local models: how nonsingular flows interact with boundary 268
6.3 Traversally generic and versal vector fields 275
6.4 On the trajectory spaces of traversing and traversally generic flows 325
6.5 The ball-based Origami theorem for traversally generic flows 349
6.6 A glimpse of holography 354

7. Combinatorics of Tangency: the Stratified Spaces of Real Polynomials 361

7.1 In this chapter ... 361
7.2 Bifurcations of the real polynomial divisors: the universal combinatorics of tangency for traversing flows 361
7.3 On stratified spaces 372
7.4 Spaces of real univariate polynomials, stratified by the combinatorial types of their real divisors 372
7.5 Combinatorics of multi-tangent trajectories 387

8. Complexity of Shadows and Traversing Flows in Terms of the Simplicial Volume 405

8.1 In this chapter ... 405
8.2 On spines and cospines 407
8.3 Shadows of manifolds with boundary 411
8.4 Simplicial semi-norms, amenable localization, and complexity of shadows 413
8.5 Complexity of traversally generic vector fields and the simplicial semi-norms 440
8.6 Complexity of the fundamental group as a lower bound for the complexity of traversing flows 464

9. The Burnside Ring-valued Morse Formula for Equivariant Vector Fields 469

9.1 In this chapter ... 469
9.2 G-equivariant vector fields and Morse stratifications 470
9.3 Equivariant Morse formula for vector fields 475
9.4 Polynomial vector fields in polynomial domains 482

9.5 Gottlieb's and Gauss-Bonnet's equivariant formulas 486
9.6 Speculations about differently flavored Burnside rings . . 489

Bibliography 491

PART 1

Flows in $2D$ and $3D$

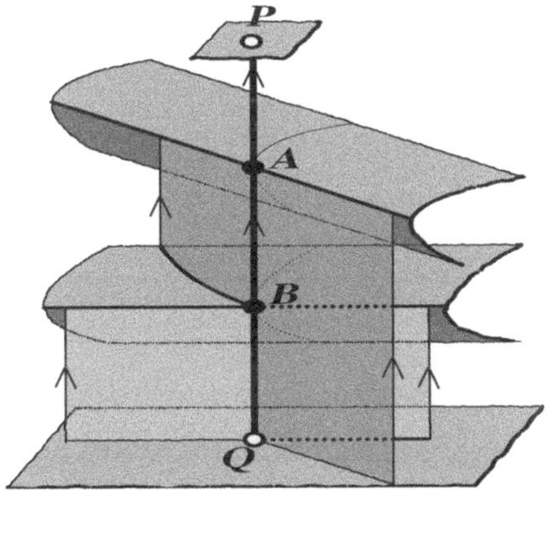

Chapter 1

A Prelude in 2D: Flows in Flatland

1.1 In this chapter ...

This chapter is an informal introduction into the philosophy and some results of the rest of the book. Most of the results from Chapter 1 are contained in [K5].

We use here the relatively simple environment of gradient flows on smooth compact and connected surfaces with boundary to illustrate the main ideas and to spell out the statements of key theorems of the book, as they manifest themselves in $2D$. The complete proofs of these results will be delivered in the chapters to come, unless the results are specific to dimension two.

The remarkable convergence of topological, geometrical, and analytical approaches to the study of closed surfaces is widely recognized by the practitioners for more than a century. We will exhibit a similar convergence of different investigative approaches to vector flows on surfaces *with boundary.*

1.2 On Morse Theory on surfaces with boundary and beyond ...

"Morse Theory", the classical book of John W. Milnor [Mi3], starts with the canonical picture of a Morse function $f : T^2 \to \mathbb{R}$ on a 2-dimensional torus T^2 (see Figure 1.1). It is portrayed as the height function f on the torus T^2, residing in the space $\mathbb{R}^3$. The height f has four *critical points*: a, b, c, and d so that

$$f(a) > f(b) > f(c) > f(d).$$

Recall that a point z is called critical if the differential df of f vanishes at z. In the vicinity of each critical point z, the torus admits a pair of local coordinate functions, say x and y, so that locally the function f acquires the form

$$f(x, y) = f(0, 0) \pm x^2 \pm y^2,$$

where the signs may form four possible combinations.

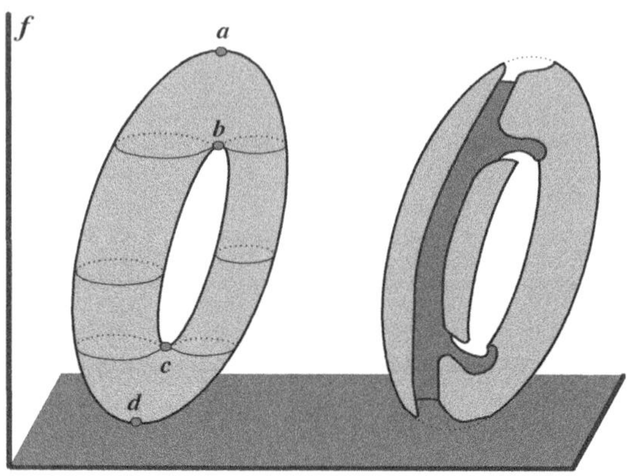

Fig. 1.1 A Morse function f on a 2-dimensional torus T^2 and its nonsingular restriction to the complement X of a disk $D^2 \subset T^2$. Note the curved geometry of the boundary loop ∂X which "remembers" the nature of f-critical points a, b, c, d.

We call a vector field v, tangent to T^2, *gradient-like* if $df(v) > 0$ everywhere outside of the set $Cr(f)$ of critical points.

If the torus is "slightly slanted" with respect to the vertical coordinate f in $\mathbb{R}^3$, then the following picture emerges. The majority of downward trajectories of the f-gradient flow $\{\Phi_t : T^2 \to T^2\}_{t\in\mathbb{R}}$ that emanate from a, asymptotically reach d. There are two trajectories that asymptotically link a to b, and two trajectories that link a to c. No (unbroken) trajectory asymptotically connects b to c.

Perhaps, a more transparent depiction of the gradient flow $\{\Phi_t\}_{t\in\mathbb{R}}$ is given in Figure 1.2. where the torus is shown in terms of its fundamental domain, the square. To form T^2, the opposite sides of the square are identified in pairs.

The Morse Theory is concerned with the *sets of constant level* $\{f^{-1}(\alpha)\}_{\alpha\in\mathbb{R}}$ and the sets *below constant level* $\{f^{-1}((-\infty, \alpha))\}_{\alpha\in\mathbb{R}}$. The main observation is that the topology of these sets is changing in an essential way only when the rising value α crosses the critical values

$$Cr(f) = \{f(a), f(b), f(c), f(d)\}.$$

When all critical values are distinct, each such "critical crossing" results in an *elementary surgery* on the set $\{f^{-1}((-\infty, \alpha))\}_{\alpha \in \mathbb{R}}$, where α is just below a critical value $\alpha_\star \in Cr(f)$. For a small $\epsilon > 0$, an elementary surgery

$$f^{-1}((-\infty, \alpha_\star - \epsilon)) \Rightarrow f^{-1}((-\infty, \alpha_\star + \epsilon))$$

attaches the *handle* $f^{-1}((\alpha_\star - \epsilon, \alpha_\star + \epsilon))$ to the set $f^{-1}((-\infty, \alpha - \epsilon))$. Eventually, when α rises above $f(a)$, the entire topology of torus T^2 is captured by a sequence of these elementary surgeries.

From a different angle, the knowledge of how the critical points a, b, c, d interact via the trajectories of the Φ_t-flow is also sufficient for reconstructing the surface T^2 as Figure 1.2 suggests (see [C] for the general results about such reconstructions).

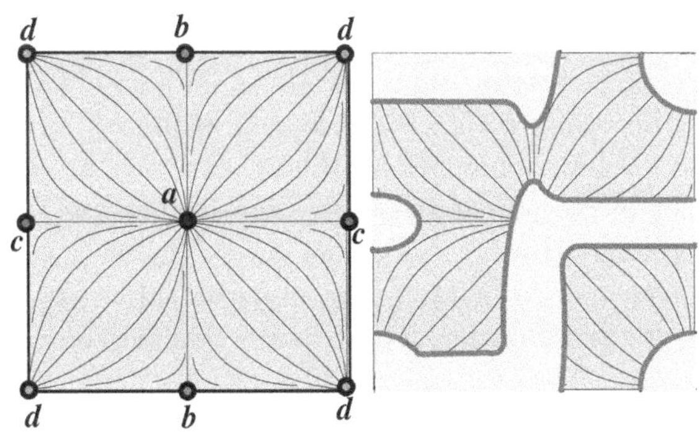

Fig. 1.2 The gradient flow of the Morse function $f : T^2 \to \mathbb{R}$ from Figure 1.1 and its restriction to the complement of a disk in the torus. The torus is represented by the square, its fundamental domain.

Note that, in the vicinity of each critical point, the gradient flow exhibits *discontinuity*: small changes in the initial position of a point z, residing in the vicinity of a critical point, result in significant differences in the position of $\Phi_t(z)$ for big positive/small negative values of t (see Figure 1.3). In fact, this discontinuity of the gradient flow, expressed in terms of the *stable* and *unstable manifolds* of critical points (see [Mi]), captures the topology of the surface (as the left diagram in Figure 1.2 suggests)!

The *space of trajectories* $\mathcal{T}(v)$ is constructed by declaring equivalent any two points that reside on the same trajectory. As a result of gradient flow discontinuity, the space $\mathcal{T}(v)$ is pathological (non-separable) (see Figure 1.3).

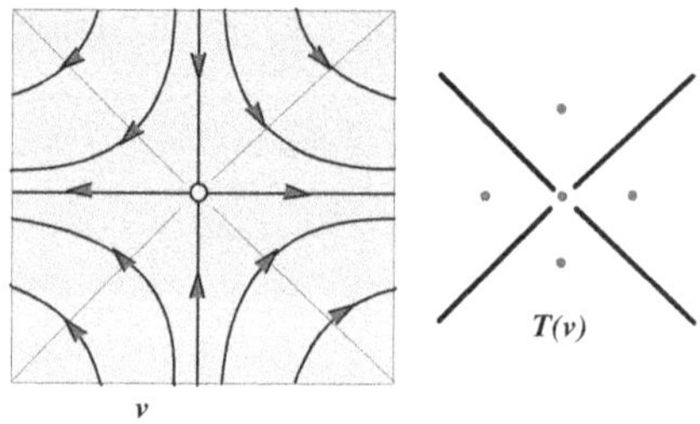

Fig. 1.3 A gradient flow v in the vicinity of a singular point and a very schematic picture of its (non-separable) trajectory space $\mathcal{T}(v)$.

The Morse theory of smooth functions and their gradient flows on compact manifolds X *with boundary* has been studied (under general boundary conditions) in the pioneering papers of Morse [Mo] and of Morse and Van Schaack [MS]. See [Ba] for the case X being a ball.

The Morse Theory on manifolds with boundary may be viewed as a very special instance of the Morse Theory on stratified spaces (the two strata ∂X and X form the stratification). The latter was developed by Goresky and MacPherson in [GM]-[GM2]. See also [Ha], [La] for the important constructions of the Morse-Smale differential complex on manifolds with conners.

If the Morse function $f : X \to \mathbb{R}$ is assumed to be locally constant on ∂X, then its gradient flow interacts with the boundary in a constrained way. In such case, the *relative topology* of the pair $(X, \partial X)$ can be captured in the ways analogous to the previous description of the Morse Theory on torus.

In the book, we offer a different philosophy for the Morse Theory on compact manifolds X *with boundary.* To formulate it, let us revisit our favorite closed surface, the torus. By deleting from T^2 small disks, centered on the points of the critical set $Cr(f)$, we manufacture a surface X whose boundary is a disjoint union of four circles. Evidently, $f : X \to \mathbb{R}$ has no critical points at all. Still it has a nontrivial topology! Can this topology be reconstructed from some data, provided by the critical-point-free f and its non-vanishing gradient-like field v? An experienced reader would notice that the restriction $f| : \partial X \to \mathbb{R}$ has critical points (maxima and minima), some of which interact *along* the boundary (with the help of a gradient-like field v^∂, tangent to ∂X). However, it is quite clear that these interactions are not sufficient for a reconstruction of the topology of X! In fact, a reconstruction of the surface X becomes possible if one introduces additional interactions between the points of $Cr(f|_{\partial X})$ that occur "through the bulk X" and are defined with the help of both vector fields v and v^∂. This observation has been explored by a number of authors, but it is not the world view that we are promoting here...

To dramatize further the situation we are facing, let us place four small disks, centered on the critical points of $f : T^2 \to \mathbb{R}$, into a single open disk D^2 and form $X = T^2 \setminus D^2$ (see Figure 1.2). Again, $f| : X \to \mathbb{R}$ has no critical points, the gradient field $v|_X \neq 0$, but its topology of X is nontrivial. This time, the boundary ∂X of the punctured torus X is just a single circle! Let us keep this challenge in mind.

Let us revisit the question that guides us: Can one propose a "Morse Theory" that is not centered on critical points, but instead focuses on the geometry of the boundary in relation to the gradient flow? The answer is affirmative. It relies on the following observation. Typically, in the vicinity of ∂X, the v-trajectories are interacting with the boundary in a number of very particular and stable ways: they are either transversal to ∂X, or are tangent to it in a *concave* or *convex* fashion[1] (see Figure 1.4). So the boundary ∂X may be "wiggly" with respect to the flow. This geometry of the v-flow in connection to the boundary is the crucial ingredient for reconstructions of X in terms of the flow (see Chapter 6, in particular, Theorem 6.12).

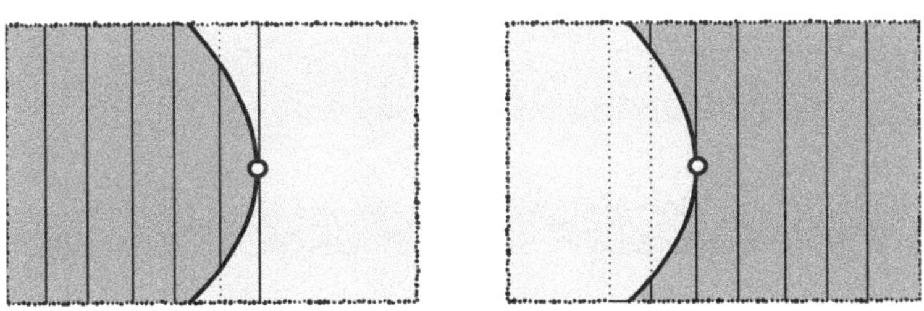

Fig. 1.4 A boundary generic field v in the vicinity of a point from $\partial_2^- X(v)$ (on the left) and in the vicinity of a point from $\partial_2^+ X(v)$ (on the right). The surface X is the shaded region, and the trajectories are shown as vertical lines.

In the vicinity of a *concave* tangency point, the v-flow is *discontinuous* in the same sense as the gradient flow is discontinuous in the vicinity of its

[1]It is possible to have a field v for which some trajectories will be *cubically* tangent to the boundary, but the majority of vector fields v avoid such cubic tangencies.

critical point: in time, close initial points become distant. In this context, the divergence of initially close points occurs due to very different *travel times* available to them; unlike the infinite travel time for the gradient flows of the Morse theory on *closed* surfaces, in the case of the nonsingular gradient flows on surfaces with boundary, every point exits the surface in a finite time. In particular, the surface is *not* flow-invariant! And again, these discontinuities of the flow reflect the topology of the surface. Let us clarify this point.

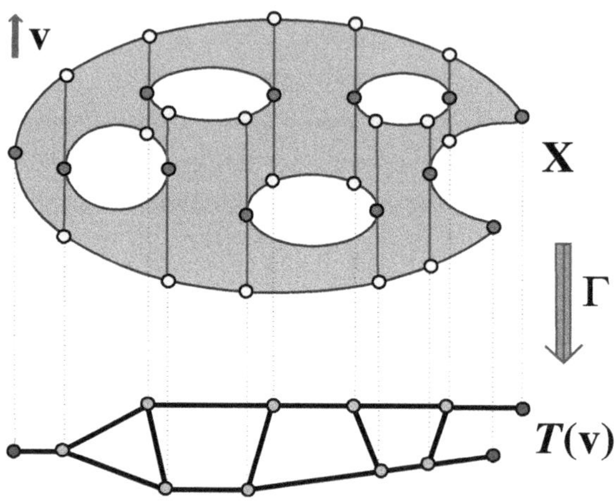

Fig. 1.5 The map $\Gamma : X \to \mathcal{T}(v)$ for a traversally generic (vertical) field v on a disk with 4 holes. The trajectory space $\mathcal{T}(v)$ is a graph whose verticies are of valencies 1 and 3.

Figure 1.5 shows a gradient flow v on a surface $X \subset \mathbb{R}^2$, the disk with 4 holes. The nonsingular function $f : X \to \mathbb{R}$ is the vertical coordinate in $\mathbb{R}^2$. Each v-trajectory is either a closed segment, or a singleton. By collapsing each trajectory to a point, we create a quotient space $\mathcal{T}(v)$ of trajectories. Since the flow trajectories are closed segments or singletons, this time, the trajectory space $\mathcal{T}(v)$ is "decent", a finite graph with verticies of valency 1 or 3 only. The verticies of valency 3 correspond to the points on ∂X where the boundary is concave with respect to the flow, and the univalent verticies to the points on ∂X where the flow is convex.

The obvious map $\Gamma : X \to \mathcal{T}(v)$ cellular. Moreover, because the fibers of Γ are contractable, Γ is a *homotopy equivalence*. In particular, the fundamental groups $\pi_1(X)$ and $\pi_1(\mathcal{T}(v))$ are isomorphic with the help of Γ. So the trajectory spaces of generic non-vanishing vector fields v of the gradient type on connected surfaces X with boundary deliver 1-dimensional *homotopy theoretical models* of X (see Corollary 7.1 for the multidimensional analogues of this fact).

1.3 Vector fields and Morse stratifications on surfaces

Let v be a smooth vector field on a compact connected surface X with boundary, such that $v|_{\partial X} \neq 0$. Following [Mo], we consider the closed locus $\partial_1^+ X(v)$, where the field v is pointing inside X, and the closed locus $\partial_1^- X(v)$, where it points outside. The intersection

$$\partial_2 X(v) =_{\mathsf{def}} \partial_1^+ X(v) \cap \partial_1^+ X(v)$$

is the locus where v is *tangent* to the boundary ∂X. Points $z \in \partial_2 X(v)$ come in two flavors: by definition, $z \in \partial_2^+ X(v)$ when $v(z)$ points inside the locus $\partial_1^+ X(v)$, otherwise $z \in \partial_2^- X(v)$. To achieve some uniformity of notations, put $\partial_0^+ X =_{\mathsf{def}} X$ and $\partial_1 X =_{\mathsf{def}} \partial X$.

Definition 1.1. We say that a vector field v on a compact surface X is *boundary generic* if:

- $v|_{\partial X} \neq 0$,
- $v|_{\partial X}$, viewed as a section of the normal 1-dimensional (quotient) bundle $T(X)|_{\partial X}/T(\partial X)$, is transversal to its zero section at the points of the locus $\partial_2 X(v)$.

  We denote by $\mathcal{B}^\dagger(X)$ the space of all boundary generic fields on a compact surface X, equipped with the C^∞-topology. ♢

In particular, for a boundary generic v, the loci $\partial_1^\pm X(v)$ are finite unions of closed intervals and circles, residing in ∂X; the loci $\partial_2^\pm X(v)$ are finite unions of points, residing in ∂X (see Figure 1.5).

Let $\chi(Z)$ denote the Euler number of a space Z. Recall that $\chi(Z)$ is the alternating sum of dimensions of the homology spaces $\{H_i(Z; \mathbb{R})\}_i$.

We denote by $\beta_1(X) =_{\mathsf{def}} \dim_{\mathbb{R}}(H_1(X; \mathbb{R}))$ the first Betti number of a space X.

Since, for a connected surface X with boundary, $H_2(X; \mathbb{R}) = 0$,

$$\chi(X) = 1 - \beta_1(X).$$

For a closed connected surface, $\chi(X) = 2 - \beta_1(X)$.

Given a vector field v with isolated zeros, we can associate an integer $\mathsf{ind}_x(v)$ with each zero x of v. This integer is the *degree* of the map which, crudely speaking, takes each point z on a small circle C_x with its center at x to the unit vector $v(z)/\|v(z)\|$. Then we define $\mathsf{Ind}(v)$, the global index of v, as the sum $\sum_{\{x\in \text{ zeros of } v\}} \mathsf{ind}_x(v)$.

The Morse formula [Mo], in the center of our investigation, computes the index $\mathsf{Ind}(v)$ of a given boundary generic vector field v on a surface X as the alternating sum of the Euler numbers of the Morse strata $\{\partial_j^+ X(v)\}_{0\leq j\leq 2}$:

$$\mathsf{Ind}(v) = \chi(X) - \chi(\partial_1^+ X(v)) + \chi(\partial_2^+ X(v)). \tag{1.1}$$

In the case of a connected surface X with boundary, $\chi(X) = 1-\beta_1(X)$, and this formula reduces to

$$\begin{aligned}\mathsf{Ind}(v) &= 1 - \beta_1(X) - \#\{\mathsf{arcs}\ in\ \partial_1^+ X(v)\} + \#\{\partial_2^+ X(v)\} \\ &= 1 - \beta_1(X) + \frac{1}{2}\Big(\#\{\partial_2^+ X(v)\} - \#\{\partial_2^- X(v)\}\Big).\end{aligned}$$

In particular, if $v \neq 0$ in X, then $\mathsf{Ind}(v) = 0$, and we get

$$\frac{1}{2}\Big(\#\{\partial_2^+ X(v)\} - \#\{\partial_2^- X(v)\}\Big) = \beta_1(X) - 1, \tag{1.2}$$

where the RHS of the equality (1.2) is the topological invariant $|\chi(X)|$ of X. In contrast, the cardinality $\#\{\partial_2^+ X(v)\}$ depends on v.

Lemma 1.1. *Let a surface X be formed by removing k open disks from a closed orientable surface Y, the sphere with g handles. Then, for any non-vanishing boundary generic field v on X,*

$$\#\{\partial_2^+ X(v)\} \geq 4g - 4 + 2k.$$

Moreover, $\#\{\partial_2^+ X(v)\} = 4g - 4 + 2k$ only when $\#\{\partial_2^- X(v)\} = 0$.

Proof. The Euler number is additive under gluing surfaces along their boundary components. Therefore, if k disks are removed from Y, the sphere with g handles, then $\chi(X) = 2 - 2g - k$. Then the Morse formulas (1.1) and (1.2) imply

$$\#\{\partial_2^+ X(v)\} \geq 4g - 4 + 2k$$

for any non-vanishing v. Moreover,

$$\#\{\partial_2^+ X(v)\} = 4g - 4 + 2k$$

if and only if $\#\{\partial_2^- X(v)\} = 0$—the main feature of the *boundary concave* fields (see Definition 1.2). □

In particular, for any non-vanishing boundary generic field v on a torus with a single hole, $\#\{\partial_2^+ X(v)\} \geq 2$ (cf. Figure 1.4).

Recall that an *immersion* is a smooth map of manifolds, whose differential has the trivial kernel.

Consider a smooth map $\alpha : X \to \mathbb{R}^2$, which is an immersion in the vicinity of ∂X. Any such α gives rise to the Gauss map $G : \partial X \to S^1$, defined by the formula $G(x) = \alpha_*(\tau_x)/\|\alpha_*(\tau_x)\|$, where τ_x is the tangent vector to ∂X at x. The direction of τ_x is consistent with the preferred orientation of ∂X, induced by the preferred orientation of X .

Let $\hat{v}$ be a non-vanishing constant field on $\mathbb{R}^2$. Since the kernel of the differential of $D\alpha : TX \to T\mathbb{R}^2$ is trivial along ∂X, the field $\hat{v}$ defines a vector field $\tilde{v} = \alpha^*(\hat{v})$ on X in the vicinity of ∂X. The pull-back field $\tilde{v}$ extends to a vector field v on X (possibly with zeros).

Then the degree of the Gauss map is given by the classical formula ([Hopf])

$$\deg(G) = \chi(X) - \mathsf{Ind}(v).$$

When $\alpha : X \to \mathbb{R}^2$ is an immersion everywhere, the pull-back field $v = \alpha^*(\hat{v}) \neq 0$ everywhere. In such case, $\mathsf{Ind}(v) = 0$, and, for a connected X with $\partial X \neq \emptyset$, we get

$$\deg(G) = \chi(X) = 1 - \beta_1(X).$$

So, for an immersion α, we get a new interpretation of formula (1.2):

$$\deg(G) = \chi(X) = \frac{1}{2}\Big(\#\{\partial_2^- X(v)\} - \#\{\partial_2^+ X(v)\}\Big). \tag{1.3}$$

This global-to-local formula has another classical geometrical interpretation. Let $\mathsf{g} = \alpha^*(\mathsf{g_E})$ be the Riemannian metric on X, the pull-back of the Euclidean metric $\mathsf{g_E}$ on $\mathbb{R}^2$. Let K_ν denote the normal curvature of ∂X with respect to g. Then

$$\deg(G) = \frac{1}{2\pi}\int_{\partial X} K_\nu \, d\mathsf{g},$$

which leads to another pleasing global-to-local connection:

$$\frac{1}{\pi}\int_{\partial X} K_\nu \, d\mathsf{g} = \#\{\partial_2^- X(v)\} - \#\{\partial_2^+ X(v)\}.$$

In particular, for a connected orientable surface X of genus g with a *single* boundary component,

$$\chi(X) = 1 - 2g = \frac{1}{2}\Big(\#\{\partial_2^- X(v)\} - \#\{\partial_2^+ X(v)\}\Big). \tag{1.4}$$

So the number of v-trajectories γ in X that are tangent to ∂X, but are not singletons (they correspond to the points of $\partial_2^+ X(v)$), as a function of genus g, grows at least as fast as $4g - 2$!

On the other hand, when ∂X is connected, by the Whitney index formula [W1], the degree of $G : \partial X \to S^1$ can be also calculated as $\mu + N^+ - N^-$, where $N^\pm$ denotes the number of positive/negative self-intersections of the curve $\alpha(\partial X) \subset \mathbb{R}^2$, and $\mu = \pm 1$. Here is a brief description of the rule by which the self-intersections acquire polarities. Let $p \in \alpha(\partial X)$ be a point where the coordinate function $y : \mathbb{R}^2 \to \mathbb{R}$ attends its minimum on the curve $\alpha(\partial X)$. If the tangent vector τ_p at p, which defines the orientation of $\alpha(\partial X)$, is ∂_x, then we put $\mu = +1$; if $\tau_p = -\partial_x$, then $\mu = -1$. Starting at p and moving in the direction of τ_p, we visit each self-intersection a twice and in a particular order. The first visitation defines a tangent vector $\tau_1(a)$, the second visitation defines a tangent vector $\tau_2(a)$. When the ordered pair $(\tau_1(a), \tau_2(a))$ defines the clockwise orientation of the xy-plane, then we attach "$-$" to a. Otherwise, the polarity of a is "$+$".

As a result, we get a somewhat mysterious connection between the self-intersections of ∂X under immersions $\alpha : X \to \mathbb{R}^2$ and the tangency patterns of the flows in X that are the α-pull-backs of non-vanishing flows in the plane.

Theorem 1.1. *Let $\hat{v}$ be a non-vanishing vector field in the plane $\mathbb{R}^2$. Let X be a genus g connected orientable surface with a connected boundary. Consider an immersion $\alpha : X \to \mathbb{R}^2$ such that the loop $\alpha(\partial X)$ has transversal self-intersections only. Assume that the pull-back $v = \alpha^*(\hat{v})$ is a boundary generic field on X. Then*

$$\frac{1}{2}\Big(\#\{\partial_2^+ X(v)\} - \#\{\partial_2^- X(v)\}\Big) = N^+ - N^- \pm 1 = 2g - 1,$$

and

$$\frac{1}{2}\Big(\#\{\partial_2^+ X(v)\} - \#\{\partial_2^- X(v)\}\Big) + 2 \leq N^+ + N^-,$$

the latter inequality being sharp by an appropriate choice of α.

Proof. The first formula is the result of combining the Whitney formula [W1] for $\deg(G)$ with formulas (1.3), (1.4).

By a theorem of Guth [Gu], for any immersion $\alpha : X \to \mathbb{R}^2$, the total number of self-intersections of the loop $\alpha(\partial X)$ admits an estimate

$$N^+ + N^- \geq 2g + 2.$$

Moreover, this lower bound is realized by an immersion $\alpha : X \to \mathbb{R}^2$! Therefore, by formula (1.4), the Guth's inequality is transformed into

$$N^+ + N^- \geq 2 + \frac{1}{2}\Big(\#\{\partial_2^+ X(v)\} - \#\{\partial_2^- X(v)\}\Big).$$

Moreover, for some optimal immersion α,

$$N^+ + N^- = 2 + \frac{1}{2}\Big(\#\{\partial_2^+ X(v)\} - \#\{\partial_2^- X(v)\}\Big) = 2 - \frac{1}{2\pi}\int_{\partial X} K_\nu \, dg.$$

□

When a surface X is oriented and a field v is boundary generic, then the points from $\partial_2^+ X(v)$ come in two *new flavors*: "$\oplus, \ominus$". By definition, a point $a \in \partial_2^+ X(v)$ has the polarity "$\oplus$" if the orientation of $T_a X$, determined by the pair $(\nu_a, v(a))$, where ν_a is the inner normal to ∂X, agrees with the preferred orientation of X. Otherwise, the polarity of a is defined to be "$\ominus$".

Thus, for each choice of orientation of X (and hence of ∂X) we get a partition

$$\partial_2^+ X(v) = \partial_2^{+,\oplus} X(v) \coprod \partial_2^{+,\ominus} X(v).$$

Switching the orientation of X switches the second polarities in the partition.

1.4 Concavity and complexity of vector flows in 2D

Definition 1.2. We say that a boundary generic vector field v is *boundary convex* if $\partial_2^+ X(v) = \emptyset$. We say that a boundary generic v is *boundary concave* if $\partial_2^- X(v) = \emptyset$ (see Figure 1.4). ◇

The existence of a boundary convex field puts severe restrictions on the topology of surface.

Lemma 1.2. *If a compact connected surface X with boundary admits a non-vanishing* boundary convex *gradient-like vector field v, then X is either a disk D^2, or an annulus A^2.*

Proof. The convexity of the field v implies that X admits a $(-v)$-directed continuous retraction on the locus $\partial_1^+ X(v)$. Since X is connected, it follows that $\partial_1^+ X(v)$ is connected as well. Thus, $\partial_1^+ X(v)$ is either a circle, or a segment. In the first case, X is diffeomorphic to an annulus $S^1 \times [0, 1]$; in the second case, X is diffeomorphic to a disk D^2. □

In contrast, the boundary concave non-vanishing gradient fields are plentiful. For example, consider the radial vector field v on an annulus A^2. Delete from A^2 any number of convex disjoint disks and restrict v to the resulting 2-disk with holes. This construction implies that any disk with holes admits a non-vanishing boundary concave gradient-like vector v.

Many other surfaces admit concave fields as well. For example, consider a Morse function $f : Y \to \mathbb{R}$ on a closed surface Y and its gradient field v. Then removing small convex (in the local Morse coordinates) balls, centered on the critical points, from Y produces a *boundary concave* non-vanishing gradient field on X. In particular, if Y is a sphere with g handles, then one can find a Morse function with $2g + 2$ critical points (see Figure 1.1). So the surface X, obtained from Y by removing $2g + 2$ balls, admits a non-vanishing concave gradient-like field v.

In fact, by Theorem 1.5, any connected orientable surface with boundary, but the disk, admits a boundary concave non-vanishing gradient field!

We view the integer $c^+(v) =_{\mathsf{def}} \#(\partial_2^+ X(v))$ as a *measure of complexity* of the v-flow, subject to the condition $\mathsf{Ind}(v) = 0$ or, alternatively, subject to the condition $v \neq 0$ in X.

We define the *complexity* of a compact connected surface X with boundary as the minimum

$$c^+(X) =_{\mathsf{def}} \min_{v \neq 0} \{c^+(v)\},$$

where v runs over all non-vanishing boundary generic fields on X.

By varying v within different spaces of fields, one may consider a variety of such minima; non-vanishing fields and non-vanishing gradient-like fields are the two most important cases. So we introduce the *gradient complexity*

$$gc^+(X) =_{\mathsf{def}} \min_{v \neq 0 \text{ of the gradient type}} \{c^+(v)\},$$

where v runs over all non-vanishing gradient-like fields on X.

Evidently $gc^+(X) \geq c^+(X)$.

Let M denote the Möbius band. In Section 1.6, we will show that $gc^+(M) = 1$, while $c^+(M) = 0$, so the two notions of complexity are different!

When $c^+(v) \leq N$, then, by Lemma 1.1, $4g + 2k - 4 \leq N$, which confines $(g, k) \in \mathbb{Z}_+^2$ to a triangle, whose area grows as a quadratic function of N. Hence, in terms of the complexity $c^+(v)$, we can restate the Lemma 1.1 as follows.

Corollary 1.1. *Let X be a connected compact surface with boundary. Let v be a boundary generic vector field on X, subject to the condition $\mathsf{Ind}(v) = 0$.*

Then the complexity of v satisfies the inequality

$$c^+(v) \geq 2 \cdot \beta_1(X) - 2 = -2 \cdot \chi(X).$$

When $\chi(X) \leq 0$, this inequality turns into the equality $c^+(v) = -2 \cdot \chi(X)$ if and only if v is boundary concave.

As a result, for any natural N, there are finitely many connected compact surfaces of bounded complexity $c^+(X) \leq N$. In fact, the number of such surfaces (counted up to a homeomorphism) grows as a quadratic function in N. ♢

For example, for any non-vanishing boundary concave field v on a torus with a single hole, $\#\{\partial_2^+ X(v)\} = 2$. In fact, the constant field v, being restricted to the complement to a convex disk, is boundary concave and has the property $\#\{\partial_2^+ X(v)\} = 2$. Thus $c^+(X) = 2$. By Theorem 1.5, $gc^+(X) = 2$ as well.

Lemma 1.1 leads immediately to the following claim.

Corollary 1.2. *Let X be a sphere with g handles and k holes, where $g, k \geq 1$. If X admits a non-vanishing boundary concave field v, then $\#\{\partial_2^+ X(v)\} = 4g - 4 + 2k$.* ♢

Given a compact surface X with boundary, we form its *double* $DX =_{\mathsf{def}} X \cup_{\partial X} X$ by attaching two copies of X along their boundaries. Note that $\chi(DX) = 2 \cdot \chi(X)$. Therefore, $\chi(X) < 0$ if and only if $\chi(DX) < 0$.

Recall that any closed orientable surface with a negative Euler number admits a metric of constant negative curvature -1. So if $\chi(X) < 0$, then DX admits such hyperbolic metric.

Let $\mathrm{Vol}(DX)$ denote the hyperbolic volume of DX, and let $\mathrm{Vol}(\Delta^2)$ denote the volume of an ideal hyperbolic triangle Δ^2 in the hyperbolic plane $\mathbf{H}^2$. In fact, $vol(\Delta^2) = \pi$.

In $2D$, a remarkable convergence of topology and geometry takes place. In the spirit of this convergence, since $\chi(DX) = -\mathrm{Vol}(DX)/\pi$, Corollary 1.1 admits a more geometric reformulation:

Theorem 1.2. *Let v be a non-vanishing boundary generic vector field on a compact connected and orientable surface X with boundary. Assume that $\chi(X) < 0$*[2]*. Then the complexity of the v-flow satisfies the inequality:*

$$c^+(v) \;\geq\; \mathrm{Vol}(DX)/\pi.$$

[2]This excludes the disk and the annulus.

Moreover, $c^+(v) = \mathrm{Vol}(DX)/\pi$ *if and only if* v *is boundary concave.* ◇

Chapter 8 is devoted to multidimensional generalizations of Theorem 1.2.

1.5 On spaces of vector fields

Definition 1.3. We say that a non-vanishing vector field v on a compact surface X is *traversing* if all its trajectories are closed segments or singletons[3]. ◇

Each trajectory γ of a traversing field v must reach the boundary both in positive and negative times: otherwise γ is not homeomorphic to a closed interval.

We denote by $\mathcal{V}_{\mathsf{trav}}(X)$ the space of all traversing fields on a given compact surface X.

We denote by $\mathcal{V}_{\mathsf{grad}}(X)$ the space of all gradient-like fields on X, and by $\mathcal{V}_{\neq 0}(X)$ the space of all non-vanishing fields on X. All these spaces are equipped with the C^∞-topology.

The next lemma claims that v is traversal if and only if it is non-vanishing and gradient-like [K1]. The proof of this lemma can be found in the arguments that lead to its multidimensional version, Corollary 5.8.

Lemma 1.3. *For any compact connected surface X with boundary,*

$$\mathcal{V}_{\mathsf{trav}}(X) = \mathcal{V}_{\mathsf{grad}}(X) \cap \mathcal{V}_{\neq 0}(X).$$

◇

The surfaces X and vector fields v we consider all are smooth. We can add an external collar to X to form a diffeomorphic surface $\hat{X} \supset X$ and to extend v to a smooth field $\hat{v}$ on $\hat{X}$. Let $\hat{\gamma}$ be a $\hat{v}$-trajectory (or rather its germ) through a point x of ∂X. We can talk about *the order of tangency* of two smooth curves, $\hat{\gamma}$ and ∂X, at $x \in \hat{\gamma} \cap \partial X$ in $\hat{X}$ (see Definition 1.5). We say that the tangency of $\hat{\gamma}$ to ∂X is *simple* if its order of tangency is 2. When the two curves are transversal at x, we say that the order of tangency is 1. In fact, these notions depend only on (X, v) and not on the extension $(\hat{X}, \hat{v})$.

Definition 1.4. A traversing vector field v on a compact surface X is called *traversally generic*, if two properties are valid:

[3]It easy to see that the ends of these segments, as well as the singletons, reside in ∂X.

(1) when a trajectory γ is tangent to the boundary ∂X, then the tangency is simple,

(2) no v-trajectory γ contains more then one simple point of tangency to ∂X[4].

We denote by $\mathcal{V}^\ddagger(X)$ the space of all traversally generic vector fields on a compact surface X. $\diamondsuit$

As the name suggests, the traversally generic fields are typical among all traversing fields; furthermore, a perturbation of any traversally generic field is traversally generic. This is the content of the next theorem. Its validation requires an involved argument, which even in $2D$ resists a significant simplification. In Chapter 6, we will prove Theorem 6.7, a multidimensional generalization of the following theorem.

Theorem 1.3. *For any compact connected surface X with boundary, the space $\mathcal{V}^\ddagger(X)$ of traversally generic fields is open and dense in the space $\mathcal{V}_{\mathsf{trav}}(X) = \mathcal{V}_{\mathsf{grad}}(X) \cap \mathcal{V}_{\neq 0}(X)$ of traversing vector fields.* $\diamondsuit$

1.6 On the graph-theoretical approach to the concavity of traversing fields in 2D

Recall that the two complexities satisfy the inequality $c^+(X) \leq gc^+(X)$. So, we pose a natural question.

Question 1.1. Are there compact connected surfaces X with boundary for which $c^+(X) < gc^+(X)$? $\diamondsuit$

Now we will take a closer look at the graph-theoretical models of the boundary concave and traversally generic vector fields in 2D. In what follows, it is essential that the trajectory spaces $\mathcal{T}(v)$ of traversally generic vector fields v are finite graphs whose verticies have valency 1 and 3 only (see Figure 1.5). Moreover, for a traversally generic boundary concave vector field v, all the verticies of $\mathcal{T}(v)$ have valency 3.

Let G be a finite connected trivalent graph with a verticies. We denote by βG its barycentric subdivision: each edge e of G is divided by a new vertex v_e, its barycenter. We consider the finite set $\mathsf{Tri}(G)$ of all *colorings* of the edges of βG with *three* colors so that, at each vertex of G, exactly three distinct colors are applied. Thus, $\#\mathsf{Tri}(G) = 6^a$.

[4]In particular, a traversally generic v is boundary generic.

Theorem 1.4. *Let G be a finite connected trivalent graph. Each coloring $\alpha \in \mathsf{Tri}(G)$ produces (in a canonical way) a compact connected surface $X(G,\alpha)$ with boundary. The surface $X(G,\alpha)$ admits a traversally generic* concave *vector field $v = v(G,\alpha)$. The cardinality of the locus $\partial_2^+ X(G,\alpha)(v)$ is the number of verticies in G.*

Conversely, every connected surface with boundary, which admits a traversally generic concave vector field, can be produced in this way.

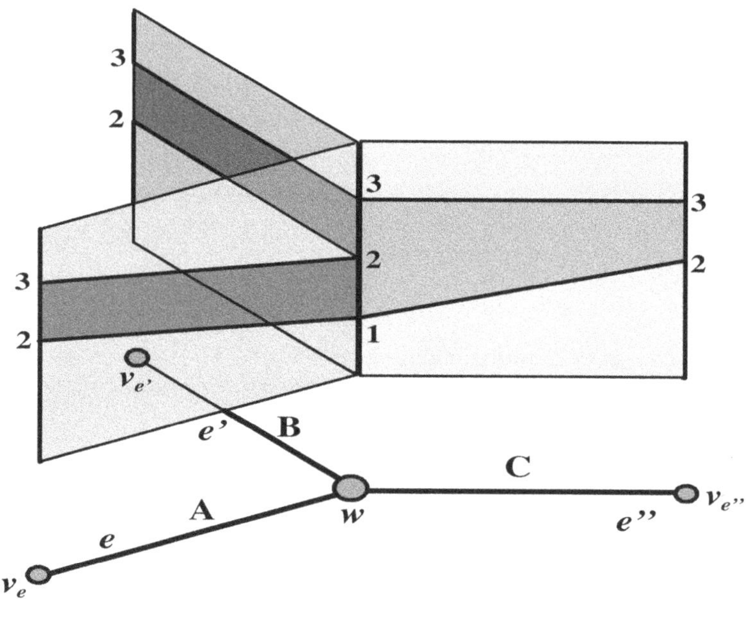

Fig. 1.6 Constructing the surface $X(G,\alpha)$ in the vicinity of a trivalent vertex $w \in G$. The verticies $v_e, v_{e'}, v_{e''} \in \beta G$ are the barycenters of the three edges e, e', e'' of G that contain w.

Proof. Let $\mathsf{A}, \mathsf{B}, \mathsf{C}$ denote three distinct colors, and $\mathcal{P} = \{\mathsf{A}, \mathsf{B}, \mathsf{C}\}$ denotes the entire pallet.

Consider a 2-dimensional space $Z = G \times (0, 4)$. It has singularities in the form of binders of the 3-page open books (see Figure 1.6). The binders correspond to the vertices of G.

First, employing a given coloring α, we will construct a piecewise linear surface $\hat{X}(G,\alpha) \subset Z$. The vector field on $\hat{X}(G,\alpha)$ will be induced by the product structure in Z.

For each edge $e \subset G$ and its barycenter $v_e \in \beta G$, we place the interval $v_e \times [2,3] \subset Z$ over v_e. Let $\hat{e}$ be half of the interval $e \subset G$, bounded two verticies $v_e \in \beta G$ and $w \in G$. Over $\hat{e}$, we place a strip $E \subset Z$. Its construction depends on the color attached to the interval $[v_e, w]$:

- if the color of $[v_e, w]$ is A, then we link the vertex $v_e \times 2$ with the vertex $w \times 1$ by a line in the rectangle $R = [v_e, w] \times (0, 4)$, and the vertex $v_e \times 3$ with the vertex $w \times 2$ by another line in R;
- if the color of $[v_e, w]$ is B, then we link by a line in R the vertex $v_e \times 2$ with the vertex $w \times 2$, and the vertex $v_e \times 3$ with the vertex $w \times 3$ by another line;
- if the color of $[v_e, w]$ is C, then we link by a line in R the vertex $v_e \times 2$ with the vertex $w \times 1$, and the vertex $v_e \times 3$ with the vertex $w \times 3$ by another line.

By definition, $E(e, w)$ is the strip in $[v_e, w] \times (0, 4)$, bounded by the two lines whose construction has been described above. Thanks to the monotonicity of the bijections

$$A : \{2,3\} \to \{1,2,3\},\ B : \{2,3\} \to \{1,2,3\},\ C : \{2,3\} \to \{1,2,3\}$$

that correspond to the colors A, B, C, the lines that bound the strips $E(e, w)$ do not intersect. We denote by $\hat{X}(G, \alpha)$ the union of all such strips.

The local model of each binder implies that $\hat{X}(G, \alpha)$ is indeed piecewise-linear surface, embedded in the singular space Z. Inside Z, one can smoothen the sharp edges of the boundary $\partial\hat{X}(G, \alpha)$ in order to get a smooth surface $X(G, \alpha)$ (we challenge the reader to visualize this smoothing in the vicinity of point $w \times 2$ from Figure 1.6). The restriction of the product structure in Z to its subspace $X(G, \alpha)$ produces a smooth non-vanishing vector field $v(G, \alpha)$ on $X(G, \alpha)$. Its trajectories (the vertical lines in Z) will be simply tangent to $\partial X(G, \alpha)$ exactly at the points of the type $w \times 2$, where w runs over the set of verticies of G. By Theorem 5.1, this vector field $v(G, \alpha)$ is of the gradient type.

Conversely, any traversally generic and concave vector field v on a connected compact surface X with boundary produces a map $\Gamma : X \to \mathcal{T}(v)$, where the space of trajectories is a finite trivalent graph. Its verticies are in 1-to-1 correspondence with the points of the locus $\partial_2^+ X(v)$.

As a point v_e in the open edge e of the graph $\mathcal{T}(v)$ approaches a vertex w, the intersection of the v-trajectory $\gamma = \Gamma^{-1}(v_e)$ with the boundary ∂X

defines a bijection of the v-ordered set $\gamma \cap \partial X$ of cardinality 2 to a v-ordered set $\Gamma^{-1}(w) \cap \partial X$ of cardinality 3, the orders being respected by the bijections. This determines one of three colors A, B, C we attach to the half-edge $[v_e, w]$. Therefore the geometry of the flow determines a tricoloring of the graph $\beta\mathcal{T}(v)$. □

Corollary 1.3. *The cardinality of the set $\partial_2^+ X(v)$ is even for any boundary concave and traversally generic vector field v. If $\partial_2^+ X(v) \neq \emptyset$, the v-trajectories that pass through the points of $\partial_2^+ X(v)$ divide X in the domains whose number is divisible by 3.*

Proof. For any boundary concave and traversally generic vector field v, the graph $\mathcal{T}(v)$ is trivalent. Its verticies are in 1-to-1 correspondence with the points of $\partial_2^+ X(v)$. Since for any trivalent graph,

$$3 \times \{\text{the number of verices}\} = 2 \times \{\text{the number of edges}\},$$

we conclude that the number of vericies in $\mathcal{T}(v)$ must be even and the number of edges is divisible by 3. This conclusion translates into the claim of the corollary. □

The next theorem provides an answer to Question 1.1.

Theorem 1.5.

- *For any compact connected surface X with boundary, but the disk,*
$$c^+(X) = 2\beta_1(X) - 2,$$
where $\beta_1(X)$ denotes the first Betti number of X.
- *For any compact connected surface X, but the disk and the Möbius band, the gradient complexity is given by the same formula:*
$$gc^+(X) = 2\beta_1(X) - 2.$$
- *Any surface X, but the disk and the Möbius band, admits a boundary concave traversing[5] vector field.*
- *In contrast, the disk D does not admit a boundary concave traversing vector field. However, $c^+(D) = gc^+(D) = 0$.*
- *The Möbius band M does not admit a boundary concave traversing vector field either. However, M does admit a concave and convex non-vanishing vector field. In fact, $c^+(M) = 0$, but $gc^+(M) = 1$.*

[5]even traversally generic

- *The annulus is the only compact connected surface that admits both concave and convex traversing vector fields.*

Proof. Recall the topological classification of closed surfaces. Any connected orientable closed surface is either a sphere S, or a torus T, or a connected sum of several tori T. Similarly, any connected non-orientable closed surface is a projective plane P, or a connected sum of several projective planes P. Evidently, any compact connected orientable surface *with boundary* can be obtained from a closed connected surface by deleting one or several open disks. Therefore, any compact connected orientable surface with boundary is either a disk D, or an annulus A, or a punctured torus $Q = T \setminus D^\circ$, or a boundary connected sum of such building blocks. Similarly, any compact connected non-orientable surface with boundary is either a Möbius band M, or a boundary connected sum of a several Möbius bands, or a connected sum of these surfaces with one or several annuli A.

By Corollary 1.1, for any compact connected surface X with boundary,

$$gc^+(X) \geq c^+(X) \geq 2\beta_1(X) - 2.$$

If X admits a *boundary concave* non-vanishing vector field v, then by the same corollary, $c^+(X) = c^+(v) = 2\beta_1(X) - 2$. In particular, if X admits a boundary concave *traversing* field, then

$$gc^+(X) = c^+(X) = 2\beta_1(X) - 2.$$

Now, let us examine the behavior of the two complexities, $c^+(\sim)$ and $gc^+(\sim)$, under the connected sum operation $\#_\partial$.

Given two boundary generic vector fields v_1 and v_2 on X_1 and X_2, respectively, there exists a boundary generic vector field w on $X_1\#_\partial X_2$ such that the cardinalities $| \sim |$ of the concave and convex loci satisfy the equations:

$$\begin{aligned} |\partial_2^+(X_1\#_\partial X_2)(w)| &= |\partial_2^+(X_1)(v_1)| + |\partial_2^+(X_2)(v_2)| + 2, \\ |\partial_2^-(X_1\#_\partial X_2)(w)| &= |\partial_2^-(X_1)(v_1)| + |\partial_2^-(X_2)(v_2)|. \end{aligned} \tag{1.5}$$

Indeed, we may attach a 1-handle H to the locus

$$\partial_1^- X_1(v_1) \coprod \partial_1^+ X_2(v_2)$$

so that H has a *neck* with respect to the extension w. Such vector field w contributes *two* new points to the locus $\partial_2^+(X_1\#_\partial X_2)(w)$ (this pair is residing in the neck of H) and no new points to $\partial_2^-(X_1\#_\partial X_2)(w)$. Of course, this construction fails when the set $\partial_1^- X_1(v_1)$ or $\partial_1^+ X_2(v_2)$ is empty; however, one may change v_1 to $-v_1$, or v_2 to $-v_2$ to arrange for $\partial_1^- X_1(-v_1) \neq \emptyset$ and/or $\partial_1^+ X_2(-v_2) \neq \emptyset$. These flips of the vector fields do no affect the loci $\partial_2^\pm X_1$ and $\partial_2^\pm X_2$.

So, using (1.5), the construction of the vector field w validates the inequalities

$$\begin{aligned} c^+(X_1 \#_\partial X_2) &\leq c^+(X_1) + c^+(X_2) + 2, \\ gc^+(X_1 \#_\partial X_2) &\leq gc^+(X_1) + gc^+(X_2) + 2. \end{aligned} \tag{1.6}$$

Note that when both vector fields v_1 and v_2 are boundary concave, then w is boundary concave as well. As a result, the set $\mathcal{C}$ of compact connected surfaces[6] that admit non-vanishing boundary concave vector fields is closed under the boundary connected sum operation $\#_\partial$! Similarly, its subset $\mathcal{CT}$, formed by compact connected surfaces that admit boundary concave and traversing vector fields, is closed under $\#_\partial$.

We will see soon that $\mathcal{C}$ is formed by all surfaces, with the exception of the disk D. At the same time, $\mathcal{CT}$ contains all compact connected surfaces with boundary, with the exception of D and M.

Let us consider the beginning of the list of compact connected surfaces with boundary, arranged according the increasing values of their first Betti numbers:

$$\begin{array}{lllll} \beta_1(X) = 0; & = 1; & = 2; & \dots & \\ D; & A,\ M; & Q,\ A\#_\partial A,\ A\#_\partial M,\ M\#_\partial M; & \dots & (1.7) \end{array}$$

Let us now compute the complexities of the surfaces D, A, Q, the basic building blocks in the representation (as boundary connected sums) of orientable connected surfaces with boundary, and the complexities of the basic blocks M, A, which participate in the representation of non-orientable connected surfaces with boundary.

Note that $c^+(D) = 0 = gc^+(D)$, since D admits a convex traversing flow. By formula (1.2), D does not admit a boundary concave non-vanishing vector field.

Also $c^+(A) = 0 = gc^+(A)$, the latter equality being delivered by the radial boundary concave and convex traversal vector field.

We claim that $c^+(Q) = 2 = gc^+(Q)$. Indeed, since $\beta_1(Q) = 2$, by Corollary 1.1, we get $c^+(Q) \geq 2$. On the other hand, there exists a trivalent graph G_Q with an appropriate tricoloring and exactly *two* verticies

[6] considered up to diffeomorphisms

such that, applying the construction from Theorem 1.4 (see Figure 1.6), we produce a traversally generic vector field $v = v(G_Q, \alpha)$ on the surface $X(G_Q, \alpha) = Q$ with the cardinality 2 locus $\partial_2^+ X(G_Q, \alpha)(v)$. As a result, both complexities of Q equal to 2.

The Möbius band M is different. We notice that M admits a non-vanishing vector field v with a single closed trajectory—the core of the Möbius band—and transversal to the boundary ∂M. Thus, $c^+(M) = 0$.

To compute $gc^+(M)$, consider a graph G_M (pictorially, "$* - \bullet$**O**") with a single vertex "$\bullet$" of valency 3 and a single vertex "$*$" of valency 1. Thus G_M is the circle "**O**" to which the segment "$* - \bullet$" is attached.

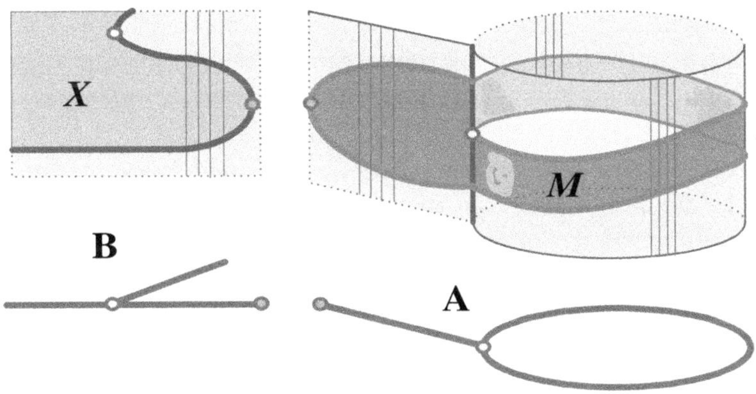

Fig. 1.7 The embedding $\alpha_M : M \to G_M \times [0, 4]$ of the Möbius band M, the right diagram A. Both diagrams A and B show a stage in the construction of the vector field $v^\#$ on $M \#_\partial X \subset G_{X \#_\partial M} \times [0, 4]$.

The construction from Theorem 1.4 (see Figure 1.7, diagram A), produces a remarkable embedding $\alpha_M : M \to G_M \times [0, 4]$! So we conclude that M admits a traversally generic vector field v_M (not a concave one!) with $\partial_2^+ M(v_M)$ being a singleton ($\partial_2^- M(v_M)$ is a singleton as well). As a result, $gc^+(M) \leq 1$.

On the other hand, any traversally generic vector field v on M must produce the graph $\mathcal{T}(v)$ which is homotopy equivalent to a circle, the homotopy type of M. If $\partial_2^+ M(v) = \emptyset$, this graph $\mathcal{T}(v)$ has no trivalent verticies, in which case, $\mathcal{T}(v)$ is homeomorphic to a circle. So $M \to \mathcal{T}(v)$ must be a fibration whose fibers (the v-trajectories) are segments. Moreover, thanks to the vector field v, this fibration is orientable, a contradiction with the non-orientability of M. Therefore, we conclude that $gc^+(M) = 1$, while $c^+(M) = 0$.

We claim that, if X admits a *concave* traversing vector field v, then so does $M\#_\partial X$. Starting with v, we deform it in the vicinity of a point $x \in \partial X \setminus \partial_2 X(v)$ to a new vector field $\tilde{v}$, so that a new pair of points $a \in \partial_2^+ X(\tilde{v})$ and $b \in \partial_2^- X(\tilde{v})$ is introduced. It is done so that $\partial_2^- X(\tilde{v}) = b$ and $\partial_2^+ X(\tilde{v}) = \partial_2^+ X(v) \cup a$. This x-localized deformation generates an edge $e \subset \mathcal{T}(\tilde{v})$ that terminates at a vertex $\tilde{b}$. Now we will join the graphs $\mathcal{T}(\tilde{v})$ and G_M by identifying the vertex $\tilde{b}$ with the vertex $* \in G_M$ of valency one (see Figure 1.7). The result is a new trivalent graph $G_{X\#_\partial M}$. Thanks to Theorem 1.4, the latter graph will produce a *concave* traversing vector field $v^\#$ on $M\#_\partial X \subset G_{X\#_\partial M} \times [0, 4]$. (Note that the construction of $v^\#$ is quite different from the more general construction of the extension w across the handle H). The existence of $v^\#$ validates the claim.

Let $K =_{\mathsf{def}} M\#_\partial M$, $L =_{\mathsf{def}} A\#_\partial M$, and $N =_{\mathsf{def}} A\#_\partial A$. In fact, Theorem 1.4 and Corollary 1.1 imply that Q, K, L, and N are the only connected surfaces of the gradient complexity 2 that admit boundary concave traversally generic vector fields. Indeed, just start with the tree "$\rangle \bullet - \bullet \langle$" with two trivalent verticies and consider the ways in which one can identify its four leaves in pairs. Then consider all admissible tricolorings of the resulting graphs G. These cases will deliver the model tricolored graphs G_Q, G_K, G_L, G_N.

Therefore $gc^+(X) = c^+(X) = 2$ only for $X = Q, K, L, N$. Moreover, for these four surfaces, $gc^+(X) = c^+(X) = 2\beta_1(X) - 2$.

Now the rest of the surfaces (of complexities higher than 2) can be obtained by applying repeatedly the connected sum operations $X \Rightarrow X\#_\partial T$, $X \Rightarrow X\#_\partial M$, and $X \Rightarrow X\#_\partial A$ to the surfaces $X = Q, K, L, N$ of complexity/gradient complexity 2.

So the surfaces Q, A admit a boundary concave traversing vector fields. Therefore, using the construction of the extension w across the handle H, we validate the following claim: if X admits a boundary concave traversing vector field, so do the surfaces $X\#_\partial A$ and $X\#_\partial Q$. We know already that if X admits a boundary concave traversing vector field, so does $X\#_\partial M$. Hence, by induction, all surfaces, with the exception of D and M, admit a boundary concave traversing vector field.

Recall that if X admits a boundary concave traversing vector field v, then $c^+(X) = gc^+(X) = 2\beta_1(X) - 2$. Therefore we proved this formula for all surfaces, but D and M. □

1.7 Combinatorics of tangency for traversing flows in 2D

Let us pick an extension $\hat{X}$ of a given compact surface X by adding an external collar to X. Let $\hat{v}$ be an extension of a given field v into $\hat{X}$. Pick a smooth auxiliary function $z : \hat{X} \to \mathbb{R}$ such that:

- 0 is a regular value of z,
- $z^{-1}(0) = \partial X$,
- $z^{-1}((-\infty, 0]) = X$, (1.8)

Definition 1.5. Let $\hat{\gamma}$ be a $\hat{v}$-trajectory through a point $x \in \partial X$. We say that $\hat{\gamma}$ has the *order/multiplicity of tangency* k to ∂X at x, if $\mathcal{L}_{\hat{v}}^{\{j\}}(z) = 0$ for all $j < k$, and $\mathcal{L}_{\hat{v}}^{\{k\}}(z) \neq 0$ at x[7]. Here $\mathcal{L}_{\hat{v}}^{\{j\}}(z)$ denotes the jth iterated $\hat{v}$-directional derivative of the function z. ◇

Given a traversally generic vector field v on a compact connected surface X, we will attach the combinatorial pattern $(1,1)$ to a typical v-trajectory $\gamma \subset X$ that corresponds to the edges of the graph $\mathcal{T}(v)$, the pattern (121) to the trajectories that correspond to the trivalent verticies of $\mathcal{T}(v)$, and the pattern (2) to the univalent verticies (see Figure 1.5). In fact, the numbers 1 and 2 in these patterns reflect the order of tangency of the curves $\hat{\gamma}$ and ∂X at the points of $\gamma \cap \partial X$ (in the sense of Definition 1.5). On a given compact surface X, for *traversally generic* fields v no other patterns (say, like (1221) or (13)) occur. In $2D$, this conclusion follows from Definition 1.5.

The lemma below is another way to state this fact. Its proof, relying on the Malgrange Preparation Theorem [Mal], can be found in Chapter 6.

Lemma 1.4. *Let v be a traversally generic field on X. Extend (X, v) to a pair $(\hat{X}, \hat{v})$. In the vicinity of each v-trajectory γ, there exist special local coordinates (u, x) in $\hat{X}$ and a real polynomial $P(u, x)$ of degree 2 or 4 such that:*

- *each $\hat{v}$-trajectory is given by the equation $\{x = const\}$,*
- *the boundary ∂X is given by the polynomial equation $\{P(u, x) = 0\}$,*
- *X is given by the polynomial inequality $\{P(u, x) \leq 0\}$.*

[7]This is equivalent to saying that the $(k-1)$-st jet at x of $z|_\gamma$ vanishes, but the k-th jet does not.

The polynomial $P(u, x)$ takes three canonical forms:

(1) $u(u-1)$, which corresponds to the combinatorial pattern (11),
(2) $u^2 - x$, which corresponds to the combinatorial pattern (2),
(3) $u\big((u-1)^2 + x\big)(u-2)$, which corresponds to the pattern (121). $\diamondsuit$

To summarize, at $\partial_2 X(v)$ the order of tangency is 2; the trajectories through $\partial_2^+ X(v)$ have the combinatorial tangency pattern (121), and through $\partial_2^- X(v)$ the combinatorial tangency pattern (2). The rest of trajectories have the pattern (11).

We denote by $\mathbf{\Omega}^\bullet_{'\langle 1]}$ the partially ordered set whose elements are $(11), (2), (121)$ and the order is defined by $(11) \succ (2)$ and $(11) \succ (121)$. This combinatorics does not look impressive. However, in higher dimensions, traversally generic fields on $(n+1)$-manifolds with boundary generate a very rich and interesting partially ordered finite list $\mathbf{\Omega}^\bullet_{'\langle n]}$ of combinatorial tangency patters. The poset $\mathbf{\Omega}^\bullet_{'\langle n]}$ is *universal* in each dimension $n+1$. The posets $\mathbf{\Omega}^\bullet_{'\langle n]}$ are discussed in detail in Chapter 7.

Chapter 2

Vector Fields, Morse Stratifications, and Gradient Spines of 3-folds

2.1 In this chapter ...

We would like to acknowledge the fundamental influence of the works of Ishii ([I], [I1]), Matveev ([Ma]), and Benedetti-Petronio ([BP], [BP1]) on the worldview developed in Chapters 2, 3, and 4. Most of the results of these chapters can be found in [K].

In the environment of vector flows on 3-dimensional manifolds with boundary, many constructions and ideas of the book will make their first appearance. Some of them are not specific to dimension 3; in Chapters 5-6, more analytical arguments will establish their validity in all dimensions. However, in $3D$, the pictorial validations in this chapter will provide the reader with more transparency and intuition.

Given a vector field v on a compact 3-fold X with boundary ∂X, such that $v \neq 0$ along ∂X, we generate the Morse stratifications $\{\partial_j^{\pm} X(v)\}_{0 \leq j \leq 3}$ of X. As in the case of vector fields on surfaces, these strata reflect the tangency patterns of the v-trajectories to the boundary ∂X.

In Theorem 2.1, we present *The Morse Law of Vector Fields*—an elegant formula (discovered in 1929 by M. Morse [Mo]) that computes the index of a vector field v in terms of the Euler numbers of the strata $\{\partial_j^{+} X(v)\}_{0 \leq j \leq 3}$.

Corollary 2.1 describes the considerable freedom in changing the shape of the Morse strata $\{\partial_j^{+} X(v)\}$ by deforming the field v within the space of a gradient-like fields of a given function $f : X \to \mathbb{R}$.

Let v_1 denote the orthogonal projection of v on ∂X. In Theorem 2.2, Theorem 2.3, and Corollary 2.6, we begin to develop the "holographic tune" of this book; for instance, we pose the question:

To what extend the v_1-flow on the boundary ∂X regulates the v-flow in the manifold X?

Along the same lines, we formulate another basic question :

What are the implications for the topology of X that result from imposing a priori certain restrictions on the Morse stratification $\{\partial_j^{\pm} X(v)\}_{0 \leq j \leq 3}$ *of some/any field v?*

While addressing these questions, we reinforce another already familiar from Chapter 1 idea—the *convexity* and *concavity* of the v-flow in connection to the boundary ∂X can be expressed and measured in terms of the Morse stratification.

For a given 3-fold X, we also discuss the notion of its *spine* $K \subset X$ (see Definition 2.3). Spines are 2-dimensional simplicial subcomplexes of X on which X collapses. It turns out (see Theorem 2.5), that any *traversally generic* (see Definition 2.5) vector field v on X canonically gives rise to a spine $K(v)$. We call such spines *gradient spines*.

In Theorem 2.4, we prove that traversally generic vector fields form an open and dense set in the space of all gradient-like fields on a 3-fold X (Chapter 6 contains the proof of a similar result for the multi-dimensional gradient flows).

Our Origami Theorem 2.6 claims that any X has a gradient spine which is the image of a 2-disk under a folding map which is 3-to-1 at most.

Let us formulate the overarching idea of this chapter: the complexity of a non-vanishing gradient v-flow on X can be measured in terms of the *singularities* of the corresponding gradient spine $K(v)$.

In Section 2.6, we study some technical generalizations of gradient spines, the so called *abstract gradient spines,* $\vec{\mathsf{Y}}$*-spines, and branched oriented spines.* It turns out (see Chapter 3), that all these notions of spines are very close.

Any gradient spine $K(v)$ allows for a reconstruction of the ambient X. Specifically, in Theorem 2.7, we show that if two 3-folds, X_1 and X_2, have homeomorphic gradient spines, the homeomorphism $h : K(v_1) \to K(v_2)$ preserving the, so called, *TN-markers* of the spines, then X_1 is diffeomorphic to X_2.

2.2 Morse stratifications on 3-folds with boundary

Let X be a compact connected smooth 3-manifold with a nonempty boundary ∂X. Let $f : X \to \mathbb{R}$ be a generic smooth function. Then f has nondegenerate critical points in the interior of X; moreover, its restriction to the boundary ∂X may also presumed to be a Morse function.

Let v be a smooth *gradient-like* vector field for f. For such fields, the property $df(v) > 0$ is valid away from the f-critical points.

Instead of working with such pairs (f, v), we could pick a Riemannian metric g on X and to choose $v = \nabla_g f$, the gradient field. As the lemma below testifies, for $v \neq 0$, both points of view are equivalent, but we prefer the first one.

Lemma 2.1. *Let $f : X \to \mathbb{R}$ be a smooth function on a 3-fold and v a smooth vector field such that $df(v) > 0$ everywhere. Then there exists a smooth Riemannian metric g on X such that the gradient $\nabla_g f = v$.*

Proof. We denote by $T(X)$ the tangent bundle of X. In a given metric g, $\nabla_g f$—the gradient of f—is defined by the formula $\langle \nabla_g f, w \rangle_g = df(w)$ for all tangent vectors $w \in T(X)$. Here $\langle \sim, \sim \rangle_g$ denotes the g-generated scalar product.

At each point $x \in X$, the condition $df(\sim) = 0$ defines a plane $H_{x,f}$ in the tangent space $T_x(X)$, and the condition $df(\sim) \geq 0$ defines a positive cone $C^+_{x,f} \subset T_x(X)$. By the definition of a gradient-like field, $v \in \mathrm{int}(C^+_{x,f})$. Note that $T_x \approx H_{x,f} \oplus L_{x,v}$, where $L_{x,v}$ is the 1-dimensional space, spanned by v. We choose a metric g so that the plane H_f is orthogonal to the line L_v and $\|v\|_g = 1$. Then for any $w = \alpha v + \beta u$, where $\alpha, \beta \in \mathbb{R}$ and $u \in H_{x,f}$, we get $\langle v, w \rangle_g = df(w)$. Thus $v = \nabla_g f$. □

The singularities of $f : \partial X \to \mathbb{R}$ come in two flavors: *positive* and *negative*. At a positive singularity, the field v is directed inward of X, and at a negative singularity—outward. This distinction between positive and negative critical points of $f|_{\partial X}$ depends only on f, not on v.

At a positive singularity and in an appropriate coordinate system $\{x_1, x_2, x_3\}$ with $\{x_1 = 0\}$ defining ∂X and $x_1 > 0$ — the interior of X, the function f can be written in the form

$$f(x) = c + x_1 + a_2 x_2^2 + a_3 x_3^2,$$

where c and $a_2 \neq 0, a_3 \neq 0$ being real constants. At a negative singularity,

$$f(x) = c - x_1 + a_2 x_2^2 + a_3 x_3^2.$$

Let $\Sigma_1^{\pm}$ be the finite set of positive/negative singularities of $f|_{\partial X}$ and let Σ_0—the finite set of f-singularities in the interior of X.

We denote by $X_{\leq c}$ the set $\{x \in X |\ f(x) \leq c\}$.

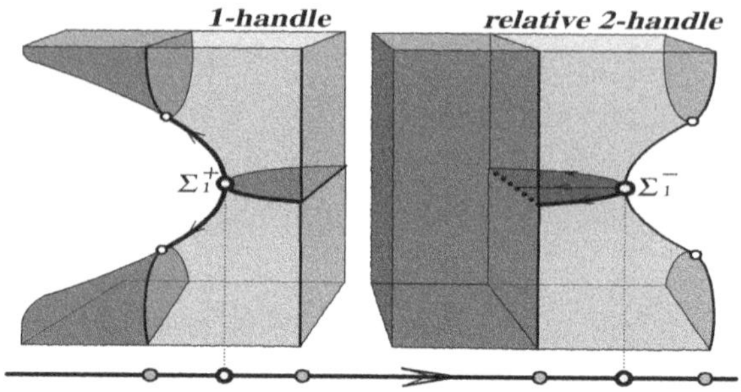

Fig. 2.1 A positive singularity Σ_1^+ of index 1 and a negative singularity Σ_1^- of index 2 of a function $f : \partial X \to \mathbb{R}$ on the boundary of a solid X. The gradient-like field v is horizontal. On the left, crossing the $f|_{\partial X}$-critical value of a *positive* critical point results in attaching 1-handle, while crossing the critical value of a *negative* critical point has no effect on the topology of X.

Crossing a critical value $c_\star$ of a positive singularity causes the topological type of $X_{\leq c}$ to change, while crossing $c_\star$ of a negative singularity has no effect on the topology of $X_{\leq c}$, as illustrated in Figure 2.1.

The exact notion of a *boundary generic* vector field v on a manifold with boundary will be introduced in Definition 5.1. Meanwhile, in the $3D$-context, it will suffice to know some basic properties of boundary generic fields. For instance, the locus L, where a boundary generic vector field is *tangent* to the boundary ∂X, is a 1-dimensional smooth submanifold of ∂X. The locus L divides ∂X into two domains: $\partial^+ X$, where v is directed inward of X, and in $\partial^- X$, where it is directed outward.

Morse noticed that, for a generic vector field v, the tangent locus L inherits a structure in relation to $\partial^+ X$ analogous to that of ∂X in relation to X ([Mo]). To explain this point, we need to revise our notations in a way which will be amenable to the recursive definitions below.

Let $\partial_0 X =_{\mathsf{def}} X$, and $\partial_1 X =_{\mathsf{def}} \partial X$. Denote by $\partial_2 X \subset \partial_1 X$ the locus L where v is tangent to $\partial_1 X$. For a boundary generic v, $\partial_2 X$ divides $\partial_1 X$ into the domain $\partial_1^+ X$, where v is directed inward X, and the domain $\partial_1^- X$, where v is outward X. Evidently, $\partial_1^\pm X \supset \Sigma_1^\pm$. Consider the set $\partial_3 X$, where v is tangent to $\partial_2 X$. The set $\partial_3 X$ divides $\partial_2 X$ into the set $\partial_2^+ X$, where v is directed inward of $\partial_1^+ X$, and the set $\partial_2^- X$, where v is directed outward of $\partial_1^+ X$. Finally, $\partial_3 X = \partial_3^+ X \coprod \partial_3^- X$, where v is directed inward of $\partial_2^+ X$ at the points of $\partial_3^+ X$ (to visualize the geometry of v in connection to all these sets, see Figures 2.2, 2.3).

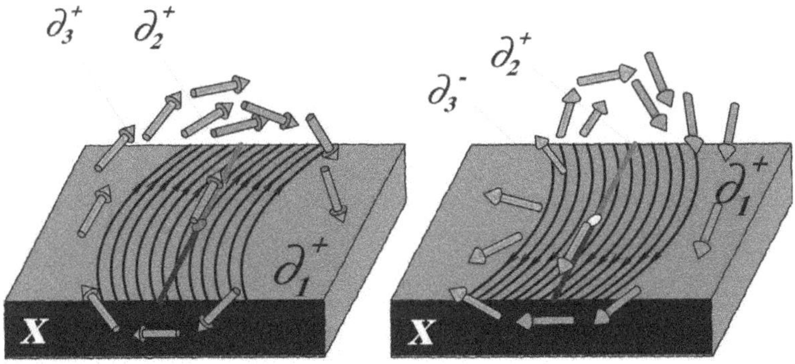

Fig. 2.2 The patterns of two fields, v (the $3D$-arrows) and v_1 (the "parabolic flow"), in the vicinity of a point from $\partial_3^+ X$ (on the left) and a point from $\partial_3^- X$ (on the right).

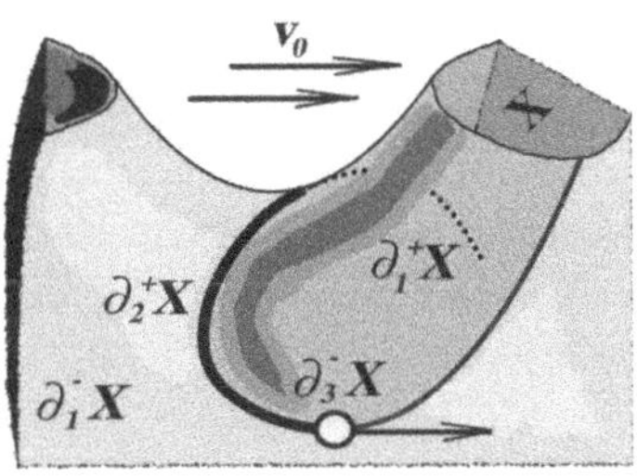

Fig. 2.3 A more realistic picture of the boundary $\partial_1 X$ in the vicinity of a point from $\partial_3^- X$ in relation to the horizontal vector field v_0.

Definition 5.1 describes accurately the notion of boundary generic vector field in any dimension. In Chapters 2, 3, and 4, we adopt the following "working definition".

Definition 2.1. A smooth vector field v on X, such that $v|_{\partial X} \neq 0$, is called *boundary generic* if all the strata $\{\partial_j X\}_{1 \leq l \leq 3}$ are regularly embedded smooth manifolds.

We call a pair (f, v) *boundary generic* if v is boundary generic and all the restrictions $f|_{\partial_j X}$ are Morse functions. $\diamondsuit$

Most of the time, the second property will be irrelevant, but when we need it, we will assume its validity. When we say that a Riemannian metric

g on X is "boundary generic", we imply that the pair $(f, \nabla_g f)$ is boundary generic.

At some point in the future, additional general position requirements will be imposed on boundary generic vector fields v or on boundary generic pairs (f, v) (see Definition 2.5). We call this subclass of vector fields v "*traversally generic*". Such vector fields will play the central role in our enterprise.

Let $\Sigma_j^{\pm} \subset \partial_j^{\pm} X$ denote the critical set of the function $f|_{\partial_j X}$.

With some generic metric g in place, let v_j be the orthogonal projection of v onto the tangent to $\partial_j X$ subspace. Let n_j denote the normal vector field to $\partial_j X$ inside $\partial_{j-1} X$. It is chosen so that it points inside $\partial_{j-1}^{+} X$. Away from the singularity locus Σ_j, we get $df(v_j) > 0$.

For a vector field v_k as above on $\partial_k X$ with isolated singularities $\{x_\star \in \Sigma_k \subset \mathrm{int}(\partial_k X)\}$, we denote by $\mathsf{ind}_{x_\star}(v_k)$ its *local index* at $x_\star$. Let

$$\mathsf{Ind}^{+}(v_k) =_{\mathsf{def}} \sum_{x_\star \in \Sigma_k^{+}} \mathsf{ind}_{x_\star}(v_k).$$

Then, according to [Mo], the following two sets of relations are equivalent.

Theorem 2.1 (The Morse Law of Vector Fields).
For any boundary generic vector field $v =_{\mathsf{def}} v_0$ and a generic metric g on a compact 3-fold X, the following formulas hold:

- $\chi(\partial_k^{+} X) = \mathsf{Ind}^{+}(v_k) + \mathsf{Ind}^{+}(v_{k+1})$[1]
- $\mathsf{Ind}^{+}(v_k) = \sum_{j=k}^{3} (-1)^j \chi(\partial_j^{+} X)$,

where $k \in [0, 3]$. ♢

There is another useful degree-type invariant of v. The set $\partial_2 X \subset \partial_1^{+} X$ carries two nonzero vector fields: the field $v = v_1$ tangent to $\partial_1 X$ along $\partial_2 X$, and the field n_2 that is normal to $\partial_2 X$ in $\partial_1 X$ and points inside $\partial_1^{+} X$. The second field trivializes the oriented tangent bundle of $\partial_1 X$ along $\partial_2 X$. With this trivialization of the bundle $T(\partial_1 X)|_{\partial_2 X}$ in place, v defines a continuous map $h : \partial_2 X \to S^1$. We view h as an element in the one-dimensional oriented bordism group $\Omega_1(S^1)$ of the circle. This group splits as $\Omega_1(pt) \oplus \Omega_0(pt) \approx \Omega_0(pt)$ (see [CF]), i.e., an element $h : M^1 \to S^1$ in $\Omega_1(S^1)$ is determined by the degree $\deg(h) = [h^{-1}(pt)] \in \mathbb{Z}$.

This degree can be easily computed in terms of the cardinalities of the cuspidal sets $\partial_3^{+} X$ and $\partial_3^{-} X$.

[1]By definition, $\mathsf{Ind}^{+}(v_3) = \#(\Sigma_3^{+})$, and $\mathsf{Ind}^{+}(v_4) = 0$.

Lemma 2.2. *The number* $\frac{1}{2}\big(\#(\partial_3^+X) - \#(\partial_3^-X)\big)$ *is equal to the degree of the map* $h : \partial_2 X \to S^1$.

Proof. Each loop γ in $\partial_2 X$ either: **(1)** entirely belongs to one of the two sets, ∂_2^+X or ∂_2^-X, or **(2)** the arcs of γ that belong to ∂_2^+X and to ∂_2^-X alternate (see Figure 2.7). In the first case, the contribution of γ to $\deg(h)$ is zero. In the second case, the contribution of each arc with the ends of opposite polarity is also zero. Each arc with two positive ends contributes a rotation of v by $+\pi$, while each arc with two negative ends contributes a rotation by $-\pi$. Hence the total rotation of $v_1 = v_0$ along $\partial_2 X$ is

$$\pi \cdot \big(\#(\partial_3^+X) - \#(\partial_3^-X)\big) = 2\pi \cdot deg(h).$$

□

In Corollary 4.7, a more refined count of the cusps from $\partial_3 X$ will produce a very different formula for the degree of $h : \partial_2 X \to S^1$.

Let $f : X \to \mathbb{R}$ be a *nonsingular* function and v its gradient-like vector field. For each point $x \in X$ we get a local map $p : X \to \mathbb{R}^2$, defined in the vicinity of the v-trajectory γ_x through x. Let us outline the construction of p. We add an external collar W to X and extend the Morse data (f, v) into $\hat{X} =_{\mathsf{def}} X \cup W$ so that $d\hat{f}(\hat{v}) > 0$ in $\hat{X}$. We denote by $(\hat{f}, \hat{v})$ this extension. We form a $\hat{v}$-*adjusted* neighborhood $U_x \subset \hat{X}$ of γ_x that consists of segments of $\hat{v}$-trajectories. The $\hat{v}$-flow defines a surjection $p_x : U_x \to V_x$ onto a neighborhood V_x of x in the hypersurface $\hat{f}^{-1}(f(x))$.

We consider the restriction $p_x : U_x \cap \partial_1 X \to V_x$, where $x \in \partial_1 X$, to the boundary $\partial_1 X$. According to Whitney [W], generic (*stable*) smooth maps $\mathbb{R}^2 \to \mathbb{R}^2$ have only folds (see (2.1)) and cusps (see (2.2)) as their singularities. Therefore, for boundary generic Morse data (f, v) and every point $x \in \partial_1 X \setminus \partial_2 X$, the map $p_x : U_x \cap \partial_1 X \to V_x$ is a surjection, for $x \in \partial_2 X \setminus \partial_3 X$, p_x is a folding along an arc in $\partial_2 X$, and at $x \in \partial_3 X$, p_x is a cusp map with $p_x(\partial_2 X)$ being the cuspidal curve. Note that along ∂_2^+X, $p_x : X \to V_x$ is locally onto, while along ∂_2^-X, it is not!

It is especially easy to visualize the stratification $\{\partial_j X\}$ when X is embedded or immersed in $\mathbb{R}^3$ and f is a pull-back of a generic linear function on $\mathbb{R}^3$. In such a case, a global surjection $p : X \to \mathbb{R}^2$ is available. Now, the locus $\partial_2 X$ can be identified with the folds of the map $p : \partial_1 X \to \mathbb{R}^2$, and the locus $\partial_3 X$ with its cusps.

Let (u_1, u_2) be local coordinates on $\partial_1 X$, centered on a point $x \in \partial_1 X$. As we deform a nonsingular vector field v within generic one-parameter families, the local structure of the projections $p_x : \partial_1 X \to \mathbb{R}^2$ can be described in terms of few canonical forms. Two of them, the *fold*

$$F(u_1, u_2) = (u_1^2,\ u_2) \tag{2.1}$$

and the *cusp*

$$F(u_1, u_2) = (u_1^3 + u_1 u_2,\ u_2), \tag{2.2}$$

describe locally the stable singularities[2] of a smooth map F from a plane to a plane. The *Swallow Tail* t-parameter family

$$F_t(u_1, u_2) = (u_1^4 + t u_1^2 + u_1 u_2,\ u_2)$$

describes a cancellation of two cusps[3]. It will play a significant role in Chapter 4.

2.3 Surgery on Morse stratifications of 3-folds

Let X be a connected compact 3-manifold X with boundary. Given a smooth function $f : X \to \mathbb{R}$ with isolated singularities, we can construct a new smooth function with *no singularities* in X. The construction is similar to its $2D$-analogue, shown in Figure 1.1. We just need to cut out of X a number of tunnels. Each tunnel starts at the boundary $\partial_1 X$ and has a dead end which "engulfs" the isolated f-singularity to be eliminated. We denote by T the interior of the tunnels. Then f, being restricted to $X \setminus T$ (which is diffeomorphic to X), is nonsingular, and its perturbation can be assumed to be of the Morse type on the new boundary $\partial(X \setminus T) \approx \partial X$ and still nonsingular in $X \setminus T \approx X$.

Lemma 2.3. *Let X be a compact connected 3-fold with boundary $\partial_1 X$. Let $f : X \to \mathbb{R}$ be a smooth function with no singularities in a regular neighborhood N of $\partial_1 X$. Denote by v its gradient-like field. Let $\gamma \subset \partial_1^{\pm} X$ be a simple path that connects two points from $\partial_2 X$ and has an empty intersection with the critical set Σ_1 of $f : \partial_1 X \to \mathbb{R}$.*

Then it is possible to deform v in N to a new f-gradient-like vector field $\tilde{v}$ for which the new set $\partial_1^{\pm} X$ will be obtained from the original one by the 1-surgery along γ. Outside of N, $v = \tilde{v}$.

A similar statement holds for any boundary generic vector field v.[4]

[2]They come, respectively, from the universal unfolding of the A_2-singularity $f(u_1) = u_1^2$ and A_3-singularity $f(u_1) = u_1^3$.

[3]It is the universal unfolding of the codimension 1 singularity of a mapping $\mathbb{R}^2$ to $\mathbb{R}^2$ coming from the universal two-parameter unfolding of A_4-singularity.

[4]not necessarily of the gradient type

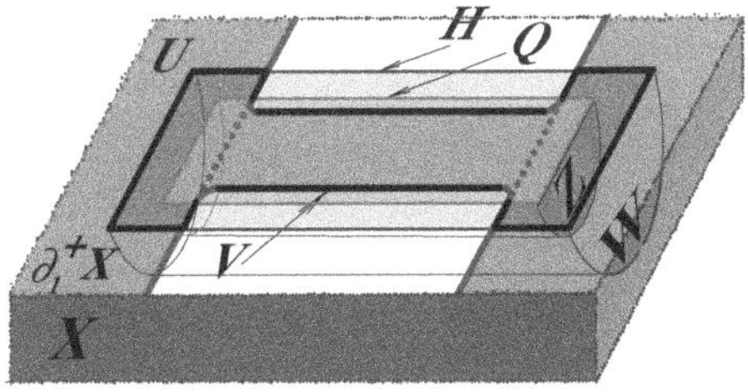

Fig. 2.4 Performing 1-surgery on the locus $\partial_1^+ X$.

Proof. By Lemma 2.1, we can pick a metric g on X in which $v = \nabla_g f$ in the vicinity of $\partial_1 X$. Let v_1 be the orthogonal projection of v on the tangent bundle $T(\partial_1 X)$. The idea is to perform surgery on $\partial_1^+ X$, which is induced by a homotopy of the field $v - v_1$, while keeping f and v_1 fixed. We start with "1-surgery" on the fields along a rectangular band H which connects two arcs, say, $A \subset \partial_2 X$ and $B \subset \partial_2 X$. The band, with the exception of small neighborhoods of A and of B, resides in $\partial_1^+ X$ (alternatively, in $\partial_1^- X$), as shown in Figure 2.4. The band avoids the singularities of $f|_{\partial_1 X}$, so that $v_1 \neq 0$ everywhere in H. We denote by Q a smaller band which is contained in H (see Figure 2.4). By smoothening the corners of Q inside H, we manufacture a 1-handle $V \subset H$ so that $V \cup \partial_1^+ X$ is a smooth domain.

Let n denote the interior normal to $\partial_1 X$ in X. We represent the field v as $v_1 + h \cdot n$, where the smooth function $h : \partial_1 X \to \mathbb{R}$ is positive in the interior of $\partial_1^+ X$ and negative in the interior of $\partial_1^- X$. In fact, we can assume that 0 is a regular value of h.

At each point $x \in N \supset \partial_1 X$, the differential df picks a particular open half-space $\mathsf{T}^+_{f,x}$ in the tangent space T_x, so that the vector field $v \in \mathsf{T}^+_f$. Along the boundary $\partial_1 X$, another family of half-spaces is available: let T^+_x denote the set of tangent vectors at $x \in \partial X$ that point inside X.

Away from the singularities of $f|_{\partial_1 X}$, the positive cone $\mathsf{T}^+_{f,x} \cap T^+_x$ is open.

Consider a smooth function $\tilde{h} : H \to \mathbb{R}$ which satisfies the following properties:

(1) $\tilde{h}^{-1}([0, +\infty)) = (V \cup \partial_1^+ X) \cap H$,
(2) zero is a regular value of $\tilde{h}$, and $\tilde{h}^{-1}(0) = \partial(V \cup \partial_1^+ X) \cap H$,
(3) the field $v_1 + \tilde{h} \cdot n \in \mathsf{T}^+_f$.

The last property can be achieved by starting with any $\tilde{h}$, subject to

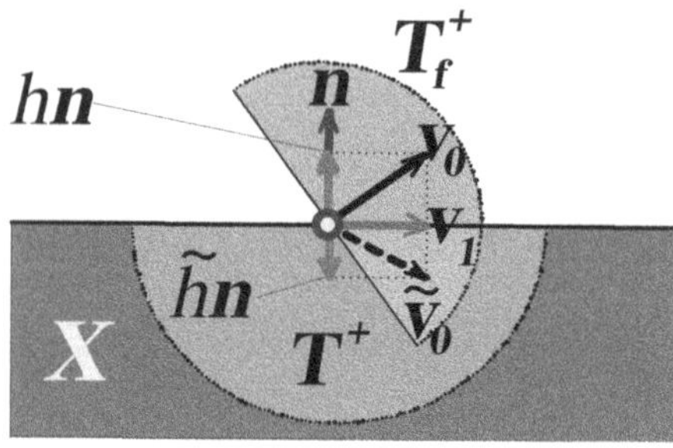

Fig. 2.5 Changing the field $v = v_1 + h \cdot n$ at a point $x \in \partial_1^- X(v)$ into a field $\tilde{v} = v_1 + \tilde{h} \cdot n$ for which $x \in \partial_1^+ X(\tilde{v})$.

(1) and (2), and then rescaling it by a positive smooth function a, so that

$$v_1 + (a\tilde{h})n \in \mathsf{T}_f^+ \cap T^+$$

(see Figure 2.5). At each point, the existence of an appropriate $a > 0$ follows from the fact that, away from Σ_1, $v_1 \in \mathsf{T}_f^+$. Because the cone $\mathsf{T}_f^+ \cap T^+$ is open and the domain H is compact, the global existence of such a scaling factor a, by a partition-of-unity argument, follows from its existence at each point of H.

We extend the field $v_1 + \tilde{h} \cdot n$ inside X to get a smooth f-gradient-like field w in a small regular neighborhood W of V in X. We use a smooth partition of unity $1 = \alpha + \beta$, subordinate to the cover W, $X \setminus Z$, where the half-cylinder neighborhood $Z \subset W$ is shown in Figure 2.4. The function α vanishes in Z and β in $X \setminus W$. Now consider the field $\tilde{v} =_{\mathsf{def}} \alpha v + \beta w$. Since $T_{f,x}^+$ is convex, $\tilde{v} \in \mathsf{T}_f^+$. Moreover, in V, $\tilde{v}$ points inside X and, in $H \setminus V$, outside X. Also, outside W, $\tilde{v} = v$.

Finally, for a fixed f, the set of all f-gradient-like fields is open and convex. Hence, any modification of a f-gradient-like field can be obtained by its deformation (linear homotopy).

The arguments for boundary generic (non-gradient) vector fields v are similar and actually simpler. □

Corollary 2.1. *Let X be a compact 3-fold with boundary $\partial_1 X$. Let $f : X \to \mathbb{R}$ be a smooth function with no singularities in a regular neighborhood N of $\partial_1 X$. Denote by v its gradient-like field. Let $\{\partial_1 X_\alpha\}_\alpha$ denote connected components of $\partial_1 X$. Put $\partial_1^\pm X_\alpha =_{\mathsf{def}} \partial_1^\pm X \cap \partial_1 X_\alpha$.*

Then there is a deformation of v in N such that, for the new f-gradient-like field, both sets, $\partial_1^+ X_\alpha$ and $\partial_1^- X_\alpha$, are nonempty, and $\partial_1^+ X_\alpha$ is homeomorphic to any given domain in $\partial_1 X_\alpha$ with a nonempty complement.

In particular, for a given generic f and all α's, there exists a gradient-like field v such that one of the two properties is valid:

- *$\partial_1^+ X_\alpha$ is homeomorphic to a disk,*
- *$\partial_1^+ X_\alpha$ and $\partial_1^- X_\alpha$ are homeomorphic surfaces.*

A similar statement is valid in a category of boundary generic vector fields.

Proof. If $\partial_1 X_\alpha = \partial_1^+ X_\alpha$ (or $\partial_1 X_\alpha = \partial_1^- X_\alpha$), then we can pick a point $x \in \partial_1 X$ where $v_1 \neq 0$. Employing an argument depicted in Figure 2.5, we can deform the field v in the vicinity of x so that, with respect to the modified gradient-like field, $x \in \partial_1^- X$ ($x \in \partial_1^+ X$, correspondingly). Thus we may assume that both sets $\partial_1^- X_\alpha$ and $\partial_1^+ X_\alpha$ are nonempty.

Now, by one-surgery on $\partial_1^+ X_\alpha$ and/or $\partial_1^- X_\alpha$, we can change the topology of $\partial_1^+ X_\alpha$ at will, as long as we keep the singular sets $\Sigma_{1,\alpha}^\pm$ of both polarities nonempty. No matter how we change the sets $\partial_1^\pm X_\alpha$, we must keep $\Sigma_{1,\alpha}^+$ inside $\partial_1^+ X_\alpha$ and $\Sigma_{1,\alpha}^-$ inside $\partial_1^- X_\alpha$. In particular, we can deform the field v so that $\partial_1^+ X_\alpha$ is a 2-disk or, say, to insure that $\partial_1^+ X_\alpha$ is homeomorphic to $\partial_1^- X_\alpha$.

Note that surgery on $\partial_1^+ X$ may change the sets $\partial_3^\pm X$! □

2.4 Morse strata, convexity, and concavity

Definition 2.2. Given a boundary generic vector field v on a manifold X with boundary and an integer $s \in [1, 3]$, we say that v is *boundary s-convex* (*boundary s-concave*), if $\partial_s^+ X = \emptyset$ ($\partial_s^- X = \emptyset$)[5]. If $\partial_2^+ X = \emptyset$ ($\partial_2^- X = \emptyset$), we simply say that v is *boundary convex* (*boundary concave*). ♢

An existence of a boundary convex gradient-like field v has strong topological implications. Let S be a surface with boundary. We denote by $\mathcal{L}(S)$ a smooth 3-manifold with boundary, obtained from the product $S \times [-1, 1]$ by rounding its corners $\partial S \times \{\pm 1\}$ and by replacing the cylindrical band $S \times [-1, 1]$ with a "curved parabolic" one, as shown in Figure 2.6. By its construction, $\mathcal{L}(S) \subset S \times [-1, 1]$. The projection $\mathcal{L}(S) \to [-1, 1]$ defines a nonsingular function f. The vertical field v in $S \times [-1, 1]$ is of the f-gradient

[5]Evidently, if $\partial_s X = \emptyset$, then v is both boundary s-convex and s-concave.

type. With respect to it, $\partial_2^- \mathcal{L}(S) = \partial S$. We call the triple $(\mathcal{L}(S), f, v)$ a *lens*, based on S.

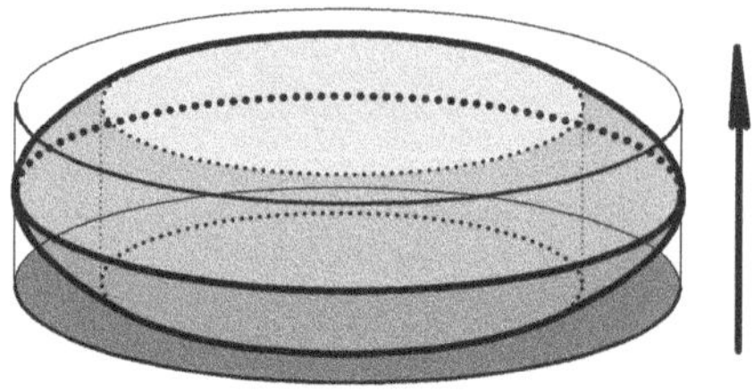

Fig. 2.6 A lens $\mathcal{L}(S)$ is convex with respect to the vertical vector field v: $\partial_2^+ \mathcal{L}(S)(v) = \emptyset$. The set $\partial_2^- \mathcal{L}(S)(v)$ is the equator of the lens.

Lemma 2.4. *If a connected* 3*-fold* X *with boundary admits a boundary convex nonsingular gradient-like vector field* v*, then* X *is diffeomorphic to a lens* $\mathcal{L}(\partial_1^+ X)$ *with the base* $\partial_1^+ X$*. Conversely, any* 3*-fold* X*, homeomorphic to a product of a compact surface with a segment, admits a boundary convex nonsingular gradient-like vector field.*

If an acyclic[6] *connected* 3*-fold* X *admits a boundary convex nonsingular gradient-like vector field, it is diffeomorphic to a* 3*-ball.*

Proof. If a boundary convex vector field v exists, we use the down flow, directed by $-v$, to produce a desired retraction of X onto $\partial_1^+ X$. We can rescale the field v by a functional factor λ to make sure that $\lambda v|_{\partial_1^+ X} = 0$. Here is the a crucial observation which validates the retraction: if $\partial_2^+ X = \emptyset$, the $(-v)$-flow is uninterrupted and continuous, so that close points in X hit $\partial_1^+ X$ at close locations. Each point has a closed segment for its trajectory, except for points in $\partial_2^- X$. For them, the trajectory is a singleton. As a result, X is diffeomorphic to the lens $\mathcal{L}(\partial_1^+ X)$ (cf. [BP], Proposition 4.2.2).

Note that the boundary convexity of v on a connected X implies that $\partial_1^+ X$ is connected.

In fact, all the handlebodies X can be obtained via the lens construction, based on an orientable compact surface, the "appropriate half" of ∂X.

When X is $\mathbb{Z}$-acyclic, a homological argument, based on the Poincaré duality, shows that $\partial_1 X$ is a homological sphere and thus is diffeomorphic to S^2. Therefore, the homology disk $\partial_1^+ X$ must be a contractible domain

[6]The homology $H_*(X; \mathbb{Z}) = 0$ for all $* > 0$.

in S^2, that is, a 2-disk. Thus, the manifold X must be shaped as a lens, one hemisphere of which is that disk. □

Figure 2.7 shows the typical behavior of a vector field v_1 in a neighborhood of $\partial_2 X$. The arcs of $\partial_2^+ X$ come in three flavors: (1) type A is bounded by a pair of points from $\partial_3^+ X$, (2) type B is bounded by a pair of points from $\partial_3^- X$, and (3) type C is bounded by a pair of mixed polarity.

This time, we play our convexity game in the dimension 2, not 3 (see Figure 2.5). At points of $\partial_3^- X$ the field v_1 in $\partial_1^+ X$ (the projection of v on the boundary) is *boundary convex* with respect to $\partial_2 X$; at the points of $\partial_3^+ X$, it is *boundary concave*.

According to Theorem 4.5, it is possible to deform v (in the space of nonsingular gradient-like vector fields) so that $\partial_3 X = \emptyset$. In other words, *there are no topological obstructions to the boundary 3-convexity and 3-concavity of traversing vector fields*! On the way to establishing this crucial fact, we need to perform 0-surgery on the finite set $\partial_3 X \subset \partial_2^+ X$.

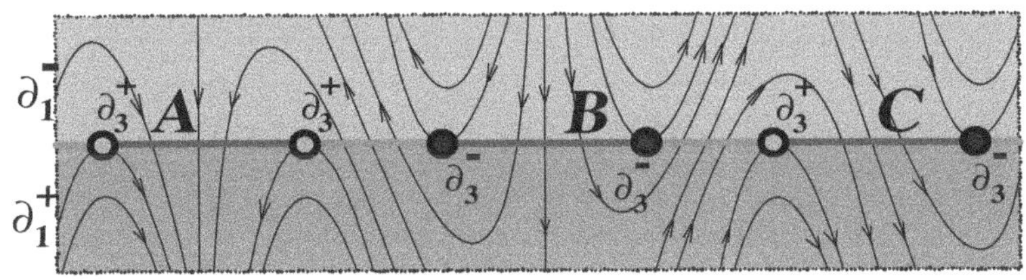

Fig. 2.7 The arcs from $\partial_2^+ X$ are of the three types: A, B, and C.

Lemma 2.5. *Let $f : X \to \mathbb{R}$ be a smooth function. Consider a vector field v on X such that $df(v) > 0$ along $\partial_1 X$. Let $C \subset \partial_2^+ X$ be an arc with one end $a \in \partial_3^+ X$, the other end $b \in \partial_3^- X$, and no other points of $\partial_3 X$ in the interior of C. Assume also that $f|_C$ has no critical points.*[7]

Then we can deform v in the vicinity of $C \subset X$ in such a way that:

- *with respect to the new f-gradient-like field $\tilde{v}$, the strata $\partial_1^+ X$ and $\partial_2 X$ remain the same as for v,*
- *the arc C changes its polarity (from being in $\partial_2^+ X$ to being in $\partial_2^- X$), and the points a, b are eliminated from the set $\partial_3 X$.*

[7]See Figure 2.7, arc C. Note that the absence of critical points of $f|_C$ implies that $f(a) < f(b)$.

Proof. Using Lemma 2.1, we choose a metric g in the vicinity of C such that $v = \nabla_g f$. Let v_1 and v_2 be the orthogonal projections of v on the tangent to $\partial_1 X$ and $\partial_2 X$ spaces, respectively.

The argument below is analogous to the one in Lemma 2.3 (see Figure 2.5). However, this time, we will deform the field v_1, while preserving both the normal to $\partial_1 X$ component $v - v_1$ and the tangent to $\partial_2 X$ component $v_2 \neq 0$ in the vicinity of C. Because the direction of the normal component $v - v_1$ remains the unchanged, the stratum $\partial_1^+ X$ and its boundary $\partial_2 X$ will be preserved, but the arc C will change its polarity from being in $\partial_2^+ X$ to being in $\partial_2^- X$.

Let us sketch the validation of these claims. As in Lemma 2.3, we take advantage of the fact that $v_2 \neq 0$ belongs to the open positive cone $\mathsf{T}^+_{x,f}$, where $x \in C$ and $\mathsf{T}^+_{x,f}$ is defined by the condition $df(\sim) > 0$. By a partition-of-unity argument, this allows, in the end of the deformation, to flip the direction of the component $v_1 - v_2$, while keeping, in the vicinity of C, the deformed component $\tilde{v}_1$ inside the half-space $\mathsf{T}^+_{x,f}$.

Finally, the vector field $\tilde{v} =_{\mathsf{def}} (v - v_1) + \tilde{v}_1 + v_2$, defined in the vicinity of C, does the job. As a result, the arc C becomes incorporated in the new $\partial_2^- X$ and its ends disappear from the set $\partial_3 X$. □

Corollary 2.2. *For any compact smooth 3-fold X, there exist a smooth boundary generic vector field v and a smooth function f on X such that $df(v) > 0$ in the vicinity of ∂X. The restriction of f to each locus $\partial_j X$, $j \in [0, 2]$, is a Morse function. Moreover, each arc $L \subset \partial_2 X$ that connects a point x, where $f|_{\partial_2 X}$ attains its minimum, to a consecutive point y, where $f|_{\partial_2 X}$ attains its maximum, has:*

- *a single point from $\partial_3 X$, which belongs to $\partial_3^- X$, provided $x \in \Sigma_2^+$ and $y \in \Sigma_2^-$,*
- *a single point from $\partial_3 X$, which belongs to $\partial_3^+ X$, provided $x \in \Sigma_2^-$ and $y \in \Sigma_2^+$,*
- *no points from $\partial_3 X$ when both $x, y \in \Sigma_2^\pm$. In this case, the polarity of the arc L is the same as the polarity of its ends x and y in Σ_2.*

The set of such pairs (f, v) is open in the space of all pairs, subject to the property $df(v) > 0$ in the vicinity of ∂X.

Proof. Let $f : X \to \mathbb{R}$ be a smooth function with no critical points residing in ∂X. Let v be its gradient vector field; thus $df(v) > 0$ in the vicinity of ∂X. By the $3D$-version of Theorem 5.2, we may smoothly perturb v to a boundary generic vector field, while preserving the "open" property $df(\sim) > 0$ along ∂X. By a smooth perturbation of f, we may assume that each function $\{f|_{\partial_j X}\}_{j\in[0,2]}$ is of the Morse type. So the maxima and the minima of $f : \partial_2 X \to \mathbb{R}$ alternate. By Lemma 2.5, we can eliminate all the arcs $M \subset \partial_2 X$ of type C (see Figure 2.7) between consecutive points $a, b \in \partial_3 X$, provided $v_2|_M \neq 0$ (recall that the polarities of a and b are opposite).

Note that $v_2 \neq 0$ in the interior of each arc L as in the hypotheses of Corollary 2.2: no critical points of $f|_L$ are present the interior of L. Therefore no arc M (with its two ends from $\partial_3 X$) of the types A and B can be contained in L. Indeed, at the two ends of M, the vectors $v = v_2$ are directed in opposite directions (see the case A and B in Figure 2.7) —a contradiction with the property $v_2|_L \neq 0$. On the other hand, the arcs of type C have been eliminated. As a result, L must contain a single point x from $\partial_3 X$. Its polarity is determined by whether v_2, the gradient of $f|_{\partial_2 X}$, points inside or outside of the arc $M \subset \partial_2^+ X$, where $x \in \partial M$.

Finally, we notice that the properties of the just constructed pair (f, v), described in Corollary 2.2, are open under smooth perturbations of pairs. □

The Morse formula for the vector fields (Theorem 2.1) helps to link the topology of $\partial_1^+ X$ with the distribution of arcs from $\partial_2^+ X$ and points from $\partial_3^\pm X$ along its boundary $\partial_2 X$.

The next lemma is similar in spirit to Theorem 4.8 from [E1]. That theorem is a very special case of the general surgery theory of folding maps (see [E1], [E2]). However, it seems that we cannot apply Eliashberg's results directly: our v-generated local folding maps $p_x : \partial_1 X \to \hat{f}^{-1}(f(x))$ have a 2-manifold target space only *locally*; in fact, a natural target space in our setting is a 2-dimensional CW-complex, typically with singularities.

Lemma 2.6. *Let X be a compact 3-fold with a non-vanishing boundary generic vector field v. Then the v-generated Morse stratification $\{\partial_k^+ X\}_{0\le k\le 3}$ of X has the following properties:*

- $\chi(\partial_1^+ X) - \chi(\partial_1^- X) = \#(\partial_3^+ X) - \#(\partial_3^- X) = 2 \cdot \deg(h : \partial_2 X \to S^1)$, *where the map h has been defined prior to Lemma 2.2.*
- $\chi(\partial_1^+ X) = \chi(X) + [\#(\partial_3^- X) - \#(\partial_3^+ X)]/2$
$= \chi(X) - \deg(h : \partial_2 X \to S^1)$.

Proof. Since $v \neq 0$ in X, the index $\mathsf{Ind}(v) = 0$. By the Morse formula (see Theorem 2.1),

$$\chi(X) - \chi(\partial_1^+ X) + \chi(\partial_2^+ X) - \chi(\partial_3^+ X) = 0.$$

Thus,

$$\chi(\partial_1^+ X) = \chi(X) + \chi(\partial_2^+ X) - \chi(\partial_3^+ X).$$

Loops in $\partial_2^+ X$ do not contribute to the Euler number $\chi(\partial_2^+ X)$, so $\chi(\partial_2^+ X) = \#(\text{arcs in } \partial_2^+ X)$. Hence,

$$\chi(\partial_1^+ X) = \chi(X) + \#(\text{arcs in } \partial_2^+ X) - \#(\partial_3^+ X).$$

Note, that the C-arcs and their ends do not contribute to the difference

$$\#(\text{arcs in } \partial_2^+ X) - \#(\partial_3^+ X)$$

—such arcs have a single end in $\partial_3^+ X$. On the other hand, the B-arcs do not contribute to $\#(\partial_3^+ X)$ and contribute -1 to the difference

$$\#(\text{arcs in } \partial_2^+ X) - \#(\partial_3^+ X).$$

Finally, the A-arcs contribute $+1$ to the difference

$$\#(\text{arcs in } \partial_2^+ X) - \#(\partial_3^+ X).$$

Therefore the difference $\chi(\partial_2^+ X) - \#(\partial_3^+ X)$ is equal to

$$\#(B\text{-arcs}) - \#(A\text{-arcs}) = \frac{1}{2}\big[\#(\partial_3^- X) - \#(\partial_3^+ X)\big].$$

By Lemma 2.2, the latter number is $-\deg(h : \partial_2 X \to S^1)$.

Recall that (see [Hopf]) for any 3-fold X,

$$\chi(X) = \frac{1}{2}\chi(\partial_1 X) = \frac{1}{2}\big[\chi(\partial_1^+ X) + \chi(\partial_1^- X)\big].$$

Replacing $\chi(X)$ with $\frac{1}{2}\big[\chi(\partial_1^+ X) + \chi(\partial_1^- X)\big]$ in the formulas above, leads to the desired relation

$$\chi(\partial_1^+ X) - \chi(\partial_1^- X) = \#(\partial_3^+ X) - \#(\partial_3^- X).$$

□

Combining Lemma 2.6 with Corollary 2.2, we get the following implication.

Corollary 2.3. $\#(\partial_3^+ X) = \#(\partial_3^- X)$ *if and only if*

$$\chi(\partial_1^+ X) = \chi(\partial_1^- X) = \chi(X).$$

When $\partial_1 X$ is connected, by deforming v, we can arrange for $\partial_1^+ X$ to be a 2-disk (see Corollary 2.1). For any such choice of v, we get

$$\#(\partial_3^+ X) - \#(\partial_3^- X) = 2 - 2 \cdot \chi(X).$$

◇

Corollary 2.4. *If a connected 3-fold X with $\chi(X) > 0$ admits a nonsingular function f with boundary 3-convex Morse data (f, v), then the restriction $f|_{\partial_1^+ X}$ must have at least $\chi(X)$ local extrema.*

Proof. The hypothesis $\partial_3^+ X = \emptyset$ implies that only B-arcs could be present in $\partial_2^+ X$. The *positive* contribution to $\chi(\partial_1^+ X)$ comes from the components of $\partial_1^+ X$, shaped as disks or spheres. If a sphere $S^2 \subset \partial_1^+ X$, then $\partial_2 X \cap S^2 = \emptyset$; the restriction of f to S^2 must have two extrema at least.

We divide the disk-shaped components of $\partial_1^+ X$ into two types:
(1) the disks D^2 with no B-arcs in their boundary (so that $\partial D^2 \subset \partial_2^\pm X$),
(2) the rest of the disks.

Any disk D^2 of the first type has the property $D^2 \cap \partial_3 X = \emptyset$ (recall that $\partial_3^+ X = \emptyset$ by the hypotheses). Since the field v is transversal to the boundary of such disk D^2 and since $df(v) > 0$ along ∂D^2, D^2 must contain at least one local extremum of $f|_{\partial_1^+ X}$.

Any disk of the second type contains at least one B-arc. Since $\partial_3^+ X = \emptyset$, from the proof of Lemma 2.6 we get $\chi(\partial_1^+ X) = \chi(X) + \#(B\text{-arcs})$. Now the lemma follows from writing down $\chi(\partial_1^+ X)$ as $\sum_\alpha 2 + \sum_\beta 1 + N$, the sum of Euler numbers of all the spheres plus the sum of Euler numbers of all the disks of the first type plus the Euler number N of the rest of components of $\partial_1^+ X$. Each disk of the second type has one B-arc at least so that the difference between the number of such B-arcs and the Euler number of the union of the second type disks is nonnegative. Therefore,

$$\sum_\alpha 2 + \sum_\beta 1 = \chi(X) + \#(\text{some } B\text{-arcs}) + \tilde{N},$$

where $\tilde{N} \geq 0$ is minus the Euler number of all components in $\partial_1^+ X$, distinct from spheres and disks. Thus $\sum_\alpha 2 + \sum_\beta 1 \geq \chi(X)$, where the LHS of this inequality represents the number of *some* local extrema in $\partial_1^+ X$. □

Corollary 2.5. *For every connected 3-fold X with a connected boundary and Euler number χ, there exist non-vanishing vector field v on X such that $\partial_1^+ X$ is a 2-disk. For such v, we get:*

$$\#(\partial_3^+ X) \geq 2\chi - 2, \quad \#(\partial_3^- X) \geq 2 - 2\chi.$$

As a result, when $\chi > 1$, the disk cannot be boundary convex with respect to the field v_1, the orthogonal projection of v on the boundary $\partial_1 X$. As χ grows, the disk $\partial_1^+ X$ becomes more "wavy".

Similarly, when $\chi < 1$, the disk cannot be boundary concave with respect to v_1, that is, $\partial_3^- X \neq \emptyset$.

Proof. By Corollary 2.1, appropriate deformations of v in the space of f-gradient-like fields will produce $\partial_1^+ X \approx D^2$. In view of Corollary 2.3, the claim follows. $\square$

The next theorem shows that the boundary convexity of Morse data (f, v) is equivalent to the possibility of special *ordering* of the $f|_{\partial_1 X}$-critical points by the magnitude of their critical values, and thus fails in general! However, if we formally attach the index $i+1$ to each $f|_{\partial_1 X}$-critical point $x \in \Sigma_1^-$ of "classical" index i, then the new self-indexing of $f|_{\partial_1 X}$ becomes possible (see [La]).

The next theorem shows that the v-flow on X and a v_1-flow on $\partial_1 X$ are strongly correlated. This correlation is an instance of the *holography phenomenon.* In particular, the boundary convexity or convexity of v can be detected in terms of v_1.

In fact, in Theorem 5.7, we will generalize the claims of Theorem 2.2 below for the gradient-like flows in any dimension; so we postpone the proof of Theorem 2.2 until we reach Chapter 5.

Theorem 2.2. *Let X be a compact 3-fold with boundary, and let $f : X \to \mathbb{R}$ be a smooth function whose restriction to ∂X is a Morse function. Let v be a boundary generic vector field on X such that $df(v) > 0$ in the vicinity of ∂X. We choose a Riemannian metric g in the vicinity of ∂X so that $v = \nabla_g f$. Consider the vector field v_1 on ∂X, the g-orthogonal projection of v on the tangent to $\partial_1 X$ planes.*

If $\partial_2^\pm X = \emptyset$, then there is no ascending trajectory $\gamma : \mathbb{R} \to \partial_1 X$ of the vector field v_1 such that

$$\lim_{t \to -\infty} \gamma(t) \in \Sigma_1^\mp \quad \text{and} \quad \lim_{t \to +\infty} \gamma(t) \in \Sigma_1^\pm.$$

Conversely, if no such v_1-trajectory γ exists, then it is possible to deform the given f-gradient vector field v to a new boundary convex/boundary concave *f-gradient vector field $\tilde{v}$.*

In particular, if $f(\Sigma_1^\pm) < f(\Sigma_1^\mp)$ as sets, then X admits boundary convex/concave Morse data (f, v).

$\diamondsuit$

Combining Theorem 2.2 and Lemma 2.3, we get (see Figure 2.6) the following claim.

Theorem 2.3. *A connected 3-fold X with boundary is a diffeomorphic to a product (with the corners being rounded) of a compact surface with a closed segment if and only if* one *of the following properties is valid:*

- *X admits a smooth non-vanishing boundary convex gradient-like vector field v.*
- *X admits a smooth nonsingular function $f : X \to \mathbb{R}$ whose restriction to the boundary $\partial_1 X$ is Morse and such that no* ascending *trajectory of the $(f|_{\partial_1 X})$-gradient field v_1 on $\partial_1 X$ links in $\partial_1 X$ a singularity from Σ_1^- to a singularity from Σ_1^+. Here v_1 is a g-orthogonal projection on $T(\partial_1 X)$ of the gradient vector field $v = \nabla_g(f)$, where g is a Riemannian metric in the vicinity of $\partial_1 X$.* ♢

In view of the remarkable validation of the Geometrization Conjecture by Perelmann [P1], [P2], the proposition below must be regarded just as an illustration. It shows which advances towards the Poincaré Conjecture are possible by the modest means of the finite-dimensional Morse Theory alone. We get the following criteria for recognizing the standard 3-balls in terms of Morse data (cf. Corollary 5.9, dealing with high dimensional flows):

Corollary 2.6. *A connected compact* acyclic *3-fold X is diffeomorphic to a 3-ball if and only if* one *of the following properties is valid:*

(1) X admits a smooth nonsingular function f whose restriction to the boundary $\partial_1 X$ is of the Morse type and such that no ascending *trajectory of the $(f|_{\partial_1 X})$-gradient field v_1 on $\partial_1 X$ links (in $\partial_1 X$) a singularity from Σ_1^- to a singularity from Σ_1^+.*

(2) X admits a smooth non-vanishing and boundary convex gradient-like vector field v.

Proof. Evidently, the standard 3-ball admits Morse data (f, v) with the properties described in (1)—(3).

By a homological argument, ∂X must be a sphere S^2, provided that X has the homology of a point. By Theorems 2.2 and 2.3, properties (1) and (2) are equivalent. Property (2) implies that X is a 3-ball. Indeed, since $\partial_1^+ X$ and X are homotopy equivalent, $\partial_1^+ X$ must be acyclic in all positive dimensions. Therefore $\partial_1^+ X$ must be a 2-ball, which, by Lemma 2.4, implies that X is a 3-ball. □

Example 2.1. Let X be the Poincaré homology 3-sphere from which a smooth 3-disk is removed. It follows from the Theorem 2.3 that, for any smooth nonsingular function $f : X \to \mathbb{R}$, there is an ascending v_1-trajectory which links a singularity from Σ_1^- to a singularity from Σ_1^+. $\diamondsuit$

2.5 Cascades, 2D-spines, and concavity

Until now, when dealing with 3-folds X, we were considering them as smooth manifolds. Now we will need to use also some structures that rely on triangulations of X. Let us give a brief overview how these different structures interact.

A priori a 3-fold X, may be equipped with one of the following competing geometric structures:

- *topological* ("TOP" for short),
- *piecewise-linear* ("PL" for short),
- *piecewise-differentiable* ("PD" or "PDIFF" for short),
- *differentiable* ("DIFF" for short).[8]

Traditionally, the structures are defined in terms of open charts, covering X, and restrictions on the changes of coordinates that occur at their intersections.

A posteriori, in dimension three, all these categories are *equivalent* (for example, see [Th], Section 3.9, for an informal and illuminating discussion)! For instance, any compact 3-dimensional TOP-manifold admits unique PL- and DIFF-structures (see [Mu], Chaper II).

In the 1950's Edwin E. Moise and R. H. Bing proved that topological 3-manifolds are always triangulable by an essentially unique (up to PL-equivalence) triangulation (see [Moi]).

Let P be a locally compact simplicial complex (polyhedron), embedded in a finite-dimensional real affine space V, so that each simplex in P is a convex closure of its vertexes.

A piecewise linear map from a subset U of an affine space into another affine space is, by definition, a restriction on U of a simplicial map, defined on some polyhedron $P \supset U$.

A *triangulation* of a topological space X is a homeomorphism $T : P \to X$, where the source space P is a polyhedron.

[8]These notations are reserved for the pseudogroup of CAT-homeomorphisms between open sets in $\mathbb{R}^n$; however, PDIFF, the piecewise-differentiable homeomorphisms of open sets in $\mathbb{R}^n$ do not form a pseudogroup, although they do *generate* some pseudogroup!

Now let us assume that the simplicial polyhedron P is a PL-*manifold.* Recall that K is a PL-manifold of dimension $n+1$ if and only if the link of each k-simplex in K is homeomorphic to a $(n-k)$-sphere. Furthermore, K is a PL-manifold if the link of each vertex in K is homeomorphic to a n-sphere.

A PL-*structure* on a topological manifold X is represented by a homeomorphism $h : P \to X$, where P is a PL-manifold. Two such homeomorphisms, $h : P \to X$ and $h' : P' \to X$, represent the same PL-structure if there exists a PL-homeomorphism $f : P' \to P$ [9] such that $h' = f \circ h$.

When we talk about a *triangulation* T of a PL-manifold X, we have in mind a *piecewise linear triangulation*, that is, a homeomorphism $T : P' \to X$ which defines a PL-structure on X, equivalent to the one defined by the given homeomorphism $h : P \to X$.

A *differentiable (or smooth) triangulation* of a differentiable manifold X is represented by a homeomorphism $T : P \to X$ which is a diffeomorphism on each open simplex in P. The smooth structures of open simplexes are induced from the smooth structure on the ambient affine space V in which P resides. The, so called, PDIFF-triangulations form a busy bridge between the PL- and DIFF-worlds.

In fact, any smooth 3-fold admits a smooth triangulation (see, for example, [Th], Theorem 3.10.2).

As a default, when a vector field on X is mentioned, X is assumed to be a smooth manifold.

Let P be a simplicial complex and $K \subset P$ its subcomplex. For the time being, we adopt a working definition of a *regular neighborhood* $\mathcal{N}(K, P)$: it is the star of K in the second barycentric subdivision $\beta^2 P$ of P, that is, the union of all simplexes in $\beta^2 P$ that have a nonempty intersection with K.

Now we are in position to introduce the notion of *spine* for a manifold with boundary. This notion is central for many constructions and results from Sections 2–4 and beyond (in Chapter 8, we will revisit the notion of spines for manifolds of any dimension). There are several competing definitions of spines (cf. [Ma], Theorem 1.1.7). We will adopt the one that suits our goals better.

Definition 2.3. Let X be a compact manifold with boundary. Let P be a compact polyhedron (simplicial complex), and $K \subset P$ a subcomplex

[9] After appropriate subdivisions of P and P', f becomes an isomorphism of simplicial complexes.

of codimension 1 at least. We assume that there is a homeomorphism $h : P \to X$ which defines a PL-structure on X.

- We say that the set $h(K) \subset X$ is a *spine* of X, if there is a simplicial map $\pi : \partial P \to K$ which is the identity on $\partial P \cap K$ and such that the mapping cylinder

$$\mathsf{cyl}(\pi) =_{\mathsf{def}} \big((\partial P \times [0,1]) \coprod K\big)/\{q \times \{0\} \sim \pi(q), \text{ where } q \in \partial P\},$$

  equipped with its natural simplicial structure, admits a simplicial homeomorphism $H : \mathsf{cyl}(\pi) \to P$, so that the restriction of H to $\partial P \times \{1\}$ is the identity map.
- We say that $h(K) \subset X$ is a *piecewise smooth spine* of a smooth manifold X, if it is a spine in the sense of the previous bullet, with both homeomorphisms $h : P \to X$ and $h \circ H : \mathsf{cyl}(\pi) \to X$ being smooth triangulations. ♢

Remark 2.1. Note that $h(K)$ is a strong deformation retract of X and that $X \setminus h(K)$ is homeomorphic to the product $\big(\partial X \setminus (h(K) \cap \partial X)\big) \times (0,1]$. ♢

Perhaps, that the most common definition of spine $K \subset X$ is given in terms of *elementary collapses* of cells or simplexes in which the ambient space X is divided. Let us review the relevant basic notions.

Let P be a simplicial complex (a polyhedron) and σ a simplex/cell in P. An elementary collapse "$\sigma \searrow \delta\sigma$" is a piecewise linear retraction $r : \sigma \to \delta\sigma$ onto a portion $\delta\sigma$ of the boundary $\partial\sigma$. The complementary portion $\sigma' =_{\mathsf{def}} \overline{\partial\sigma \setminus \delta\sigma} \neq \emptyset$ must be again a simplex or a cell. Moreover, σ is the only simplex/cell in P that contains the interior of σ'. We call such pair $\sigma' \subset \sigma$ *free*. For each free pair $\sigma \supset \sigma'$, we replace P with a smaller simplicial complex $Q =_{\mathsf{def}} P \setminus (\sigma^\circ \cup \sigma'^\circ)$ and write "$P \searrow Q$".

We notice that the polyhedron P from Definition 2.3 is collapsible onto K: indeed, simplexes σ' from $\partial P \times \{1\}$ can be used as the free faces for elementary collapses of the cells $\sigma =_{\mathsf{def}} \sigma' \times [0,1] \subset \mathsf{cyl}(\pi) \approx P$. Thus X is collapsible onto $h(K)$.

The converse is also true: if a PL-manifold P collapses onto its subcomplex K, then P is PL-homeomorphic to a mapping cylinder of a PL-map $\pi : \partial P \to K$ (for example, see [Ma], Theorem 1.1.7). In such a case, $h(K)$ is a spine of X in the sense of Definition 2.2 (where $h : P \to X$ is a homeomorphism which defines a PL-structure in X).

Here is a sketch of an argument that validates this claim.

To simplify the notations, from now and on, we do not distinguish religiously between $K \subset P$ and its image $h(K) \subset X$, the spine.

Let $X_1 \subset X$ both be n-dimensional PL-manifolds such that $X \setminus \mathrm{int}(X_1)$ is a PL-ball B^n whose boundary is divided into two complementary $(n-1)$-balls, $D^{n-1} \subset \partial X$ and $B^{n-1} \subset \partial X_1$. Then $X \searrow X_1$ with D^{n-1} serving as the free portion through which the elementary collapse is executed. In such a case, we say that $X \searrow X_1$ is an *elementary shelling*. A sequence of elementary shellings is called a *shelling*.

If X shells to X_1, then there exists a simplicial homeomorphism $h : X \to X_1$ which is an identity outside any given neighborhood of $X \setminus X_1$ in X (see [RS], Lemma 3.25).

Next crucial step is to prove that, if $K_1 \subset K \subset X$ is a pair of polyhedra and $K \searrow K_1$, then a regular neighborhood $\mathcal{N}(K, X)$ shells to a regular neighborhood $\mathcal{N}(K_1, X)$ ([RS], Theorem 3.26). In particular, if $X \searrow K$, then X shells to $\mathcal{N}(K, X)$. Thus X and $\mathcal{N}(K, X)$ are simplicially homeomorphic via a homeomorphism that is an identity in a smaller regular neighborhood $\mathcal{N}_\beta(K, X)$ of K.

Finally, we notice that any regular neighborhood $\mathcal{N}(K, X)$ can be canonically identified with a mapping cylinder of a simplicial map $\pi_K : \partial\mathcal{N}(K, X) \to K$. Let us glance at the construction of π_K. Its construction can be localized to n-dimensional simplicies Δ in which X is subdivided. Let $Q =_{\mathsf{def}} K \cap \Delta$. Consider $\mathcal{N}(Q, \Delta) = \mathcal{N}(K, X) \cap \Delta$, the star of Q in the second barycentric subdivision $\beta^2\Delta$ of Δ. Let us denote by $Q^\vee$ the set of simplexes in the first barycentric subdivision $\beta\Delta$ that have an empty intersection with Q. Then any simplex $\tilde{\Delta} \subset \beta\Delta$ is a join of a unique simplex $\Delta' \subset \beta Q$ with a unique simplex $\Delta'' \subset Q^\vee$. Therefore, each point $x \in \tilde{\Delta} \setminus (Q^\vee \cap \tilde{\Delta})$ has a unique line $l_x \subset \tilde{\Delta}$ that passes through x and Δ'. Since $\mathcal{N}(Q, \Delta) \cap Q^\vee = \emptyset$, the construction $x \Rightarrow l_x \cap Q$ defines a simplicial map $\pi_Q : \mathcal{N}(Q, \Delta) \to Q$ which gives $\mathcal{N}(Q, \Delta)$ a structure of the mapping cylinder $\mathsf{cyl}(\pi_Q : \partial\mathcal{N}(Q, \Delta) \to Q)$. This local construction of π_Q is automatically coherent across various n-simplexes Δ.

Therefore, the notion of a spine $K \subset X$, based on a collapse $X \searrow K$, and the one, given in Definition 2.2, are equivalent.

Remark 2.2. Any locally compact simplicial complex K embeds in $\mathbb{R}^{2\dim(K)+1}$, where its regular neighborhood $\mathcal{N}(K)$ is a PL-manifold whose spine is K. As a result, there are no restrictions on the topology of K coming from the fact that K is a spine of a *manifold* X whose dimension is

$2\dim(K) + 1$ at least. However, if a spine K has a low codimension in an ambient manifold X, its local topology could be rather restricted; this is especially pronounced in when $\dim(K) = \dim(X) - 1$, the most important case for us (see Figure 2.14).

To illustrate this observation, consider a *nontrivial* PL-bundle $E \to S^k$ over a sphere S^k with a disk fiber D^k (for example, take $E \to S^1$ to be the Möbius band bundle). Let $K = E \cup_{S^k} D^{k+1}$, the disk D^{k+1} being attached to the zero section $\sigma \approx S^k$ of E with the help of a homeomorphism $\phi : \partial D^{k+1} \to \sigma$. Then K cannot be a spine of any $(2k+1)$-dimensional manifold X: otherwise the nontrivial bundle E would have to extend over the contractible space D^{k+1}, an impossible task. $\diamondsuit$

Our next goal is to use "generic" non-vanishing gradient-like vector fields v on X in order to construct quite special spines of X. We call such special spines *gradient*.

For a given boundary generic v, consider locus $\partial_2^+ X$. It is comprised of a finite number of disjoint arcs or loops $\{E_j\}$. For each connected curve $E_j \subset \partial_2^+ X$, consider the set of points in X which can be reached from E_j by moving down along the trajectories of $-v$. Denote by W_j the closure in X of this set. We call such a set W_j a *waterfall* (see Figure 2.10) The union $\cup_j W_j$ of all waterfalls is called a *cascade* and is denoted by $\mathcal{C}(\partial_2^+ X)$ (see Figure 2.11). We denote by $\mathcal{C}(\partial_3^- X)$ the finite union of the downward trajectories through the points of $\partial_3^- X$. We call these trajectories *free*.

Figures 2.9, 2.11, and 2.12 show mechanisms by which generic waterfalls are created, as well as their typical shapes.

Definition 2.4. Let M, N be smooth manifolds. A smooth map $f : M \to N$ is called *stable* if there is an open neighborhood $\mathcal{O}(f)$ of f in the space $C^\infty(M, N)$ of smooth maps (equipped with the Whitney topology) such that, for any $g \in \mathcal{O}(f)$, there exist smooth diffeomorphisms $\psi : M \to M$ and $\phi : N \to N$ so that $\phi \circ f = g \circ \psi$. $\diamondsuit$

A reader-friendly account of the properties of stable maps can be found in the book of Golubitsky and Guillemin [GG], Chapters III—V. For now, we need to know only that the stable maps form an open set in the space $C^\infty(M, N)$.

Before proceeding to the construction of *gradient spines* in Theorem 2.5 below, we need to establish some general properties of vector fields, suitable for the production of spines. The next definition is a very special case of

Definition 6.3 that introduces the notion of *traversally generic* vector fields on manifolds with boundary.

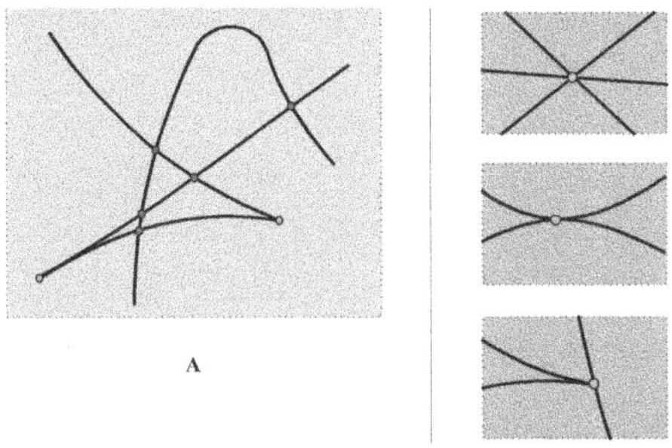

Fig. 2.8 Diagram A shows a traversally generic pattern of the locus $\partial_2^+ X$, being projected to a plane, locally transversal to the v-flow. Diagrams B, C and D show typical patterns of the projection in which the traversing genericity is violated.

Definition 2.5. We say that a boundary generic non-vanishing gradient-like field v on X is *traversally generic* if:

- for any two distinct points $x, y \in \partial_2^+ X$ that belong to the same v-trajectory γ, the surfaces of the waterfalls W_x and W_y are *transversal* at each point of $\gamma \cap W_x \cap W_y$,
- locally, no three waterfalls share a point,
- for each point $x \in \partial_3 X$, the v-trajectory γ_x through x has the property $\gamma_x \cap \partial_2 X = x$; that is, each $x \in \partial_3 X$ belongs locally to a single waterfall.

(See Figures 2.8 and 6.2 which illustrate the scope of this definition.)

Here is an *equivalent* and more economic definition, which is also tailed to dimension three. We embed X properly into a slightly bigger manifold $\hat{X}$ and extend v smoothly into $\hat{X}$. We denote this extension by $\hat{v}$. Let $U \subset \hat{X}$ be an open neighborhood of a v-trajectory $\gamma \subset X$, consisting of $\hat{v}$-trajectories, and S a transversal section of the flow in U. We say that a non-vanishing gradient-like field v on X is *traversally generic* if, for any v-trajectory γ and a pair $U \supset S$ as above, the $\hat{v}$-flow-governed projection $\partial_1 X \cap U \to S$ is *stable*. $\diamondsuit$

The next theorem is a special case of Theorem 6.7 which deals with the multi-dimensional traversally generic vector fields.

Theorem 2.4. *Let X be a smooth connected and compact 3-fold with boundary.*

- *For any non-vanishing gradient-like vector field v on X, there exists a C^∞-small perturbation of v that produces a traversally generic vector field.*
- *The traversally generic fields form an open set in the space of all non-vanishing gradient-like vector fields on X.*

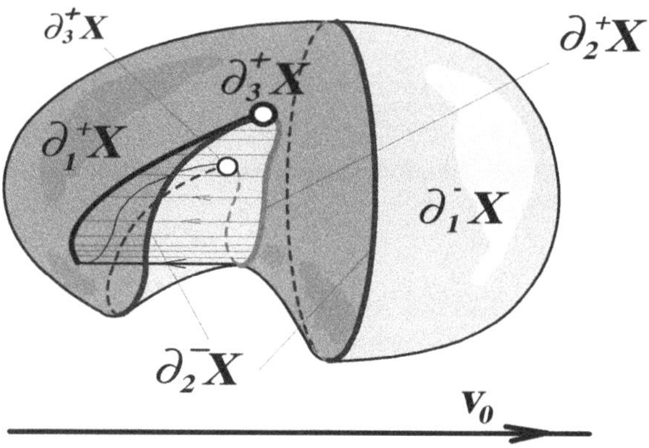

Fig. 2.9 Killing the 1-cycle in the annulus $\partial_1^+ X$ (the dark area) by attaching to it the 2-cell $\mathcal{C}(\partial_2^+ X)$. As a result, the spine $\partial_1^+ X \cup \mathcal{C}(\partial_2^+ X)$ of the 3-ball X is produced.

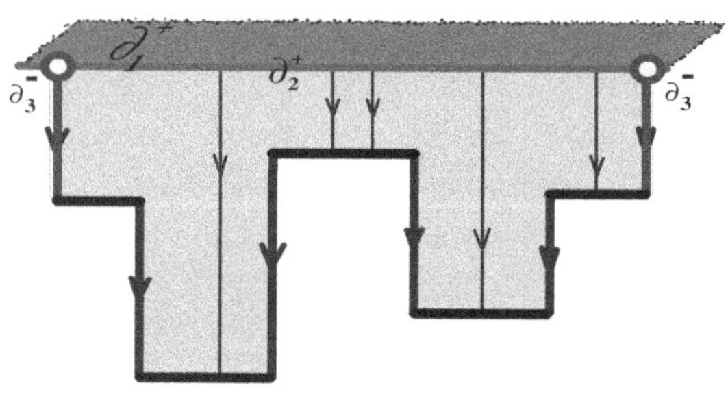

Fig. 2.10 A waterfall of an A-arc in $\partial_2^+ X$.

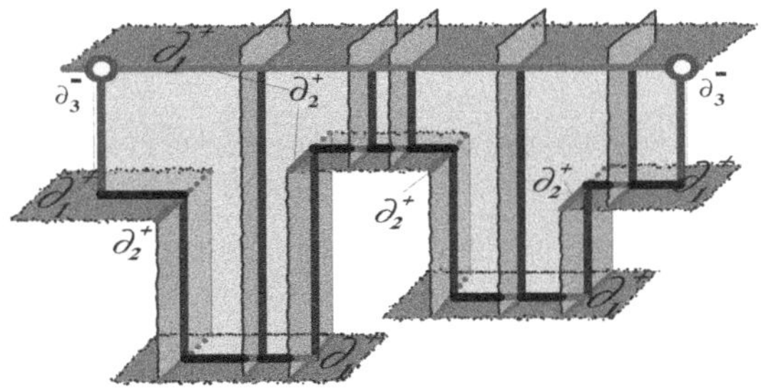

Fig. 2.11 The same waterfall in a cascade.

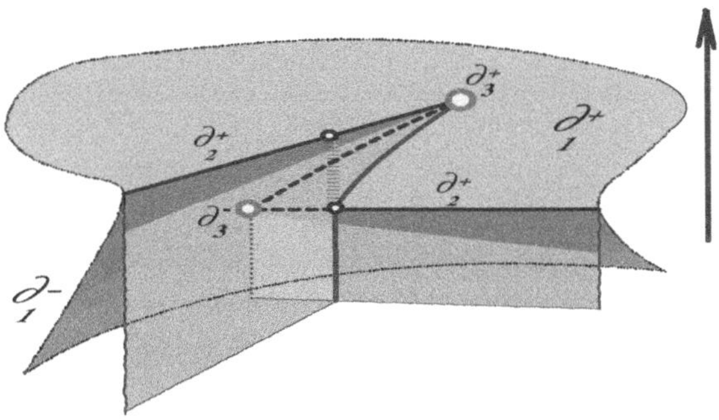

Fig. 2.12 A cascade of two arcs from $\partial_2^+ X$ with two cusps from $\partial_3^+ X$ and $\partial_3^- X$. Note the single trajectory shared by the two waterfalls.

Proof. We need to show that, for a vector field on a 3-fold, being traversally generic is an "open and dense property" in the space of all non-vanishing vector fields of the gradient type. This fact can be established by rather standard, but involved transversality arguments. Let us just sketch the arguments, which will be presented in their full generality in the proof of Theorem 6.7, the high-dimensional analogue of Theorem 2.4.

We start with a non-vanishing boundary generic vector field v of the gradient type. Then we extend v into a larger manifold $\hat{X}$ so that the embedding $X \subset \hat{X}$ is proper, and the extension $\hat{v}$ of v remains non-vanishing and of the gradient type.

Consider a special finite open cover $\mathcal{U}$ of X and its refinement $\mathcal{V}$ such that:

- each element $U_\alpha \in \mathcal{U}$ is an open regular neighborhood of some $\hat{v}$-trajectory in $\hat{X}$ and an union of $\hat{v}$-trajectories[10],
- each U_α is equipped with a smooth and transversal section S_α of the $\hat{v}$-flow, so that U_α is diffeomorphic to the product $S_\alpha \times (0,1)$, where the diffeomorphism is taking each $\hat{v}$-trajectory in U_α onto the fiber of the projection $S_\alpha \times (0,1) \to S_\alpha$,
- each section S_α is diffeomorphic to a 2-disk,
- each S_α contains properly an open 2-disk T_α so that the $\hat{v}$-trajectories through T_α form an open set $V_a \subset \hat{X}$, and the collection $\mathcal{V} =_{\mathsf{def}} \{V_\alpha\}_\alpha$ is an open cover of X as well.

The correspondence $\gamma \to \gamma \cap S_\alpha$ defines a projection $p_\alpha(\hat{v}) : U_\alpha \to S_\alpha$ whose fibers are $\hat{v}$-trajectories in the interior of $\hat{X}$.

Recall that the stable smooth maps $\pi : \mathbb{R}^2 \to \mathbb{R}^2$ have fold-type and cusp-type singularities only (see [W]). In local coordinates, they are given by the already familiar formulas: $(x, y) \to (x, y^2)$ and $(x, y) \to (x, y^3 - xy)$, correspondingly. Moreover, the π-images of the folding loci are curves with transversal intersections, and no triple intersections are allowed. The π-images may have isolated cusps, each cusp point locally belongs to a single (cuspidal) curve. Such stable maps $\pi : \mathbb{R}^2 \to \mathbb{R}^2$ are open and dense in the space of all smooth maps $\phi : \mathbb{R}^2 \to \mathbb{R}^2$ (note that the property of stable maps being dense fails in high dimensions (see [GG], Fig. 4 on page 163)).

If $\hat{v}$ is already a traversally generic field with respect to $\partial_1 X$, by Definition 2.5, the maps $p_\alpha(\hat{v}) : \partial_1 X \cap U_\alpha \to S_\alpha$ must have only the fold and cusp type singularities, so that for all α's the following properties are valid:

(1) $p_a(\hat{v})$ projects the folding loci $\partial_2 X \setminus \partial_3 X$ onto transversally intersecting curves in S_α with no triple intersections,
(2) all the $p_a(\hat{v})$-images of points from $\partial_3 X$ (the cusp points) are distinct,
(3) no point of the pure stratum $\partial_2 X \setminus \partial_3 X$ projects to a cusp point in S_α.

(2.3)

In other words, the property of $\hat{v}$ being traversally generic can be formulated just in terms of the projections $\{p_\alpha(\hat{v}) : \partial_1 X \cap U_\alpha \to S_\alpha\}_\alpha$ for *any* choice of sections $\{S_\alpha\}_\alpha$ as above. In such formulation, by the properties of the cover $\mathcal{V}$, we can use even smaller sections $\{T_\alpha \subset S_\alpha\}_\alpha$ as local devices for probing the traversally generic vector fields.

[10]We call such neighborhoods *v-adjusted.*

A C^∞-small perturbation $\hat{v}'$ of any given non-vanishing gradient-like vector field $\hat{v}$ on $\hat{X}$ produces a C^∞-small perturbations $\{p_\alpha(\hat{v}')\}_\alpha$ of the maps $\{p_\alpha(\hat{v})\}_\alpha$, where each map $p_\alpha(\hat{v}')$ is restricted to the surface $\delta V_\alpha =_{\mathsf{def}} \partial_1 X \cap V_\alpha$. (In fact, by the choice of the cover $\mathcal{V}$, each surface δV_α can be made diffeomorphic to a disjoint union of open 2-disks.)

By the choice of $\mathcal{V}$, the tubular solids $V_\alpha =_{\mathsf{def}} (p_\alpha(\hat{v}))^{-1}(T_\alpha)$ form a finite cover of X and thus of $\partial_1 X$. The perturbations $\hat{v}'$ of $\hat{v}$ have been chosen so small that S_α still remains transversal to all $\hat{v}'$-trajectories through δV_α. So we are dealing with the maps $p_\alpha(\hat{v}) : \delta V_\alpha \to T_\alpha \subset S_\alpha$ and their perturbations $p_\alpha(\hat{v}') : \delta V_\alpha \to S_\alpha$ (the perturbed map $p_\alpha(\hat{v}')$ may have a smaller domain than the map $p_\alpha(\hat{v}) : U_\alpha \to S_\alpha$, but is still targeting the same section S_α).

Using the stable properties (1)–(3) from (2.3) of the maps $\{p_\alpha(\hat{v})\}_\alpha$, we conclude that, for $\hat{v}'$ sufficiently close to $\hat{v}$, the maps $\{p_\alpha(\hat{v}')\}$ must still poses properties (1)–(3), which imply that the field $\hat{v}'$ is traversally generic in each set, formed by the $\hat{v}'$-trajectories through the surface δV_α. Note that this is the point where the arguments for flows on 3-folds are much simpler than the parallel arguments for high-dimensional flows that lead to Theorem 6.7.

Since $\{\delta V_\alpha\}_\alpha$ form a cover of $\partial_1 X$, each $\hat{v}'$-trajectory γ' hits some surface δV_α. So the perturbed maps $\{p_\alpha(\hat{v}') : \delta V_\alpha \to S_\alpha\}_\alpha$ insure that $\hat{v}'$ has the desired properties (2.3) in the vicinity of each trajectory γ', and thus $\hat{v}'$ is traversally generic everywhere. Therefore, the traversally generic vector fields form an open set in the space of all non-vanishing gradient-like vector fields on X.

Now we are going to show that the traversally generic vector fields form a dense set in the space of all non-vanishing gradient-like vector fields on X.

First, we perturb a given non-vanishing gradient-like vector field v to insure that the Morse strata $\{\partial_j X\}_{1 \leq j \leq 3}$ for the perturbed field will be closed manifolds, and that $\{\partial_j^\pm X\}_{1 \leq j \leq 3}$ will be their compact submanifolds; in other words, we perturb v to a non-vanishing gradient-like and *boundary generic* vector field. This can be arranged by first perturbing v to v_1 so that, as a section $\bar{v}_1$ of the quotient 1-bundle $T(X)/T(\partial_1 X)|_{\partial_1 X}$, $\bar{v}_1$ is transversal to the zero section. The set $\partial_2 X(v_1)$ is the locus where the section v_1 vanishes. Then, by an additional perturbation v_2 of v_1, which does not violate the previous transversality of sections and has the property $\partial_2 X(v_2) = \partial_2 X(v_1)$, it is possible to arrange that v_2, viewed as a section $\bar{v}_2$ of the quotient 1-bundle $T(\partial_1 X)/T(\partial_2 X(v_1))|_{\partial_2 X(v_1)}$, will be transversal to the zero section. Now $\partial_3 X(v_3)$ is the locus where the section $\bar{v}_2$ vanishes.

Let us denote by v' the perturbed vector field. It satisfies this hierarchical 3-staged transversality. Note that any field sufficiently close to v' still satisfies the same 3-staged transversality. So we may assume that $\hat{v}'$ is boundary generic, non-vanishing, and of the gradient type.

We employ again the special covers $\mathcal{U}, \mathcal{V}$ of X (whose elements are built out of $\hat{v}'$-trajectories) and the transversal sections $\{S_\alpha\}$ of the $\hat{v}'$-flow in the tubular sets $\{U_\alpha \supset V_\alpha\}_\alpha$. We will perturb $\hat{v}'$ gradually by moving from one element of the cover $\mathcal{U}$ to the next, each sequential perturbation being supported in a single tube $U_\alpha \subset \hat{X}$.

Let B be the set of indices that enumerate some elements of the finite cover $\mathcal{U}$. Assume that $\hat{v}'$ is such that for each $\beta \in B$, the map $p_\beta(\hat{v}') : \delta V_\beta \to S_\beta$ already has the properties that are listed in (2.3). Take any index $\alpha \notin B$. Then we perturb the field $\hat{v}'$ in U_α so that the perturbation $\hat{v}''$ will preserve the desired properties (2.3), already achieved for the maps $\{p_\beta(\hat{v}') : \delta V_\beta \to S_\beta\}_{\beta \in B}$, and thus for the maps

$$\{p_\beta(\hat{v}') : \delta V_\alpha \cap \delta V_\beta \to S_\beta\}_{\beta \in B}.$$

In addition, we will seek properties (2.3) for the map $p_\alpha(\hat{v}'') : V_\alpha \to S_\alpha$.

Let us sketch how the desired perturbation $\hat{v}''$ of $\hat{v}'$ in U_α can be constructed; for more details the reader should examine the proof of Theorem 6.7.

Consider the connected components $\{\delta V_{\alpha,i}\}_i$ of δV_α, each of which is a 2-disk $D_{\alpha,i}$, mapped by $p_\alpha(\hat{v}')$ to S_α. For a narrow tube V_α, the maps $p_\alpha(\hat{v}') : D_{\alpha,i} \to S_\alpha$ can be of three types: a folding map, a cuspidal map, or a surjection. The fact that these three types cover all the local possibilities for $p_\alpha(\hat{v}')$ follows from the 3-staged transversality of $\hat{v}'$ with respect to the Morse stratification (see Lemma 6.1). Therefore, each map $p_\alpha(\hat{v}') : D_{\alpha,i} \to S_\alpha$ automatically has properties (2.3).

Next, we separate the components $\{\delta V_{\alpha,i}\}_i$ by smooth transversal sections $S_{\alpha,i}$ of the $\hat{v}'$-flow in U_α so that $\delta V_\alpha \cap (\coprod_i S_{\alpha,i}) = \emptyset$. The desired perturbation $\hat{v}''$ is supported in the vicinity of $\coprod_i S_{\alpha,i}$ and does not affect the vector field in the vicinity of δV_α. It is delivered by the action on $\hat{v}'$ of sufficiently small and *correlated* diffeomorphisms ("twists") $\{\phi_{\alpha,i} : Z_{\alpha,i} \to Z_{\alpha,i}\}_i$, where $Z_{\alpha,i}$ denotes a small regular neighborhood of the "separating" section $S_{\alpha,i}$. This neighborhood $Z_{\alpha,i}$ is diffeomorphic to the cylinder $S_{\alpha,i} \times [-\epsilon, +\epsilon]$ and is built out of segments of $\hat{v}'$-trajectories. The diffeomorphisms $\{\phi_{\alpha,i}\}_i$ preserve the field $\hat{v}'$ along the boundaries $\{\partial Z_{\alpha,i}\}_i$, so that the fields $\hat{v}'' =_{\mathsf{def}} (D\phi_{\alpha,i})_*(\hat{v}')$ in $\{Z_{\alpha,i}\}_i$ and the field

$\hat{v}'$ in $\hat{X} \setminus (\coprod_i Z_{\alpha,i})$ match smoothly along the boundaries $\{\partial Z_{\alpha,i}\}_i$.

The purpose of the twists $\{\phi_{\alpha,i}\}_i$ is to place the $p_\alpha(\hat{v}'')$-images of the folding curves and cusp points from $\{\delta V_{\alpha,i}\}_i$ in general position in S_α (see Figure 2.8, diagrams B, C, D), that is, to assure properties (2.3) for the map $p_\alpha(\hat{v}'')$.

Let us stress that, as we perturb $\hat{v}'$ to $\hat{v}''$, we make sure that the desired and *stable* properties of maps $\{p_\beta(\hat{v}') : \delta V_\beta \cap \delta V_\alpha \to S_\beta\}_{\beta \in B}$ are preserved for $\{p_\beta(\hat{v}'') : \delta V_\beta \cap \delta V_\alpha \to S_\beta\}_{\beta \in B}$.

This completes the inductive step $B \Rightarrow B \cup \alpha$ of the argument, as well as the entire proof. □

The following theorem describes how any traversally generic vector field v canonically produces a *gradient spine.* Unlike the spines in the first bullet of Definition 2.3, the gradient spines are descendants of the smooth structure on X, and not objects, defined in terms of the simplicial/cellular structure on X. This difference in the nature of the two notions generates some tension to be resolved. Although the gradient spines $K \subset X$ are produced *canonically* by an appropriate vector field, the retraction $r : X \to K$—the key ingredient of the general spines' definition—is far from being canonical. This is another focal point of the tension.

Theorem 2.5. *Let X be a compact connected 3-fold with boundary, and v a traversally generic vector field on X. Then the 2-dimensional set*

$$K(v) =_{\mathsf{def}} \partial_1^+ X \ \cup \ \mathcal{C}(\partial_2^+ X)$$

is a spine of X. Here the cascade $\mathcal{C}(\partial_2^+ X)$ is a union of downward $(-v)$-trajectories through the points of the locus $\partial_2^+ X$.

Proof. First we will show that the set $K = \partial_1^+ X \cup \mathcal{C}(\partial_2^+ X)$ is a subcomplex of some smooth triangulation of X. As we move towards this goal, we will not distinguish between the triangulations, as homeomorphisms with polyhedra as sources, and the images of such homeomorphisms in the target spaces.

Let $\mathcal{H}$ be the pattern in the surface $\partial_1^+ X$, generated by the curves that form the set $\partial_2 X$ and the curves that form the set $\partial_1^+ X \cap \mathcal{C}(\partial_2^+ X)$ (the two sets share the locus $\partial_2^+ X$). The first diagram in Figure 2.20 shows an example of such a pattern $\mathcal{H}$.

We denote by Σ the union of all the cusp points from $\partial_3^+ X$, the flow-generated projections of the cusp points from $\partial_3^- X$, and the trivalent vertexes of $\mathcal{H}$. The latter are produced by the self-intersection points of the

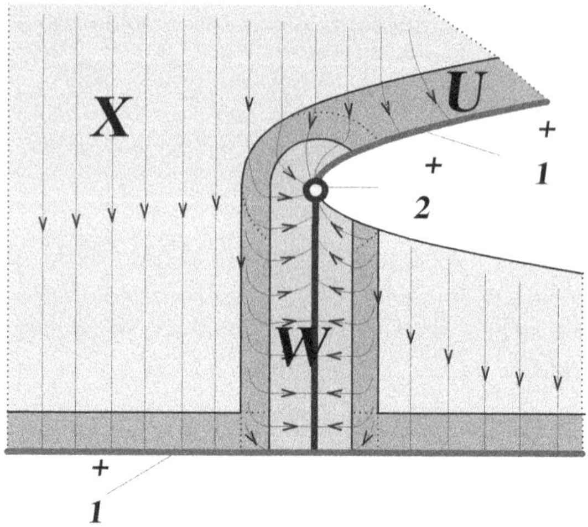

Fig. 2.13 A typical section of the gradient spine $K = \partial_1^+ X \cup \mathcal{C}(\partial_2^+ X)$ by a plane, transversal to $\partial_2^+ X$. The arrows indicate a retraction $r : X \to K$.

projections of $\partial_2^+ X$ on a transversal to the flow section (as shown in Figure 2.8, diagram A); each transversal crossing "X" contributes the trivalent configuration "T" to $\mathcal{H}$ by dropping one of the four rays that form the cross.

Choose a smooth triangulation T of $\partial_1^+ X$ such that the points of Σ are among the vertexes of T and the closures of smooth arcs from $\mathcal{H} \setminus \Sigma$ are unions of edges from T. Moreover, by refining T, if necessarily, we can assume that the map $\pi : \partial_2^+ X \to \mathcal{H}$, defined by the downward flow according to the formula

$$\pi(x) =_{\mathsf{def}} \gamma_x \cap (\partial_1^+ X \setminus \partial_2^+ X),$$

where $x \in \partial_2^+ X$ and γ_x denotes the trajectory through x, is simplicial.

We construct a simplicial 2-complex $\mathcal{T}(v)$ from T—the smoothly triangulated surface $\partial_1^+ X$—by identifying each pair of edges e, e' in T such that $\pi(e) = e'$. This identification generates a triangulation of $\mathcal{T}(v)$ and the simplicial quotient map $T \to \mathcal{T}(v)$ of the two 2-complexes. We may think of $\mathcal{T}(v)$ as *the space of v-trajectories*.

Consider the map $\Gamma : X \to \mathcal{T}(v)$ that takes each $x \in X$ to a point $\Gamma(x)$ in $\mathcal{T}(v)$ that is represented by the set $\gamma_x \cap \partial_1^+ X$ (of cardinality 2 at most).

For each simplex $\Delta \subset \mathcal{T}(v)$, consider its preimage $\mathbf{\Delta} =_{\mathsf{def}} \Gamma^{-1}(\Delta)$ under the map Γ. For each Δ in $\mathcal{T}(v)$ and its interior Δ°, the space $\mathbf{\Delta}^\circ =_{\mathsf{def}}$

$\Gamma^{-1}(\Delta^\circ)$ is either diffeomorphic to the prism $\Delta^\circ \times [0,1]$ or, in the case of $\Delta \subset \partial_2^- X$, to Δ° itself.

Note that the intersection $\mathbf{\Delta} \cap \partial_1^+ X$ is already triangulated by T, so that $\Gamma : \mathbf{\Delta} \cap \partial_1^+ X \to \Delta$ is a simplicial map. It is possible to extend this triangulation T_Δ of $\mathbf{\Delta} \cap \partial_1^+ X$ to a triangulation $\mathbf{T}_{\mathbf{\Delta}}$ of $\mathbf{\Delta}$ in such a way that $\Gamma : \mathbf{\Delta} \to \Delta$ will be a simplicial map. By an inductive argument, these extensions $\mathbf{T}_{\mathbf{\Delta}}$ can be correlated across all simplexes Δ: first, we start with the 0-skeleton $\mathcal{T}^{(0)}(v)$ of $\mathcal{T}(v)$ (for which the obvious extension is provided by T), then we proceed to the 1-skeleton $\mathcal{T}^{(1)}(v)$ of $\mathcal{T}(v)$, and finally, to the 2-skeleton $\mathcal{T}^{(2)}(v)$ of $\mathcal{T}(v)$. At each stage of the induction, we pick a point $a_\Delta \in \Delta^\circ \times \{0\} \subset \mathbf{\Delta}$ and use a cone with the apex at a_Δ over the already manufactured triangulation of the boundary $\partial\mathbf{\Delta}$ to subdivide $\mathbf{\Delta}$. This refinement must be accompanied by using the point $\Gamma(a_\Delta) \in \Delta^\circ$ as an apex from which a similar subdivision of Δ is performed.

Note that this particular way of extending the triangulation from $\Gamma^{-1}(\mathcal{T}^{(0)}(v))$ to $\Gamma^{-1}(\mathcal{T}^{(1)}(v))$ and then to $\Gamma^{-1}(\mathcal{T}^{(2)}(v))$ is designed so that the $(-v)$-generated down flow governs an elementary collapse of each block $\mathbf{\Delta}$ onto the polyhedron

$$\mathbf{S}(\Delta) =_{\mathsf{def}} \mathbf{\Delta} \setminus \mathsf{cone}(\Delta^\circ \times \{1\}),$$

where $a_\Delta \in \Delta^\circ \times \{0\}$ serves as the apex of $\mathsf{cone}(\Delta^\circ \times \{1\})$. The top $\Delta \times \{1\}$ of $\mathbf{\Delta}$ is a "free face" residing in $\partial_1^- X$, while the union of the polyhedra $\mathbf{S}(\Delta)$ contains K.

By the very construction of the triangulation $\mathbf{T} =_{\mathsf{def}} \{\mathbf{T}_{\mathbf{\Delta}}\}$, the map Γ : $\mathbf{T} \to \mathcal{T}(v)$ is simplicial, and the space $K = \partial_1^+ X \cup \mathcal{C}(\partial_2^+ X)$ is a subcomplex of X in the triangulation $\mathbf{T}$.

Now it is possible to collapse the polyhedron $\mathbf{T}$ onto the spine K. We start the process by collapsing each 3-dimensional block $\mathbf{\Delta}$ onto $\mathbf{S}(\Delta)$ along the fibers of $\Gamma : \mathbf{\Delta} \to \Delta$. This leaves the faces of $\mathsf{cone}(\partial\Delta \times \{1\})$ free. Next, we use these free faces to collapse each 3-simplex $\sigma \subset \mathbf{S}(\Delta)$ with a free face onto the union of its faces that do not belong to $\mathsf{cone}(\partial\Delta \times \{1\})$. These elementary collapses continue until we collapse each 3-dimensional block $\mathbf{\Delta}$ onto the 2-complex $\delta\mathbf{\Delta} =_{\mathsf{def}} \partial\mathbf{\Delta} \setminus (\Delta^\circ \times \{1\})$. Then, using a similar recipe for 2-dimensional blocks $\mathbf{\Delta} \subset \mathbf{T}$, we collapse the 2-dimensional complex

$$\delta\mathbf{T} =_{\mathsf{def}} \bigcup_{\{\Delta |\, \dim(\Delta)=2\}} \delta\mathbf{\Delta}$$

onto K, starting with the free edges that reside in $\partial_1^- X^\circ$. Finally, the edges of $\delta\mathbf{T}$ that have free vertexes from $\partial_1^- X^\circ$ are collapsed. These collapses

$\mathbf{T} \searrow \delta\mathbf{T} \searrow K$ terminate with K. Therefore, using Theorem 1.1.7 from [Ma], we conclude that $K = \partial_1^+ X \cup \mathcal{C}(\partial_2^+ X)$ is a spine of X in the sense of Definition 2.3. □

Definition 2.6. For a given vector field v on X, we introduce an equivalence relation " $\sim_v$ " among points of X: two points are defined to be equivalent if they belong to the same v-trajectory. The quotient space $\mathcal{T}(v) =_{\mathsf{def}} X/\sim_v$, equipped with the quotient topology, is called the *trajectory space* of the field v. ◇

Remark 2.3. If v vanishes in $\mathrm{int}(X)$, then $\mathcal{T}(v)$ is a non-separable space. In contrast, for a traversing boundary generic v, the space $\mathcal{T}(v)$ can be given the structure of a two-dimensional CW-complex, which is homotopy equivalent to X. In fact, for a traversing boundary generic v on a 3-fold X, the obvious maps $\Gamma : X \to \mathcal{T}(v)$ and $\Gamma^\partial : \partial_1^+ X \cup \mathcal{C}(\partial_2^+ X) \to \mathcal{T}(v)$ both are *quasi-fibrations* (see [May] for the definition). Thus, for such v, the gradient spine can be also regarded as a *homotopy substitute* for the space of v-trajectories. ◇

Let v be a traversing vector field on X. Consider the map $p : \partial_2 X \to \partial_1^+ X$ that takes each point $x \in \partial_2 X$ to the point $p(x) \in \gamma_x \cap \partial_1^+ X$ which lies below x on the v-trajectory γ_x through x; when x is the lowest point in $\gamma_x \cap \partial_1^+ X$, we put $p(x) = x$. Typically, the map p is discontinuous.

Let $F : \partial_1^+ X \to \mathbb{R}$ be a continuous function such that $F(p(x)) < F(x)$ for any $x \in \partial_2 X$, provided $p(x) \neq x$. In the proof of Lemma 5.6, we will construct such a smooth F for any *traversing*[11] vector field v. For a given vector field v, by the linear homotopy, the space of such functions F is contractible.

Let $\mathcal{I}_F(x)$ denote the segment $[F(p(x)), F(x)] \subset \mathbb{R}$.

We introduce an equivalence relation "$\sim_\gamma$" among the points of $\partial_2 X$: by definition, $x \sim_\gamma y$ if they belong to the same v-trajectory. Let $\mathcal{T}_2$ be the quotient space $\partial_2 X/\sim_\gamma$, and $\Gamma : \partial_2 X \to \mathcal{T}_2$ the quotient map. Evidently, $\Gamma(x) = \Gamma(p(x))$. For a boundary generic v, the space $\mathcal{T}_2$ is a finite graph. Its verticies τ have the property $\#(\Gamma^{-1}(\tau)) > 1$. The rest of points $\eta \in \mathcal{T}_2$ belong to the interior of edges.

[11]equivalently, for any non-vanishing gradient-like vector field

With the help of F, we construct an *embedding* $\alpha_F : \partial_2 X \to \mathcal{T}_2 \times \mathbb{R}$. By definition, $\alpha_F(x) =_{\mathsf{def}} (\Gamma(x), F(x))$ (see Figure 6.6). Clearly, α_F is a 1-to-1 map.

Consider the subspace $\mathcal{C}_F(v) \subset \mathcal{T}_2 \times \mathbb{R}$, comprising all the points of the form $(\Gamma(x), t)$, where $x \in \partial_2 X$ and $t \in \mathcal{I}_F(x)$. The singular surface $\mathcal{C}_F(v)$ is the candidate for modeling the cascade $\mathcal{C}(\partial_2^+ X)$.

We introduce the quotient space

$$\mathcal{E}_F(v) =_{\mathsf{def}} \mathcal{C}_F(v) \coprod \partial_1^+ X \big/ \{(\Gamma(x), F(x)) \sim i(x);\ (\Gamma(x),\ F(p(x)) \sim p(x)\},$$

where $x \in \partial_2 X$, $i : \partial_2 X \to \partial_1^+ X$ stands for the obvious inclusion, and $p : \partial_2 X \to \partial_1^+ X$ is the map defined above.

Lemma 2.7. *Let v be a traversing boundary generic vector field on a compact connected 3-fold X with boundary. For any auxiliary function $F : \partial_1^+ X \to \mathbb{R}$ as above, the space $\mathcal{E}_F(v)$ is homeomorphic to the gradient spine $K(v)$. Thus, the topological type of the gradient spine $K(v)$ depends only on the map $p : \partial_2 X \to \partial_1^+ X$.*

Proof. A priori, the topological type of the space $\mathcal{E}_F(v)$ depends only on the map $p : \partial_2 X \to \partial_1^+ X$ and the auxiliary function $F : \partial_1^+ X \to \mathbb{R}$. In fact, the topological type of $\mathcal{E}_F(v)$ does not depend on a particular choice of F. Indeed, consider another such continuous function $G : \partial_1^+ X \to \mathbb{R}$. Any v-trajectory γ generates an ordered sequence of points $\{x_i\}_{i \in [1,q]} = \gamma \cap \partial_1^+ X$. For such a sequence, there exists a piecewise linear homeomorphism $\Phi^\gamma_{F,G} : \mathbb{R} \to \mathbb{R}$ that maps the ordered monotone sequence $\{F(x_i)\}_i$ to the ordered monotone sequence $\{G(x_i)\}_i$. By definition, the slope of $\Phi^\gamma_{F,G}$ is equal to the fraction $(G(x_2) - G(x_1))/(F(x_2) - F(x_1))$ for all $x \leq x_1$ and to the fraction $(G(x_q) - G(x_{q-1}))/(F(x_q) - F(x_{q-1}))$ for all $x \geq x_q$.

The homeomorphism $\Phi^\gamma_{F,G}$ takes the interval $\mathcal{I}_F(x)$, $x \in \partial_2 X$, to the interval $\mathcal{I}_G(x)$.

To construct the desired homeomorphism $\Psi_{F,G} : \mathcal{C}_F(v) \to \mathcal{C}_G(v)$, we first apply $\Phi^\gamma_{F,G}$ to each line $\tau \times \mathbb{R}$, where $\tau = \Gamma(\gamma)$ is a *vertex* of $\mathcal{T}_2$ (so the cardinality of the Γ-fiber over τ exceeds 1). We denote by V_τ a small regular neighborhood (a star) of τ in $\mathcal{T}_2$. For each edge $[\tau, \tau'] \subset V_\tau$, we interpolate between $\Phi^\gamma_{F,G}$ and $\Phi^{\gamma'}_{F,G}$. Let $\Gamma(\gamma) = \tau$ and $\Gamma(\gamma') = \tau'$. We consider a typical point $\tau_\theta \in [\tau, \tau']$, given by the formula $(1-\theta)\cdot\tau + \theta\cdot\tau'$, where $\theta \in [0,1]$. Let γ_θ be the v-trajectory such that $\Gamma(\gamma_\theta) = \tau_\theta$. Note that $\gamma_\theta \cap \partial_2 X$ is a singleton for all $\theta \neq 0$.

The interpolating homeomorphism of the number line is defined in two stages. First, consider the linear interpolation

$$\tilde{\Phi}^{\theta}_{F,G} =_{\mathsf{def}} (1-\theta)\cdot \Phi^{\gamma}_{F,G} + \theta \cdot \Phi^{\gamma'}_{F,G}.$$

Then we compose $\tilde{\Phi}^{\theta}_{F,G}$ with the affine map Ψ^{θ} that takes the closed interval $\tilde{\Phi}^{\theta}_{F,G}(F(\gamma_\theta))$ to the closed interval $G(\gamma_\theta)$. We denote by $\Psi^{\gamma_\theta}_{F,G}$ this composition. We apply $\Psi^{\gamma_\theta}_{F,G}$ to each line $\{\tau_\theta\}\times\mathbb{R}$, where $\tau_\theta \in V_\tau$, and for all stars $\{V_\tau\}_\tau$.

For any point $\eta \in \mathcal{T}_2 \setminus \coprod_\tau V_\tau$, we use the original homomorphism $\Phi^{\gamma}_{F,G} : \{\eta\}\times\mathbb{R} \to \{\eta\}\times\mathbb{R}$, where $\Gamma(\gamma) = \eta$. This completes the construction of the homeomorphism $\Psi_{F,G} : \mathcal{C}_F(v) \to \mathcal{C}_G(v)$. It respects the equivalence relations in $\mathcal{C}_F(v)\coprod \partial_1^+X$ and in $\mathcal{C}_G(v)\coprod\partial_1^+X$ that define the quotients $\mathcal{E}_F(v)$ and $\mathcal{E}_G(v)$, respectively. Therefore, $\Psi_{F,G}$ gives rise to a homeomorphism $\tilde{\Psi}_{F,G} : \mathcal{E}_F(v) \to \mathcal{E}_G(v)$.

Let $\mathcal{T}(v)$ denote the space of v-trajectories in X. By a similar argument, utilizing a smooth function $F^\bullet : X \to \mathbb{R}$ such that $dF^\bullet(v) > 0$ (see the proof of Lemma 5.6, validating the existence of $F^\bullet$), we embed X in $\mathcal{T}(v)\times\mathbb{R}$. This embedding is given by the formula $A_{F^\bullet}(x) =_{\mathsf{def}} (\Gamma(\gamma_x), F^\bullet(x))$. In particular, $A_{F^\bullet}$ can be applied to $K(v) \subset X$.

Let $F =_{\mathsf{def}} F^\bullet|_{\partial_2 X}$. It is easy to see that the space $\mathcal{E}_F(v)$, whose topological type depends on p only, is homeomorphic to the image $A_{F^\bullet}(K(v))$. The homeomorphism $\mathcal{E}_F(v) \to A_{F^\bullet}(K(v))$ is given by applying the map $A_{F^\bullet}\coprod Id : \partial_1^+X\coprod\mathcal{C}_F(v) \to A_{F^\bullet}(K(v))$. Thus, the topological type of the gradient spine $K(v)$ depends only on the map $p : \partial_2 X \to \partial_1^+ X$. □

The 2-cells in which the gradient spine K is subdivided admit a preferred orientation, induced by the preferred orientation of ∂X, so that they form an integral 2-chain $[K]$. Its algebraic boundary $\partial[K]$ consists of the 1-chain $\partial_2^- X \cup \mathcal{C}(\partial_3^- X)$, together with the singularity locus $s(K)$ of K. This locus is comprised of curves, forming the intersection $\partial_1^+X\cap\mathcal{C}(\partial_2^+X)$, and the the segments of trajectories, shared by pairs of waterfalls. In short, each edge from the support of $\partial[K]$ contributes to the cycle $\partial[K]$ with multiplicity ±1 (and not ±3 !)—an important property, which the gradient spines share with the so called *branched spines* (see Corollary 3.1.7 in [BP] and Definition 2.11). Such spines were originally studied in [GR], [GR1], and [GR2].

Remark 2.4. Note that changing the function $f : X \to \mathbb{R}$ to $-f : X \to \mathbb{R}$ and its gradient-like field v to $-v$ exchanges the strata $\partial_1^+X \Leftrightarrow \partial_1^-X$, $\partial_3^+X \Leftrightarrow \partial_3^-X$, but keeps the strata $\partial_2^+X, \partial_2^-X$ fixed (recall that, in the

Morse theory on closed manifolds, the change $f \Rightarrow -f$ leads to a form of the Poincaré duality). Therefore, the gradient spines $K(f, v)$ and $K(-f, -v)$ "complement" each other in X: they share $\partial_2 X$ and their cascades $\mathcal{C}(f, v)$ and $\mathcal{C}(-f, -v)$ complement each other in the set spanned by all the trajectories through $\partial_2^+ X$. Evidently, the topologies of $K(f, v)$ and $K(-f, -v)$ could be radically different. However, their *complexities* (see Definition 3.1) are equal! ◊

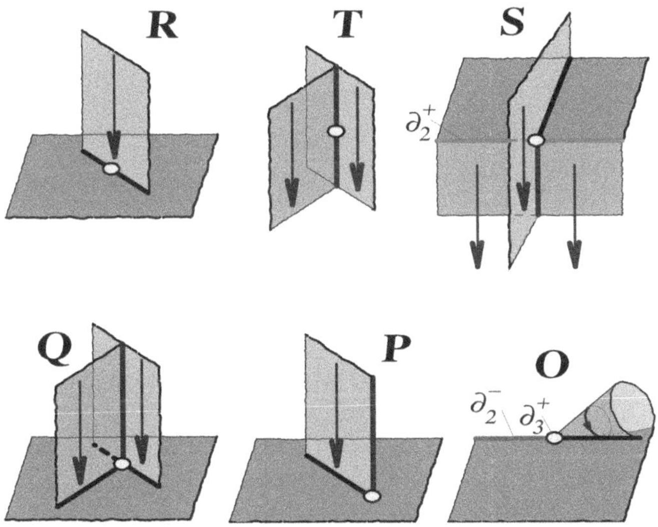

Fig. 2.14 Singularity types of traversally generic gradient spines.

By its very construction, any gradient spine $K = \partial_1^+ X \cup \mathcal{C}(\partial_2^+ X)$, produced by a traversally generic vector field, has a very particular stratified local geometry. In fact, as abstract cellular 2-complexes, gradient spines are *simple* spines in the sense of [Ma]. However, in our context (as well as in the case of branched spines), *the smooth structure of* $\partial_1^+ X$ and of the waterfalls *breaks down the* 3-*fold symmetry* of simple combinatorial spines: speaking metaphorically, the smoothness of the flow distinguishes between the shape "Y" and the shape"T".

The stars of points in K are shown in Figure 2.14. We will call them by their letter labels: "R", "T", "S", "Q", "P", and "O".

The six types (1) – (6) of links of points from K are depicted in Figure

2.21.

A generic point from the set $\partial_2^- X \cup \mathcal{C}(\partial_3^- X)$ has a neighborhood, shaped as a half-disk, and has a link of type (2). The singular set $s(K)$ of K consists of the points of types R, T, and S, all having links of type (5). The T-type is produced by the trajectories that belong to two distinct waterfalls or to two branches of the same waterfall. The R-points are generic to loci where waterfalls hit the ground $\partial_1^+ X$. The S-points are generated when a waterfall transversally hits an arc from $\partial_2^+ X$. They are hybrids of T- and R-types. Topologically R-, T-, and S-types are indistinguishable. The Q-type is generated where two waterfalls hit the ground. Stars of Q-points are shaped as a union of a disk with a half-disk with a quoter-disk, and their links are of type (6). The Q-type singularities are isolated in K. The singularities of P-type are located where a free $(-v)$-trajectory through $\partial_3^- X$ hits the ground. Hence, they are in 1-to-1 correspondence with points of $\partial_3^- X$. A neighborhood of a P-singularity is a union of a disk with a quoter-disk which share a common radius; its link is of type (4). Finally, the singularities of the O-type are just points from $\partial_3^+ X$. They also have stars shaped as cones over a circle union a radius (see Figure 2.14). Topologically the O- and P-types are the same.

Next, we prove that 3-folds have gradient spines which are rather special *"origami"*, folded from a 2-disk (see Figure 2.15). This result is very similar to Theorem 1.2 from the Ishii paper [I], where an origami is built from a, so called, *normal pair*. However, one technical difference is evident: in [I], the v-flow is transversal to the normal disk, while in our construction, the gradient flow is not necessarily transversal to $\partial_1^+ X$ along its boundary $\partial_2 X$.

For the multi-dimensional generalization of the following Theorem 2.6 see Theorem 6.11.

Theorem 2.6 (The Origami Theorem for 3-folds).

The gradient spine $K =_{\mathsf{def}} \partial_1^+ X \cup \mathcal{C}(\partial_2^+ X)$ of a 3-fold X, produced with the help of any traversally generic vector field v, is an image under a cellular map $\Phi : S \to K$ of a cellular 2-complex S, homeomorphic to the surface $\partial_1^+ X$. The map Φ is 1-to-1 in the interiors of the 2-cells in which S is subdivided, 2-to-1 at most on the 1-skeleton without the vertices, and 3-to-1 at most on the set of vertices. The local geometry of Φ can be described by the four identification patters in Figure 2.16.

Moreover, any 3-fold X has a gradient spine K which is an image of a 2-disk D^2 under a cellular map $\Phi : D^2 \to K$ which is 1-to-1 in the interiors of the 2-cells in D^2, 2-to-1 at most on the 1-skeleton of D^2 without the vertices, and 3-to-1 at most on the set of vertices.

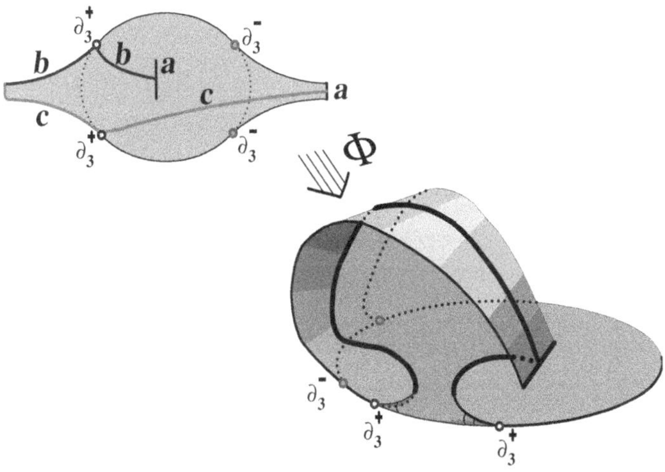

Fig. 2.15 An "origami" map $\Phi : D^2 \to K$ with a collapsable 2-complex K. The pairs of arcs in D^2, identified by Φ, are marked with the same letter: a, b, or c.

Proof. One can resolve the gradient spine $K = \partial_1^+ X \cup \mathcal{C}(\partial_2^+ X)$ along its singular set $s(K)$ so that the resulting space is a cellular 2-complex

$$S =_{\mathsf{def}} \partial_1^+ X \bigcup_{\partial_2^+ X} \Big(\coprod_j W_j\Big).$$

It is obtained from $\partial_1^+ X$ by attaching individual waterfalls $\{W_j\}$ along the loops and arcs, forming $\partial_2^+ X$. As a result, S is homeomorphic to the surface $\partial_1^+ X$. In particular, when $\partial_1^+ X$ is a 2-disk, so is the resolution S.

This resolution $K \Rightarrow S$ can be realized by performing cuts of K of several types. First, at each singularity of any type, but the T-type (see Figures 2.14, 2.17), the cut separates the cascade from the "ground" surface $\partial_1^+ X$. Then, at each singularity of the T-type (see Figure 2.18), the cut separates the preferred "half-waterfall" from the adjacent waterfall; the cut is performed along a segment of the trajectory that both waterfalls share.

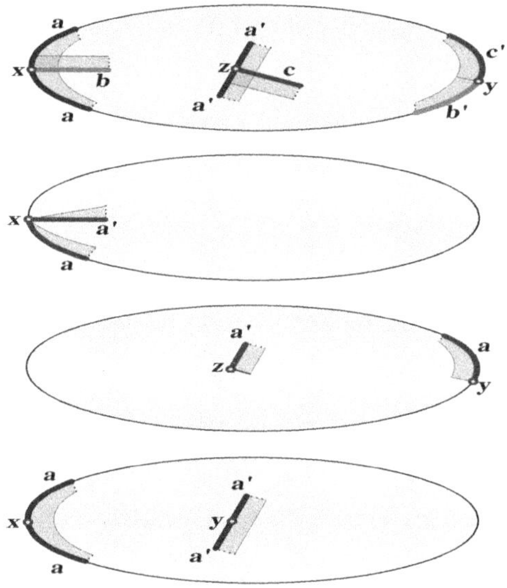

Fig. 2.16 The four local identification patterns for the origami that produces gradient spines. The pairs of letters (a, a'), (b, b'), and (c, c') mark the edges to be identified.

The shape of the waterfalls $\{W_j\}$ depends on the type of the arc or loop in $\partial_2^+ X$ from which W_j falls. Recall that, after an appropriate change of the vector fields, the arcs come in three flavors: A, B and C (see Figure 2.7). The A-type waterfall has two "free" edges, which are not affected by the gluing that produces K. The edges emanate from the two ends $x, y \in \partial_3^- X$ of an A-arc. The B-type waterfall has no "free" edges at all. It falls from a B-arc whose ends are in $\partial_3^+ X$. The C-type waterfall has one "free" edge, which emanates from a point in $\partial_3^- X$—the end of the arc. Figure 3.9 depicts the topological models of A, B, and C types in vicinity of an arc τ, where they hit the ground $\partial_1^+ X$.

By Corollary 2.1, for appropriate Morse data, we can assume that $\partial_1^+ X$ is a disk.

Evidently, there is a mapping $\Phi : S \to K$ which glues the complex K back. The map Φ is 1-to-1 on the compliment to the 1-skeleton of S, it is 2-to-1 at most on the interior of its edges, and 3-to-1 at most on its vertices (see Figures 2.16–2.19). With the local topology of K being restricted by the list in Figure 2.14, one can verify that Figure 2.16 lists all the possible gluing patterns for Φ. □

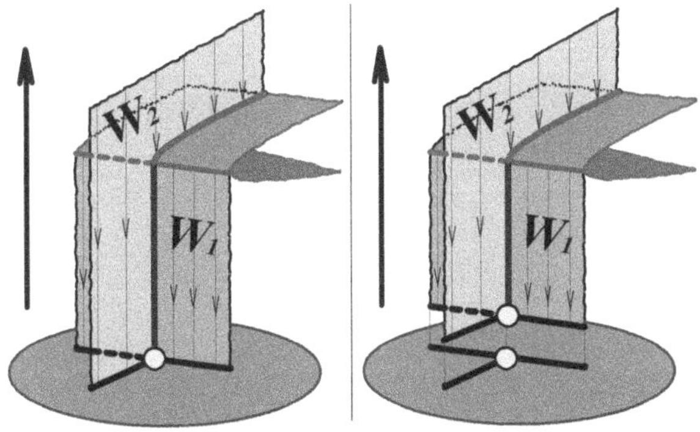

Fig. 2.17 A singularity of the Q-type and its resolution.

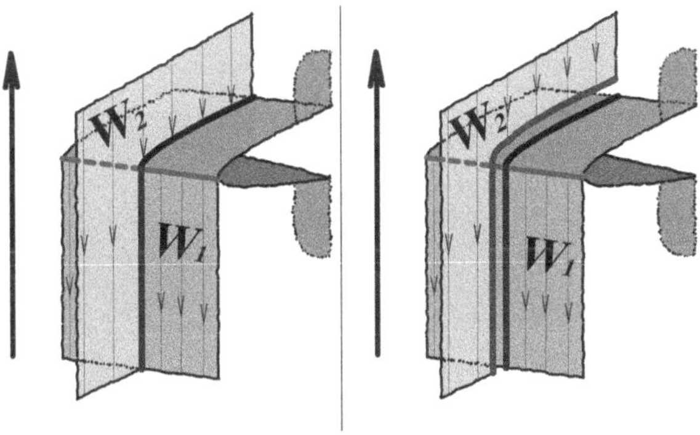

Fig. 2.18 A singularity of the T-type and its resolution.

Example 2.2. Globally, the gluing patterns of $\Phi : S \to K$ can be quite intricate, as shown in Figure 2.20. It depicts an example of a map $\Phi : D^2 \to K$ as in Theorem 2.6 with a contractible, but not collapsable K. The second diagram in Figure 2.20 is obtained from the first one by elementary collapses, performed through the free trajectories. The third diagram suggests contracting the dark shaded disk to a point, thus ignoring the subtleties of the Φ-gluing pattern inside the disk, but keeping the presentation of the fundamental group $\pi_1(K) \approx 1$. $\diamondsuit$

Lemma 2.8. *Let X be a connected 3-fold with boundary. Then any traversally generic vector field v, such that the locus $\partial_1^+ X$ is a 2-disk, gives rise*

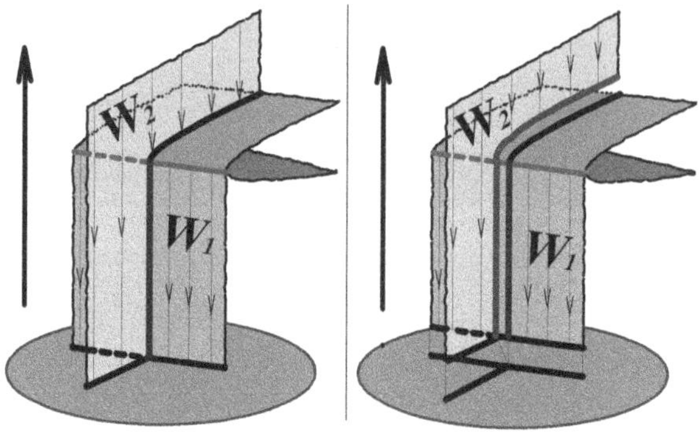

Fig. 2.19 A complete resolution of singularities.

to a finite presentation of the fundamental group $\pi_1(X)$ *(as is described in the proof below).*

Any traversally generic vector field v *(with no restrictions on the topology of the surface* $\partial_1^+ X$*), gives rise to a finite presentation of the fundamental group* $\pi_1(X/\partial_1^+ X)$.

Proof. If, for a traversally generic v, the locus $\partial_1^+ X$ is a 2-disk, the gradient spine $K = K(v)$ produces a *finite presentation* of the fundamental group $\pi_1(X) \approx \pi_1(X/\partial_1^+ X)$. Its generators are the oriented v-trajectories that are shared by pairs of waterfalls, or by two branches of the same waterfall (two waterfalls may share a number of trajectories), together with the free trajectories that emanate from $\partial_3^- X$ (each waterfall has at most two free trajectories). Let $\{\omega_{i\beta}\}_\beta$ be the set of shared trajectories in W_i and let $\{\alpha_{i\delta}\}_\delta$ be the set of free trajectories.

Since $\partial_1^+ X$ is contractible, the homotopy class of each loop $\gamma \subset K$ in the quotient space $K/\partial_1^+ X$ is characterized by its "traces" in the waterfalls (see Figures 2.11 and 2.20). In each waterfall W_i, the trace of γ can be replaced by a word in the alphabet $\{\omega_{i\beta}\}_\beta$ and $\{\alpha_{i\delta}\}_\delta$. The relations are produced by looking independently at each waterfall, decorated with its shared trajectories.

If we cut W_i along the $\{\omega_{i\beta}\}_\beta$'s, it will break into a number of polygons. Each polygon has certain shared and free trajectories in its boundary, and the rest of the boundary consists of arcs that belong to $\partial_1^+ X$. A free trajectory belongs to a single polygon, and each polygon has two free trajectories

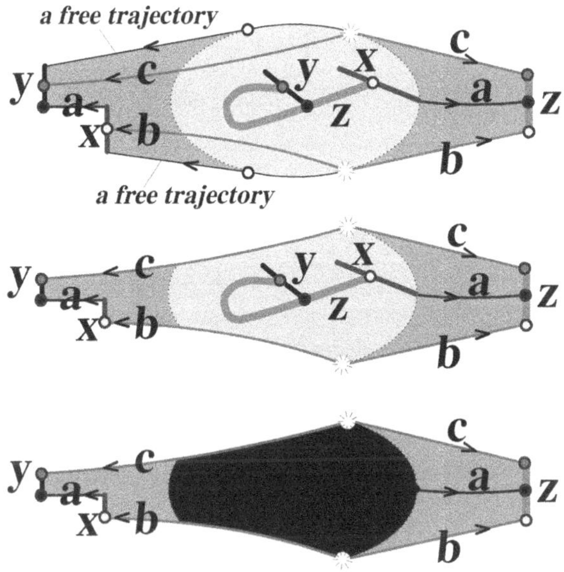

Fig. 2.20 The combinatorics of an origami map $\Phi : D^2 \to K$ which gives rise to a non-collapsable 2-dimensional complex K with $\pi_1(K) = 1$, presented as $\{a, b, c \mid ba = c^{-1},\, ab^{-1} = 1,\, ac^{-1} = 1\}$.

at most. Each polygon is oriented and contributes a single relation: moving along its oriented boundary produces a word in the alphabet that contains the free and shared trajectories and their inverses (the arcs of the boundary of the polygon, were it is attached to the disk $\partial_1^+ X$ are ignored). In fact, when a polygon contains a free trajectory, we can delete it from the list of generators and the polygon itself from the list of relations. Eventually, this will eliminate all free trajectories with the exception of the pairs that belong to a single waterfall, free of shared trajectories in its *interior*. One of the free trajectories in each of such pairs can be dropped from the list of generators, and its polygon from the list of relations.

Thus, v determines the Q^{**}*-class* (as defined in [Me]) of the $\pi_1(X)$-presentation. By [Ra], this Q^{**}-class is a topological invariant of X.

For any locus $\partial_1^+ X$, similar considerations lead to a presentation of the fundamental group $\pi_1(X/\partial_1^+ X)$. □

2.6 Abstract gradient spines

The geometry of gradient spines $K = K(v)$ provides us with a particular way of orienting their 2-cells. This orientation is induced by a preferred orientation of ∂X and is spread along the cascade by the $(-v)$-flow. Furthermore, not only the surfaces in $K^\circ =_{\mathsf{def}} K \setminus s(K)$ acquire a *preferred orientation*, but they also have *preferred sides* in the ambient X, the sides picked by the inner normals to $\partial_1^+ X$ in X. The inner normals uniquely extend by continuity over each waterfall W and pick its preferred side in the ambient 3-fold X.

These distinct properties of gradient spines can be captured in the notion of an *abstract gradient spine* (see Definitions 2.7 - 2.10). We start with a few preliminary definitions.

Definition 2.7. A *simple polyhedron* is a compact 2-dimensional polyhedron K such that each of its points has a link[12], homeomorphic to one of the six shapes in Figure 2.21[13], and the corresponding stars from Figure 2.14.

Type (1) represents the 2-dimensional set of generic points in K. Types (2) and (5) represent the 1-dimensional sets of points in K, while types (3), (4), and (6) correspond to isolated points (the "most singular" locus in K). ◇

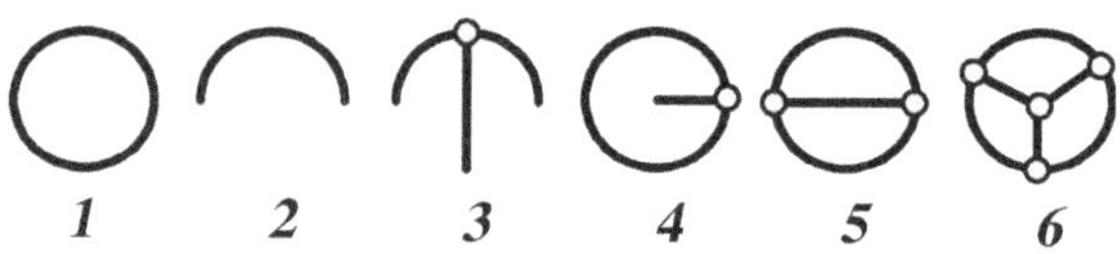

Fig. 2.21 Link types of points in simple polyhedra.

The six types admit a partial order "$\succ$", induced by the inclusions of closures of the corresponding strata of K. For example,

$(1) \succ (2),\ (1) \succ (5),\ (2) \succ (3),\ (2) \succ (4),\ (5) \succ (3),\ (5) \succ (4),\ (5) \succ (6).$

The gradient spines, produced by traversally generic vector fields, are simple polyhedra. Typically, they admit 2-collapses; thus the need for considering link types $(2), (3)$, and (4).

[12]the boundary of a star associated to a point

[13]Actually, type (3) does not appear in generic gradient complexes, but is present in their 1-parameter deformations.

The "topological boundary" $s_\bullet(K)$ of K is formed by points with links of type (2), degenerating into types (3) and (4).

The graph $s_\dagger(K)$ is formed by the points with links of type (5), degenerating into types (3), (4), and (6). In the vicinity of a point of type (5), K looks as an open book with 3 pages. Thus a generic point from a curve from $s_\bullet(K)$ belongs locally to the boundary a single surface in K, while a generic point from the curve from $s_\dagger(K)$ belongs to the boundaries of three surfaces. Let

$$s(K) =_{\mathsf{def}} s_\bullet(K) \cup s_\dagger(K).$$

We denote by $ss(K)$ the finite set of points of types (3), (4), and (6). The bivalent vertices from the finite set

$$ss_{\bullet\dagger}(K) =_{\mathsf{def}} s_\bullet(K) \cap s_\dagger(K)$$

are of type (4), and the rest of vertices from $ss(K)$ are 4-valent and of type (6). We also denote by $Q(K)$ the finite set, formed by the points of type (6).

Recall that a simple spine K is called *special* in [Ma] or *standard* in [BP] if the stratification $ss(K) \subset s(K) \subset K$ gives K a structure of cellular 2-complex, i.e., if $K^\circ =_{\mathsf{def}} K \setminus s(K)$ is a disjoint union of open 2-cells, and $s(K)^\circ =_{\mathsf{def}} s(K) \setminus ss(K)$ is a disjoint union of open 1-cells.

If K is a simple *special* spine of X, whose points have local models of types (1), (5), and (6), then X can be *uniquely reconstructed* just from a regular neighborhood $\mathcal{N}(s(K)) \subset K$ of the singular set $s(K)$ ([Ma], Theorem 1.1.17). For the reconstruction to work, it is important that $K \setminus s(K)$—the disjoint union of disks—does not support nontrivial line bundles. For any gradient spine K, the bundle, normal to $K \setminus s(K)$ in X, is also trivial. This property of gradient spines K will permit a unique reconstruction of X from K as well (see Theorem 2.7).

For a simple polyhedron K, in order to distinguish intrinsically the T-shaped configurations of three pages from the Y-shaped ones, we use a particular *system of markers*, placed along the edges of the graph $s_\dagger(K)$. The marker is a short segment, emanating from a generic point $x \in s_\dagger(K)$. It is transversal to the graph $s_\dagger(K)$ and is contained in one of the three surfaces ("pages") that join at x. We call such segments *T-markers.* A T-marker m, the vertical leg of letter T, tells us that the two pages that do not contain m are thought "to form 180° angle" in the ambient X.

In the world of, so called, *branched* spines K, the local geometry of K along $s_\dagger(K)$ also picks one page out of three (see [BP] and Definition 2.11).

It suffices to place a single T-marker at each edge or loop of the graph $s_\dagger(K)^\circ =_{\mathsf{def}} s_\dagger(K) \setminus ss(K)$: by continuity, the marker spreads along the edge or the loop until it reaches an isolated singularity from $ss(K)$. There the pattern, formed by the markers, must satisfy an additional constraint, which we will describe below.

The *N-marker* is attached to generic points $x \in s_\dagger(K)$ and is contained in one of the two pages that do not carry a T-marker. Informally, one can think of the pages with T-markers as labeling the waterfalls. The N-markers reside both in the "ground" $\partial_1^+ X$ and in the waterfalls. They can be regarded as substitutes for the *preferred normals* (in X) to the oriented pages that contains the T-markers.

For the gradient spines $K = \partial_1^+ X \cup \mathcal{C}(\partial_2^+ X)$, the field v provides us with a T-marker at points of $\partial_1^+ X \cap s(K)$. For points on a trajectory, shared by two waterfalls, the T-marker resides in the page, complementary to the two pages that "form 180° angle". For each waterfall W, the vectors, tangent to $\partial_1^+ X$ and pointing to the preferred side of W, produce the rest of N-markers.

For a subcomplex $A \subset K$, let $\mathcal{N}(A)$ denote a regular neighborhood in K of A.

Effectively, the T and N-markers generate a particular *coloring* with three colors of the "beams" that form the space $\mathcal{N}(s_\dagger(K)) \setminus \mathcal{N}(ss(K))$; the colors of the three pages, which share an edge, being distinct.

A single pair of TN-markers approaches each singularity of types (3) or (4) from Figure 2.21. In fact, the markers distinguish between the spines of type (4) in the vicinity of a point from $\partial_3^+ X$ (O-type in Figure 2.14) and a point y, where a trajectory through $\partial_3^- X$ hits the ground $\partial_1^+ X$ (P-type in Figure 2.14): in the first case, moving along the loop in (4) in the direction of N, we return from the direction marked by T; in the second case, moving along the loop in the direction of N, we return from the unmarked direction.

Four pairs of TN-markers approach each singularity x of the type (6) from four different directions. The link L_x of x in K is shown in Figure 2.22, A and C. The markers reflect the "*plane* union *half-plane* union *quoter-plane* geometry" (as seen from the North Pole). Among the six edges of L_x, there is a single edge α with two T-markers and a single edge β with two N-markers, α and β sharing a vertex. The rest of the markers are determined by Figure 2.22, A and C. In the figure, configuration B is obtained from configuration A by opening the quoter-plane (the higher waterfall) into a half-plane and bending the ground plane (along the bold line) so that its

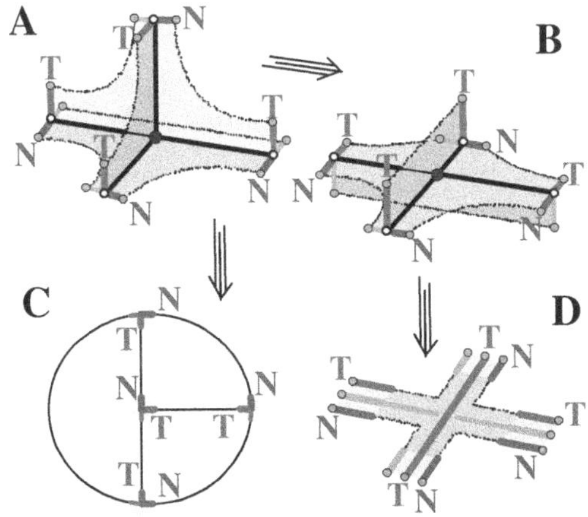

Fig. 2.22 TN-markers in the vicinity of a Q-singularity and the plane pattern they generate. The patterns in diagrams A and B are topologically equivalent. Diagram C is a view of configuration A from the top, diagram D is a view of configuration B from the top.

two halves form the right angle. We have introduced diagrams B and D just to make some of our drawings more in the spirit of *the plane calculus of* 4*-valent graphs*, which has been developed in [I], [I1], [BP], [Ko].

The TN-pattern in the link L_x, $x \in Q(K)$, is completely determined by an ordered pair of edges (α, β) that share a vertex in the complete graph Γ_4 with four vertices. Hence there are $4 \times C_3^2 = 12$ ways of attaching a pattern of TN markers as in Figure 2.22 to Γ_4. On the other hand, there are $(3!)^4 = 1296$ ways to mark the four Y-shaped beams that join at the singularity x. So, the majority of the four-beam patterns will not match the local geometry of Q-singularity.

Definition 2.8. An *abstract* TN*-polyhedron* is a simple polyhedron K with T and N-markers along the edges of the graph $s_\dagger(K)$. We insist that, at each vertex x, whose link is a complete graph in four vertices, these markers satisfy the pattern from Figure 2.22 or its mirror image.

Vertexes $x \in ss(K)$, whose links $L(x)$ are of type (4) from Figure 2.21, admit two patterns of the TN-markers. They reflect the difference between the local shapes of spines of types P and O from Figure 2.14: in the P-case, moving along the loop in $L(x)$ in the direction of the N-marker, we return via the unmarked branch, while in the O-case, moving along the loop in $L(x)$

in the direction of the N-marker, we return via the T-marked branch.[14] $\diamondsuit$

Given an abstract TN-polyhedron K, using the T-markers, we can cut it open along $s_{\dagger}(K)$, so that locally each T-marked page is separated from the rest. For the resolution to work, it is important that the TN markers at the isolated singularities will obey the combinatorial rules depicted in Figure 2.22. The result of this T-resolution is a compact surface $res_T(K)$ (not necessarily connected). It is equipped with the canonical map $\Phi : res_T(K) \to K$.

Definition 2.9. An abstract TN-polyhedron K is said to be *oriented* if its resolution $res_T(K)$ is an oriented compact surface.[15] $\diamondsuit$

Definition 2.10. An *abstract gradient spine* of a 3-fold X is an abstract oriented TN-polyhedron which is a spine of X (in the sense of Definition 2.3). $\diamondsuit$

Remark 2.5. In particular, for any traversally generic vector field v on X, the gradient spine $K(v) = \partial_1^+ X(v) \cup \mathcal{C}(\partial_2^+ X)$ is an abstract gradient spine.

At the same time, not any "natural" spine is an abstract gradient spine. For example, consider the 3-fold $(\mathbb{R}P^3)^\circ =_{\mathsf{def}} \mathbb{R}P^3 \setminus \mathrm{int}(D^3)$. Under the obvious embedding $\mathbb{R}P^2 \subset (\mathbb{R}P^3)^\circ$, $K = \mathbb{R}P^2$ is a "natural" spine for $(\mathbb{R}P^3)^\circ$; however, K is non-orientable in the sense of Definition 2.8 and thus is not an abstract gradient spine! In contrast, for any traversally generic v on $X = (\mathbb{R}P^3)^\circ$, the subcomplex $K(v) = \partial_1^+ X(v) \cup \mathcal{C}(\partial_2^+ X)$ is an abstract gradient spine. We encourage the reader to construct such gradient spine explicitly. $\diamondsuit$

Next, we introduce local models of isolated singularities for abstract gradient spines. They are motivated by the local geometry of gradient spines.

Let D^3_+ be the half-ball $\{x^2 + y^2 + z^2 \leq 1,\ z \geq 0\}$ with the equatorial disk $D^2 =_{\mathsf{def}} \{z = 0\}$. Let $\mathcal{Q} \subset D^3_+$ be the union of D^2 with the half-disk

$$D^2_+ =_{\mathsf{def}} \{y = 0,\ z \geq 0\} \cap D^3_+$$

[14] This observation represents an intrinsic distinction between the type (4) vertexes; it is analogous to distinction between points of $\partial_3^+ X$ and of $\partial_3^- X$ for gradient spines.

[15] Note that for a gradient spine $K \subset X$, the resolution $res_T(K)$ is homeomorphic to $\partial_1^+ X$ and therefore is oriented by the inner normal to ∂X in X, provided that X itself is oriented.

and with the quoter-disk

$$D^2_{++} =_{\mathsf{def}} \{x = 0,\ y \geq 0,\ z \geq 0\} \cap D^3_+.$$

We think of the pair $D^2 \subset \partial D^3_+$ as mimicking the portion of $\partial_1^+ X \subset X$.

Up to PL-homeomorphisms of D^3_+ that keep D^2 invariant, there is a unique way to embed the complete graph Γ_4 from Figure 2.22, C, (without the TN-markers) in the sphere ∂D^3_+ in such a way that the "circle" in Γ_4 is mapped onto ∂D^2 and the rest of the graph into $\partial D^3_+ \setminus D^2$.

For abstract gradient spines K, the vicinities of the O- and P-type singularities also admit preferred embeddings in D^3_+. In the case of a P-singularity (an analogue of a cusp point $x \in \partial_3^- X$), the regular neighborhood $U_K(x)$ of x in K can be identified with the polyhedron

$$\mathcal{P} =_{\mathsf{def}} D^2 \cup D^2_{++} \subset D^3_+.$$

In the case of O-singularity (an analogue of a cusp $x \in \partial_3^+ X$), the regular neighborhood $U_K(x)$ can be identified with the polyhedron

$$\mathcal{O} =_{\mathsf{def}} (D^2 \setminus B^2) \cup D^2_+ \subset D^3_+,$$

where B^2 denotes one of the two 90° sectors in which the ray $D^2 \cap D^2_{++}$ divides the half-disk

$$\{x^2 + y^2 \leq 1,\ z = 0,\ y \geq 0\} \subset D^2.$$

In this model, the half-disk D^2_+ imitates the waterfall, the v-flow being modeled by the circular flow in the interior of D^2_+[16]. However, this identification

$$(U(x), U_K(x)) \approx (D^3_+, \mathcal{O}),$$

where $U(x)$ denotes a regular neighborhood of x in X, is only a piecewise differential homeomorphism; it violates the local smooth geometry of the of the gradient spine. Indeed, the angle between D^2_+ and D^2 along the ray $D^2_+ \cap B^2$ is 90°, while in the vicinity of a cusp $x \in \partial_3^+ X$ the waterfall W is tangent to the portion of $\partial_1^+ X$ along the arc in $\partial_2^+ X$ from which it falls down (so the angle between $\partial_1^+ X$ and W is 180° along the arc $\partial_2^+ X$). In fact, we can improve the model $\mathcal{O} \subset D^3_+$ by smoothening $\mathcal{O}$ along the edge $D^2_+ \cap B^2$.

Each of the model embeddings $\mathcal{Q}, \mathcal{O}, \mathcal{P} \subset D^3_+$ admits two distinct N-markings that pick the preferred side of the "waterfall" (they are mirror images of each other); the T-markings are uniquely determined by the

[16]A better model is the field $w \neq 0$ in D^2_+, whose trajectories interpolate between the unit semicircle $\{x^2 + z^2 = 1,\ z \geq 0\}$ and the parabola $\{z = -x^2\}$ through the origin.

geometry of the embeddings. In what follows, we fix one of the two possible choices for the TN-markers in $\mathcal{Q}, \mathcal{O}, \mathcal{P}$.

For the 3-dimensional gradient flows, the proof of the next theorem is similar to the proofs in [Ma], Lemma 1.1.15 and Theorem 1.1.17 (however, the arguments in [Ma] apply only to, so called, *special* spines).

Theorem 2.7. *If two compact connected orientable smooth 3-folds, X_1 and X_2, have homeomorphic abstract gradient spines, K_1 and K_2, so that the homeomorphism $h : K_1 \to K_2$ respects the TN-markers, then the manifolds X_1 and X_2 are diffeomorphic.*

Proof. Note that any homeomorphism $h : K_1 \to K_2$ of spines must respect the stratifications in K_1 and K_2 by the topological types of links of their points (depicted in Figure 2.21). In particular, any given homeomorphism $h : K_1 \to K_2$ maps the finite set $ss(K_1)$ to the finite set $ss(K_2)$, and the graph $s(K_1)$ to the graph $s(K_2)$.

Thanks to the preservation of markers, h discriminates between the isolated singularities of types O and P in Figure 2.14, whose links are homeomorphic. Indeed, according to Definition 2.8, the patterns of TN-markers for the O-singularity and the P-singularity are different.

Now we need to digress a bit. Recall that the famous *Hauptvermutung Conjecture* suggests that any two triangulations of a triangulable space have a common refinement. Stated differently, any two simplicial complexes which are homeomorphic, are homeomorphic via a homeomorphism that is a simplicial map. Using the Reidemeister torsion, Milnor gave an example of two 6-dimensional complexes that are homeomorphic, but combinatorially distinct [Mi]. However, the Hauptvermutung Conjecture is valid in small dimensions [Moi]. We will need here a very special case of this fact: given two compact triangulated surfaces, S_1 and S_2, any homeomorphism $h : S_1 \to S_2$ is isotopic to (can be approximated by) a simplicial homeomorphism (with respect to appropriately refined triangulations).

By definition of a spine (see the beginning of Section 2.5 for the relevant definitions), there exists a triangulation $\mathbf{T}_i$ of X_i $(i = 1, 2)$ such that the abstract gradient spine $K_i \subset X_i$ is a subcomplex of $\mathbf{T}_i$.

Any given homeomorphism $\tilde{h} : K_1 \to K_2$ can be approximated by a PL-homeomorphism $h : K_1 \to K_2$ which is a simplicial map in the triangulations T_1 and T_2 of the two spines, induced by the ambient triangulations $\mathbf{T}_1$ and $\mathbf{T}_2$ (perhaps, appropriately refined) of X_1 and X_2. Indeed, $\tilde{h}$

automatically takes $ss(K_1)$ to $ss(K_2)$, while preserving the types of singularities. Then we refine T_1 and T_2 and approximate $\tilde{h}$ so that the modified $\tilde{h} : s(K_1) \to s(K_2)$ is a simplicial map of the two graphs.

Using the T-markers, we resolve each spine K_i into a compact surface $S_i = res_T K_i$ (as shown in Figure 2.19). The T-markers help to lift the PL-homeomorphism $\tilde{h} : s(K_1) \to s(K_2)$ to a PL-homeomorphism $\tilde{h} : \partial S_1 \to \partial S_2$. Finally, we fix $\tilde{h} : \partial S_1 \to \partial S_2$ and approximate $\tilde{h} : S_1 \to S_2$ by a PL-homeomorphism $h : S_1 \to S_2$, simplicial with respect to the lift of refined smooth triangulations T_1 and T_2.

Now we can assume that $h : K_1 \to K_2$ is a marker-preserving simplicial map with respect to the two triangulations, induced by the ambient triangulations $\mathbf{T}_1$ of X_1 and $\mathbf{T}_2$ of X_2.

In what follows, the words "regular neighborhood" of a simplicial subcomplex $K \subset X$ mean the union of simplexes from the second barycentric subdivision of a triangulation $\mathbf{T}$ of X that have a nonempty intersection with K.

Let U_x be the $\mathbf{T}_1$-amenable regular neighborhood of a point $x \in ss(K_1)$ in X_1. Then there exists a PL- homeomorphism α_x of the pair $(U_x, U_x \cap K_1)$ with one of the three model pairs: $(D^3_+, \mathcal{Q})$, $(D^3_+, \mathcal{O})$, or $(D^3_+, \mathcal{P})$, so that α_x preserves the TN markers. Let $V_{h(x)}$ be a regular neighborhood of $h(x)$ in X_2, formed by using the triangulation $\mathbf{T}_2$ of X_2. Then, similarly, there exists a marker-preserving PL-homeomorphism $\beta_{h(x)}$ of the pair $(V_{h(x)}, V_{h(x)} \cap K_2)$ with the same (of the three) model pair. So the PL-homeomorphism $\hat{h}_x =_{\mathsf{def}} (\beta_{h(x)})^{-1} \circ \alpha_x$ maps the pair $(U_x, U_x \cap K_1)$ onto the pair $(V_{h(x)}, V_{h(x)} \cap K_2)$ and preserves the TN markers.

Since $h : K_1 \to K_2$ is a simplicial (and marker-preserving) map with respect to T_1 and T_2, we get $h(U_x) = V_{h(x)}$ for all $x \in ss(K_1)$.

The composite PL-homeomorphism $\beta_{h(x)} \circ h \circ \alpha_x$ of the model 2-complexes $\mathcal{Q}, \mathcal{O}$, or $\mathcal{P}$ preserves the markers and, via the cone structure, extends to PL-homeomorphism of the ambient ball D^3_+.

With the help of such extensions, we can construct a PL-homeomorphism

$$\hat{h} : K_1 \cup \left(\cup_{x \in ss(K_1)} U_x \right) \to K_2 \cup \left(\cup_{x \in ss(K_1)} V_{h(x)} \right)$$

of stratified spaces, which takes K_1 to K_2 and preserves their TN-markings.

In a similar way, after "fattening" $h \Rightarrow \hat{h}$ of h in the vicinity of $ss(K_1)$, we can extend $\hat{h}$ into a regular neighborhood $\mathcal{N}(s(K_1), X_1)$ of $s(K_1)$ in X_1, while respecting the markings in the source and the target. To accomplish this goal we need the following "tubular" models of spines:

- $(D^1 \cup D^1_+) \times [0,1] \subset D^2_+ \times [0,1]$,
- $(D^1 \cup D^1_+) \times S^1 \subset D^2_+ \times S^1$,
- $D^1_+ \times [0,1] \subset D^2_+ \times [0,1]$,
- $D^1_+ \times S^1 \subset D^2_+ \times S^1$,

each model being embedded into a solid cylinder $D^2_+ \times [0,1]$ or a solid torus $D^2_+ \times S^1$. Here $D^0 =_{\mathsf{def}} (0,0,0)$, $D^1 =_{\mathsf{def}} D^2 \cap D^2_+$, and $D^1_+ =_{\mathsf{def}} D^2_+ \cap D^2_{++}$. The first pair models the abstract gradient spines K_i the vicinity of an arc or a loop from the singular sets $s_\dagger(K_i)$, $(i = 1,2)$, the second pair models the abstract gradient spines K_i the vicinity of an arc or a loop from $s_\bullet(K_i)$. The NT-markers of the spines in these tubular models are suggested by their geometry. When the ends of these tubes are attached to the pairs $(D^3_+, \mathcal{O})$, $(D^3_+, \mathcal{P})$, $(D^3_+, \mathcal{Q})$ along their boundaries, the markers guide the attachment.

Finally, the third, most problematic, "fattening" $h \Rightarrow \hat{h}$ of the modified homeomorphism $h : K_1 \to K_2$ occurs in a regular neighborhood W_i of the surface

$$K_i^{\odot} =_{\mathsf{def}} K_i \setminus \mathcal{N}(s(K_i), K_i),$$

residing in $X_i \setminus \mathcal{N}(s(K_i), X_i)$. The fattening is possible because the surfaces from $K_i^{\odot}$ are oriented (see Definition 2.9), and so are the ambient 3-folds X_i, which results in the normal bundles $\nu(K_i^{\odot}, X_i)$ being PL-trivial. This crucial observation makes it possible to extend the previously constructed PL-homeomorphism

$$\nu(K_1^{\odot}, X_1)|_{\partial K_1^{\odot}} \cup K_1^{\odot} \to \nu(K_2^{\odot}, X_2)|_{\partial K_2^{\odot}} \cup K_2^{\odot}$$

to a PL-homeomorphism

$$W_1 =_{\mathsf{def}} \nu\big(K_1^{\odot},\ X_1 \setminus \mathcal{N}(s(K_1), X_1)\big) \to W_2 =_{\mathsf{def}} \nu\big(K_2^{\odot},\ X_2 \setminus \mathcal{N}(s(K_2), X_2)\big).$$

Therefore we managed to construct a PL-homeomorphism $\mathcal{N}(K_1, X_1) \to \mathcal{N}(K_2, X_2)$ of regular neighborhoods.

Recall that any regular neighborhood of a spine K in a PL-manifold X is PL-homeomorphic to X (for example, see [Ma], Theorem 1.1.7). Thus, X_1 is PL-homeomorphic to X_2. However, in dimension 3, any PL-homeomorphism $h : X_1 \to X_2$ of smoothly triangulated manifolds admits an approximation by a diffeomorphism (see [Th], Proposition 3.10.7 and Theorem 3.10.8). Therefore, X_1 is diffeomorphic to X_2. □

Lemma 2.9. *Let $G_4(c)$ denote the number of connected regular 4-valent graphs with c vertices, taken up to a homeomorphism. Then the number of* *connected* special *abstract gradient spines, whose points have links of the types* (1), (5), *and* (6) *from Figure 2.21 does not exceed* $G_4(c) \cdot 12^c$.[17] *In turn,* $G_4(c)$ *can be crudely estimated from above by the number of elements in the symmetric group* S_{4c} *that move every symbol in* $(1, 2, 3, \dots, 4c)$.

Proof. Recall that a spine K is special if $K \setminus s(K)$ is a disjoint union of 2-cells, and $s(K) \setminus ss(K)$ is a disjoint union of 1-cells. A special spine K, with all its points modeled after types (1), (5), and (6), is completely determined by a regular neighborhood $\mathcal{U} \subset K$ of its one-skeleton $s(K)$, a regular 4-valent graph Γ. In order to reconstruct K from $\mathcal{U}$, we just attach a 2-disk to every circular component of $\partial\mathcal{U}$.

There are $G_4(c)$ such graphs Γ. Each edge $\gamma \subset \Gamma$ is a core of a beam $B_\gamma \subset \mathcal{U}$ with a Y-shaped section.

Let $V_\star$ denote a star of a Q-singularity (see Figure 2.22). Let V be the disjoint union of c copies of the star $V_\star$. With the TN-markers on both ends $a, b \in \partial V$ of each beam in place, intrinsically, there is a unique way to attach the beam B_γ to V, so that the TN-markers are consistent along the beam. Therefore *any* TN-pattern as in Figure 2.22, assigned to each copy of $V_\star$ in V, will determine the rules for attaching the beams, and thus the reconstruction of K. Since V supports 12^c TN-patterns, the total number of special abstract gradient spines does not exceed $G_4(c) \cdot 12^c$.

Let us fix a graph Γ as above. Then $\mathcal{U}$ is determined by assigning, for each edge γ of Γ, a pairings between the two Y-shaped plugs in V that correspond to γ. For each γ, there are six pairings. Since the number of edges in Γ is $2c$, the total number of non-homeomorphic $\mathcal{U}$'s is bounded from above by 6^{2c}. This number $2^{2c} \cdot 3^{2c}$ should be compared with the estimate $12^c = 2^{2c} \cdot 3^c$ for the number of neighborhoods $\mathcal{U}$ that are consistent with the abstract gradient spine structure.

Crudely, $G_4(c)$ can be estimated from above by considering the number of ways in which c crosses "X" can be connected to form a regular 4-valent graph. Each pairing of the $4c$ free edges is a permutation from the permutation group S_{4c}, which moves all the $4c$ symbols. $\square$

For a simple spine K, its nonsingular part $K^\circ =_{\mathsf{def}} K \setminus s(K)$ is a union of orientable surfaces. Consider the set of integral 2-chains $C_2^\square(K)$ that are combinations of the fundamental cycles of connected surfaces that form K°, the coefficients in the combinations being ± 1. Also consider the set of

[17]By Theorem 3.3, the same number $G_4(c) \cdot 12^c$ gives an upper bound on the number of irreducible with no essential annuli 3-folds.

1-chains $C_1^{\square}(K)$ that are combinations of the fundamental cycles of arcs or loops that form $s(K)^\circ =_{\mathsf{def}} s(K) \setminus ss(K)$, the coefficients in the combinations being ± 1.

Following [GR], [GR1], and [BP], we introduce the following notion:

Definition 2.11. An *oriented branching* on a simple complex K is a 2-chain $\alpha \in C_2^{\square}(K)$ such that its boundary $\partial\alpha \in C_1^{\square}(K)$. $\diamondsuit$

In other words, an oriented branching is a special choice of orientations for each of the connected components of K°. Note that, for an arbitrary $\alpha \in C_2^{\square}(K)$, some edges of $s(K)$ could be present in $\partial\alpha$ with multiplicity ± 3. The oriented branching structure denies this possibility.

Next, we introduce the notion of $\vec{\mathsf{Y}}$-*structure* for spines $K \subset X$. This *extrinsic* definition, which resembles the intrinsic notion of branched spines, will play a significant role in Chapters 3 and 4.

Consider a configuration $\mathcal{Y}$ in $\mathbb{R}^3$ of three distinct half-planes that share a line l (an open book with three pages and the binder l). For any nonzero vector $w \in \mathbb{R}^3$, not parallel to l, consider a linear surjection $p_w : \mathbb{R}^3 \to \mathbb{R}^2$ with the kernel generated by w. There are two possibilities for the map $p_w : \mathcal{Y} \to \mathbb{R}^2$:

(1) generic points in $\mathbb{R}^2$ have preimages of cardinalities one and two, and p_w is onto, or

(2) the cardinalities of preimages are zero, one, and three (p_v is not onto).

We attach symbol "$\vec{\mathsf{Y}}$" to the first situation and symbol "$\vec{\mathsf{W}}$" to the second one.[18] The $\vec{\mathsf{Y}}$-configurations are generated when w and $-w$ point into distinct chambers in which the three half-planes divide $\mathbb{R}^3$.

At each point $x \in s_\dagger(K)^\circ$ of a simple spine $K \subset X$, the linearization of the three surfaces that join at x generates a configuration $\mathcal{Y}_x$ in the tangent space T_xX. At each point $x \in Q(K)$, the linearization of the four surfaces that meet at x generates a 2-complex $\mathcal{X}_x$, which divides T_xX into four pyramids. We will prefer configurations $\mathcal{X}_x$ for which the vectors w and $-w$ point into distinct pyramids. Such configurations are said to be of the $\vec{\mathsf{X}}$-type. For them, the fibers of the projection $p_w : \mathcal{X}_x \to \mathbb{R}^2$ will be of cardinality 1, 2, and 3.

Definition 2.12. Let X be an oriented 3-fold. We say that a spine $K \subset X$ is a $\vec{\mathsf{Y}}$-*spine* if there exists a vector field w, a section of $T(X)$ along the

[18]The shapes of the two letters are mimicking the desired properties of the half-plane configuration with respect to a horizontal w.

graph $s_\dagger(K) \subset X$, which is transversal to each of the surfaces that form K and join along $s_\dagger(K)$. Moreover, for each $x \in s_\dagger(K)^\circ$, the three-page configuration $\mathcal{Y}_x \subset T_xX$ is of the $\vec{\mathsf{Y}}$-type, and, for each $x \in Q(K)$, the four-page configuration $\mathcal{X}_x$ is of the $\vec{\mathsf{X}}$-type with respect to the vector $w(x)$.

In addition, we require that the global orientations of surfaces that form K° will agree with their local orientations induced by w and the preferred orientation of X in the vicinity of $s_\dagger(K)$[19]. ♢

Example 2.3. Consider a union Y of three radii in a disk D^2, Y being symmetric under the rotation ϕ on the angle $2\pi/3$. Let X be the mapping torus of $\phi : D^2 \to D^2$ and K the mapping torus of $\phi : \mathsf{Y} \to \mathsf{Y}$. Since there is ϕ-equivariant retraction of $D^2 \to \mathsf{Y}$, K is a simple spine of X, but not a $\vec{\mathsf{Y}}$-spine. ♢

Lemma 2.10. *For any traversally generic v, the gradient spine $K = K(f, v)$ in X is a $\vec{\mathsf{Y}}$-spine.*

Proof. For a traversally generic v, away from $\partial_2^+ X$, the waterfalls and the ground $\partial_1^+ X$ are transversal along their mutual intersections/self-intersections. At a generic point $x \in s_\dagger(K)$, two out of three pages of K form an angle 180°. Thus, for an open and dense set of vectors $w(x) \in T_xX$, the configuration $\mathcal{Y}_x$ is of the $\vec{\mathsf{Y}}$-type.

Consider the two pages S_T and S_N of K at x that are marked with the T and N-markers. Let $\mathcal{C}_x \subset T_xX$ be the interior of the convex closure of the two half-spaces, tangent to S_T and S_N at x. Since the tangent half-space of the unmarked page forms 180°-angle with the the tangent half-space of the N-marked page, it is never in $\mathcal{C}_x$. Thus the positive cone $\mathcal{C}_x$ picks a unique chamber among the three chambers in which K divides X in the vicinity of $x \in s_\dagger(K)^\circ$.

At each Q-singularity x, the spine K divides the star of x in X into four pyramids. For a gradient spine K, the NT-markers pick one pyramid P_x out of these four pyramids: its triangular base is built out of the three edges that are marked with TT, NN and TN-markers in Figure 2.22, diagram C. This choice is consistent with the choice of chambers $\{\mathcal{C}_{x_i}\}_{1 \leq i \leq 4}$ of the four beams that merge at x.

[19]This requirement is redundant for surfaces whose boundary meets $s_\dagger(K)$ along connected curves.

We already noticed that, for vectors $w(x) \in \mathcal{C}_x$, the configuration $\mathcal{Y}_x$ is of the $\vec{\mathsf{Y}}$-type. Utilizing the convexity of the cones $\{\mathcal{C}_x\}_{x \in s_\dagger(K)}$, by a partition of unity argument, we conclude that there is a vector field w along the graph $s_\dagger(K)$ such that w belongs to the NT-preferred chamber along $s_\dagger(K)^\circ$ and to the preferred pyramid P at the points of $Q(K) \subset ss(K)$. At the same time, $-w$ does not belong to P. Moreover, w (which has a nonzero projection on the N-marked normal to the waterfalls) extends to a field which is transversal to K along $s_\bullet(K)$ as well. As Figure 2.13 and Figure 3.1 testify, w points into the half-spaces, picked by the inner normals to $\partial_1^+ X$, which are spread by continuity along the waterfalls. Thus, w locally induces the same orientation of the surfaces in the spine, as the inner normals to $\partial_1^+ X$ do. □

Lemma 2.11. *For an oriented 3-fold X, the notions of oriented branched spine and of $\vec{\mathsf{Y}}$-spine are equivalent.*

Proof. Assume that K is an oriented branched spine (see Definition 2.11). Since X is oriented, the orientation of each component S of K° picks a particular normal ν_S to S in X. For any $x \in s_\dagger(K)$, an oriented branching $\alpha \in C_2^\square(K)$ on K, picks a preferred surface S_x out of the three oriented surfaces that join at x. Here is the recipe for choosing S_x: if γ is the edge of $s_\dagger(K)$ through x, locally, there are exactly two surfaces, say S_1 and S_2, such that γ is present in ∂S_1 and ∂S_2 with the same sign. Then γ is present in ∂S_x with the opposite sign.

Let U be a small regular neighborhood of $s_\dagger(K)$ in X, equipped with a Riemannian metric. For some $x \in s_\dagger(K)$, both normals ν_{S_x} and $-\nu_{S_x}$ may belong to the same locally-defined chamber in $X \setminus K$. For those x, we employ an isotopy of X to "widen" the angle between S_1 and S_2, so that, in the end of the isotopy, the vectors ν_{S_x} and $-\nu_{S_x}$ will reside in different chambers. Such angle-adjusting isotopy is the identity on the complement to U and on $S_x \cap U$. It is smooth in $X \setminus s_\dagger(K)$. As a result, S_x, S_1, S_2 will form a $\vec{\mathsf{Y}}$-configuration with respect to the normal field ν_{S_x}. In a sense, ν_{S_x} is a piece-wise smooth surrogate of the desired field $w(x)$ under construction.

By Proposition 3.1.6 in [BP], any oriented branching extends to $Q(K) \subset ss(K)$; that is, for any $x \in Q(K)$, the preferred chambers of the four Y-beams that approach x "share" a vector $w(x) \in T_xX$, transversal to K. More accurately, the infinitesimal parallel shifts of this vector $w(x)$ along the four edges $\gamma_1, \gamma_2, \gamma_3, \gamma_4 \subset s_\dagger(K)$ that emanate from x belong to the preferred chambers $\mathcal{C}_{\gamma_1}, \mathcal{C}_{\gamma_2}, \mathcal{C}_{\gamma_3}, \mathcal{C}_{\gamma_4}$ of the beams, while the infinitesimal shifts of $-w(x)$ do not belong to the preferred chambers. Next, we smoothly interpolate between the parallel shifts of $\{w(x)\}_{x \in Q(K)}$ in the vicinity of $Q(K)$ and the fields ν_{S_x} along the edges of $s_\dagger(K)$. This interpolation gives the desired field w along $s(K)$.

Conversely, if K admits a $\vec{\mathsf{Y}}$-structure, then K is an oriented branched spine. Indeed, the field w, transversal to the pages of K along $s(K)$, picks a coherent orientation of each component S of K° along its boundary ∂S. According to Definition 2.12, this localized orientation of S is global. Thus, the field w generates an element $\alpha \in C_2^{\square}(K)$. Because of the very nature of the $\mathcal{Y}_x$-configurations, α has the required property $\partial\alpha \in C_1^{\square}(K)$. □

Chapter 3

Concavity and Complexity of Traversing Flows on 3-folds

3.1 In this chapter ...

The main results in this chapter have their origins in [K].

Inspired by the Matveev notion of *combinatorial complexity* $c(X)$ of 3-folds X ([Ma]), we introduce and study its analogue, the *gradient complexity* $gc(X)$. It is defined as the *minimal number of double-tangent trajectories* that any traversally generic vector field v on X may have.

The inequality $c(X) \leq gc(X)$ is of the fundamental importance. It enables to convert any estimate of $c(X)$ from below into a lower estimate of $gc(X)$. In particular, we get such lower estimates of $gc(X)$ in terms of:

- the homology groups $H_1(X; \mathbb{Z})$ (Theorem 3.5),
- the complexity of the fundamental group $\pi_1(X)$ (Theorem 3.7),
- the volumes of punctured hyperbolic X (Theorem 3.6).

Most of the discussion in Section 3.2 revolves around the relation and tension between generic simple spines K of X, descendants of the PL-structure on X, and the gradient spines $K(v)$, produced by traversally generic v-flows on X, descendants of the smooth structure. In fact, the symmetry of generic spines is broken by the gradient and abstract gradient spines; thus the gap between the two complexities, $c(X)$ and $gc(X)$. Theorem 3.11 gives a handle on the size of this gap. At the same time, Theorem 3.10, claims the equality of various types of complexities based on different kinds of spines (branched oriented, abstract gradient, and $\vec{\mathsf{Y}}$-spines), all sharing similar braking of the S_3-symmetry.

Finally, the last Section 3.3 is dedicated to structural descriptions of 3-folds X with $gc(X) \leq 1$ (see Theorem 3.12 and Theorems 3.14 - 3.17).

3.2 Combinatorial and gradient complexities of 3-folds

Following [Ma], we introduce the notion of complexity for simple 2-dimensional polyhedra and 3-dimensional manifolds.

Definition 3.1.

- The *combinatorial complexity* $c(K)$ of a simple 2-dimensional polyhedron K is defined to be the cardinality of the set $Q(K)$, formed by points of type (6) in Definition 2.7.
- The *combinatorial complexity* of a compact connected 3-fold X is defined as the minimum of the combinatorial complexities of its simple spines. ◇

For traversally generic vector fields v (see Definition 2.5), this notion of $c(K)$ can be applied to gradient spines $K =_{\mathsf{def}} K(v) = \partial_1^+ X \cup \mathcal{C}(\partial_2^+ X)$.

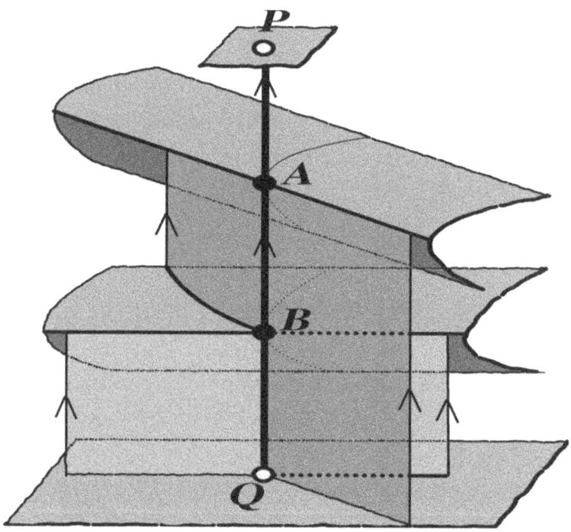

Fig. 3.1 For a traversally generic v, any v-trajectory QP, tangent to $\partial_1 X$ at A and B, generates a singularity at Q of type (6) from Figure 2.21.

Here and on, by a *trajectory* of a vector field v we mean a v-integral curve that does not admit a continuation in X. For example, AB in Figure 3.1 is not a trajectory, while PQ is.

Figure 3.1 leads to a key observation: for a traversally generic vector field v, the Q-singularities of the gradient spine $\partial_1^+ X \cup \mathcal{C}(\partial_2^+ X)$ are in 1-to-1 correspondence with segments $[BQ]$ of the v-trajectories that are shared by either a pair of waterfalls or, locally, by two branches of the same waterfall. The number of such shared segments in a cascade can be given another, less technical, interpretation. We notice that, for a traversally generic vector field v, a shared segment $[BQ]$ corresponds to a unique pair of distinct points $A, B \in \partial_2^+ X$ that are linked by a v-trajectory $[PQ]$. In turn, such trajectories are exactly the ones that link distinct points $A, B \in \partial_1 X$ and that are *tangent* to $\partial_1 X$ at A and B. Indeed, $\partial_2 X$ is the locus where v is tangent to $\partial_1 X$, and points of $\partial_2^- X$ do not "communicate" through the bulk X. We call such trajectories $[PQ]$ *double-tangent.*

A traversally generic field v provides us with an *oriented tangle* $\mathcal{T}an(v) \subset X$ of segments $[AB]$ of double-tangent trajectories which connect pairs of points in $\partial_2 X$, the orientation of $\mathcal{T}an(v)$ being induced by v.

When $\partial_1^+ X$ is a disk, then $\mathcal{T}an(v)$ produces a coupling of points in its circular boundary $\partial_2 X$. This tangle is not unlike the one naturally associated with a smooth plane curve having transversal crossings only (cf. [Ar]). However, in the case of a traversally generic field on X, the orientation of the strings $\{[AB]\}$ tells us which local branch of $\partial_2^+ X$ is "above" and which is "below"; in terms of a smooth function $f : X \to \mathbb{R}$, subject to the constraint $df(v) > 0$, this is the property $f(A) > f(B)$.

Question 3.1. *How does the tangle $\mathcal{T}an(v)$ change, as v deforms in the space of non-vanishing gradient-like fields?*

Is the topological type of the tangle $\mathcal{T}an(v)$ constant for each path-connected component of the space of traversally generic vector fields? ◇

We will attempt to answer these questions in Chapter 4.

Motivated by [Ma], we propose the following two definitions.

Definition 3.2. The *complexity* $c(v)$ of a traversally generic field v on X is defined to be the number of the double-tangent v-trajectories. ◇

We have seen that any $x \in Q(K)$ corresponds to a trajectory γ, tangent to $\partial_1 X$ at a pair of points $A, B \in \partial_2^+ X$. Since v orients γ, the vector field helps to break the symmetry between the tangency points A and B.

For traversally generic vector fields v on an oriented 3-fold X and their gradient spines $K(v)$, the isolated singularities of the Q-type (equivalently, the double-tangent trajectories of v) admit two kinds of orientations: the *tangential* and the *normal*. Let us describe the two types.

The 3-fold X is oriented; so its boundary, with the help of the inner normals, acquire a preferred orientation. The preferred orientation of $\partial_1^+ X \subset \partial_1 X$, again with the help of the inner normals, induces a preferred orientation of $\partial_2 X = \partial(\partial_1^+ X)$. Consider a vector w_A, *tangent* to $\partial_2^+ X$ at A, and a vector w_B, *tangent* to $\partial_2^+ X$ at B, the directions of both vectors agreeing with the preferred orientation of $\partial_2 X$. The $(-v)$-flow spreads w_A and w_B and produces a normal frame $(\tilde{w}_B, \tilde{w}_A)$ along the trajectory $\gamma = [P, Q]$. At $x = Q$, we project the frame $(\tilde{w}_B(x), \tilde{w}_A(x))$ on the tangent space $T_Q(\partial_1 X)$ along the direction of the vector $-v(Q)$. The orientation, induced by the image of the frame $(\tilde{w}_B(x), \tilde{w}_A(x))$ under this projection, may agree or disagree with the preferred orientation of $\partial_1^+ X$ at Q. In the first case, the *tangential polarity* of the spine singularity Q and of the trajectory γ through it is defined to be positive ("+"), in the second case, it is negative ("−"). We denote by $Q(K)^+$ and $Q(K)^-$, respectively, the sets of positively and negatively tangentially polarized points in the singular set $Q(K) \subset K$.

Now consider a vector ν_A, *normal* to $\partial_1 X$ at $A \in \partial_2^+ X$ and pointing inside X, and a vector ν_B, also *normal* to $\partial_1 X$ at $B \in \partial_2^+ X$ and pointing inside X. The $(-v)$-flow spreads ν_A and ν_B and produces a normal frame $(\tilde{\nu}_B, \tilde{\nu}_A)$ along the trajectory $\gamma = [P, Q]$. At $x = Q$, we project the frame $(\tilde{\nu}_B(x), \tilde{\nu}_A(x))$ on the tangent space $T_Q(\partial_1 X)$ along the direction of the vector $-v(Q)$. Again, the orientation, induced by the image of the frame $(\tilde{\nu}_B(x), \tilde{\nu}_A(x))$ under this projection, may agree or disagree with the preferred orientation of $\partial_1^+ X$ at Q. In the first case, the *normal polarity* of the singularity Q and the trajectory γ is defined to be positive "+", in the second case, it is negative "−".

Reversing the preferred orientation of X reverses the preferred orientations of $\partial_1^+ X$ and of $\partial_2 X$; so the frame (w_B, w_A) is replaced by $(-w_B, -w_A)$ which is similarly oriented. Therefore, the tangential orientation of Q flips (because the orientation of $T_Q(\partial_1 X)$ flips). Similarly, reversing the preferred orientation of X does not affect the frames $(\tilde{\nu}_B(x), \tilde{\nu}_A(x))$; however it does reverse the orientation of $\partial_1^+ X$. Therefore, the normal orientation of Q and γ also flips.

Comparing the tangential and normal orientations of a particular double-tangent trajectory γ or of the corresponding singularity Q, we see that they may agree or disagree; in the first case, we attach the polarity "$\oplus$" to γ, in the second case, the polarity "$\ominus$". Note that these new polarities do not depend on the preferred orientation of X. Moreover, the polarities $\oplus, \ominus$ are well-defined even for a non-orientable X!

We denote by $Q(K)^{\oplus}$ and $Q(K)^{\ominus}$, respectively, the sets of positively and negatively polarized points in the singular set $Q(K) \subset K$.

Definition 3.3. The *polarized complexity* $c^{\oplus}_{\ominus}(v)$ of a traversally generic field v on X is defined to be the difference between the number of positively polarized and negatively polarized double-tangent trajectories. In other words, $c^{\oplus}_{\ominus}(v) = \#Q(K)^{\oplus} - \#Q(K)^{\ominus}$ ◊

We are going to show that the polarized complexity $c^{\oplus}_{\ominus}(v)$ can be given another interpretation, which has the flavor of a "self-linking number" for the 1-cycle $\partial[K]$ in X.

By Lemma 2.10, the gradient spine $K = K(v)$ has a preferred vector field w along $s(K)$ which gives K its $\vec{\mathsf{Y}}$-structure. This field is transversal to the surfaces forming K and points inside X along $\partial_1^+ X$ and into the preferred side of each waterfall from the cascade $\mathcal{C}(\partial_2^+ X)$.

Recall that the surfaces that form the gradient spine K acquire orientations, induced by the orientation of $\partial_1 X$. Thus we get the fundamental 2-chain $[K] \in C_2^{\square}(K)$. We view the 1-cycle $\partial[K] \in C_1^{\square}(K)$ with support in $s(K)$ as an oriented graph. Employing a preferred field vector field w along $s(K)$, we push $\partial[K]$ a bit in the direction of w. We denote by $\partial[K]_w$ the perturbed graph. By the construction of w (see the proof of Lemma 2.10), $\partial[K]_w$ belongs to the positive cone, spanned by the markers T and N along $s(K) \setminus ss(K)$. The 2-chain $[K]$ and the 1-cycle $\partial[K]_w$ are in a general position in X, and their intersection points occur only in the vicinity of $Q(K)$, a single intersection per each point of $Q(K)$ (see Figure 2.22, diagram A). Therefore,

$$c(v) =_{\mathsf{def}} c(K) = \#(\partial[K]_w \cap [K]).$$

We claim that an *algebraic* count $\partial[K]_w \circ [K]$ of intersection points in $\partial[K]_w \cap [K]$ also makes sense[1], provided that X is oriented. Let us explain this observation. Each vertex of the oriented graph $\partial[K]$, whose multiplicity exceeds two, is a Q-singularity. It has valency four: a pair of incoming and a pair of outgoing edges. Therefore, at each $x \in Q(K)$, there are exactly two *oriented* resolutions of the oriented graph $\partial[K]$ into a pair of arcs. One of the resolutions is the boundary of the oriented resolved

[1] Recall that, in general, the Intersection Theories on spaces with singularities of codimension one fail miserably.

surface $res_T(K) \subset X$ (see Figure 2.19, the right diagram). We denote by $\overline{res}(\partial[K])$ the other resolution of $\partial[K]$. Let $\overline{res}(\partial[K])_w$ denote its w-shift.

Now consider the algebraic intersection number $\overline{res}(\partial[K])_w \circ res_T(K)$ of the oriented curve $\overline{res}(\partial[K])_w$ with the oriented surface $res_T(K)$. It is easy to see that the points of $\overline{res}(\partial[K])_w \circ res_T(K)$ and of $\partial[K]_w \cap [K]$ are in 1-to-1 correspondence. It requires more effort to check that the standard orientation, assigned to each point of $\overline{res}(\partial[K])_w \cap res_T(K)$, is positive if and only if the corresponding singularity $x \in Q(K)$ has positive polarity "$\oplus$". Recall that the preferred sides (labeled with the N-markers) of surfaces from $K \setminus s(K)$ that join at x depend only on v (and not on the orientation of $\partial_1 X$); and so does the construction of field w, essentially determined by the geometry of the cascade. Thus one needs to consider only the interaction of a particular w-shift with the four localized choices for the the orientations of the two waterfalls. We leave the rest of this verification to the reader, who should keep in mind that the normal orientation of the double-tangent trajectory γ through x is determined by the geometry of the cascade, while its tangent orientation depends on the local orientation of the two branches of $\partial_2^+ X$ (equivalently, on the orientation of the two waterfalls) that intersect γ.

So we define the *self-linking number* of the graph $\partial[K] \subset X$ by the formula

$$lk_w(\partial[K], \partial[K]) =_{\mathsf{def}} \overline{res}(\partial[K])_w \circ res_T(K).$$

By standard homological reasoning, $lk_w(\partial[K], \partial[K])$ depends only on the oriented graph $\partial[K]$ and the preferred field w along it; in turn, up to an isotopy of X, the field w depends only on the TN-marked germ of K in X at $s(K)$.

Combining Definitions 3.2 and 3.3 with the arguments above, we get:

Theorem 3.1. *For a traversally generic vector field v and its gradient spine $K = \partial_1^+(X) \cup \mathcal{C}(\partial_2^+ X)$, carrying the preferred vector field w along the singular set $s(K)$, the complexity and the polarized complexity of v can be calculated in terms of the embedding $K \subset X$ by the formulas:*

$$c(v) = \#(\partial[K]_w \cap [K]) \tag{3.1}$$

$$c_{\ominus}^{\oplus}(v) = lk_w(\partial[K], \partial[K]). \tag{3.2}$$

Hence, the number of double-tangent v-trajectories is greater or equal than the absolute value of the linking number $lk_w(\partial[K], \partial[K])$. ◇

We can refine the complexity of the v-flow and view it as an ordered *pair* of nonnegative integers $(\#[Q(K)^{\oplus}], \#[Q(K)^{\ominus}])$, where $K = K(v)$.

The construction that validates Corollary 3.2 implies that once a pair $(c_{\oplus}, c_{\ominus})$ has been realized by some vector field v, then $(c_{\oplus}+1, c_{\ominus})$ and $(c_{\oplus}, c_{\ominus}+1)$ can be realized as well.

Contemplating about formula (3.2), we realize that it makes perfect sense for any simple $\vec{\mathsf{Y}}$-spine $K \subset X$ (see Definition 2.12).

Definition 3.4. Let X be a compact connected 3-fold with boundary. Its *gradient complexity*, $gc(X)$, is the minimum, taken over all traversally generic vector fields v, of the number of double-tangent v-trajectories.

Alternatively, we can define the gradient complexity of X by the formula:

$$gc(X) =_{\mathsf{def}} \min_{\{K,w\}} \#(\partial[K]_w \cap [K]),$$

where $K = K(v)$ runs over gradient spines, equipped with the preferred vector fields w along the singular loci of K. $\diamondsuit$

Definition 3.5. Let X be a compact connected 3-fold with boundary. Its $\vec{\mathsf{Y}}$-*complexity*, $c_{\vec{\mathsf{Y}}}(X)$, is the minimum, taken over all $\vec{\mathsf{Y}}$-spines K in X, of the combinatorial complexities $c(K)$.

Alternatively, we can define the $\vec{\mathsf{Y}}$-complexity of X by the formula:

$$c_{\vec{\mathsf{Y}}}(X) =_{\mathsf{def}} \min_{\{K,w\}} \#(\partial[K]_w \cap [K]),$$

where K runs over $\vec{\mathsf{Y}}$-spines in X, equipped with the preferred vector fields w along the singular loci of K. $\diamondsuit$

By Lemma 2.11 and the definitions above,

$$gc(X) \geq c_{\vec{\mathsf{Y}}}(X) \geq c(X).$$

In fact, Theorem 3.10 implies that $gc(X) = c_{\vec{\mathsf{Y}}}(X)$.

However, in general, $gc(X) > c(X)$. For example, for the punctured lens space $L^{\circ}_{3,1} =_{\mathsf{def}} L_{3,1} \setminus D^3$, one gets $c(L^{\circ}_{3,1}) = 0$, while $gc(L^{\circ}_{3,1}) \geq 2$ (see the proof of Theorem 3.15).

Since, for any handlebody X, there exists a vector field so that $\partial_2^+ X = \emptyset$, it follows that the gradient complexity of handlebodies is zero.

In general, computing $c(X)$ is hard; it is much easier to estimate it from above or below. For instance, if X admits a triangulation, consisting of n tetrahedrons, then $c(X) \leq n$ ([Ma], Proposition 2.1.6).

Now, let us review briefly some standard vocabulary from the field of 3-dimensional topology.

Definition 3.6.

- An embedded compact surface $\Sigma \subset X$ is called *proper*, if $\partial\Sigma = \Sigma \cap \partial X$.
- An embedded compact surface $\Sigma \subset X$ is called *incompressible*, if any loop in Σ that is contractible in X is also contractible in Σ.
- A 3-fold X is *irreducible*, if any embedded sphere $S^2 \subset X$ bounds a 3-ball; in such a case, S^2 is called *nonessential.*
- X is called *prime* if no essential embedded 2-sphere separates it[2].
- We say that X is *boundary irreducible* if, for any proper 2-disk $(D, \partial D) \subset (X, \partial X)$, the loop ∂D bounds a 2-disk in ∂X.
- A proper annulus $(A, \partial A) \subset (X, \partial X)$ is called *parallel to* ∂X, if there is an embedding $\mathbf{T} \subset X$ of the solid torus $\mathbf{T}$, so that the annulus $A \subset \partial\mathbf{T}$, while the complementary annulus $\partial\mathbf{T} \setminus A^\circ$ is contained in ∂X. The core loop of $\mathbf{T}$ must be homotopic to a loop from ∂A.
- A proper annulus $(A, \partial A) \subset (X, \partial X)$ is called *nonessential*, if either A is parallel to ∂X, or the core loop of A is contractible in X. The rest of annuli are called *essential.* ◇

Remark 3.1. If X is irreducible and boundary irreducible, then, for any proper 2-disk $(D, \partial D) \subset (X, \partial X)$, there is a 3-ball $\mathbf{B} \subset X$ so that its boundary $\partial\mathbf{B}$ consists of D and the complementary disk $\partial\mathbf{B} \setminus D^\circ \subset \partial X$.

We notice that, if one deletes a 3-ball from an irreducible 3-fold, then the resulting manifold with boundary is boundary irreducible and has no essential annuli. ◇

As the next theorem testifies, for boundary irreducible 3-folds with no essential annuli, we can reduce generic $\vec{\mathsf{Y}}$-spines to *special* ones without compromising their combinatorial complexity (cf. Theorem 2.2.4 in [Ma]).

Theorem 3.2. *Let X be a connected compact and boundary irreducible with no essential annuli 3-fold, and let $K \subset X$ be a simple spine such that the surface $K^\circ =_{\mathsf{def}} K \setminus s(K)$ is orientable. Then X has another simple spine K' such that:*

[2]For example, $S^2 \times S^1$ is prime.

- *$K' \setminus s(K')$ is a collection of 2-disks,*
- *$s(K')$ is a connected graph with vertices of multiplicity 1 and 4 only,*
- *$c(K') \leq c(K)$,*
- *when K is a $\vec{\mathsf{Y}}$-spine, so is K',*
- *when K is an abstract gradient spine, so is K',*
- *if $c(K) > 0$, then K' is a special spine.*

Proof. Let $K \subset X$ be a simple (in the sense of Definition 2.7) spine. In the proof, we use the notations "$s(K), ss(K), s_\dagger(K), s_\bullet(K), s_{\bullet\dagger}(K)$", which have been introduced in the text that follows Definition 2.7.

By Definition 2.3, X is PL-homeomorphic to the mapping cylinder of a cellular map $p : \partial_1 X \to K$. We denote by $r : X \to K$ the retraction, induced by the structure of the mapping cylinder. Since X is a manifold, for any $x \in K \setminus s(K)$, the preimage $r^{-1}(x)$ is PL-homeomorphic to a closed interval, and, for any $x \in s_\bullet(K)$, $r^{-1}(x)$ is homeomorphic to a singleton.

We pick an open regular neighborhood $\mathcal{N}$ of the closed 1-dimensional set $s_\dagger(K)$, formed by the points whose links in K are of the main type (5) and its limit types (3)[3], (4), and (6) from Figure 2.21. See also Figure 3.9 for the depiction of $\mathcal{N}$ in the vicinity of an arc from $s_\dagger(K)$.

Let S be a connected component of the compact surface $K \setminus \mathcal{N}$. The boundary ∂S of S consists of the set $\partial_\dagger S =_{\mathsf{def}} \partial(\mathcal{N}) \cap S$ and its complement $\partial_\bullet S =_{\mathsf{def}} \overline{\partial S \setminus \partial_\dagger S}$, each of the two sets being a union of simple arcs and loops.

When $K \setminus \mathcal{N}$ is a *closed* surface, then $s(K) = \emptyset$, and X is PL-homeomorphic to a segment bundle over the surface K.

Consider a simple closed path $\gamma \subset S$ such that: γ is not contractible in S, and the restriction of the normal line bundle $\nu(S, X)|_\gamma$ is trivial (for the gradient spines, $\nu|_\gamma$ is automatically trivial). In fact, such a loop γ always exists, unless S is homeomorphic to D^2, S^2, or $\mathbb{R}P^2$. Since S is presumed to be orientable, D^2 and S^2 are the only exceptions. If $S = S^2$, then $s(K) = \emptyset$. In such a case, $K = S^2$ and $X = S^2 \times [0, 1]$.

We denote by A the annulus $r^{-1}(\gamma)$. Its boundary resides in $\partial_1 X$. Since no essential annuli are permitted, either **(1)** A is parallel to the boundary $\partial_1 X$, or **(2)** γ is contractible in X.

---

[3]For the gradient spines, these are the cusps from $\partial_3^+ X$ and the loci, where the trajectories through $\partial_3^- X$ hit $\partial_1^+ X$.

In the first case, γ must divide S. Assume, for a moment, that it does not divide S. Then we can find a loop $\delta \subset S$ which intersects with A at a singleton; this would imply, contrary to the hypothesis, that the annulus A is essential. Indeed, by the property of A being parallel to $\partial_1 X$, there is a solid torus $\mathbf{T} \subset X$ whose boundary is divided into A and the complementary annulus $A' \subset \partial_1 X$; so, for homological reasons, no closed loop $\delta \subset X$ intersects with $\partial\mathbf{T}$ at a singleton.

As we delete $\mathbf{T}^\circ$ from X, we do not change the topological type of X, but we do change K to a new complex $K' = K \setminus (K \cap \mathbf{T}^\circ)$. Due to the construction of A as $r^{-1}(\gamma)$ and due to the fact that A divides X, deleting $\mathbf{T}^\circ$ preserves the mapping cylinder structure: we notice that $X' = r^{-1}(K')$ is still the mapping cylinder of $p : \partial_1 X \setminus A' \to K'$. Indeed, no point $x \in \partial_1 X \setminus A'$ is mapped by r to a point $r(x) \in K \cap \mathbf{T}^\circ$; this would imply that the segment $[x, r(x)]$ from the mapping cylinder $\mathsf{cyl}(r)$ must intersect A, a set entirely built out of r-preimages of points from K.

Therefore, K' is a new spine of $X' \approx X$. Now $\gamma \subset s_\bullet(K') = K' \cap \partial_1 X'$ (so that each point of γ has a link in K' of type (2) from Figure 2.21).

Next, we consider the second case, when $\gamma \subset S$ is nullhomotopic in X. Then, by Dehn's Lemma [De], each of the two loops γ_1 and γ_2, comprising the boundary of A and residing in $\partial_1 X$, bounds an embedded disk in X. Since X is boundary irreducible, γ_1 bounds a disk $D_1 \subset \partial_1 X$ and γ_2 bounds a disk $D_2 \subset \partial_1 X$. We push D_2 slightly inside X so that the loop ∂D_2 slides along A. Let us denote by D_2' the pushed-in disk.

Let us consider the proper 2-disk $D \subset X$, formed by D_2' and the portion of A between $\partial D_2'$ and γ_1. Its boundary $\partial D_1 = \gamma_1$. Note that $D \cap K = \gamma$. Recalling that X is boundary irreducible, we conclude that D bounds a 3-ball $\mathbf{B} \subset X$, whose boundary is $D \cup D_1$. Thus γ must divide S into $S'' =_{\mathsf{def}} \mathbf{B} \cap S$ and its complement $S' =_{\mathsf{def}} (X \setminus \mathbf{B}^\circ) \cap S$; at the same time, γ divides K into $K'' =_{\mathsf{def}} \mathbf{B} \cap K$ and its complement $K' =_{\mathsf{def}} (X \setminus \mathbf{B}^\circ) \cap K$.

We denote by A' the subannulus of A, bounded by γ and $\partial D_2'$. Since the ball $\mathbf{B}$ is collapsible on the disk $A' \cup D_2' \subset \partial\mathbf{B}$, the complex $K^\sharp =_{\mathsf{def}} K' \cup A' \cup D_2'$ is a spine of $X \setminus \mathbf{B}^\circ \approx X$. In fact, we have replaced S by a new surface $S^\sharp =_{\mathsf{def}} S' \cup A' \cup D_2'$, in which γ is evidently contractible (the surface $S^\sharp$ is shaped as a hat with a brim).

Let us summarize the outcome: any closed loop γ in S with $\nu|_\gamma$ being trivial, can be a source for one of the two spine modifications, described above. By modifications of the second type, we can insure that each S is *incompressible* in X. Then, by modifications of the first type, we can insure that all the nontrivial homotopy classes of simple loops in S will be represented in the portion $\partial_\bullet S$ of its boundary, and thus will be disjoint from $s_\dagger(S)$. Therefore, we can assume that the surface S *has no handles* (i.e., $S \neq T^2 \# S_1$ for any surface S_1) and that any nontrivial loop in S is homotopic to a loop, contained in $\partial_\bullet S$ (unless $S = S^2$, in which case $X = S^2 \times [0,1]$). Thus, S must be either (1) a disk, or (2) a disk with a number of holes.

Next, we claim that any simple path γ, whose boundary $\partial\gamma \subset \partial_\bullet S$, must separate S. Assume, to the contrary, that such non-separating γ does exist. Since $r^{-1}(x)$ is a segment or a singleton for all $x \in S$, $r^{-1}(\gamma)$ is a disk $D \subset X$ whose boundary $\partial D \subset \partial_1 X$. Since X is boundary irreducible, there exists a 3-ball $\mathbf{B} \subset X$, whose boundary $\partial\mathbf{B}$ comprises D union with another disk $D' \subset \partial_1 X$. The intersection $K \cap \mathbf{B}$ bounds γ (on one side), contrary to the hypothesis about γ. Therefore, $\partial_\bullet S$ contains no more than *one* loop from ∂S: otherwise a non-separating path γ as above must exist. On the other hand, we have already arranged for any nontrivial loop in S to be homotopic to a loop from $\partial_\bullet S$. Thus, S is either an annulus, or a disk.

If S in a modified spine is an annulus, then one of its boundary loops must lie in $\partial_\bullet S$, and the other one in $\partial_\dagger S$. So we can collapse S onto $\partial_\dagger S$ and further onto $s_\dagger(K)$, thus simplifying K further. Therefore, we managed to construct a spine K with all the components of S being homeomorphic to a disk. Some of these disks S could have the property $\partial_\bullet S \neq \emptyset$, in which case they also can be collapsed, again simplifying the spine.

The moment we arranged for K° to be a disjoint union of 2-disks, the graph $s(K)$, the singular locus of K, becomes connected. Suppose, to the contrary, that $s(K) = s'(K) \coprod s''(K)$, where $s'(K), s''(K) \neq \emptyset$. Consider regular neighborhoods $\mathcal{N}'$ and $\mathcal{N}''$ of $s'(K)$ and $s''(K)$ in K, respectively. Then the boundaries $\partial\mathcal{N}'$ and $\partial\mathcal{N}''$ each is a disjoint union of loops. In order to form K, we attach disks to $\partial\mathcal{N}'$ and $\partial\mathcal{N}''$; however, this will lead to a disconnected K, clearly a contradiction.

The vertices of $s(K)$ are the isolated singularities of K of types (3), (4), and (6) from Figure 2.21. Types (3) and (4) have a single edge of $s_\dagger(K)$ of type (5) and at least one free edge of type (2) that terminates there; type (6) is a 4-valent vertex. After all the collapses, the free edges will disappear and $s(K) = s_\dagger(K)$ will become a regular 4-valent graph.

Let us examine how the spine modifications above have affected the given abstract gradient or $\vec{\mathsf{Y}}$-structures of the original spine. Let w be a preferred vector field along $s(K)$. Note that all we did amounts to deleting from X a number of relative 3-balls $\mathbf{B}$ and solid tori $\mathbf{T}$. In all cases, but one, the effect on a spine K was deleting its portion $K \cap \mathbf{B}^\circ$ or $K \cap \mathbf{T}^\circ$. Evidently, these operations neither increase the complexity of K, nor destroy the orientations of S and the TN-markers (in the case when K is an abstract gradient spine). In the case of $\vec{\mathsf{Y}}$-spines, the cuts do not change the fact that w along $s(K)$ is transversal to each surface S and that the configurations $\mathcal{Y}_x \subset T_xX$ are of the $\vec{\mathsf{Y}}$-type with respect to w. The only less trivial case occurred when we replaced K with $K^\sharp =_{\mathsf{def}} K' \cup A' \cup D'_2$, but again, the procedure could only eliminate a portion of $s(K)$ which was *disjoint* from the rest and thus did not affected the transversality of w to K or the TN-markers. Note that the orientation of S' uniquely spreads over the cup $A' \cup D'_2$.

The only simplifying move that could harm the abstract gradient structure (specifically, the correlated orientations of the pages) of a given spine is the collapse of some of the disks or annuli S in the very end of the game (see Remark 3.2). So, if we want to keep the abstract gradient structure of K, we should stop there. □

Remark 3.2. Note that elementary expansions and collapses of abstract gradient (or oriented branched) spines K are very different from the elementary expansions and collapses of K, viewed just as the underlying 2-dimensional complexes. The orientations of 2-cells can prevent us from executing some collapses, which non-oriented complexes would support. Also, we need to pay close attention to the choice of NT-markers, an integral part of the abstract gradient complex structure.

For example, take a plane on which a circular fence is erected. The fence divides the plane into two domains, the disk and its exterior. If the orientations of these two domains disagree, we cannot collapse the fence, if we want to keep the oriented branched structure.

Because the T-resolution of an abstract gradient spine leads to an oriented surface, we can always collapse a fence marked with T. In fact, an unmarked fence also can be collapsed. ◇

The inequality $gc(X) \geq c(X)$ can be combined with a number of results about the combinatorial complexity $c(X)$, thus delivering lower bounds on the number of double-tangent trajectories.

Matveev proved that, up to a homeomorphism, there are only finitely many compact boundary irreducible 3-folds X that have no essential proper annuli and with a bounded combinatorial complexity $c(X)$ (see [Ma], Theorem 2.2.5). We can be a bit more specific, as we establish the following similar result for the gradient complexity $gc(X)$.

Theorem 3.3. *Let X be a compact connected 3-fold X with boundary. For any traversally generic vector field v, the number of double-tangent v-trajectories is greater than or equal to the combinatorial complexity $c(X)$.*

The number of topological types of boundary irreducible X with no essential annuli and gradient complexity $gc(X) = c$ does not exceed $\Gamma_4(c) \cdot 12^c$, where $\Gamma_4(c)$ is the number of topological types of four-valent connected graphs with c vertices at most[4].

In particular, there are no more than $\Gamma_4(c) \cdot 12^c$ distinct orientable closed hyperbolic 3-folds Y such that $gc(X) = c$, where $X = Y \setminus \mathrm{int}(D^3)$.

Proof. Since we have established that gradient spines of traversally generic vector fields are special kind of simple spines, and since for gradient spines the number of double-tangent v-trajectories is equal to the number of their singularities of the type Q (see the arguments centered on Figure 3.1), the first claim is clear.

In order to prove the second claim, consider a gradient spine $K(v) \subset X$ with $c(v) = gc(X) = c$. Then by Theorem 3.2, we can simplify $K(v)$ to a *special* abstract gradient spine K with c copies of Q-singularities at most. Examining the constructions that lead to the proof of Theorem 3.2, we see that deleting 3-balls and solid tori, adjacent to $\partial_1 X$, did not change the combinatorics of the special abstract gradient spine in the vicinity of the remaining Q-singularities. After collapsing all its 2-cells, whose boundary touches the topological boundary $s_\bullet(K)$ of K, we could eliminate some of the Q-singularities and transform the Y-shaped beams that connect them into I-shaped ones. We conclude that, in the end, $s(K) = s_\dagger(K)$ must be a connected regular 4-valent graph with c vertices at most. By Lemma 2.8, the number of such abstract gradient spines has a upper boundary $\Gamma_4(c) \cdot 12^c$. By Lemma 1.1.15 in [Ma], any special simple spine $K \subset X$ determines the topological type of its ambient X (and thus its PL- and smooth types).

Finally, the last claim is based on the property of the hyperbolic $X = Y \setminus \mathrm{int}(D^3)$ being boundary irreducible and with no essential annuli ([Ma], Lemma 2.6.7). □

[4]Crudely, the number of regular 4-valent graphs with c vertices is less than $(4c-1)!!$, where $k!!$ denotes the product of all odd numbers that do not exceed k.

In Chapter 2, we have seen how the v-flow through the "bulk" X and the v_1-flow along its boundary $\partial_1 X$ are intimately linked. Here v_1 denotes the orthogonal projection (in the v-adjusted Riemannian metric) of v on the tangent to $\partial_1 X$ planes. This connection allows for partial reconstruction of certain properties of (X, v) just in terms of the boundary data $(\partial_1 X, v_1)$, an instance of general *holography phenomenon*, which we are going to investigate briefly in Chapter 8.

Meanwhile, we get the following interpretation of Theorem 2.2.

Corollary 3.1. *Let (f, v) and $(f|_{\partial_1 X}, v_1)$ be Morse data on X and $\partial_1 X$, where $v \neq 0$ is a traversally generic f-gradient field in some metric on X, and v_1 denotes the orthogonal projection of v on $T(\partial_1 X)$.*

If there is no ascending v_1-trajectory that connects in $\partial_1 X$ a point of Σ_1^- to a point of Σ_1^+, then $gc(X) = 0$.

Stated differently, if $gc(X) \neq 0$, then for any traversally generic v, there is an ascending v_1-trajectory that connects in $\partial_1 X$ a point of Σ_1^- to a point of Σ_1^+.

Proof. By Theorem 2.2, the absence of an ascending v_1-trajectory γ that connects in $\partial_1 X$ a point in Σ_1^- to a point in Σ_1^+ implies the convexity of the Morse data (f, v), that is, $\partial_2^+ X = \emptyset$. On the other hand, if $\partial_2^+ X = \emptyset$, then no v-trajectory γ contains a pair of points $x, y \in \partial_1 X$, where γ is tangent to $\partial_1 X$. So $gc(X) = 0$.

If $gc(X) \neq 0$, then no traversally generic vector field is convex. Then by Theorem 2.2, we get the second claim. □

Given 3-folds X' and X'' with boundary, two types of connected sum operations are available: $X' \# X''$ and $X' \#_\partial X''$. In the first construction, we delete a 3-ball from the interior of each of the two manifolds and then attach a 1-handle $S^2 \times D^1$ to the boundaries of the balls. In the second construction, a 1-handle $D^2 \times D^1$ is attached to the boundary $\partial_1 X' \coprod \partial_1 X''$ (so the boundary $\partial_1 X' \coprod \partial_1 X''$ is subjected to 1-surgery as well).

Theorem 3.4. *The gradient complexity is a semi-additive invariant:*

$$gc(X' \# X'') \leq gc(X') + gc(X''),$$

$$gc(X' \#_\partial X'') \leq gc(X') + gc(X'').$$

As a result, attaching 1-handles to the interior of X or to its boundary does not increase the gradient complexity $gc(X)$. Also deleting 3-balls from *the interior of X does not change its gradient complexity. Therefore, only 2-surgery on X has the potential to increase its gradient complexity.*

Proof. Take any two compact 3-folds X', X'' with boundary. Unlike the "honest" additivity, the *semi*-additivity

$$gc(X' \#_\partial X'') \leq gc(X') + gc(X'')$$

is easy to validate. Let the Morse pair (f', v') delivers $gc(X')$ and the Morse pair (f'', v'') delivers $gc(X'')$. Assume that f'' attains its minimum at $a \in \partial_1^+ X''$ and f' attains its maximum at $b \in \partial_1^- X'$. By adding a positive constant to f'', we may assume that $\min(f'') > \max(f')$.

By perturbing v'' a bit (see Theorem 2.4), we can arrange that the cascade in X'' has an empty intersection with a small disk $D_a^2 \subset \partial_1^+ X''$, centered at a, without changing the original topology of the gradient spine K''. Similarly, by perturbing v' if necessary, we can pick a sufficiently small disk $D_b^2 \subset \partial_1^- X'$ so that the $(-v')$-trajectories through D_b^2 do not intersect the cascade in X'. Then we attach a 1-handle $H \approx D^2 \times [0,1]$ to $X' \coprod X''$ at a and b and extend the Morse data from the top disk D_a^2 and the bottom disk D_b^2 into H. The neck $\gamma = \partial D^2 \times \{\frac{1}{2}\}$ belongs to the set $\partial_2^+(X' \# X'')$ and the annulus $\partial D^2 \times [\frac{1}{2}, 1]$ to the set $\partial_1^+(X' \#_\partial X'')$. Thus, the cylindrical waterfall streaming from γ does not intersect with the cascade in X' and hits $\partial_1^+ X'$ without producing new shared trajectories. By the choice of D_a^2, the cascade in X'' does not fall through H. Thus $gc(X' \#_\partial X'') \leq gc(X') + gc(X'')$.

Deleting the interior of a 3-ball $\mathbf{B}$ from the interior of X does not change its gradient complexity. Indeed, if (f, v) delivers $gc(X)$, then we pick a sufficiently small convex ball $\mathbf{B}$ whose center lies on a v-trajectory γ that is not in the cascade. Then $\partial\mathbf{B}$ is concave with respect to v, and the trajectories tangent to ∂B are separated from the old cascade. As a result, they do not contribute to the set of double-tangent trajectories of v in $X \setminus \mathbf{B}^\circ$. Hence $gc(X \setminus \mathbf{B}^\circ) \leq gc(X)$.

Consider v that delivers $gc(X \setminus \mathbf{B}^\circ)$. Since $\mathcal{T}(v)$ is connected, there is a trajectory γ which connects a point $x \in \partial\mathbf{B}$ to a point in $y \in \partial_1 X$ and is transversal at x and y to the boundary of $X \setminus \mathbf{B}^\circ$. Drilling a narrow tunnel $W \subset X \setminus \mathbf{B}^\circ$, centered on γ and with a concave bottleneck, produces a manifold $\hat{X}$ homeomorphic to X (we need to smoothen $\hat{X}$ at both ends of the tunnel). This can be done in such a way that $X \setminus \mathbf{B}^\circ$ and $\hat{X}$ will share the same cascade. Hence, $gc(X \setminus \mathbf{B}^\circ) \geq gc(\hat{X}) = gc(X)$. Therefore $gc(X \setminus \mathbf{B}^\circ) = gc(X)$.

Now $X'\#X''$ can be obtained by the following operations: **(1)** delete 3-balls $\mathbf{B}', \mathbf{B}''$ from X' and X'', respectively, **(2)** form a connected sum

$$(X' \setminus (\mathbf{B}')^\circ)\#_\partial(X'' \setminus (\mathbf{B}'')^\circ)$$

by attaching a 1-handle $H =_{\text{def}} D^2 \times [0,1] \subset S^2 \times [0,1]$ to $\partial\mathbf{B}' \cup \partial\mathbf{B}''$, and **(3)** attach a 3-ball to the boundary of $(S^2 \times [0,1]) \setminus H$. By the previous arguments, these operations lead to the inequality $gc(X'\#X'') \leq gc(X') + gc(X'')$.

Let $\mathbf{T}$ denote a solid torus. Since $gc(\mathbf{T}) = 0$, we have $gc(X\#_\partial\mathbf{T}) \leq gc(X)$ and $gc(X\#\mathbf{T}) \leq gc(X)$.

Therefore, only 2-surgery on X has the potential to increase its gradient complexity. □

Remark 3.3. In dimension 2, the analogue of gradient complexity is the number

$$gc(X) =_{\text{def}} \min_{v \text{ being traversally generic}} \{\#\partial_2^+ X(v)\}.$$

We have seen in Chapter 1 that this number is not additive under the connected sum of surfaces (see the proof of Theorem 1.5 and (1.6)).

To establish the conjectured *additivity* of $gc(\sim)$ under the connected sum operations in $3D$ seems to be much harder. The additivity of $c(X)$ is a nontrivial fact, which relies on Haken's theory of normal surfaces ([Ma], Theorem 2.2.9). Based on that theorem, we notice that the defect $gc(X) - c(X)$ is also semi-additive under the connected sum operations. ◇

Corollary 3.2. *Let X be a compact connected smooth 3-fold with boundary. Then, for any non-negative integer n, there exists a traversally generic flow on X with $gc(X) + n$ double-tangent trajectories.*

Proof. Consider the unit sphere $S^2 = \{x^2 + y^2 + z^2 = 1\}$ and its equator $E = S^2 \cap \{z = 0\}$.

There exists a simple smooth curve $\mathcal{L} \subset S^2$ such that $\mathcal{L}$ is transversal to the vertical field $v^\star =_{\text{def}} \partial_z$ and projects orthogonally onto a curve in the xy-plane, shaped as the figure "∞". So the projection $\mathcal{L} \to \infty$ is a regular map. We may choose $\mathcal{L}$ so that $E \cap \mathcal{L}$ is a pair of points, say, a and b. The union $\mathcal{L} \cup E$ divides S^2 into four connected regions. In fact, one can find this remarkable curve $\mathcal{L}$ on every tennis ball!

Let $\mathbf{B}$ be a solid, obtained from the round ball $\mathbf{D}^3 = \{x^2+y^2+z^2 \leq 1\}$ by making a narrow smooth grove (indentation) along $\mathcal{L}$. The section of $\mathbf{B}$ by each plane Π, which contains the z-axis, is a 2-disk D_Π. With respect to the vector field $v^\star|_{D_\Pi}$, the disk has two concave tangency points, a' and b', and two convex tangency points, a'' and b'', so that $z(a'') > z(a')$ and $z(b'') < z(b')$. Moreover, a' and a'' are in the vicinity of one point $A \in \mathcal{L} \cap \Pi$, and b' and b'' are in the vicinity of another point $B \in \mathcal{L} \cap \Pi$.

Then the locus $\partial_2^+\mathbf{B}(v^\star)$ is a simple loop in $\partial\mathbf{B}$, which is C^∞-close to $\mathcal{L}$. At the same time, the locus $\partial_2^-\mathbf{B}(v^\star)$ is a disjoint union of two simple closed curves, $\mathcal{C}_+$ and $\mathcal{C}_-$. The curve $\mathcal{C}_-$ resides above (in the z-direction) $\mathcal{L}$ and above the equatorial plane $\{z = 0\}$, while the curve $\mathcal{C}_+$ below $\mathcal{L}$ and below the equatorial plane. In fact, the locus $\partial_2^-\mathbf{B}(v^\star)$ is obtained from $E \cup \mathcal{L}$ by resolving the two cross-intersections at the points $a, b \in E \cap \mathcal{L}$. The vector field $v^\star$ is traversal to the pairs of arcs that locally resolve each crossing.

Since $\partial_2^+\mathbf{B}(v^\star)$ is C^∞-close to $\mathcal{L}$, its projection on the xy-plane is still regular and its image is still diffeomorphic to figure ∞. Because the projection of $\partial_2^+\mathbf{B}(v^\star)$ on the xy-plane has a single self-intersection, $v^\star$ has a single double-tangent trajectory in $\mathbf{B}$.

Next, we use a traversally generic vector field v that delivers $gc(X)$. We form a connected sum $X\#_\partial\mathbf{B}$, equipped with the field that extends v and $v^\star = \partial_z$ across a 1-handle, as is done in the proof of Theorem 3.4. Let $\tilde{v}$ denote this extension. Evidently, the number of new double-tangent $\tilde{v}$-trajectories in $X \approx X\#_\partial D^3$ is $gc(X)+1$.

Recycling this construction proves the claim. □

Each time we have a lower bound on the combinatorial complexity $c(X)$, for any traversally generic gradient-like flow, the number of double-trajectories must be at least $c(X)$ (Theorem 3.3). Fortunately, [Ma] contains a complete list of *punctured* closed irreducible 3-manifolds of combinatorial complexity ≤ 6. To validate this list is a labor-intense accomplishment, facilitated my a computer.

We use the notation "X°" for the result of deleting a 3-ball from a given closed 3-fold X. For example, using computations in [Ma], pages 77 and 461-462, for punctured elliptic manifolds (they are orbit spaces of the classical finite groups $\mathsf{P}_{24}, \mathsf{P}_{48}, \mathsf{P}_{120}$, that act freely on the 3-sphere) we get:

$$gc((S^3/\mathsf{P}_{24})^\circ) \geq 4, \quad gc((S^3/\mathsf{P}_{48})^\circ) \geq 5, \quad gc((S^3/\mathsf{P}_{120})^\circ) \geq 5.$$

For $2 \leq n \leq 6$ and the quaternion unit groups Q_{4n},

$$gc((S^3/\mathsf{Q}_{4n})^\circ) \geq n.$$

For punctured lense spaces $L^\circ_{p,q}$, one has:

$$\begin{aligned} gc(L^\circ_{4,1}) \geq 1,\ gc(L^\circ_{5,2}) \geq 1; \\ gc(L^\circ_{5,1}) \geq 2,\ gc(L^\circ_{7,2}) \geq 2,\ gc(L^\circ_{8,3}) \geq 2; \\ gc(L^\circ_{6,1}) \geq 3,\ gc(L^\circ_{9,2}) \geq 3,\ gc(L^\circ_{10,3}) \geq 3, \\ gc(L^\circ_{11,3}) \geq 3,\ gc(L^\circ_{12,5}) \geq 3,\ gc(L^\circ_{13,5}) \geq 3, \end{aligned} \tag{3.3}$$

etc.

Reinterpreting Theorem 2.6.2 in [Ma] (see also [MP]), one gets a lower homological bound on the number of double-tangent trajectories of traversally generic flows on closed 3-folds with a number 3-balls being deleted.

Theorem 3.5. *Let X be a 3-fold, obtained from a closed irreducible and orientable 3-fold Y, different from the lens space $L_{3,1}$, by removing a number of 3-balls. Then any traversally generic flow on X has at least*

$$2 \cdot \log_5 |\mathrm{tor}(H_1(X;\mathbb{Z})| + \mathrm{rk}(H_1(X;\mathbb{Z})) - 1$$

double-tangent trajectories. Here $|\mathrm{tor}(H_1(X;\mathbb{Z})|$ denotes the order of the torsion subgroup $\mathrm{tor}(H_1(X;\mathbb{Z})) \subset H_1(X;\mathbb{Z})$. ◇

For closed hyperbolic X, both $c(X^\circ)$ and $gc(X^\circ)$ exhibit at least a linear growth as functions of the hyperbolic volume $\mathrm{Vol}(X)$ ([K]). This result (Theorem 3.6 below) should be compared with results of Gromov [Gr1], which have a similar flavor, but a different and broader context. In [Gr1], the objects of study, closest to ours, are the folding maps $F : X \to \mathbb{R}^2$ from a closed 3-fold X to the plane, but the goals of investigations are similar: to estimate from below (in terms of topological invariants of X) the number of self-crossings for the F-image of the 1-dimensional folding loci $\Sigma_F \subset X$.

Theorem 3.6. *Let X be a compact 3-fold, obtained form a closed hyperbolic 3-fold Y by removing a number of 3-balls. Let $\mathrm{Vol}(Y)$ denote the hyperbolic volume of Y, and $\mathrm{Vol}(\Delta^3)$ the hyperbolic volume of a regular ideal tetrahedron in the hyperbolic space $\mathbf{H}^3$.*

Then, any traversally generic flow v on X has at least $\mathrm{Vol}(Y)/\mathrm{Vol}(\Delta^3)$ double-tangent trajectories.

Proof. The statement follows from [Ma], Lemma 2.6.7 and Corollary 2.6.8, coupled with the inequality $c(X) \leq gc(X)$. □

Theorems 8.9 and 8.10 from Chapter 8, may be regarded as multidimensional analogues of Theorem 3.6.

Definition 3.7. Let G be a finitely presented group. Consider the relations in a particular presentation of G. The *length of a relation* is the number of generators and their inverses that are present in the relation.

The *complexity of a presentation* is defined to be the sum of lengths of all its relations (for example, the complexity of the presentation in Figure 2.20 is $3 + 2 + 2 = 7$).

We define $c(G)$, the *presentation complexity* of G, to be the minimum complexity among all G-presentations.

Theorem 3.7 below is a generalization of Proposition 2.6.6 from [Ma] which claims $c(X) \geq -1 + \frac{1}{3}c(\pi_1(X))$ for any punctured closed *irreducible* orientable 3-fold X, different from $S^3, \mathbb{R}P^3$, and $L_{1,3}$. In fact, we can drop the assumption of irreducibility of X.

Theorem 3.7. *For any compact connected 3-fold X with boundary and a traversally generic vector field v on X,*

- $gc(v) \geq \frac{1}{3}\, c\big(\pi_1(X/\partial_1^+X(v))\big)$;
- *if, in addition, $\partial_1^+X(v)$ is homeomorphic to a 2-disk, then*
$$gc(v) \geq \frac{1}{3}\, c\big(\pi_1(X)\big);$$
- *if the boundary $\partial_1 X$ is a union of 2-spheres, then also*
$$gc(X) \geq \frac{1}{3}\, c\big(\pi_1(X)\big);$$
- *the group $\pi_1\big(X/\partial_1^+X(v)\big)$ has a finite presentation with not more than $gc(v)$ generators. Under the hypotheses of the second or third bullets, the group $\pi_1(X)$ has a finite presentation with the number of generators that does not exceed $gc(X)$.*

Proof. Let us return to the description of the presentation of $\pi_1(X/\partial_1^+X)$, given by a traversally generic gradient cascade and described in the proof of Lemma 2.8. Each waterfall is a union of several polygons Π that are bounded by segments of double-tangent trajectories, free trajectories that stream from the locus ∂_3^-X, and segments that belong to ∂_1^+X. The interiors of the polygons are disjoint (see Figure 3.1).

So the generators of $\pi_1(X/\partial_1^+X)$ are the shared segments of double-tangent trajectories from the cascade (like the segment AB from Figure 3.1), together with some free trajectories of waterfalls that have two free trajectories and no shared segments at all. Each waterfall of this kind contributes a single "free" generator and no relations.

Therefore, the number of generators of $\pi_1(X/\partial_1^+X)$ does not exceed $gc(v)$, the number of double-tangent trajectories, plus half of the cardinality of the locus ∂_3^-X. However, by Theorem 4.6, if a traversally generic v has $gc(v)$ double-tangent trajectories, then there exists another traversally generic vector field v' such that $gc(v') = gc(v)$ and $\partial_3X(v') = \emptyset$. Therefore $\pi_1(X/\partial_1^+X)$ admits a presentation with $gc(v)$ generators at most. In particular, if a traversally generic v delivers $gc(X)$, then there exists another traversally generic vector field v' such that $gc(v') = gc(X)$ and $\partial_3X(v') = \emptyset$. Therefore $\pi_1(X/\partial_1^+X)$ admits a presentation with $gc(X)$ generators at most.

We notice that each shared segment in a cascade belongs to *three* polygons Π (see Figure 2.11) at most. Thus, it is present in the relations *three* times at most. As a result, the complexity of the presentation of $\pi_1(X/\partial_1^+X)$, induced by any traversally generic cascade, does not exceed three times the number of shared segments of the v-trajectories, or, stated differently, the number of double-tangent trajectories of the v-flow.

When $\partial_1^+X(v)$ is a disk, $\pi_1(X/\partial_1^+X) \approx \pi_1(X)$, and the same estimates of the number of relations and generators apply.

In view of these arguments, we need to clarify only the statement in the third bullet, which deals with the case of a simply-connected boundary ∂_1X. In such a case, the fundamental group $\pi_1(X/\partial_1^+X, \bar{x}_0)$ is a free product of the fundamental group $\pi_1(X, x_0)$ with the free group F_m in m generators, where $m =_{\mathsf{def}} |\pi_0(\partial_1^+X, x_0)| = |\pi_0(\partial_1^+X)| - 1$. Let us justify this claim.

The space $X \cup_{\partial_1^+X} \mathsf{cone}(\partial_1^+X)$ can serve as a homotopy equivalent model of the quotient X/∂_1^+X. Pick a point $x_0 \in \mathrm{int}(X)$ and connect it by smooth paths $\{\gamma_\beta\}_\beta$ to each connected component $\partial_1^+X_\beta$ of the set ∂_1^+X, so that different paths share only x_0. Let Z be a small open regular neighborhood of the graph $\bigcup_\beta \gamma_\beta$ in X. Consider a small open regular neighborhood W of ∂_1^+X in the cone $\mathsf{cone}(\partial_1^+X)$ so that $W \approx \partial_1^+X \times [0, \epsilon)$.

Let $A =_{\mathsf{def}} Z \cup \mathsf{cone}(\partial_1^+X)$, $B =_{\mathsf{def}} X \cup W$, and $C =_{\mathsf{def}} Z \cup W$. Then A, B, C are path-connected and each one contains the base point x_0. Moreover, $A \cap B = C$ and $A \cup B = X \cup_{\partial_1^+X} \mathsf{cone}(\partial_1^+X)$. Now we can apply the van Kampen Theorem (see [Hat], Theorem 1.20) to conclude that

$$\pi_1(A \cup B, x_0) = \pi_1(A, x_0) *_{\pi_1(C,\, x_0)} \pi_1(B, x_0).$$

Let Γ be the suspension of a finite set of cardinality $m+1$. Since A is homotopy equivalent to the graph Γ, we conclude that $\pi_1(A, x_0) \approx \mathsf{F}_m$, the free group in m generators. By the construction of C and A, the natural homomorphism $\pi_1(C, x_0) \to \pi_1(A, x_0)$ is trivial. Using the hypotheses about the simply-connected nature of $\partial_1 X$, for any choice of $x_0' \in \partial_1^+ X$, the composition

$$\pi_1(\partial_1^+ X, x_0') \to \pi_1(\partial_1 X, x_0') \to \pi_1(X, x_0')$$

is trivial. Thus $\pi_1(C, x_0) \to \pi_1(B, x_0)$ must be trivial as well. Therefore the balanced product structure of $\pi_1(A, x_0) *_{\pi_1(C, x_0)} \pi_1(B, x_0)$ reduces to the free product $\pi_1(A, x_0) * \pi_1(B, x_0)$. As a result,

$$\pi_1(X/\partial_1^+ X, \bar{x}_0) = \mathsf{F}_m * \pi_1(X, x_0).$$

We consider an optimal traversally generic vector field v, whose cascade contains $gc(X)$ double-tangent trajectories. Each double-tangent trajectory is present no more than 3 times in the relations of the presentation $\mathcal{P}_v$ of $\pi_1(X/\partial_1^+ X, x_0)$, generated by the cascade $\mathcal{C}(\partial_2^+ X)(v)$. Thus $c(\mathcal{P}_v) \leq 3 \cdot gc(X)$.

Let $f_1, \dots f_m$ be some free generators of F_m. The epimorphism

$$\rho : \pi_1(X/\partial_1^+ X, x_0) \approx \mathsf{F}_m * \pi_1(X, x_0) \to \pi_1(X, x_0)$$

is defined by sending each free generator to f_j to 1. With the help of ρ, the presentation $\mathcal{P}_v$ of $\pi_1(X/\partial_1^+ X, x_0)$ produces a presentation $\mathcal{P}_v'$ of $\pi_1(X, x_0)$. Since ρ maps any set of generators to a set of generators, and any set of relations to a set of relations, $c(\mathcal{P}_v) \geq c(\mathcal{P}_v')$. Therefore,

$$3 \cdot gc(X) \geq c(\mathcal{P}_v) \geq c(\mathcal{P}_v') \geq c(\pi_1(X, x_0)).$$

□

We know only one mechanism that occasionally gives *upper* bounds of the gradient complexity $gc(X)$, where X is produced from a closed manifold Y by deleting a number of balls. The estimate is given in terms of Morse functions on Y. We will explore it in some generality in Theorem 5.10 and Corollary 5.12. Let us glance at how this mechanism works in dimension three.

Let Y be a closed 3-fold with a self-indexing Morse function $f : Y \to \mathbb{R}$, which has a single minimum. So we assume that the Morse index $\lambda(x) =$

$f(x)$ for all f-critical points $x \in \Sigma_f$. Then f and its generic gradient-like field v give rise to a presentation $\mathcal{P}_v$ of $\pi_1(Y)$: its generators are in 1-to-1 correspondence with the f-critical points x of Morse index one and its relations are generated by some of the f-critical points y of Morse index two.

For a given gradient-like field v, we denote by $W^s(x), W^u(x)$ the stable and the unstable manifolds of a critical point x. We assume that v is such that the manifolds $W^s(x)$ and $W^u(y)$ are in general position in Y at each point where they intersect (this is the famous Morse-Smale transversality property [Sm]).

Let $x, y \in \Sigma_f$ be such that $\lambda(y) = 2$ and $\lambda(x) = 1$. Each relation $\{R_y = 1\}$ of the presentation $\mathcal{P}_v$ is obtained by moving along the boundary $\delta_y =_{\mathsf{def}} W^u(y) \cap f^{-1}(1.5)$ of the unstable disk

$$W^u(y) \cap f^{-1}[1.5,\, 2]$$

and recording oriented transversal intersections (in the surface $f^{-1}(1.5)$) of the loop δ_y with the stable disks

$$\big\{W^s(x) \cap f^{-1}[1,\, 1.5]\big\}_{\{x \in \Sigma_f \,|\, \lambda(x)=1\}}$$

in the order they appear along δ_y. This generates a word R_y in the alphabet $\{x, x^{-1}\}_{\{x \in \Sigma_f \,|\, \lambda(x)=1\}}$ and thus a presentation $\mathcal{P}_v$.

Consider the number

$$c_M(Y) =_{\mathsf{def}} \min_{\{f,v\}} c(\mathcal{P}_v),$$

the minimum of presentation complexities being taken over all Morse pairs (f, v) as above (the subscript "M" in $c_M(Y)$ stands for "Morse").

Equivalently, we can think of $c_M(Y)$ as the minimal number of trajectories, connecting critical points of Morse index 1 to critical points of Morse index 2, that a gradient-like field v of a self-indexing Morse function $f : Y \to \mathbb{R}$ with a single minimum might have, provided that v satisfies the Morse-Smale transversality condition. This invariant of Y is not easy to compute precisely; however, each specific pair (f, v) will deliver an upper bound of $c_M(Y)$.

Evidently,

$$c_M(Y) \geq c(\pi_1(Y)).$$

Let (f, v) be the Morse data such that $c(\mathcal{P}_v) = c_M(Y)$. We denote by $\mu(Y)$ the minimal cardinality of the critical set Σ_f for such $c_M(Y)$-realizing Morse pairs (f, v). Clearly, $\mu(Y) \geq 2$.

Theorem 3.8. *Let Y be a connected closed 3-fold. Let X has been obtained from Y by removing m open 3-balls, where $m \geq \mu(Y)$. Then*

$$\frac{1}{3} \cdot c(\pi_1(Y)) \leq gc(X) \leq 4 \cdot c_M(Y).$$

Moreover, for any gradient-like vector field v on Y that satisfies the Morse-Smale transversality condition and realizes $c_M(Y)$, we get:

$$gc(v|_X) \leq 4 \cdot c_M(Y), \text{ and } gc^{\oplus}_{\ominus}(v|_X) = 0.$$

Proof. Note that $\pi_1(X) \approx \pi_1(Y)$. Thus, in view of Theorem 3.7, we need to validate only the inequality $gc(X) \leq 4 \cdot c_M(Y)$.

Consider (f, v) as above and such that $c(\mathcal{P}_v) = c_M(Y)$. Pick a Riemannian metric g on Y, which is Euclidean in the Morse coordinates in the vicinity of each f-critical point. Assume that $v = \nabla_g f$ in the vicinity of the critical set Σ_f and that, for any pair of critical points $x, y \in \Sigma_f$, the stable and the unstable manifolds $W^s(x)$ and $W^u(y)$ are transversal at the points where they intersect.

As usual, v produces a handle decomposition of Y with a single 0-handle, a number of 1-handles which correspond to h-critical points of Morse index 1, and a number of 2-handles which correspond to the critical points of Morse index 2.

Since $W^s(x)$ and $W^u(y)$ are transversal, the v-trajectories that connect a critical x of Morse index 1 to a critical y of Morse index 2 are in 1-to-1 correspondence with the points of the intersection

$$W^u(y) \cap f^{-1}(1.5) \cap W^s(x).$$

As we just mentioned, this geometry generates a presentation $\mathcal{P}(v, Y)$ of $\pi_1(Y)$, whose generators are the critical points $\{x \in \Sigma_f | \lambda(x) = 1\}$ and whose relations are labelled with the critical points $\{y \in \Sigma_f |\ \lambda(x) = 2\}$. Each relation is obtained by moving along the loop $W^u(y) \cap f^{-1}(1.5)$ and writing down the symbol $x^{\pm}$ each time we encounter an intersection

$$W^u(y) \cap f^{-1}(1.5) \cap W^s(x).$$

With the manifolds $\{W^u(y)\}$, $\{W^s(x)\}$, and Y being given arbitrary orientations, the sign of the exponent $x^{\pm}$ depends on the sign of the intersection of the oriented curve $W^u(y) \cap f^{-1}(1.5)$ with the oriented curve $f^{-1}(1.5) \cap W^s(x)$ in the oriented surface $f^{-1}(1.5)$. Evidently, the complexity $c(\mathcal{P}(v, Y))$ is the total number of the trajectories that link critical points of index 1 with critical points of index 2. So, the number of such trajectories is $c(\pi_1(Y))$ at least.

Now let us delete from Y some small geodesic balls $\{\mathbf{D}^3_z\}_{z\in\Sigma(f)}$, centered on the critical points of f. Their size is picked so that, in the vicinity of each critical point of indices 1 and 2, the variation of f in the balls is less than $\epsilon < 0.5$. Their number can be assumed to be $\mu(Y)$.

We denote by X the complement to the balls in Y and restrict (f, v) to X. Then each critical point x of index 1 contributes an equatorial annulus $A_x \subset \partial\mathbf{D}^3_x$, and each critical point y of index 2 contributes a pair of "polar" disks $B^\pm_y \subset \partial\mathbf{D}^3_y$ to the set $\partial^+_1 X$ (see Figure 3.2). The boundaries of A_x and $B^\pm_y$ form the set $\partial^+_2 X$. Moreover, $\partial^+_1 X$ consists of $\{A_x\}_x$ and $\{B^\pm_y\}_y$, together with the sphere $S =_{\mathsf{def}} f^{-1}(0.5)$, centered on the point $f^{-1}(0)$ of the absolute minimum 0.

The cascade $\mathcal{C}(v) = \mathcal{C}(\partial^+_2 X)$ falls from the union of loops

$$\Big(\coprod_x \partial(A_x)\Big) \coprod \Big(\coprod_y \partial(B^\pm_y)\Big)$$

on the "ground" $S \coprod (\coprod_x A_x))$. Note that this particular choice of Morse data (f, v) on X has the two properties: **(1)** $\partial^-_2 = \emptyset$, so the field v is *concave*, and **(2)** $\partial_3 X = \emptyset$ so that v is also 3-*convex* and 3-*concave*! All these observations are based on the quadratic nature of h in the Morse coordinates.

We denote by $\mathsf{W}^\pm_x$ the two waterfalls that stream down from the two loops $\partial B^\pm_x$, and by $\mathsf{W}^\pm_y$ the two waterfalls that stream down from the two loops that form ∂A_y. The portions of trajectories, shared by $\mathsf{W}^\pm_x$ and $\mathsf{W}^\pm_y$, are in 1-to-1 correspondence with the trajectories γ that are tangent to both spheres $\partial\mathbf{D}^3_x$ and $\partial\mathbf{D}^3_y$. Where they reach the ground sphere S, the Q-singularities of the gradient spine $K(v) = \partial^+_1 X \cup \mathcal{C}(v)$ are generated. Moreover, every Q-singularity of $K(v)$ is generated in this way by an appropriate pair of critical points x, y.

We also introduce the sets $\mathsf{V}(z) \subset Y$ that are formed by all the v-trajectories that are tangent to the sphere $\partial\mathbf{D}^3_z$, where $z \in \Sigma(f)$.

Let us examine what happens to the trajectories that are tangent to both spheres, $\partial\mathbf{D}^3_y$ and $\partial\mathbf{D}^3_x$, as the balls $\mathbf{D}^3_y$ and $\mathbf{D}^3_x$ shrink.

As the radius of $\mathbf{D}^3_y$ approaches zero, the two loops $\mathsf{V}(y) \cap f^{-1}(1.5)$ become very close in the C^∞-topology to the loop $W^u(y) \cap f^{-1}(1.5)$ which separates them. Similarly, as the radius of $\mathbf{D}^3_x$ approaches zero, the two loops $\mathsf{V}(x) \cap f^{-1}(1.5)$ become very close in the C^∞-topology to the loop $W^s(x) \cap f^{-1}(1.5)$ which separates them.

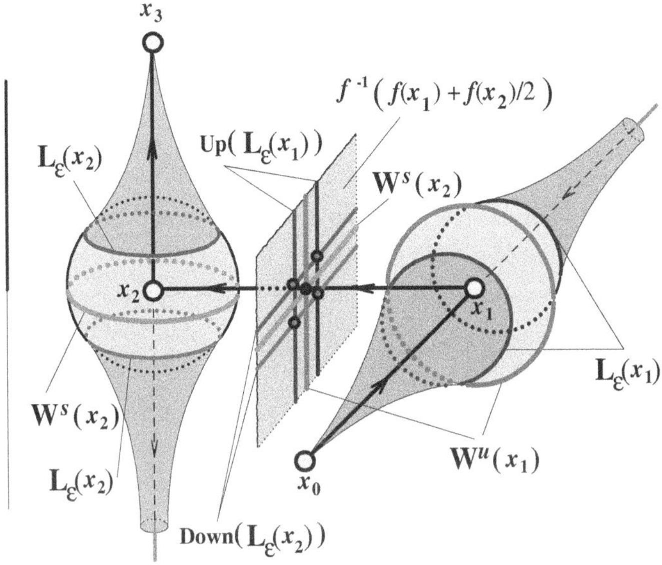

Fig. 3.2 In the diagram, $\{x_i\}_{i\in[0,3]}$ are critical points of indicies i of a Morse function f. $W^s(x_i)$ and $W^u(x_i)$ denote the stable and unstable manifolds of x_i, and $L_\epsilon(x_i)$ the loci where the f-gradient flow is tangent to the ϵ-spheres, centered on x_i. Each trajectory that connects x_1 to x_2 generates *four* trajectories that connect $L_\epsilon(x_1)$ to $L_\epsilon(x_2)$. The plane represents the constant level surface $f^{-1}\big(\frac{1}{2}(f(x_1)+f(x_2))\big)$.

Therefore, using that the intersections of the loop $W^u(y)\cap f^{-1}(1.5)$ with the loop $W^s(x)\cap f^{-1}(1.5)$ are transversal, each point a from

$$W^u(y)\cap f^{-1}(1.5)\cap W^u(x)$$

corresponds to *four* points from

$$\mathsf{V}(y)\cap f^{-1}(1.5)\cap \mathsf{V}(x),$$

all of which reside in the vicinity of a (see Figure 3.2). All these 4-tuples are exactly the Q-singularities of the spine in X, generated by v.

Therefore

$$gc(v) = 4\cdot \#\{\big(\cup_x\, W^s(x)\big)\cap f^{-1}(1.5)\cap \big(\cup_y\, W^u(y)\big)\}$$

—the length of the presentation $\mathcal{P}(v,Y)$ of the group $\pi_1(Y)\approx\pi_1(X)$. Recall that v was chosen so that $\mathcal{P}(v,Y)=c_M(Y)$. Hence,

$$gc(X)\leq gc(v) = 4\cdot c_M(Y).$$

Employing Theorem 3.4, we conclude that the same inequalities hold for any other manifold $\tilde{X}$ obtained from X by deleting any additional 3-balls (i.e., by deleting $m \geq \mu(Y)$ balls from Y).

We notice that the two "parallel" loops from $\mathsf{V}(y) \cap f^{-1}(1.5)$ have opposite orientations (induced by the orientation of the polar cups $B_y^{\pm} \subset \partial_1^+ X$). Similarly, the two "parallel" loops from $\mathsf{V}(x) \cap f^{-1}(1.5)$ have opposite orientations (their orientations are induced by the orientation of the annulus $A_x \subset \partial_1^+ X$). Hence, each 4-tuple as above contains exactly two intersection points of each polarity. As a result, the Q-singularities of the spine occur in pairs of *opposite* polarities $\oplus, \ominus$. Therefore, for the $c_M(Y)$-realizing field v, the polarized complexity $gc_{\ominus}^{\oplus}(v|_X) = 0$. □

For example, we can apply these considerations to the closed 3-folds that admit a genus 1 Heegaard splittings: such manifolds are obtained by gluing two solid tori along their boundaries. With the exception of $S^2 \times S^1$ and S^3, they are the lens spaces $L_{p,q}$.

Let us pick a matrix $A \in \mathsf{SL}_2(\mathbb{Z})$ whose first row is $(p, -r)$ and the second one is (q, s), so that the determinant $qr + ps = 1$. We employ A to form a closed manifold Y_A from the disjoint union of two solid tori, $\mathbf{T}_1$ and $\mathbf{T}_2$. Their boundaries, T_1 and T_2, are the 2-tori. They admit meridian-longitude bases (μ_1, λ_1) and (μ_2, λ_2), where the loop μ_i bounds an embedded disk in $\mathbf{T}_i$. The two tori T_1 and T_2 are identified with the help of a diffeomorphism $\mathcal{A}$, induced by the linear isomorphism $\tilde{\mathcal{A}} : \mathbb{R}^2 \to \mathbb{R}^2$ with the matrix $A \in \mathsf{SL}_2(\mathbb{Z})$. The gluing diffeomorphism $\mathcal{A} : T_1 \to T_2$ maps the meridian μ_1 to the simple geodesic loop $p \cdot \mu_2 + q \cdot \lambda_2$. The image $\mathcal{A}(\mu_1)$ determines the topological type of the resulting manifold

$$Y_A =_{\mathsf{def}} \mathbf{T}_1 \cup_{\{\mathcal{A}: T_1^2 \to T_2^2\}} \mathbf{T}_2.$$

So the topology of Y_A depends only on p and q. For $p \geq 2$ and $0 < q < p$ (p and q being relatively prime), we get the lens space $L_{p,q}$.

Corollary 3.3. *Let X be a 3-fold, obtained from the lens space $L_{p,q}$ by removing four 3-balls. Then,*

$$gc(X) \;\leq\; 4 \cdot \min\{q,\, p - q,\, q',\, p - q'\},$$

where $qq' = \pm 1 \mod p$, and $q,\, q' \in [1,\, p-1]$.

Proof. The corollary is obtained by constructing a Morse function

$$f : L_{p,q} = \mathbf{T}_1 \cup_{\{\mathcal{A}: T_1^2 \to T_2^2\}} \mathbf{T}_2 \to \mathbb{R},$$

induced by a Morse function $h : \mathbf{T}_1 \coprod \mathbf{T}_2 \to \mathbb{R}$ such that:

(1) h is constant on $T_1 \coprod T_2$,

(2) h has a single minimum and a single critical point of index 1 in $\mathbf{T}_2$,

(3) h has a single maximum and a single critical point of index 2 in $\mathbf{T}_1$.

Here is a way of producing the desired function $h : \mathbf{T}_i \to \mathbb{R}$. Cut the solid tori $\mathbf{T}_i$ open to get solid cylinders $\mathbf{C}_i$ of hight 1 and with the base $\{x^2 + y^2 \leq 1\}$. Pick a smooth periodic function $\phi : \mathbb{R} \to [0, 0.4]$ with period 1, a single simple maximum in the interval $[0, 1]$, located at 0.5, and a single simple minimum 0, located at 0 (we choose ϕ so that $\phi''(0) > 0$ and $\phi''(0.5) < 0$). Let

$$h_1(x, y, z) =_{\text{def}} (1 - x^2 - y^2)\phi(z) + (x^2 + y^2).$$

Evidently, it is periodic in z with period 1 and is identically 1 on the boundary $\{(x, y, z)|\ x^2 + y^2 = 1\}$. A straightforward computation shows that the only (non-degenerated) critical points of $h_1 : \mathbf{C}_1 \to \mathbb{R}$ are: $a = (0, 0, 0)$ of Morse index 0, and $b = (0, 0, 0.5)$ of Morse index 1. The function $h_2 : \mathbf{T}_2 \to \mathbb{R}$, in the new-old coordinates x, y, z, is defined by the formula $h_2 = 2 - h_1$. It is identically 1 on the boundary of the torus and has a single simple local maximum d and a single critical point c of Morse index 2. In fact, $h_1 \coprod h_2 : \mathbf{T}_1 \coprod \mathbf{T}_2 \to \mathbb{R}$ induces a function $f : L_{p,q} \to \mathbb{R}$ with the desired properties.

For any f-gradient-like flow v on $L_{p,q}$ the unstable disk W_c^u can be used to define the meridian $\mu_2 =_{\text{def}} W_c^u \cap T_2$. Similarly, the stable disk W_b^s can be used to define the meridian $\mu_1 =_{\text{def}} W_b^s \cap T_1$. Thus the simple loop $\mathcal{A}(\mu_1) = p \cdot \mu_2 + q \cdot \lambda_2$ has q points of intersection with the meridian μ_2. These points are labeling the v-trajectories that connect b to c.

Now, by applying to f the construction in Theorem 3.7, we get that $gc(L_{p,q}) \leq 4q$. Since $L_{p,q}$ and $L_{p,q'}$ are diffeomorphic, when $q' \equiv \pm q^{\pm 1}$ mod p (see [Re]), the claim follows. □

3.3 Which spines are gradient?

Next, we address a natural question: Which spines $K \subset X$ are produced in the form $\partial_1^+ X \cup \mathcal{C}(\partial_2^+ X)$ by traversally generic fields v? In other words, *which spines are gradient?*

It turns out that any $\vec{\mathsf{Y}}$-spine and, thus by Lemma 2.11, any *branched* spine K can be *approximated* by a gradient spine without compromising its combinatorial complexity $c(K)$.

Theorem 3.9. *Let K be a simple $\vec{\mathsf{Y}}$-spine*[5] *in a smooth compact connected 3-fold X with boundary. Then, there exist a nonsingular smooth function $f : X \to \mathbb{R}$ and its traversally generic gradient-like vector field v, so that the gradient spine $K(v) \subset X$ has the following properties:*

- *K is PL-homeomorphic to a 2-complex, obtained from $K(v)$ by elementary collapses of some of its 2-cells,*
- *$c(K(v)) = c(K)$.*

Assume that a traversally generic Morse pair (f', v') in X is such that the vector field v', being restricted to $s(K)$, gives the simple spine K its $\vec{\mathsf{Y}}$-spine structure. Then, for each $\epsilon > 0$, exist new Morse pair (f, v), as in the first two bullets, that satisfies two additional properties:

- *f and f' coincide, when restricted to K,*
- *the f-controlled size of all elementary 2-collapses, in which the collapse $K(v) \searrow K$ decomposes, does not exceed ϵ.*[6]

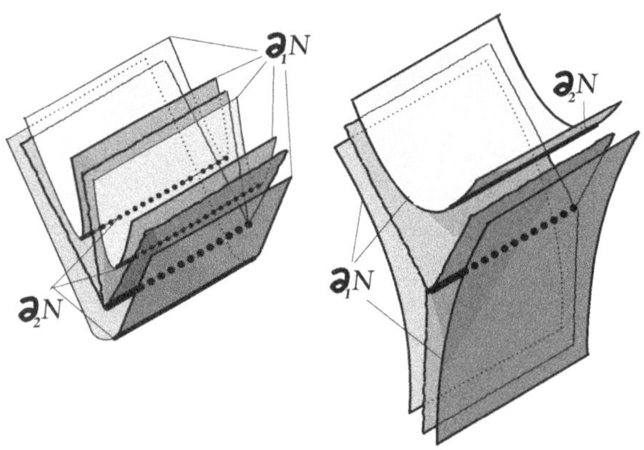

Fig. 3.3 A smooth regular neighborhood $\mathcal{N}$ of a spine K, the 3-page open book. The vector field v is orthogonal to the plane of drawing Π. Consider the projection $p : \partial_1\mathcal{N} \to \Pi$ along $-v$. The locus $\partial_2\mathcal{N}$, or rather its the p-image, is shown as the three bold curves in the left diagram (the $\mathcal{W}$-configuration) and as the single bold curve in the right one (the $\mathcal{Y}$-configuration). Note that $p(\partial_2\mathcal{N})$ is a *simple* curve, so no double-tangent v-trajectories are present in $\mathcal{N}$.

[5] see Definition 2.12.

[6] i.e. the f-images of the collapsing 2-cells are ϵ-small.

Proof. Let $K \subset X$ be a simple $2D$-subcomplex of a smooth triangulation of X. Let w be a vector field along $s(K)$ that delivers the $\vec{\mathsf{Y}}$-structure to K. We start with a construction of w-adjusted Morse pair (f, v) by extending w from the graph $s(K)$ to a smooth field $v \neq 0$ in an open neighborhood of $s(K)$ in X. Since $s(K)$ is 1-dimensional and w is transversal to the pages of K along $s(K)$, for a sufficiently small $\epsilon > 0$, the v-flow ϕ_t will have the property $\phi_t(s(K)) \cap \phi_{t'}(s(K)) = \emptyset$ for all $0 \leq t < t' \leq \epsilon$. Hence, the set $L =_{\mathsf{def}} \{\phi_t(x)\}_{x \in s(K),\, 0 \leq t \leq \epsilon}$ can be given a product structure $H : L \approx s(K) \times [0, \epsilon]$, where H is a piecewise differential homeomorphism of the stratified spaces. We define a function $\hat{f} : L \to \mathbb{R}$ as the composition of H with the projection on $[0, \epsilon]$.

We approximate $\hat{f} : L \to \mathbb{R}$ by a smooth function $\tilde{f} : X \to \mathbb{R}$ with isolated singularities and such that $d\tilde{f}(v) > 0$ in the vicinity of $s(K)$. Applying a smooth isotopy to $K \subset X$, which is fixed on $s(K)$, we may assume that the singularities of $\tilde{f}$ in X are not located on K. Thus they can be removed by finger moves $X \Rightarrow X \setminus U$, where an open set $U \subset X \setminus K$ is a disjoint union of dead end tunnels; each tunnel originates at some point from $\partial_1 X$ and is confined to $X \setminus K$. The restriction f of $\tilde{f}$ to the resulting manifold $X \setminus U \approx X$ is a nonsingular smooth function. Its gradient field v has the property $v|_{s(K)} = w$ and gives K its $\vec{\mathsf{Y}}$-spine structure.

We consider a small open regular neighborhood $\mathcal{N}$ of $K \subset X$. The Regular Neighborhood Theorem (see [Hud], Theorems 2.11, 2.16) implies that K is a spine for $\mathcal{N}$ (see Definition 2.3 and Remark 2.1) and that there is a PL-homeomorphism $g : \mathcal{N} \to X$ which is an identity on K. In dimension three, we can smoothen $\mathcal{N}$ and approximate g by a diffeomorphism (see [Th], Proposition 3.10.7 and Theorem 3.10.8). Abusing notations, we still denote the smoothen versions of $\mathcal{N}$ and g by the same symbols.

Using the smooth product structure in $\mathcal{N} \setminus K$, we can construct a smooth function $F : \mathcal{N} \to [0, 1]$ such that:

- $F^{-1}(0) = K$,
- $F^{-1}(1) = \partial\mathcal{N}$,
- the interval $(0, 1]$ is the set of regular values for F,
- K is the critical set of F.

The minus gradient-like flow of F defines a collapse $\mathcal{N} \searrow K$.

Next, we pick $\epsilon > 0$ so small that $\mathcal{N}_\epsilon =_{\mathsf{def}} F^{-1}([0, \epsilon])$ interacts with the v-flow as is depicted in Figure 3.3 the right diagram, Figure 3.4, and Figure 3.7, diagrams 1 and 2. This depiction—a linearization at a point from $s_\dagger(K)$ of the surfaces that form K—is based on the properties of v being in general position with respect to the surfaces K° and giving K its $\vec{\mathsf{Y}}$-structure.

Consider now the stratification of the given simple spine K by points, whose links are shown in Figure 2.21, diagrams (5), (6), (1), (2), (4).

Let us take a closer look at the interaction of v with the boundary $\partial_1\mathcal{N}_\epsilon =_{\mathsf{def}} \partial\mathcal{N}_\epsilon$:

(1) in the vicinity of $s_\dagger(K) \setminus Q(K)$,
(2) in the vicinity of $Q(K)$,
(3) along the loops of v-tangency, which are located in K°,
(4) in the vicinity of $s_\bullet(K) \setminus s_{\bullet\dagger}(K)$,
(5) in the vicinity of $s_{\bullet\dagger}(K)$.

These five cases correspond to the five link types of points from K, listed in the order in which they appear in Figure 2.21. The third case is exceptional: it does not reflect the local geometry of the spine K, only the way in which the field v, under construction, interacts with $K \setminus s(K)$; we will see how to take care of this anomaly during the course of the proof.

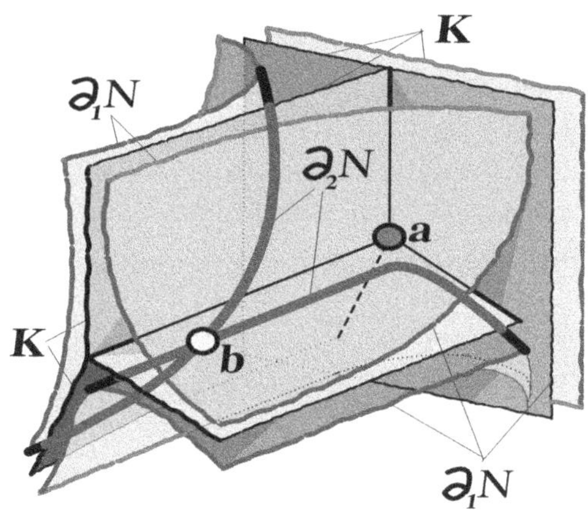

Fig. 3.4 Let Π be the plane of the drawing, orthogonal to the vector field v. The picture shows a regular neighborhood $\mathcal{N}$ of the spine in the vicinity of its Q-singularity a. Note the intersection point b of the two bold curves—the p-images of the folds of the projection $p : \partial_1\mathcal{N} \to \Pi$ along $-v$. Point b is the p-image of the unique double-tangent v-trajectory in $\mathcal{N}$.

In the first case (see the three-page open book in Figure 3.3, the right diagram, and its 2D-section in Figure 3.7, diagram (2)), the gradient spine $K^\sharp_\epsilon =_{\mathsf{def}} K(\mathcal{N}_\epsilon, v)$, generated by the vector field v on $\mathcal{N}_\epsilon$, is locally homeomorphic to the given simple spine K. This conclusion depends heavily on the $\vec{\mathsf{Y}}$-nature of the field. Moreover, because of this nature, for a sufficiently small ϵ, the image of the fold is a simple arc—no double-tangent trajectories are present in the vicinity of $s_\dagger(K) \setminus Q(K)$ (see Figure 3.3, the right diagram). Let us compare diagrams (2) and (3) in Figure 3.7: diagram (2) reflects the fact that v is of the $\vec{\mathsf{Y}}$-type, while in diagram (3) shows the vector field of the $\vec{\mathsf{W}}$-type.

In the second case (see Figure 3.4, where $\mathcal{N}_\epsilon$ is abbreviated to $\mathcal{N}$, and Figure 3.8), the surface $\partial_1\mathcal{N}_\epsilon$ and the field v in the vicinity of the Q-singularity are transversal. At Q, the vectors $v, -v$ point into two (out of four) chambers, in which X is divided by K. Each of the other two chambers contains the folding locus $\partial_2 X$, shown as the two bold arcs in Figure 3.4. The transversal intersection b of the two arcs is the image of a single double-tangent v-trajectory for the v-generated gradient spine $K^\sharp_\epsilon$ of $\mathcal{N}_\epsilon$.

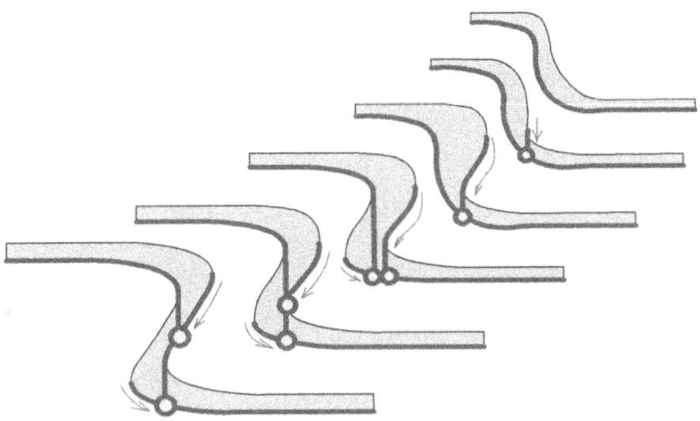

Fig. 3.5 Slicing the gradient spine $K^\sharp_\epsilon \subset \mathcal{N}_\epsilon$ by a 1-parameter family of v-invariant "parallel" planes in the vicinity of a cusp. The arrows indicate elementary collapses of 2-cells in $K^\sharp_\epsilon$. The collapses result in a 2-complex, PL-homeomorphic to the given K.

The third case (see Figure 3.7, diagram (1), and Figure 3.5), the behavior of v with respect to $\partial\mathcal{N}_\epsilon$ and to K° is similar to the behavior of a boundary generic field with respect to $\partial_1 X$ along $\partial_2 X$ (in other words, the v-guided projections of $\partial\mathcal{N}_\epsilon$ and of K° on local transversal sections of the flow have the fold and cusp singularities only). Consider a smooth arc

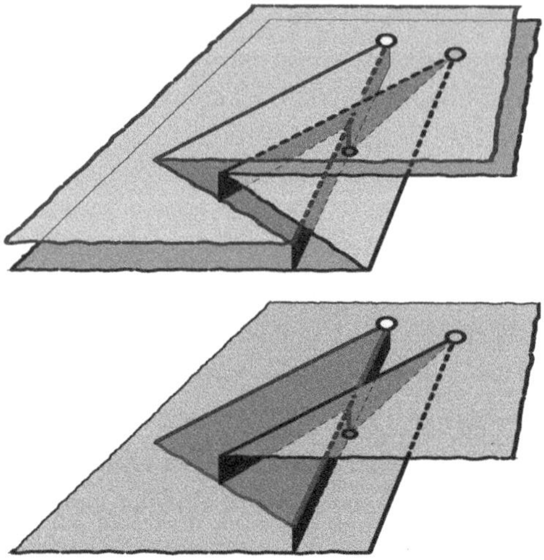

Fig. 3.6 A regular neighborhood $\mathcal{N}_\epsilon$ of a cusp from $\partial_3(K^\circ, v)$ (the upper diagram) and its gradient spine $K^\sharp_\epsilon \subset \mathcal{N}_\epsilon$ with a single Q-singularity (the lower diagram).

$L \subset K^\circ$, were v is tangent to the surface K° and transversal to L. In the vicinity of a point $x \in L$, the neighborhood $\mathcal{N}_\epsilon$ and the flow are represented, up to a diffeomorphism, by the product of the first diagram in Figure 3.7 with a segment. We consider the portion of the cascade $K^\sharp_\epsilon$, generated in $\mathcal{N}_\epsilon$ by v and marked with a dotted line. Collapsing this portion produces a 2-complex, which is locally PL-homeomorphic to the original K.

Points $x \in L$, where v is tangent to L, require additional attention. There K° and v have local geometry, similar to the geometry of $\partial_1 X$ in the vicinity of a cusp point from $\partial_3 X$ (see Figures 3.5 and 3.6). Figure 3.5 demonstrates how $K^\sharp_\epsilon$ collapses onto the given K. We notice that each cusp from $\partial_3(K^\circ, v)$ contributes a single Q-singularity to the spine $K^\sharp_\epsilon$ (i.e., one double-tangent trajectory to $\mathcal{N}_\epsilon$, as shown in Figure 3.6), a Q-singularity which has no counterpart in the original K! Fortunately, according to Theorem 4.6, we can modify v away from a neighborhood of $s(K)$ (where v is transversal to $K \subset X$ and gives it its $\vec{\mathsf{Y}}$-structure) so that, for the modified vector field, the field-guided projections of K° onto a transversal to the flow local sections will have no cusps! This cusp-eliminating deformation of v, supported in the vicinity of $K \setminus (\mathcal{N}_\epsilon(s(K)) \cap K)$, is not a small one. It can be performed prior to considering cases (1), (2) and (4), (5) from the above list of five items.

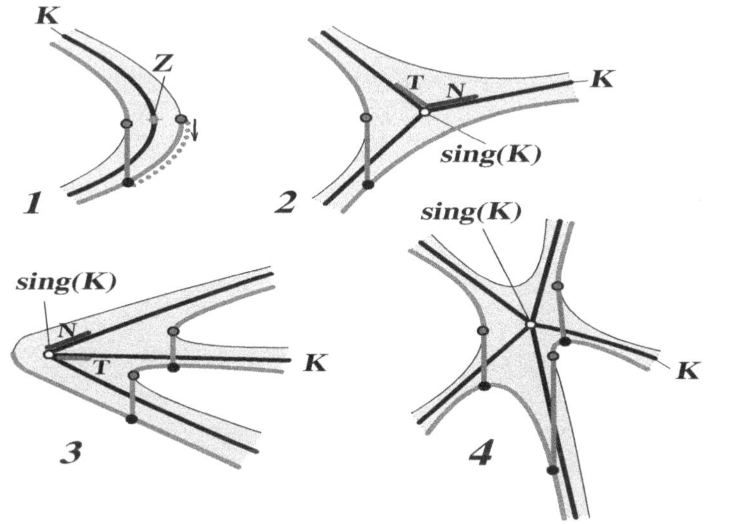

Fig. 3.7 Sections of a spine K and of the associated gradient spine $K^\sharp_\epsilon \subset \mathcal{N}_\epsilon$ by a v-invariant plane, transversal to the singular set $s(K)$. Diagrams 3 and 4 illustrate complications for the shape of $K^\sharp_\epsilon$, arising when v is of $\vec{\mathsf{W}}$-type, or when $K \subset X$ is not simple.

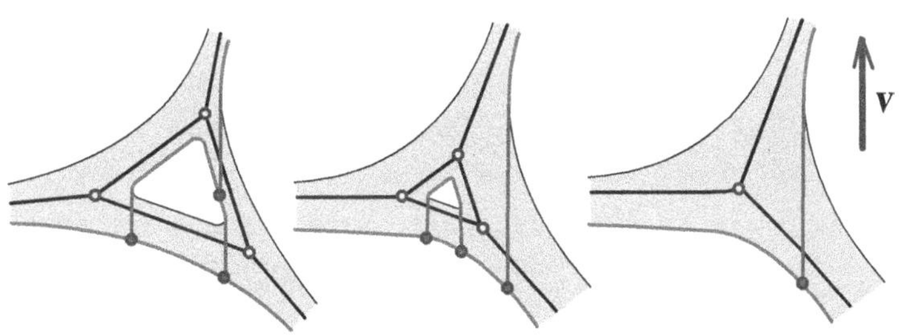

Fig. 3.8 Sections of the gradient spine $K^\sharp_\epsilon \subset \mathcal{N}_\epsilon$ by a generic family of parallel surfaces in vicinity of a Q-singularity. The surfaces are invariant under the v-flow.

Simpler cases (4) and (5) (cf. Figure 3.5) are treated similarly. None of them contributes Q-singularities to the gradient spine of the neighborhood $\mathcal{N}_\epsilon$.

Therefore, for the appropriate v, we get $gc(\mathcal{N}_\epsilon, v) = c(K^\sharp_\epsilon) = c(K)$.

Next, we use the diffeomorphism $g_\epsilon^{-1} : X \to \mathcal{N}_\epsilon$ to transplant the previously constructed Morse pair (f, v) from $\mathcal{N}_\epsilon$ to X. Evidently, the transplanted pair and their gradient spine still have all the desired properties.

We leave to the reader to verify that the size of elementary collapses, as measured in terms of the f-variation, can be made arbitrary small. The argument uses the uniform continuity of the smooth functions F and f, together with the fact that v is in general position with respect to K. These imply that, for a sufficiently small $\epsilon' > 0$, there is $\epsilon > 0$ so that in $\mathcal{N}_\epsilon =_{\mathsf{def}} F^{-1}([0, \epsilon])$ the variation of f along each v-trajectory does not exceed ϵ' (see Figure 3.7, diagram 1). Moreover, the f-controlled size of elementary collapses does not exceed the f-controlled size of the trajectories inside $\mathcal{N}_\epsilon$. This property is preserved under the diffeomorphism g_ϵ^{-1}. □

Definition 3.8. The *branched oriented complexity* $boc(X)$ of a 3-fold X is the minimum of the combinatorial complexities $\{c(K)\}$, where K varies over all simple spines of X that admit a branched orientation (as in Definition 2.11). ◇

Theorem 3.9 has an important implication:

Theorem 3.10. *For any compact connected 3-fold X with boundary,*

$$boc(X) = c_{\vec{\mathsf{Y}}}(X) = gc(X).$$

Proof. Let $K \subset X$ be a simple $\vec{\mathsf{Y}}$-spine with $c(K) = c_{\vec{\mathsf{Y}}}(X)$—the optimal number of Q-singularities. By Theorem 3.9, there exists a traversally generic field v on X such that $gc(v) = c_{\vec{\mathsf{Y}}}(X)$. Thus $gc(X) \leq c_{\vec{\mathsf{Y}}}(X)$. On the other hand, by Lemma 2.10, any gradient spine is a $\vec{\mathsf{Y}}$-spine, a wider category of spines. Hence, examining the definitions of the two complexities, we get $gc(X) = c_{\vec{\mathsf{Y}}}(X)$. By Lemma 2.11, the notions of oriented branched spine and of the $\vec{\mathsf{Y}}$-spine are equivalent. So $boc(X) = c_{\vec{\mathsf{Y}}}(X)$. Thus we have shown that $gc(X)$ is equal to $boc(X)$ as well. □

Theorem 3.11. *For any 3-fold X, which is a boundary connected sum (or a connected sum) of boundary irreducible with no essential annuli 3-folds $\{X_k\}_k$ with $c(X_k) > 1$, the following inequality is valid:*

$$c(X) \leq gc(X) \leq 6 \cdot c(X).^{7}$$

[7]Note that, for the lens space $L_{8,3}$, both Theorem 3.11 and Corollary 3.3 give the same upper boundary 12 for its gradient complexity $gc(L_{8,3})$.

Proof. By Theorem 3.2 (cf. [Ma] Theorem 2.2.4), any boundary irreducible with no essential annuli X_k has a *special* spine K_k of complexity $c(X_k)$ at most, provided $c(X_k) > 1$. According to [BP], Theorem 3.4.9, for any special spine K_k with $c(K_k) > 1$, there exists a sequence of not more than $5 \cdot c(K_k)$ oriented Matveev-Piergallini moves (see [Ma] Figure 1.12, for the definition of these moves) which convert K_k into an *oriented branched* spine. This Benerdetti-Petronio Theorem is based on an intricate argument, utilizing the calculus of oriented branched fat graphs, developed in [BP]. The sequence of moves is depicted in [BP], Figure 3.19. Furthermore, each move increases the combinatorial complexity of the modified spine by one. Thus, these moves result in a branched spine of complexity $6 \cdot c(X_k)$ at most. Next, by Lemma 2.11, any branched spine is a $\vec{\mathsf{Y}}$-spine. By Theorem 3.9, it admits an approximation by a gradient spine of the same complexity. Therefore $gc(X_k) \leq 6 \cdot c(X_k)$.

Now the claim follows from the semi-additivity of $gc(\sim)$ (Theorem 3.4) and the additivity of $c(\sim)$ under connected sums/boundary connected sums of 3-folds ([Ma], Theorem 2.2.9), the later nontrivial property being based on the Haken theory of normal surfaces. □

Example 3.1. Since $c(L^\circ_{5,1}) = 2$ and $c((S^3/\mathrm{Q}_8)^\circ) = 2$ (see [Ma], Table A.9 on page 461), by Theorem 3.11, we get:

$$gc\big(L^\circ_{5,1}\#\,(S^3/\mathrm{Q}_8)^\circ\big) \leq 6\big[c(L^\circ_{5,1}) + c((S^3/\mathrm{Q}_8)^\circ)\big] = 6(2+2) = 24.$$

However, using Corollary 3.3, this estimate can be improved:

$$gc\big(L^\circ_{5,1}\#\,(S^3/\mathrm{Q}_8)^\circ\big) \leq gc(L^\circ_{5,1}) + gc((S^3/\mathrm{Q}_8)^\circ) \leq 4 + 6\cdot 2 = 16.$$

Therefore, we get

$$4 = c\big(L^\circ_{5,1}\#\,(S^3/\mathrm{Q}_8)^\circ\big) \leq gc\big(L^\circ_{5,1}\#\,(S^3/\mathrm{Q}_8)^\circ\big) \leq 16.$$

Thus, any traversally generic flow on $L^\circ_{5,1}\#\,(S^3/\mathrm{Q}_8)^\circ$ has 4 double-tangent trajectories at least, and there exists a traversally generic flow with 16 double-tangent trajectories at most. ◇

Let $\sigma(p,q)$ denote the sum of all partial quotients in the expansion of p/q as a *regular continued fraction*.

Corollary 3.4. *For the connected sum/boundary connected sum of punctured lens spaces* $L^\circ_{p_i,q_i}$ *with* $(p_i,q_i) \neq (3,1),(4,1),(5,2)$, *one gets*

$$gc(\#_i\, L^\circ_{p_i,q_i}) \leq 6 \cdot \sum_i \big(\sigma(p_i,q_i) - 3\big).$$

Proof. The only punctured lens spaces with the combinatorial complexity less than 2 are: $L_{3,1}, L_{4,1}$, and $L_{5,2}$ (see [Ma], Table A.9 on the page 461). Since $c(L^\circ_{p,q}) \leq \sigma(p,q) - 3$ (see [Ma], page 77), using Theorem 3.11, we get the desired upper estimate. □

3.4 The 3-folds of gradient complexity 0 and 1

Since every handlebody X admits a convex gradient flow, $gc(X) = 0$. By Theorem 3.4, deleting several open 3-balls from X produces a new manifold X' also with zero gradient complexity. Note that $\pi_1(X)$ is a free group.

It seems natural to expect that any X with $gc(X) = 0$ is a domain in $\mathbb{R}^3$, a handlebody with holes. However, Theorem 3.12 below claims that there are other examples of 3-folds X with $gc(X) = 0$ and rather complex fundamental groups! We are going to describe a combinatorial scheme that generates all of them. At the same time, when $\partial_1 X$ is just a union of 2-spheres and $gc(X) = 0$, then the topology of X is very special indeed (Theorem 3.13); in particular, $\pi_1(X)$ is free. Moreover, $gc(X) = 0$ is definitely a stronger requirement on X than the constraint $c(X) = 0$!

We start building a combinatorial scheme that captures faithfully the topology of abstract gradient spines of zero complexity. The scheme is cumbersome, so the reader may skip its construction and go directly to the list (3.4) of properties that describe the scheme and to Theorem 3.12 that employs the list.

Consider an oriented 3-fold X which admits a simple abstract gradient spine K (see Definition 2.10) with the singular set $Q(K) = \emptyset$. Let $p : X \searrow K$ be the collapsing map, defined in terms of the PL-map $p^\partial = p| : \partial X \to K$ and the PL-homeomorphism $h : \mathsf{cyl}(p^\partial) \to X$ (see Definition 2.3). In this section, we do not distinguish between $h(K)$ and K; so we treat K as a subset of X.

Since $Q(K) = \emptyset$, the singularity set $s_\dagger(K)$ [8] consists of a number of disjoint simple loops and arcs. The orientation of X, coupled the NT-induced orientation of the normal to $s_\dagger(K)$ bundle, picks a preferred orientation of the arcs and loops in $s_\dagger(K)$. As a part of the abstract gradient spine structure, the surfaces in $K^\circ = K \setminus s_\dagger(K)$ are also oriented.

Let $\mathcal{N}^\dagger_K =_{\mathsf{def}} \mathcal{N}(s_\dagger(K), K)$ be a closed regular neighborhood of the graph $s_\dagger(K)$ in K.

[8] Generic points of $s_\dagger(K)$ have links of the type (5) from Figure 2.21.

First, let us consider the case (we call it "pattern A") when a connected component of $s_\dagger(K)$ is a simple loop ω. Because $\mathcal{N}_K^\dagger$ admits TN-markings, in the vicinity of each loop ω, the boundary $\partial\mathcal{N}_K^\dagger$, consists of three simple loops, parallel to ω and residing on the boundary of the solid torus $\mathbf{T}_\omega$, the regular neighborhood of ω in X. Let

$$\mathcal{N}_\omega =_{\mathsf{def}} \mathcal{N}_K^\dagger \cap \mathbf{T}_\omega \ \text{ and } \ \partial\mathcal{N}_\omega =_{\mathsf{def}} \mathcal{N}_K^\dagger \cap \partial\mathbf{T}_\omega.$$

Recall that ω contributes with the multiplicity ± 1 to the boundary of the fundamental 2-chain $[K] \in C_2(K, s(K); \mathbb{Z})$, so that the orientation of one of the three loops from $\partial\mathcal{N}_\omega$ disagrees with the orientations of the other two.

Next, we concentrate on the role of singular points in $ss(K)$ of the type (4) from Figure 2.21 (for the gradient spines, these correspond to cusps from $\partial_3 X$). Figure 3.9 shows the three patterns of regular neighborhoods $\mathcal{N}_\tau =_{\mathsf{def}} \mathcal{N}(\tau, K) \subset \mathcal{N}_K^\dagger$ of an *arc* τ from the locus $s_\dagger(K)$ (cf. Figure 2.14). These three patterns, which we will refer to as "B, C, D", reflect the fundamental assumption $Q(K) = \emptyset$. The TN-markers along the bold arc

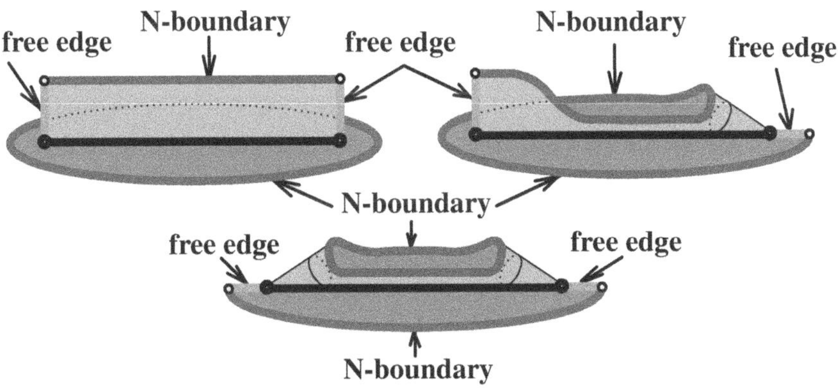

Fig. 3.9 The types B, C, and D of a regular neighborhoods in K of an arc τ (the bold segment) from the locus $s_\dagger(K)$.

τ are not shown, but suggested by the shapes: the T-marker is normal to the oval domains, the N-marker points towards the viewer.

In pattern B, the set $\partial\mathcal{N}_\tau \subset \partial\mathcal{N}_K$, that is shared by $\mathcal{N}_\tau$ and the closure of its complement $K \setminus \mathcal{N}_\tau$ (marked as "N-boundary" in Figure 3.9), is a loop union with an arc; in pattern C, it is a single arc; in pattern D, it is again a loop union with an arc. Each of the three patterns has two *free arcs* that belong to $s_\bullet(K)$, the set of points in K whose links are of the types (2) and (4) from Figure 2.21. We notice the quadrangular bands $H_\tau \subset \mathcal{N}_\tau$ that are bounded by a pair of free arcs, as shown in Figure 3.9, patterns B and D.

Each regular neighborhood $\mathcal{N}_\tau$ is the intersection of the regular neighborhood $\mathcal{N}(\tau, X)$ of the arc τ in X with the spine K. Since X is a PL-manifold, $\mathcal{N}(\tau, X)$ is a PL-homeomorphic to a 3-ball $\mathbf{B}_\tau$. The graph $\partial\mathcal{N}_\tau$ resides in its boundary.

As before, let $p : X \to K$ be the retraction, defined by giving X the mapping cylinder structure of the map $p^\partial : \partial X \to K$. Let

$$K^\odot =_{\mathsf{def}} K \setminus \mathrm{int}(\mathcal{N}_K^\dagger).$$

Because $p^{-1}(K^\odot) \subset X$ inherits a mapping cylinder structure, $K^\odot$ is an abstract gradient spine of the 3-fold $X^\odot =_{\mathsf{def}} p^{-1}(K^\odot)$.

The spine K is obtained from the 2-complex

$$\mathcal{N}_K^\dagger = \big(\coprod_\omega \mathcal{N}_\omega\big) \coprod \big(\coprod_\tau \mathcal{N}_\tau\big)$$

by attaching compact connected and oriented surfaces $\{S_\kappa\}_\kappa$ to $\partial\mathcal{N}_K^\dagger$ along some disjoint loops and arcs in their boundaries $\{\partial S_\kappa\}_\kappa$.

Let us record the information how to assemble K from the building blocks $\{S_\kappa\}$, $\{\mathcal{N}_\omega\}$, $\{\mathcal{N}_\tau\}$ by introducing a *bookkeeping graph* $\mathbf{\Gamma}$. The verticies of $\mathbf{\Gamma}$ represent the arcs and loops that form $s(K)$, as well as the connected components of the surfaces that form $K \setminus s(K)$. As we progress towards this goal, we will add some necessary decorations to $\mathbf{\Gamma}$ (see Figure 3.10).

We start now with a lengthy description of the bookkeeping graph $\mathbf{\Gamma}$. Abusing notations, we denote by the same symbol "ω" a loop $\omega \subset s_\dagger(K)$ and the vertex of $\mathbf{\Gamma}$ that represents it. Similarly, we denote by the same symbol "τ" an arc $\tau \subset s_\dagger(K)$ and the vertex of $\mathbf{\Gamma}$ that represents it. Finally, each connected surface $S_\kappa \subset K \setminus s(K)$ corresponds to a unique vertex $\kappa \in \mathbf{\Gamma}$.

The boundary ∂S_κ is a disjoint union of several loops. Some of these loops have an empty intersection with the neighborhood $\mathcal{N}_K^\dagger$; we call them "*free*". Others are attached to $\mathcal{N}_K^\dagger$ along arcs or/and loops, residing in $\partial\mathcal{N}_K^\dagger$. Each surface S_κ is represented in $\mathbf{\Gamma}$ by a unique vertex κ. We denote by Π the set of such vertexes.

A few edges $\{\gamma_{\kappa,i}\}_i$ emanate from each vertex κ. Each edge $\gamma_{\kappa,i}$ is "plugged" in one of the three "connectors" (see Figure 3.9 for type (2) connector):

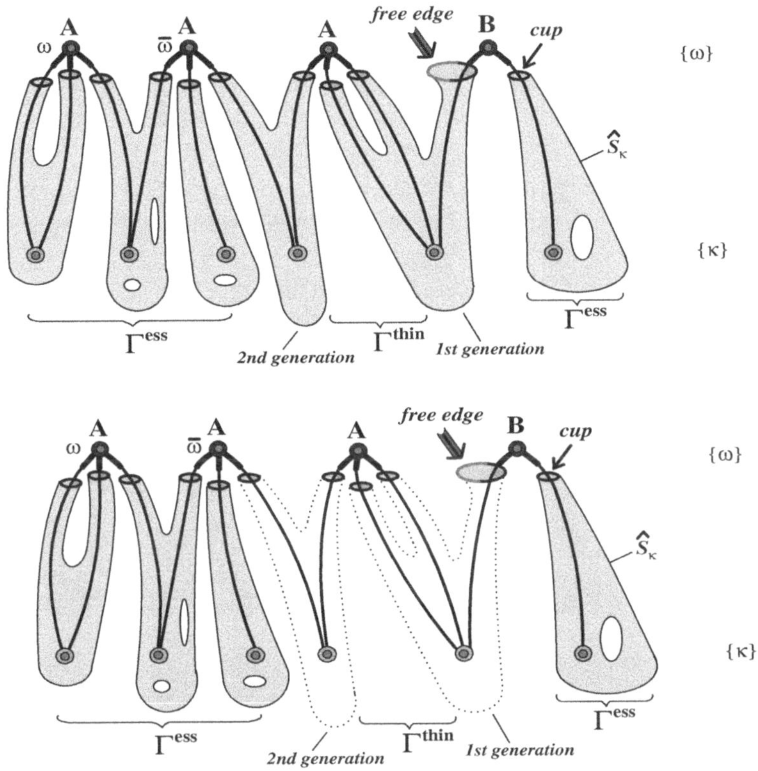

Fig. 3.10 A gradient spine K of zero complexity (the figure shows just the manifold portion $K^{\odot}$ of K) and the graph $\mathbf{\Gamma}$ that organizes the combinatorial information about K. It has three connectors of the type A and one connector of the type B. A free arc in the vicinity of this B-connector serves as the locus, where a collapse of K on its subcomplex originates. The surfaces in K whose union admits a collapse onto 1-dimensional graph are marked as "thin", the rest of the surfaces are marked as "ess".

(1) con_ω—a germ of the star of a trivalent vertex $\omega \in \mathbf{\Gamma}$ that represents a regular neighborhood $\mathcal{N}_\omega$ of a loop $\omega \subset s_\dagger(K)$,

(2) con_τ—a germ of the star of a bivalent or univalent vertex $\tau \in \mathbf{\Gamma}$ that represents a regular neighborhood $\mathcal{N}_\tau$ of an arc $\tau \subset s_\dagger(K)$,

(3) con_ϕ—a germ of the star of an univalent vertex $\phi \in \mathbf{\Gamma}$ that represents a regular neighborhood in K of a free loop or arc ϕ in $\partial S_\kappa \cap \partial_\bullet K$.

Such connectors come in different shapes: type A represents the neighborhood $\mathcal{N}_\omega$ of a loop $\omega \subset s_\dagger(K)$. Since $\partial\mathcal{N}_\omega$ consists of tree loops, the connector con_ω has tree edges ("sockets") to which the edges $\{\gamma_{\kappa,i}\}_{\kappa,i}$ are attached. We decorate two of these tree edges with letters N and T to record the NT-markings of the abstract gradient spine K.

Type B and D connectors con_τ represent the neighborhoods $\mathcal{N}_\tau$ of types B and D from Figure 3.9. For such $\mathcal{N}_\tau$, the boundary $\partial\mathcal{N}_\tau$ is a disjoint union of a loop with an arc; thus con_τ has two sockets which are marked appropriately "loop" and "arc".

If an arc τ is of type C, then $\partial\mathcal{N}_\tau$ is a single arc; it is represented by a vertex τ with a single edge-socket.

Finally, we also may have vertexes $\phi \in \mathbf{\Gamma}$ of type F that represent free loops or arcs from $\coprod_\kappa \partial S_\kappa$, residing in $\partial_\bullet K$. Each ϕ has a single socket.

The resulting graph $\mathbf{\Gamma}$ (see Figure 3.10) has the following properties (they faithfully reflect the definition of an abstract gradient spine K):

- $\mathbf{\Gamma}$ is a connected graph,
- each socket of each connector is occupied by an edge $\gamma_{\kappa,i}$ that emanates from some vertex $\kappa \in \Pi$.

We aim to separate the surfaces $\{S_\kappa\}$ into two types: "*thin*" and "*essential*". Crudely speaking, the union of essential surfaces with $\mathcal{N}_K^\dagger$ cannot be collapsed on its subcomplex; while the union of thin surfaces with $\mathcal{N}_K^\dagger$ collapses on its 1-dimensional subcomplex.

With this goal in mind, consider a subgraph $\mathbf{\Gamma}'$ of $\mathbf{\Gamma}$ that has the following two properties:

- Each vertex $\kappa \in \mathbf{\Gamma}' \cap \Pi$ belongs only to the edges of $\mathbf{\Gamma}$ that are attached to the connectors con_ω of type A and/or to the connectors con_τ of types B and D. Moreover, the edges are plugged only in the sockets of con_τ that are labeled "loops".
- For each connector con_ω of type A, there are at least two edges of $\mathbf{\Gamma}'$ that are plugged in the connector. They retain the NT markers.

Let $\mathbf{\Gamma}^{\mathsf{ess}}$ be the *maximal subgraph* of $\mathbf{\Gamma}$ that possess these two properties of $\mathbf{\Gamma}'$. Note that $\mathbf{\Gamma}^{\mathsf{ess}}$ may not be a connected graph. We call each connected component of $\mathbf{\Gamma}^{\mathsf{ess}}$ a *"cluster"*.

We partition the set Π into $\Pi^{\mathsf{ess}} =_{\mathsf{def}} \Pi \cap \mathbf{\Gamma}^{\mathsf{ess}}$ and $\Pi^{\mathsf{thin}} =_{\mathsf{def}} \Pi \setminus \Pi^{\mathsf{ess}}$. So the surfaces $\{S_\kappa\}_{\kappa\in\Pi}$ that form $K^\odot$ come in two flavors: "thin" and "essential". The thin surfaces correspond to vertexes $\kappa \in \Pi^{\mathsf{thin}}$, and the essential surfaces correspond to vertexes $\kappa \in \Pi^{\mathsf{ess}}$ (see Figure 3.10).

We also partition the arcs τ and loops ω into two categories, "thin" and "ess": by definition, τ or ω belong to ess, if an essential surface S_κ has a nonempty intersection with $\mathcal{N}_\tau$ or $\mathcal{N}_\omega$; the rest of arcs or loops belong to the thin kind.

Similarly, we divide the spine K into a union of two subcomplexes:

$$K^{\mathsf{ess}} =_{\mathsf{def}} \Big(\bigcup_{\kappa \in \Pi^{\mathsf{ess}}} S_\kappa \Big) \cup \Big(\bigcup_{\omega \in \mathbf{\Gamma}^{\mathsf{ess}}} \mathcal{N}_\omega \Big) \cup \Big(\bigcup_{\tau \in \mathbf{\Gamma}^{\mathsf{ess}}} \mathcal{N}_\tau \Big),$$

$$K^{\mathsf{thin}} =_{\mathsf{def}} \Big(\bigcup_{\kappa \in \Pi^{\mathsf{thin}}} S_\kappa \Big) \cup \Big(\bigcup_{\omega \in \mathbf{\Gamma}^{\mathsf{thin}}} \mathcal{N}_\omega \Big) \cup \Big(\bigcup_{\tau \in \mathbf{\Gamma}^{\mathsf{thin}}} \mathcal{N}_\tau \Big).$$

Since each neighborhood $\mathcal{N}_\omega$ or $\mathcal{N}_\tau$ belongs either to thin or ess categories, but not to both, the 2-complex $K^{\mathsf{thin}} \subset K$ is the closure of $K \setminus K^{\mathsf{ess}}$. Chasing definitions, we see that $K^{\mathsf{thin}} \cap K^{\mathsf{ess}} \subset \partial \mathcal{N}_\dagger^K$.

The complex K^{thin} admits a collapse on a graph $G^{\mathsf{thin}} \subset K^{\mathsf{thin}}$, its 1-dimensional skeleton. Here is a description of the collapse $K^{\mathsf{thin}} \searrow G^{\mathsf{thin}}$, executed in stages.

We say that a connected surface $S_\kappa \subset K^\odot$ is *thin in the first generation*, if ∂S_κ has an arc or a loop that belongs to $s_\bullet(K) = \partial_\bullet K$ (such an arc or a loop may be used as the origin of the collapse); in other words, if there is a point in ∂S_κ whose link in K is of type (2) from Figure 2.21. Note that any S_κ, whose boundary shares an *arc* with $\partial \mathcal{N}_K^\dagger$, is automatically thin in the first generation (see Figure 3.9)! If a surface S_κ is not thin in the first generation, then $\partial S_\kappa \cap \partial_\bullet K = \emptyset$.

We say that a surface S_κ is *thin in the second generation*, if S_κ is not thin in the first generation, but ∂S_κ has a loop in some $\partial \mathcal{N}_\omega$, so that the other two loops from $\partial \mathcal{N}_\omega$ belong to the surfaces that are thin in the first generation (see Figures 3.10 and 3.11 that show a surface, thin in the second generation). A surface is *thin in the third generation* if it is not in the first and second generations, but ∂S_κ has a loop in some $\partial \mathcal{N}_\omega$ so that the other two loops from $\partial \mathcal{N}_\omega$ belong to the surfaces that are thin in the first two generations. This recursive definition can be recycled to introduce *thin multigenerational surfaces.*

We call the surfaces that are thin in all generations simply "*thin*". A surface S_κ is called "*essential*" if it is not thin. In fact, S_κ being essential is equivalent to the property $S_\kappa \subset K^{\mathsf{ess}}$. Similarly, thin surfaces S_κ are characterized by the property $S_\kappa \subset K^{\mathsf{thin}}$.

For each S_κ that is thin in the first generation, using the free portions of ∂S_κ, we can collapse S_κ onto a 1-dimensional graph $G_\kappa \subset S_\kappa$, its spine. Some loops or arcs from $\partial S_\kappa \cap \partial \mathcal{N}_K^\dagger$ may be incorporated into G_κ.

As we execute the collapse $S_\kappa \searrow G_\kappa$, we erase the open star $St(\kappa) \subset \mathbf{\Gamma}$ of the vertex κ from the bookkeeping graph $\mathbf{\Gamma}$. After all such collapses have been performed and the corresponding stars $St(\kappa)$ have been erased, we will be left with surfaces S_κ whose boundaries in their entirety consist of *loops* that are only present either in the boundaries $\{\partial\mathcal{N}_\omega\}_\omega$, or in $\{\partial\mathcal{N}_\tau\}_\tau$ of the types B and D, or in both.

Among the loops ω, some may have neighborhoods $\mathcal{N}_\omega$ with the following property: $\partial\mathcal{N}_\omega$ contains two loops β', β'' that belong to thin in the first generation surfaces $S_{\kappa'}$ and $S_{\kappa''}$. In such a case, the surface S_κ, that is attached to the third loop $\beta \subset \partial\mathcal{N}_\omega$, is thin in the second generation.

The neighborhood $\mathcal{N}_\omega$ (an open book with three pages) is the union of three annuli, $A' \supset \beta'$, $A'' \supset \beta''$, and $A \supset \beta$ which share the loop ω. Using the product structure in the annuli, the collapses $S_{\kappa'} \searrow G_{\kappa'} \supset \beta'$, $S_{\kappa''} \searrow G_{\kappa''} \supset \beta''$, extend further to collapses

$$S_{\kappa'} \cup A' \searrow G^\star_{\kappa'} \supset \omega, \quad S_{\kappa''} \cup A'' \searrow G^\star_{\kappa''} \supset \omega,$$

where $G^\star_{\kappa'} \subset S_{\kappa'} \cup A'$ and $G^\star_{\kappa''} \subset S_{\kappa''} \cup A''$ are graphs, homeomorphic to $G_{\kappa'}$ and $G_{\kappa''}$, respectively.

Since ω has free edges in $S_\kappa \cup A \approx S_\kappa$, it is possible to extend the first generation collapses to a second generation collapse

$$S_\kappa \cup S_{\kappa'} \cup S_{\kappa''} \cup \mathcal{N}_\omega \searrow G^\star_\kappa \cup G^\star_{\kappa'} \cup G^\star_{\kappa''}.$$

In parallel, we eliminate the corresponding star $St(\kappa)$ from the modified $\mathbf{\Gamma}$.

This algorithm can be repeated again and again until no further collapses of 2-cells and 1-cells in K are possible. In the end, we will be left with possibly disconnected subgraph $\mathbf{\Gamma}^{\mathsf{ess}}$ of $\mathbf{\Gamma}$, whose vertexes $\kappa \in \Pi \cap \mathbf{\Gamma}^{\mathsf{ess}}$ are connected only to vertexes whose connectors are of types A, B and D; moreover, each vertex ω of type A is of valency 3 and 2 in $\mathbf{\Gamma}^{\mathsf{ess}}$ (it had valency 3 in $\mathbf{\Gamma}$), and each edge leading to vertexes τ of types B and D is labeled "loop". Furthermore, $\mathbf{\Gamma}^{\mathsf{ess}}$ inherits the NT-markers from $\mathbf{\Gamma}$ which decorate the connectors con_ω (reflecting the NT-markers of K).

Eventually, we will collapse K onto a subcomplex $G \cup K^{\mathsf{ess}}$, where G is a 1-subcomplex whose intersection with K^{ess} is a finite set. This collapse $K \searrow G \cup K^{\mathsf{ess}}$ maps K^{thin} onto G and is an identity on K^{ess}.

Let $X^{\mathsf{ess}} =_{\mathsf{def}} p^{-1}(K^{\mathsf{ess}})$ and $H =_{\mathsf{def}} p^{-1}(K^{\mathsf{thin}})$. Since

$$K^{\mathsf{thin}} \cap K^{\mathsf{ess}} \subset \partial\mathcal{N}^\dagger_K \subset K^\odot$$

and since $p^{-1}(\partial\mathcal{N}^\dagger_K) \to \partial\mathcal{N}^\dagger_K$ is a trivial fibration whose fiber is a segment, both sets X^{ess} and H are submanifolds of X (with corners along $H \cap X^{\mathsf{ess}} \cap \partial_1 X$).

Since H is a mapping cylinder of the map $p : H \cap \partial_1 X \to K^{\mathsf{thin}}$, it collapses on K^{thin}. The latter has a 1-dimensional spine G. Thus the 3-fold H collapses further on the graph G, its 1-dimensional spine. According to [RS] (see also Theorem 1.1.7 in [Ma]), the regular neighborhood $\mathcal{N}(G, H)$ is PL-homeomorphic to H. Therefore H is a union of *handlebodies*. It is attached to X^{ess} along a number of disjoint disks, residing in $\partial X^{\mathsf{ess}}$. Thus X is obtained from X^{ess} by attaching a number of solid 1-handles to $\partial X^{\mathsf{ess}}$.

Next, let us investigate the topology of X^{ess} and its spine K^{ess}. First, we consider a typical connected component X' of X^{ess} and its spine K', the corresponding connected component of K^{ess}.

When $\mathcal{N}_\tau$ is of types B and D, we notice that any loop $\gamma \subset \partial\mathcal{N}_\tau$ bounds a 2-disk $D^2_\tau \subset \mathcal{N}_\tau$. We modify each essential surface S_κ by *cupping* every such loop γ in ∂S_κ with the disc D^2_τ. The boundary of the resulting surface $\hat{S}_\kappa$ is attached only to loops from $\{\partial\mathcal{N}_\omega\}_{\omega\in\Gamma^{\mathsf{ess}}}$ only. Note that the cupping procedure $S_\kappa \Rightarrow \hat{S}_\kappa$ is a "1-to-1 operator".

Recall that, for each neighborhood $\mathcal{N}_\omega$, where $\omega \in \Gamma^{\mathsf{ess}}$, two or three loops from $\partial\mathcal{N}_\omega$ are adjacent to the $\hat{S}_\kappa$'s.

We will employ a modified graph $\mathbf{\Gamma}'$ to record the information that is needed to assemble K' from the blocks $\{\hat{S}_\kappa\}$ and $\{\mathcal{N}_\omega\}$. Each surface $\hat{S}_\kappa$ corresponds to a vertex $\kappa \in \Pi' =_{\mathsf{def}} \mathbf{\Gamma}' \cap \Pi^{\mathsf{ess}}$, and each loop $\omega \subset K^{\mathsf{ess}}$ corresponds to a vertex ω of $\mathbf{\Gamma}'$. Now the valency of each vertex ω in $\mathbf{\Gamma}'$ is either 2 or 3. To distinguish between these two types of ω's, we will reserve the notation "ω" for the vertexes/loops of valency 3, and will use the notation "$\bar{\omega}$" for the vertexes/loops of valency 2 (see Figure 3.10).

So the vertices of $\mathbf{\Gamma}'$ are divided into three groups: $\{\kappa\}$, $\{\omega\}$, and $\{\bar{\omega}\}$. The edges of $\mathbf{\Gamma}'$ record the way the components of $\partial\hat{S}_\kappa$ are attached to the loops from $\{\partial\mathcal{N}_\omega\}$ and $\{\partial\mathcal{N}_{\bar{\omega}}\}$.

We denote by $\gamma_{\kappa,\omega}$ a typical edge that links a vertex κ with a vertex ω, and by $\gamma_{\kappa,\bar{\omega}}$ a typical edge that links κ with $\bar{\omega}$. There are also some "fake" edges $\gamma(\bar{\omega}, \phi)$ that record the the 3-paged geometry of some neighborhoods $\mathcal{N}_{\bar{\omega}}$; they represent "bridges to nowhere", the memories of the last thin generation...

We require that $\mathbf{\Gamma}'$ will satisfy the following (regrettably long) list of properties:

(1) $\mathbf{\Gamma}'$ is a connected graph,
(2) its vertexes are divided into four groups: $\{\kappa\}, \{\omega\}, \{\bar{\omega}\}$, and $\{\phi\}$, the cardinalities of the sets $\{\bar{\omega}\}$ and $\{\phi\}$ being equal,

(3) each vertex $\bar{\omega}$ is linked by a single edge $\gamma_{\bar{\omega},\phi}$ to a unique vertex $\phi = \phi(\bar{\omega})$,
(4) each vertex ω is linked by edges $\{\gamma_{\kappa,\omega}\}$ only to the vertices from the set $\{\kappa\}$, so that the valency $\mu(\omega) = 3$,[9]
(5) each vertex $\bar{\omega}$ is linked by two edges to the vertices from the set $\{\kappa\}$, so that the valency $\mu(\bar{\omega}) = 3$,
(6) the markers N, T, and E are used to decorate and order cyclically "$\rightsquigarrow N \rightsquigarrow T \rightsquigarrow E \rightsquigarrow$" the three edges that join at each vertex ω or $\bar{\omega}$,
(7) a number $\epsilon(\gamma_{\kappa,\omega}) = \pm 1$ [10] is assigned to each edge $\gamma_{\kappa,\omega}$ so that, for each ω, the relation $\sum_\kappa \epsilon(\gamma_{\kappa,\omega}) = \pm 1$ is valid (in fact, there is a bit of flexibility in choosing the collection of units $\{\epsilon(\gamma_{\kappa,\omega})\}_{\kappa,\omega}$: for each ω, we may switch synchronically the signs of all $\{\epsilon(\gamma_{\kappa,\omega})\}_\kappa$),
(8) a number $\epsilon(\gamma_{\kappa,\bar{\omega}}) = \pm 1$ is assigned to each edge $\gamma_{\kappa,\bar{\omega}}$,
(9) an integer $\chi(\kappa) \leq 2$ ("the Euler number of the surface $\hat{S}_\kappa$") is attached to each vertex κ.

(3.4)

It turns out that the data in (3.4) allow for a reconstruction of K' and, thanks to Theorem 3.12 and Theorem 2.7, for a reconstruction of X'.

Next proposition describes all orientable 3-folds of gradient complexity zero in terms of the combinatorial structures as in (3.4).

Theorem 3.12. *Let X be a compact connected orientable 3-fold with boundary, such that $gc(X) = 0$.*

- *Then X is obtained by attaching solid 1-handles to the boundary of a disjoint union $\coprod X'$ of connected orientable 3-folds X' with boundaries. Each 3-fold X' is either homeomorphic to a product of a closed surface S with a closed interval, or has an abstract gradient spine K' that is uniquely determined by a graph $\mathbf{\Gamma}'$, decorated with combinatorial data as in (3.4). Moreover, the graph $\mathbf{\Gamma}'$ determines X', up to a diffeomorphism.*
- *If X has no essential annuli, then each $\mathbf{\Gamma}'$ is a tree, and each surface S_κ from $K' \setminus s_\dagger(K')$ is a 2-sphere with a number of holes, so that $\chi(\kappa) = 2 - \mu(\kappa)$, where $\mu(\kappa)$ denotes the valency of the vertex $\kappa \in \mathbf{\Gamma}'$.*

[9]Some κ may be linked to a particular ω or $\bar{\omega}$ by more than one edge; so the notation "$\gamma_{\kappa,\omega}$" is a bit misleading.
[10]This assignment imitates the branched orientation of the spine.

- *On the other hand, each graph* $\mathbf{\Gamma}'$, *subject to (3.4), gives rise to a unique orientable* 3*-fold* X' *of gradient complexity zero. Its Euler number* $\chi(X')$ *is given by the RHS of formula (3.5), and its fundamental group has a presentation, described by formulas (3.6) - (3.8).*
- *If the boundary* $\partial_1 X$ *is simply-connected, then the fundamental group* $\pi_1(X)$ *is free or trivial.*

Proof. Since $gc(X) = 0$, for an appropriate traversally generic vector field v on X, the gradient spine $K(v)$ has no Q-singularities[11]. Therefore X has an abstract gradient spine $K =_{\mathsf{def}} K(v)$ such that its combinatorial complexity $c(K)$ is zero. Of course, for any subcomplex $K' \subset K$, $c(K') = 0$ as well. So the conclusions of the discussion that preceded the formulation of Theorem 3.12 do apply to K and its subcomplexes. We keep the notations that have been introduced in the course of these discussions, except for one: in what follows, the *cupped* surface $\hat{S}_\kappa$ will be denoted "S_κ".

We continue to investigate the 3-fold X', a connected component of X^{ess}, with its abstract gradient spine K'. It may happen that $s_\dagger(K') = \emptyset$. In such a case, K' is homeomorphic to a closed surface S_κ (for some index κ). Then X' is homeomorphic to $S_\kappa \times [0,1]$, and, by definition, $\mathbf{\Gamma}'$ is just a singleton κ.

As before, put $\mathcal{N}^\dagger_{K'} =_{\mathsf{def}} \mathcal{N}(s_\dagger(K'), K')$.

When $s_\dagger(K') \neq \emptyset$, let us restate how K' gives rise to a graph $\mathbf{\Gamma}'$ as in (3.4). Recall that $res_T(K')$ denotes the T-induced resolution of K'. We fix an orientation for each loop ω and $\bar{\omega}$ in $s_\dagger(K')$ and an orientation for each connected surface $S_\kappa \subset (K')^\odot =_{\mathsf{def}} K' \setminus \mathcal{N}^\dagger_{K'}$, consistent with the orientation of $(K')^\odot \subset res_T(K')$, the latter being a part of the abstract gradient spine structure on K' (see Definition 2.10).

The connected components $\{\beta\}$ of ∂S_κ correspond to the edges $\{\gamma_{\kappa,\beta}\}$ of $\mathbf{\Gamma}'$ that emanate from the vertex κ and terminate at some vertex ω or $\bar{\omega}$. So the number of components in $\partial\hat{S}_\kappa$ is the valency $\mu(\kappa)$ of the vertex κ in $\mathbf{\Gamma}'$.

In the induced orientation, each loop β in the boundary ∂S_κ is attached to a loop ω or $\bar{\omega}$ in $s_\dagger(K')$ via either an orientation-preserving homeomorphism, in which case $\epsilon(\gamma_{\kappa,\omega}) = 1$, or via an orientation-reversing homeomorphism, in which case $\epsilon(\gamma_{\kappa,\omega}) = -1$. Similar conventions apply when β is identified with a loop $\bar{\omega}$. The TN-markers on K' induce a branched

[11] points whose links in $K(v)$ are of type (6) from Figure 2.21

orientation of K' (see Definition 2.11). They decorate the edges of $\mathbf{\Gamma}'$ and enforce the equality $\sum_\kappa \epsilon(\gamma_{\kappa,\omega}) = \pm 1$ for each vertex (loop) ω.

We notice that the topological type of S_κ determines and is determined by the Euler number $\chi(\kappa) =_{\mathsf{def}} \chi(S_\kappa)$ and the number $\mu(\kappa)$ of connected components in ∂S_κ. Let us denote by $\hat{g}_\kappa$ the number of handles (the "genus") in S_κ. Then

$$\chi(\kappa) = 2 - 2\hat{g}_\kappa - \mu(\kappa).$$

Therefore K' produces a graph $\mathbf{\Gamma}'$ as in (3.4).

Conversely, the decorated graph $\mathbf{\Gamma}'$ provides an unambiguous assembly instructions for the abstract gradient spine K': the numbers $\chi(\kappa)$ and $\mu(\kappa)$ (see item (9) in (3.4)) determine the topological type of the surface S_κ; the edges, emanating from the vertex κ, tell to which neighborhoods $\mathcal{N}_\omega$ or/and $\mathcal{N}_{\bar{\omega}}$ the components $\{\beta_{\kappa,i}\}$ of ∂S_κ should be attached; the *TNE*-markers (item (6) in (3.4)) help to distinguish between the three pages from $\mathcal{N}_\omega$ (the three edges of $\mathbf{\Gamma}'$ that meet at the vertex ω); finally, the numbers $\{\epsilon(\gamma_{\kappa,\omega})\}$, $\{\epsilon(\gamma_{\kappa,\bar{\omega}})\}$ (items (7) and (8) in (3.4)) tell whether the attaching diffeomorphisms $\psi_{\kappa,i} : \partial S_{\kappa,i} \to \omega$ or $\bar{\psi}_{\kappa,i} : \partial S_{\kappa,i} \to \bar{\omega}$ of the two circles preserve or reverse their orientations.

In turn, by Theorem 2.7, the abstract gradient spine K' gives rise to a unique 3-fold X'. Therefore the decorated graph $\mathbf{\Gamma}'$ in (3.4) determines the smooth type of X'.

By its very construction, the oriented branched complexity of such X' is zero. By Theorem 3.10, $gc(X') = 0$ for any component X' of X^{ess}. Since X is obtained from $\coprod X'$ by attaching solid 1-handles to its boundary $\coprod \partial X'$, according to Theorem 3.4, $gc(X) = 0$ as well.

This proves the first bullet of the theorem and a part of the third bullet.

We continue validating the other claims of the theorem.

The neighborhood $\mathcal{N}^\dagger_{K'}$ is a mapping cylinder of a map $\pi : \partial\mathcal{N}^\dagger_{K'} \to s_\dagger(K')$. As we attach the surfaces $\{S_\kappa\}$ to $\mathcal{N}^\dagger_{K'}$ along their boundaries to form K', the map π identifies the vertices and edges from $\coprod_\kappa \partial S_\kappa$ either *in triples* (in the case of a loop $\omega \subset s_\dagger(K')$) or in *pairs* (in the case of a loop $\bar{\omega} \subset s_\dagger(K')$). Therefore, in a triangulation of $\coprod_\kappa S_\kappa$, the number of edges that have been identified in K' via π is the same as the number of the vertexes that have been identified in K'. As a result, the Euler numbers $\{\chi(S_\kappa)\}$ add up to produce the Euler number $\chi(K')$. Since K' is a spine of X', we get

$$\chi(X') = \sum_{\kappa} \big(\chi(\kappa) - \mu(\kappa)\big) = \sum_{\kappa} \big(2 - 2\hat{g}_\kappa - \mu(\kappa)\big)$$
$$= 2\sum_{\kappa}(1 - \hat{g}_\kappa) \; - 2l - 3k, \tag{3.5}$$

where k is the number of loops ω and l is the number of loops $\bar{\omega}$ in $s_\dagger(K')$. So the total number of edges, emanating from the vertexes $\{\kappa\}$ in $\mathbf{\Gamma}'$, is $3k + 2l = \sum_\kappa \mu(\kappa)$.

We notice that $1 - \hat{g}_\kappa > 0$ only when $\hat{g}_\kappa = 0$. Therefore, if $\chi(X') \gg 0$, formula (3.5) implies that, among the surfaces $\{S_\kappa\}_\kappa$, there are many disks.

Recall that, for any odd-dimensional manifold X, $\chi(\partial_1 X) = 2 \cdot \chi(X)$. Thus $\partial_1 X'$ must be a closed oriented surface with the Euler number

$$\chi(\partial_1 X') = 4\sum_{\kappa}(1 - \hat{g}_\kappa) - 4l - 6k.$$

If $\partial_1 X'$ is a union of q copies of spheres S^2, then $\chi(X') = q$. In such a case,

$$q = 2\sum_{\kappa}(1 - \hat{g}_\kappa) - 2l - 3k.$$

The previous formulas imply:

Corollary 3.5. *Let X be a 3-fold which has an abstract gradient spine $K \subset X$ such that $c(K) = 0$*[12]*. Let $s(K) = s_\dagger(K)$ be a union of k loops. Then*

$$\chi(X) \equiv k \mod 2, \;\text{ and }\; \chi(X) \equiv \sum_{\kappa}(g_\kappa - 1) \mod 3,$$

where g_κ is the number of handles in a typical surface $S_\kappa \subset K \setminus s(K)$.

In particular, if $\partial_1 X$ is a union of q spheres, then $q \equiv k \mod 2$, and $q \equiv \sum_\kappa (g_\kappa - 1) \mod 3$. ◊

We return to the proof of Theorem 3.12. Assume for now that the original 3-fold X has no essential annuli. Since X is obtained from $X^{\mathsf{ess}} \supset X'$ by attaching solid 1-handles to the boundary $\partial X^{\mathsf{ess}} \supset \partial X'$, the natural homomorphism $\pi_1(X') \to \pi_1(X)$ is a monomorphism. Therefore, if the core γ of an annulus $(A, \partial A) \subset (X', \partial X')$ is contractible in X, it must be contractible in X' as well.

[12]then, by Theorem 3.10, $gc(X) = 0$.

Now assume that an annulus $(A, \partial A) \subset (X', \partial X')$ is parallel to the boundary of X with the help of a solid torus $\mathbf{T} \subset X$, whose core γ is parallel to each of the two loops that form ∂A. Let us show that $(A, \partial A)$ is parallel to the boundary of X' as well.

Let $A_\star =_{\mathsf{def}} \mathbf{T} \cap \partial X$. By definition, $\partial \mathbf{T} = A \cup A_\star$. The annuli $A_\star$ and A share common boundary, which resides in $\partial X'$.

Consider all the solid 1-handles $\{H_i \approx D_i \times [0,1]\}_i$ that were attached to $X' \subset X^{\mathsf{ess}}$ in order to form X. Put $\delta H_i = \partial D_i \times [0,1]$. By an isotopy of A in X', we may assume that $H_i \cap \partial A = \emptyset$.

Since $\partial A \subset \partial X'$ and $\partial A = \partial A_\star$, we conclude that either **(1)** $A_\star \cap \delta H_i = \emptyset$, or **(2)** $A_\star \supset \delta H_i$, in which case, $\mathbf{T} \supset H_i$. In the second case, the core γ of $\mathbf{T}$ must represent a nontrivial loop in the quotient space X/X^{ess} (which has a homotopy type of a 1-dimensional complex). However, γ is homotopic to a loop from $\partial A \subset X'$, a contradiction. Therefore, $A_\star \subset \partial X'$ and $\mathbf{T} \subset X'$. So we conclude that A is parallel to $\partial X'$.

Hence, when a connected component X' of X^{ess} has an essential annulus, then so does X. Therefore the hypotheses of the theorem imply that each connected component X' of X^{ess} has no essential annuli.

Recall that each edge $\gamma_{\kappa,i} \subset \mathbf{\Gamma}'$ stands for a particular component of ∂S_κ. Let the loop $\beta =_{\mathsf{def}} \beta_{\kappa,i} \subset \partial S_\kappa$ be that component. If the 1-chain $\gamma_{\kappa,i}$ (or its multiple) can be closed to form a 1-cycle μ in $\mathbf{\Gamma}'$, then X' has an essential annulus $p^{-1}(\beta)$, where $p : X' \searrow K'$ is the collapsing map that gives X' the structure of the mapping cylinder $\mathsf{cyl}(p : \partial X' \to K')$. Indeed, the loop $\mu \subset \mathbf{\Gamma}'$ helps to construct a loop $\tilde{\mu} \subset X'$ which has a nonzero algebraic intersection number with the annulus $p^{-1}(\beta)$. Therefore the graph $\mathbf{\Gamma}'$ (as in (3.4)), produced with the help of the abstract gradient spine K' (in particular, subject to the constraint $c(K') = 0$), must be a *tree* for any 3-fold X' with no essential annuli.

Similarly, if X' has no essential annuli, no surface $S_\kappa \subset K'$ has a handle. Indeed, for any pair of loops $\tau, \eta \in S_\kappa$ that intersect transversally at a singleton, the annulus $p^{-1}(\tau)$ must be essential. Moreover, τ and η can be pushed into the boundary $\partial_1 X'$, where they still will intersect at a singleton: recall that $p : p^{-1}(S_\kappa) \to S_\kappa$ is a trivial fibration with a closed segment for the fiber. Therefore, when X' has no essential annuli, each surface S_κ must be a 2-sphere with a number of 2-disks being removed. Hence its "genus" $\hat{g}_\kappa = 0$. This proves the second bullet of the theorem.

For any oriented loop $\beta \subset \partial S_\kappa$, both loops, forming the boundary of the annulus $p^{-1}(\beta)$, reside in $\partial X'$ and are evidently homotopic in X' to β.

On the other hand, the loop β is homotopic to the core loop $\omega \subset s_\dagger(K')$, where $\partial\mathcal{N}_\omega \supset \beta$, or to its inverse ω^{-1}, the loop ω carrying the opposite orientation. Similar observations are valid for loops $\bar{\omega}$ such that $\partial\mathcal{N}_{\bar{\omega}} \supset \beta$. Therefore, each loop $\omega \subset s_\dagger(K')$ is in the image of the inclusion-induced homomorphism $\pi_1(\partial_1 X', x_0) \to \pi_1(X', x_0)$ for an appropriate choice of base point $x_0 \in \partial_1 X'$. When $\partial_1 X'$ is connected, a single base point x_0 can serve all loops in $s_\dagger(K')$.

Now let us investigate the special case when $\partial_1 X = \coprod S^2$, the union of spheres. Again, in such case, no surface $S_\kappa \subset (K')^\odot$ contains a handle. We have noticed already that any two loops in S_κ, which intersect at a singleton, can be pushed into the boundary $\partial_1 X = \coprod S^2$, where they still must share only a singleton, a clear impossibility. So, each surface S_κ must be a 2-sphere with a number of holes.

Since $\partial_1 X = \coprod S^2$, each loop $\omega \subset s_\dagger(K)$ is homotopic to a loop in some ∂S_κ, and thus can be homotoped to a loop in $\partial_1 X$. As a result, all ω's and $\bar{\omega}$'s are contractible in X. Since X is obtained from X^{ess} by attaching solid 1-handles to X^{ess}, we conclude that any loop $\omega \subset s_\dagger(X^{\mathsf{ess}})$ must be also contractible in X^{ess}. Similar conclusion holds for all loops $\bar{\omega} \subset s_\dagger(X^{\mathsf{ess}})$. Thus any loop from $s_\dagger(K')$ is contractible in K', provided that $\partial_1 X$ is simply-connected.

We notice that the fundamental group $\pi_1(K')$, where K' is a connected component of K^{ess} such that $s_\dagger(K') \neq \emptyset$, is generated by:

- the basic 1-cycles $\{\gamma \in H_1(\mathbf{\Gamma}'; \mathbb{Z})\}$ in the graph $\mathbf{\Gamma}'$, being lifted to K' (these are "free" generators),
- the loops $\{\omega\}$ and $\{\bar{\omega}\}$ that form $s_\dagger(K')$,
- the oriented components $\{\beta_{\kappa,i}\}_i$, of the boundaries $\{\partial S_\kappa\}_\kappa$,
- the basic loops $\{a_{\kappa,1}, b_{\kappa,1}, \dots, a_{\kappa,\hat{g}_\kappa}, b_{\kappa,\hat{g}_\kappa}\}_\kappa$, residing in the surfaces S_κ's, where $2\hat{g}_\kappa = 2 - \chi(\kappa) - \mu(\kappa)$.

Evidently, these generators are subject to the following relations:

(1) For each κ and any ordering $o =_{\mathsf{def}} (\beta_{\kappa,1}, \beta_{\kappa,2}, \dots, \beta_{\kappa,\mu(\kappa)})$ of the connected components $\{\beta_{\kappa,i}\}_i$ (i.e., of the edges that emanate from κ),

$$\mathcal{R}_{\kappa,o} =_{\mathsf{def}} \Big\{ \prod_{r=1}^{\mu(\kappa)} \beta_{\kappa,r} = \prod_{l=1}^{\hat{g}_\kappa} [a_{\kappa,l},\, b_{\kappa,l}] \Big\}, \tag{3.6}$$

where $[\sim, \sim]$ denotes the commutator (just consider the fundamental $(4\hat{g}_\kappa)$-gon with $\mu(\kappa)$ holes).

(2) For each vertex ω and the triple of edges $\gamma_{\kappa_1,\omega}, \gamma_{\kappa_2,\omega}, \gamma_{\kappa_3,\omega}$ that terminate at ω,

$$\mathcal{T}_\omega =_{\mathsf{def}} \left\{\beta_{\kappa_s,\omega} = \omega^{\epsilon(\gamma_{\kappa_s},\omega)}\right\}, \text{ where } s = 1, 2, 3. \tag{3.7}$$

(3) For each vertex $\bar{\omega}$ and the pair of edges $\gamma_{\kappa_1,\bar{\omega}}, \gamma_{\kappa_2,\bar{\omega}}$ that terminate at $\bar{\omega}$,

$$\mathcal{T}_{\bar{\omega}} =_{\mathsf{def}} \left\{\beta_{\kappa_s,\bar{\omega}} = \bar{\omega}^{\epsilon(\gamma_{\kappa_s},\bar{\omega})}\right\}, \text{ where } s = 1, 2. \tag{3.8}$$

In the case of $\partial_1 X = \coprod S^2$, according to the previous arguments, all the loops $\{\omega, \bar{\omega} \subset s_\dagger(K')\}$, are contractible in K', and since each $\hat{g}_\kappa = 0$, all the generators $\{a_{\kappa,1}, b_{\kappa,1}, \dots, a_{\kappa,\hat{g}_\kappa}, b_{\kappa,\hat{g}_\kappa}\}$ disappear. This forces $\pi_1(X') \approx \pi_1(\Gamma')$ to be *free* or *trivial*. Therefore $\pi_1(X)$, being a free product of the groups $\{\pi_1(X'_\alpha)\}_\alpha$ with a free group, whose genesis is in the handlebody H that is attached to $\coprod_\alpha X'_\alpha$ to form X, is a free or a trivial group.

For a closed 3-fold Y, let $Y^\circ =_{\mathsf{def}} Y \setminus D^3$. According to Matveev's classification Table 9.1 on page 407 in [Ma], the only *closed* oriented 3-folds Y with $c(Y^\circ) = 0$ are the connected sums of the following *prime* blocks: **(1)** spheres S^3, **(2)** products of spheres $S^2 \times S^1$, **(3)** projective spaces $\mathbb{R}P^3$, and **(4)** lens spaces $L_{3,1}$.

As usual, $gc(Y^\circ) = 0$ implies $c(Y^\circ) = 0$. Given that $\partial Y^\circ = S^2$, by the previous arguments, we conclude that $\pi_1(X)$ is free or trivial. So the presence of the $\mathbb{R}P^3$'s and $L_{3,1}$'s in the connected sums is ruled out: their fundamental groups have torsion elements that will survive in $\pi_1(Y^\circ)$. In particular, any generic gradient flow on $(\mathbb{R}P^3)^\circ$ and $(L_{3,1})^\circ$ must have at least one double-tangent trajectory[13].

Therefore, if $\partial_1 X = \coprod S^2$ and $gc(X) = 0$, X must be a connected sum of 3-balls with a number of 3-folds $S^2 \times S^1$, a claim of Theorem 3.14 below.

In fact, any such a connected sum

$$X = D^3 \# \dots \# D^3 \# (S^1 \times S^2) \# \dots \# (S^1 \times S^2)$$

is of zero gradient complexity, provided $\partial X \neq \emptyset$. Indeed, consider a Morse function $h : S^2 \to \mathbb{R}$ with only two critical points, a and b, and a Morse function $g : S^1 \to \mathbb{R}$ with only two critical points, c and d. We may assume that $h(a) = 0$, $h(b) = 2$ and $g(c) = 0$, $g(d) = 3$. Consider the Morse function $f = h + g$ on $S^1 \times S^2$. Let M be the 3-fold, produced from $S^2 \times S^1$ by removing four small balls, centered on the critical points of f.

[13]Soon we will show that any gradient flow these 3-folds must have two double-tangent trajectories at least.

By an argument as in the proof of Theorem 3.8, applied to f, we construct a gradient spine of M. The boundary ∂M consists of four 2-spheres, two of which are $f^{-1}(1)$ and $f^{-1}(4)$. Since $f(b \times c) > f(d \times a)$, the f-critical point of index 1 lies above the f-critical point of index 2. Hence, these two critical points are not linked by a gradient trajectory in $S^2 \times S^1$. As a result, the waterfalls in M, generated in the vicinity of these two points, do not intersect each other. They hit the sphere $f^{-1}(1) \subset \partial_1 M$ along four disjoint simple loops. Therefore, $gc(M) = 0$. By Theorem 3.4, $gc(f^{-1}([1,4])) = gc(f^{-1}([0,4]) = gc(M) = 0$. Since any connected sum X is assembled from the blocks of the types $f^{-1}([1,4])$ and $f^{-1}([0,4])$, the semi-additivity of $gc(\sim)$ under connected sums (Theorem 3.4) implies the claim.

We notice that Exercise 5.3 from [He], being strengthen by the validated Poincaré Conjecture [P1], [P2], [MT], leads to the following pleasing theorem.

Theorem 3.13. *Let X be a compact connected 3-fold, whose boundary is a disjoint union of connected surfaces $\Sigma_1, \dots, \Sigma_q$.*

If $\pi_1(X)$ is free of rank r, then X is the connected sum

$$X = H_1 \# \dots \# H_q \# E_1 \# \dots \# E_s,$$

where H_i is a handlebody with $(1 - \frac{1}{2}\chi(\Sigma_i))$ handles, E_j is a 2-sphere bundle over S^1 (trivial when X is orientable), and

$$s = r - \sum_{i=1}^{q} \Big(1 - \frac{1}{2}\chi(\Sigma_i)\Big).$$

In particular, when the boundary is a union of q copies of S^2, then

$$X = D^3_1 \# \dots \# D^3_q \# E_1 \# \dots \# E_s.$$

◊

If we combine Theorem 3.13 with the recently reached conclusion "$\pi_1(X)$ is free for any 3-fold X with a simply-connected boundary and such that $gc(X) = 0$", we could avoid using in our arguments the classification results from [Ma].

This completes the proof of Theorem 3.12, as well as the proof of Theorem 3.14 below (the latter theorem does not use on the validity of the Poincaré Conjecture!). □

Theorem 3.14. *Let X be a connected oriented 3-fold with a simply-connected boundary. If $gc(X) = 0$, then X is a connected sum of several 3-balls with several copies of $S^2 \times S^1$.*

In particular, if $\partial_1 X = S^2$, $\pi_1(X) = 0$, and $gc(X) = 0$, then X is a 3-ball. ◇

Remark 3.4. In fact, the group presentations as in (3.6) - (3.8) allow for a great variety of fundamental groups $\pi_1(X)$. Although we do not have a good conjecture, describing all these groups intrinsically, such a description seems feasible. ◇

Example 3.2. Let us consider few basic examples of 3-folds X with $gc(X) = 0$. They are generated by quite simple abstract gradient spines K with $s(K) = s_\dagger(K)$ being a single loop ω. We intend to show that the fundamental groups of such 3-folds X may deviate from the class of free groups; they have quite interesting presentations.

There are three possibilities for $\mathbf{\Gamma} =_{\mathsf{def}} \mathbf{\Gamma}'$ as in (3.4): it has a single vertex ω and either **(1)** three vertices $\kappa_1, \kappa_2, \kappa_3$, or **(2)** two vertices κ_1, κ_2, or **(3)** one vertex κ_1.

In the first case, the graph $\mathbf{\Gamma}$ is a trivalent tree, and $\pi =_{\mathsf{def}} \pi_1(X)$ is generated by the sets of generators

$$\{a_1, b_1, \dots, a_{g_1}, b_{g_1}\},\ \{c_1, d_1, \dots, c_{g_2}, d_{g_2}\},\ \{e_1, f_1, \dots, e_{g_3}, f_{g_3}\},$$

$$\omega,\ \beta_{\kappa_1},\ \beta_{\kappa_2},\ \beta_{\kappa_3},$$

subject to the of relations as in (3.6) - (3.7). Expelling ω and β's from this set of relations, we get just two relations:

$$\prod_{i=1}^{g_1}[a_i, b_i] = \prod_{j=1}^{g_2}[c_j, d_j] = \Big(\prod_{k=1}^{g_3}[e_k, f_k]\Big)^{-1}. \tag{3.9}$$

Note that $\pi/[\pi, \pi]$ is free abelian of rank $2(g_1 + g_2 + g_3)$.

In the second case, $\mathbf{\Gamma}$ contains a single base loop f which contains the vertex κ_1. According to (3.6) - (3.7), the fundamental group $\pi =_{\mathsf{def}} \pi_1(X)$ is generated by the following sets of generators:

$$\{a_1, b_1, \dots, a_{g_1}, b_{g_1}\},\ \{c_1, d_1, \dots, c_{g_2}, d_{g_2}\},\ f,\ \omega,\ \beta_{\kappa_1,1}, \beta_{\kappa_1,2},\ \beta_{\kappa_2,1}.$$

They are subject to the following two distinct sets of relations:

$$\prod_{i=1}^{g_1}[a_i, b_i] \;=\; \beta_{\kappa_1,1}\beta_{\kappa_1,2} \;=\; \beta_{\kappa_1,2}\beta_{\kappa_1,1},$$
$$\beta_{\kappa_1,1} = \omega, \quad \beta_{\kappa_1,2} = \omega,$$
$$\prod_{i=1}^{g_1}[c_i, d_i] = \beta_{\kappa_2,1}, \quad \beta_{\kappa_2,1} = \omega^{-1}, \tag{3.10}$$

or

$$\prod_{i=1}^{g_1}[a_i, b_i] \;=\; \beta_{\kappa_1,1}\beta_{\kappa_1,2} \;=\; \beta_{\kappa_1,2}\beta_{\kappa_1,1},$$
$$\beta_{\kappa_1,1} = \omega, \quad \beta_{\kappa_1,2} = \omega^{-1},$$
$$\prod_{i=1}^{g_1}[c_i, d_i] = \beta_{\kappa_2,1}, \quad \beta_{\kappa_2,1} = \omega. \tag{3.11}$$

Expelling ω and β's form (3.10), we get that $\pi_1(X)$ is generated by $\{a_1, b_1, \dots, a_{g_1}, b_{g_1}\}$, $\{c_1, d_1, \dots, c_{g_2}, d_{g_2}\}$, and f, subject to a single relation:

$$\prod_{i=1}^{g_1}[a_i, b_i]\Big(\prod_{i=1}^{g_1}[c_i, d_i]\Big)^2 \;=\; 1 \tag{3.12}$$

(which is similar to the two relations from (3.9)).

In contrast, expelling ω and β's form (3.11), we get that $\pi_1(X)$ is generated by $\{a_1, b_1, \dots, a_{g_1}, b_{g_1}\}$, $\{c_1, d_1, \dots, c_{g_2}, d_{g_2}\}$, and f, subject to a single relation

$$\prod_{i=1}^{g_1}[a_i, b_i] \;=\; 1. \tag{3.13}$$

In both cases, $\pi/[\pi, \pi]$ is free abelian of rank $2(g_1 + g_2) + 1$.

In the third case, $\mathbf{\Gamma}$ contains two basic loops f_1, f_2. Similar considerations imply that the fundamental group π is freely generated by $\{a_1, b_1, \dots, a_{g_1}, b_{g_1}\}$ together with f_1, f_2.

In all these descriptions of π, we have assumed that $g_1, g_2, g_3 > 0$. Numerous exceptions to this assumption lead to groups π of a slightly different kind. For example, when $\mathbf{\Gamma}$ is as in the second case and $g_1 = 0$, we get a group π with the 2-torsion element $\prod_{j=1}^{g_2}[c_j, d_j]$; however, it is trivial in the homology $\pi/[\pi, \pi]$. ♢

Next, we will show that there is no orientable 3-fold X with a simply-connected boundary such that $gc(X) = 1$. In contrast, the combinatorial complexity does attain value one: in fact, $c(L^\circ_{4,1}) = 1$ and $c(L^\circ_{5,2}) = 1$. Based on partial evidence (see Corollary 3.3 and Theorem 3.8), we propose:

Conjecture 3.1. *Let X be a compact connected 3-fold with a simply-connected boundary. If $gc(X) > 0$, then $gc(X) \geq 4$.* ◊

Theorem 3.15. *If an orientable compact connected 3-fold X with simply-connected boundary admits a traversally generic flow with a* single *double-tangent trajectory, then X is a connected sum of a number of 3-balls and products $S^2 \times S^1$. As a result, $gc(X) = 0$.*

In other words, no orientable 3-fold X with a simply-connected boundary has the gradient complexity one.

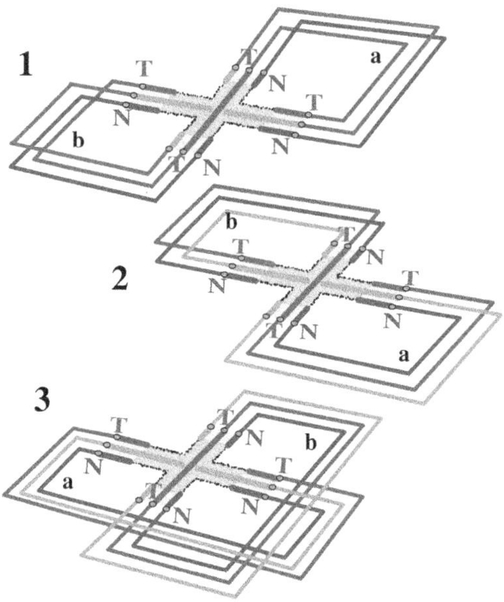

Fig. 3.11 The three patterns for the graphs $\partial(\mathcal{N}(s_\dagger(K), K))$, corresponding to abstract gradient spines K with $c(K) = 1$, and a connected singular set $s(K) \approx$ "∞" (not shown in the figure). The intersections of the curves should be ignored.

Proof. Recall that the combinatorial complexity $c(Y)$ of a closed 3-fold Y is defined as $c(Y \setminus D^3)$.

Assume that $gc(X) = 1$. Then $c(X) \leq 1$. If X has a simply-connected boundary, one can cup each sphere in the boundary of X with a 3-ball, thus producing a closed oriented 3-fold $\hat{X}$. By Proposition 2.1.5 [Ma] the number of deleted 3-balls does not affect the combinatorial complexity; thus $c(\hat{X}) =_{\mathsf{def}} c(\hat{X} \setminus D^3) = c(X) \leq 1$.

The 3-fold $\hat{X}$ admits a decomposition into a connected sum of closed orientable irreducible 3-folds $\{\hat{X}_\alpha\}_\alpha$. Let $X_\alpha =_{\mathsf{def}} \hat{X}_\alpha \setminus D^3$.

By the additivity of combinatorial complexity under connected sums (see [Ma], Theorem 2.2.9), only one of the irreducible summands has complexity 1, the rest must be of complexity 0. According to the classification in [Ma], Table A.9, the only closed irreducible orientable 3-folds of combinatorial complexity 1 are the lens spaces $L_{4,1}$ or $L_{5,2}$. Therefore either $c(\hat{X}) = 0$, or $\hat{X}$ is a connected sum of $L_{4,1}$ or $L_{5,2}$ with a closed manifold of combinatorial complexity zero. We are going to show that the assumption $gc(X) = 1$ contradicts to these options for the decomposition of $\hat{X}$ into irreducible pieces.

By the semi-additivity of the gradient complexity under connected sums (see Theorem 3.4), at least one of the irreducible 3-folds $\{\hat{X}_\alpha\}_\alpha$ has the property $gc(X_\alpha) = 1$. Let us concentrate on this irreducible 3-fold $\hat{X}_{\alpha^\star}$. Put $X_\star =_{\mathsf{def}} \hat{X}_{\alpha^\star} \setminus D^3$.

Let $K \subset X_\star$ be an abstract gradient spine of the combinatorial complexity $c(K) = 1$. A regular neighborhood $\mathcal{N}(a_\dagger) =_{\mathsf{def}} \mathcal{N}(a_\dagger, K)$ of its Q-singularity $a_\dagger$ is depicted in Figure 2.22. In the vicinity of $a_\dagger$, there are four arcs from $s_\dagger(K)$ that emanate from $a_\dagger$. If one of these arcs γ terminates at a point $b \in ss(K)$ of type P from Figure 2.14, then, in the vicinity of b, the topological boundary $s_\bullet(K)$ of K has a free arc that may be used to collapse K along γ onto a subcomplex $K' \subset K$ such that $s_\dagger(K')$ does not contain $\text{int}(\gamma)$ (note that $a_\dagger \in s_\dagger(K')$), and hence $c(K') = 0$. This contradicts the assumption that the branched oriented complexity $boc(X) = gc(X) = 1$ (see Definition 3.8 and Theorem 3.10). Similarly, if γ terminates at a point $b \in ss(K)$ of type O from Figure 2.14, then again a free arc in $s_\bullet(K)$ can be used to initiate a collapse $K \searrow K'$, which will result in an abstract gradient spine K' of zero complexity. Therefore, each arc $\gamma \subset s_\dagger(K) \cap \mathcal{N}(a_\dagger)$ must join in $s_\dagger(K)$ with another arc $\gamma' \subset s_\dagger(K)$, emanating from $a_\dagger$, that is, the component of $s_\dagger(K)$ that contains $a_\dagger$ must be homeomorphic to the figure "∞".

Examining Figure 2.22, we notice that there are only three (up to the mirror symmetry) possible combinatorial patterns for connecting *in pairs* the four 2-dimensional beams (their sections are shaped as "Y", or rather as "T") that join in the vicinity of $a_\dagger$ to form $\mathcal{N}(\infty, K)$. The three patterns must produce the neighborhood $\mathcal{N}(s_\dagger(K), K)$ that will admit an oriented branching in the sense of Definition 2.11 (equivalently, will admit TN-markings which are consistent along the core ∞ and will match as in Figure 2.22). These three patterns are shown in Figure 3.11. The figure does not show the singular set $\infty \subset s_\dagger(K)$. A regular neighborhood $\mathcal{N}(\infty, X_\star)$ of the graph ∞ in $X_\star$ is a handlebody whose core is the graph ∞ and whose boundary is a surface Σ of genus 2. The three graphs $\{\partial\mathcal{N}(\infty, K)\}$ in Figure 3.11 should be imagined as residing in Σ. They are collections of simple loops (ignore the self-intersections in Figure 3.11). In fact, the number of loops from $\partial\mathcal{N}(\infty, K)$ in the three patterns from Figure 3.11 are: $1, 2$, and 3, respectively. At this stage, the simple disjoint loops that form $\partial\mathcal{N}(\infty, K)$ have "rigid" patterns in relation to the basic longitudes $\tilde{a}, \tilde{b}$ of the surface Σ, while their winding in the direction of the basic meridians $\tilde{c}, \tilde{d}$ of Σ (the meridians bound disks in $\mathcal{N}(\infty, X_\star)$) is flexible.

Since $\hat{X}_{\alpha_\star}$ is irreducible, $X_\star$ is an irreducible and boundary irreducible 3-fold with no essential annuli. Therefore, by Theorem 3.9, $X_\star$ must have an abstract gradient spine $K_\star$ such that $\mathcal{N}(\infty, K_\star) = \mathcal{N}(\infty, K)$ and $K_\star$ is obtained by attaching 2-disks and annuli (with one of their two boundary components being left free) to the boundary $\partial\mathcal{N}(\infty, K_\star)$. When the annuli are present, $K_\star$ can be collapsed onto a spine of the oriented branched complexity zero. Therefore, we need to consider only the case when 2-disks are attached to $\mathcal{N}(\infty, K_\star)$ along its boundary. Then the following three outcomes are possible:

(1) $|\pi_0(\partial\mathcal{N}(\infty, K_\star))| = 1$, and $\Pi =_{\mathsf{def}} \pi_1(X_\star)$ is generated by two generators, a and b, subject to a single relation $\{a[a,b]b^{-1} = 1\}$ (so that $H_1(X_\star) \approx \mathbb{Z}$);
(2) $|\pi_0(\partial\mathcal{N}(\infty, K_\star))| = 2$, and $\pi_1(X_\star) = \{a, b \,|\, aba^{-2}b = 1,\ b = 1\} \approx 1$;
(3) $|\pi_0(\partial\mathcal{N}(\infty, K_\star))| = 3$, and $\pi_1(X_\star) = \{a, b \,|\, [a,b] = 1, a = 1, b = 1\} \approx 1$.

(3.14)

Here a, b can be identified with the two generators that span freely the group $\pi_1(\infty) \approx \mathsf{F}_2$.

In the first case, a single 2-handle is attached to the surface $\Sigma = \partial\mathcal{N}(\infty, X)$ of genus two so that $\pi_1(X_\star) \approx \Pi$. In the second case, two 2-handles are attached to Σ so that $X_\star$ is simply-connected. In the third case, three 2-handles are attached so that $X_\star$ is again simply-connected.

Since the loop pattern $\partial\mathcal{N}(\infty, K_\star)$ is rigid in $\mathcal{N}(a_\dagger, X_\star)$, the vicinity of $a_\dagger$ in $X_\star$ (see Figure 2.22, A), the pattern is determined up to Dehn twists of each of the two solid 1-handles that are attached to the 3-ball $\mathcal{N}(a_\dagger, X_\star)$ to form the handlebody $\mathcal{N}(\infty, X_\star)$. On the other hand, the topological type of the manifold $X_\star$, obtained by attaching solid 2-handles along the loops $\partial\mathcal{N}(\infty, K_\star)$ to the handlebody $\mathcal{N}(\infty, X_\star)$, does not depend of such Dehn twists. Therefore, the topology of $X_\star$ is uniquely determined by the three patterns in (3.14). As a result, in the case (2), $X_\star$ is homeomorphic to D^3, and, in the case (3), $X_\star$ is homeomorphic to $S^2 \times [0, 1]$.

Recall that $gc(D^3) = 0$ and $gc(S^2 \times [0, 1]) = 0$, although by the proof of Corollary 3.2, these manifolds indeed admit traversally generic fields with a single double-tangent trajectory! The only other closed irreducible manifolds of combinatorial complexity zero are the projective space $\mathbb{R}P^3$ and the lens space $L_{3,1}$ (see [Ma], Table A.9). However $gc(\mathbb{R}P^{3\circ}) \neq 1$ and $gc(L^\circ_{3,1}) \neq 1$, since the fundamental groups of these 3-folds are different from the ones in (3.14). Hence $gc(\mathbb{R}P^{3^\circ}) > 1$ and $gc(L^\circ_{3,1}) > 1$. This leaves us with a single irreducible and boundary irreducible with no essential annuli candidate for $X_\star$ of gradient complexity 1, namely, the first case in (3.14).

We claim that this unique candidate $X_\star$ indeed has the gradient complexity 1. By its very construction, $X_\star$ has an abstract gradient (equivalently, a $\vec{\mathsf{Y}}$-) spine $K_\star$ with the property $c(K_\star) = 1$. By Theorem 3.9, $X_\star$ admits a vector field $v_\star$ such that $gc(v_\star) \leq 1$. However, $gc(v_\star) \neq 0$, since according to Theorem 3.12 (see also the presentations in (3.6)-(3.9)), the fundamental groups of irreducible and boundary irreducible with no essential annuli 3-folds of the gradient complexity 0 are free or trivial.

Remark 3.5. It would be interesting to construct a traversally generic vector field $v_\star$ with a single double-tangent trajectory on the 3-fold $X_\star$ *explicitly*, as we did for D^3 and $S^2 \times [0, 1]$ (see Corollary 3.2). ◇

To complete the proof of Theorem 3.15, let us revisit the case of orientable 3-folds X with a simply-connected boundary. Table A.9 from [Ma] implies that $L_{4,1}$ and $L_{5,2}$ are the only two new possibilities for prime blocks in X with $c(X) = 1$; and again, they both fail to be of gradient complexity 1 because their fundamental groups do not match with the ones in (3.14) (supported by the three TN-marked patterns in Figure 3.11).

Therefore we conclude that $gc(X) \neq 1$ for any orientable 3-fold X with a simply-connected boundary!

So, if an orientable X with a simply-connected boundary admits a traversally generic field with a single double-tangent trajectory, it must also admit a traversally generic field with no double-tangent trajectories! By Theorem 3.14, such X is a connected sum of 3-balls and products $S^2 \times S^1$. □

Theorem 3.16. *Let X be a connected oriented irreducible and boundary irreducible with no essential annuli 3-fold.*

If $gc(X) = 1$, then X is obtained from the handlebody $\mathbf{H}$, whose boundary is a surface Σ of genus 2, by attaching one solid 2-handle in accordance with the longitudinal pattern (1) in (3.14). The topological type of X is determined by this construction. In particular, the fundamental group $\pi_1(X)$ is isomorphic to the group

$$\Pi =_{\mathsf{def}} \{a, b|\ a[a, b]b^{-1} = 1\}.$$

Also $H_1(X; \mathbb{Z}) \approx \mathbb{Z}$ and $H_2(X; \mathbb{Z}) \approx 0$.

Proof. Basically, the argument is already contained in the proof of Theorem 3.15. There we have shown that any X of gradient complexity 1 has a gradient spine K such that $s_\dagger(K) \supset \infty$ (recall that each gradient spine is, in particular, an abstract gradient spine and a $\vec{\mathsf{Y}}$-spine). Thus $\partial\mathcal{N}(\infty, K)$ must be as in Figure 3.11.

By Theorem 3.2, we can find a new abstract gradient spine $K_\star$ so that: (1) $c(K_\star) \leq c(K) = 1$, and (2) $K_\star^\odot =_{\mathsf{def}} K_\star \setminus \mathcal{N}(s_\dagger(K_\star), K_\star)$ is a disjoint union of 2-disks. Note that $c(K_\star) = 0$ would contradict to the assumption $gc(X) = 1$ since $boc(X) = gc(X)$ by Theorem 3.10; thus $c(K_\star) = 1$.

Some of the disks from $K_\star^\odot$ must be attached to $\mathcal{N}(\infty, K_\star)$ along the curves $\partial\mathcal{N}(\infty, K_\star)$ so that a connected subcomplex of $K_\star$ is formed. Since $K_\star$ is connected, it follows that $K_\star$ *is* the union of these disks with the regular neighborhood $\mathcal{N}(\infty, K_\star)$ and that $s_\dagger(K_\star) = \infty$. Therefore we can use the three models from (3.14), together with the TN-markers, to describe $K_\star$ faithfully.

By Theorem 2.7, the marked abstract gradient spine $K_\star$ determines the topological type of X.

By Theorem 3.9, the abstract gradient spine admits an approximation by a gradient spine of equal complexity. Therefore, $gc(X) \leq 1$ for any choice of simple curves in Σ, subject to (3.14). On the other hand, in the proof of Theorem 3.15, we have shown that $gc(X) \neq 0$. Therefore, remarkably, there exists a *unique* connected oriented irreducible and boundary irreducible with no essential annuli 3-fold of gradient complexity 1.

Combining the description of $\pi_1(X)$ with the homological long exact sequence of the pair $(X, \partial X)$ and taking into consideration the Poincaré duality, we get that ∂X is a 2-torus. □

Corollary 3.6. *Let X be a connected sum of a connected oriented irreducible and boundary irreducible with no essential annuli 3-fold $X^\star$ with a 3-fold X'.*

If $gc(X^\star) = 1$, then the fundamental group $\pi_1(X)$ is isomorphic to the free product $\Pi * \pi_1(X')$, where Π is the group as in Theorem 3.16. ♢

Theorem 3.17. *Let X be a connected oriented 3-fold with boundary. Let $\mathbf{H}$ be a handlebody whose boundary is a surface of genus 2, equipped with basic longitudes a, b and the meridians c, d.*

If $gc(X) = 1$, then one of the three options is available:

- *There exists an annulus $(A, \partial A) \subset (X, \partial X)$ so that cutting X along A divides X into the handlebody $\mathbf{H}$ and a 3-fold X' of gradient complexity 0. The core of A is a simple loop whose longitude pattern*[14] *in $\partial\mathbf{H}$ is $a[a, b]b^{-1}$.*
- *There exist two disjoint annuli $(A_1, \partial A_1), (A_2, \partial A_2) \subset (X, \partial X)$ so that cutting X along $A_1 \coprod A_2$ divides X into the handlebody $\mathbf{H}$ and a 3-fold X' of gradient complexity 0. The core of A_1 is a simple loop whose longitude pattern in $\partial\mathbf{H}$ is $aba^{-2}b$, and the core of A_2 is a simple loop whose longitude pattern is b.*
- *There exist three disjoint annuli $(A_1, \partial A_1), (A_2, \partial A_2), (A_3, \partial A_3) \subset (X, \partial X)$ so that cutting X along $A_1 \coprod A_2 \coprod A_3$ divides X into the handlebody $\mathbf{H}$ and a 3-fold X' of gradient complexity 0. The core of A_1 is a simple loop whose longitude pattern in in $\partial\mathbf{H}$ is $[a, b]$, the core of A_2 a simple loop whose longitude pattern is a, and the core of A_3 a simple loop whose longitude pattern is b.*

Proof. By the familiar arguments from the proofs of Theorem 3.15 and Theorem 3.16, X has an abstract gradient spine K with a single Q-singularity $a_\dagger$ so that $s(K) \supset \infty$, where $a_\dagger$ is the vertex of valency 4 in

[14] By the "longitude pattern" we mean any word w in the alphabet $\{a, b, c, d\}$ such that letting $c = 1$ and $d = 1$ produces a given word in $\{a, b\}$.

the graph ∞. The regular neighborhood $\mathcal{N}(\infty, K)$ is described by the three combinatorial patterns in Figure 3.11.

Let $p : X \to K$ be the retraction map as in Definition 2.3. Then $\mathbf{H} =_{\mathsf{def}} p^{-1}(\mathcal{N}(\infty, K))$ is a handlebody, obtained from the 3-ball $p^{-1}(\mathcal{N}(a_\dagger, K))$ by attaching two solid 1-handles. The annuli in the theorem form the set $p^{-1}(\partial\mathcal{N}(\infty, K))$.

Evidently, the manifold $X' = X \setminus \mathbf{H}$ has an abstract gradient spine $K' = K \setminus \mathcal{N}(\infty, K)$ such that $c(K') = 0$. By Theorem 3.9, the abstract gradient spine admits an approximation by a gradient spine of equal complexity. Thus $gc(X') = 0$. The rest of the claim that describes the longitudinal patterns of the annuli cores follows from (3.14). $\square$

Chapter 4

Deformations of Traversing Flows in 3D and Modifications of Gradient Spines

4.1 In this chapter...

We will study the effect of deforming non-vanishing gradient-like vector fields v in $3D$ on the gradient spines $K(v)$ they generate. From a different angle, we aim to describe the path connected components of the space of traversally generic vector fields on a given 3-fold. We claim only a partial progress towards this goal. Our approach is influenced heavily by [BP] and [Ar].

Since the gradient spines $K(v)$ in our inquiry are just tools to study the gradient-like flows, we cannot restrict our considerations only to, so called, *special* ("*standard*" in the terminology of [BP]) spines. Recall that these are spines K for which $s(K) \setminus ss(K)$ is a disjoint union of open arcs and $K \setminus s(K)$ is a disjoint union of open disks. There is nothing natural about the flows that produce special gradient spines. In this context, there is little value in developing some combinatorial calculus of special gradient spines, based on 1-dimensional graphs (appropriately decorated) as it is done in [BP]. In terminology of [BP1], the required version of the "*Combed Calculus*" is not local. The reader may view this section as an attempt to build an analogue of the Benerdetti-Petronio Combed Calculus of concave traversing flows (cf. [BP], Figure 4.7, and [BP1]) for the category of traversally generic flows.

We introduce a number elementary localized transformations of gradient spines $K(v) \subset X$. Some of them, denoted α, β and γ, resemble the three basic Reidemester moves for the plane diagrams of links [Re1]. Others, deal with diagrams, formed by plane curves with cusps (see Figure 4.1). All these elementary moves have semi-algebraic models, described in the list (4.4).

In Definition 4.1, we introduce an equivalence relation among traversally generic vector fields on a given 3-fold X with boundary. This equivalence relation is realized by special homotopies of traversing vector fields on X. We call such homotopies *truly generic*. The equivalence relation resembles the notion of quasi-isotopy.

If two traversally generic vector fields are linked by a truly generic homotopy, they define the same Spin^c-structure on X. We conjecture that the converse is true.

In Theorem 4.4, we prove that, if traversally generic vector fields v_0 and v_1 are linked by a truly generic homotopy, the spine $K(v_1)$ can be obtained from the spine $K(v_0)$ by a sequence of the elementary moves from the list (4.4).

Theorem 4.6 claims that any traversally generic vector field v can be deformed into a traversally generic vector field v' such that $\partial_3 X(v') = \emptyset$; moreover, the gradient complexity $gc(v') = gc(v)$. In particular, $gc(X)$ is realizable by traversally generic vector fields that are 3-concave and 3-convex.

Finally, we use basic properties of the elementary moves to coorient the walls of various chambers in the space $\mathcal{V}^\ddagger(X)$ of traversally generic vector fields on X, as well as in a variety of similar spaces like $\mathcal{V}^\ddagger_{\mathsf{fold}}(X)$—the space of traversally generic fields v, subject to the property "$\partial_3 X(v) = \emptyset$". We describe how piercing the walls of chambers in these spaces affects the gradient complexity $gc(v)$ and the polarized gradient complexity $gc^\oplus_\ominus(v)$ of a vector field v (see Theorem 4.7, Theorem 4.8, and Corollary 4.9).

4.2 On the combinatorics of 3D and 4D traversing vector flows

We start with a few observations about the combinatorics of traversally generic flows on the 3D- and 4D-manifolds with boundary. These observations will be reinforced, generalized, and validated in Chapters 5, 6, and 7.

Let v be a boundary generic vector field on a compact connected 3-fold X with boundary. Each trajectory γ of such a vector field generates a combinatorial pattern ω_γ, a finite sequence of natural numbers. The numbers are in one-to-one correspondence with points $x \in \gamma \cap \partial X$, the v-orientation of γ ordering the points. Each number reflects the multiplicity of tangency between γ and ∂X at the corresponding point $x \in \gamma \cap \partial X$. The sequence ω_γ either consists of a single even number, or starts and terminates with an odd number, the rest of the numbers being even. We denote by $\mathbf{\Omega}^\bullet$ the set of all such sequences ω.

Here is the sublist of combinatorial patterns ω_γ, available for *traversally generic* (see Definition 2.5) vector fields on a compact connected 3-folds:

$$\mathbf{\Omega}^\bullet_{'\langle 2]} =_{\mathsf{def}} \big\{(11);\ (2), (121);\ (31), (1221), (13)\big\}. \tag{4.1}$$

A boundary generic vector field, which is not traversally generic, may generate other combinatorial types: say, the types (12221), and (321). If the field is not even boundary generic, it may generate more combinatorial types: like the type (141).

Given a finite sequence $\omega = (\omega_1, \omega_2, \dots, \omega_q)$ of natural numbers, we introduce a non-negative integer $|\omega|' =_{\mathsf{def}} \sum_{i=1}^{q}(\omega_i - 1)$. We call $|\omega|'$ the *reduced norm* of ω. For a traversally generic v, we may think of $|\omega_\gamma|'$ as the *codimension* of the set of v-trajectories γ that share the combinatorial type ω_γ in the space of all trajectories on X. Note that in (4.1), the semicolons separate ω's by the increasing values $0, 1, 2$ of their reduced norms; thus the bizarre notation $\mathbf{\Omega}^\bullet_{'\langle 2]}$ for the list in (4.1).

We notice that, for a boundary generic v, the combinatorial patterns ω_γ of the v-trajectories γ that intersect the locus $\partial_2 X(v) \setminus \partial_3 X(v)$ must contain some $\omega_i = 2$. Similarly, the combinatorial patterns ω_γ of the v-trajectories γ that intersect the locus $\partial_3 X(v)$ must contain some $\omega_i = 3$.

Consider the extended list of combinatorial types ω:

$$\mathbf{\Omega}^\bullet_{'\langle 3]} =_{\mathsf{def}} \big\{\Omega^\bullet_{'\langle 2]};\ (4), (321), (123), (141), (12221)\big\}. \tag{4.2}$$

They arise as tangency patterns of trajectories of traversally generic vector fields v on 4-dimensional manifolds Y with boundary. In fact, $\mathbf{\Omega}^\bullet_{'\langle 3]}$ is a partially ordered set, depicted in Figure 7.2.

Each pattern ω defines a set of points $\partial_1 Y(v, \omega) \subset \partial_1 Y(v)$: it is generated by the intersections of the v-trajectories $\gamma \subset Y$ of the combinatorial type ω with the boundary $\partial_1 Y(v)$. Similarly, each ω defines a set of points $\partial_1^+ Y(v, \omega) \subset \partial_1^+ Y(v)$. For example, each v-trajectory γ of the combinatorial type (12221) hits $\partial_1^+ Y$ at four points, three of which belong to the locus $\partial_2^+ Y(v)$. The points, forming the intersection $\gamma \cap \partial_1 Y$, are ordered as they appear along the v-oriented trajectory γ.

Note that every γ, except for the ones of the main type (11), hits the points from the locus $\partial_2 Y(v)$. So we will concentrate on the intersections of γ with $\partial_2 Y(v)$. We denote by $\partial_2 Y(v, \omega)$ the locus $\{\gamma \cap \partial_2 Y(v)\}_\gamma$, where γ's are of the combinatorial type ω. Similarly, $\partial_2^\pm Y(v, \omega)$ stands for the locus $\{\gamma \cap \partial_2^\pm Y(v)\}_\gamma$, where γ's are of the combinatorial type ω.

4.3 One-parameter families of traversing flows on 3-folds

Consider the space $\mathcal{V}_{\mathsf{trav}}(X)$ of all traversing (equivalently, non-vanishing gradient-like) vector fields on X and its subspace $\mathcal{V}^\dagger(X) = \mathcal{B}^\dagger(X) \cap \mathcal{V}_{\mathsf{trav}}(X)$, formed by boundary generic traversing vector fields. Informally, we will also interpret $|\omega|'$ as the codimension, in the space $\mathcal{V}^\dagger(X)$, of the subspace $\mathcal{V}^\dagger(X, \omega)$, formed by the vector fields v that have trajectories of the combinatorial type ω.

For a smooth compact connected 3-fold X with boundary, put $Y =_{\mathsf{def}} X \times [0,1]$, $\delta Y =_{\mathsf{def}} \partial X \times [0,1]$, and $X_t =_{\mathsf{def}} X \times \{t\}$.

Given a continuous path $\tau : [0,1] \to \mathcal{V}_{\mathsf{trav}}(X)$, we form the vector field $v(x,t) =_{\mathsf{def}} \tau(t)(x)$ on Y. Let $v_t = \tau(t)$ denote the restriction of v to the slice X_t. By its construction, v_t is tangent to X_t.

We say that a path τ is *smooth*, if the vector field v is smooth on Y. Any continuous path τ can be approximated by a smooth path $\tilde{\tau} : [0,1] \to \mathcal{V}_{\mathsf{trav}}(X)$ with the same end points $\tau(0)$ and $\tau(1)$. Therefore we may deal only with smooth paths $\tilde{\tau}$.

For $\omega \in \mathbf{\Omega}^\bullet_{'\langle 4]}$ and $j \in [1,4]$, we consider the locus

$$\partial_j^\pm Y(v,\omega) =_{\mathsf{def}} \bigcup_{\{\gamma |\ \omega_\gamma = \omega\}} \gamma \cap \partial_j^\pm Y(v).$$

Because v is tangent to all X_t's, it is possible to check (see the proof of Lemma 6.12) that $\partial_j^\pm Y(v) = \bigcup_{t\in[0,1]} \partial_j^\pm X_t(v_t)$.

The next definition introduces an equivalence relation among the traversally generic vector fields on a given 3-fold X. A priori, this equivalence relation is more refined then the equivalence relation, defined by the obvious map $\mathcal{V}^\ddagger(X) \to \pi_0(\mathcal{V}^\ddagger(X))$. However, it has an advantage of allowing a tight control of the modifications of shapes of gradient spines within the equivalence class.

Definition 4.1. Let X be a compact connected 3-fold X with boundary and let $Y = X \times [0,1]$. Put $X_t = X \times \{t\}$.

We say that a continuous path $\tau : [0,1] \to \mathcal{V}_{\mathsf{trav}}(X)$ is *trully generic*, if:

- For all, but finitely many $t \in (0,1)$, the vector field $v_t =_{\mathsf{def}} \tau(t)$ on X_t is traversally generic. We denote by $\Theta_\tau \subset (0,1)$ the finite set of exceptional values.
- The vector field $v(x,t) =_{\mathsf{def}} v_t(x)$ on Y is *traversally generic* with respect to the portion δY of the boundary ∂Y.

- Each slice X_t that contains a v_t-trajectory γ of a combinatorial type $\omega \in \mathbf{\Omega}^\bullet_{'\langle 3]} \setminus \mathbf{\Omega}^\bullet_{'\langle 2]}$ is transversal to the surface $\partial_2 Y(v)$ and to the curve $\partial_3 Y(v)$ at every point of their intersection $\gamma \cap \partial_2 Y(v)$ or $\gamma \cap \partial_3 Y(v)$.
- For each $\omega \in \mathbf{\Omega}^\bullet_{'\langle 2]}$, the restriction $\pi_\omega : \partial_2 Y(v, \omega) \to [0,1]$ of the obvious projection $\pi : Y \to [0,1]$ to the locus $\partial_2 Y(v, \omega)$ is a Morse function.
- The critical values of these functions $\{\pi_\omega\}_{\omega \in \mathbf{\Omega}^\bullet_{'\langle 2]} \setminus \mathbf{\Omega}^\bullet_{'\langle 1]}}$ are all distinct and are contained in the finite set Θ_τ. ♢

Remark 4.1. By Lemma 6.12, any smooth path $\tau : [0,1] \to \mathcal{V}^\dagger(X)$ produces a boundary generic and traversing vector field v on $Y = X \times [0,1]$.

Again, by Lemma 6.12, any smooth path $\tau : [0,1] \to \mathcal{V}^\ddagger(X)$ produces a traversally generic vector field v on Y. In particular, any such path is truly generic in the sense of Definition 4.1. Therefore, $t \in \Theta_\tau$ from that definition only if $\tau(t) \notin \mathcal{V}^\ddagger(X)$. ♢

Consider a triple $v_0, v_1, v_2 \in \mathcal{V}^\ddagger(X)$. We notice that if v_0 can be connected to v_1 by a truly generic path, and if v_1 can be connected to v_2 by a truly generic path, then v_0 can be connected to v_2 by a truly generic path. So we may introduce a *truly generic path component* of a given vector field $v \in \mathcal{V}^\ddagger(X)$. Thus each truly generic path component is an open subset of $\mathcal{V}^\ddagger(X)$, comprising perhaps many path components of $\mathcal{V}^\ddagger(X)$.

Let us denote by $\pi_0^\bullet(\mathcal{V}^\ddagger(X))$ the set of truly generic path components of the space $\mathcal{V}^\ddagger(X)$. Evidently, we got a natural surjective map

$$\pi_0(\mathcal{V}^\ddagger(X)) \to \pi_0^\bullet(\mathcal{V}^\ddagger(X)).$$

Let us give a more conceptual interpretation of the traversing flows v_t on X_t that correspond to the exceptional $t \in \Theta_\tau$, an interpretation which works well only in small dimensions[1].

Let L be a linear subspace of dimension n in $\mathbb{R}^{n+k}$. Consider a proper *stable* (see Definition 2.4) smooth map $F : M \to \mathbb{R}^{n+k}$ from an open $(n+k)$-manifold M to the vector space $\mathbb{R}^{n+k}$. Assume that F is transversal to L. Then $f =_{\mathsf{def}} F| : F^{-1}(L) \to L$ is a smooth map from a n-dimensional manifold $N =_{\mathsf{def}} F^{-1}(L)$ to the n-dimensional vector space $L \approx \mathbb{R}^n$. Of course, $f : N \to L$ may not be stable. Still, f is not completely random.

We say that a given smooth map $f : N \to L$ is *versally k-stable*[2] if it is produced in this way by some stable map $F : M \to \mathbb{R}^{n+k}$.

[1]We assume that the relevant dimensions do not exceed 9.

[2]Perhaps, this is not the conventional use of the term "versal".

Definition 4.2. Let X be a smooth compact connected n-manifold with boundary. We say that a traversing vector field $v \in \mathcal{V}_{\mathsf{trav}}(X)$ is *versally k-stable* if any v-trajectory γ has a neighborhood $U \subset X$ and a flow section S (as in Definition 2.5) so that the flow-governed projection $p : \partial_1 X \cap U \to S$ is a versally l-stable map, where $l \leq k$.

We denote by $\mathcal{V}^{k-\mathsf{st}}_{\mathsf{trav}}(X)$ the space of all versally k-stable traversing vector fields on X. ◇

By definition, if a vector field is versally $(k-1)$-stable, then it is versally k-stable. Put

$$^\circ\mathcal{V}^{k-\mathsf{st}}_{\mathsf{trav}}(X) =_{\mathsf{def}} \mathcal{V}^{k-\mathsf{st}}_{\mathsf{trav}}(X) \setminus \mathcal{V}^{(k-1)-\mathsf{st}}_{\mathsf{trav}}(X).$$

This space is formed by traversing vector fields $v \in \mathcal{V}^{k-\mathsf{st}}_{\mathsf{trav}}(X)$ that have at least one v-trajectory γ, such that the local projection $p : \partial_1 X \cap U_\gamma \to S_\gamma$ is versally k-stable, but not versally $(k-1)$-stable.

Remark 4.2.

- By Definition 2.5, and the Whitney Theorem [W] (about the stable maps of a plane to itself), the traversally generic fields on 3-folds are exactly the ones that are versally 0-stable!
- By Theorem 6.7, for a 3-fold X, the space $\mathcal{V}^{0-\mathsf{st}}_{\mathsf{trav}}(X) = \mathcal{V}^{\ddagger}(X)$ is open and *dense* in the space $\mathcal{V}_{\mathsf{trav}}(X)$. Thus we may regard the difference $\mathcal{V}_{\mathsf{trav}}(X) \setminus \mathcal{V}^{0-\mathsf{st}}_{\mathsf{trav}}(X)$ as "the boundary" of the open domain $\mathcal{V}^{0-\mathsf{st}}_{\mathsf{trav}}(X)$ in $\mathcal{V}_{\mathsf{trav}}(X)$. This boundary is partitioned into the spaces $\{^\circ\mathcal{V}^{k-\mathsf{st}}_{\mathsf{trav}}(X)\}_{k\geq 1}$ and their "pathological" complement $\mathcal{V}^{\infty}_{\mathsf{trav}}(X)$ of infinite codimension. So we will treat $^\circ\mathcal{V}^{(k+1)-\mathsf{st}}_{\mathsf{trav}}(X)$ as the "main portion" of the boundary of $^\circ\mathcal{V}^{k-\mathsf{st}}_{\mathsf{trav}}(X)$. ◇

In small dimensions ($\dim(Y) \leq 4$), for a traversally generic vector field v on Y, in the vicinity of each v-trajectory γ, the flow generated local projections $\{p_\gamma : \delta Y \cap U_\gamma \to S_\gamma\}_\gamma$ are stable maps.

These local projections are in the core of Definition 6.1 that introduces the notion of traversally generic vector fields v on $(n+1)$-manifolds. In a more technical jargon, for a traversally generic v on the 4-dimensional manifold Y, the local projections $\{p_\gamma\}_\gamma$ have only singularities of the Boardman types S_1, S_{11}, S_{111}, and the maps $\{p_\gamma\}_\gamma$ are transversal to the Boardman strata in the appropriate jet spaces (i.e., they satisfy the *normal crossing condition*) (see [Bo]).

Therefore, by Definition 4.1, the restrictions

$$\{p| : \delta Y \cap U_\gamma \cap X_t \to S_\gamma \cap X_t\}_\gamma$$

are versally 1-stable. So a truly generic path $\tau : [0,1] \to \mathcal{V}_{\mathsf{trav}}(X)$ is, in fact, a path in the space $\mathcal{V}^{1-\mathsf{st}}_{\mathsf{trav}}(X)$, which intersects the subspace ${}^\circ\mathcal{V}^{1-\mathsf{st}}_{\mathsf{trav}}(X)$, the main portion of the boundary of $\mathcal{V}^{0-\mathsf{st}}_{\mathsf{trav}}(X)$, at finitely many points.

Recall that $\mathcal{V}^{0-\mathsf{st}}_{\mathsf{trav}}(X)$ is open and dense in the space $\mathcal{V}_{\mathsf{trav}}(X)$. Treating ${}^\circ\mathcal{V}^{2-\mathsf{st}}_{\mathsf{trav}}(X)$ as a "codimension 2 subspace" in the space in $\mathcal{V}_{\mathsf{trav}}(X)$, we speculate that τ can be perturbed so that the new path $\tilde{\tau}$ will miss the closure of the subspace ${}^\circ\mathcal{V}^{2-\mathsf{st}}_{\mathsf{trav}}(X)$. Therefore, we are compelled to state the following conjecture.

Conjecture 4.1. *Any two traversally generic vector fields v_0 and v_1 (i.e., any two points $v_0, v_1 \in \mathcal{V}^{0-\mathsf{st}}_{\mathsf{trav}}(X) = \mathcal{V}^{\ddagger}(X)$) that are connected by continuous path $\tau : [0,1] \to \mathcal{V}_{\mathsf{trav}}(X)$ can be also connected by a smooth truly generic path $\tilde{\tau} : [0,1] \to \mathcal{V}^{1-\mathsf{st}}_{\mathsf{trav}}(X)$ (as in Definition 4.1).* ◇

The challenge of proving the conjecture lies the *integrability* of the field $v = v_\tau$ on Y, the integral being the function $\pi : Y \to [0,1]$. By Theorem 6.7, we may perturb any traversing v so that the perturbed vector field $\tilde{v}$ is traversally generic with respect to δY, however, $\tilde{v}$ may be not integrable with respect to any function $\tilde{\pi} : Y \to \mathbb{R}$.

4.4 Local algebraic models of traversally generic 4*D*-flows and their 3*D*-slices

Now, for each $\omega = (\omega_1, \dots, \omega_q) \in \mathbf{\Omega}^\bullet_{'\langle 3]}$, we will construct a model smoothly stable map

$$F_\omega : \coprod_{i=1}^{q} (\mathbb{R}^3_i, 0_i) \longrightarrow (\mathbb{R}^3, 0),$$

where the rank of the differential $DF_\omega(0_i)$ at $0_i \in \mathbb{R}^3_i$ is $4 - \omega_i$. These maps $\{F_\omega\}_{\omega \in \mathbf{\Omega}^\bullet_{'\langle 3]}}$ will serve as *models* of the flow-generated local projections $\{p_\gamma : \partial_1 Y \cap U_\gamma \to S_\gamma\}_\gamma$ for traversally generic vector fields on 4-dimensional manifolds Y with boundary (here the combinatorial type of the core trajectory $\gamma \subset U_\gamma$ is ω).

Let us form the polynomial

$$P(u, \vec{x}) =_{\mathsf{def}} \prod_{i=1}^{q} \Big((u-i)^{\omega_i} + \sum_{k=0}^{\omega_i - 2} x_{i,k}(u-i)^k\Big), \tag{4.3}$$

were $(u, \{x_{i,k}\})$ are among the coordinates in $\mathbb{R}^4$ (depending on ω, the total number of the coordinates $\vec{x} =_{\mathsf{def}} \{x_{i,k}\}$ may be less than 3, and the degree of $P(u, \vec{x})$ in u has the range $\{2, 4, 6, 8\}$). Here u is the preferred coordinate, so that ∂_u models the traversally generic vector field on the 4-manifold $Z_\omega =_{\mathsf{def}} \{P(u, \vec{x}) \leq 0\}$. The latter will serve as a model for Y from Definition 4.1.

We also consider the hypersurface $H_\omega =_{\mathsf{def}} \{P(u, \vec{x}) = 0\}$, the model for δY. Let ℓ denote the u-coordinate line in $\mathbb{R}^4$. We notice that the ∂_u-trajectory ℓ intersects the hypersurface H_ω exactly at the points $\{(i, \vec{0})\}_{i \in [1,q]}$, and the multiplicity of tangency of ℓ to H_ω at $(i, \vec{0})$ is ω_i. So the combinatorial tangency pattern ω_ℓ of ℓ is the given sequence ω.

Let $F : \mathbb{R}^4 \to \mathbb{R}^3$ be the obvious projection along the u-direction. Let F_ω denote the restriction of F to the hypersurface H_ω.

We will concentrate on a sufficiently small open cylindrical neighborhood V of the line $\ell \subset \mathbb{R}^4$. The radius of the cylinder V is chosen so small, that the nonsingular hypersurface $H_\omega \cap V$ is diffeomorphic to the disjoint union of $q = |\omega| - |\omega|'$ open 3-balls $\{B_i\}$. We identify each B_i with the space $\mathbb{R}^3_i$.

Now let $F_\omega : H_\omega \cap V \to W$ be the u-directed projection on the open 3-ball $W =_{\mathsf{def}} \{u = 0\} \cap V \approx \mathbb{R}^3$, the transversal section of the ∂_u-flow and the base of the cylinder V.

Let T_i denote the F_ω-image of $\partial_{\omega_i} Z_\omega(\partial_u)$, localized to the vicinity of the point $(i, 0) \in l \cap H_\omega$. In Chapter 6, we will verify that the restrictions $\{F_\omega : \partial_{\omega_i} B_i(\partial_u) \to W\}_i$ are immersions. In fact, T_i is an affine subspace of $\mathbb{R}^3$ that is spanned by the set of $\vec{x}$-coordinate vectors, complementary to the coordinates $\{x_{i,k}\}_k$. So the vector spaces $\{\mathsf{T}_i\}_{\{i |\, \omega_i > 1\}}$ are automatically in general position in W. As a result, the vector field ∂_u is traversally generic in the vicinity of the line ℓ.

For $\omega_i > 1$, T_i may be a point, a line, or a plane. Here is a list of all possible generic configurations $\coprod_{\{i |\, \omega_i > 1\}} \mathsf{T}_i \subset \mathbb{R}^3$. For $\omega = (2), (121)$, the configuration $\bigcup_{\{i |\, \omega_i > 1\}} \mathsf{T}_i$ is a single plane in $\mathbb{R}^3$. For $\omega = (31), (13)$, the configuration $\bigcup_{\{i |\, \omega_i > 1\}} \mathsf{T}_i$ is a single line in $\mathbb{R}^3$. For $\omega = (1221)$, the configuration $\bigcup_{\{i |\, \omega_i > 1\}} \mathsf{T}_i$ is a a pair of transversal planes in $\mathbb{R}^3$. For $\omega = (4), (141)$, the configuration $\bigcup_{\{i |\, \omega_i > 1\}} \mathsf{T}_i$ is a singleton. For $\omega = (321), (123)$, the configuration $\bigcup_{\{i |\, \omega_i > 1\}} \mathsf{T}_i$ is a line transversal to a plane. Finally, for $\omega = (12221)$, the configuration $\bigcup_{\{i |\, \omega_i > 1\}} \mathsf{T}_i$ is a union of three mutually transversal planes.

This completes the description of the ω-dependent model map F_ω :

$\coprod_{i=1}^{q}(\mathbb{R}_i^3, \vec{0}_i) \to (\mathbb{R}^3, \vec{0})$.

Now we aim to describe the way in which a generic ∂_u-invariant $3D$-foliation in $\mathbb{R}^4$ interacts with the map F_ω. Given a nonsingular smooth function $L : W \to \mathbb{R}$, we consider the 2-dimensional foliation $\mathcal{L}$ in $W \approx \mathbb{R}^3$, whose leaves are $\{\mathcal{L}_c =_{\mathsf{def}} L^{-1}(c)\}_{c \in \mathbb{R}}$. The $3D$-slices $\{F^{-1}(\mathcal{L}_c) \cap Z_\omega\}_{c \in \mathbb{R}}$ will model the foliation $\{X_t\}_{t \in [0,1]}$ in $Y = X \times [0, 1]$.

We pick L so that, in the vicinity of the origin $\vec{0} \in \mathbb{R}^3$, each surface $\mathcal{L}_c$ is in general position with the configuration $\bigcup_{\{i|\,\omega_i > 1\}} \mathsf{T}_i \subset \mathbb{R}^3$. To achieve this, it suffices to choose the plane $\mathsf{L}_0 \subset \mathbb{R}^3$, tangent to the surface $\mathcal{L}_0$ at the origin, to be in general position with respect to $\bigcup_{\{i|\,\omega_i > 1\}} \mathsf{T}_i$ and then to shrink the radius of the cylinder V, if necessary.

Although the configuration $\bigcup_{\{i|\,\omega_i > 1\}} \mathsf{T}_i$ is generic in $\mathbb{R}^3$, the *affine* spaces $\{\mathsf{T}_{\{i|\,\omega_i > 1\}} \cap \mathsf{L}_0\}_i$ *may not be in general position* in the plane L_0! Thus ∂_u may not be traversally generic in the $3D$-slice $F^{-1}(\mathcal{L}_0) \subset \mathbb{R}^4$.

By its construction, the plane L_0 is in general position with the configuration $\bigcup_{\{i|\,\omega_i > 1\}} \mathsf{T}_i$. Therefore, by examining the list of all possible configurations above, we conclude that the configuration $\bigcup_{\{i|\,\omega_i > 1\}}(\mathsf{T}_i \cap \mathsf{L}_c)$ of affine spaces *is* in general position in L_c, provided $c \neq 0$. Therefore, for a sufficiently small $|c| \neq 0$, the submanifolds $\{F_\omega(\partial_{\omega_i} B_i(\partial_u)) \cap \mathcal{L}_c\}_i$ are in general position in $\mathcal{L}_c$. In other words, ∂_u is traversally generic in each $3D$-slice $Z_\omega \cap F_\omega^{-1}(\mathcal{L}_c)$, where $|c| \neq 0$ is small enough.

Since, when $\omega_i < 4$, the space T_i is transversal to each surface $\{\mathcal{L}_c\}_{c \in (-\epsilon, \epsilon)}$, the map F_ω is transversal to the foliation $\{\mathcal{L}_c\}_c$ in a sufficiently narrow cylinder $V \supset \ell$. As a result, the source space H_ω of F_ω acquires the pull-back $2D$-foliation $\{F_\omega^{-1}(\mathcal{L}_c)\}_{c \in (-\epsilon, \epsilon)}$.

Let $\{v_t\}_{t \in [0,1]}$ be a truly generic family (see Definition 4.1) of vector fields. Recall that, by Lemma 2.7, the gradient spine K_t of $X_t \subset Y$ is determined by the v_t-flow generated map $p_t : \partial_2^+ X_t(v_t) \to \partial_1^+ X_t(v_t)$. So, passing to the local models, we aim to understand how the gradient spines $\{K_c^\omega \subset F^{-1}(\mathcal{L}_c) \cap Z_\omega\}_{c \in \mathbb{R}}$, generated by ∂_u, evolve in the L-generated $3D$-slices $\{F^{-1}(\mathcal{L}_c)\}_{c \in \mathbb{R}}$ of the model $(Z_\omega, H_\omega, \partial_u)$. The model gradient spine K_c^ω in the slice can be recovered from the ∂_u-induced projection

$$p_c^\omega : \partial_2 Z_\omega(\partial_u) \cap F_\omega^{-1}(\mathcal{L}_c) \to \partial_1^+ Z_\omega(\partial_u) \cap F_\omega^{-1}(\mathcal{L}_c)$$

that takes each point $x \in \partial_2 Z_\omega(\partial_u) \cap F_\omega^{-1}(\mathcal{L}_c)$ to the point $y \in \partial_1^+ Z_\omega(\partial_u) \cap F_\omega^{-1}(\mathcal{L}_c)$ that resides just below x in the ∂_u-trajectory γ_x through x (i.e., $u(y) < u(x)$). As before, if x is the lowest point in $\gamma_x \cap H_\omega$, we define $p_c^\omega(x) = x$.

Alternatively, each map p_c^ω, and thus each spine K_c^ω, is determined by the F_ω-image of the 1-dimensional locus $\partial_2 Z_\omega(\partial_u) \cap F_\omega^{-1}(\mathcal{L}_c)$ in the surface $\mathcal{L}_c \approx \mathbb{R}^2$, together with the depiction[3] of the F_ω-image in $\mathcal{L}_c$ of the germs of the components of the surface $\partial_1^+ Z_\omega(\partial_u) \cap F_\omega^{-1}(\mathcal{L}_c)$ that are locally bounded by the corresponding arcs of $\partial_2 Z_\omega(\partial_u) \cap F_\omega^{-1}(\mathcal{L}_c)$. The layers-components of the surfaces $\partial_1^+ Z_\omega(\partial_u) \cap F_\omega^{-1}(\mathcal{L}_c)$, suspended over the surface $\mathcal{L}_c$, are locally ordered by the the function $u : \partial_1^+ Z_\omega(\partial_u) \cap F_\omega^{-1}(\mathcal{L}_c) \to \mathbb{R}$. As a result, the local branches of the curves from $F_\omega(\partial_2 Z_\omega(\partial_u)) \cap \mathcal{L}_c$ are also ordered in the vicinity of their intersection points. In the vicinity of a point $F_\omega(x)$, where $x \in \partial_3^+ Z_\omega(\partial_u)$, the germ of $\partial_1^+ Z_\omega(\partial_u) \cap F_\omega^{-1}(\mathcal{L}_c)$ may be ramified over $\mathcal{L}_c$.

We will use these conventions to simplify the depiction of any gradient spine K_c^ω by planar graphs with some layered shadings of the domains, adjacent to the curves that form the graphs (see Figure 4.1).

Thus, for an appropriate choice of $\omega \in \mathbf{\Omega}^\bullet_{'\langle 3]}$ and of the foliation $\{\mathcal{L}_c\}_c$ in $W \approx \mathbb{R}^3$, our next goal is to list and describe the elementary changes of localized patterns of the plane contour curves that appear in the 1-parameter families

$$\{(\bigcup_{i \in [1,q]} F_\omega(\partial_2 B_i(\partial_u))) \cap \mathcal{L}_c\}_{c \in (-\epsilon, \epsilon)},$$

where the 3-balls $\{B_i\}_i$ are the connected components of $\partial_1 Z_\omega(\partial_u)$ in the vicinity of the line $F^{-1}(0)$, and of the associated layerings

$$\{\partial_1^+ B_i(\partial_u) \cap F_\omega^{-1}(\mathcal{L}_c)\}_{i \in [1,q]}$$

of the domains $F_\omega(\partial_1^+ Z_\omega(\partial_u)) \cap \mathcal{L}_c$, adjacent to the curves.

4.5 Algebraic models of elementary surgery for truly generic deformations of traversing $3D$-flows

For each $\omega \in \mathbf{\Omega}^\bullet_{'\langle 3]} \setminus \mathbf{\Omega}^\bullet_{'\langle 2]}$, any generic choice of the foliation $\{\mathcal{L}_c\}_c$ will generate a distinct elementary modification, as we compare the slice of the ∂_u-flow in Z_ω by the hypersurfaces $F^{-1}(\mathcal{L}_{-\epsilon})$ and $F^{-1}(\mathcal{L}_\epsilon)$. For $\omega \in \mathbf{\Omega}^\bullet_{'\langle 2]}$, a slightly different mechanism is available. It is based on a more careful correlated choice of the foliation $\{\mathcal{L}_c\}_c$. That choice requires not only to pick the tangent plane L_0 to the surface $\mathcal{L}_0$ generically, but also the 2-jet of the surface $\mathcal{L}_0$.

[3]shading like in Figure 4.1 and Figure 4.4, the lower diagrams

We present now a long list of elementary moves that arise from the combinatorial types $\omega \in \boldsymbol{\Omega}^{\bullet}_{'\langle 3]}$ and the generic 2-jets or 1-jets of surfaces $\mathcal{L}_0 \subset \mathbb{R}^3$ at the origin. First, we will describe these moves algebraically.

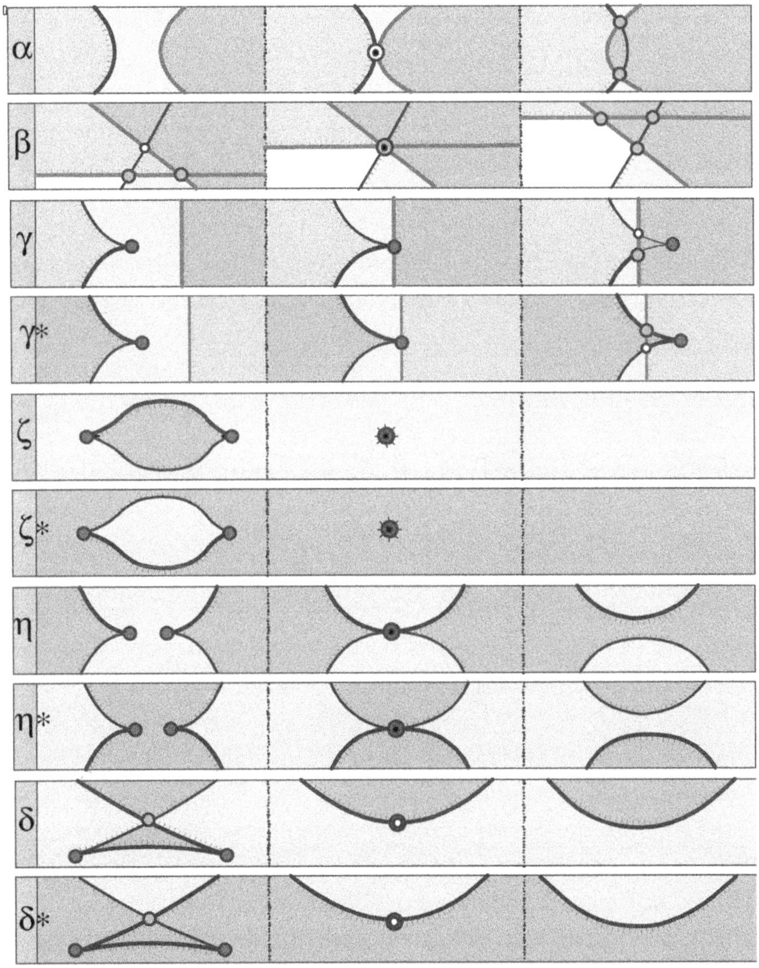

Fig. 4.1 *The elementary moves, depicted as a change in the projections of the loci* $\partial_2 X_t(v_t) \subset \partial_1^+ X_t(v_t)$ *on the local traversal section of the* v_t*-flow. The passages from the left to the right columns and back show the moves, while the middle diagrams depict the exceptional patterns.*

Reference (4.4) marks the following bulleted list of moves. (4.4)

- **The α-move.** (The exceptional plane curves pattern is shown in Figure 4.1 and in Figure 2.8, diagram C. See also Figure 4.3.)
  The α-move is an analogue of the type II Redemeister move for planner loop diagrams.

  Let $\omega = (1221) \in \mathbf{\Omega}^\bullet_{'\langle 2]}$, and let $(u, x_{2,0}, x_{3,0}, y)$[4] be coordinates in $\mathbb{R}^4$. We form the polynomial

  $$P(u, x_{2,0}, x_{3,0}) =_{\mathsf{def}} (u-1)\big((u-2)^2 + x_{2,0}\big)\big((u-3)^2 + x_{3,0}\big)(u-4).$$

  The hypersurface $H_{(1221)} \subset \mathbb{R}^4$ is given by the polynomial equation $P(u, x_{2,0}, x_{3,0}) = 0$, and the domain $Z_{(1221)} \subset \mathbb{R}^4$ by the inequality $P(u, x_{2,0}, x_{3,0}) \leq 0$. The projection F is defined by the formula $(u, x_{2,0}, x_{3,0}, y) \to (x_{2,0}, x_{3,0}, y)$. The two planes, T_2 and T_3, represent the images of $\partial_2^+ Z_{(1221)}(\partial_u) \cap V$ under the map F. They are in general position in $\mathbb{R}^3$.

  The foliation $\mathcal{L}$ is produced with the help of the function

  $$L(x_{2,0}, x_{3,0}, y) =_{\mathsf{def}} \ell(x_{2,0}, x_{3,0}) - y^2,$$

  where $(x_{2,0}, x_{3,0}, y)$ are coordinates in $\mathbb{R}^3$ and $\ell(x_{2,0}, x_{3,0})$ is a generic linear function, such that the degree-one term of the quadratic polynomial $\ell(-(u-2)^2, -(u-3)^2)$ does not vanish. In particular, we may choose $\ell(x_{2,0}, x_{3,0}) = x_{2,0} + x_{3,0}$.

  Put $\mathcal{L}_c =_{\mathsf{def}} L^{-1}(c)$. The surface $\mathcal{L}_0$ is tangent to the y-axis, the F-image of the set of ∂_u-trajectories of the combinatorial type (1221). As c crosses 0, the α-move $(\mathsf{T}_2 \cup \mathsf{T}_3) \cap \mathcal{L}_{-\epsilon} \Longrightarrow (\mathsf{T}_2 \cup \mathsf{T}_3) \cap \mathcal{L}_{\epsilon}$ is executed.

- **The β-move.** (The exceptional plane curves pattern is shown in Figure 4.1 and in Figure 2.8, diagram B. See also Figure 4.4.)
  The β-move is an analogue of the type III Redemeister move.

  Let $\omega = (12221) \in \mathbf{\Omega}^\bullet_{'\langle 3]} \setminus \mathbf{\Omega}^\bullet_{'\langle 2]}$. Let $(u, x_{2,0}, x_{3,0}, x_{4,0})$ be the coordinates in $\mathbb{R}^4$. Then

  $$P(u, x_{2,0}, x_{3,0}, x_{4,0}) =_{\mathsf{def}}$$

  $$(u-1)\big((u-2)^2 + x_{2,0}\big)\big((u-3)^2 + x_{3,0}\big)\big((u-4)^2 + x_{4,0}\big)(u-5).$$

[4] The strange choice of the notations "$x_{2,0}, x_{3,0}$" for the variables is needed to make our notations consistent trough a long list of ω's.

The hypersurface $H_{(12221)} \subset \mathbb{R}^4$ is given by the equation

$$P(u, x_{2,0}, x_{3,0}, x_{4,0}) = 0.$$

The map $F_{(12221)} : H_{(12221)} \to \mathbb{R}^3$ drops the u-coordinate. The plane $\mathsf{T}_2 \subset \mathbb{R}^3$ is given by the equation $\{x_{2,0} = 0\}$, the plane $\mathsf{T}_3 \subset \mathbb{R}^3$ by the equation $\{x_{3,0} = 0\}$, and the plane $\mathsf{T}_4 \subset \mathbb{R}^3$ by the equation $\{x_{4,0} = 0\}$. The union of these three coordinate planes is the $F_{(12221)}$-image of the folding loci $\partial_2^+ Z(\partial_u) \cap V$.

We pick a generic linear function $L(x_{2,0}, x_{3,0}, x_{4,0})$ so that the degree-one term of the quadratic polynomial $L(-(u-2)^2, -(u-3)^2, -(u-4)^2)$ does not vanish. In particular, we may pick

$$L(x_{2,0}, x_{3,0}, x_{4,0}) = x_{2,0} + x_{3,0} + x_{3,0}.$$

Let $\mathcal{L}_c =_{\mathsf{def}} L^{-1}(c)$. The passage from the configuration $(\mathsf{T}_2 \cup \mathsf{T}_3 \cup \mathsf{T}_4) \cap \mathcal{L}_{-\epsilon} \subset \mathcal{L}_{-\epsilon}$ to the configuration $(\mathsf{T}_2 \cup \mathsf{T}_3 \cup \mathsf{T}_4) \cap \mathcal{L}_\epsilon \subset \mathcal{L}_\epsilon$ represents the β-move. The exceptional configuration $(\mathsf{T}_2 \cup \mathsf{T}_3 \cup \mathsf{T}_4) \cap \mathcal{L}_0$ is formed by three lines through the origin.

- **The γ-move.** (The exceptional curves pattern is shown in Figure 2.8, diagram D, and Figure 4.1, and the cascades in Figure 4.8, the upper diagram.)

  The γ-move is an analogue of the type II Redemeister move for plane curves with cusps.

  Let $\omega = (321) \in \mathbf{\Omega}^\bullet_{\prime\langle 3]} \setminus \mathbf{\Omega}^\bullet_{\prime\langle 2]}$. Let $(u, x_{1,0}, x_{1,1}, x_{2,0})$ be the coordinates in $\mathbb{R}^4$. Then

  $$P(u, x_{1,0}, x_{1,1}, x_{2,0}) =_{\mathsf{def}}$$

  $$\big((u-1)^3 + x_{1,1}(u-1) + x_{1,0}\big)\big((u-2)^2 + x_{2,0}\big)(u-3).$$

  The line $\mathsf{T}_1 \subset \mathbb{R}^3$ is given by the equations $\{x_{1,0} = 0, x_{1,1} = 0\}$, and the plane $\mathsf{T}_2 \subset \mathbb{R}^3$ by the equation $\{x_{2,0} = 0\}$; so they are in general position in $\mathbb{R}^3$.

  Consider the surface $\Sigma \subset H_{(321)}$, given by the two polynomial equations

  $$\{Q(u, x_{1,0}, x_{1,1}) = 0,\ \frac{\partial}{\partial u} Q(u, x_{1,0}, x_{1,1}) = 0\},$$

  where

  $$Q(u, x_{1,0}, x_{1,1}) =_{\mathsf{def}} (u-1)^3 + x_{1,1}(u-1) + x_{1,0}.$$

  The $F_{(321)}$-image of the surface Σ is the product of the cuspidal curve $\mathcal{C}$ in the $(x_{1,0}, x_{1,1})$-plane *times* the $x_{2,0}$-coordinate line λ. The intersection $\Sigma \cap V$ is a portion of the locus $\partial_2 Z_{(321)}(\partial_u)$. Thus the $F_{(321)}$-image of $\partial_2 Z_{(321)}(\partial_u)$ is the union of the coordinate plane T_2 with the surface $F_{(321)}(\Sigma) = (\mathcal{C} \times \lambda)$ (note that $\mathsf{T}_2 \cap F_{(321)}(\Sigma) = \mathcal{C}$).

Let $L(x_{1,0}, x_{1,1}, x_{2,0})$ be a linear function such that its zero plane $\mathcal{L}_0$ is in general position with the configuration $\mathsf{T}_1 \bigcup \mathsf{T}_2$. In particular, we may pick the linear function

$$L(x_{1,0}, x_{1,1}, x_{2,0}) =_{\mathsf{def}} x_{1,1} - x_{2,0}.$$

The passage from the configuration $(\mathsf{T}_2 \cup (\mathcal{C} \times \lambda)) \cap \mathcal{L}_{-\epsilon} \subset \mathcal{L}_{-\epsilon}$ to the configuration $(\mathsf{T}_2 \cup (\mathcal{C} \times \lambda)) \cap \mathcal{L}_{\epsilon} \subset \mathcal{L}_{\epsilon}$ represents the γ-move.

- **The $\gamma^\star$-move.** (The exceptional curves pattern is shown in Figure 2.8, diagram D, and Figure 4.1, and the cascades in Figure 4.8, the lower diagram.)

  It is an analogue of the type II Redemeister move for plane curves with cusps.

  This move is very similar to the γ-move, but based on $\omega = (123) \in \mathbf{\Omega}^\bullet_{\prime\langle 3]} \setminus \mathbf{\Omega}^\bullet_{\prime\langle 2]}$. The corresponding polynomial

$$P(u, x_{1,0}, x_{1,1}, x_{2,0}) =_{\mathsf{def}}$$
$$(u-1)\big((u-2)^2 + x_{2,0}\big)\big((u-3)^3 + x_{3,1}(u-3) + x_{3,0}\big).$$

  The line $\mathsf{T}_3 \subset \mathbb{R}^3$ is given by the equations $\{x_{3,0} = 0, x_{3,1} = 0\}$ and the plane $\mathsf{T}_2 \subset \mathbb{R}^3$ by the equation $\{x_{2,0} = 0\}$.

  This time, the portion of the locus $\partial_2 Z_{(123)}(\partial_u)$ in the vicinity of the point $(2, \vec{0})$ is below the the portion of locus $\partial_3 Z_{(123)}(\partial_u)$ in the vicinity of $(3, \vec{0})$. Therefore the geometry of the cascades is different from the geometry of the cascades in the γ-move, although the plane curves patterns are similar.

  Let $L(x_{1,0}, x_{1,1}, x_{2,0})$ be a linear function such that its zero plane $\mathcal{L}_0$ is in general position with the configuration $\mathsf{T}_2 \bigcup \mathsf{T}_3$.
  The passage from the configuration $(\mathsf{T}_2 \cup (\mathcal{C} \times \lambda)) \cap \mathcal{L}_{-\epsilon} \subset \mathcal{L}_{-\epsilon}$ to the configuration $(\mathsf{T}_2 \cup (\mathcal{C} \times \lambda)) \cap \mathcal{L}_{\epsilon} \subset \mathcal{L}_{\epsilon}$ represents the $\gamma^\star$-move.

- **The δ-move (*the Swallow Tail cusp cancellation*).** (Figure 2.12 shows the geometry of the slice before the cancellation. See Figure 4.1 and Figure 4.6, diagrams 5 and 6.)

  The δ-move is an analogue of the type I Redemeister move for plane curves with cusps.

  Let $\omega = (4) \in \mathbf{\Omega}^\bullet_{\prime\langle 3]} \setminus \mathbf{\Omega}^\bullet_{\prime\langle 2]}$ and let $(u, x_{1,0}, x_{1,1}, x_{1,2})$ be the coordinates in $\mathbb{R}^4$. Then

$$P(u, x_{1,0}, x_{1,1}, x_{1,2}) =_{\mathsf{def}} (u-1)^4 + x_{1,2}(u-1)^2 + x_{1,1}(u-1) + x_{1,0}.$$

We pick the linear function $L(x_{1,0}, x_{1,1}, x_{1,2})$ so that, for small $|c|$, its constant value foliation $\{\mathcal{L}_c\}_c$ is transversal to each stratum of the Swallow Tail surface, the $F_{(4)}$-image of the locus $\{P = 0, \frac{\partial}{\partial u}P = 0\}$. In particular, we may choose

$$L(x_{1,0}, x_{1,1}, x_{1,2}) =_{\mathsf{def}} x_{1,2}.$$

The slice $F_\omega(\partial_2 Z(\partial_u)) \cap \mathcal{L}_{-\epsilon}$ represents a smooth curve with a single transversal self-intersection and two cusps, the three singular points being the vertices of curvilinear triangle. As a negative c crosses 0, the triangle shrinks, and the slice $F_\omega(\partial_2 Z(\partial_u)) \cap \mathcal{L}_\epsilon$ becomes a simple nonsingular curve.

- **The $\delta^\star$-move.** (See Figure 4.1 and Figure 4.6, diagrams 1 and 2.) *The $\delta^\star$-move is an analogue of the type I Redemeister move for plane curves with cusps.*
  This move is very similar to the δ-move, but based on $\omega = (141) \in \mathbf{\Omega}^\bullet_{'\langle 3]} \setminus \mathbf{\Omega}^\bullet_{'\langle 2]}$. Therefore the geometry of the cascades is different from the geometry of the cascades in the δ-move, although the plane curves patterns of the projection $F_{(141)}$ are similar. Let the polynomial

  $$P(u, x_{2,0}, x_{3,1}, x_{3,0}) =_{\mathsf{def}}$$

  $$= (u-1)\big((u-2)^4 + x_{2,2}(u-2)^2 + x_{2,1}(u-2) + x_{2,0}\big)(u-3).$$

  The foliation $\{\mathcal{L}_c\}_c$ is defined by the linear function $L(x_{2,0}, x_{2,1}, x_{2,2})$ as in the case of the δ-move. In particular, we may choose

  $$L(x_{2,0}, x_{2,1}, x_{2,2}) =_{\mathsf{def}} x_{2,2}.$$

- **The ζ-move (*the cusp cancellation via* 0*-surgery*).** (See Figure 4.1, Figure 4.6, diagrams 3 and 4, and Figure 4.5).
  Let $\omega = (31) \in \mathbf{\Omega}^\bullet_{'\langle 2]}$ and let $(u, x_{1,0}, x_{1,1}, y)$ be the coordinates in $\mathbb{R}^4$. Then

  $$P(u, x_{1,0}, x_{1,1}) =_{\mathsf{def}} = \big((u-1)^3 + x_{1,1}(u-1) + x_{1,0}\big)(u-2).$$

  We consider a smooth function

  $$L(x_{1,0}, x_{1,1}, y) =_{\mathsf{def}} \ell(x_{1,0}, x_{1,1}) + y^2,$$

  where $x_{1,0}, x_{1,1}, y$ are coordinates in $\mathbb{R}^3$, and the 1-jet of the function ℓ has the $x_{1,1}$-term with nonzero coefficient. In particular, we may choose

  $$L(x_{1,0}, x_{1,1}, y) =_{\mathsf{def}} x_{1,1} + y^2.$$

$L(x_{1,0}, x_{1,1}, y)$ defines the foliation $\{\mathcal{L}_c\}_c$.
The slice $F_\omega(\partial_2 Z(\partial_u)) \cap \mathcal{L}_{-\epsilon}$ is a smooth loop with two cusps; the exceptional slice $F_\omega(\partial_2 Z(\partial_u)) \cap \mathcal{L}_0$ is the singleton, while the slice $F_\omega(\partial_2 Z(\partial_u)) \cap \mathcal{L}_\epsilon$ is empty.

- **The $\zeta^\star$-move (*the cusp cancellation via* 0-*surgery*).** (See Figure 4.1, Figure 4.6, diagrams 3 and 4, and Figure 4.5.)
Let $\omega = (13) \in \mathbf{\Omega}^\bullet_{'\langle 2]}$. The plane curves patterns are similar to the ones in the ζ-move, but the cascades are different. Let

$$P(u, x_{1,0}, x_{1,1}) =_{\mathsf{def}}= (u-1)\big((u-2)^3 + x_{2,1}(u-2) + x_{2,0}\big)$$

and let

$$L(x_{2,0}, x_{2,1}, y) =_{\mathsf{def}} \ell(x_{2,0}, x_{2,1}) + y^2,$$

where $x_{2,0}, x_{2,1}, y$ are coordinates in $\mathbb{R}^3$, and the 1-jet of the function ℓ has the $x_{2,1}$-term with nonzero coefficient. In particular, we may choose

$$L(x_{2,0}, x_{2,1}, y) =_{\mathsf{def}} x_{2,1} + y^2.$$

The function $L(x_{2,0}, x_{2,1}, y)$ defines the foliation $\{\mathcal{L}_c\}_c$. The slices $F_\omega(\partial_2 Z(\partial_u)) \cap \mathcal{L}_{\pm\epsilon}$ are as in the ζ-move.

- **The η-move (*the cusp cancellation via* 1-*surgery*).** (See Figure 4.1, Figure 4.6, diagrams 1 and 2, and Figure 4.5.)
Let $\omega = (31)$ and

$$P(u, x_{1,0}, x_{1,1}) =_{\mathsf{def}}= \big((u-1)^3 + x_{1,1}(u-1) + x_{1,0}\big)(u-2).$$

and let

$$L(x_{1,0}, x_{1,1}, y) =_{\mathsf{def}} \ell(x_{1,0}, x_{1,1}) - y^2,$$

where $x_{1,0}, x_{1,1}, y$ are coordinates in $\mathbb{R}^3$, and the 1-jet of the function ℓ has the $x_{1,1}$-term with nonzero coefficient. In particular, we may choose

$$L(x_{1,0}, x_{1,1}, y) =_{\mathsf{def}} x_{1,1} - y^2.$$

The function $L(x_{1,0}, x_{1,1}, y)$ defines the foliation $\{\mathcal{L}_c\}_c$.
The slice $F_\omega(\partial_2 Z(\partial_u)) \cap \mathcal{L}_{-\epsilon}$ is a disjoint union of two smooth arcs; the slice $F_\omega(\partial_2 Z(\partial_u)) \cap \mathcal{L}_0$ is a union of two cuspidal curves that share a cusp at the origin, while the slice $F_\omega(\partial_2 Z(\partial_u)) \cap \mathcal{L}_\epsilon$ is a disjoint union of two curves, each one with its own cusp.

- **The $\eta^\star$-move (*the cusp cancellation via* 1*-surgery*).** (See Figure 4.1, Figure 4.6, diagrams 1 and 2, and Figure 4.5.) It is similar to the η-move: the plane curves patterns are similar to the ones in the ζ-move, but the cascades are different.

  The $\eta^\star$-move is based on $\omega = (13) \in \mathbf{\Omega}^\bullet_{'\langle 2]}$ and the polynomial
  $$P(u, x_{2,0}, x_{2,1}) =_{\mathsf{def}}= (u-1)\big((u-2)^3 + x_{2,1}(u-2) + x_{2,0}\big).$$
  Let
  $$L(x_{2,0}, x_{2,1}, y) =_{\mathsf{def}} \ell(x_{2,0}, x_{2,1}) - y^2,$$
  where $x_{2,0}, x_{2,1}, y$ are coordinates in $\mathbb{R}^3$, and the 1-jet of the function ℓ has the $x_{2,1}$-term with nonzero coefficient. In particular, we may choose
  $$L(x_{2,0}, x_{2,1}, y) =_{\mathsf{def}} x_{2,1} - y^2.$$
  The function $L(x_{1,0}, x_{1,1}, y)$ defines the foliation $\{\mathcal{L}_c\}_c$.

  The slices $F_\omega(\partial_2 Z(\partial_u)) \cap \mathcal{L}_{\pm\epsilon}$ are the same as in the η-move.

Finally, we have reached the end of the elementary surgery list.

4.6 How truly generic deformations of traversing vector fields affect the gradient spines

Prior to engaging with Theorem 4.1 below, let us describe in some details the geometry of the loci $\{\partial_2 Y(v,\omega)\}_{\omega \in \mathbf{\Omega}^\bullet_{'\langle 3]}}$ and $\{\partial_2^{\pm} Y(v,\omega)\}_{\omega \in \mathbf{\Omega}^\bullet_{'\langle 3]}}$ for *any* traversally generic vector field v on an arbitrary compact connected and smooth 4-dimensional manifold Y with boundary (examine Figure 4.2 for guidance). This description is based on the partial order in $\mathbf{\Omega}^\bullet_{'\langle 3]}$ and on the local models of traversally generic vector fields, the focus of Chapter 6. The poset $\mathbf{\Omega}^\bullet_{'\langle 3]}$ and its "multidimensional" generalizations are described in detail in Chapter 7 (see Figure 7.2).

First of all, locally, each set $\partial_2 Y(v,\omega)$ is a regularly embedded submanifold of $\partial_1 Y$ of codimension $|\omega|'$. The locus $\partial_3 Y(v)$ is a union of simple arcs and loops that separate $\partial_2 Y(v)$ into two compact surfaces, $\partial_2^+ Y(v)$ and $\partial_2^- Y(v)$. For any $\omega \in \mathbf{\Omega}^\bullet_{'\langle 3]} \setminus \mathbf{\Omega}^\bullet_{'\langle 2]}$ (equivalently, for any ω with $|\omega|' = 3$), the set $\partial_2^+ Y(v,\omega)$ is finite. For any such ω, but $\omega = (12221)$, the points of $\partial_2^+ Y(v,\omega)$ belong the union of curves that form $\partial_3 Y(v)$. The finite locus $\partial_2^+ Y(v, (12221))$ is a union of several ordered triples, which belong to the interior of $\partial_2^+ Y(v)$.

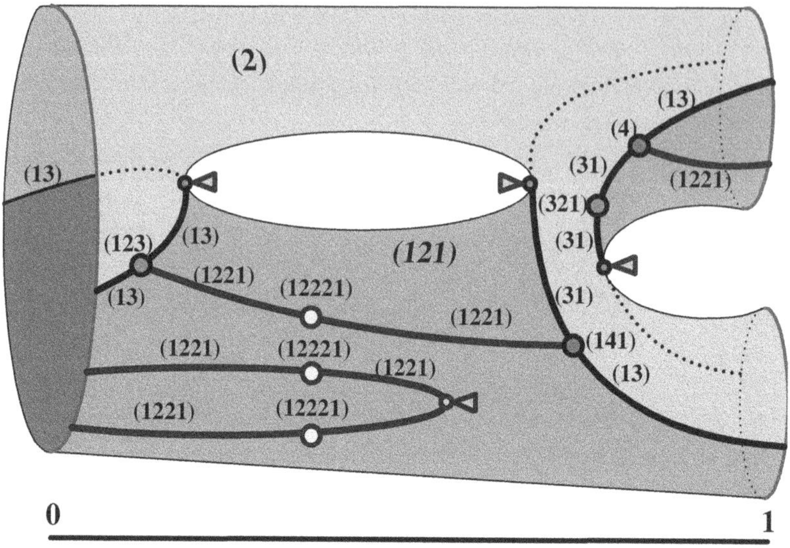

Fig. 4.2 *A typical stratification of the surface* $\partial_2 Y(v)$ *by the loci* $\{\partial_2 Y(v,\omega)\}_{\omega \in \mathbf{\Omega}^{\bullet}_{'\langle 3]}}$ *for a traversally generic* v *on a* 4*-manifold* Y*. The critical points of the Morse functions* $\pi_\omega : \partial_2 Y(v,\omega) \to [0,1]$ *are marked by the symbols "* $\lhd$ *" and "* $\rhd$ *".*

For any ω such that $|\omega|' = 2$, $\partial_2^+ Y(v,\omega)$ is a union of simple open arcs and loops. In particular, the closure of $\partial_2^+ Y(v,(13))$ is the locus $\partial_3^+ Y(v)$, while the closure of $\partial_2^+ Y(v,(31))$ is the locus $\partial_3^- Y(v)$. Each point from $\partial_2^+ Y(v,(4))$ or from $\partial_2^+ Y(v,(141))$ locally separates $\partial_3 Y(v)$ into two arcs that belong to $\partial_3^+ Y(v)$ or to $\partial_3^- Y(v)$, respectively. The points from $\partial_2^+ Y(v,(123))$ belong to the interior of $\partial_3^+ Y(v)$, while the points from $\partial_2^+ Y(v,(321))$ belong to the interior of $\partial_3^- Y(v)$.

At each point of $\partial_2^+ Y(v,(12221))$, two local arcs from $\partial_2^+ Y(v,(1221))$ meet (recall that the points of $\partial_2^+ Y(v,(12221))$ occure in triples). Each point from $\partial_2^+ Y(v,(123))$ belongs to the closures of two local arcs from $\partial_2^+ Y(v,(13))$ and one local arc from $\partial_2^+ Y(v,(1221))$. Each point from $\partial_2^+ Y(v,(321))$ belongs to the closures of two local arcs from $\partial_2^+ Y(v,(31))$ and one local arc from $\partial_2^+ Y(v,(1221))$.

Now let us go back to the truly generic families of vector fields on 3-folds. As before, put $Y =_{\mathsf{def}} X \times [0,1]$, $\delta Y =_{\mathsf{def}} \partial X \times [0,1]$, and $X_t = X \times \{t\}$.

Let $v(x,t) =_{\mathsf{def}} v_t(x)$. So $\{v_t\}_{t\in[0,1]}$ defines a path $v : [0,1] \to \mathcal{V}_{\mathsf{trav}}(X)$. At the same time, v is a vector field on Y, tangent to the fibers of the obvious projection $\pi : Y \to [0,1]$.

Lemma 4.1. *Let X be a compact connected 3-fold with boundary, and $\tau : [0,1] \to \mathcal{V}_{\mathsf{trav}}(X)$ a continuous path which generates the smooth vector field $v(x,t) =_{\mathsf{def}} \tau(t)(x)$ on Y.*

- *If, for some $T \in (0,1)$, the vector field v_T is boundary generic on X_T, then the vector field v on Y is boundary generic with respect to δY in an open neighborhood $V_{T,\epsilon} = \bigcup_{t\in(T-\epsilon,\, T+\epsilon)} X_t$ of X_T in Y.*
- *If, for some $T \in (0,1)$, the vector field v_T is traversally generic on X_T then:*
  (1) *the vector field v on Y is* traversally generic *with respect to δY in an open neighborhood $V_{T,\epsilon}$ of X_T in Y,*
  (2) *the $\mathbf{\Omega}^{\bullet}_{\prime\langle 2]}$-stratified topological type of the gradient spine*
  $$K_t = \partial_1^+ X_t(v_t) \cup \mathcal{C}(\partial_2^+ X_t(v_t))$$
  is constant for all $X_t \subset V_{T,\epsilon}$.

Proof. The first bullet and the first statement of the second bullet will be validated in Lemma 6.12.

The validation of the second statement is based on the previous description of traversally generic vector fields on arbitrary 4-dimensional manifolds.

Let γ_x denotes the v-trajectory through $x \in Y$. We denote by $p : \partial_2 Y(v) \to \partial_1^+ Y(v)$ the map that assigns to each point $x \in \partial_2 Y(v)$ the point of the intersection $\gamma_x \cap \partial_1^+ Y(v)$ below x. When x is the lowest point in the intesection, we define $p(x) = x$. Similarly, $p_t : \partial_2 X_t(v_t) \to \partial_1^+ X_t(v_t)$ is the restriction of p to $\partial_2 X_t(v_t)$. (See Figure 2.16 for the possible local patterns of p_t.)

Let $I_\epsilon(T) =_{\mathsf{def}} (T-\epsilon,\, T+\epsilon)$.

By the first bullet, v is traversally generic in some neighborhood $V_{T,\epsilon}$. Thus, by the proof of Lemma 6.12, for each $\omega \in \mathbf{\Omega}^{\bullet}_{\prime\langle 2]}$, the projection $\pi_\omega : \partial_2 Y(v,\omega) \to [0,1]$ is an submersion along $X_T \cap \partial_2 Y(v,\omega)$. So there exists a smaller $\epsilon > 0$ so that π_ω is an submersion on $\partial_2 Y(v,\omega) \cap V_{T,\epsilon}$. Therefore the curves $C(\omega) =_{\mathsf{def}} \partial_2 Y(v,\omega) \cap V_{T,\epsilon}$, forming the three loci

$$\partial_2 Y(v,(13)) \cap V_{T,\epsilon},\quad \partial_2 Y(v,(31)) \cap V_{T,\epsilon},\quad \partial_2 Y(v,(1221)) \cap V_{T,\epsilon},$$

are nonsingular and transversal to each slice $X_t \subset V_{T,\epsilon}$. Similarly,

$$\pi_2 : \partial_2 Y(v) \cap V_{T,\epsilon} \to I_\epsilon(T)$$

is a regular function on the surface $\partial_2 Y(v) \cap V_{T,\epsilon}$, and thus π_2 is a fibration.

Since $\pi_1 : \delta Y \to [0,1]$ is a fibration and $\partial_2 Y(v) = \partial(\partial_1^+ Y(v))$, it follows that $\pi_1 : \partial_1^+ Y(v) \cap V_{T,\epsilon} \to I_\epsilon(T)$ is a fibration as well. Indeed, using the product structure in $\partial_2 Y(v) \cap V_{T,\epsilon}$ we construct a non-vanishing vector field w, tangent to $\partial_2 Y(v) \cap V_{T,\epsilon}$ and transversal to $\partial_1 X_T$. Then we use the product structure in $\partial_1 Y(v) \cap V_{T,\epsilon}$ to extend w to a vector field $\tilde{w}$ on $\partial_1^+ Y(v) \cap V_{T,\epsilon}$ in such way that $\tilde{w}$ is transversal to $\partial_1^+ X_T(v_T)$. Srinking ϵ if necessary, we conclude that $\tilde{w}$ defines a product structure in $\partial_1^+ Y(v) \cap V_{T,\epsilon}$. In particular, for $t \in I_\epsilon(T)$, $\partial_1^+ X_t(v_t)$ is diffeomorphic to $\partial_1^+ X_T(v_T)$.

The following arguments may also require adjusting $\epsilon > 0$; abusing notations, we denote by "ϵ" the results of all the appropriate adjustments.

For $\omega = (1221), (13)$, the surface $Y(v, \omega) \subset Y$ (which consists of the v-trajectories of the combinatorial type ω), in addition to its intersections $C(\omega)$ with the surface $\partial_2 Y(v)$, has also transversal intersection with the space $\mathrm{int}(\partial_1^+ Y(v))$. Similarly, for $\omega = (1221), (31)$, the surface $Y(v, \omega)$ has transversal intersection with the space $\mathrm{int}(\partial_1^- Y(v))$. These intersections are smooth curves $C_\dagger^+(\omega) \subset \partial_1^+ Y(v) \cap V_{T,\epsilon}$ and $C_\dagger^-(\omega) \subset \partial_1^- Y(v) \cap V_{T,\epsilon}$ that "mirror" the curves $C(\omega) \subset \partial_2 Y(v)$. In other words, $C_\dagger^+(\omega) = p(C(\omega))$ for all $\omega \in \mathbf{\Omega}^\bullet_{'\langle 2]} \setminus \mathbf{\Omega}^\bullet_{'\langle 1]}$.

Now we construct a new vector field w on the surface $V_{T,\epsilon} \cap \partial_2 Y(v)$. The w-flow will give a product structure $\partial_2(X_T)(v_T) \times I_\epsilon(T)$ to the surface. First, we define w to be tangent to all the curves $\{C(\omega)\}_{\omega \in \mathbf{\Omega}^\bullet_{'\langle 2]} \setminus \mathbf{\Omega}^\bullet_{'\langle 1]}}$. Then we extend w smoothy to a non-vanishing field, tangent to $\partial_2 Y(v) \cap V_{T,\epsilon}$ and transversal to the fibers $\{\partial_1 X_t\}_{t \in I_\epsilon(T)}$. Abusing notations, we still call it w. We change w by a conformal factor so that $(D\pi_2)^*(w)$ is the unit vector field in the interval $I_\epsilon(T)$.

For $t \in I_\epsilon(T)$, the w-flow defines a diffeomorphism $\Psi_2(t) : \partial_2 X_T(v_T) \to \partial_2 X_t(v_t)$, which preserves the stratifications of both 1-dimensional manifolds by the combinatorial types $(2), (121); (1221), (13), (31)$. Moreover, $\Psi_2(t)$ takes each v_T-ordered pair $\gamma \cap \partial_2^+ X_T(v_T)$, where γ is a v_T-trajectory of the combinatorial type (1221), to a v_t-ordered pair $\Psi(\gamma) \cap \partial_2^+ X_t(v_t)$.

For all $t \in I_\epsilon(T)$, the maps $p_t : \partial_2 X_t(v_t) \to \partial_1^+ X_t(v_t)$ are 1-to-1 and continuous *away* from the finite locus $\partial_2 X_t(v_t, (1221))$; at the locus $\partial_2 X_t(v_t, (1221))$, the order of each pair $\gamma \cap \partial_2^+ X_t(v_t)$ is preserved by $\Psi_2(t)$. Therefore the map

$$\Psi_2^\#(t) =_{\mathsf{def}} p_t \circ \Psi_2(t) \circ (p_T)^{-1} : p_T(\partial_2 X_T(v_T)) \to p_t(\partial_2 X_t(v_t))$$

is a well-defined homeomorphism of the two graphs. Moreover, the family

$\{\Psi_2^{\#}(t)\}_{t \in I_\epsilon(T)}$ defines a product structure in the 2-dimensional polyhedron $p(\partial_2 Y(v) \cap V_{T,\epsilon}) \subset \partial_1^+ Y(v)$. This product structure generates a continuous non-vanishing vector field $w^{\#}$ on $p(\partial_2 Y(v) \cap V_{T,\epsilon})$. It is tangent to δY and transversal to the slices $\{\partial_1 X_t\}$. The differential of $\pi : \delta Y \to [0,1]$ maps $w^{\#}$ to the unit vector field on $[0,1]$.

Examining the definition of p and using that v is traversally generic in $V_{T,\epsilon}$, we see that a point $y \in p(\partial_2 Y(v)) \cap \partial_2 Y(v) \cap V_{T,\epsilon}$, if and only if,

$$y \in \big[\partial_2 Y(v,(2)) \cup \partial_2 Y(v,(31)) \cup \partial_2 Y(v,(1221))\big] \cap V_{T,\epsilon}.$$

So, for a traversally generic vector field v_t, the intersection $p_t(\partial_2 X_t(v_t)) \cap \partial_2 X_t(v_t)$ consists of points from the strata

$$\partial_2 X_t(v_t,(2)),\ \partial_2 X_t(v_t,(31)),\ \partial_2 X_t(v_t,(1221)).$$

By the construction of $\Psi_2^{\#}(t)$, at each point $y \in p_t(\partial_2 X_t(v_t)) \cap \partial_2 X_t(v_t)$, the product structure in $p(\partial_2 Y(v)) \cap V_{T,\epsilon}$ agrees with the product structure in $\partial_2 Y(v) \cap V_{T,\epsilon}$, defined by w. Thus $w^{\#} = w$ at the points from the intersection $\big(p(\partial_2 Y(v)) \cap \partial_2 Y(v)\big) \cap V_{T,\epsilon}$. We denote by $w^\dagger$ the vector field on $\big(p(\partial_2 Y(v)) \cup \partial_2 Y(v)\big) \cap V_{T,\epsilon}$, obtained by "pasting" w and $w^{\#}$.

Finally, we extend $w^\dagger$ to a non-vanishing vector field $w^\bullet$ on the 3-fold $\delta Y \cap V_{T,\epsilon}$, so that $w^\bullet$ is transversal to the 2D-slices $\partial_1 X_t(v_t)$. By the construction of w, the sets $\partial_2 Y(v)$, $\partial_2^{\pm} Y(v)$ are invariant under the $w^\bullet$-flow. Thus the set $\partial_1^+ Y(v)$ is invariant under the $w^\bullet$-flow as well. As a result, the $w^\bullet$-flow produces homeomorphisms

$$\Psi_1^\bullet(t) : \big(\partial_1^+ X_T(v_T),\ p_T(\partial_2 X_T(v_T))\big) \to \big(\partial_1^+ X_t(v_t),\ p_t(\partial_2 X_t(v_t))\big)$$

of the pairs. Moreover, by its construction, $\Psi_1^\bullet(t)$ commutes with the maps p_T and p_t.

According to Lemma 2.7, each gradient spine K_t is determined by the surface $\partial_1^+ X_t(v_t)$ and the map $p_t : \partial_2 X_t(v_t) \to \partial_1^+ X_t(v_t)$. Therefore, $\Psi_1^\bullet(t)$ induces a homeomorphism $\Phi(t) : K_T \to K_t$ of spines, which preserves the $\mathbf{\Omega}^\bullet_{'\langle 2]}$-stratifications in both polyhedra. Moreover, $\Phi(t)$ respects the TN-markers, delivered by the geometry of the gradient spines. □

The second claim in the second bullet of Lemma 4.1 leads to:

Corollary 4.1. *For a connected 3-fold X with boundary, the gradient complexity function $gc : \mathcal{V}^\ddagger(X) \to \mathbb{Z}_+$ is constant on each path-connected component of the space $\mathcal{V}^\ddagger(X)$.* ◇

The next lemma claims that, for a *truly generic* family of vector fields $\{v_t\}_t$, any change in the topology of the Morse stratification of $\{\partial_j^{\pm} X(v_t)\}_j$ must involve some modifications of the cubic loci $\partial_3 X(v_t)$. See Figure 4.6 supporting this observation.

Lemma 4.2. *Let* $v = \{v_t\}_{t \in [0,1]}$ *be a truly generic family of smooth traversing vector fields on a compact connected 3-fold* X *with boundary. Then the Morse function* $\pi_2 =_{\mathsf{def}} \pi| : \partial_2 Y(v) \to [0,1]$ *has no critical points in the open surface* $\partial_2 Y(v,(2)) \cup \partial_2 Y(v,(121))$.

Proof. If $x_\star \in \partial_2 Y(v,(2))$ is a critical point of π_2, then the topology of $\partial_2^- X_t(v_t) \subset \partial_2 Y(v)$ must change via an elementary surgery, as t crosses the critical value $t_\star = \pi_2(x_\star)$. Since the field v is traversally generic, any v-trajectory γ that intersects $\partial_2 Y(v)$ and resides in the vicinity of $\gamma_{x_\star}$ is of the combinatorial type (2). Using that $\partial(\partial_1^+ X_t(v_t)) = \partial_2^- X_t(v_t)$ in the vicinity of $x_\star$, this change in $\partial_2^- X_t(v_t)$ forces the $x_\star$-localized topological type and even the homology type of $\partial_1^+ X_t(v_t)$ to change as well. The cascade $\mathcal{C}(\partial_2^+ X(v_t))$, being separated from the locus where the change of $\partial_1^+ X_t(v_t)$ takes place, is not affected. Therefore, the topological and even the homological type of the gradient spine K_t must change, as t crosses the critical value t_*. Since K_t is a spine of $X_t \approx X$ no such changes are possible.

Let us make the previous argument a bit more formal. By Definition 4.1, there is an open interval $I_{\epsilon_\star}(t_\star) =_{\mathsf{def}} (t_\star - \epsilon_\star, t_\star + \epsilon_\star)$ with the property $I_{\epsilon_\star}(t_\star) \cap \Theta_v = t_\star$. Consider the neighborhood $U =_{\mathsf{def}} \pi^{-1}(I_{\epsilon_\star}(t_\star))$ of $X_{t_\star}$ in Y. Then, using that $x_\star \in \partial_1^- Y(v)$, there exists a neighborhood $V =_{\mathsf{def}} V(x_\star) \times I_{\epsilon_\star}(t_\star) \subset U$, where $V(x_\star)$ is a neighborhood of $x_\star$ in $X_{t_\star}$ so small that $V \cap \partial_2^+ Y(v) = \emptyset$ and $V \cap \partial_2 Y(v)$ admits Morse coordinates (centered at $x_\star$) for the function π_2.

Consider the integral relative homology $H_*(V \cap \partial_1^+ X_t,\, \partial V \cap \partial_1^+ X_t)$. Then the rank of $H_*(V \cap \partial_1^+ X_t,\, \partial V \cap \partial_1^+ X_t)$ changes by ± 1, as t crosses $t_\star$. As the same time, $H_*(K_t \setminus V,\ K_t \cap \partial V)$ does not change (recall that $K_t = \partial_1^+ X_t(v_t) \cup \mathcal{C}(\partial_2^+ X_t(v_t)))$. By the excision arguments, the rank of $H_*(K_t)$ must change.

However, K_t is a gradient spine of $X_t = X$. So no homology change in $H_*(K_t)$ (or, by Theorem 2.7, no topological change) is possible. Hence π_2 has no critical points in $\partial_2 Y(v,(2))$.

Now consider the case when $x_\star \in \partial_2 Y(v,(121))$ is a critical point of the Morse function π_2. Then, in the vicinity of $x_\star$, the topology of $\partial_2^+ X_t(v_t) \subset$

$\partial_2 Y(v)$ must change via an elementary surgery, as t crosses the critical value $t_\star$. Since the vector field v is traversally generic, any v-trajectory γ that intersects $\partial_2 Y(v)$ and resides in the vicinity of $\gamma_{x_\star}$ is of the combinatorial type (121). Using that $\partial(\partial_1^+ X_t(v_t)) = \partial_2^+ X_t(v_t)$ in the vicinity of $x_\star$, this change in $\partial_2^+ X_t(v_t)$ forces the homology type of the spine K_t in the vicinity of the cascade $\mathcal{C}(\partial_2^+ X_t(v_t))$ to change.

Let us justify this claim. Again, by Definition 4.1, there is an open interval $I_{\epsilon_\star}(t_\star) \subset [0,1]$ with the property $I_{\epsilon_\star}(t_\star) \cap \Theta_v = t_\star$. Consider the neighborhood $U =_{\mathsf{def}} \pi^{-1}(I_{\epsilon_\star}(t_\star))$ of $X_{t_\star}$ in Y. Then there exists a neighborhood $V = V(x_\star) \times I_{\epsilon_\star}(t_\star) \subset U$, where $V(x_\star)$ is a neighborhood in $X_{t_\star}$ of the segment $[x_\star, y_\star] \subset \mathcal{C}(\partial_2^+ X_t(v_t))$ of the $v_\star$-trajectory $\gamma_{x_\star}$. Here $y_\star$ is the point in $\gamma_{x_\star} \cap \partial_1^+ X_t(v_t)$ that resides below $x_\star$. We pick $V(x_\star)$ so narrow that $V \cap \partial_2^- Y(v) = \emptyset$ and $V \cap \partial_2 Y(v)$ admits some Morse coordinates for the function $\pi_2 : \partial_2 Y(v) \to [0,1]$.

Consider the integral relative homology $H_*(V \cap K_t,\ \partial V \cap K_t)$. Then the rank of $H_*(V \cap K_t,\ \partial V \cap K_t)$ changes by ± 1, as t crosses $t_\star$. In fact, if $i_\star$ is the Morse index at $x_\star$ of π_2, then a new absolute $(2 - i_\star)$-cycle is generated as a result of crossing. As the same time, $H_*(\partial_1^+ X_t \setminus V,\ \partial_1^+ X_t \cap \partial V)$ does not change. So, by the excision arguments, the rank of $H_*(K_t)$ must change, as t crosses $t_\star$.

However, K_t is a gradient spine of $X_t = X$. So, again, no homology change in $H_*(K_t)$ is possible. Hence π_2 has no critical points in $\partial_2 Y(v, (121))$. □

Corollary 4.2. *Let $v = \{v_t\}_{t \in [0,1]}$ be a truly generic family of smooth traversing vector fields on a compact connected 3-fold X with boundary.*

Then the critical points of the Morse function $\pi_2 : \partial_2 Y(v) \to [0,1]$ are located in the interiors of the edges of the graph

$$\bigcup_{\{\omega \in \mathbf{\Omega}^\bullet_{\langle 3]} \setminus \mathbf{\Omega}^\bullet_{\langle 1]}\}} \partial_2 Y(v, \omega).$$

In other words, all the critical points of π_2 are among the critical points of the function π_2, being restricted to the curves $\partial_2 Y(v, (1221))$, $\partial_2 Y(v, (13))$, $\partial_2 Y(v, (31))$. ◇

We are in position to state the first main result of this section.

Theorem 4.1. *Let $v = \{v_t\}_{t \in [0,1]}$ be a truly generic*[5] *family of smooth traversing vector fields on a compact connected 3-fold X with boundary.*

[5]in the sense of Definition 4.1

Let

$$K_t =_{\mathsf{def}} \partial_1^+ X(v_t) \bigcup \mathcal{C}(\partial_2^+ X(v_t))$$

denote the gradient spine of X_t*, generated by* v_t*. Then the following claims are valid:*

- *The* 2*-complex* K_1 *can be obtained from the* 2*-complex* K_0 *by a finite sequence of elementary moves and their inverses, described in the local models from the list (4.4) .*
- *The* β*-move is the result of the parameter* t *crossing a special value* $t_\star \in \Theta_v$ *for which* $v_{t_\star}$ *has trajectories of the combinatorial type* (12221)*. The* $\gamma, \gamma^\star$*-moves are the result of* t *crossing a special value* $t_\star \in \Theta_v$ *for which* $v_{t_\star}$ *has trajectories of the combinatorial types* (321) *or* (123)*, respectively.*

  The $\delta, \delta^\star$*-moves are the result of* t *crossing a special value* $t_\star \in \Theta_v$ *for which* $v_{t_\star}$ *has trajectories of the combinatorial type* (4) *or* (141)*, respectively.*
- *The* α*-move corresponds to* t *crossing a critical value* $t_\star \in \Theta_v$ *of the Morse function* $\pi| : Y(v, (1221)) \to [0,1]$ *on the curve* $Y(v, (1221)) \subset \partial_2 Y(v)$*.*

  The ζ, η*-moves correspond to* t *crossing a critical value* $t_\star \in \Theta_v$ *of Morse the function* $\pi| : Y(v, (31)) \to [0,1]$ *on the curve* $Y(v, (31)) \subset \partial_3 Y(v)$*.*

  The $\zeta^\star, \eta^\star$*-moves correspond to* t *crossing a critical value* $t_\star \in \Theta_v$ *of the Morse function* $\pi| : Y(v, (13)) \to [0,1]$ *on the curve* $Y(v, (31)) \subset \partial_3 Y(v)$*.*

Proof. We denote by $\Theta_v \subset (0,1)$ the finite set of exceptional (in the sense of Definition 4.1) points. By Lemma 4.1, the topology of the gradient spines K_t may change only when t crooses an exceptional value $t_\star$, for which $v_{t_\star}$ either:

(1) fails to be traversally generic because $v_{t_\star}$ has trajectories of the combinatorial types from $\Omega^\bullet_{'\langle 3]} \setminus \Omega^\bullet_{'\langle 2]}$ (such ω's are forbidden for traversally generic fields on any 3-fold X), or

(2) fails to be traversally generic because $t_\star$ is a critical value of the Morse function $\pi_2 : \partial_2(Y, \omega) \to [0,1]$ for some $\omega \in \mathbf{\Omega}^\bullet_{'\langle 2]}$ (although the combinatorial types of all $v_{t_\star}$-trajectories belong to the "permitted" set $\mathbf{\Omega}^\bullet_{'\langle 2]}$).

Motivated by this dichotomy, we organize in the first family the elementary changes in the topology of K_t that occur when t crosses a special value $t_\star \in \Theta_v$, for which $v_{t_\star}$ has trajectories of combinatorial types $\omega \in \mathbf{\Omega}^\bullet_{'\langle 3]} \setminus \mathbf{\Omega}^\bullet_{'\langle 2]}$. The first family contains $\beta, \gamma, \gamma^\star$, and $\delta, \delta^\star$-moves and their inverses. For them, the slice $X_{t_\star}$ is transversal to the the loci $\partial_2 Y(v), \partial_3 Y(v)$.

The second family of moves deals with the "permitted" types $\omega \in \mathbf{\Omega}^\bullet_{'\langle 2]}$, but the exceptional slice $X_{t_\star}$ is tangent to the loci $\partial_2 Y(v), \partial_3 Y(v)$. The second family contains the $\alpha, \zeta, \zeta^\star, \eta, \eta^\star$-moves and their inverses.

According to Definitions 4.1 and 4.2, the truly generic vector field v on Y is traversally generic and thus boundary generic with respect to δY. So the loci $\{\partial_j Y(v)\}_{j\in[2,4]}$ and $\{\partial_j^\pm Y(v)\}_{j\in[1,4]}$ are regularly embedded submanifolds of δY. Using that $Y = X \times [0,1]$ and that, for all t, $v|_{X_t} = v_t$, it is possible to verify that $\{\partial_j^\pm Y(v) \cap X_t = \partial_j^\pm X(v_t)\}_{j\in[1,4]}$ (see the proof of Lemma 6.12). However, for some exceptional values of $t \in \Theta_v$, these intersections may not be submanifolds of ∂X_t.

By Definition 4.1, for each $\omega \in \mathbf{\Omega}^\bullet_{'\langle 3]}$, the restriction $\pi_\omega : \partial_2 Y(v,\omega) \to [0,1]$ of $\pi : Y \to [0,1]$ to the locus $\partial_2 Y(v,\omega)$ is a Morse function with finitely many critical points.

Also, by Definition 4.1, for all $t \in [0,1] \setminus \Theta_v$, the field v_t is traversally generic on X_t, and t is the regular value of the Morse functions

$$\{\pi_\omega : \partial_2 Y(v,\omega) \to [0,1]\}_{\omega\in\Omega^\bullet_{'\langle 3]}}.$$

Thus, by Lemma 2.7 and Lemma 6.12, for a regular $t \notin \Theta_v$, the property of v_t being traversally generic on X_t implies that v has to be traversally generic in the vicinity of $X_t \subset Y$. In particular, if $\partial_4 Y(v) \cap X_t \neq \emptyset$, then v_t is not traversally generic; so such $t \in \Theta_v$. This means that all fibers X_t that have a v_t-trajectory of the combinatorial types $(12221), (123), (321), (4), (141)$ from the set $\mathbf{\Omega}^\bullet_{'\langle 3]} \setminus \mathbf{\Omega}^\bullet_{'\langle 2]}$ are labeled by exceptional $t \in \Theta_v$.

We denote by $\Theta_v^{[3]} \subset \Theta_v$ the subset, formed by such exceptional t's. We denote by $\Theta_v^{[2]} \subset \Theta_v$ the set of critical values of the Morse functions

$$\big\{\pi_2 : \partial_2 Y(v,\omega) \to [0,1]\big\}_{\omega\in\mathbf{\Omega}^\bullet_{'\langle 2]}\setminus\mathbf{\Omega}^\bullet_{'\langle 1]}},$$

where $\mathbf{\Omega}^\bullet_{'\langle 2]} \setminus \mathbf{\Omega}^\bullet_{'\langle 1]} = \{(1221), (31), (13)\}$. Finally, we denote by $\Theta_v^{[1]} \subset \Theta_v$ the set of critical values of the Morse functions

$$\big\{\pi_2 : \partial_2 Y(v,\omega) \to [0,1]\big\}_{\omega\in\mathbf{\Omega}^\bullet_{'\langle 1]}\setminus\mathbf{\Omega}^\bullet_{'\langle 0]}},$$

where $\mathbf{\Omega}^\bullet_{'\langle 1]} \setminus \mathbf{\Omega}^\bullet_{'\langle 0]} = \{(121), (2)\}$.

Remarkably, by Lemma 4.2 and Corollary 4.2, $\Theta_v^{[1]} \subset \Theta_v^{[2]}$; in particular, the changes in the topology of $\partial_2 X_t(v_t)$ and $\partial_3 X_t(v_t)$ are *synchronized.* Therefore, by Lemma 4.1, we need to examine only the changes in the shapes of the gradient spines K_t, as t crosses the exceptional values from the set $\Theta = \Theta_v^{[3]} \coprod \Theta_v^{[2]}$. Each such crossing will generate an elementary move, described in the local models from the list in (4.4) and their inverses. Let us justify this claim.

Let I denote a closed segment. We add a collar $\partial X \times I$ to X, thus producing a larger manifold $\hat{X}$ that properly contains X. Then we form the manifold $\hat{Y} = \hat{X} \times [0, 1]$, which properly contains Y. We extend v smoothly to a vector field w on $\hat{Y}$ and project orthogonally the extension w on the fibers $\{\hat{X}_t\}$. We denote that projection by $\hat{v}$. By shrinking the collar if necessary, we may assume that $\hat{v}$ is still non-vanishing and of the gradient type.

Since $\hat{v}$ is traversally generic with respect to δY, by Lemma 6.4, each v-trajectory γ in the complement to $\hat{X}_0 \coprod \hat{X}_1$ has a neighborhood $U \subset \hat{Y}$ such that the U admits special coordinates $(u, \vec{x}, \vec{y}) : U \to \mathbb{R}^4$ in which δY is given by a polynomial equation $P(u, \vec{x}) = 0$, and Y by a polynomial inequality $P(u, \vec{x}) \leq 0$, where the polynomial $P = P_\omega$ is as in (4.3). As before, we denote by Z_ω the domain $\{P_\omega \leq 0\}$ in $\mathbb{R}^4$ and by H_ω its boundary $\{P_\omega = 0\}$. The polynomial P_ω is determined by the combinatorial type $\omega =_{\mathsf{def}} \omega_\gamma \in \mathbf{\Omega}^\bullet_{'\langle 3]}$, and therefore the entire local model $(Z_\omega, H_\omega, \partial_u)$ depends only on ω.

Freezing the coordinates $(\vec{x}, \vec{y})$ defines the v-trajectories; thus the field ∂_u represents v in U.

The truly generic vector field $\hat{v}$ on $\hat{Y}$ has a very special property: it is tangent to the $3D$-foliation $\hat{\mathcal{M}}_\pi =_{\mathsf{def}} \{\hat{X}_t\}_{t \in [0,1]}$ in $\hat{Y}$ (i.e., $\hat{v}$ has a first integral, the function $\pi : \hat{Y} \to [0, 1]$). We denote by $\mathcal{N}_\omega$ the image of the foliation $\hat{\mathcal{M}}_\pi$ under the coordinate map $(u, \vec{x}, \vec{y}) : U \to \mathbb{R}^4$. Since the coordinate map takes $\hat{v}$-trajectories to ∂_u-trajectories, each leaf of $\mathcal{N}_\omega$ consists of ∂_u-trajectories. We stress that the foliation $\mathcal{N}_\omega$ is not canonical in $\mathbb{R}^4$: it does not depend on ω alone. However, we will need only the 1-jet or the 2-jet of the nonsingular functions that determine the foliation.

Consider $S =_{\mathsf{def}} \{u = 0\}$, a transversal section of the ∂_u-flow. Then $\mathcal{N}_\omega$ is determined by a 2-dimensional foliation $\mathcal{L}_\omega$ in $S \approx \mathbb{R}^3$, such that $\mathcal{N}_\omega = F^{-1}(\mathcal{L}_\omega)$. In the vicinity of the origin, $\mathcal{L}_\omega$ depends on a choice of an appropriate submersion $L_\omega : \mathbb{R}^3 \to \mathbb{R}$ whose constant level sets form $\mathcal{L}_\omega$.

In what follows, we treat the local model $(Z_\omega, H_\omega, \partial_u, L_\omega)$ from the list (4.4) as a *germ* in the vicinity of the u-coordinate line ℓ_u.

For any $\omega \in \mathbf{\Omega}^\bullet_{'\langle 3]} \setminus \mathbf{\Omega}^\bullet_{'\langle 2]}$, the $\vec{y}$-coordinates are absent in the model $(Z_\omega, H_\omega, \partial_u, L_\omega)$. For such ω, by the third bullet of Definition 4.1, the 3D-leaf $(L_\omega \circ F)^{-1}(0)$ is transversal to the surface $\partial_2 Z_\omega(\partial_u)$ at any point $(i, \vec{0})$, where the line ℓ_u intersects with the hypersurface H_ω. The leaf $(L_\omega \circ F)^{-1}(0)$ is also transversal to the curve $\partial_3 Z_\omega(\partial_u)$ at any point $(i, \vec{0}) \in \ell_u \cap H_\omega$.

Note that a particular choice of the germ (at the origin) of the submersion $L_\omega : \mathbb{R}^3 \to \mathbb{R}$ in list (4.4) is immaterial, as long as $(L_\omega \circ F)^{-1}(0)$ is transversal to $\partial_2 Z_\omega(\partial_u)$ and $\partial_3 Z_\omega(\partial_u)$ at each point $(i, \vec{0})$: different choices of L_ω lead to the diffeomorphic modifications

$$(Z_\omega, H_\omega, \partial_u)|_{(L_\omega \circ F)^{-1}(-c)} \Rightarrow (Z_\omega, H_\omega, \partial_u)|_{(L_\omega \circ F)^{-1}(c)},$$

where $c > 0$, and thus to homeomorphic modifications in the germs of the corresponding gradient spines.

We can reformulate this transversality of $\mathcal{N}_\omega$ to $\partial_2 Z_\omega(\partial_u)$ and $\partial_3 Z_\omega(\partial_u)$ in terms of the foliation $\mathcal{L}_\omega$ and the images $F_\omega(\partial_2^+ Z_\omega(\partial_u))$ and $F_\omega(\partial_3 Z_\omega(\partial_u))$.

Consider the surface $\mathcal{D}^2_\omega \subset \mathbb{R}^3$, the F_ω-image of the nonsingular surface $\partial_2 Z_\omega(\partial_u)$. It is the union of real discriminant varieties and contains $\mathcal{D}^1_\omega =_{\mathsf{def}} F_\omega(\partial_3 Z_\omega(\partial_u))$, the union of real algebraic curves. Finally, $\mathcal{D}^0_\omega$, the F_ω-image of $\partial_4 Z_\omega(\partial_u)$, is the origin.

For $\omega = (12221)$, $\mathcal{D}^2_\omega$ is the union of three coordinate planes, and $\mathcal{D}^1_\omega = \emptyset$. For $\omega = (123), (321)$, the surface $\mathcal{D}^2_\omega$ is the coordinate plane $\{x_2 = 0\}$ union with the product of a cuspidal curve $\mathcal{C}$ in the x_0x_1-plane and the x_2-coordinate line λ, while $\mathcal{D}^1_\omega = \lambda$. For $\omega = (4), (141)$, $\mathcal{D}^2_\omega$ is the Swallow Tail surface with its singular locus $\mathcal{D}^1_\omega$ being the union of three curves that emanate from the origin: they represent the ∂_u-trajectories of the combinatorial types $(13), (31)$, and (1221) (see Figure 7.3).

So, for a truly generic v and for any $t_\star \in \Theta^{[3]}$, the leaves of the corresponding foliation $\mathcal{L}_\omega$ are transversal to the surface $\mathcal{D}^2_\omega \setminus \mathcal{D}^1_\omega$ and to the curve $\mathcal{D}^1_\omega$ in the vicinity of the origin.

This concludes our analysis of the $\beta, \gamma, \gamma^\star, \delta, \delta^\star$-moves.

In contrast, by the fourth bullet from Definition 4.1, for a critical $t \in \Theta^{[2]}$, the loci $\partial_2 Y(v)$ or $\partial_3 Y(v)$ are quadratically tangent to X_t. Therefore, for $t \in \Theta^{[2]}$, the leaf $\mathcal{L}_0 = L_\omega^{-1}(0)$ is quadratically tangent to $\mathcal{D}^2_\omega$ or/and to $\mathcal{D}^1_\omega$. In fact, by Lemma 4.2, if the 3-fold X_t is quadratically tangent to the surface $\partial_2 Y(v)$, then it is automatically tangent to the curve $\partial_3 Y(v)$.

With the exception of the α-move, for all moves from the second family, the converse is true: if X_t is simply tangent to $\partial_3 Y(v)$, then it is automatically tangent to $\partial_2 Y(v)$ in such a way that $\pi_2 : \partial_2 Y(v) \to [0,1]$ is a Morse function in the vicinity of the tangency point.

In order to validate this observation, we notice that each move from the set $\{\zeta, \zeta^\star, \eta, \eta^\star\}$ is based on a ω with one of its components $\omega_i = 3$. Let P denote the cubic factor of the u-polynomial P_ω. We write each nonsingular function $L_\omega(y, x_0, x_1)$ from the sublist of (4.4) that describes $\zeta, \zeta^\star, \eta, \eta^\star$-moves as $L_\omega = \pm y^2 + \ell(x_0, x_1)$. Here x_0 denotes the free term and x_1 denotes the coefficient of the first degree term in the $(u-i)$-polynomial P. The 1-jet of the function $\ell(x_0, x_1)$ contains x_1 with a nonzero coefficient; also $\ell(0,0) = 0$.

Consider the pair of equations $\{P(u,x) = 0, \frac{\partial}{\partial u} P(u,x) = 0\}$ that defines the locus $\partial_2 Z_\omega(\partial_u)$ in the vicinity of $(i, \vec{0})$. This pair of equations allows to expres x_1 as $-3(u-i)^2$, while x_0 is a function divisible by $(u-i)^3$. Thus $(u-i, y)$ are smooth coordinates on the surface $\partial_2 Z_\omega(\partial_u)$ in the vicinity of $(i, \vec{0})$. Substituting these coordinates in the function $L_\omega(y, x_0, x_1)$, we get that the restriction of $L_\omega \circ F_\omega$ to the surface $\partial_2 Z_\omega(\partial_u)$ can be written in the form $Q(u-i, y) + (u-i)^3 R(u-i)$, where Q is a non-degenerate quadratic form in two variables $u-i$, y, and $R(u-i)$ is a polynomial in $u-i$.

So the Hessian of this function is non-degenerated at the origin. As a result, $L_\omega \circ F_\omega : \partial_2 Z_\omega(\partial_u) \to \mathbb{R}$ has a critical point of the Morse type at $(i, 0)$. Therefore, for a truly generic v, the function $\pi_2 : \partial_2 Y(v) \to [0,1]$ must have a critical point of the Morse type at any point of tangency between X_t and $\partial_3 Y(v)$.

In contrast, the α-moves correspond to special $t \in \Theta_v$ for which X_t is simply tangent to the curve $\partial_2 Y(v, (1221))$, but t is a regular value for $\pi_2 : \partial_2 Y(v) \to [0,1]$.

This completes the proof of Theorem 4.1. □

Next, we take a second look at the morphing geometry of the gradient spines, generated by the 3D traversing flows. Now we will pay a close attention to the evolutions of cusps.

Temporarily, put $v = v_t$, where $t \in [0,1]$. In the notations below, we drop the dependence of the Morse strata $\{\partial_j^\pm X(v)\}_j$ on v.

Recall that, when X is oriented, new *polarities* "$\oplus$" and "$\ominus$" can be given to the isolated singularities from the locus $\partial_3 X = \partial_3^+ X \cup \partial_3^- X$.

For each point $x \in \partial_3 X$, consider an arc $\delta \subset \partial_2 X$ which contains x. If the orientation of δ, defined by the tangent vector $v(x)$, agrees with the orientation of δ, induced by the preferred orientation of $\partial_2^+ X$ (the latter is defined by the preferred orientation of $\partial_1^+ X$ and the inner normal $\nu(\partial_2 X, \partial_1^+ X)$), then we assign the polarity "$\oplus$" to x; otherwise, we assign to it the polarity "$\ominus$".

The polarities $\oplus, \ominus$ divide $\partial_3 X$ into two sets, $\partial_3^{\oplus} X$ and $\partial_3^{\ominus} X$. Therefore, the set $\partial_3 X$ is subdivided into *four* disjoint subsets:

$$\partial_3^{+\oplus} X,\ \partial_3^{-\oplus} X,\ \partial_3^{+\ominus} X,\ \partial_3^{-\ominus} X.$$

Put

$$\partial_3^A X =_{\mathsf{def}} \partial_3^{+\oplus} X \cup \partial_3^{-\ominus} X \ \text{ and } \ \partial_3^B X =_{\mathsf{def}} \partial_3^{-\oplus} X \cup \partial_3^{+\ominus} X.$$

As we deform v and watch the transformations of the corresponding gradient spines $K(v) \subset X$, we will follow the evolution of polarities, attached to the Q-singularities (see the discussion prior to Definition 3.3) and to the points from $\partial_3 X$.

As in the proof of Theorem 4.1, let us reexamine the traversally generic flow-generated projections of the folding loci $\partial_2^+ X$ on a flow section S (as in Figure 2.8, diagram A).

The flow-generated local projections $p : \partial_1 X \to S$, where $S \approx \mathbb{R}^2$ denotes the transversal local section of the flow, the curve patterns $p(\partial_2 X)$ become exceptional in a number distinct ways, as shown in Figure 4.1, the middle column diagrams.

In fact, Theorem 4.1 testifies that, for *truly generic* families $\{v_t\}_t$, the following list of violations, parallel to the list (4.4), is indeed complete.

- In the deformation, described by the α-move, momentarily, the flow governed projection $p(\partial_2 X) \subset S$ has two local nonsingular branches that share a singleton x, so that the two tangent lines to the branches at x coincide. In other words, there exists a trajectory γ that is tangent to $\partial_1 X$ at two distinct points, and the two local waterfalls that contain segments of γ are tangent along the portion of γ that they share. The combinatorial type $\omega_\gamma = (1221)$ alone does not reflect this special geometry!

  The two Q-singularities of spines, generated in an α-move, have opposite polarities $\oplus$ and $\ominus$. So α-moves and their inverses do not change the *polarized* complexity $c_{\ominus}^{\oplus}(\sim)$ (see Definition 3.3) of the gradient spine (equivalently, of the traversally generic vector field).

  If we forget the TN-markers (which break the symmetry between the left and right dome-shaped surfaces in Figure 4.3, (b)), then the α-moves make sense for any simple (unmarked) 2-complex K.

We denote by $\mathcal{V}^{\asymp}_{\mathsf{trav}}(X)$ the set of traversing vector fields on X that have such a trajectory γ.

- In the deformation, described by the β-move, momentarily, the flow governed projection $p(\partial_2 X) \subset S$ has three local nonsingular branches that share a singleton. At that point, the branches are transversal in pairs. In other words, there exists a trajectory γ that is tangent to $\partial_1 X$ at tree distinct points and the three local waterfalls that contain segments of γ are transversal in pairs. The combinatorial type $\omega_\gamma = (12221)$ reflects faithfully the move.

  We notice that the β-move increases by one the number of double-tangent trajectories in the cascades. The β-moves come in different flavors, depending on the orientations and coorientations of the three folds from $\partial_2^+ X$, the folding loci being *ordered* by v (Figure 4.4 shows only one of the possible flavors). In fact, there are essentially two types of the β-moves: as the number of double-tangent trajectories increases by one, that new trajectory may have polarity $\oplus$ or $\ominus$; the other two double-tangent trajectories keep their polarities. So, as a result of the β-move, the polarized complexity $c^{\oplus}_{\ominus}(\sim)$ of the flow/spine changes by ± 1.
  The β-moves have a direct analogue in the category of simple spines, where they are called *the Matveev-Piergallini moves* [Ma].

  We denote by $\mathcal{V}^{*}_{\mathsf{trav}}(X)$ space of traversing vector fields that have such triple-tangent trajectory γ.

- In the case of $\gamma, \gamma^\star$-moves, $p(\partial_2 X) \subset S$, momentarily, has two local branches that share a singleton x. The first branch is a cuspidal curve $\mathcal{C}$ with the cusp at x, the second branch is nonsingular at x; the "tangent" ray of the the curve $\mathcal{C}$ is transversal to the tangent line of the nonsingular branch at x. The trajectory γ through x is simply tangent to $\partial_1 X$ at one point and cubically tangent to $\partial_1 X$ at a different point. The combinatorial type ω_γ of such trajectory is (321) or (123).

  The γ-move increases the number of double-tangent trajectories by 1; the new trajectory may have polarity $\oplus$ or $\ominus$.

  We denote by $\mathcal{V}^{)\prec}_{\mathsf{trav}}(X)$ the space of traversing vector fields that have such trajectory γ.

- In the case of $\delta, \delta^\star$-moves, $p(\partial_2 X) \subset S$, momentarily, cancels two cusps and one transversal intersection point. The exceptional trajectory γ is tangent to $\partial_1 X$ with multiplicity 4 at a singleton. The two combinatorial types $\omega_\gamma = (141), (4)$ reflect faithfully the moves.

We denote by $\mathcal{V}_{\mathsf{trav}}^{\prec\times\succ}(X)$ the space of traversing vector fields that have such trajectory γ.

- In the case of $\zeta, \zeta^\star$-moves, the property of the locus $\partial_2 X$ being a 1-dimensional manifold is violated momentarily. At the special moment, the locus $\partial_2 X$ has a singleton for its component. As one loop from $\partial_2 X$ degenerates into a singleton, a loop from $p(\partial_2 X) \subset S$ degenerates into a singleton and then disappears as well. In the process, the two cusps of $p(\partial_2 X)$ are cancelled.

  We denote by $\mathcal{V}_{\mathsf{trav}}^{\overset{\prec\succ}{\smile}}(X)$ the space of traversing vector fields that have a singleton as a part of the locus $\partial_2 X$.
- In the case of $\eta, \eta^\star$-moves, the property of the locus $\partial_2 X$ being a 1-dimensional manifold is also violated momentarily. In the end, the two cusps of $p(\partial_2 X)$ are cancelled.

  We denote by $\mathcal{V}_{\mathsf{trav}}^{\overset{\succ\prec}{\frown}}(X)$ the space of traversing vector fields that have a pair of intersecting arcs as a part of the locus $\partial_2 X$.

Of course, there are other special configurations of the v-flow-governed projections of $\partial_2 X$ on local sections of the flow, but they typically occur in the families of traversing fields that depend on *two* parameters at least.

Inspired by Arnold's treatment in [Ar] of the spaces of immersed plane curves, we are compelled to think of the subspaces

$$\mathcal{V}_{\mathsf{trav}}^{*}(X),\ \mathcal{V}_{\mathsf{trav}}^{\asymp}(X),\ \mathcal{V}_{\mathsf{trav}}^{)\prec}(X),\ \mathcal{V}_{\mathsf{trav}}^{\prec\times\succ}(X),\ \mathcal{V}_{\mathsf{trav}}^{\overset{\prec\succ}{\smile}}(X),\ \mathcal{V}_{\mathsf{trav}}^{\overset{\succ\prec}{\frown}}(X)$$

as codimension one "walls" that divide the space $\mathcal{V}_{\mathsf{trav}}^{1\text{-st}}(X) \subset \mathcal{V}_{\mathsf{trav}}(X)$ into infinitely many "chambers".

In part, the next theorem is a reformulation of Theorem 4.1 in terms of the spines.

Theorem 4.2. *Let $v = \{v_t\}_{t\in[0,1]}$ be a truly generic family of traversing vector fields on a compact connected 3-fold X with boundary. Let K_t denote the gradient spine of X_t, generated by v_t. Then the following claims are valid:*

- *The $\delta, \delta^\star, \zeta, \zeta^\star, \eta, \eta^\star-$ moves result in elementary expansions and collapses of 2-cells of the gradient spines K_t.*
- *The $\alpha, \delta, \delta^\star$-moves do not change the topological type of the locus $\partial_1^+ X_t(v_t)$. The $\zeta, \zeta^\star$-moves result in 0-surgery on $\partial_1^+ X_t(v_t)$, while the $\eta, \eta^\star$-moves result in 1-surgery on $\partial_1^+ X_t(v_t)$.*

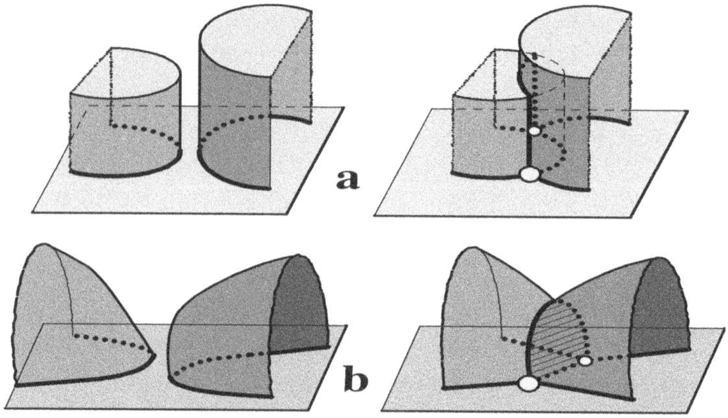

Fig. 4.3 *The α-move shown as: (a) a change in the shape of the cascades, and (b) as a change in the shape of the 2-spines .*

- *The $\beta, \gamma, \gamma^\star, \delta^\star$-moves change the combinatorial complexity $c(K_t)$*[6] *by ± 1, the α-moves by ± 2, the $\delta, \zeta, \zeta^\star, \eta, \eta^\star$-moves do not affect $c(K_t)$.*
- *In the deformation process, the cusps from $\partial_3 X_t(v_t)$ can cancel in pairs as shown in Figure 4.6, diagrams 1-6, and their mirror images.*
- *The change in topology of $\partial_1^+ X_t(v_t)$ is accompanied by cancelations of pairs of points from $\partial_3 X_t(v_t)$ with the opposite polarity $(\oplus, \ominus)$.*
- *The cancellations of pairs with the opposite polarity $(+, -)$ do not change the topology of $\partial_1^+ X_t(v_t)$.*
- *The pairs of points from $\partial_3 X_t(v_t)$, sharing the same first and second polarities, cannot be cancelled.*

Proof. Thanks to Theorem 4.1 and Lemma 4.1, we only need to describe the changes in the shapes of the cascades, generated by the elementary moves in (4.4). By the proof of Lemma 4.1, the shape of a gradient spine and its cascade depend soley on the v_t-flow-denerated map $p_t : \partial_2 X(v_t) \to \partial_1^+ X(v_t)$. Therefore, the proof of Theorem 4.2 amounts to analysing the changing patterns (as c crosses 0) of the maps

$$p_\omega(c) : \partial_2 Z_\omega(\partial_u) \cap [(L_\omega \circ F_\omega)^{-1}(c)] \to \partial_1^+ Z_\omega(\partial_u) \cap [(L_\omega \circ F_\omega)^{-1}(c)],$$

for the models $\{(Z_\omega, H_\omega, \partial_u, L_\omega)\}_{\omega \in \mathbf{\Omega}^\bullet_{'\langle 3]}}$ from the list (4.4).

[6]the number of trajectories of the combinatorial type (1221) in X_t

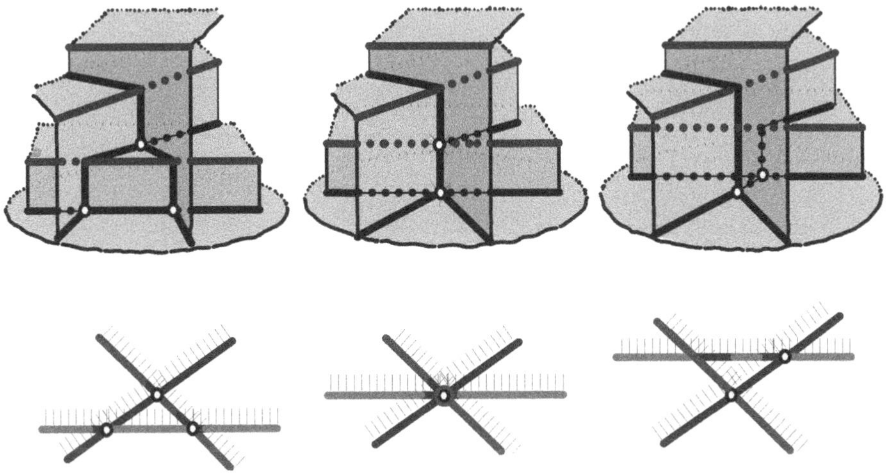

Fig. 4.4 *A typical β-move, shown as the change (from left to right) in the shape of the cascades (the upper diagrams) and the change of the localized v-projection of $\partial_2^+ X$ on $\partial_1^+ X$ (the lower diagrams). As the lower diagrams show, the lines in the triangular configurations are not only ordered by their height above the plane of projection, but are also equipped with normals that point to the half-planes, imitating the plateaus of $\partial_1^+ X$.*

These maps $\{p_\omega(c)\}_{c\in(-\epsilon,\epsilon)}$, the localized analogues of the maps $\{p_t\}_{t\in[0,1]}$, are determined by the maps $\{F_\omega|_{(L_\omega\circ F_\omega)^{-1}(c)}\}_{c\in(-\epsilon,\epsilon)}$ in a somewhat subtle way. The subtlety is due to the fact that the target space of the continuous map $F_\omega|_{(L_\omega\circ F_\omega)^{-1}(c)}$ is a transversal section of the ∂_u-flow, while the target space of the discontinous map $p_\omega(c)$ is the space $\partial_1^+ Z_\omega(\partial_u) \cap (L_\omega \circ F_\omega)^{-1}(c)$.

For example, for $\omega = (1221)$ and $c \neq 0$, the $F_\omega|_{(L_\omega\circ F_\omega)^{-1}(c)}$-image of $\partial_2 Z_\omega(\partial_u) \cap [(L_\omega \circ F_\omega)^{-1}(c)]$ is a pair of transversal arcs "X", while its $p_\omega(c)$-mage is a disjoint union of a ray "I" with the "T"-shaped configuration of three rays (so that the F_ω-image of "I $\coprod$ T" is a cross "X").

See Figure 4.1 for the local patterns of $F_\omega(\partial_2 X(v_t))$ and for somewhat cryptic depictions of $p_\omega(\partial_2 X(v_t))$ in terms of the shading.

In the rest of the proof, we will shuttle freely between (X_t, v_t) and its local models $\{(Z_\omega, H_\omega, \partial_u, L_\omega)|_{(L_\omega\circ F_\omega)^{-1}(c)}\}_{\omega\in\mathbf{\Omega}^\bullet_{'\langle 3]}}$, while preferring in general a much simpler notation "(X_t, v_t)".

We start with $\eta, \eta^\star, \zeta, \zeta^\star, \delta, \delta^\star$-moves, responsible for the surgery of the Morse stratification $\partial_1^\pm X(v_t) \supset \partial_2^\pm X(v_t) \supset \partial_3^\pm X(v_t)$. The relation be-

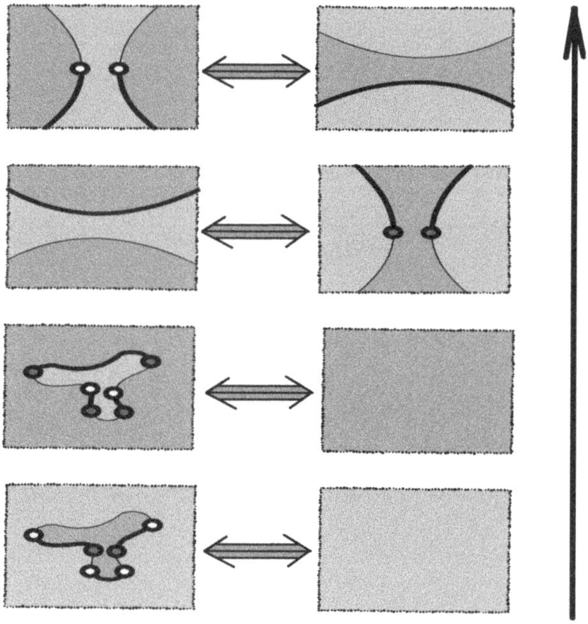

Fig. 4.5 Elementary surgery on $\partial_1^+ X$ (the dark domains) and its effect on $\partial_2^+ X$ (bold arcs) and on $\partial_3^+ X$ (bold dots).

tween these strata before and after such an elementary move is a result of a direct computation of the Morse stratifications for the analytic models $\{(Z_\omega, H_\omega, \partial_u, L_\omega)|_{(L_\omega \circ F_\omega)^{-1}(c)}\}_{\omega \in \mathbf{\Omega}^\bullet_{'\langle 3]}}$. The computations involve the partial u-derivatives ∂_u, $(\partial_u)^2$ of the polynomials P_ω that determine the hypersurface H_ω.

Based on these computations, Figure 4.6 depicts elementary surgery on the loci $\partial_3^\pm X(v_t)$, $\partial_2^\pm X(v_t)$ and $\partial_1^+ X(v_t)$, as t crosses an exceptional value $t_\star \in \Theta_v$. The surgery is supported in the vicinity of a point $x_\star \in \partial_1 X$ such that $t_\star = \pi(x_\star)$. The figure shows the localized projection of the Morse strata on the plane, tangent to $\partial_1 X$ at $x_\star$. In the figure, the field $v_{t_\star}$ is vertical at $x_\star$.

Diagrams 1-2 in Figure 4.6 depict 1-surgery (the $\eta, \eta^\star$-moves), and diagrams 3-4 depict 0-surgery and 2-surgery on $\partial_1^+ X(v_t)$ (the $\zeta, \zeta^\star$-moves). In each diagram, the curves from $\partial_2 X(v_t)$ are oriented; so the points from $\partial_3 X(v_t)$ acquire four flavors: $(+, \oplus), (+, \ominus), (-, \oplus), (-, \ominus)$.

Reversing the orientation of $\partial_2 X(v_t)$ flips the polarities $\oplus \Leftrightarrow \ominus$. Such flip leads to another four diagrams (not shown in Figure 4.6) that are the mirror images of the diagrams 1-4. They complete the elementary surgery list.

The first polarities $(+, -)$ of the canceling singularities from $\partial_3 X(v_t)$ are the same, and the second polarities $(\oplus, \ominus)$ are opposite. The orientation of $\partial_2 X(v_t)$ prevents pairs of the types $(+, \oplus), (+, \oplus)$ or of the types $(-, \ominus), (-, \ominus)$ from cancellation (see diagrams 7-8). Again, we think about moves depicted in Figure 4.6 as localized events (occurring at a point of the boundary $\partial_1 X$) whose effect on the gradient spine propagates inside X with the help of the v_t-flows.

We notice that each pair of canceling singularities from $\partial_3 X(v_t)$ has one point in the locus

$$\partial_3^A X(v_t) = \partial_3^{+\oplus} X(v_t) \cup \partial_3^{-\ominus} X(v_t)$$

and the other in

$$\partial_3^B X(v_t) = \partial_3^{-\oplus} X(v_t) \cup \partial_3^{+\ominus} X(v_t).$$

Hence, the difference $\#(\partial_3^A X(v_t)) - \#(\partial_3^B X(v_t))$ stays *invariant* under the transformations from Figure 4.6. Perhaps, by Conjecture 4.1 and Theorem 4.5, this difference is always zero!

Diagrams 5-6 from Figure 4.6 show another generic mechanism by which the singularities from $\partial_3 X(v_t)$ cancel, a mechanism which has no effect on the topology of $\partial_1^+ X(v_t)$, but which modifies $\partial_2^+ X(v_t)$. This type of cancellation is generated by the universal Swallow Tail family (the $\delta, \delta^\star$-moves), described by the formula that follows (2.2) and whose instance is depicted in Figure 2.12.

Again, reversing the orientations of the arcs from $\partial_2 X(v_t)$ will produce two more diagrams, not shown in Figure 4.6. This time, the cancellation occurs among the points of opposite first (i.e., "$+, -$") and same second (i.e., "$\oplus, \ominus$") polarities.

Let us first examine the effect of 1-surgery—the $\eta, \eta^\star$-moves (see the 1st and the 2nd moves in Figure 4.6) on the shape of the cascades. Figure 4.7 shows what happens with cascades, as two cusps merge (as in Figure 4.6, diagrams 1-2). Viewed in terms of the local projections $p_x : \partial_1 X(v_t) \to S_x$, where S_x is a local transversal section of the v_t-flow, this transformation is generated via a generic one-parameter family of mirror-symmetric merging cusps.

The upper diagram in Figure 4.7 depicts the case when the "groovings" are located in $\partial_1^+ X(v_t)$, and the lower diagram when they are in $\partial_1^- X(v_t)$.[7] In both diagrams, the merging cusp points a and b (or c and d) have opposite polarities "$\oplus, \ominus$" (see also diagrams 1-2 from Figure 4.6); in the upper diagram, they belong to the set $\partial_3^+ X(v_t)$, in the lower one, to the set $\partial_3^- X(v_t)$.

In the upper-right diagram, consider the arc-shaped band, marked with a dotted curve and residing in $\partial_1^+ X(v_t)$ (most of the band is in the rear and thus is invisible). In order to get the upper-left diagram, we collapse a 2-cell (a "trapezoid") in that band onto a portion of its boundary. In the lower diagram, the cusp points c and d of opposite second polarities belong to the set $\partial_3^- X(v_t)$ and are assumed to be very close to each other. Therefore, the $(-v_t)$-trajectories through c and d will hit the same plato in $\partial_1^+ X(v_t)$ (shown as the lower disk)—the typical cascade in the 1-parameter family is assumed to be traversally generic. Locally, the spine in the left-lower diagram can be obtained from this plato by two elementary expansions of 2-cells, while the spine in the right-lower diagram by a single elementary expansion.

The spirit of considerations for the $\zeta, \zeta^\star$ -moves, delivering elementary 0-surgery and 2-surgery of $\partial_1^+ X(v_t)$, is similar. It is illustrated in Figure 2.9, which portrays a small indent in a round 3-ball D^3 and its effect on the spine (the indent generates a hole in the set $\partial_1^+(D^3)$ of the original round ball).

The change in the gradient spines, produced by $\delta, \delta^\star$-moves is shown Figure 2.12 (see also diagram 6 in Figure 4.6). Again, the cascades before and after the Swallow Tail cancellation are linked by elementary 2-collapses or expansions. Note that the $\delta^\star$-move adds a new trajectory of the type (1221), while δ-move adds a new "virtual trajectory" of the type (22), residing in the local model "outside" X.

In short, we conclude that the changing topology of the Morse triple $(\partial_1^+ X(v_t), \partial_2^+ X(v_t), \partial_3^+ X(v_t))$ via surgery (via the $\zeta, \zeta^\star, \eta, \eta^\star, \delta, \delta^\star$-moves) induces only 2-*dimensional expansions and collapses* of the corresponding gradient spines $K(v_t)$.

Next, let us examine how the gradient spines are affected by deformations $\{v_t\}_t$ that do not change the topology of the Morse triple $(\partial_1^+ X(v_t), \partial_2^+ X(v_t), \partial_3^+ X(v_t))$. These are the $\alpha, \beta, \gamma, \gamma^\star$-moves and their

[7] In the first case, the f-controlled size of collapses and expansions is small, while in the second case, it can be quite big—a localized change in convexity of the boundary $\partial_1 X$ causes a distributed change in the shape of cascades.

inverses. Evidently, under such deformations, the corresponding spine can change only via changing interactions of waterfalls among themselves and with the "ground" $\partial_1^+ X(v_t) \setminus \partial_2 X(v_t)$. We analyze these interactions by considering the $(-v)$-governed projections of $p_t : \partial_2^+ X(v_t) \to \partial_1^+ X(v_t)$ (see Figure 2.8 and Figure 2.16), while, in the vicinity of each trajectory, keeping track of which branch of $\partial_2^+ X(v_t)$ is "above" and which is "below", as determined by v.

To simplify the notations below, we suppress the dependence of the waterfalls on the "time" parameter t (equivalently, on the parameter c in the local models). Let us consider:

(i) two waterfalls, W_1 and W_2, which fall from two arcs $C_1, C_2 \subset \partial_2^+(X)$ so that, locally, C_1 is above C_2, or

(ii) three waterfalls, W_1 W_2, and W_3, which fall from three arcs $C_1, C_2, C_3 \subset \partial_2^+ X$ so that, locally, C_1 is above C_2, and C_2 is above C_3.

- In the α-move, initially, W_1 and W_2 are disjoint in the vicinity of a given trajectory $\gamma_\star$. In the deformation process, C_2 touches tangentially W_1 at a single point $x_\star$ (see diagram C in Figure 2.8 for the exceptional $v_{t_\star}$-directed projection of the two arcs) and then intersects W_1 at a pair of nearby points y and z, as shown in Figure 4.1, (a). In such a the case, the topology of the cascade is changing as in Figure 4.1, (a), and its complexity changes by $+2$. The $(\oplus, \ominus)$-polarities of the two Q-singularities, generated by the α-move, are opposite.
- In the β-move, the ordered and cooriented projections of three branches of $\partial_2^+ X$ form a triangular configuration that deforms into a new triangular configuration. Momentarily, a triple-tangent trajectory $\gamma_\star$ is created (see Figure 4.4 and diagram B in Figure 2.8 for the exceptional $v_{t_\star}$-directed projection of the three arcs). As a result of the β-move, the complexity of the gradient spine jumps by one; moreover, the polarized complexity also changes by ± 1.
- In the $\gamma, \gamma^\star$-moves, initially, W_1 and W_2 are disjoint in the vicinity of a given trajectory $\gamma_\star$. As the the field v_t changes, C_2 can pierce W_1 transversally at a single cusp point $x_\star$ (see Figure 4.6 and diagram D in Figure 2.8 for the exceptional $v_{t_\star}$-directed projection of the two arcs). In the end, a new double-tangent trajectory is created; thus the spine complexity jumps by 1, moreover, the polarized complexity also changes by ± 1. Note that γ-moves come in two flavors: γ_+ (the upper diagram in Figure 4.6) that employed a point of $\partial_3^+ X$, and $\gamma^\star$ (the lower diagram in Figure 4.6) that employs a point of $\partial_3^- X$.

(4.5)

We notice that the moves in list (4.5) closely resemble the generic moves that describe deformations of plane curves in [Ar] (although, in our context, the curves are not immersed: they may have cusps and, in that regard, resemble the plane projections of Legendrian links). □

4.7 Philips' Theorem and the spaces of vector fields in 3D

Next, we consider the space $\mathcal{F}(X)$ of smooth *nonsingular* functions on a smooth compact oriented manifold X with boundary. In other words, $\mathcal{F}(X)$ is the space of smooth *submersions* $f : X \to \mathbb{R}$ from the manifold X to the manifold $\mathbb{R}$. Let $\mathcal{V}_{\neq 0}(X)$ denote the space of smooth non-vanishing vector fields on X.

Philips' remarkable Theorem B [Ph] is pivotal for the rest of this chapter. It claims that, for an open smooth manifold M and a smooth manifold N, the differential

$$D : \mathsf{Sub}(M, N) \to \mathsf{Epi}(TM, TN)$$

is a weak homotopy equivalence. Here $\mathsf{Sub}(M, N)$ denotes the space of submersions, and $\mathsf{Epi}(TM, TN)$ the space of bundle epimorphisms.

We will need a special case of Theorem B, when $N = \mathbb{R}$. By picking a Riemannian metric g on M, we may interpret $\mathsf{Epi}(TM, T\mathbb{R})$ as the space of non-vanishing vector fields on M. In order to construct the corresponding vector field, we consider the kernel of a given epimorphism $A \in \mathsf{Epi}(TM, T\mathbb{R})$ and take the vector field that is orthogonal to the distribution $\ker A$ and mapped by A to the unit vector field.

This leads to the following claim.

Theorem 4.3 (Phillips). *For a fixed Riemannian metric g on a connected compact*[8] *smooth manifold X with boundary, the gradient map*

$$\nabla_g : \mathcal{F}(X) \to \mathcal{V}_{\neq 0}(X)$$

is a weak homotopy equivalence. ◇

Corollary 4.3. *Let X be a smooth connected compact manifold with boundary. Let g be a Riemannian metric on X.*

[8]Actually, the Phillips' Theorem is stated for an *open* manifold X, but the proof relies on the existence of a handle decomposition of X with no handles of top dimension $\dim(X)$; so the compact connected manifold with boundary fits the argument well.

- *If two non-vanishing gradient vector fields on X can be connected by a path in the space of non-vanishing vector fields $\mathcal{V}_{\neq 0}(X)$, then they also can be connected by a path in the space $\nabla_g(\mathcal{F}(X))$ of non-vanishing* gradient *vector fields.*
- *Similarly, if two non-vanishing gradient-like vector fields on X can be connected by a path in the space of non-vanishing fields $\mathcal{V}_{\neq 0}(X)$, then they also can be connected by a path in the space $\mathcal{V}_{\mathsf{grad}}(X) \cap \mathcal{V}_{\neq 0}(X)$ of non-vanishing* gradient-like *vector fields.*

Proof. According to Theorem 4.3, the ∇_g-induced map

$$(\nabla_g)_* : \pi_0(\mathcal{F}(X)) \to \pi_0(\mathcal{V}_{\neq 0}(X))$$

is a bijection, which proves the first bullet.

Let us validate the second bullet. By Lemma 2.1, if v_0 and v_1 are non-vanishing gradient-like vector fields, there exist Riemannian metrics g_0 and g_1 and smooth functions f_0 and f_1 on X such that $\nabla_{g_0}(f_0) = v_0$ and $\nabla_{g_1}(f_1) = v_1$. Since the space of Riemannialn metrics on X is path-connected, the two maps $\nabla_{g_0} : \mathcal{F}(X) \to \mathcal{V}_{\neq 0}(X)$ and $\nabla_{g_1} : \mathcal{F}(X) \to \mathcal{V}_{\neq 0}(X)$ are homotopic via a homotopy $\nabla_{g_t} : \mathcal{F}(X) \to \mathcal{V}_{\neq 0}(X)$, where $t \in [0, 1]$.

Let $\mathcal{F}_0$ and $\mathcal{F}_1$ be the two path-connected components of the space $\mathcal{F}(X)$ that contain f_0 and f_1, respectively. Similarly, let $\mathcal{V}_0$ and $\mathcal{V}_1$ be the path-connected components of the space $\mathcal{V}_{\neq 0}(X)$ that contain v_0 and v_1, respectively.

We intend to show that, if $\mathcal{V}_0 = \mathcal{V}_1$, then $\mathcal{F}_0 = \mathcal{F}_1$. Thanks to the homotopy ∇_{g_t}, the path-connected sets $\nabla_{g_1}(\mathcal{F}_1)$ and $\nabla_{g_0}(\mathcal{F}_1)$ belong to the same path-connected component $\mathcal{V}_1$ of $\mathcal{V}_{\neq 0}(X)$ that contains v_1.

Now assume that $v_0 = \nabla_{g_0}(f_0)$ and $v_1 = \nabla_{g_1}(f_1)$ belong to the same connected component $\mathcal{V}_0 = \mathcal{V}_1$ of $\mathcal{V}_{\neq 0}(X)$. Then both $\nabla_{g_0}(\mathcal{F}_1)$ and $\nabla_{g_1}(\mathcal{F}_1)$ are contained in $\mathcal{V}_0$.

On the other hand, by the Phillips Theorem,

$$(\nabla_{g_t})_* : \pi_0(\mathcal{F}(X)) \to \pi_0(\mathcal{V}_{\neq 0}(X))$$

is a bijection for all t, including $t = 0$. Since $\nabla_{g_0}(\mathcal{F}_1)$ and $\nabla_{g_0}(\mathcal{F}_0)$ belong to the same component $\mathcal{V}_0$, we conclude that $\mathcal{F}_0 = \mathcal{F}_1$. It follows that f_0 and f_1 can be connected by a path f_t in the space of submersions $\mathsf{Sub}(X, \mathbb{R}) =_{\mathsf{def}} \mathcal{F}(X)$.

So, if two gradient-like fields v_0 and v_1 belong to the same path-connected component of the space $\mathcal{V}_{\neq 0}(X)$, they also can be connected by a path $\{\nabla_{g_t}(f_t)\}_{t\in[0,1]}$ in the space $\mathcal{V}_{\mathsf{grad}}(X) \cap \mathcal{V}_{\neq 0}(X)$ of gradient-like nonsingular vector fields. □

Let $\tilde{\nabla}(f) =_{\mathsf{def}} \nabla(f)/\|\nabla(f)\|$, the normalized gradient of f.

When X is an oriented connected 3-fold with boundary, the tangent bundle of X is trivial. With the help a trivialization, the space of non-vanishing vector fields $\mathcal{V}_{\neq 0}(X)$ fibers over the space $C^\infty(X, S^2)$ of C^∞-maps from X to the sphere S^2 with a contractible fiber $\mathbb{R}_+$. Therefore, in 3D, Philips' Theorem 4.3 reduces to the following proposition.

Corollary 4.4. *Let X be a compact connected oriented Riemannian 3-fold with boundary. Fix a trivialization $\tau : TX \to X \times \mathbb{R}^3$ of the tangent bundle TX. Then the τ-induced normalized gradient map*

$$\tilde{\nabla} : \mathcal{F}(X) \to C^\infty(X, S^2)$$

is a weak homotopy equivalence, that is, the map $\tilde{\nabla}$ induces homotopy group isomorphisms

$$\tilde{\nabla}_* : \pi_n(\mathcal{F}(X)) \approx \pi_n(C^\infty(X, S^2))$$

for all $n \geq 0$. As sets,

$$\pi_0(\mathcal{F}(X)) \overset{\nabla_*}{\approx} \pi_0(\mathcal{V}_{\neq 0}(X)) \approx H^2(X; \mathbb{Z}),$$

and

$$\pi_1(\mathcal{F}(X), f_0) \overset{\nabla_*}{\approx} \pi_1(\mathcal{V}_{\neq 0}(X), v_0) \approx H^1(X; \mathbb{Z}) \oplus H^2(X; \mathbb{Z}),$$

where $v_0 =_{\mathsf{def}} \nabla(f_0)$.

Proof. In order to prove the corollary, we need to explain just the validity of the last two bijections. With a trivialization $\tau : TX \to X \times \mathbb{R}^3$ being fixed, any vector field $v \neq 0$ gives rise to the normalized map $v_\tau : X \to S^2$, whose homotopy class $[v_\tau] \in \pi^2(X)$, the 2nd *cohomotopy* set. Evidently $[v_\tau]$ depends only on the homotopy class $[v] \in \pi_0(\mathcal{V}_{\neq 0}(X))$ and the homotopy class of the trivialization τ.

Since X is 3-*coconnected*[9], by the Hopf Theorem (see Theorem 11.5, [Hu]), the natural map

$$Hopf : \pi^2(X) \to H^2(X; \mathbb{Z})$$

[9]That is, $H^i(X; G) = 0$ for all $i \geq 3$ and any coefficient group G.

is a bijection (in fact, an isomorphism). We denote by $h(v)$ the element $Hopf([v_\tau]) \in H^2(X;\mathbb{Z})$[10]. Thus, $h(v)$ detects the element $[v] \in \pi_0(\mathcal{V}_{\neq 0}(X))$.

By Theorem 4.3, any smooth map $v_\tau : X \to S^2$ is homotopic to a map of the form $\tilde{\nabla} f : X \to S^2$ for some submersion $f : X \to \mathbb{R}$. Hence the desired bijection

$$\pi_0(\mathcal{F}(X)) \approx \pi_0(\mathcal{V}_{\neq 0}(X)) \approx H^2(X;\mathbb{Z}).$$

Similarly, with the help of τ, any loop $\gamma : S^1 \to \mathcal{V}_{\neq 0}(X)$ generates a map $\gamma_\tau : X \times S^1 \to S^2$, whose homotopy class $[\gamma_\tau] \in \pi^2(X \times S^1)$. If two loops, γ_0 and γ_1, are homotopic in $\mathcal{V}_{\neq 0}(X)$, then $[(\gamma_0)_\tau] = [(\gamma_1)_\tau]$ in $\pi^2(X \times S^1)$.

Dealing with loops in $\mathcal{V}_{\neq 0}(X)$ through a fixed base point $v_\star$ amounts to fixing the map $\gamma_\tau| : X \times \{\star\} \to S^2$, where $\star$ denotes a base point in S^1. Thus the elements of $\pi_1(\mathcal{V}_{\neq 0}(X), v_\star)$ map bijectively to the set of elements from $\pi^2(X \times S^1)$, whose restrictions in $\pi^2(X \times \{\star\})$, under the obvious embedding $i : X \times \{\star\} \subset X \times S^1$, are fixed.

As a result, the elements of the fundamental group $\pi_1(\mathcal{F}(X), f_\star) \approx \pi_1(\mathcal{V}_{\neq 0}(X), v_\star)$, where $v_\star =_{\mathsf{def}} \tilde{\nabla}(f_\star)$, can be identified with the elements of the set $\pi^2(X \times S^1)$ that map to $h(v_\star)$ under the composition

$$\pi^2(X \times S^1) \xrightarrow{Hopf} H^2(X \times S^1;\mathbb{Z}) \xrightarrow{i^*} H^2(X;\mathbb{Z}).$$

Obstructions to linking any such pair of maps, $(\gamma_0)_\tau : X \times S^1 \to S^2$ and $(\gamma_1)_\tau : X \times S^1 \to S^2$, by a homotopy lie in the group

$$\oplus_{j \geq 0} H^j(X \times S^1, X \times \{\star\};\ \pi_j(S^2)).$$

Since X is 3-coconnected, $\pi_2(S^2) \approx \mathbb{Z} \approx \pi_3(S^2)$, and

$$(\gamma_0)_\tau|_{X \times \{\star\}} = (v_\star)_\tau = (\gamma_1)_\tau|_{X \times \{\star\}},$$

via the suspension isomorphisms, these obstructions lie in $H^1(X;\mathbb{Z}) \oplus H^2(X;\mathbb{Z})$. With a little more work, one can show that any element of $H^1(X;\mathbb{Z}) \oplus H^2(X;\mathbb{Z})$ is realizable as an obstruction to homotopy between the appropriate two maps $X \times S^1 \to S^2$ as above. □

Another invariant $e(v) \in H^2(X;\mathbb{Z})$ of non-vanishing vector fields v on an orientable 3-fold X is available. Its definition is independent of the choice of a trivialization of τ of TX. Consider an oriented 2-dimensional vector bundle η on X, formed by the planes transversal to v. The orientation of η is induced by the orientation of X and by v. Let $e(v) \in H^2(X;\mathbb{Z})$ be

---

[10] The choice of $h(v)$ is equivalent to the choice of a Spin^c-structure on X.

the Euler class of η. The element $e(v)$ is invariant under homotopies of v through non-vanishing fields. The isomorphism class of η is described by the homotopy class of the classifying map $E(v) : X \to \mathsf{BSO}(2) = \mathbb{CP}^\infty$. Since X is 3-dimensional and the 3-skeleton of $\mathbb{CP}^\infty$ is a 2-sphere, $E(v)$ is homotopic to a map with S^2 as a target. The homotopy class of $E(v)$ is detected by the Euler class of the bundle η. The relation between $e(v)$ and $h(v)$ is well-known. It is described by the lemma below, whose proof is left to the reader.

Lemma 4.3. *The homotopy class of the map $E(v) : X \to S^2$ is twice the homotopy class of the map $h(v) : X \to S^2$. Thus, $e(v) = 2h(v)$.* ◇

Theorem 4.4. *Let X be a connected compact 3-fold with boundary. Let v_0 and v_1 be a pair of traversally generic vector fields on X.*

- *Then v_0 and v_1 are connected by a continuous path*
$$\tau : [0,1] \to \mathcal{V}_{\mathsf{trav}}(X) = \mathcal{V}_{\mathsf{grad}}(X) \cap \mathcal{V}_{v \neq 0}(X)$$
if and only if $h(v_0) = h(v_1)$[11].
- *If v_0 and v_1 are connected by a truly generic path*[12] *in the space $\mathcal{V}^{1-\mathsf{st}}_{\mathsf{trav}}(X) \subset \mathcal{V}_{\mathsf{trav}}(X)$, then there exists a finite sequence of elementary 2-expansions, 2-collapses (induced by the $\delta, \delta^\star, \zeta, \zeta^\star, \eta, \eta^\star$-moves and their inverses), and $\alpha, \beta, \gamma, \gamma^\star$-moves and their inverses which transforms the gradient spine $K(v_0)$ into the gradient spine $K(v_1)$.*
- *Assuming the validity of Conjecture 4.1, we speculate that if $h(v_0) = h(v_1)$ for a pair v_0 and v_1 of traversally generic vector fields on X, then there exists a finite sequence of elementary 2-expansions, 2-collapses, and $\alpha, \beta, \gamma, \gamma^\star$-moves and their inverses which transforms the gradient spine $K(v_0)$ into the gradient spine $K(v_1)$.*

Proof. By their definition, traversally generic vector fields are gradient-like and nonsingular. By Corollary 4.3, if $h(v_0) = h(v_1)$, then v_0 and v_1 are connected by a continuous path in the space $\mathcal{V}_{\mathsf{trav}}(X)$. By the definition of the invariant $h(\sim)$, the converse is obvious.

The second claim of the theorem is a reformulation of Theorems 4.1 and 4.2.

[11] Both fields define equivalent Spin^c-structures on X.
[12] See Definition 4.1.

Conjecture 4.1 suggests that if v_0 and v_1 are connected by a continuous path in the space $\mathcal{V}_{\mathsf{trav}}(X)$, then they may be connected by a trully generic path. Therefore, by the first claim, if $h(v_0) = h(v_1)$, then v_1 and v_2 are connected by a trully generic path, provided the validity of Conjecture 4.1. By the second claim of the theorem, there exists a finite sequence of elementary 2-expansions, 2-collapses, and $\alpha, \beta, \gamma, \gamma^\star$-moves and their inverses, which transforms the gradient spine $K(v_0)$ into the gradient spine $K(v_1)$.

□

The first bullet of Theorem 4.4 leads to

Corollary 4.5. *Let X be a connected compact 3-fold with boundary. If $H^2(X;\mathbb{Z}) = 0$, then the space $\mathcal{V}_{\mathsf{trav}}(X)$ is path-connected.* ◇

We would like to sort out what happens to gradient spines $K(v_t)$ when, in the process of deforming a traversing vector field v_t, the value of $h(v_t) \in H^2(X;\mathbb{Z})$ jumps, as the result of $v_{t_\star}$ developing singularities and thus exiting the space $\mathcal{V}_{\mathsf{trav}}(X)$. It suffices to analyze how the gradient spines change in any collection of 1-parameter deformations $\{v_t\}$, such that the variations of $\{h(v_t)\}$ span the entire $H^2(X;\mathbb{Z})$.

We add an exterior collar to $\partial_1 X$ and denote by $\hat{X} \supset X$ the extended manifold. Let us fix a Riemannian metric on $\hat{X}$. We will consider 1-parameter families of Morse functions $\{f_t\}_{t\in[0,1]}$ on $\hat{X}$ and the families of their gradient fields $\{v_t =_{\mathsf{def}} \nabla(f_t)\}_{t\in[0,1]}$. In particular, we are going to examine how the f_t-critical points "traveling through X in $\hat{X}$" affect the evolution of gradient spines $K(v_t)$ (see Figure 4.9) Here we assume that f_0 and f_1 are nonsingular in X and that the "traveling" critical points $x_t^\star$ are of the Morse type. We will arrange for the f_t-critical points to trace some arcs that represent generators of the group $H_1(X, \partial X;\mathbb{Z})$.

With this goal in mind, consider a homotopy $\{f_t\}_{t\in[0,1]}$ of Morse functions on X so that the singular set

$$\Sigma_{\{f_t\}} =_{\mathsf{def}} \{(x,t) \in X \times [0,1] \,\big|\, x \text{ is a critical point of } f_t\}$$

consists of a collection $\tilde{J}$ of arcs in $X \times (0,1)$ with the endpoints in $\partial X \times (0,1)$. Let J be the image of $\tilde{J}$ under the obvious projection onto X. By an isotopy of $\tilde{J}$ in $X \times (0,1)$, we may assume that $J \subset X$ is a union of disjoint embedded arcs $\{J_\alpha\}$ with the end points in ∂X. These arcs are oriented by the increasing parameter $t \in (0,1)$.

Let us explain the images in Figure 4.9. The figure depicts the effect on the shape of the gradient spine $K(\nabla f_t)$ of a single f_t-critical point $x_t^\star$

(of index one or two) tracing an oriented arc J in X. We assume that the gradient spine $K(\nabla f_t)$ is well-defined at least for those $t \in [0,1]$ for which $\nabla f_t \neq 0$ in X. In fact, the figure shows an isotopy $\phi_t : X \to \hat{X}$ of the manifold X against the background of a "stationary" Morse function $f : \hat{X} \to \mathbb{R}$ with a critical point $x^\star$, its "stationary" gradient field $\hat{v}$ in $\hat{X}$, and a trivialization $\beta : T\hat{X} \to \hat{X} \times \mathbb{R}^3$ of the tangent bundle. The isotopy ϕ_t is supported in a cylinder $C =_{\mathsf{def}} D^2 \times \hat{J} \subset \hat{X}$ whose core $\hat{J}$ is the unstable manifold of $x^\star$ (the upper diagrams) and the stable manifold of $x^\star$ (the lower diagrams). In C, the isotopy ϕ_t is arranged to be rotationally symmetric with respect to the axis $\hat{J}$.

Let $f_t =_{\mathsf{def}} f \circ \phi_t$. When we isotop X inside of $\hat{X}$, the induced trivialization $\beta|_{D\phi_t(TX)}$ of TX changes by a homotopy, which is constant outside C. As long as $v_t =_{\mathsf{def}} \hat{v}|_{\phi_t(X)} \neq 0$, this deformation does not change the class of the map $\tilde{\nabla} f_t : X \to S^2$ in $\pi^2(X)$. Note that, when the critical point $x^\star$ of $f : \hat{X} \to \mathbb{R}$ is inside $\phi_t(X)$, the map $\tilde{\nabla} f_t =_{\mathsf{def}} \nabla f_t / \|\nabla f_t\|$ is only well-defined in $X \setminus \phi_t^{-1}(x^\star)$.

For all t such that $v_t \neq 0$ in X, the preimage of a base point $* \in S^2$ under the map $\tilde{\nabla} f_t$ is the locus $J_t \subset \phi_t^{-1}(\hat{J})$, where the field v_t is vertical with respect to the trivialization β and points up (see the upper-left diagram in Figure 4.9). In the upper diagrams, the Morse index $i(x^\star) = 2$, and J_0 is the intersection of $\phi_0(X)$ with one of the two unstable trajectories of v, emanating from the critical point $x^\star \in \hat{X}$. The preimage J_1 is empty for the upper-right diagram.

The case of $i(x^\star) = 1$ is similar: the gradient ∇f flips its direction (relative to the one shown the upper diagrams). In the lower-left diagram, the locus, where the field v_0 is vertical and points upwards, is empty; in the lower-right diagram, it is represented by the arc J_1, the intersection of $\phi_1(X)$ with one of the two stable trajectories of v, emanating from the critical point $x^\star \in \hat{X}$. The v-induced orientations of the arcs $J_0, J_1 \subset \hat{J}$ in the upper and lower diagrams are opposite.

By the next lemma, in the case of Morse index $i(x^\star) = k$, the Poincaré-dual of the variation $h(v_1) - h(v_0)$ is given by the formula $(-1)^k[J]$, were the curve J is oriented by the direction of motion of the critical point $x_t^\star$.

Lemma 4.4. *Let $\{f_t\}_{t\in[0,1]}$ be a generic family of Morse functions on an oriented 3-fold X, and $\{v_t\}_{t\in[0,1]}$ a family of their gradients. Assume that f_0, f_1 are nonsingular.*

Then the cohomology class $h(v_1) - h(v_0) \in H^2(X;\mathbb{Z})$ is Poincaré dual of the relative 1-cycle $\sum_\alpha (-1)^{i_\alpha}[J_\alpha]$, where $i_\alpha = 1, 2$ denotes the Morse *index of the f_t-critical point $(x^\star_\alpha)_t$ that traces the oriented arc $(J_\alpha, \partial J_\alpha) \subset (X, \partial X)$.*

When all the indices $\{i_\alpha\}$ are even, then $h(v_1) - h(v_0) = \Phi^*[S^2]$, where $\Phi : X \to S^2$ is the map, defined by the Thom-Pontryagin construction on the oriented union of arcs $J =_{\mathsf{def}} \Phi^{-1}(*)$. Here $* \in S^2$ is a regular value of the map Φ, and $[S^2]$ denotes the fundamental cohomology class.

Proof. Let $\{f_t\}_{t\in[0,1]}$ be a family of Morse functions on X, which satisfies the lemma hypotheses. Since all the singularities $\{\Sigma_{f_t}\}$ are of the Morse type and thus stable, the family avoids the birth and death of critical points, unless, as t changes, they "enter" or "exit" X trough its boundary ∂X.

Put $Y =_{\mathsf{def}} X \times [0,1]$. Consider the 3-bundle τ on Y, tangent to the fibers of the obvious projection $Y \to [0,1]$, and its trivialization $\beta : \tau \approx Y \times \mathbb{R}^3$. The gradient-like fields $\{v_t\}_{t\in[0,1]}$ define a section w of τ, which is transversal to the zero section and vanishes on the set

$$\tilde{J} =_{\mathsf{def}} \bigcup_{t\in[0,1]} \Sigma_{f_t} \times \{t\}.$$

Since the births and deaths of critical points are forbidden in the family, this locus $\tilde{J}$ is a disjoint union of *arcs* whose ends are attached to ∂Y; moreover, since $\Sigma_{f_0} = \emptyset = \Sigma_{f_1}$ by the hypotheses, the ends of $\tilde{J}$ are attached to $\delta X =_{\mathsf{def}} \partial X \times (0,1) \subset \partial Y$.

By a small isotopy of Y, which preserves the fibers of the projection $Y \to [0,1]$, we can perturb the family $\{f_t\}_{t\in[0,1]}$ so that J, the projection of $\tilde{J}$ on X, is a disjoint union of simple curves whose ends reside in ∂X.

The section w and the trivialization β of the bundle τ give rise to a well-defined map $\Phi : Y \setminus \tilde{J} \to S^2$.

We denote by U a small regular neighborhood of $\tilde{J}$ in Y, such that all the fibers U_t of the projection $U \subset Y \to [0,1]$ are homeomorphic to a disjoint union of 3-disks $\{U_{t,\alpha}\}_\alpha$, centered on the critical point $\{x^\star_{t,\alpha}\}_\alpha$ (the disks got truncated, as the critical points approach $\partial_1 X \times [0,1]$).

For a generic w, we may assume that $\Phi : Y \setminus U \to S^2$ is transversal to the base point $* \in S^2$.

Let Z be the surface $\Phi^{-1}(*) \cap (Y \setminus U)$, equipped with the orientation ,induced by the orientations of S^2 and Y.

Consider the portion of the boundary $\delta Z =_{\mathsf{def}} \Phi^{-1}(*) \cap \{\delta Y \setminus (U \cap \delta Y)\}$.

Since the index of the gradient field in the vicinity of each Morse-type critical point $x^\star_{t,\alpha}$ is given by the formula

$$\deg\left(\tilde{\nabla} f_t : \partial U_{t,\alpha} \to S^2\right) = (-1)^{i(x^\star_\alpha)},$$

the boundary of the 2-chain $Z \subset Y \setminus U$ is the 1-cycle

$$\partial Z = (\tilde{\nabla} f_1)^{-1}(*) - (\tilde{\nabla} f_0)^{-1}(*) - \sum_{\alpha} (-1)^{i_\alpha} \tilde{J}'_\alpha + \delta Z.$$

Here the arc $\tilde{J}'_\alpha \subset \partial U$ is parallel in U to $\tilde{J}_\alpha$ and is oriented by the increasing parameter t, and $\delta Z \subset \delta Y$.

Since δZ is contained in δY, we conclude that the relative 1-homology class vanishes in the group $H_1(Y, \delta Y; \mathbb{Z}) \approx H_1(X, \partial X; \mathbb{Z})$:

$$(\tilde{\nabla} f_1)^{-1}(*) - (\tilde{\nabla} f_0)^{-1}(*) - \sum_{\alpha} (-1)^{i_\alpha} \tilde{J}'_\alpha = 0.$$

Hence, $h(\tilde{\nabla} f_1) - h(\tilde{\nabla} f_0)$ is the Poincaré-dual in X of the relative 1-cycle $\sum_\alpha (-1)^{i_\alpha} [J_\alpha]$. □

Let $K = K(v)$ be a gradient spine of X. Imagine cutting a disk D^2 out of an oriented surface $K^\circ =_{\mathsf{def}} K \setminus s(K)$, flipping the orientation of D^2, and gluing it back. This would be a clear violation of the branched orientation in K°. To protect the consistency of the branched orientation, let us erect a cylindrical wall $W = S^1 \times [0, 1]$ along the cut and orient W so that the singularity locus $\partial D^2 = S^1 \times \{0\}$ contributes with the multiplicity ± 1 to the boundary of the oriented 2-chain $[K^\circ \cup W]$. This results in adding a disjoint oriented circle ∂D^2 to the old singularity set $s_\dagger(K)$. Although $K^\circ \cup W$ is collapsible on K°, within the category of *branched* spines, we cannot collapse the wall on its circular base ∂D^2: the reversed orientation of D^2 prevents the collapse.

According to our usual recipe, the geometry of the downward flow in Figure 4.9, left diagrams, prescribes two ways of marking the new gradient spine $K \cup W$ with the TN-markers. When $i(x^\star) = 2$, we mark the inner normal of $\partial D^2 \subset D^2$ with T, and place the marker N either in W along $S^1 \times \{0\}$ so that it points towards $S^1 \times \{1\}$, or N marks the exterior normal of $\partial D^2 \subset D^2$. When $i(x^\star) = 1$, we mark the outer normal of $\partial D^2 \subset D^2$ with T, and place the marker N either in W along $S^1 \times \{0\}$, or N marks the interior normal of $\partial D^2 \subset D^2$.

We call such changes in the shape of (abstract) gradient spines *mushroom flips* (the name is inspired by the shape of the spine in the lower-left diagram from Figure 4.9). The mushroom flips are reminiscent of, so called, *Pontryagin Moves* (see Definition 6.2.1 in [BP]).

They are geometrical manifestations of jumps of the Spin^c-structure on X, defined by varying non-vanishing vector fields v. The first type of flip occurs when a critical point of index 2 is traversing X, the second type corresponds to a critical point of index 1 (see Figure 4.9).

Corollary 4.6. *Let v be a traversally generic vector field on a compact oriented 3-fold X with boundary, and $h \in H^2(X;\mathbb{Z})$ a given cohomology class. Then there exists another traversally generic vector field v' on X, such that:*

- *$v' = v$ outside a tubular neighborhood U of the disjoint union of several simple arcs $\{J_\alpha \subset X\}_\alpha$ whose ends reside in ∂X,*
- *$h(v') = h(v) + h$,*
- *the gradient spine $K(v')$ is obtained from the gradient spine $K(v)$ by a finite sequence of mushroom flips, all supported in $K(v) \cap U$,*
- *the combinatorial complexities $c(K(v'))$ and $c(K(v))$ are equal.*

Proof. Lemma 4.4 describes the variations of the Spin^c-structure on X in terms of the variations in the shape of gradient spines $K(v_t)$ that occur when critical points $\{x^\star_{t,\alpha}\}$ of index $i_\alpha = 1$ or 2 are traversing X along some oriented arcs $\{J_\alpha\}$, the union of arcs representing a given generator λ of $H^1(X, \partial X;\mathbb{Z})$. The arcs, representing λ, may be chosen in general position with respect to a given gradient spine $K(v_0)$. That is, they are transversal to $K(v_0)^\circ =_{\mathsf{def}} K(v_0) \setminus s(K(v_0))$ at finitely many points and have an empty intersection with the singular set $s(K(v_0))$. Furthermore, if an oriented arc J_α hits the cascade $\mathcal{C} =_{\mathsf{def}} \mathcal{C}(\partial_2^+ X(v_0)) \subset K(v_0)$ transversally at a point x, then we can replace J_α with two oriented arcs J'_α and J''_α such that $J'_\alpha \cup J''_\alpha$ is homologous to J_a in $H_1(X, \partial X;\mathbb{Z})$, and $(J'_\alpha \cup J''_\alpha) \cap \mathcal{C} = (J_a \cap \mathcal{C}) \setminus x$. To construct the new arcs, use the downward v_0-trajectory γ_x through x (without loss of generality, we can assume that γ_x is not contained in a double-tangent trajectory) and a rectangular band $B \subset X$ with the core γ_x, the band being transversal to the waterfall W that contains x.

For a traversing vector field v_0, we get $\partial_1^- X(v_0) \neq \emptyset \neq \partial_1^+ X(v_0)$. Employing this fact, by a somewhat similar construction, we can find a representative of $\lambda \in H^1(X, \partial X;\mathbb{Z})$ so that that each arc J_α hits $\partial_1^+ X(v_0)$ at a single point x. For example, if we have an arc J_α with two ends, $x, y \in \partial_1^+ X$, then we pick a path δ, connecting a point $z \in \mathrm{int}(J_\alpha)$ to a point $u \in \partial_1^- X$, and form a narrow rectangular band B with the core δ. We use two arcs in ∂B to replace J_a with two arcs J'_α and J''_α, each having the desired property. Therefore, we can assume that each arc J_α representing λ either is oriented in accordance with the vector $v_0(x)$, where $x = J_\alpha \cap \partial_1^+ X$, or opposite to it. In the first case, we send a critical point of an even index to trace J_α, in the second case, we send a critical point of an odd index, as shown in Figure 4.9. The result is a new submersion $f_1 : X \to \mathbb{R}$ and its gradient field v_1, such that $h(v_1) - h(v_0)$ is Poincaré-dual to λ. By a small deformation of v_1 inside the space of f_1-gradient fields, employing Theorem 2.4, we may assume that v_1 traversally generic.

Thus we need to describe the transformation $K(v_0) \Rightarrow K(v_1)$ only the case where each arc J_α hits $\partial_1^+ X(v_0)$ transversally at a singleton x_α and is disjointed from all the cascades $\mathcal{C} \subset K(v_0)$. Each intersection x_α of this type will be responsible for one mushroom flip as in Figure 4.9.

These mushroom flips, being disjointed from the singularity set $s(K(v_0))$, do not affect the combinatorial complexity of the original spine.

So, for a given traversally generic vector field v_0 on X, the Poincaré dual h of any given class $\lambda \in H_1(X, \partial X; \mathbb{Z})$ can be realized in the form $h(v_1) - h(v_0)$ for an appropriate traversally generic v_1. The spines $K(v_0)$ and $K(v_1)$ of equal complexity are linked by a finite sequence of mushroom flips. □

Assuming the validity of Conjecture 4.1 and using Theorem 4.4, the previous argument motivates the following conjecture.

Conjecture 4.2. *Let v_0 and v_1 be two traversally generic vector fields on a compact oriented 3-fold X with boundary. Then there exists a sequence of mushroom flips, followed by a sequence of 2-expansions, 2-collapses, $\alpha, \beta, \gamma, \gamma^\star$-moves and their inverses, which transforms the gradient spine $K(v_0)$ into the gradient spine $K(v_1)$.* ♢

In [K], this conjecture was mistakenly stated as a theorem.

4.8 Eliminating cusps by surgery on traversing vector fields, while preserving their complexity

The next Theorem 4.5 and Theorem 4.6, form a bridge to some results of Eliashberg ([E1], [E2]) about the *folding maps* of surfaces. We will describe these results and their strong generalizations in Chapter 5.

Theorem 4.5. *Let $S \subset X$ be a connected compact and two-sided surface, regularly embedded into an open oriented 3-fold X. Let v be a vector field on X which is non-vanishing in the vicinity of S, boundary generic with respect to S, and transversal to S along its boundary ∂S.*

Then there exists a deformation of v, which is supported in the vicinity of S, is fixed in the vicinity of ∂S, and such that the new deformed vector field v' on X is boundary generic with respect to S, and the new cuspidal set $\partial_3 S(v')$ is empty.

If v is a gradient-like vector field on X, then so is v'. If $v \neq 0$ in X then so is v'. Moreover, the entire deformation of v takes place in the space of non-vanishing vector fields of the gradient type. ♢

Before proving Theorem 4.5, we need to adjust the familiar construction of the Morse stratification to the hypotheses of the theorem.

Since X is orientable and S is two-sided, the surface S is orientable. Since S has a preferred side in X, it can be divided, with the help of v, into two complementary domains, S^+ and S^-, which share a common portion $L =_{\mathsf{def}} \partial S^+ \cap \partial S^-$ of their boundaries. By definition, in S^+, the field v points into the preferred side of S and is tangent to S along the locus L. Since v is transversal to S along ∂S, for a generic v, the locus L is a collection of smooth *loops*. The preferred orientation of S^+, together with the normals $\nu(L, S^+)$, induces a particular orientation on L. The 1-submanifold L is divided by the cusp locus $C =_{\mathsf{def}} \partial_3 S(v)$ into two portions, L^+ and L^-. Along L^+, v points inside S^+.

As in the very beginning of Section 4.2, the points from C acquire four flavors: $(+, \oplus), (-, \ominus), (-, \oplus), (+, \ominus)$. The first polarity $\{+, -\}$ reflects the fact that v at a point of C points inside or outside of L^+. The second polarity $\{\oplus, \ominus\}$ tells us whether, at the points of C, the field agrees or disagrees with the orientation of L, induced by the orientation of $S^+ \subset S$. We denote by C^A the points of the two flavors $(+, \oplus), (-, \ominus)$, and by C^B of the other two flavors $(-, \oplus), (+, \ominus)$.

The proof of Theorem 4.5 relies on the next three lemmas.

Lemma 4.5. *Under the hypotheses of Theorem 4.5, $\#(C^A) = \#(C^B)$. In other words, for a boundary generic with respect to S vector field v,*

$$\#(\partial_3^{+\oplus} S) - \#(\partial_3^{+\ominus} S) + \#(\partial_3^{-\ominus} S) - \#(\partial_3^{-\oplus} S) = 0. \qquad (4.6)$$

Proof. We already remarked that L must be a disjoint union of simple loops. Each loop L_i from L either has no cusps, or the arcs from L^+ and L^- alternate along L_i. For an arc from L^+, examining the four possible flavors attached to its end points, a and b, we see that exactly one of the two polarities of a and b must be different. Therefore, if $a \in C^A$, then

$b \in C^B$. Since the arcs from L^+ and L^- alternate along each loop from L, we get $\#(C^A) = \#(C^B)$. This validates formula (4.6). □

Lemma 4.6. *Let two cusp points $a, b \in C$ have the following flavors:*

(1) *either* $(+, \oplus)$ *and* $(+, \ominus)$, *or*

(2) $(-, \oplus)$ *and* $(-, \ominus)$.

In the first case, assume that a and b can be connected by a simple path $\gamma \subset S^+$; in the second case, assume that $\gamma \subset S^-$.

Then a and b can be cancelled via a deformation of v as in Figure 4.6 diagrams 1-2 (via the $\eta, \eta^\star$-moves that perform 1-surgery on $S^\pm$). The deformation is the identity away from a regular neighborhood of the path γ in X.

If $v \neq 0$ in X, then so is the deformed field v'. If, for a smooth function $f : X \to \mathbb{R}$, $df(v) > 0$ on a compact $K \supset S$, then $df(v') > 0$ on K. In fact, if v is non-vanishing and of the gradient type on X, so are all the intermediate vector fields.

Proof. Consider two cusps, a and b, (as in Figure 4.6, diagrams 1-2), and a path $\gamma \subset S$ as in the hypotheses of the lemma. We may assume that $v = \nabla(f)$ for an appropriate metric and a smooth function $f : X \to \mathbb{R}$ in the vicinity of γ in X. Then we apply Lemma 2.3 (see Figure 2.4) to construct a homotopy (which is supported in the vicinity of γ) between v and a new non-vanishing field v' with the desired properties. The construction of the homotopy is contained in the proof of Lemma 2.3 and is illustrated in Figure 2.5. By the proof of Lemma 2.3, the deformed field v' has the property $df(v') > 0$ in the vicinity of γ in X.

If, for a global function $f : X \to \mathbb{R}$, v has the property $df(v) > 0$ on a compact $K \subset X$ that contains S, then again, by the proof of Lemma 2.3, the deformed field v' has the property $df(v') > 0$ in K.

If we do not care about constructing a gradient-like v', here is an alternative, singularity theory-flavored proof, perhaps, a bit more transparent. The two cusps a and b are mirror images of each other in the sense that there are ambient coordinates x, y, z in the vicinity of path $\gamma \subset X$ so that:

(1) the path γ can be given by $\{y = z = 0\}$ with x ranging from a to b,
(2) the surface S is given by the equation $y = z^3 + (x - a)z$ at the cusp $(a, 0, 0)$, and by the equation $y = z^3 - (x - b)z$ at the cusp $(b, 0, 0)$,

(3) in the vicinity of γ, the gradient field v is of the form

$$\frac{\partial g}{\partial x}\,\partial_x + \frac{\partial g}{\partial y}\,\partial_y + \partial_z,$$

where $(a,0)$ and $(b,0)$ are critical points of an appropriate Morse function $g(x,y)$.

The surface S is transversal to v along $\mathrm{int}(\gamma)$, and the space of all germs of two-sided oriented surfaces with this property has the homotopy type of a 0-dimensional sphere (there is no topological obstruction to making the surface standard along γ). So, we can cancel the two cusps by embedding a neighborhood of γ into the standard model and using the η or $\eta^\star$-deformation as in Figure 4.6, diagrams 1-2, and Figure 4.7. □

Lemma 4.7. *If two consecutive cusps, a and b, along a loop from the tangecy locus $L \subset S$ are as in Lemma 4.6, then a, b can be cancelled, so that the new vector field $v' \neq 0$ has a tangency locus $L' \subset S$ that is the result of 0-surgery on L; the loci $S^\pm$ are subjected to 1-surgery. The deformation of v into v' is supported in an arbitrary small neighborhood of the arc $[a,b] \subset L$.*

If v is a non-vanishing gradient-like vector field on X, so is v' and all the intermediate vector fields.

Proof. In the case of polarities $(+,\ominus)$ and $(+,\oplus)$, there is a path γ along the surface that links a and b inside of S^+. In the other case, γ that links a and b can be found inside of S^-. This path γ exists in an arbitrarily small neighborhood of the arc $[a,b] \subset L$. By Lemma 4.6, the cancellation of the two cusps is possible. □

Lemma 4.8. *Let $S \subset X$ and v be as in Theorem 4.5.*

For any point $x_\star \in L^+$, there is a deformation (the $\delta, \delta^\star$-moves) v' of v, supported in the vicinity of $x_\star$ in X, so that two new consecutive cusps of the types $(+,\oplus)$ and $(-,\oplus)$, or of the types $(+,\ominus)$ and $(-,\ominus)$, are introduced to the new v'-tangency locus L'.

If v is a non-vanishing gradient-like vector field on X, so is v' and all the intermediate vector fields.

Proof. The argument is suggested by Figure 4.6, the passage from right to left in diagrams 5 and 6 and their mirror images (not shown in the figure). Pick a point $x_\star \in L$, where $v(x_\star)$ is transversal to the tangency locus L.

Consider a localized deformation of the normal to S component of v that realizes the modification of the set $\partial_1^+ X$, as depicted in the diagrams 5 and 6 in Figure 4.6, while keeping the tangential component of the field fixed. As in the proof of Lemma 2.3 (see Figure 2.5), this deformation of v can be realized so that the new field still remains gradient-like for the same function $f : X \to \mathbb{R}$ that served the original field v.

The analytical models of such vector field deformations are delivered by the $\delta, \delta^\star$-moves from the list in (4.4).

Here is another sketch of the desired deformation. Consider the polynomial

$$P_t(x, y, u) =_{\mathsf{def}} u^4 + tu^2 + xu + y,$$

where t is a real parameter.

We introduce local coordinates (x, y, u) in the vicinity of the point $x_\star \in L$, so that the x-axis is tangent to the curve L at $x_\star$. In fact, these coordinates can be chosen so that locally S is given by the equation $P_\epsilon(x, y, u) = 0$, $\epsilon > 0$, and the locus L by the pair of equations: $\{P_\epsilon(x, y, u) = 0,\ \frac{\partial}{\partial u} P_\epsilon(x, y, u) = 0\}$, while the field v becomes ∂_u.

The Swallow Tail deformation is given by projecting the surfaces

$$\Sigma_t =_{\mathsf{def}} \{P_t(x, y, u) = 0\} \subset \mathbb{R}^3_{xyu}$$

on the xy-plane along the u-axis. The folding locus of the projection is given by the pair of equations:

$$\{P_t(x, y, u) = 0,\ \frac{\partial}{\partial u} P_t(x, y, u) = 0\}.$$

As the parameter t changes from negative to positive, the two cusps[13] of the projection $p_t : \Sigma_t \to \mathbb{R}^2$ cancel at $(0, 0)$. See Figure 4.6, diagrams 5-6, and Figure 2.12 depicting the Swallow Tail surface for negative values of t. The reverse process generates a pair of cusps of the desired polarities: $(+, \oplus)$ and $(-, \oplus)$, or $(+, \ominus)$ and $(-, \ominus)$.

Consider the isotopy $\{\alpha_t : \mathbb{R}^3 \to \mathbb{R}^3\}_{t \in [-\epsilon, \epsilon]}$, defined by the formula $\alpha_t(x, y, u) =_{\mathsf{def}} (x, y - 2tu^2, u)$. Note that α_t takes the surface Σ_t to the surface Σ_{-t}. It is possible to localize α_t so that the new isotopy $\tilde{\alpha}_t$ will be supported in an appropriately chosen neighborhood U of the origin. That U must contain a neighborhood V which, in turn, contains the two cusps of the projection $\Sigma_{-\epsilon} \to \mathbb{R}^2_{xy}$. Moreover, the restriction $\tilde{\alpha}_\epsilon|_V = \alpha_\epsilon|_V$. The folding locus of $\tilde{\alpha}_\epsilon(\Sigma_\epsilon) \to \mathbb{R}^2_{xy}$ in U is similar to the folding locus of $\Sigma_{-\epsilon} \to \mathbb{R}^2_{xy}$ in U. We leave the further details to the reader. □

[13]They are symmetric with respect to y-axis and have the u-coordinates $\pm(-t/6)^{\frac{1}{2}}$ and the x-coordinates $\pm 8(-t/6)^{\frac{3}{2}}$.

Proof of Theorem 4.5. Take any two consecutive cusps a, b along L so that the connecting arc $[a, b]$ lies in L^+. Say, a is of type $(+, \oplus)$. Then b is either of type $(+, \ominus)$ or $(-, \oplus)$. In the first case, according to Lemma 4.6, a and b can be cancelled by deforming the vector field in the vicinity of S. In the second case, by Lemma 4.7, we can introduce two cusps $c, d \in [a, b]$ of types $(+, \ominus)$ and $(-, \ominus)$, respectively. Then, by Lemma 4.6, a and c can be cancelled, as well as d and b. The same argument works in the case when a is of any other type than $(+, \oplus)$. Continuing in this way, all cusps can be eventually eliminated by an appropriate deformation of the vector field.

Note that the property $\partial_3 X(v) = \emptyset$ is stable (see Definition 2.4) under perturbations of v: a map that is C^∞-close to a folding map is again a folding map.

If the original v is gradient-like in X, then, according to Lemmas 4.6 - 4.8, so are the deformed vector field and all the intermediate vector fields. □

Theorem 4.6. *Let X be a connected oriented compact 3-fold with boundary. Any traversally generic vector field v on X can be deformed, within the space $\mathcal{V}_{\mathsf{trav}}(X)$, into a new traversally generic vector field v', so that the new locus $\partial_3 X(v') = \emptyset$, and $gc(v') = gc(v)$. The vector fields v and v' define the same Spin^c-structure on X.*

In other words, there are no topological obstructions to the 3-convexity and 3-concavity, even within the space of traversally generic vector fields with a fixed number *of double-tangent trajectories.*

Proof. In order to prove Theorem 4.6, we need to examine carefully the arguments in the proof of Theorem 4.5, where the role of surface S is played by $\partial_1^+ X(v)$ *union a narrow collar* $\partial_2 X(v) \times [0, \epsilon)$, residing in $\partial_1^- X(v)$.

Since we are going to deform v on multiple occasions, for the rest of the proof, we will use the neutral notation "$\partial_j X$" instead of more accurate notations "$\partial_j X(v)$" or "$\partial_j X(v_t)$".

Employing Theorem 4.5, we will always construct new fields that are traversing (equivalently, gradient-like and non-vanishing) on X.

Consider all the arcs $[a, b] \subset \partial_2^- X \subset \mathsf{int}(S)$ with the different first polarities (i.e., "$+, -$") of the adjacent cusps a and b. Then, by the argument from Lemma 4.5, the second polarities (i.e., "$\oplus, \ominus$") of a, b must agree. For every such arc $[a, b]$, by the $\delta, \delta^\star$-moves, we introduce a pair of cusps $c, d \in [a, b]$, as in the proof of Theorem 4.5 and Lemma 4.8 (see Figure

2.12). Because the arc $[a, b] \subset \partial_2^- X$, the waterfall, falling from $[c, d]$—an arc in the new locus $\partial_2^+ X$— is "protected" by the old $\partial_1^- X$ and is *disjoint* from the rest of the old waterfalls. As a result, the original gradient complexity is not affected by the introduction of the new cusps c, d (contrast this with Figure 2.12, where the interior of X is facing the viewer and the gradient complexity increases by 1 in the process of creating the Swallow Tail Singularity; in our case, the the interior of X is *behind* the Swallow Tail surface in the figure).

Introducing c, d also has no affect on the degree

$$deg(h) = \frac{1}{2}\left[\#(\partial_3^+ X) - \#(\partial_3^- X)\right]$$

of the map $h : \partial_2 X \to S^1$ from Lemma 2.2.

Thus, we can assume that every arc from $\partial_2^- X$ is bounded by cusps of *opposite* second polarity.

Now, for each arc $[a, b] \subset \partial_2^- X$, we pick a path $\gamma \subset \partial_1^{\pm} X$ which connects two canceling cusps a, b (as in Lemmas 4.6 and 4.7) and resides in the vicinity of $[a, b]$. The new portion of the waterfall that is generated after the cancelation of cusps is also localized to the vicinity of $[a, b]$. Hence it is again "protected" by $\partial_1^- X$ and separated from the old waterfalls that existed before the cancellation (see Figure 4.7, the upper diagram). Therefore, canceling a, b by the $\eta, \eta^\star$-moves, again, does not change the number of double-tangent trajectories of the original field. Because the first polarities of a and b agree, each cancelation does change the degree of the map h by one. Since $deg(h) = 0$ for vector fields with no cusps, we will need at least $deg(h_v)$ cancellations to get from the given vector field v to a traversing boundary generic vector field v' with $\partial_3 X(v') = \emptyset$.

According to Theorem 2.4, by a C^∞-small perturbation of v' we can transform it into a traversally generic field. A sufficiently small perturbation of v' will preserve the stable property $\partial_3 X(v') = \emptyset$.

All these deformations take place in the space $\mathcal{V}_{\mathsf{trav}}(X)$.

Unfortunately, we do not know how to insure that the intermediate vector fields $\{v_t\}_{t \in [0,1]}$ in our arguments will be traversally generic, except perhaps at finitely many instances $t_\star$, where $v_{t_\star}$ needs to be a slice of a traversally generic vector field w on $X \times [0, 1]$. In other words, we do not know how to control the deformation so that the path $\{v_t\}_{t \in [0,1]}$ will be truly generic in the sense of Definition 4.1. □

Corollary 4.7. *In terms of the polarized cusps, the degree of the map*

$h : \partial_2 X \to S^1$ can be expressed as follows:

$$deg(h : \partial_2 X \to S^1) = [\#(\partial_3^{+\oplus} X) - \#(\partial_3^{-\oplus} X)] \tag{4.7}$$
$$= [\#(\partial_3^{+\ominus} X) - \#(\partial_3^{-\ominus} X)] = \chi(X) - 2\chi(\partial_1^+ X).$$

Hence, for the traversally generic fields v with fixed values of $\chi(\partial_1^+ X)$[14]*, the number*

$$\#(\partial_3^{+\oplus} X) - \#(\partial_3^{-\oplus} X) = \#(\partial_3^{+\ominus} X) - \#(\partial_3^{-\ominus} X)$$

is a topological invariant of X.

Proof. By Lemma 4.5,

$$\#(\partial_3^{+\oplus} X) - \#(\partial_3^{+\ominus} X) + \#(\partial_3^{-\ominus} X) - \#(\partial_3^{-\oplus} X) = 0.$$

Combined with Lemma 2.2, this leads to the formula for the degree of $h : \partial_2 X \to S^1$ claimed in the corollary. □

4.9 On the spaces of traversing vector fields in 3D with no cusps

Now, consider the space $\mathcal{B}_{\partial_3=\emptyset}(X)$ of boundary generic vector fields v on a compact connected 3-fold X with boundary, such that v is *transversal* to the submanifold $\partial_2 X(v) \subset \partial_1 X$. In other words, in the space of all boundary generic fields, a field $v \in \mathcal{B}_{\partial_3=\emptyset}(X)$ is characterized by the property "$\partial_3 X(v) = \emptyset$". For fields $v \in \mathcal{B}_{\partial_3=\emptyset}(X)$, the local v-directed projections p of the boundary $\partial_1 X$ on a transversal section S of the v-flow are either submersions or folding maps.

Next, we consider the space

$$\mathcal{V}_{\mathsf{fold}}(X) =_{\mathsf{def}} \mathcal{B}_{\partial_3=\emptyset}(X) \cap \mathcal{V}_{\mathsf{trav}}(X),$$

where

$$\mathcal{V}_{\mathsf{trav}}(X) = \mathcal{V}_{\mathsf{grad}}(X) \cap \mathcal{V}_{\neq 0}(X)$$

is the space of non-vanishing gradient like vector fields.

Remark 4.3. Similar spaces of smooth maps $F : M \to N$ from a given manifold M to a given manifold N, with folds only for the F-singular loci, have been studied by Eliashberg in great generality in [E1], [E2]. In the universe of traversing vector fields, we are lacking a target manifold N.

[14]For example, for v with $\partial_1^+ X$ being a 2-disk.

However, its role is played by the space $\mathcal{T}(v)$ of v-trajectories, a space which has a structure of a cellular 2-complex and is singular in general. Recall that the map $\Gamma^\partial : \partial_1 X \to \mathcal{T}(v)$ that takes each point $x \in \partial_1 X$ to a trajectory γ_x through x is a focus of study in this book. $\diamondsuit$

Let $\mathcal{V}^\ddagger_{\mathsf{folds}}(X)$ denote the subspace of $\mathcal{V}_{\mathsf{folds}}(X)$, consisting of traversally generic vector fields (see Definition 2.4) with no cusps. Note that each $v \in \mathcal{V}^\ddagger_{\mathsf{folds}}(X)$ has no triple-tangent trajectories.

Consider the function $gc : \mathcal{V}^\ddagger(X) \to \mathbb{Z}_+$ that associates the number of double-tangent trajectories with each traversally generic vector field.

Theorem 4.6 leads to the following corollary.

Corollary 4.8. *$\mathcal{V}^\ddagger_{\mathsf{folds}}(X)$ is an open and dense subset of the space $\mathcal{V}_{\mathsf{folds}}(X)$ and an open subset of the space $\mathcal{V}^\ddagger(X)$.*

The function $gc : \mathcal{V}^\ddagger_{\mathsf{folds}}(X) \to \mathbb{Z}_+$ is constant on each path-connected component of the space $\mathcal{V}^\ddagger_{\mathsf{folds}}(X)$.

The range of the function $gc : \mathcal{V}^\ddagger_{\mathsf{folds}}(X) \to \mathbb{Z}_+$ is the set of all integers that are greater than or equal to $gc(X)$. Thus,

$$gc(X) = \min_{\{v \in \mathcal{V}^\ddagger_{\mathsf{folds}}(X)\}} gc(v). \tag{4.8}$$

Proof. Since locally-defined v-directed submersions and folding maps $p : \partial_1 X \to S$ are stable under C^∞-small perturbations of v, the space $\mathcal{V}_{\mathsf{folds}}(X)$ is open in the space $\mathcal{V}_{\mathsf{trav}}(X)$ of all traversing fields. Therefore, by Theorem 2.4, $\mathcal{V}^\ddagger_{\mathsf{folds}}(X)$ is an *open and dense* subset of $\mathcal{V}_{\mathsf{folds}}(X)$ and an open subset of the space of traversally generic fields $\mathcal{V}^\ddagger(X)$.

By claim (2) from Lemma 4.1, the function $gc : \mathcal{V}^\ddagger_{\mathsf{folds}}(X) \to \mathbb{Z}_+$ is constant on each path-connected component of the space $\mathcal{V}^\ddagger_{\mathsf{folds}}(X)$.

Let v be a traversally generic field on X with $gc(v)$ double-tangent trajectories. By Theorem 4.6, the same value $gc(v)$ is realizable by a field $v' \in \mathcal{V}^\ddagger_{\mathsf{folds}}(X)$. Thus formula (4.8) follows.

By Corollary 3.2, any integer in the interval $[gc(X), +\infty)$ is the value $gc(v)$ for some traversally generic field v with the property $\partial_3 X(v) = \emptyset$. $\square$

Formula (4.8), being combined with Theorem 3.11, has an immediate implication:

Corollary 4.9. *Let a 3-fold X be as in Theorem 3.11. Then X admits a traversally generic vector field v with $6 \cdot c(X)$ double-tangent trajectories at most and such that $\partial_3 X(v) = \emptyset$.* $\diamondsuit$

Example 4.1. Let X be a punctured Poincaré sphere $(S^3/\mathsf{I}_{120})^\circ$, where I_{120} denotes the icosahedron group of order 120. Its combinatorial complexity $c((S^3/\mathsf{I}_{120})^\circ) = 5$. Then by Corollary 4.9 and Theorem 3.11, X admits a 3-convex and 3-concave traversally generic vector field v with $30 = 5 \times 6$ double-tangent trajectories at most and with 5 double-tangent trajectories at least. $\diamondsuit$

Example 4.2. Let X be a 3-fold, produced with the help of combinatorial data as in (3.4). Then Theorem 3.12 and formula (4.8) imply that X admits a 3-convex and 3-concave traversally generic vector field v with no double-tangent trajectories. $\diamondsuit$

Example 4.3. Let $X^\clubsuit$ be a 3-fold as in Theorem 3.16. Its fundamental group

$$\pi_1(X^\clubsuit) = \big\{a, b \big|\ a[a, b]b^{-1} = 1\big\}.$$

Then formula (4.8) implies that $X^\clubsuit$ admits a 3-convex and 3-concave traversally generic vector field $v^\clubsuit$ with a single double-tangent trajectory. From the gradient complexity perspective, the pair $(X^\clubsuit, v^\clubsuit)$ is "atomic" with respect to connected sums of 3-folds. $\diamondsuit$

Recall that the function $h : \mathcal{V}_{\neq 0}(X) \to H^2(X; \mathbb{Z})$ represents faithfully the set of path-connected components $\pi_0(\mathcal{V}_{\neq 0}(X))$ (see Corollary 4.4) of the space of all non-vanishing vector fields $\mathcal{V}_{\neq 0}(X)$.

Corollary 4.10. *The map* $h : \mathcal{V}^\ddagger_{\mathsf{folds}}(X) \to H^2(X; \mathbb{Z})$ *is surjective. In other words, any* Spin^c*-structure on a connected oriented 3-fold* X *with boundary is realizable by a traversally generic vector field* v *such that* $\partial_3 X(v) = \emptyset$.

Proof. Lemma 4.4 and Corollary 4.4 imply that the restriction of h to the space $\mathcal{V}_{\mathsf{trav}}(X) = \mathcal{V}_{\mathsf{grad}}(X) \cap \mathcal{V}_{\neq 0}(X)$ of non-vanishing gradient-like fields identifies the path-connected components from $\pi_0(\mathcal{V}_{\mathsf{trav}}(X))$ with the elements of $H^2(X; \mathbb{Z})$. Furthermore, by Theorem 4.6, any traversing v can be connected by a path in the space $\mathcal{V}_{\mathsf{trav}}(X)$ to a field $v' \in \mathcal{V}^\ddagger_{\mathsf{folds}}(X)$. Therefore, $h : \mathcal{V}^\ddagger_{\mathsf{folds}}(X) \to H^2(X; \mathbb{Z})$ is surjective as well. Note that $h_* : \pi_0(\mathcal{V}^\ddagger_{\mathsf{folds}}(X)) \to H^2(X; \mathbb{Z})$ is far from being injective. $\square$

The rest of the chapter is somewhat speculative, but the conjectures it contains seem to be reasonable.

We notice that no $\gamma, \gamma^\star, \delta, \delta^\star, \zeta, \zeta^\star, \eta, \eta^\star$-moves are possible within $\mathcal{V}_{\mathsf{folds}}(X)$ since they require nonempty strata $\partial_3 X$ (see Figure 4.1). Also the α-and β-moves, although consistent with the space $\mathcal{V}_{\mathsf{folds}}(X)$, do violate the traversally generic nature of vector fields $v \in \mathcal{V}^\ddagger_{\mathsf{folds}}(X)$. Therefore, we propose the following conjecture, a version of Conjecture 4.1, adjusted to the vector fields from the space $\mathcal{V}_{\mathsf{folds}}(X)$ (compare the description below with Figure 4.4).

Conjecture 4.3. *Let X be a connected compact 3-fold with boundary.*

Any two vector fields $v_0, v_1 \in \mathcal{V}^\ddagger_{\mathsf{folds}}(X)$ that are connected by a continuous path in the space $\mathcal{V}_{\mathsf{folds}}(X)$ are also connected by a smooth path $\tau : [0,1] \to \mathcal{V}_{\mathsf{folds}}(X)$ such that:

- *the vector fields $v_t = \tau(t)$ on the slices $X_t =_{\mathsf{def}} X \times \{t\}$, produce a vector field v on $Y = X \times [0,1]$ which is traversally generic with respect to $\delta Y = \partial_1 X \times [0,1]$,*
- *the field v_t is traversally generic on X_t for all, but finally many, $t \in (0,1)$. In particular, if v_t has a trajectory of the combinatorial type (12221), then t belongs to the finite exceptional set $\Theta_\tau \subset (0,1)$,*
- *the obvious projection $\pi : \partial_2 Y(v) \to [0,1]$ is a fibration,*
- *the obvious projection $\pi : \partial_2 Y(v, (1221)) \to [0,1]$ is a Morse function on the curve $\partial_2 Y(v, (1221))$ with its critical values in Θ_τ.*

The existence of such path τ implies that the gradient spines $K(v_0)$ and $K(v_1)$ are linked by a finite sequence of α, β-moves and their inverses. $\diamondsuit$

Definition 4.3. We call a path $\tau : [0,1] \to \mathcal{V}_{\mathsf{folds}}(X)$ *truly generic* if it has the properties as in Conjecture 4.3[15]. $\diamondsuit$

It follows from Theorem 2.4 that the space $\mathcal{V}^\ddagger_{\mathsf{folds}}(X)$ is locally path-connected. In fact, by Lemma 6.12, any smooth path $\tau : [0,1] \to \mathcal{V}^\ddagger_{\mathsf{folds}}(X)$ defines a truly generic family!

If v_0 can be connected to v_1 by a truly generic path, and if v_1 can be connected to v_2 by a truly generic path, then if v_0 can be connected to v_2 by a truly generic path. So we may talk about *truly generic path component* of a given vector field $v \in \mathcal{V}^\ddagger_{\mathsf{folds}}(X)$. Thus each truly generic path component is an open subset of $\mathcal{V}^\ddagger_{\mathsf{folds}}(X)$, comprising many path components of $\mathcal{V}^\ddagger_{\mathsf{folds}}(X)$.

We denote the set of such components by the symbol $\pi_0^\bullet(\mathcal{V}^\ddagger_{\mathsf{folds}}(X))$.

[15]In particular, $\tau(0), \tau(1) \in \mathcal{V}^\ddagger_{\mathsf{folds}}(X)$.

By Lemma 6.12, we get a natural surjection

$$\pi_0(\mathcal{V}^{\ddagger}_{\mathsf{folds}}(X)) \to \pi_0^{\bullet}(\mathcal{V}^{\ddagger}_{\mathsf{folds}}(X)).$$

In the spirit of Arnold's studies of the spaces of immersions of plane curves [Ar], we continue to exercise *wishful thinking* by pretending that various spaces of traversing vector fields on a given 3-fold X behave as infinite-dimensional manifolds or hypersurfaces in such manifolds. The hypersurfaces may have codimension two "singularities" in the space $\mathcal{V}_{\mathsf{trav}}(X)$ of traversing vector fields; conjecturally, perturbations of given path $\tau : [0,1] \to \mathcal{V}_{\mathsf{trav}}(X)$ can avoid them. We do not know how to justify these claims: the property of a vector field to be traversally generic or even traversal is hard to express in local terms, say in terms of avoiding certain subvarieties of jet bundles. It seems that, in order to describe the boundaries of subspaces like $\mathcal{V}^{\ddagger}_{\mathsf{folds}}(X) \subset \mathcal{V}_{\mathsf{trav}}(X)$ or $\mathcal{V}^{\ddagger}(X) \subset \mathcal{V}_{\mathsf{trav}}(X)$, one needs to deal with some serious Functional Analysis.

Undeterred by these difficulties, we conjecture that the codimension one walls that bound the chambers of $\mathcal{V}^{\ddagger}_{\mathsf{folds}}(X) \subset \mathcal{V}_{\mathsf{folds}}(X)$, are contained in the "thin" space $\mathcal{V}_{\mathsf{folds}}(X) \setminus \mathcal{V}^{\ddagger}_{\mathsf{folds}}(X)$. They consist of vector fields v with:

(1) either *triple-tangent trajectories*, or/and

(2) double-tangent trajectories γ that are locally contained in two waterfalls that are *quadratically tangent* along γ (see Figure 2.8, diagrams B and C);

(3) the rest of the fields in $\mathcal{V}_{\mathsf{folds}}(X)$, which form an even "thinner" set of codimension two.

The "triple-tangent" walls $\mathcal{W}_*(X) \subset \mathcal{V}_{\mathsf{folds}}(X)$ can be cooriented according to the the following rule: the germ of a truly generic path $\tau : [0,1] \to \mathcal{V}_{\mathsf{folds}}(X)$, which represents a β-move, defines a positive coorientation of $\mathcal{W}_*(X)$ when, as a result of the β-move, the number of double-tangent trajectories jumps up by 1, while the difference between the numbers of double-tangent trajectories of the polaritiies $\oplus$ and $\ominus$[16] changes by ± 1. This coorientation is similar in spirit but different from the one used in [Ar].

We denote by $\mathcal{W}_{\asymp}(X) \subset \mathcal{V}_{\mathsf{folds}}(X)$ the codimension one walls, comprised of vector fields with a trajectory γ, whose segment is contained in two waterfalls that are simply tangent along that segment. These walls are cooriented by the α-moves: if $\tau : [0,1] \to \mathcal{V}_{\mathsf{folds}}(X)$ is a truly generic path,

[16]equivalently, $\#[Q(K(v))^{\oplus}] - \#[Q(K(v))^{\ominus}]$

whose germ represents an α-move, then τ, the normal to the wall, is oriented so that the number of double-tangent trajectories increases (by 2) along τ.

We summarize these observations in the following *conjecture-independent* theorem.

Theorem 4.7. *The codimension* 1 *walls, bounding the path-connected components of the space* $\mathcal{V}^{\ddagger}_{\mathsf{folds}}(X) \subset \mathcal{V}_{\mathsf{folds}}(X)$, *are cooriented by the* α *and* β-*moves. Along any truly generic path* $\tau : [0,1] \to \mathcal{V}_{\mathsf{folds}}(X)$ *that pierces the walls in the direction of positive normals, the gradient complexity* $gc(\tau(t))$ *increases, as a function of* $t \in [0,1]$.

Each transversal intersection of the path τ *with the walls from the triple-tangent subspace* $\mathcal{W}_*(X)$ *increases the gradient complexity of the vector field by* 1 *and changes the polarized gradient complexity by* ± 1. *Each transversal intersection of* τ *with the walls from the subspace* $\mathcal{W}_{\asymp}(X)$ *increases the gradient complexity by* 2, *but does not affect the polarized complexity.* $\diamondsuit$

In Theorem 4.6, we have shown that any traversally generic field v can be deformed (within the space of traversing fields) into a traversally generic field with no cusps. For fields with $\partial_3 X = \emptyset$, both loci, $\partial_2^+ X$ and $\partial_2^- X$, are collections of simple oriented loops in $\partial_1 X$, the orientation being induced by the orientation of $\partial_1^+ X$.

Within the space $\mathcal{V}_{\mathsf{folds}}(X)$, no surgery on $\partial_2 X(v)$, induced by deformations of fields v, is possible (see Figures 4.5 and 4.6). Indeed, the transversality of v to $\partial_2 X(v)$ is imposed and prevents the loops from $\partial_2 X(v)$ from touching each other, or being born/annihilated. Thus, each connected component of $\mathcal{V}_{\mathsf{folds}}(X)$ has its own *oriented and polarized loop pattern*

$$\Xi_v =_{\mathsf{def}} \partial_2^+ X(v) \coprod \partial_2^- X(v) \subset \partial_1 X.$$

We may associate a dual oriented graph Ξ_v^* with Ξ_v: its vertexes correspond to the connected components of $\partial_1 X \setminus \partial_2 X$, and its edges correspond to the boundary loops that are shared by the adjacent components. The edges are oriented so that, moving in the preferred direction, one moves from a region in $\partial_1^- X$ to a region in $\partial_1^+ X$. In addition, the edges of Ξ_v^* have polarities "$\pm$" depending whether the shared loop belongs to $\partial_2^+ X$ or to $\partial_2^- X$. This structure can be regarded as a 1-cochain $\Delta_v \in C^1(\Xi_v^*; \mathbb{Z})$.

Question 4.1. *For a given connected compact 3-fold* X *with boundary, what is the minimal number of positive loops that form* $\partial_2^+ X(v)$ *for traversing vector fields* v *with* $\partial_3 X(v) = \emptyset$?

Also, for a given X, which oriented loop patterns Ξ_v in $\partial_1 X$ are realizable? $\diamondsuit$

We will revisit Question 4.1 and similar questions in Section 5.5.

Recall that by [BP], Theorem 4.4.5, any X that admits an oriented branched simple spine, admits a *concave* vector field. Thus, for such oriented X, the minimal number of negative loops is zero.

Let $\overline{\mathcal{W}_*(X)}$ stand for the closure in $\mathcal{V}_{\mathsf{folds}}(X)$ of the space $\mathcal{W}_*(X)$. Within the space $\mathcal{V}_{\mathsf{folds}}(X) \setminus \overline{\mathcal{W}_*(X)}$ no β-moves and their inverses are permitted. Based on this observation, we will construct a map from $\pi_0\big(\mathcal{V}_{\mathsf{folds}}(X) \setminus \overline{\mathcal{W}_*(X)}\big)$ to integral-valued bilinear forms over finitely generated free $\mathbb{Z}$-modules. In a sense, the forms are induced by the intersections in the v-trajectory space $\mathcal{T}(v)$ of the 1-cycles, forming $\partial_2^+ X$. Since $\mathcal{T}(v)$ may have singularities of codimension one, in general, this intersection has no obvious homological interpretation. However, within the constraints of a given chamber of the space $\mathcal{V}_{\mathsf{folds}}(X) \setminus \overline{\mathcal{W}_*(X)}$, it is well-defined.

Consider a $\mathbb{Z}$-module M_v, freely generated by the oriented loops $\{\omega_i\}_i$ that form $\partial_2^+ X(v)$, the orientation being induced from $\partial_1^+ X(v)$ (in turn, that orientation, with the help of the inner normals, is induced by the orientation of X).

Define the bilinear form $\Psi_v(\omega_i, \omega_j)$ to be the sum of ± 1's which are contributed by the double-tangent trajectories γ_{ij}, linking ω_i to ω_j. The sign, contributed by γ_{ij}, is the $\oplus/\ominus$-polarity, associated with γ_{ij}. Evidently, the form Ψ_v is preserved under the α-moves—the only admissible transformations of the gradient spine $K(v)$ within a given chamber of $\mathcal{V}_{\mathsf{folds}}(X) \setminus \overline{\mathcal{W}_*(X)}$ (see Figures 4.1 and 4.4).

These considerations lead directly to the following theorem.

Theorem 4.8. *The following objects are locally constant on the space $\mathcal{V}^{\ddagger}_{\mathsf{folds}}(X)$ of traversally generic vector fields v that are 3-concave and 3-convex:*

- *the oriented and polarized loop patterns $\Xi_v \subset \partial_1 X$,*
- *the cochain $\Delta_v \in C^1(\Xi_v^*; \mathbb{Z})$,*
- *In contrast, the bilinear form $\Psi_v : \mathsf{M}_v \otimes \mathsf{M}_v \to \mathbb{Z}$ and thus the linking number*

$$lk_w(\partial[K], \partial[K]) = \#(Q(K)^{\oplus}) - \#(Q(K)^{\ominus})$$

are locally constant on the subspace $\mathcal{V}_{\mathsf{folds}}(X) \setminus \overline{\mathcal{W}_*(X)}$ of traversing vector fields v with no triple-tangent trajectories and no cusps.

Here $K = K(v)$ denotes the gradient spine, and the linking number $lk_w(\partial[K], \partial[K])$ was introduced prior to Theorem 3.1. $\diamondsuit$

Corollary 4.11. *Let two vector fields, v_0 and v_1, belong to the space $\mathcal{V}_{\mathsf{folds}}(X) \setminus \overline{\mathcal{W}_*(X)}$*[17]*. Then any truly generic path $\tau : [0,1] \to \mathcal{V}_{\mathsf{folds}}(X)$, which connects the points v_0 and v_1, must have at least*

$$|gc^{\oplus}_{\ominus}(v_0) - gc^{\oplus}_{\ominus}(v_1)|$$

intersections with the walls $\mathcal{W}_*(X)$ of chambers, forming $\mathcal{V}_{\mathsf{folds}}(X) \setminus \mathcal{W}_*(X)$.

In other words, any truly generic 1-parameter family τ of vector fields must have at least $|gc^{\oplus}_{\ominus}(v_0) - gc^{\oplus}_{\ominus}(v_1)|$ members with triple-tangent trajectories. $\diamondsuit$

We have reached the end of our journey through the "three-dimensional part" of our enterprise. Some of the results from Chapters 1-4 will guide us through the forthcoming study of high-dimensional gradient flows, others are clearly unique to the three-dimension worlds.

[17]In particular, the claim applies to any pair $v_0, v_1 \in \mathcal{V}^{\ddagger}_{\mathsf{folds}}(X)$.

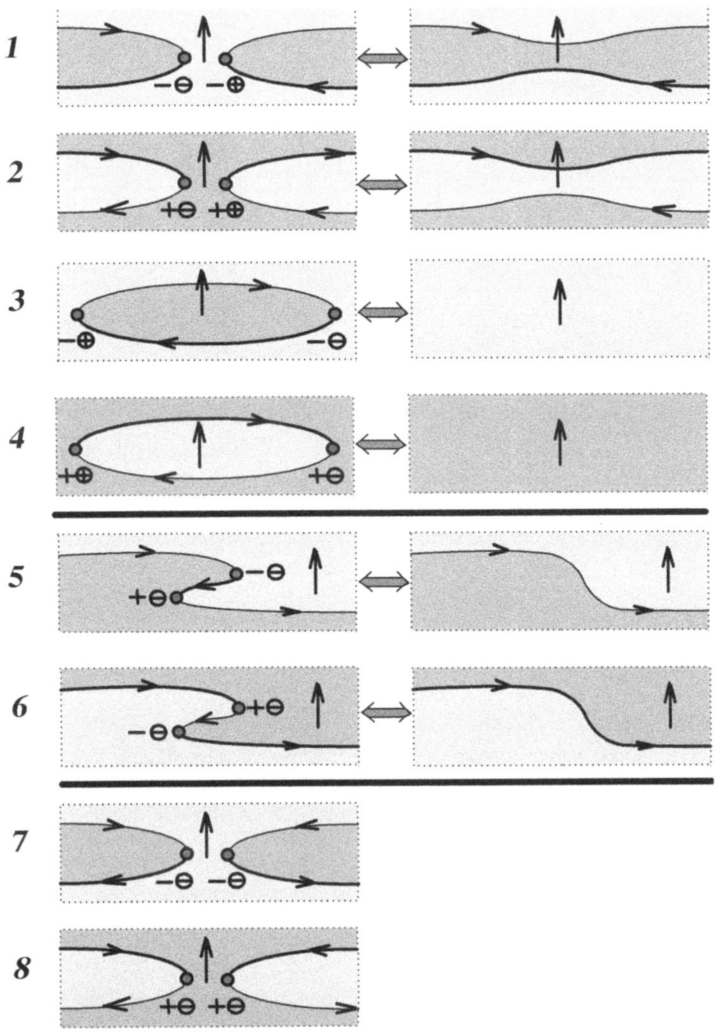

Fig. 4.6 The vector field is "almost vertical" in the vicinity of a point on the boundary $\partial_1 X$, where the surgery is supported. In diagrams 1-4 the topology of $\partial_1^+ X$ (the dark domains) changes, in diagrams 5-6 it does not; diagrams 7-8 show "impossible cancellations".

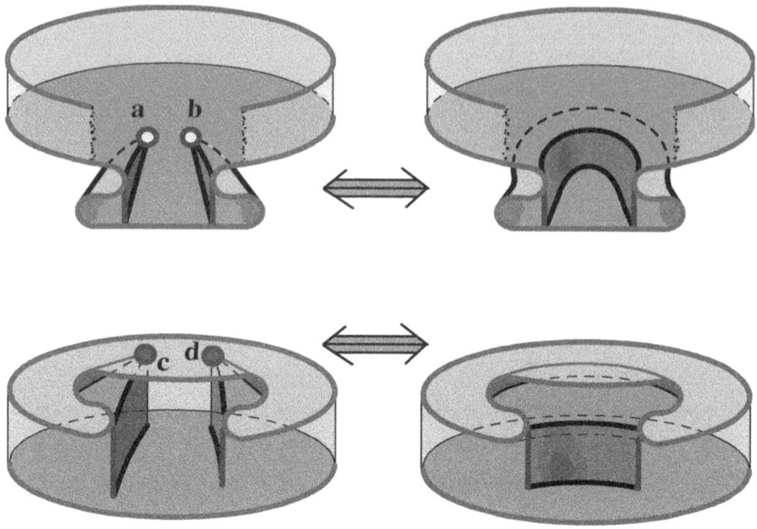

Fig. 4.7 The $\eta, \eta^\star$-moves: changing the gradient spines $\partial_1^+ X(v_t) \cup \mathcal{C}(\partial_2^+ X(v_t))$ by elementary collapses and expansions of 2-cells. The moves correspond to elementary 1-surgery on $\partial_1^+(X)(v_t)$.

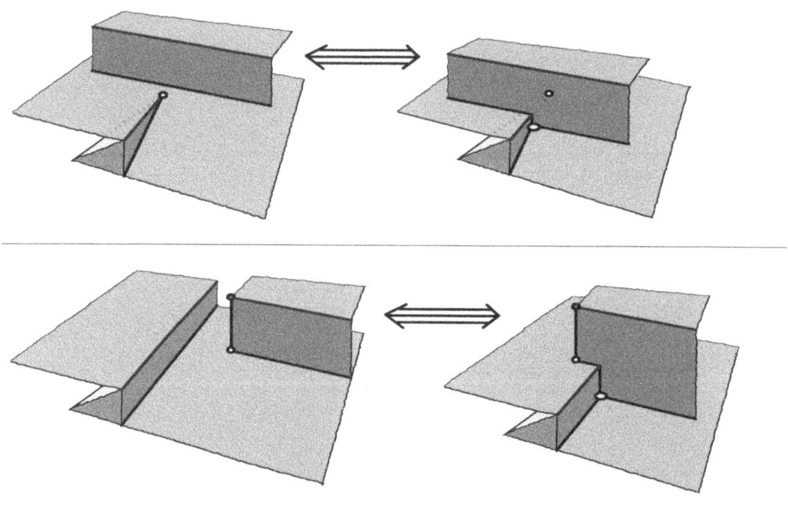

Fig. 4.8 The $\gamma, \gamma^\star$-moves: generating a new Q-singularity by deforming the cascade. Under these moves, the topology of the pair $(\partial_1^+ X, \partial_2^+ X)$ does not change, but the topology of the cascade and thus of the gradient spine does.

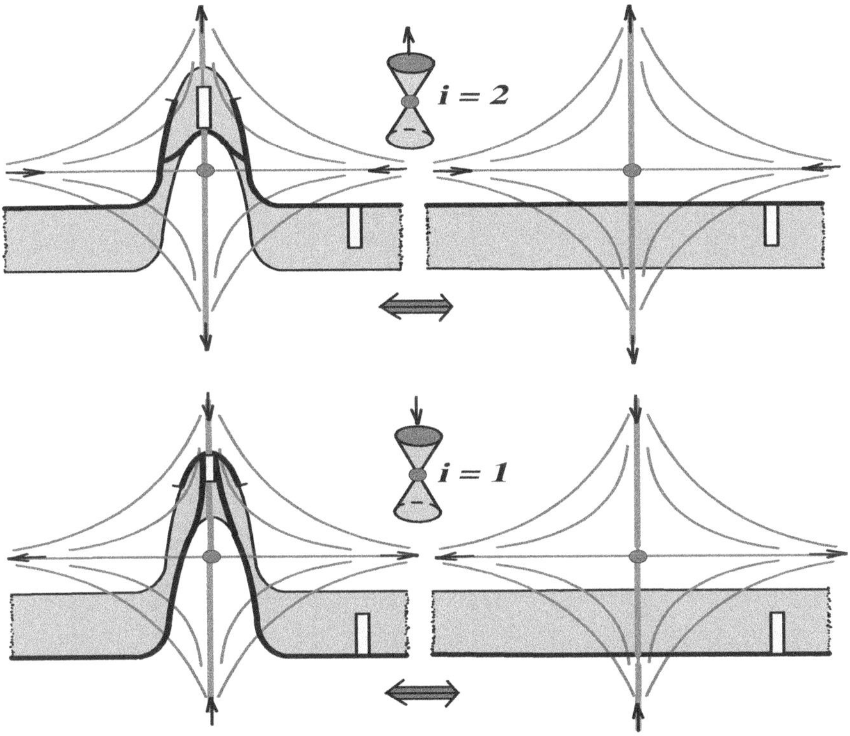

Fig. 4.9 Changing the characteristic class $h(\nabla(f_t)) \in H^2(X;\mathbb{Z})$ by the dual of the 1-cycle $(-1)^i[J]$: the deformation is shown as an isotopy of X in the ambient $\hat{X}$ against a fixed background of a particular Morse function $f : \hat{X} \to \mathbb{R}$ and its gradient field (a more realistic $3D$-picture is obtained from the flat diagrams by revolving them around the vertical axis). The upper diagram shows the case of a single f-critical point of Morse index $i = 2$, the lower one—of Morse index $i = 1$. The spine changes by the, so called, *mushroom flips*. Note the change in the orientation (shown by small normal bars) of the newly formed disk, the "mushroom head" in the lower diagram.

PART 2

High-dimensional Flows

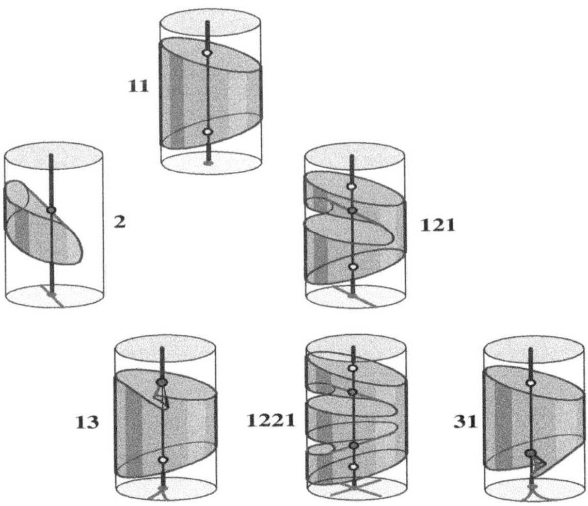

Chapter 5

Stratified Convexity and Concavity of Gradient Flows on Manifolds with Boundary

5.1 In this chapter...

We launch now our investigations of the Morse Theory of gradient flows on smooth compact $(n + 1)$-dimensional manifolds with boundary, a special case of the well-developed Morse theory on stratified spaces (see [GM], [GM1], and [GM2]). For us, however, the starting point and the source of inspiration is the 1929 paper of Morse [Mo].

As in the previous chapters, we intend to present to the reader a version of the Morse Theory in which the critical points remain behind the scene, while shaping the geometry of the boundary. Some of the concepts that animate our approach can be found in [K1] and are already familiar from Chapters 2-4, where they were adopted to the special environment of gradient flows in dimension three. These notions include *stratified convexity or concavity* of traversing flows in relation to the boundary of the manifold. That stratified concavity will serve as a measure of *intrinsic complexity* of a given manifold X with respect to any traversing flow. As we know already, both convexity and concavity of 3D-flows may have strong topological implications.

In Section 5.2, for so-called *boundary generic* fields v on X (see Definition 5.1), we explore the Morse stratification $\{\partial_j^{\pm} X(v)\}_j$ of the boundary ∂X (see formula (5.1) and [Mo]), induced by the vector field v on X.

In Section 5.3, we investigate the degrees of freedom to change this stratification by deforming a given vector field within the space of gradient-like fields (Theorem 5.5, Theorem 5.6, Corollary 5.6, and Corollary 5.7).

In Section 5.4, for vector fields on compact manifolds, we introduce the pivotal notion of boundary s-convexity/s-concavity, $s = 2, 3, \dots$ (see Definition 5.5). Then we explore some topological implications of the existence of a boundary 2-convex/2-concave traversing field on X (see Lemma 5.7, Corollary 5.9, Corollary 5.3, and Corollary 5.10).

Another central theme that will reappear in this chapter is *the holographic properties of traversing flows* on manifolds with boundary. The ultimate aim here is to reconstruct (perhaps, only partially) the bulk of the manifold and the dynamics of the flow on it from some residual structures on the boundary. Thus the name "*holography*".

Let v_1 denote the orthogonal projection of the field $v|_{\partial X}$ on the bundle $T(\partial X)$ tangent to the boundary. Occasionally, we can determine whether a given field v is convex/concave just by observing the behavior of the v_1-trajectories on the boundary $\partial_1 X$ (Theorems 5.7 and 5.8). We view the possibility of such determination as an instance of a general phenomenon, which we call "holography" (see Section 6.6, in particular, Theorem 6.12). The holography phenomenon occupies us fully in [K6], [K7], and is the focus of the forthcoming book [K13].

In Section 5.5, we employ the Eliashberg's surgery theory of folding maps [E1], [E2] to help to describe the patterns of Morse stratifications for traversing 3-concave and 3-convex fields (Theorem 5.11, Corollary 5.13 and Conjecture 5.2).

Finally, in Section 5.6, we study implications for the topology of X when it admits a concave flow with very restricted patterns of tangency to the boundary ∂X.

5.2 Revisiting the Morse stratification $\{\partial_j^+ X(v)\}$

Inspired by [Mo], we start by reintroducing some basic notions and constructions that describe the way in which generic vector fields on a compact smooth manifold interact with its boundary. Some of these are familiar to the reader from our investigations of $3D$-flows.

Let X be a compact smooth $(n+1)$-dimensional manifold with a boundary ∂X. Let v be a smooth vector field on X which does not vanish on the boundary. As a rule, we assume that X is properly contained in a $(n+1)$-dimensional manifold $\hat{X}$ and that the field v extends to a field $\hat{v}$ on $\hat{X}$ so that $v|_{\hat{X}\setminus X} \neq 0$. In fact, we always treat the pair $(\hat{X}, \hat{v})$ as a *germ* of a space and a field in the vicinity of the given pair (X, v).

Typically, we will consider vector fields only with the isolated *Morse-type* singularities (zeros), located away from the boundary. This means that v, viewed as a section of the tangent bundle $T(X)$, is transversal its zero section. In other words, in the vicinity of each singular point, there is a local system of coordinates $(x_1, \dots, x_{n+1})$ such that the field v can be represented as $v = (a_1x_1, \dots, a_{n+1}x_{n+1})$, where all $a_i \neq 0$.

To achieve some uniformity in our notations, let $\partial_0 X =_{\mathsf{def}} X$ and $\partial_1 X =_{\mathsf{def}} \partial X$.

The vector field v gives rise to a partition $\partial_1^+ X \cup \partial_1^- X$ of the boundary $\partial_1 X$ into two sets: the locus $\partial_1^+ X$, where the field is directed inward of X, and $\partial_1^- X$, where it is directed outward. We assume that v, viewed as a section of the quotient line bundle $T(X)/T(\partial X)$ over ∂X, is transversal to its zero section. This assumption implies that both sets $\partial_1^\pm X$ are compact manifolds which share a common boundary $\partial_2 X =_{\mathsf{def}} \partial(\partial_1^+ X) = \partial(\partial_1^- X)$. Evidently, $\partial_2 X$ is the locus where v is *tangent* to the boundary $\partial_1 X$.

Morse has noticed that, for a generic vector field v, the tangent locus $\partial_2 X$ inherits a similar structure in connection to $\partial_1^+ X$, as $\partial_1 X$ has in connection to X (see [Mo]). That is, v gives rise to a partition $\partial_2^+ X \cup \partial_2^- X$ of $\partial_2 X$ into two sets: the locus $\partial_2^+ X$, where the field is directed inward of $\partial_1^+ X$, and $\partial_2^- X$, where it is directed outward of $\partial_1^+ X$. Again, let us assume that v, viewed as a section of the quotient line bundle $T(\partial_1 X)/T(\partial_2 X)$ over $\partial_2 X$, is transversal to its zero section.

For generic fields, this structure replicates itself: the cuspidal locus $\partial_3 X$ is defined as the locus where v is tangent to $\partial_2 X$; $\partial_3 X$ is divided into two manifolds, $\partial_3^+ X$ and $\partial_3^- X$. In $\partial_3^+ X$, the field is directed inward of $\partial_2^+ X$, in $\partial_3^- X$, outward of $\partial_2^+ X$. We can repeat this construction until we reach the zero-dimensional stratum $\partial_{n+1} X = \partial_{n+1}^+ X \cup \partial_{n+1}^- X$ (see Figure 2.2 and 2.3 which depict these strata for fields in $3D$).

Thus a generic vector field v on X gives rise to two stratifications:

$$\begin{aligned} \partial X &=_{\mathsf{def}} \partial_1 X \supset \partial_2 X \supset \cdots \supset \partial_{n+1} X, \\ X &=_{\mathsf{def}} \partial_0^+ X \supset \partial_1^+ X \supset \partial_2^+ X \supset \cdots \supset \partial_{n+1}^+ X, \end{aligned} \tag{5.1}$$

the first one by closed submanifolds, the second one—by compact ones. Here $\dim(\partial_j X) = \dim(\partial_j^+ X) = n + 1 - j$. For simplicity, the notations "$\partial_j^\pm X$" do not reflect the dependence of these strata on the vector field v. When the field varies, we use a more accurate notation "$\partial_j^\pm X(v)$".

These considerations motivate a more formal definition.

Definition 5.1. Let X be a compact smooth $(n+1)$-dimensional manifold with boundary ∂X, and v a smooth vector field on X such that $v|_{\partial X} \neq 0$.

We say that v is *boundary generic* if v produces a filtrations of X as in (5.1) whose strata $\{\partial_j^+ X \subset \partial_j X\}_{1 \leq j \leq n+1}$ are defined inductively in j as follows:

- $\partial_0 X =_{\mathsf{def}} \partial X$, $\partial_1 X =_{\mathsf{def}} \partial X$ [1],
- v, viewed as a section of the tangent bundle $T(X)$, is transversal to its zero section,
- $v|_{\partial X} \neq 0$,
- for each $k \in [1, j]$, the v-generated stratum $\partial_k X$ is a closed smooth submanifold of $\partial_{k-1} X$,
- the field v, viewed as section of the quotient 1-bundle

$$T_k^\nu =_{\mathsf{def}} T(\partial_{k-1} X)/T(\partial_k X) \to \partial_k X,$$

  is transversal to the zero section of T_k^ν for all $k \leq j$,
- the stratum $\partial_{j+1} X$ is the zero set of the section $v \in T_j^\nu$ [2],
- the stratum $\partial_{j+1}^+ X \subset \partial_{j+1} X$ is the locus where v points inside of $\partial_j^+ X$.

We denote the space of smooth boundary generic vector fields on X by the symbol $\mathcal{B}^\dagger(X)$. ♢

Remark 5.1. Replacing v with $-v$ affects the Morse stratification according to the formula:

$$\partial_j^+ X(-v) = \partial_j^\epsilon X(v),$$

where $\epsilon = +$ when $(n+1) - j \equiv 0 \mod 2$, and $\epsilon = -$ otherwise. ♢

We will postpone the proof of the theorem below until Chapter 6 (see Morin's Theorem 6.1, or [K3], Theorem 3.4, an extension of Theorem 5.1). There we will develop the needed analytical tools.

Theorem 5.1. *Boundary generic vector fields form an open and dense subset $\mathcal{B}^\dagger(X)$ in the space $\mathcal{V}(X)$ of all smooth fields on X.* ♢

Definition 5.2. We say that a smooth vector field v on X is *of the gradient type (or gradient-like)* if, for a smooth function $f : X \to \mathbb{R}$,

- the differential df and the field v vanish on the same locus $Z \subset X$,
- the function $df(v) > 0$ in $X \setminus Z$,
- in the vicinity of Z, there exist a Riemannian metric g on X so that $v = \nabla_g f$, the gradient field of f in the metric g. ♢

[1] So $\partial_0 X$ and $\partial_1 X$—the base of induction—do not depend on v.
[2] Thus if all $\{\partial_k X\}_{\{k \leq j\}}$ are smooth manifolds, then by the transversality of the section v, so is $\partial_{j+1} X$.

Definition 5.3. A smooth function $f : X \to \mathbb{R}$ is called *Morse function* if its differential df, viewed as a section of the cotangent bundle $T^*(X)$, is transversal to the zero section. $\diamondsuit$

Recall that, for a Morse function f on a compact $(n+1)$-manifold X, the critical set $Z =_{\mathsf{def}} \{x \in X|\ df_x = 0\}$ is finite, and each point $x \in Z$ has special local coordinates $(x_1, \dots, x_{n+1})$ such that $df = \sum_{1 \leq i \leq n+1} a_i x_i dx_i$, where $a_i \neq 0$ for all i (for example, see [GG]).

Definition 5.4. Let $f : X \to \mathbb{R}$ be a smooth function and v its gradient-like vector field. We say that the pair (f, v) is *boundary generic* if the field v is boundary generic in the sense of Definition 5.1 and the restrictions of f to each stratum $\partial_j X =_{\mathsf{def}} \partial_j X(v)$ are Morse functions for all $0 \leq j \leq n$. $\diamondsuit$

Lemma 5.1. *Let V be a compact smooth manifold, and Y a smooth manifold, which is stratified by submanifolds $\{Y_j\}_j$. Let $\mathcal{Z} = \mathcal{Z}(V, Y)$ be the space of smooth maps $\Psi : V \to Y$ that are transversal to each stratum Y_j. Put $V_j^\Psi =_{\mathsf{def}} \Psi^{-1}(Y_j)$. Next consider the space $\mathcal{X} = \mathcal{X}(V, Y)$ of pairs (f, Ψ) such that $\Psi \in \mathcal{Z}$ and $f : V \to \mathbb{R}$ has the property: $\{f|_{V_j^\Psi}\}_j$ are Morse functions for all j.*

Then $\mathcal{X}$ is open and dense in the space $C^\infty(V, Y \times \mathbb{R})$.

Proof. Consider the space $(T^*V) \times Y$, where T^*V denotes the cotangent bundle of V. The property $(f, \Psi) \in \mathcal{X}$ is equivalent to the property of the section df of the bundle $T^*V \times Y \to V \times Y$ to be transversal to each (transversal) intersection of the Ψ-graph $\Gamma_\Psi \subset V \times Y$ with each stratum $V \times Y_j$. The latter property defines a open set in $C^\infty(V, Y \times \mathbb{R})$.

In order to validate density of $\mathcal{X}$ in $C^\infty(V, Y \times \mathbb{R})$, we first perturb a given map $\Psi : V \to Y$ to make it transversal to each stratum $Y_j \subset Y$, and then perturb a given function $f : V \to \mathbb{R}$ to make the section df of T^*V transversal to each manifold $V_j^\Psi = \Psi^{-1}(Y_j)$. $\square$

Theorem 5.2. *The boundary generic*[3] *Morse pairs (f, v) on a compact manifold X form an open and dense subset in the space of all smooth functions $f : X \to \mathbb{R}$ and their gradient-like fields v.*

---

[3]in the sense of Definition 5.1

Proof. By Theorem 5.1, the boundary generic fields v form an open and dense set in the space of all fields.

Let F^n be a complete flag in $\mathbb{R}^n$, formed by subspaces F_j of codimension j. In the proof of Theorem 6.5 (see also Theorem 3.4 from [K2]), for every field v, we have constructed a smooth map $\Psi^\partial(v) : \partial_1 X \to \mathbb{R}^n$ such that $\partial_j X(v) = \Psi^\partial(v)^{-1}(F_j)$. Moreover, $\Psi^\partial(v)$ is transversal to each F_j, if and only if, v is a boundary generic field. The construction of the map $\Psi^\partial(v)$ utilizes high order Lie derivatives $\{\mathcal{L}_v^j\}_{0 \leq j \leq n}$ in the direction of v of an auxiliary function $z : \hat{X} \to \mathbb{R}$, as in Lemma 6.1.

Now the property of boundary generic Morse pairs (f, v) to be open and dense in the space of all pairs follows from Lemma 5.1: just let $V = \partial_1 X$, $Y = \mathbb{R}^n$, $Y_j = F_j$, and $\Psi = \Psi^\partial(v)$ in that lemma.

For the reader's convenience, let us sketch an alternative argument that establishes just the density of boundary generic Morse pairs (f, v) in the space of all pairs. It does not rely on the construction of the map $\Psi^\partial(v)$.

We start with a pair (f, v) where $v|_{\partial X} \neq 0$ and $df(v) > 0$ at the points of the set where $v \neq 0$. By a small perturbation of f, we may assume the f is a Morse function on X and v its gradient-like field.

Let $K \supset \partial X$ be a compact regular neighborhood of ∂X in X, so small that $v|_K \neq 0$. By Theorem 5.1, we can perturb v to a new vector field $\tilde{v}$ so that $\tilde{v}$ is boundary generic in the sense of Definition 5.1 and still $\tilde{v}|_K \neq 0$.

For a given f, the condition $df(u)|_K > 0$ defines an open cone in the space of all fields u, subject to the constraint $u|_K \neq 0$. Therefore $\tilde{v}$ can be chosen both boundary generic and gradient-like for $f|_K$. When $\tilde{v}|_K$ is fixed, so are the stratifications $\{\partial_j^+ X(\tilde{v}) \subset \partial_j X(\tilde{v})\}_j$.

Next, with $\tilde{v}|_K$ being fixed, we perturb f again to a new function $\tilde{f}$ so that $d\tilde{f}(\tilde{v})|_K > 0$ and $\{f|_{\partial_j X(\tilde{v})}\}$ are Morse function for all j. The perturbation will be supported in the compact K. We start constructing $\tilde{f}$ inductively, first by adjusting it on the 1-manifold $\partial_n X(\tilde{v})$ and then moving sequentially to the strata $\partial_j X$ with lower indices j. We pick each perturbation $\tilde{f}$ so small that the open condition $d\tilde{f}(\tilde{v})|_K > 0$ is not violated. The existence of the desired j-th perturbation is based on the fact that Morse functions on a compact manifold Y (in this case, on $\partial_j X(\tilde{v})$) form an open and dense subset in $C^\infty(Y)$, the space of all smooth functions on Y with the Whitney topology. Note that since $\tilde{v}$ is tangent to $\partial_j X(\tilde{v})$ along $\partial_{j+1} X(\tilde{v})$ and $d\tilde{f}(\tilde{v})|_{\partial_{j+1} X(\tilde{v})} > 0$, the restriction $\tilde{f}|_{\partial_j X(\tilde{v})}$ has no critical points in the vicinity of $\partial_{j+1} X(\tilde{v})$. Thus we need to perturb $\tilde{f}|_{\partial_j X(\tilde{v})}$ only on a compact subset $Q_j \subset \partial_j X(\tilde{v})$ which has an empty intersection with

$\partial_{j+1}X(\tilde{v})$. This perturbation extends smoothly from Q_j to X. Eventually, we reach the upper stratum $\partial_0 X =_{\mathsf{def}} X$, thus constructing a boundary generic approximation of the given pair (f, v).

All the perturbations $(\tilde{f}, \tilde{v})$ of (f, v) we have introduced so far, but the first one, are supported in K, where $\tilde{v} \neq 0$ and $d\tilde{f}(\tilde{v}) > 0$. This proves that the boundary generic pairs form a *dense* set in the space of all pairs (f, v), where v being a f-gradient-like field, subject to the constraints: $v|_{\partial X} \neq 0$, and $f : X \to \mathbb{R}$ being a Morse function. □

For a given Morse pair (f, v), we denote by $\Sigma_j \subset \partial_j X(v)$ the set of critical points of the function $f| : \partial_j X \to \mathbb{R}$. For a boundary generic Morse pair (f, v), the finite critical set Σ_j is divided into two complementary sets: the set $\Sigma_j^+ \subset \partial_j^+ X$ of positive critical points and the set $\Sigma_j^- \subset \partial_j^- X$ of negative ones (see Figure 2.1).

Remark 5.2. Note that when $\partial_j^+ X \neq \emptyset$, it may happen that $\Sigma_j^+ = \emptyset$. However, if a component $\partial_j^+ X_\alpha$ of $\partial_j^+ X$ is a *closed* manifold, then $f : \partial_j^+ X_\alpha \to \mathbb{R}$ must have local extrema, in which case $\Sigma_j^+ \neq \emptyset$. ◇

Consider a generic vector field v and a Riemannian metric g on X. We denote by v_j the orthogonal projection of the field v on the tangent space $T(\partial_j X)$. Note that if v is a gradient field for a function $f : X \to \mathbb{R}$ in metric g, then v_j is automatically a gradient field for the restriction $f|_{\partial_j X}$ and metric $g|_{\partial_j X}$.

Take a smooth vector field v on a compact $(m+1)$-manifold Y with isolated singularities $\{y_\star \in \Sigma(v) \subset \mathrm{int}(Y)\}$. We denote by $\mathsf{ind}_{y_\star}(v)$ the localized index of the vector field v at its typical singular point $y_\star$. In a local chart, $\mathsf{ind}_{y_\star}(v)$ is crudely defined as the degree of a map $G_v : S^m_{y_\star} \to S^m$ from a small $y_\star$-centered m-sphere to the unit m-sphere. The map takes each point $a \in S_{y_\star}$ to the point $v(a)/\|v(a)\| \in S^m$.

We define the "global" index $\mathsf{Ind}(v)$ as the sum $\sum_{y_\star \in \Sigma(v)} \mathsf{ind}_{y_\star}(v)$.

For a boundary generic vector field v and a Riemannian metric g on X, we form the fields $\{v_j\}$ on $\{\partial_j X(v)\}$ and define the global index of v_j by the formula:

$$\mathsf{Ind}^+(v_j) =_{\mathsf{def}} \sum_{\{x_\star \in \Sigma_j^+\}} \mathsf{ind}_{x_\star}(v_j).$$

Let us revisit the beautiful Morse formulas [Mo]:

Theorem 5.3 (The Morse Law of Vector Fields).
For a boundary generic vector field v and a Riemannian metric on a $(n+1)$-manifold X, such that the singularities of the fields v_j are isolated for all $j \in [0, n+1]$, the following two equivalent sets of formulas hold:

$$\chi(\partial_j^+ X) = \mathsf{Ind}^+(v_j) + \mathsf{Ind}^+(v_{j+1})$$

$$\mathsf{Ind}^+(v_j) = \sum_{k=j}^{n+1} (-1)^k \chi(\partial_k^+ X), \tag{5.2}$$

where $\chi(\sim)$ stands for the Euler number of the appropriate space[4]. ◇

For vector fields with symmetry, the Morse Law of Vector Fields has an *equivariant* generalization [K9]. We devote Chapter 9 to the equivariant Morse Law of Vector Fields.

Morse formula (5.2) has an instant, but significant implication:

Corollary 5.1. *Let N be a neighborhood of the zero set of a smooth vector field v on a compact smooth $(n+1)$-manifold X, the boundary ∂N being smooth. Assume that v is boundary generic with respect to both boundaries, ∂X and ∂N. Then*

$$\mathsf{Ind}(v) = \sum_{j=0}^{n+1} (-1)^j \chi(\partial_j^+ N) = \sum_{j=0}^{n+1} (-1)^j \chi(\partial_j^+ X).$$

◇

Remark 5.3. Therefore, the numbers

$$\sum_{j=0}^{n+1} (-1)^j \chi(\partial_j^+ N) \text{ and } \sum_{j=0}^{n+1} (-1)^j \chi(\partial_j^+ X)$$

can serve as "*more* and *less localized*" *definitions* of the index invariant $\mathsf{Ind}(v)$. ◇

An interesting discussion, connected to Theorem 5.3, its social, topological, and geometrical implications, can be found in the popular paper of Daniel Gotlieb [Got]. For us, reading [Got] was the starting point of the entire project, described in this book. The paper [Got1] contains significant generalizations of [Got]. The "*Topological Gauss-Bonnet Theorem*" below is just a sample of these results.

[4]By definition, $\mathsf{Ind}^+(v_{n+1}) = |\Sigma_{n+1}^+|$ and $\mathsf{Ind}^+(v_{n+2}) = 0$.

For a given smooth map $F : X \to \mathbb{R}^{n+1}$, which is an immersion in the vicinity of $\partial_1 X$, the *Gauss map* $G : \partial_1 X \to S^n$ is produced by taking each point $x \in \partial_1 X$ to the unit vector $\nu_{F(x)}$, normal to the appropriate local branch of $F(\partial_1 X)$ at the point $F(x)$, and shifted to the origin $0 \in \mathbb{R}^{n+1}$. The normal $\nu_{F(x)}$ points inside the half-space picked by $DF(n_x)$, where n_x is outward normal to $\partial_1 X$ in X.

Theorem 5.4 (Gotlieb). *Let X be a compact smooth $(n+1)$-dimensional manifold, and $\Phi : X \to \mathbb{R}^{n+1}$ a smooth map, which is an* immersion *in the vicinity of $\partial_1 X$. Let g be a Riemannian metric on X, which, in the vicinity of $\partial_1 X$, is the pull-back $\Phi^*(g_{\mathsf{E}})$ of the Euclidean metric on $\mathbb{R}^{n+1}$. Consider a generic submersion $\ell : \mathbb{R}^{n+1} \to \mathbb{R}$ such that the composite function $f =_{\mathsf{def}} \ell \circ \Phi$ has only isolated singularities, residing in the interior of X. Let $v =_{\mathsf{def}} \nabla_g f$ be the gradient field of f*[5]. *Assume that v is boundary generic on X.*

Then the degree of the Gauss map

$$G : \partial_1 X \to \Phi(\partial_1 X) \to S^n$$

can be calculated either

(1) *by integrating over $\partial_1 X$ the normal curvature K_ν (in the metric g) of the hypersurface $\partial_1 X \subset X$, or*

(2) *in terms of the v-induced stratification*

$$\partial_1^+ X(v) \supset \partial_2^+ X(v) \supset \cdots \supset \partial_{n+1}^+ X(v)$$

by the formula[6]

$$\deg(G) = \frac{1}{\mathrm{vol}(S^n)} \int_{\partial_1 X} K_\nu \, d\mu_g = \chi(X) - \mathsf{Ind}(v)$$
$$= -\sum_{j=1}^{n+1} (-1)^j \chi(\partial_j^+ X(v)). \tag{5.3}$$

◇

Example 5.1. Let X be an orientable surface of genus g with a single boundary component. We pick an immersion $\Phi : X \to \mathbb{R}^2$ and a linear function $l : \mathbb{R}^2 \to \mathbb{R}$. Let $f : X \to \mathbb{R}$ be the composite function $f = l \circ \Phi$. If the vector field $v =_{\mathsf{def}} \nabla f$ is as in Theorem 5.4, then the statement of Theorem 5.4 is spelled out in Theorem 1.1. ◇

[5] Thus v is a *transfer* by Φ of the gradient field $\nabla_{g_{\mathsf{E}}} l$.

[6] Recall that $\mathrm{vol}(S^n) = 2\pi^{\frac{n+1}{2}} / \Gamma(\frac{n+1}{2})$.

Corollary 5.2. *Let $(n+1)$-manifolds $X \subset \hat{X}$, the map $\Phi : \hat{X} \to \mathbb{R}^{n+1}$, the metric g on $\hat{X}$, the function $\ell : \mathbb{R}^{n+1} \to \mathbb{R}$, the function $f = \ell \circ \Phi : \hat{X} \to \mathbb{R}$, and the vector field $v = \nabla_g f$ be as in Theorem 5.4. Assume that v is boundary generic relative to $\partial_1 X$.*

For each $j > 0$, consider a ϵ-small tubular neighborhood U_j of the manifold $\partial_j X(v)$ in $\hat{X}$. Then $\Phi : U_j \to \mathbb{R}^{n+1}$ is an immersion. This setting gives rise to the Gauss map $G_j : \partial U_j \to S^n$, defined by the formula $G_j(x) = D_x\Phi(\nu_x)/\|D_x\Phi(\nu_x)\|$, where $x \in \partial U_j$ and ν_x is the unit vector, inward normal to ∂U_j at x.

Then the degree of the Gauss map G_j can be calculated either

(1) *by integrating (with respect to the n-measure μ_g) over ∂U_j the normal curvature K_ν of the hypersurface $\partial U_j \subset \hat{X}$, or*

(2) *in terms of the v-induced stratum $\partial_j X(v)$ by the formula*

$$\deg(G_j) = \frac{1}{\mathrm{vol}(S^n)} \int_{\partial U_j} K_\nu \, d\mu_g = \chi(\partial_j X(v)). \tag{5.4}$$

Proof. We will apply Theorem 5.4 to the field v in U_j to conclude that

$$\deg(G_j) = \frac{1}{\mathrm{vol}(S^n)} \int_{\partial U_j} K_\nu \, d\mu_g = \chi(U_j) - \mathsf{Ind}(v).$$

Since $v \neq 0$ in U_j, $\mathsf{Ind}(v) = 0$, and the last term of this equation reduces to $\chi(U_j) = \chi(\partial_j X(v))$. □

Remark 5.4. Of course, for an odd-dimensional $\partial_j X(v)$, the Euler number $\chi(\partial_j X(v)) = 0$, and so is $\deg(G_j)$. When $\partial_j X(v)$ is even-dimensional (i.e., $n + 1 - j = 2l$), the integral in equation (5.4) can be expressed in terms of the *intrinsic* Riemannian geometry of the manifold $\partial_j X(v)$, namely, in terms of the *Pfaffian* $Pf(\Omega)$. The Pfafian is a $2l$-differential form, whose construction utilizes the curvature tensor on the manifold (see [MiS]). So, when $j = n + 1 - 2l$,

$$\deg(G_j) = \frac{1}{\mathrm{vol}(S^n)} \int_{\partial U_j} K_\nu \, d\mu_g = (2\pi)^{-l} \int_{\partial_j X(v)} Pf(\Omega) = \chi(\partial_j X(v)).$$

◇

Given a boundary generic vector field v on X, we introduce a sequence of basic degree-type invariants $\{d_k(v)\}$ which are intimately linked, via the Morse formula (5.2), to the invariants $\{\chi(\partial_j^+ X(v))\}$. For 3-folds, we are already familiar with one such invariant, namely, $\deg(h) = d_2(v)$ (see Lemma 2.2 and the discussion that precedes it).

We use a Riemannian metric g on X to produce the orthogonal projection v_j of the vector field v on the tangent subspace $T(\partial_j X(v)) \subset T(X)$.

Let $S(\partial_{k-1}X)$ be the bundle of unit $(n+1-k)$-spheres, associated with the tangent bundle of the manifold $\partial_{k-1}X$. We denote by $S_\#(\partial_{k-1}X)$ the restriction of the bundle $S(\partial_{k-1}X) \to \partial_{k-1}X$ to the subspace $\partial_k X \subset \partial_{k-1}X$.

For each k, consider two vector fields, v and the inward normal field ν_k to $\partial_k X$ in $\partial_{k-1}^+ X$, as *sections* of the sphere bundle $p_k : S_\#(\partial_{k-1}X) \to \partial_k X$ (remember, $v \neq 0$ is tangent to $\partial_{k-1}X$ along $\partial_k X$ so that $v = v_{k-1}$ along $\partial_k X$!). Assume that the sections v and ν_k are transversal in the space $S_\#(\partial_{k-1}X)$. This transversality can be achieved by a perturbation of ν_k (equivalently, by a perturbation of the metric g), supported in the vicinity of the singularity locus Σ_k^+. Indeed, the intersections occur where the field v_{k-1} is positively proportional to ν_k, that is, where $v_k = 0$. The later locus is exactly the locus Σ_k^+. That perturbation does not affect the stratification $\{\partial_j^+ X\}_j$. Assuming the transversality of the intersection, the locus $v(\partial_k X) \cap \nu_k(\partial_k X) \subset S_\#(\partial_{k-1}X)$ is zero-dimensional.

We define the integer $d_k(v)$ as the *algebraic* intersection number $v \circ \nu_k$ of two $(n+1-k)$-cycles, $v(\partial_k X)$ and $\nu_k(\partial_k X)$, in the ambient oriented manifold $S_\#(\partial_{k-1}X)$ of dimension $2(n+1-k)$.

Lemma 5.2. *For a boundary generic vector field v on a Riemannian manifold X, the following formula holds:*

$$d_k(v) = \mathsf{Ind}^+(v_k) = \sum_{j=k}^{n+1} (-1)^j \chi(\partial_j^+ X).$$

Proof. We already have noticed that the intersection set $v(\partial_k X) \cap \nu_k(\partial_k X)$ of the two sections projects bijectively under the map $p_k : S_\#(\partial_{k-1}X) \to \partial_k X$ onto the locus Σ_k^+, where the component v_k of v vanishes and v points inward of $\partial_{k-1}^+ X$. It takes a bit more work to see that the sign, attached to the transversal intersection point $a \in v(\partial_k X) \cap \nu_k(\partial_k X)$, is $(-1)^{\mathsf{ind}(p_k(a))}$, where $\mathsf{ind}(p_k(a))$ is the index (the localized degree) of the field v_k in the vicinity of its singularity $p_k(a) \in \Sigma_k^+$. Thus $d_k(v) =_{\mathsf{def}} v \circ \nu_k = \mathsf{Ind}^+(v_k)$. By the Morse Formula (5.2), the claim of the lemma follows. □

Corollary 5.3. *The integer $d_k(v) = \sum_{j=k}^{n+1}(-1)^j\chi(\partial_j^+ X)$ depends only on on the local indices of the vector field v_k at the points of the singular locus Σ_k^+.* ◇

Question 5.1. *How to compute the invariant $d_j(v)$ in the terms of Riemannian Geometry and in the spirit of Theorem 5.4 and Corollary 5.2?* ◇

For a boundary generic vector field v and a fixed Riemannian metric g on X, each manifold $\partial_j X(v)$ comes equipped with a *preferred normal framing* fr_j of the normal bundle $\nu(\partial_j X(v),\, \partial_1 X)$: just consider the unitary inward normal field ν_1 of $\partial_j X(v)$ in $\partial_{j-1}^+ X(v)$, then the unitary inward normal field ν_2 of $\partial_{j-1}X(v)$ in $\partial_{j-2}^+ X(v)$, being restricted to $\partial_j X(v)$, then the unitary inward normal field ν_3 of $\partial_{j-2}X(v)$ in $\partial_{j-3}^+ X(v)$, being restricted to $\partial_j X(v)$, and so on...

Via the Pontryagin construction [Po], this framing fr_j generates a continuous map $G_j(v, g) : \partial_1 X \to S^{j-1}$. Its homotopy class $[G_j(v, g)]$ is an element of the *cohomotopy set* $\pi^{j-1}(\partial_1 X)$. If $\partial_j X(v) = \emptyset$, then we define $G_j(v, g) : \partial_1 X \to S^{j-1}$ to be the trivial map that takes $\partial_1 X$ to the base point in S^{j-1}.

Unfortunately, as we will see soon, the homotopy class $[G_j(v, g)] = 0$! However, when $\partial_{j+1}X(v) = \emptyset$, each of the two loci $\partial_j^\pm X(v)$ is a closed manifold. Then we can apply the Pontryagin construction only to, say, $\partial_j^+ X(v)$ to get a map $G_j^+(v, g) : \partial_1 X \to S^{j-1}$. This application leads directly to the following proposition.

Corollary 5.4. *Consider a boundary generic vector field v on X such that $\partial_{j+1}X(v) = \emptyset$*[7]*. Let g be a Riemannian metric, defined in the vicinity of $\partial_1 X$ in X. Then these data give rise to continuous map $G_j^+(v, g) : \partial_1 X \to S^{j-1}$.*

The homotopy class $[G_j^+(v, g)] \in \pi^{j-1}(\partial_1 X)$ is independent of the choice of g and the homotopy class of v within the open subspace of the space $\mathcal{B}^\dagger(X)$, defined by the constraint $\partial_{j+1}X(v) = \emptyset$.

In particular, when $\partial_3 X(v) = \emptyset$, we get an element

$$[G_2^+(v)] \in \pi^1(\partial_1 X) \approx H^1(\partial_1 X; \mathbb{Z}),$$

and when $\partial_4 X(v) = \emptyset$, an element

$$[G_3^+(v)] \in \pi^2(\partial_1 X) \approx H^2(\partial_1 X; \mathbb{Z}).$$

[7] In Definition 5.5, such vector fields are called both $(j+1)$-convex and $(j+1)$-concave.

If $\partial_1 X = S^n$, we can interpret $[G_j^+(v)]$ also as an element of the homotopy group $\pi_n(S^{j-1})$. $\diamondsuit$

The elements $[G_j(v)]$ and $[G_j^+(v)]$ have another classical interpretation as elements of *oriented framed cobordism set* $\Omega^{\mathsf{fr}}_{n-j+1}(\partial_1 X)$. In fact, the pair $(\partial_j X(v), \mathsf{fr}_j)$ defines the trivial element in $\Omega^{\mathsf{fr}}_{n-j+1}(\partial_1 X)$. In contrast, if $\partial_{j+1} X(v) = \emptyset$, then the bordism class $(\partial_j^+ X(v), \mathsf{fr}_j)$ may be nontrivial.

Let us recall the definition of framed cobordisms (for example, see [Kos]). Let $M_0, M_1 \subset Y$ be oriented closed smooth m-dimensional submanifolds of a compact $(m+k)$-manifold Y, whose normal bundles $\nu(M_0, Y)$ and $\nu(M_1, Y)$ are equipped with framings fr_0 and fr_1, respectively.

We say that two pairs (M_0, fr_0) and (M_1, fr_1) define the same element in the bordism group $\Omega^{\mathsf{fr}}_m(Y)$, if there is a compact $(m+1)$-dimensional oriented submanifold $W \subset Y \times [0,1]$ whose normal bundle $\nu(W, Y \times [0,1])$ admits a framing Fr so that:

(1) $\partial W = (M_1 \times \{1\}) \coprod (-M_0 \times \{0\})$,

(2) the restriction of Fr to $M_1 \times \{1\}$ coincides with fr_1, and the restriction of Fr to $M_0 \times \{0\}$ coincides with fr_0.

Then the Pontryagin construction establishes a bijection $P : \Omega^{\mathsf{fr}}_m(Y) \to \pi^k(Y)$, where $m + k = \dim Y$. If $m < k - 1$, both sets admit the structure of abelian groups and the bijection P becomes a group isomorphism.

Now we are in a better position to explain why $[G_j(v)] = 0$.

We add a collar to X to form a larger manifold $\hat{X}$ properly containing X. We fix a product structure $\partial_1 X \times [-1,1]$ in a regular neighborhood U of $\partial_1 X$ in $\hat{X}$ so that $\partial_1 X = \partial_1 X \times \{0\}$ and $X \cap U = \partial_1 X \times [0,1]$. Then we extend the metric $g|_{\partial_1 X}$ to a product metric g on U. Consider the obvious embedding

$$\alpha : \partial_1^+ X(v) \subset \partial_1 X \times \{0\} \subset \partial_1 X \times [-1,1] \approx U.$$

Fix $j > 1$. Using the product structure in U, we can isotop $\alpha(\partial_1 X)$ in U to a regular embedding $\beta : \partial_1 X \subset U$ such that:

(1) the isotopy deforms $\alpha(\partial_j^+ X(v))$ inside $\alpha(\partial_j^+ X(v)) \times [0,1] \subset U$ to a regular embedding $\beta : \partial_j^+ X(v) \subset \alpha(\partial_j^+ X(v)) \times [0,1]$,

(2) $\beta|_{\partial_{j+1} X(v)} = \alpha|_{\partial_{j+1} X(v)}$,

(3) the hypersurfaces $\alpha(\partial_j^+ X(v))$ and $\beta(\partial_j^+ X(v))$ are orthogonal along $\alpha(\partial_{j+1} X(v))$.

We denote by μ_{j+1} the unit vector field, g-normal to $\beta(\partial_j^+ X(v))$ in $\alpha(\partial_j^+ X(v)) \times [0,1]$. Similarly, let ν_{j+1} stand for the g-normal to $\alpha(\partial_{j+1} X(v))$ in $\alpha(\partial_j^+ X(v))$ unit vector field.

By the properties (2) and (3) of β, the field ν_{j+1} on $\alpha(\partial_{j+1}X(v))$ extends smoothly to the field μ_{j+1}. In addition, we consider the images $\mu_2, \mu_3, \dots \mu_j$ of the fields $\nu_2, \nu_3, \dots \nu_j$ along $\alpha(\partial^+_{j+1}X(v))$ under the differential map $D\beta$.

So, for any $j \geq 1$, the normal framing fr_{j+1} of $\alpha(\partial_{j+1}X(v))$ in $\alpha(\partial_1 X)$ extends to a normal framing fr^β_j of $\beta(\partial^+_j X)$ in $\partial_1 X \times [0,1]$. Therefore $[G_j(v)] = 0$ as an element of the framed bordisms of $\partial_1 X$. As a result, when $\partial_{j+1}X(v) = \emptyset$, we get $[G^+_j(v)] = -[G^-_j(v)]$ in $\Omega^{\mathsf{fr}}_{n-j+1}(\partial_1 X)$ (equivalently, in $\pi^{j-1}(\partial_1 X)$).

5.3 Deforming the Morse stratification

Let X be a smooth compact $(n+1)$-manifold with boundary ∂X. A boundary generic vector field v (see Definition 5.1) gives rise to two stratifications (5.1). We are going to investigate how the stratification $\{\partial^\pm_j X(v)\}_j$ changes as a result of deforming the vector field v.

Lemma 5.3. *Let $N \subset Y$ be a locally closed submanifold of an open manifold Y, and let M a closed manifold. Consider a family of maps $\{f_t : M \to Y\}_{t\in[0,1]}$ such that each f_t is transversal to N. All the manifolds, maps, and families of maps are assumed to be smooth.*

Then all the submanifolds $\{f_t(M)\cap N\}$ are isotopic in N. In particular, the intersections $f_0(M)\cap N$ and $f_1(M)\cap N$ are diffeomorphic.

Proof. Let $F : M \times [0,1] \to Y$ be the map, defined by the family $\{f_t\}$. Thanks to the transversality hypothesis, F is transversal to N and $F^{-1}(N)$ is a submanifold of $M \times [0,1]$ whose boundary is

$$f_0^{-1}(N) \sqcup f_1^{-1}(N) \subset M \times \partial([0,1]).$$

Let w be a non-vanishing vector field on $F^{-1}(N)$, normal to each codimension 1 submanifold $f_t^{-1}(N)$ in $F^{-1}(N)$. In the construction of w, we rely on the property of each f_t being transversal to N. Since $\partial(F^{-1}(N)) = f_0^{-1}(N) \sqcup f_1^{-1}(N)$ and $w \neq 0$, each w-trajectory that originates at a point of $f_0^{-1}(N)$ must reach $f_1^{-1}(N)$ in finite time. Therefore, employing the w-flow, $F^{-1}(N)$ is diffeomorphic to $f_0^{-1}(N) \times [0,1]$, and the F-image of that product structure in $F^{-1}(N)$ defines a smooth isotopy between $f_0(M)\cap N$ and $f_1(M)\cap N$ in N. This isotopy extends to an ambient isotopy of N itself [Thom].

Note that these arguments fail in general when ether M or N have boundaries. However, under additional assumptions (such as $f_t|_{\partial M}$ being

t-independent and $f_t(M) \cap \partial N = \emptyset$), the relative versions of the lemma are valid. □

Theorem 5.5. *The smooth topological type of each stratum $\partial_j^\pm X(v)$ is constant within each path-connected component of the space $\mathcal{B}^\dagger(X)$ of boundary generic fields.*

Proof. If two boundary generic vector fields, v_0 and v_1, are connected by a continuous path $v : [0,1] \to \mathcal{B}^\dagger(X)$, then they can be connected by a path $\tilde{v} : [0,1] \to \mathcal{B}^\dagger(X)$ such that the dependence of the field $\tilde{v}(t)$ on $t \in [0,1]$ is smooth. The argument is based on the property of boundary generic fields to form an open set in the space of all fields (Theorem 5.1), the smooth partition of unity technique (which utilizes the compactness of manifold $X \times [0,1]$), and the standard techniques of approximating continuous functions with the smooth ones.

Thus it suffices to consider a smooth 1-parameter family of vector fields $v_t \in \mathcal{B}^\dagger(X)$, connecting v_0 to v_1. Since any boundary generic vector field $v_t|_{\partial_1 X}$, viewed as a section of the vector bundle $\eta_1 : TX|_{\partial_1 X} \to \partial_1 X$, is transversal its zero section, we may apply Lemma 5.3 (with $M = \partial_1 X$, N being the zero section of η, $Y = E(\eta_1)$, the total space of η_1, and $f_t = v_t$) to conclude that all the submanifolds $\{\partial_2 X(v_t)\}_t$ are isotopic in $\partial_1 X$.

Each $\partial_2 X(v_t)$ divides $\partial_1 X$ into a pair of complementary domains, $\partial_1^+ X(v_t)$ and $\partial_1^- X(v_t)$. Since their polarity $\pm$ is determined by the inward/outward direction of v_t which changes continuously with t, the ambient isotopy h_t of $\partial_1 X$ (taking $\partial_2 X(v_0)$ to $\partial_2 X(v_t)$) must take $\partial_1^+ X(v_0)$ to $\partial_1^+ X(v_t)$. The isotopy $h_t : \partial_1 X \to \partial_1 X$ extends to an isotopy $\tilde{h}_t : X \to X$.

A similar argument applies to lower strata $\partial_j^\pm X(v_t)$. Indeed, with the isotopy $h_t : \partial_1 X \to \partial_1 X$ that takes $\partial_2 X(v_0)$ to $\partial_2 X(v_t)$ in place, consider the two sections, v_0 and $(h_t^{-1})_*(v_t)$, of the bundle $\eta_2 : T(\partial_1 X)|_{\partial_2 X(v_0)} \to \partial_2 X(v_0)$, both sections being transversal to the zero section of η_2. Applying again Lemma 5.3, we conclude that the loci $\partial_3 X(v_0)$ and $h_t^{-1}(\partial_3 X(v_t))$ are isotopic in $\partial_2 X(v_0)$ (recall that these loci are exactly the transversal intersections of two sections, v_0 and $(h_t^{-1})_*(v_t)$, of η_2 with its zero section). Again, an isotopy $h'_t : \partial_2 X(v_0) \to \partial_2 X(v_0)$ that takes $\partial_3 X(v_0)$ to $h_t^{-1}(\partial_3 X(v_t))$ must take $\partial_2^+ X(v_0)$ to $h_t^{-1}(\partial_2^+ X(v_t))$. The isotopy h'_t extends to an isotopy $\tilde{h}'_t : X \to X$ which preserves the pair $\partial_2 X(v_0) \subset \partial_1 X$. So, the pairs $\partial_3 X(v_0) \subset \partial_2^+ X(v_0)$ and $\partial_3 X(v_t) \subset \partial_2^+ X(v_t)$ are diffeomorphic via the composite isotopy $\tilde{h}'_t \circ \tilde{h}_t$.

This reasoning can be recycled to prove that all the pairs $\partial_j^+ X(v_0)$ and $\partial_j^+ X(v_t)$ are diffeomorphic via a *single* isotopy of X. This argument will be presented explicitly in the proof of Theorem 6.6 (see also Theorem 3.4 from [K2]). □

Corollary 5.5. *Let X be a $(n+1)$-dimensional compact smooth manifold with boundary.*

Within each path-connected component of the space $\mathcal{B}^\dagger(X)$ of boundary generic fields, the numbers $\{d_k(v)\}_{0 \leq k \leq n}$ [8]*, as well as the numbers $\{\chi(\partial_k^\pm X(v))\}_{1 \leq k \leq n+1}$, are constant.*

Proof. The claim follows instantly from Theorem 5.5 and Lemma 5.2. □

Let X be a connected compact smooth manifold X with boundary. By deforming any given smooth function $f : X \to \mathbb{R}$ and its gradient-like field v, we can expel the isolated v-singularities from X. This can be achieved by the appropriate "finger moves" which originate at points of the boundary ∂X and engulf the isolated singularities of v. The result of these manipulations lead to the following lemma.

Lemma 5.4. *Any connected $(n+1)$-manifold X with a nonempty boundary admits a Morse function $f : X \to \mathbb{R}$ with no critical points in the interior of X and such that $f| : \partial X \to \mathbb{R}$ is a Morse function. Such functions form an open nonempty set in the space $C^\infty(X)$ of all smooth functions on X.*

As a result, the non-vanishing *gradient-like vector fields v on X form an open nonempty set in the space $\mathcal{V}(X)$ of all all vector fields on X.*

Proof. Let us sketch the main idea of the argument. Start with a Morse function $f : X \to \mathbb{R}$. Connect each critical point in the interior of X by a smooth path to a point on the boundary in such a way that a system of non-intersecting paths is generated. Then delete from X small regular neighborhoods of those paths ("we dig a system of dead-end tunnels") and restrict f to the remaining portion $X^\odot$ of X. Smoothen the entrances of the tunnels so that the boundary of $X^\odot$ will be a smooth manifold which is diffeomorphic to X. We got a nonsingular function f on $X^\odot$. A slight perturbation of f on $X^\odot$ will not introduce critical points in the interior of $X^\odot$ and will deliver a Morse function on its boundary. Indeed, recall

[8] introduced prior to Lemma 5.2

that the sets of Morse functions on X and on ∂X are open and dense in the spaces $C^\infty(X)$ and $C^\infty(\partial X)$ of all smooth functions, respectively (for example, see [GG]).

Of course $v \neq 0$ is an open condition imposed on a vector field on a compact manifold. On the other hand, if $df(v) > 0$, then any vector field v', sufficiently close to v, will have the property $df(v') > 0$. The previous arguments show that the set of gradient-like non-vanishing fields is nonempty. So it is an open nonempty subspace in the space $\mathcal{V}(X)$ of all all vector fields on X. □

Eliminating isolated critical points of a given function $f : X \to \mathbb{R}$ on a manifold with boundary is not "a free lunch" (see Figure 1.1): the elimination introduces new critical points of the restricted function $f : \partial X \to \mathbb{R}$. This is a persistent theme throughout this book:

Expelling critical points of gradient flows from a manifold X leaves crucial residual geometry on its boundary ∂X relative to the new flow.

Under the right conditions, this boundary-confined geometry allows for a reconstruction of the topology of X (see Theorem 6.12). Ideas like these will be developed elsewhere, in particular, in the forthcoming book [K13]. Meanwhile, the following lemma gives a taste of things to come.

Lemma 5.5. *Let $f : X \to \mathbb{R}$ be a Morse function with no local* extrema *in the interior of a $(n+1)$-manifold X. Then the elimination by a finger move*[9] *of each f-critical point $p_\star$ of the Morse index $i(p_\star)$ results in the introduction of $[2(n+1-i(p_\star)) - 1]$ new critical points of positive type and $2i(p_\star)+1$ new critical points of negative type for the modified function $\tilde{f} : \partial X \to \mathbb{R}$.*

Proof. Let $p_\star$ be a Morse singularity of f in the interior of X. Denote by $S_{p_\star}$ the sphere that bounds a small disk $D_{p_\star}$, centered on $p_\star$, and such that $f|_{S_{p_\star}}$ is a Morse function. Without loss of generality, we can assume that, in the Morse coordinates $\{x_i\}$, $S_{p_\star}$ is given by $\sum_{i=1}^{n+1} x_i^2 = 1$, while $f(x) = \sum_{i=1}^{n+1} a_i x_i^2$ with all the $\{a_i \neq 0\}$ being *distinct*. Then $f|_{S_{p_\star}}$ has only Morse-type singularities at the points where the coordinate axes pierce the sphere $S_{p_\star}$. With respect to the pair $(X \setminus D_{p_\star}, f)$, these points come in two flavors: positive and negative. The two types are separated by the hypersurface of the cone $C = \{\sum_{i=1}^{n+1} a_i x_i^2 = 0\}$. In the vicinity of $p_\star$, the

[9] as in the proof of Lemma 5.4

intersection $C \cap S_{p_\star}$ is exactly the locus

$$\partial_2(X \setminus D_{p_\star}) = \partial_2^+(X \setminus D_{p_\star}),$$

so that the f-gradient field v (tangent to $S_{p_\star}$ along $C \cap S_{p_\star}$) is transversal to $C \cap S_{p_\star}$, the product of two spheres. Therefore, in the vicinity of $x_\star$, $\partial_3(X \setminus D_{p_\star}) = \emptyset$!

The function $f|_{S_{p_\star}}$ has exactly $2 \cdot i(p_\star)$ critical points of the positive type and exactly $2(n+1-i(p_\star))$ critical points of the negative type. We shall denote these sets by $\Sigma_1^\mp(p_\star)$ and the two domains in which C divides $S_{p_\star}$—by $S_{p_\star}^\pm$.

Let $x \in S_{p_\star}^+$ be a local maximum of $f|_{S_{p_\star}}$. Note that it is possible to connect x to a nonsingular for $f|_{\partial X}$ point $y \in \partial X$ by a smooth path γ along which f is increasing. Indeed, any non-extendable path γ such that $df(\dot\gamma) > 0$ either approaches a critical point or reaches the boundary ∂X. By a small perturbation, we can insure that γ avoids all the (hyperbolic) critical points in the interior of X (by the hypothesis, f has no local maxima/minima in the interior of X). Thus γ can be extended until it reaches the boundary ∂X at a point y.

Drilling a narrow tunnel U, diffeomorphic to the product $\gamma \times D^n$, along γ does not change the topology of X; the function $f|_{X \setminus U}$ retains almost the same list of singularities at the boundary as the function $f|_{X \setminus D_{p_\star}}$ has: more accurately, the local maximum at $x \in S_{p_\star}^+$ disappears in $\partial(X \setminus U)$ and a *negative* critical point of index 1 of $f|_{\partial(X \setminus U)}$ appears near the y-end of the tunnel U. Thus we have modified f and have eliminated the critical point $p_\star$ in the interior of X at the cost of introducing on the boundary $2(n+1-i(p_\star)) - 1$ critical points of positive type and $2i(p_\star) + 1$ critical points of negative type. □

Soon, motivated by Lemma 5.4, we will restrict our attention to *nonsingular* functions $f : X \to \mathbb{R}$ and their gradient-like fields v—an open subset in the space of all gradient-like pairs (f, v); but for now, let us investigate a more general case.

Consider Morse data (f, v), where the field v is nonsingular along the boundary $\partial_1 X$. Extend (f, v) to $\hat{X} =_{\mathsf{def}} X \cup C$, where C is some external collar of $\partial_1 X$ so that the extension $(\hat{f}, \hat{v})$ is nonsingular in C. At each point $x \in \partial_1 X$, the $\hat{v}$-flow defines a projection p_x of the germ of $\partial_1 X$ into the germ of the hypersurface $\hat{f}^{-1}(\hat{f}(x))$.

Let $\partial_j X^\circ$ denote the *pure* stratum $\partial_j X \setminus \partial_{j+1} X$, and let $\partial_j^\pm X^\circ$ denote $\partial_j^\pm X \setminus \partial_{j+1} X$. At the points $x \in \partial_1 X^\circ$, p_x is a surjection; at the points of $x \in \partial_2 X^\circ$, it is a folding map; at the points $x \in \partial_3 X^\circ$, it is a cuspidal map. Often we will refer to points $x \in \partial_1 X$ by the smooth types of their p_x-projections.

As the theorem and the corollary below testify, for a given function $f : X \to \mathbb{R}$, we enjoy a considerable freedom to change the given Morse stratification $\{\partial_j^+ X =_{\mathsf{def}} \partial_j^+ X(v)\}$ by deforming the f-gradient-like field v (cf. Lemma 2.3 and Corollary 2.1).

Theorem 5.6. *Let X be a compact connected smooth $(n+1)$-manifold with boundary. Let $f : X \to \mathbb{R}$ be a smooth function with no critical points along ∂X, and let v be its gradient-like field. Consider a stratification*

$$X =_{\mathsf{def}} Y_0 \supset Y_1 \supset Y_2 \supset \cdots \supset Y_{n+1}$$

of X by compact smooth manifolds $\{Y_j\}$, and let S_j and S_j^∂ denote the critical sets of the restrictions $f|_{Y_j}$ and $f|_{\partial Y_j}$, respectively. Assume that the following properties are satisfied:

- $\dim(Y_j) = n + 1 - j$,
- $Y_1 \subset \partial X$ *and* $\{Y_j \subset \partial Y_{j-1}\}_{j \in [2,n+1]}$ *are regular embeddings,*
- *for each $j \leq n+1$ the functions $f|_{Y_j}$ and $f|_{\partial Y_j}$ have Morse-type critical points at the finite loci S_j and S_j^∂, respectively,*
- *at the points of S_j, $df(\nu) > 0$ and, at the points of $S_{j-1}^\partial \setminus S_j$, $df(\nu) < 0$, where ν is the inward normal to ∂Y_{j-1} in Y_{j-1}*[10]*.*

Then, within the space of f-gradient-like vector fields, there is a deformation of v into a new boundary generic gradient-like field $\tilde{v}$, such that the stratification $\{\partial_j^+ X(\tilde{v})\}_{1 \leq j \leq n+1}$ coincides with the given stratification $\{Y_j\}_{1 \leq j \leq n+1}$.

Proof. The arguments are similar to the ones employed in the proof of Corollary 2.1 and illustrated in Figure 2.5.

We pick a Riemannian metric g in a collar U of ∂X in X so that v becomes the gradient field of f. Consider auxiliary vector fields $\{v_j\}$, where v_j denotes the orthogonal projection of v on the tangent spaces of the manifold Y_j.

The construction of the desired field $\tilde{v}$ is inductive in nature, the induction being executed in increasing values of the index k. Figure 2.5 illustrates a typical inductive step.

[10] This condition is metric-independent.

Assume that $v =_{\mathsf{def}} \tilde{v}$ has been already constructed so that $\partial_j^+ X(v) = Y_j$ and $\Sigma_j^+(v) = S_j$ for all $j < k$. This assumption implies that v is tangent to Y_j exactly along its boundary ∂Y_j for all $j < k$. Along $\partial Y_{k-1} = \partial(\partial_{k-1}^+ X(v)) = \partial_k X(v)$ (and thus along $Y_k \subset \partial Y_{k-1}$), we decompose v as $v_k + \sum_{j=1}^{k-1} n_j$, where $n_j =_{\mathsf{def}} v_{j-1} - v_j$.

The idea is to modify v in the direction normal to $\partial_k X(v)$ in $\partial_{k-1} X(v)$, while keeping the rest of its components $\{n_j\}$ unchanged.

Denote by T_x the tangent space of Y_{k-1} at $x \in \partial Y_{k-1}$. Let T_x^+ be the open half-space of T_x, positively spanned by the vectors that point inside Y_{k-1}. Let $T_x^+(f)$ be half of the tangent space T_x, defined by $df(u) > 0$, where $u \in T_x$. We introduce the complementary to T_x^+ and $T_x^+(f)$ open half-spaces T_x^- and $T_x^-(f)$.

At each point $x \in Y_k$, consider the open cone $C_x^+ = T_x^+ \cap T_x^+(f)$ and, at each point $x \in \partial Y_{k-1} \setminus Y_k$, the open cone $C_x^- = T_x^- \cap T_x^+(f)$ (see Figure 2.5). These cones are nonempty, except perhaps at the points of S_{k-1}^∂, where $\pm v_{k-1}$ is anti-parallel to the inward normal ν_k of $\partial Y_{k-1} \subset Y_{k-1}$. However, at $x \in S_k$, $C_x^+ \neq \emptyset$, and at $x \in S_{k-1}^\partial \setminus S_k$, $C_x^- \neq \emptyset$ due to the last bullet in the hypotheses of the theorem. Thus, for each $x \in Y_k^\circ$, there is a number h so that the vector $u_k = v_k + h \cdot \nu_k \in C_x^+$ (this conclusion uses the the property $df(v_k) > 0$ on the set $Y_k \setminus S_k$). Similarly, for each $x \in \partial Y_{k-1} \setminus Y_k$, there is a number h so that $u_k \in C_x^-$. By the partition of unity argument, which employs convexity of the cones $C_x^\pm$, there is a smooth function $h : \partial Y_{k-1} \to \mathbb{R}$ which delivers the desired vector field u_k along ∂Y_k. In order to insure the continuity of h and u_k across the boundary $\partial Y_k \subset \partial Y_{k-1}$, we require $h|_{\partial Y_k} = 0$. Thus $u_k = v_k \neq 0$ on ∂Y_k.

Put $v' = u_k + \sum_{j=1}^{k-1} n_j$. Now, $\partial_j^+ X(v') = \partial_j^+ X(v) = Y_j$ for all $j < k$ (these strata depend on the n_j's only), and $\partial_k^+ X(v') = Y_k$ by the construction of u_k. Moreover, $\Sigma_j^+(v') = \Sigma_j^+(v) = S_j$ for all $j \leq k$. In fact, v' is tangent to Y_{k-1} along ∂Y_{k-1}. Note that this inductive argument should be modified for $k = n+1$, since $Y_{n+1} = S_{n+1}$ is 0-dimensional.

We smoothly extend v' into a regular neighborhood V of ∂Y_{k-1} in X. Abusing notations, we denote this extension by v' as well. The neighborhood V is chosen so that there $df(v') > 0$.

To complete the proof of the inductive step $k-1 \Rightarrow k$, we form the field $\tilde{v} =_{\mathsf{def}} \psi_0 v + \psi_1 v'$, where the functions $\{\psi_0, \psi_1\}$ deliver a smooth partition of unity, subordinate to the cover $\{X \setminus V, V\}$ of X. Since $df(\sim) > 0$ defines a convex cone in the space of vector fields, $\tilde{v}$ is a f-gradient-like field with the desired Morse stratification. □

Theorem 5.6 has an immediate implication:

Corollary 5.6. *Let $f : X \to \mathbb{R}$ be a Morse function and v its boundary generic gradient-like vector field with the Morse stratification $\{\partial_j^+ X(v) \subset \partial_j X(v)\}_{0 \leq j \leq n+1}$. Assume that compact codimension zero submanifolds $Y_j \subset \partial_j X(v)$ are chosen so that, for each j, $Y_j \supset \Sigma_j^+(v)$ and $Y_j \cap \Sigma_j^-(v) = \emptyset$.*

Then, within the space of f-gradient-like fields, it is possible to deform v into a new gradient-like boundary generic vector field $\tilde{v}$, such that the stratification $\{\partial_j^+ X(\tilde{v})\}_{1 \leq j \leq n+1}$ coincides with the given stratification $\{Y_j\}_{1 \leq j \leq n+1}$. Moreover, $\{\partial_j X(\tilde{v}) = \partial_j X(v)\}_{1 \leq j \leq n+1}$.

In particular, if $\Sigma_k^+(v) = \emptyset$, the claim is valid for any stratification $\{Y_j\}_{1 \leq j \leq n+1}$ as above that terminates with $Y_k = \emptyset$. ◇

The next proposition (based on Corollary 5.6) shows that, for a given Morse function $f : X \to \mathbb{R}$, by an appropriate choice of gradient-like field v, the Morse stratification $\partial_j^+ X$ can be made topologically very simple and regular: namely, each stratum $\partial_j^+ X$ is a disjoint union of $(n+1-j)$-dimensional disks. Moreover, when the boundary $\partial_1 X$ is connected and $j \in [1, n-1]$, each stratum $\partial_j^+ X$ is a just a single disk.

Corollary 5.7. *Let $f : X \to \mathbb{R}$ be a Morse function on a compact $(n+1)$ -manifold X, f being nonsingular along the boundary $\partial_1 X$. We divide the connected components $\{\partial_1 X_\alpha\}_\alpha$ of $\partial_1 X$ into two types,* A *and* B. *By definition, for type* A, *the singularity set $\Sigma_1^+(f) \cap \partial_1 X_\alpha \neq \emptyset$, and for type* B, *$\Sigma_1^+(f) \cap \partial_1 X_\alpha = \emptyset$.*

Then any f-gradient-like vector field v can be deformed, within the space of f-gradient-like vector fields, into a boundary generic vector field $\tilde{v}$ so that, for each component $\partial_1 X_\alpha$ of type A *and all $j < n$, the stratum $\partial_j^+ X(\tilde{v}) \cap \partial_1 X_\alpha$ is diffeomorphic to a disk D^{n+1-j}. At the same time, for the components of type* B *and all $j \geq 1$, the stratum $\partial_j^+ X(\tilde{v}) \cap \partial_1 X_\alpha = \emptyset$.*

In contrast, for the components of type A, *the 1-dimensional stratum $\partial_n^+ X(\tilde{v}) \cap \partial_1 X_\alpha$ is a finite union of arcs, residing in the circle $\partial_n X(\tilde{v}) \cap \partial_1 X_\alpha$. Moreover, $\chi(\partial_n^+ X(\tilde{v}))$, the number or arcs in $\partial_n^+ X(\tilde{v})$, and the number of points in $\partial_{n+1}^+ X(\tilde{v})$ are linked via the formula*

$$\left|\partial_{n+1}^+ X(\tilde{v})\right| = \chi(\partial_n^+ X(\tilde{v})) + (-1)^{n+1}\left[\mathsf{Ind}(v) - \chi(X)\right] + \frac{m}{2}\left[(-1)^{n+1} - 1\right],$$

where $\mathsf{Ind}(v) = \mathsf{Ind}(\tilde{v})$ is the index of the vector field v, and m is the number of boundary components of type A.

Proof. If $n \geq 2$, for each type A connected component $\partial_1 X_\alpha$ of $\partial_1 X$, the singularity set $\Sigma_1^+(f) \cap \partial_1 X_\alpha$ can be included in a disk $D_\alpha^n \subset \partial_1^+ X(v)$. By Corollary 5.6, we can deform v to a new f-gradient-like vector field v', so that the new stratum $\partial_1^+ X(v') \cap \partial_1 X_\alpha = D_\alpha^n$. If $n \geq 3$, then the singularity set $\Sigma^+(f|_{\partial D_\alpha^n})$ can be encapsulated in a single disk $D_\alpha^{n-1} \subset \partial D_\alpha^n$. By the same token, after still another deformation v'' of v', we can arrange for $\partial_1^+ X(v'') \cap \partial_1 X_\alpha = D_\alpha^n$ and $\partial_2^+ X(v'') \cap \partial_1 X_\alpha = D_\alpha^{n-1}$. This process repeats itself, unless the dimension of $\partial_j^+ X(\tilde{v}) \cap \partial_1 X_\alpha$ becomes one. At its final stage, $\partial_n^+ X(\tilde{v}) \cap \partial_1 X_\alpha$ consists of several arcs, which are contained in the circle ∂D_α^2.

For each type B connected component $\partial_1 X_\alpha$ of $\partial_1 X$, by a similar reasoning, we can arrange for $\partial_1^+ X(\tilde{v})_\alpha = \emptyset$. Thus, $\partial_j^+ X(\tilde{v}) \cap \partial_1 X_\alpha = \emptyset$ for all $j \geq 1$ and $\alpha \in \mathsf{B}$.

Therefore, letting $Y_j = \coprod_{\alpha \in \mathsf{A}} D_\alpha^{n+1-j}$ for all $j \in [1, n-1]$ in Corollary 5.6, we have established all the claims of the corollary, but the last one.

Since v and $\tilde{v}$ both are the gradient-like fields for the same Morse function f, their indices, $\mathsf{Ind}(v)$ and $\mathsf{Ind}(\tilde{v})$, are equal. Thus we get

$$\mathsf{Ind}(v) = \mathsf{Ind}(\tilde{v}) = \chi(X) + \frac{m}{2}\left[(-1)^{n+1} - 1\right] + (-1)^n\left[\chi(\partial_n^+ X(\tilde{v})) - \chi(\partial_{n+1}^+ X(\tilde{v}))\right],$$

where $\frac{m}{2}[(-1)^{n+1} - 1]$ is the contribution of all the disk-shaped strata $\{\partial_j^+ X(\tilde{v})\}_{1 \leq j < n}$ to the Morse formula (5.2). □

Recall that, by Corollary 2.5, for any 3-fold X and a boundary generic field $v \neq 0$ on it, we get

$$|\partial_3^+ X(v)| \geq 2\chi(X) - 2,$$

provided $\partial_1^+ X(v) \approx D^2$. Thus, as a positive $\chi(X)$ increases, the boundary of the disk $\partial_1^+ X(v)$ becomes more "wavy". For example, by attaching many 2-handles to the boundary of a 3-disk, we get a 3-fold X with a big value χ of $\chi(X)$. On such X, $|\partial_3^+ X(v)| > 2\chi - 2$ for any gradient-like field $v \neq 0$ such that $\partial_1^+ X(v) \approx D^2$.

These examples motivate the following:

Question 5.2. *For boundary generic gradient-like vector fields v with a fixed value i of the index $\mathsf{Ind}(v)$ and a disk-shaped stratification $\{\partial_j^+ X(v)\}_{1 \leq j < n}$ as in Corollary 5.7, what is the minimum $\mu(X, i)$ of $|\partial_{n+1}^+ X(v)|$?*

Is $\mu(X, i)$ semi-additive under the connected sum (boundary connected sum) operation: in other words, is the inequality

$$\mu(X_1 \# X_2, i_1 + i_2) \leq \mu(X_1, i_1) + \mu(X_2, i_2)$$

valid? ♢

Evidently, such a number $\mu(X, i)$ is an invariant of the diffeomorphism type of X.

5.4 Boundary convexity and concavity of vector fields

We are ready to introduce pivotal concepts of the *stratified convexity and concavity* for smooth vector fields on manifolds with boundary.

Definition 5.5. Given a boundary generic vector field v (see Definition 5.1), we say that v is boundary *s-convex*, if $\partial_s^+ X(v) = \emptyset$. In particular, if $\partial_2^+ X(v) = \emptyset$, we say that v is boundary 2-convex, or just *boundary convex.*

We say that v is boundary *s-concave*, if $\partial_s^- X(v) = \emptyset$. In particular, if $\partial_2^- X(v) = \emptyset$, we say that v is boundary 2-concave, or just *boundary concave.* ♢

Example 5.2. Assume that a compact manifold X, a 0-dimensional submanifold in the interior of a Riemannian manifold (Y, g), is given by an inequality $\{x : h(x) \geq 0\}$, where $h : Y \to \mathbb{R}$ is a smooth function with 0 being a regular value.

Recall that, in local coordinates $\{x^i\}$ on Y, the *Hessian* of h is given by the formula

$$Hess(h)_{ij} = \Big(\frac{\partial^2 h}{\partial x^i \partial x^j} - \sum_k \Gamma_{ij}^k \frac{\partial h}{\partial x^k}\Big) dx^i \otimes dx^j,$$

where Γ_{ij}^k are Christoffel symbols, determined by the metric g.

Then the boundary convexity of a gradient-like vector field v in X can be expressed in terms of the Hessian $Hess(h)$ by the inequality

$$\langle Hess_x(h)v(x),\ v(x)\rangle_g < 0$$

at all points x, where $v(x)$ is tangent to ∂X. If

$$\langle Hess_x(h)v(x),\ v(x)\rangle_g > 0,$$

where $v(x)$ is tangent to ∂X, then the field v is boundary concave. ♢

Example 5.3. Let f be a Morse function on a Riemannian manifold, and $v = \nabla f$ its gradient field. According to the argument in Lemma 5.5, the complement to a small convex (in the Morse coordinates, centered at a f-critical point) disk, is *boundary concave* with respect to v. In fact, the vector field v is both boundary 3-concave and 3-convex! So, if $f : Y \to \mathbb{R}$ is a Morse function on a *closed* manifold Y with a critical set Σ, then the complement X in Y to a small locally convex neighborhood of Σ admits a boundary concave $f|_X$-gradient-like vector field v with $\partial_3 X(v) = \emptyset$! $\diamondsuit$

Theorem 5.7 and Theorem 5.8 below belong to a family of results which we call "*holographic*". In such results, the intension is to reconstruct some structures on the "bulk" X (or even the space X itself) from the appropriate flow-generated structures ("observables") on its boundary ∂X (see [K4] and [K7] for a more complete realization of this line of investigations).

In Theorem 5.7, we describe how some *boundary-confined* interactions between the critical points of a given function $f : \partial_1 X \to \mathbb{R}$ of opposite polarities can serve as an indicator of the convexity/concavity of the gradient field ∇f in X (the convexity/concavity properties of the v-flow do require knowing the field in the *vicinity* of $\partial_1 X$ in X!). Recall that we are already familiar with the 3D-version of Theorem 5.7 (see Theorem 2.2).

Theorem 5.7. *Let $f : X \to \mathbb{R}$, $f| : \partial_1 X \to \mathbb{R}$ be Morse functions, and v and v_1 their gradient vector fields with respect to a Riemannian metric g on X and its restriction to $\partial_1 X$, respectively. Assume that v is boundary generic.*

- *If $\partial_2^{\pm} X(v) = \emptyset$, then there is no* ascending *v_1-trajectory $\gamma : \mathbb{R} \to \partial_1 X$, such that*

$$\lim_{t \to -\infty} \gamma(t) \in \Sigma_1^{\mp} \text{ and } \lim_{t \to +\infty} \gamma(t) \in \Sigma_1^{\pm}$$

  (both critical sets $\Sigma_1^{\pm}$ depend only on f).
- *Conversely, for a given f-gradient pair (v, v_1), if no such v_1-trajectory $\gamma \subset \partial_1 X$ exists, then one can deform (v, v_1) to a new boundary generic pair $(\tilde{v}, \tilde{v}_1)$ of the f-gradient type, so that $\partial_2^{\pm} X(\tilde{v}) = \emptyset$. Moreover, the fields v_1 and $\tilde{v}_1$ on $\partial_1 X$ can be chosen to be arbitrary close in the C^∞-topology.*

  In particular, if $f(\Sigma_1^+) < f(\Sigma_1^-)$ (as sets), then X admits a boundary generic and convex f-gradient-like field $\tilde{v}$; similarly, if $f(\Sigma_1^+) > f(\Sigma_1^-)$, then X admits a boundary generic and concave f-gradient-like field $\tilde{v}$.

Proof. First consider the convex case, that is, the relation between the property $\partial_2^+ X(v) = \emptyset$ and the absence of an ascending v_1-trajectory $\gamma : \mathbb{R} \to \partial_1 X$ that connects Σ_1^- to Σ_1^+.

Consider the function $h : \partial_1 X \to \mathbb{R}$, defined via the formula $v = v_1 + h \cdot n$, where n denotes a unitary vector field, inward normal to $\partial_1 X$ in X. Since v is boundary generic, zero is a regular value of h. Then

$$\partial_1^+ X(v) = h^{-1}([0, +\infty)), \quad \partial_1^- X(v) = h^{-1}((-\infty, 0])$$

and $\partial_2 X(v) = h^{-1}(0)$.

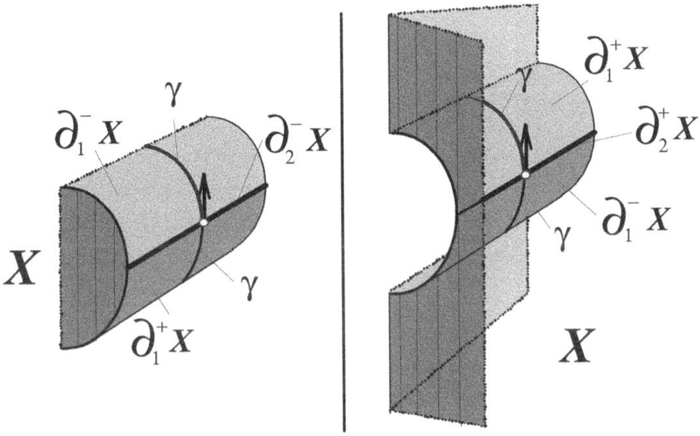

Fig. 5.1 The behavior of a gradient field v in the vicinity of a point from $\partial_2^- X(v)$ (the left diagram) and in the vicinity of a point from $\partial_2^+ X(v)$ (the right diagram), as perceived by an observer moving along an ascending v_1-trajectory $\gamma \subset \partial_1 X$.

If an ascending v_1-trajectory $\gamma \subset \partial_1 X$, which links $\Sigma_1^- \subset \partial_1^- X(v)$ with $\Sigma_1^+ \subset \partial_1^+ X(v)$, does exist, it must cross somewhere the boundary $\partial_2 X(v)$ of $\partial_1^- X(v)$. Since the field v_1 is an orthogonal projection of v on $T(\partial_1 X)$, the two fields, v and v_1, must agree at any point $x \in \partial_2 X(v)$—the locus where v is tangent to $\partial_1 X$. Thus, v_1 is the gradient of f at $x \in \gamma \cap \partial_2 X(v)$. Therefore, as $\gamma(t)$ crosses from $\partial_1^+ X(v)$ into $\partial_1^- X(v)$ at x, in its vicinity, the arc $\gamma \cap \partial_1^- X(v)$ lies *below* the arc $\gamma \cap \partial_1^+ X(v)$ (see Figure 5.1).

By the definition of the locus $\partial_2^+ X(v)$, such a crossing $x \in \gamma \cap \partial_2 X(v)$ belongs to $\partial_2^+ X(v)$. Therefore, $\partial_2^+ X(v) \neq \emptyset$, contrary to the theorem hypothesis.

On the other hand, if no such v_1-trajectory γ exists, then we claim the existence of a codimension one closed submanifold $N \subset \partial_1 X$, which separates $\partial_1 X$ in two manifolds, $A \supset \Sigma_1^+$ and $B \supset \Sigma_1^-$ ($\partial A = N = \partial B$), so that the vector field v_1, or rather its perturbation $\tilde{v}_1$, is *transversal* to N and points *outward* of A. Indeed, for each critical point $x \in \Sigma_1^+$, in the local Morse coordinates $(y_1, \dots, y_n)$ on $\partial_1 X$, consider a small closed ϵ-disk $D^n_\epsilon(x) =_{\mathsf{def}} \{\sum_k y_k^2 = \epsilon^2\}$, centered on the critical point x. Denote by $U_\epsilon(x) \subset \partial_1 X$ the closure of the union of *downward* trajectories of the v_1-flow, passing through the points of $D^n_\epsilon(x)$ (see Figure 5.2, the left diagram). Let A_ϵ be the union $\cup_{x \in \Sigma_1^+} U_\epsilon(x)$ (see Figure 5.1, the right diagram).

Since we assume that no descending v_1-trajectory γ links a point of Σ_1^+ to a point of Σ_1^-, we can choose the disks $\{D^n_\epsilon(x)\}_{x \in \Sigma_1^+}$ so small that the set Σ_1^- belongs to the complement $\partial_1 X \setminus A_\epsilon$ of A_ϵ.

For each $x \in \Sigma_1^+$, the zero cone $\{Hess_x(f|_{\partial_1 X}) = 0\}$ of the Morse function $f|_{\partial_1 X}$ separates the sphere $\partial D^n_\epsilon(x)$ into two handles, $H^-_\epsilon(x)$ and $H^+_\epsilon(x)$ (each being a product of a sphere with a disk). We denote by $H^-(x)$ the handle in $\partial D^n_\epsilon(x)$ whose spherical core is formed by the intersection of the unstable disk through x with the sphere $\partial D^n_\epsilon(x)$. Then, by definition, the set $U_\epsilon(x)$ is a collection of downward trajectories through the points of $H^-(x)$ union with $D^n_\epsilon(x)$. Note that the downward trajectories from a different set $U_\epsilon(y)$ could enter the disk $D^n_\epsilon(x)$ only through the complementary handle $H^+(x) = \partial D^n_\epsilon(x) \setminus H^-(x)$ in its boundary. As a result, $U_\epsilon(x) \cup U_\epsilon(y)$ is a manifold whose piecewise smooth boundary could have corners (see Figure 5.2, the right diagram). Similarly, A_ϵ is a domain in $\partial_1 X$ whose boundary is piecewise smooth manifold with corners.

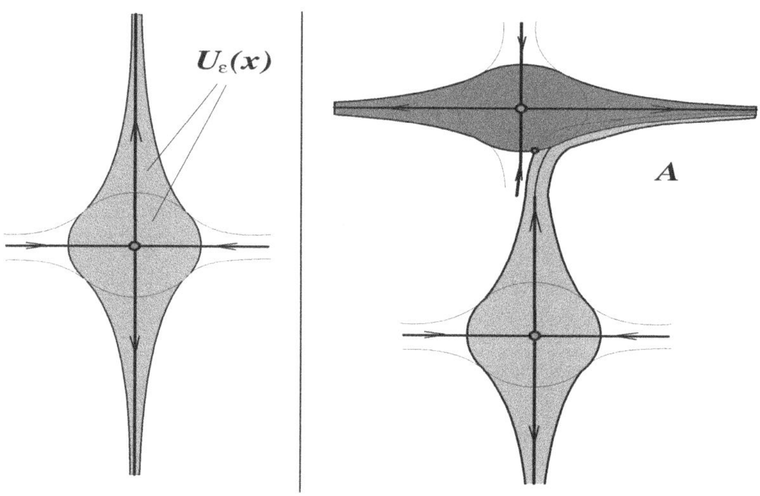

Fig. 5.2 The set $U_\epsilon(x)$ (the left diagram) and the set $A =_{\mathsf{def}} A_\epsilon$ (the right diagram). Note that the arrows point in the direction of $-v_1$.

Since A_ϵ consists of the downward trajectories of v_1, then any point $y \in \gamma_x$ that can be reached from $x \in A_\epsilon$, following the field $-v_1$ (for short, "y is below x") belongs to A_ϵ as well. Therefore the boundary ∂A_ϵ is assembled either from downward trajectories or from singletons; the singletons are contributed by some portions of $\coprod_{x \in \Sigma_1^+} \partial D^n_\epsilon(x)$, where v_1 points outside of the relevant disk $D^n_\epsilon(x)$. Thus either v_1 is tangent to ∂A_ϵ, or it points outside A_ϵ.

Away from $\Sigma_1^+ \cup \Sigma_1^-$, $v_1 \neq 0$ is of the f-gradient type. Thus, in each tangent space $T_x(\partial_1 X)$, where $x \in \partial_1 X \setminus \Sigma_1$, there is an open positive cone $\mathcal{C}_x(f)$, spanned by f-gradient type vectors, and $v_1(x) \in \mathcal{C}_x(f)$. Therefore, in the vicinity of ∂A_ϵ, we can perturb v_1 to a new field $\tilde{v}_1$ of the f-gradient type, so that $\tilde{v}_1$ points strictly outside A_ϵ and still $\tilde{v}_1(x) \in \mathcal{C}_x(f)$ for all $x \in \partial A_\epsilon$. It is possible to smoothen the boundary ∂A_ϵ so that, with respect to a new smooth boundary $\partial \tilde{A}_\epsilon$, the field $\tilde{v}_1$ still points outside $\tilde{A}_\epsilon \supset \Sigma_1^+$, the new domain bounded by $\partial \tilde{A}_\epsilon$, and $\tilde{v}_1(x) \in \mathcal{C}_x(f)$ for all $x \in \partial \tilde{A}_\epsilon$.

If, for some f-regular value c, $f(\Sigma_1^+) < c < f(\Sigma_1^-)$, then $N =_{\mathsf{def}} f^{-1}(c)$ can serve as a separator.

Let $\tilde{A} =_{\mathsf{def}} \tilde{A}_\epsilon$ and $\tilde{B} =_{\mathsf{def}} \overline{\partial_1 X \setminus \tilde{A}}$. With the separator $N = \partial \tilde{A}$ in place, consider a smooth function $\tilde{h} : \partial_1 X \to \mathbb{R}$ with the properties:

(1) zero is a regular value of $\tilde{h}$, and $\tilde{h}^{-1}(0) = N$,
(2) $\tilde{h}^{-1}((-\infty, 0]) = \tilde{A}$, $\tilde{h}^{-1}([0, +\infty) = \tilde{B}$,
(3) $\tilde{h} = h$ in a neighborhood of $\Sigma_1^+ \cup \Sigma_1^-$,
(4) $\tilde{v} =_{\mathsf{def}} \tilde{v}_1 + \tilde{h} \cdot n \in \mathcal{C}(f)$, where n is the inward normal to $\partial_1 X$ in X.

We notice that the field $\tilde{v}$ points inside X along $\tilde{A}$ and outside X along $\tilde{B}$. It also points outside $\tilde{A}$ along $N = \tilde{A} \cap \tilde{B}$. As a result, we conclude that $\partial_2^- X(\tilde{v}) = N$ and $\partial_2^+ X(\tilde{v}) = \emptyset$; in other words, $\tilde{v}$ is boundary convex. Note that $\tilde{v}_1$ can be chosen arbitrary close to v_1. Indeed, employing Theorem 5.1, we can perturb $\tilde{v}_1$ to insure its genericity with respect to the pair $(\partial_1^+ X(\tilde{v}), \partial_2 X(\tilde{v}))$, and thus the boundary genericity of $\tilde{v}$ itself.

The argument in the concave case, which deals with the relation between the property $\partial_2^- X(v) = \emptyset$ and the absence of an ascending v_1-trajectory $\gamma : \mathbb{R} \to \partial_1 X$ that connects Σ_1^+ to Σ_1^-, is analogous. We just need to switch the polarities of the relevant sets. □

Now we need to introduce a number basic notions to which we will return on many occasions in the future.

Definition 5.6. Let ω be a differential 1-form on a smooth manifold Y.

We say that a path $\gamma : [0,1] \to Y$ is *ω-positive* (*ω-negative*), if, $\omega(\dot{\gamma}(t)) > 0$ (< 0) for all values of the parameter $t \in [0,1]$. ◊

Definition 5.7. Let ω be a closed differential 1-form on a manifold Y, equipped with a Riemannian metric g. We say that a vector field v on Y is the *gradient* of ω (and denote it "$\nabla_g \omega$"), if $\omega(w) = \langle v, w \rangle_g$ for any vector field w on Y. ◊

Definition 5.8. Let ω be a differential 1-form on a manifold Y and let Σ_ω be the set of points $y \in Y$, where $\omega : T_y Y \to \mathbb{R}$ is the zero map. Assume that $\omega = df$ for some smooth function f in the vicinity of Σ_ω.

We say that a vector field v is of *ω-gradient type* if $\omega(v) > 0$ on $Y \setminus \Sigma_\omega$ and $v = \nabla_g f$ in the vicinity of Σ_ω. Here g is *some* Riemannian metric in the vicinity of Σ_ω (cf. Definition 5.2). ◊

We are in position now to formulate a generalization of Theorem 5.7 for closed differential 1-forms—another instance of somewhat weaker "holographic phenomenon", this time for fields which may not be gradient-like globally.

Theorem 5.8. *Let ω be a closed 1-form on a compact manifold X, equipped with a Riemannian metric g. Assume that ω and $\omega|_{\partial_1 X}$ have only Morse-type singularities. Let the gradient $v = \nabla_g \omega$ be a boundary generic vector field, and let $v_1 =_{\mathsf{def}} \nabla_{g|_{\partial_1 X}}(\omega|_{\partial_1 X})$.*

If $\partial_2^{\pm} X(v) = \emptyset$, then there is no ω-ascending v_1-trajectory $\gamma \subset \partial_1 X$, such that

$$\lim_{t \to -\infty} \gamma(t) \in \Sigma_1^{\mp} \text{ and } \lim_{t \to +\infty} \gamma(t) \in \Sigma_1^{\pm}.$$

Assume that there exists a codimension one submanifold $N \subset \partial_1 X$, which separates Σ_1^+ and Σ_1^- and such that the field v_1 is transversal to N and points outward/inward of the domain in $\partial_1 X$ that is bounded by N and contains Σ_1^+. Then one can deform the ω-gradient vector fields (v, v_1) to a new boundary generic pair $(\tilde{v}, \tilde{v}_1)$ of the ω-gradient type so that $\partial_2^{\pm} X(\tilde{v}) = \emptyset$.

Proof. The $(\omega|_{\partial_1 X})$-gradient fields v_1 on $\partial_1 X$ are characterized by the property $\omega(v_1) > 0$, valid on the locus where $\omega|_{\partial_1 X} \neq 0$. Usually, in this setting, we do not have a natural choice for the wall $N \subset \partial_1 X$ which would separate the singularities of opposite polarities $\Sigma_1^+ = \Sigma_1^+(\omega)$ and $\Sigma_1^- = \Sigma_1^-(\omega)$ and such that the field v_1 would be transversal to N.

It seems unlikely that the absence of an ascending v_1-trajectory that links Σ_1^- with Σ_1^+ alone is sufficient to guarantee the existence of a separator N. However, in the presence of such separator N, the arguments are identical with the ones employed in the proof of Theorem 5.7. □

Remark 5.5. In Theorems 5.7 and 5.8, the partition $\Sigma_1^+ \coprod \Sigma_1^-$ of the singular set Σ_1 must satisfy some basic relations:

$$\sum_{x\in\Sigma_1^+} \mathsf{ind}_x(v_1) + \sum_{x\in\Sigma_1^-} \mathsf{ind}_x(v_1) = 0, \text{ when } n+1 \equiv 0 \mod 2,$$

$$\sum_{x\in\Sigma_1^+} \mathsf{ind}_x(v_1) + \sum_{x\in\Sigma_1^-} \mathsf{ind}_x(v_1) = 2\cdot\chi(X), \text{ when } n+1 \equiv 1 \mod 2.$$

These relations reflect the fact that $\chi(\partial_1 X) = 0$ when $n+1 \equiv 0 \mod 2$, and $\chi(\partial_1 X) = 2\cdot\chi(X)$ when $n+1 \equiv 1 \mod 2$. ♢

Given a Riemannian metric g on a smooth $(n+1)$-manifold X, let us recall the definition of the *Hodge Star Operator*

$$*_g : \bigwedge^p T^*(X) \to \bigwedge^{n+1-p} T^*(X).$$

For $p = 1$, formula (5.5) below provides a definition in terms of local coordinates (see [Ca]).

Pick a local basis $\alpha =_{\mathsf{def}} (\alpha_1, \dots, \alpha_{n+1})$ of 1-forms in $T^*(X)$ and consider the associated basis

$$\alpha^\vee =_{\mathsf{def}} (\dots,\ (-1)^{k+1}\alpha_1 \wedge \dots \overset{\vee^{(k)}}{} \dots \wedge \alpha_{n+1},\ \dots)$$

of $\bigwedge^n T^*(X)$, where $1 \leq k \leq n+1$ and the symbol "$\vee^{(k)}$" stands for omitting the k-th form α_k from the product $\alpha_1 \wedge \dots \wedge \alpha_{n+1}$.

Assume that, in the dual to α basis α^* of $T(X)$, the metric g is locally given by a symmetric matrix $\mathsf{g} = (g_{ik})$. Then the matrix G of the $*_g$-operator in the bases α, $\alpha^\vee$ is given by the formula

$$\mathsf{G} = \sqrt{\det(\mathsf{g})}\cdot \mathsf{g}^{-1}, \tag{5.5}$$

whence $\det(\mathsf{G}) = (\det(\mathsf{g}))^{\frac{n-1}{2}}$.

The *co-derivative* δ_g of a differential p-form α is a form defined by the formula

$$\delta_g(\alpha) =_{\mathsf{def}} (-1)^{(p+1)\dim X+1}(*_g \circ d \circ *_g)(\alpha).$$

By definition, the Laplace operator $\Delta_g =_{\mathsf{def}} d\delta_g + \delta_g d$.

In this book, by a *harmonic form* β we mean a closed ($d\beta = 0$) and co-closed differential form ($\delta_g\beta = 0$). Every such form is a solution of the Laplace equation $\Delta_g\beta = 0$. However, on manifolds with boundary, there may be solutions of the Laplace equation that are not closed and co-closed!

Definition 5.9. A closed differential 1-form ω on a compact manifold Y is called *intrinsically harmonic* if there exists a Riemannian metric g on Y such that the form $*_g(\omega)$ is *closed*[11]. $\diamondsuit$

Example 5.4. Let Y be a closed smooth manifold and $H : Y \to S^1$ a smooth map with isolated Morse-type singularities. Consider the closed 1-form $\omega =_{\mathsf{def}} H^*(d\theta)$, the pull-back of the canonic 1-form $d\theta$ on the circle S^1. Assume that one of the H-fibers, $F_0 = H^{-1}(*)$, is connected. Then ω is intrinsically harmonic [FKL] (see also [F]). $\diamondsuit$

Let Σ_ω denote the singularity set of a closed 1-form ω on a compact manifold Y. We assume that $\Sigma_\omega \subset \mathrm{int}(Y)$.

By Calabi's Proposition 1 [Ca], ω is intrinsically harmonic if and only if, through every point $y \in Y \setminus \Sigma_\omega$, there is a ω-*positive* path γ which either is a loop, or which starts and terminates at the boundary ∂Y.

Theorem 5.9. *Let ω be a closed 1-form on a Riemannian manifold X, such that its singularity set $\Sigma_\omega \subset \mathrm{int}(X)$. Assume that $\omega|_{\partial_1 X}$, the restriction of ω to $T(\partial_1 X)$, is a co-closed form.*

Then the gradient field $v = \nabla\omega$ in X is not *boundary convex or boundary concave (that is, $\partial_2^+ X(v) \neq \emptyset$* and *$\partial_2^- X(v) \neq \emptyset$). Thus, if $\partial_2 X(v)$ is connected, then $\partial_3 X(v) \neq \emptyset$.*

Proof. We abbreviate $\partial_j^\pm X(v)$ to $\partial_j^\pm X$ and $*_{g|_{\partial_1 X}}$ to $*_\partial$. Here $*_\partial$ is the $*$-operator on the boundary ∂X with respect to the restriction of the given Riemannian metric g to ∂X.

If $*_\partial(\omega|_{\partial_1 X})$ is a closed $(n-1)$-form on $\partial_1 X$, then by the Stokes' Theorem,

$$\int_{\partial_2 X} *_\partial(\omega|_{\partial_1 X}) = \int_{\partial_1^+ X} d(*_\partial(\omega|_{\partial_1 X})) = 0.$$

However, for a concave/convex gradient field $v = \nabla\omega$, the $(n-1)$-form $*_\partial(\omega|_{\partial_1 X})$, being restricted to $\partial_2 X$, is proportional to the volume form of

[11]This requirement is equivalent to $\delta_g\omega = 0$.

$\partial_2 X$ with negative/positive functional coefficient. Indeed, at the points of $\partial_2^+ X$, the angle between v and the normal n to $\partial_2 X$ in $\partial_1^+ X$ is acute, while it is obtuse at the points of $\partial_2^- X$ (see Figure 5.1). Therefore, $\int_{\partial_2 X} *_\partial(\omega|_{\partial_1 X}) \neq 0$ when either $\partial_2^+ X = \emptyset$ or $\partial_2^- X = \emptyset$. The resulting contradiction proves that $\partial_2^+ X \neq \emptyset$ and $\partial_2^- X \neq \emptyset$.

Therefore, when $\partial_2 X$ is connected, then $\partial_2^+ X$ and $\partial_2^- X$ must share the common nonempty boundary $\partial_3 X$—the gradient field v must have cubic tangency loci. □

Example 5.5. Let X be a compact connected smooth manifold and $H : X \to S^1$ a smooth map with isolated Morse-type singularities. Consider the closed 1-form $\omega = H^*(d\theta)$ on X, the pull-back of the canonic 1-form $d\theta$ on the circle S^1. Assume that one of the fibers of the map $H : \partial_1 X \to S^1$ is connected. Then there exists a metric g^∂ on ∂X (and thus a metric g on X) such that the form $\omega^\partial =_{\mathsf{def}} H^*(d\theta)|_{\partial_1 X}$ is co-closed in $\partial_1 X$ ([Ca], [K10]). Consider the gradient field $v = \nabla_g(\omega)$. Then by Theorem 5.9, $\partial_2^+ X(v) \neq \emptyset$ and $\partial_2^- X(v) \neq \emptyset$ for any Riemannian metric g whose restriction to $\partial_1 X$ "harmonizes" ω^∂. ◇

Let us formally introduce here an already familiar class of traversing vector fields.

Definition 5.10. A non-vanishing vector field v on a compact manifold X is called *traversing* if each v-trajectory is either a closed segment whose both ends reside in ∂X, or a singleton which belongs to ∂X.

We denote by $\mathcal{V}_{\mathsf{trav}}(X)$ the space of traversing vector fields on X. ◇

Remark 5.6. The definition excludes fields with zeros in X (they will generate trajectories that are homeomorphic to open or semi-open intervals) and fields with closed trajectories. Note that all gradient-like fields of nonsingular functions are traversing, but the gradient-like fields of nonsingular closed 1-forms may be not traversing! ◇

Lemma 5.6. *Any traversing vector field is of the gradient type.*

Proof. Let v be a traversing field on a compact $(n+1)$-dimensional X. We extend the pair (X, v) to a pair $(\hat{X}, \hat{v})$ so that X is properly contained in $\hat{X}$ and $\hat{v} \neq 0$.

For every v-trajectory $\gamma^\star \subset X$ we choose a local transversal compact section $S_{\gamma^\star} \subset \text{int}(\hat{X})$ of the $\hat{v}$-flow so that $S_{\gamma^\star}$ is diffeomorphic to a n-dimensional ball with its center at the singleton $\gamma^\star \cap S_{\gamma^\star}$. We denote by $\tilde{U}_{\gamma^\star}$ the union of $\hat{v}$-trajectories through $S_{\gamma^\star}$.

We can adjust the section $S_{\gamma^\star}$ so that the set $\tilde{U}_{\gamma^\star}$ will contain a compact cylinder $\hat{U}_{\gamma^\star} \approx S_{\gamma^\star} \times [-a_{\gamma^\star}, b_{\gamma^\star}]$, where $a_{\gamma^\star}, b_{\gamma^\star}$ are positive constants (which depend on $\gamma^\star$), with the following properties:

(1) $\hat{U}_{\gamma^\star} \supset \tilde{U}_{\gamma^\star} \cap X$,
(2) for any $\hat{v}$-trajectory $\tilde{\gamma}$ through $S_{\gamma^\star}$, the intersection $\hat{\gamma} =_{\mathsf{def}} \tilde{\gamma} \cap \hat{U}_\gamma$ is a closed segment,
(3) the point $\tilde{\gamma} \cap S_{\gamma^\star}$ belongs to the interior of segment $\hat{\gamma}$.

Then the collection $\mathcal{U} =_{\mathsf{def}} \{\hat{U}_{\gamma^\star} \cap X\}_{\gamma^\star}$ forms a cover of X. Since $X \subset \mathsf{int}(\hat{X})$ is compact, we can choose a finite subcover $\mathcal{U}' \subset \mathcal{U}$ of X.

For each $\hat{U}_{\gamma^\star} \cap X \in \mathcal{U}'$ and the corresponding section $S_{\gamma^\star}$, we produce a smooth function $\phi_{\gamma^\star} : \hat{U}_{\gamma^\star} \to \mathbb{R}$ by integrating the vector field $\hat{v}$ and using $S_{\gamma^\star}$ as the initial value locus for the integration. More accurately, let

$$\psi_{\hat{\gamma}} : [-a_{\gamma^\star}, b_{\gamma^\star}] \to \hat{U}_{\gamma^\star}$$

be the parametrization of a typical trajectory $\hat{\gamma} \subset \hat{U}_{\gamma^\star}$, such that

$$\frac{d}{dt}\psi_{\hat{\gamma}}(\tau) = \hat{v}(\psi_{\hat{\gamma}}(\tau))$$

for all $\tau \in [-a_{\gamma^\star}, b_{\gamma^\star}]$ and $\psi_{\hat{\gamma}}(0) = \hat{\gamma} \cap S_{\gamma^\star}$. This bijective parametrization introduces a smooth product structure

$$\Phi : \hat{U}_{\gamma^\star} \approx S_{\gamma^\star} \times [-a_{\gamma^\star}, b_{\gamma^\star}]$$

by the formula $\Phi(x) =_{\mathsf{def}} \big(\hat{\gamma}_x \cap S_{\gamma^\star}, (\psi^{-1}_{\hat{\gamma}_x}(x)\big)$.

We define a smooth function $\hat{\phi}_{\gamma^\star} : \hat{U}_{\gamma^\star} \to \mathbb{R}$ by the formula $x \to \psi^{-1}_{\hat{\gamma}_x}(x)$ and denote it (quite appropriately) by the symbol $\int^x_{S_{\gamma^\star}} \hat{v}$.

Let $\chi_{\gamma^\star} : S_{\gamma^\star} \to \mathbb{R}_+$ be a smooth non-negative function that vanishes only on the boundary $\partial S_{\gamma^\star}$ together with all its partial derivatives. Let $\tilde{\chi}_{\gamma^\star} : \hat{U}_{\gamma^\star} \to \mathbb{R}_+$ denote the composition of the $\hat{v}$-directed projection $\pi_{\gamma^\star} : \hat{U}_{\gamma^\star} \to S_{\gamma^\star}$ with the function $\chi_{\gamma^\star}$. Since $\tilde{\chi}_{\gamma^\star}$ vanishes on $\partial\hat{U}_{\gamma^\star} \cap X$ together with all its partial derivatives, $\tilde{\chi}_{\gamma^\star}$ extends on X to produce a smooth function $\hat{\chi}_{\gamma^\star} : X \to \mathbb{R}_+$, constant along the $\hat{v}$-trajectories and with the support in $\hat{U}_{\gamma^\star} \cap X$.

Now consider the smooth function

$$f(x) =_{\mathsf{def}} \sum_{\hat{U}_{\gamma^\star} \in \mathcal{U}'} \hat{\chi}_{\gamma^\star}(x)\Big(\int^x_{S_{\gamma^\star}} \hat{v}\Big). \tag{5.6}$$

It is well-defined on X. Let us compute its v-directional derivative:

$$\mathcal{L}_v(f(x)) = \sum_{\hat{U}_{\gamma^\star} \in \mathcal{U}'} \mathcal{L}_v\Big[\hat{\chi}_{\gamma^\star}(x) \cdot \Big(\int_{S_{\gamma^\star}}^{x} \hat{v}\Big)\Big]$$

$$= \sum_{\hat{U}_{\gamma^\star} \in \mathcal{U}'} \hat{\chi}_{\gamma^\star}(x)\, \mathcal{L}_v\Big(\int_{S_{\gamma^\star}}^{x} \hat{v}\Big) > 0. \tag{5.7}$$

Let us clarify formula (5.7). By the very definition of $\hat{\chi}_{\gamma^\star}$, it is constant on each $\hat{v}$-trajectory, so that $\mathcal{L}_v(\hat{\chi}_{\gamma^\star}) = 0$. Also, $\hat{\chi}_{\gamma^\star} > 0$ in $\mathrm{int}(\hat{U}_{\gamma^\star})$. At the same time, $\mathcal{L}_v\big(\int_{S_{\gamma^\star}}^{x} \hat{v}\big) > 0$, since $\frac{d}{dt}\psi_{\hat{\gamma}} = \hat{v}(\psi_{\hat{\gamma}}) \neq 0$ and $\psi_{\hat{\gamma}}$ increases in the direction of v. Finally, each $x \in X$ belongs to the interior of some set $\hat{U}_{\gamma^\star}$. Therefore, $df(v) = \mathcal{L}_v(f) > 0$, so v is a gradient-like field for v. □

Corollary 5.8. *Let X be a smooth compact connected manifold with boundary.*

Then $\mathcal{V}_{\mathsf{trav}}(X)$—the space of traversing vector fields on X—is nonempty and coincides with the intersection $\mathcal{V}_{\mathsf{grad}}(X) \cap \mathcal{V}_{\neq 0}(X)$, where $\mathcal{V}_{\mathsf{grad}}(X)$ denotes the space of gradient-like fields, and $\mathcal{V}_{\neq 0}(X)$ the space of all non-vanishing fields on X.

Proof. By definition, any traversing field v on X does not vanish. By Lemma 5.6, v must be of the gradient type. Thus

$$\mathcal{V}_{\mathsf{trav}}(X) \subset \mathcal{V}_{\mathsf{grad}}(X) \cap \mathcal{V}_{\neq 0}(X).$$

On the other hand, for a compact X with a gradient-like $v \neq 0$, each v-trajectory γ_x through $x \in \mathrm{int}(X)$ must reach the boundary in both finite positive and negative times (since it is controlled by some Lyapunov function f). As a result,

$$\mathcal{V}_{\mathsf{trav}}(X) = \mathcal{V}_{\mathsf{grad}}(X) \cap \mathcal{V}_{\neq 0}(X).$$

It remains to show that $\mathcal{V}_{\mathsf{trav}}(X) \neq \emptyset$. By Lemma 5.4, $\mathcal{V}_{\mathsf{grad}}(X) \cap \mathcal{V}_{\neq 0}(X) \neq \emptyset$, which implies that $\mathcal{V}_{\mathsf{trav}}(X) \neq \emptyset$. □

Recall that there are simple topological obstructions to the existence of a boundary convex *traversing* vector field v on a given manifold X. The next lemma testifies that the existence of such v imposes severe restrictions on the topology of X.

Lemma 5.7. *A connected compact $(n+1)$-manifold X with boundary admits a* boundary convex *traversing*[12] *vector field v, if and only if, X is diffeomorphic to a product of a connected compact n-manifold and a segment, the corners of the product being smoothly rounded.*

Proof. Indeed, if such convex v exists, $\partial_1^+ X =_{\mathsf{def}} \partial_1^+ X(v)$ must be a deformation retract of X: just use the down flow to produce the retraction. Therefore, when $\partial_2^+ X(v) = \emptyset$, then X is homeomorphic to the quotient space $\{(\partial_1^+ X) \times [0,1]\}/\sim$, where the equivalence relation "$\sim$" is defined by collapsing each segment $\{x \times [0,1]\}_{x \in \partial_2^- X}$ to a point. If we round the corners of the product, we will get a diffeomorphism between X and the "lens-shaped" space $\{(\partial_1^+ X) \times [0,1]\}/\sim$ (see Figure 2.6).

On the other hand, any product $Y \times [0,1]$, whose conners $\partial Y \times \partial([0,1])$ being rounded, admits a traversing vector field of the desired boundary convex type. □

Let us review briefly some results form the glorious days of classical Differential Topology.

Let X be a smooth compact *contractible* $(n+1)$-manifold. Its boundary must be a homology sphere. For $n \geq 6$, the h-cobordism theorem [Sm] implies that any such X with a *simply-connected* boundary ∂X is diffeomorphic to the standard $(n+1)$-ball. In fact, every smooth homology 4-sphere is diffeomorphic to the boundary of a smooth compact *contractible* 5-manifold X ([Ke], [Sm]).

For $n = 2$, by Theorem 3.13, based on the Perelman's solution [P1], [P2] of the Thurston's geometrization conjecture, X is diffeomorphic to the standard 3-ball.

The hard case ($n = 3$) of a 4-dimensional smooth fake ball X is wide open. Recall that Mazur [Maz] and Poenaru [Poe] gave examples of contractible compact smooth 4-manifolds that are not diffeomorphic to the standard 4-ball. Their boundaries are homology 3-spheres. In particular, these boundaries may be the Brieskorn homology spheres (see [Bri] for the definition) $\Sigma(2,5,7), \Sigma(3,4,5)$, and $\Sigma(2,3,13)$ [AkKi]. However, the Poincare homology sphere does not bound a contractible smooth 4-manifold because the Rochlin invariant [Rokh] provides an obstruction.

We notice that any such contractible X admits a boundary generic and traversing vector field.

---

[12] equivalently, a non-vanishing gradient-like vector field

Corollary 5.9. *Any smooth compact* acyclic *4-manifold X with boundary, which either*

- *admits a boundary* convex *traversing vector field, or*
- *admits a nonsingular function $f : X \to \mathbb{R}$ such that $f(\Sigma_1^+(f|_{\partial_1 X})) < f(\Sigma_1^+(f|_{\partial_1 X}))$ as sets,*

is diffeomorphic to the standard ball D^4.

Proof. By Lemma 5.7, X is diffeomorphic to a product of a 3-fold $Y = \partial_1^+ X(v)$ with $[0,1]$, the corners of the product being rounded. If $\partial Y = \emptyset$, then the 3-cycle $[Y]$ prevents X from being acyclic. So $\partial Y \neq \emptyset$.

By Theorem 3.13, any compact connected 3-fold Y with boundary is a connected sum of handlebodies and products $S^1 \times S^2$, unless $Y = D^3$. The acyclicity of X implies the acyclicity of Y and thus excludes the handlebodies and products $S^1 \times S^2$ in the decomposition of Y. So Y must be the standard 3-ball. As a result, X is diffeomorphic to the standard 4-ball.

In particular, any smooth compact contractible 4-manifold X that admits a boundary convex traversing vector field is diffeomorphic to the standard ball D^4.

Now we deal with the case, described in the second bullet. We denote by $v = \nabla f$ the gradient field of $f : X \to \mathbb{R}$ and by $v_1 = \nabla_{\partial_1 X} f$ its orthogonal projection on $\partial_1 X$. When f is a surjection, then v is traversing. If X admits a surjection f as in the second bullet, then by Theorem 5.7, we may deform v into a new f-gradient-like traversing vector field $\tilde{v}$ and its orthogonal projection $\tilde{v}_1$ so that no $\tilde{v}_1$-ascending trajectory of $\tilde{v}_1$ may cross from $\partial_1^- X(\tilde{v})$ into $\partial_1^+ X(\tilde{v})$. This implies that $\partial_2 X(\tilde{v}) = \partial_2^- X(\tilde{v})$, the field $\tilde{v}$ is boundary convex. So, by the previous argument, X is the standard ball D^4. □

We notice that, in general, $H_n(X; \mathbb{Z}) \neq 0$ is an *obstruction* to finding a *boundary convex* traversing vector field v on a $(n+1)$-dimensional manifold X with a connected boundary.

Corollary 5.10. *Let X be a smooth connected compact $(n+1)$-manifold with boundary, which admits a boundary* convex *traversing vector field.*

If $H_n(X; \mathbb{Z}) \neq 0$, then X is diffeomorphic to the product $Y \times [0,1]$, where Y is a closed *manifold.*

In particular, no connected X, with the property $H_n(X;\mathbb{Z}) \neq 0$ and whose nonempty boundary ∂X has a number of connected components different from two, admits a boundary convex traversing vector field.

Proof. If such boundary convex traversing field v exists, $\partial_1^+ X$ must be a deformation retract of X. Therefore, for a connected X, $\partial_1^+ X$ must be connected as well.

On the other hand, if $\partial(\partial_1^+ X) = \partial_2 X \neq \emptyset$, then the connected $\partial_1^+ X$ must be of a homotopy type of a $(n-1)$-dimensional complex. In such a case, the groups $H_n(\partial_1^+ X;\mathbb{Z}) \approx H_n(X;\mathbb{Z})$ must vanish.

Thus when $H_n(X;\mathbb{Z}) \neq 0$ and v is boundary convex, the only remaining option is $\partial(\partial_1^+ X) = \partial_2 X = \emptyset$—the manifold $\partial_1^+ X$ is closed. In such a case, X is a product of a connected closed n-manifold with an interval; so the boundary ∂X must be the union of *two* diffeomorphic components. □

As with the case of boundary convex traversing vector fields, perhaps, there are topological obstructions to the existence of a boundary *concave* traversing vector field on a given manifold? Presently, the contours of the universe of such obstructions are murky. By Theorem 1.5, we know only that the disk D^2 does not admit a non-vanishing boundary concave vector field and that the Möbius band M^2 does not admit a boundary concave traversing vector field.

Lemma 5.8. *If a boundary generic vector field v on an even-dimensional compact orientable manifold X is boundary concave, then its index*

$$\mathsf{Ind}(v) = \chi(X) + \chi(\partial_1^+ X) = \chi(X) + \frac{1}{2}\cdot\chi(\partial_2 X).$$

Thus, for all boundary concave vector fields v with a fixed value of index $\mathsf{Ind}(v)$, the Euler number $\chi(\partial_2 X) = \chi(\partial_2^+ X)$ is topological invariant.

If a boundary generic vector field v on an odd-*dimensional compact orientable manifold X is boundary concave, then its index*

$$\mathsf{Ind}(v) = \chi(X) - \chi(\partial_1^+ X).$$

Thus, for all boundary concave vector fields v with a fixed value of index $\mathsf{Ind}(v)$, the Euler number $\chi(\partial_1^+ X)$ is topological invariant.

Proof. For a boundary concave vector field v, $\partial(\partial_1^+ X) = \partial_2 X = \partial_2^+ X$. Therefore, the Morse formula (5.2) reduces to the equation

$$\mathsf{Ind}(v) = \chi(X) - \chi(\partial_1^+ X) + \chi(\partial_2 X). \tag{5.8}$$

Recall that, for any orientable *odd*-dimensional manifold Y, $\chi(\partial Y) = 2 \cdot \chi(Y)$. Therefore, when $\dim(X) \equiv 0 \mod 2$, we get $2 \cdot \chi(\partial_1^+ X) = \chi(\partial_2 X)$. Thus formula (5.8) transforms into

$$\mathsf{Ind}(v) = \chi(X) + \chi(\partial_1^+ X) = \chi(X) + \frac{1}{2} \cdot \chi(\partial_2 X).$$

For an odd-dimensional X, the closed manifold $\partial_2 X$ is odd-dimensional, so $\chi(\partial_2 X) = 0$. Therefore $\mathsf{Ind}(v) = \chi(X) - \chi(\partial_1^+ X)$. □

Corollary 5.11. *Let X be a 4-dimensional compact connected oriented smooth manifold with boundary.*

If $\chi(X) < 0$, then for any boundary generic concave *vector field v on X of index 0, the locus $\partial_2 X = \partial_2^+ X$ contains at least $|\chi(X)|$ two-dimensional spheres.*

Proof. Since $\partial_2 X$ is a closed orientable 2-manifold, its Euler number is positive only if $\partial_2 X$ contains sufficiently many 2-spheres. By Lemma 5.8, $\chi(\partial_2 X) = -2 \cdot \chi(X) > 0$. Therefore $\partial_2 X$ contains at least $|\chi(X)|$ two-dimensional spheres. □

In Chapter 1, we have seen that the disk D^2 does not admit a non-vanishing concave field. In contrast, D^3 does admit a boundary generic concave non-vanishing vector field: just take the obvious action of the circle S^1 on the sphere S^3 and the Hopf vector field v it generates. Then consider the restriction of v to the northern hemisphere $D^3_+ \subset S^3$. We transplant v from D^3_+ to the unitary disk $D^3 \subset \mathbb{R}^3$, centered at the origin. We can describe the transplanted $\tilde{v}$ informally as the sum of the velocity field of the solid D^3, spinning around the z-axis, with the solenoidal field of the loop $L = \{x^2 + y^2 = 4/9,\ z = 0\}$. However, this field $\tilde{v}$ is not of the traversing type: it has closed trajectories (residing in the solid torus $\{dist(\sim, L) \leq 1/3\}$).

These observations encourage us to formulate the following simple looking conjecture.

Conjecture 5.1. *The standard ball D^{n+1} does not admit a traversing boundary concave vector field.* ◊

Thanks to Theorem 5.7, to prove Conjecture 5.1 will suffice to show that, for any boundary generic traversing vector field v on D^{n+1}, the vector field v_1 (the orthogonal projection of v on S^n) has an ascending trajectory δ

that links a point of Σ_1^+ to a point of Σ_1^-. Indeed, the existence of such δ implies that $\partial_2^+ X(v) \neq \emptyset$.

5.5 Morse stratifications of the boundary 3-convex and 3-concave vector fields

The construction of a boundary concave field on a 2-disk with several holes (see Figure 1.5) admits a simple generalization which leads to a mass production of manifolds with traversing boundary generic 3-convex and 3-concave vector fields. Such vector fields have trajectories of the combinatorial types

$$\{(11), (2), (121), \dots, (12\dots 21)\}$$

only. The construction uses a finite *ordered* collection of codimension 0 submaniflods of a given manifold Y to create *cavities* in the product manifold $Y \times [0,1]$. So we will call it "*the cavity construction*".

We consider a closed n-manifold Y. Let $\{Z_i \subset Y\}_{1 \leq i \leq s}$ be compact submanifolds also of dimension n. Let $W =_{\mathsf{def}} Y \times [0,1]$. We pick s disjointed close intervals $\{I_i\}_i$ in the interval $[0,1]$. Then we form the product $U_i =_{\mathsf{def}} Z_i \times I_i$. By rounding the corners of U_i, we get a $(n+1)$-manifold $V_i \subset U_i$ so that each segment $z \times I_i$, where $z \in \mathrm{Int}(Z_i)$, hits V_i along a closed segment, and each segment $z \times I_i$, where $z \in \partial(Z_i)$, hits V_i along a singleton.

Let us form the manifold $X =_{\mathsf{def}} W \setminus \coprod_i \mathrm{int}(V_i)$. Its boundary consists of two copies of Y together with the disjoint union of ∂V_i (they are the doubles of Z_i's). The obvious vertical field v on W, being restricted to X, is boundary concave. In fact, $\partial_1^+ X(v) = Y \times \{0\} \coprod A$, where $A \approx \coprod_i Z_i$, and $\partial_2^+ X(v) \approx \coprod_i \partial Z_i$.

Example 5.6. Let $S^4 \odot D^4$ denote the space of the unitary 4-disk bundle η over the 4-sphere, whose Euler class is $2[S^4]$. Consider the smooth 8-dimensional manifold Y^8, obtained by plumbing on the twisted products $\{S_i^4 \odot D_i^4\}_{1 \leq i \leq 8}$ so that the intersection form in $H_4(Y^8; \mathbb{Z})$ realizes the canonical non-degenerated integral bilinear form

$$\mathsf{E}_8 : \mathbb{Z}^8 \otimes \mathbb{Z}^8 \to \mathbb{Z}$$

of index 8. The boundary ∂Y^8 is the famous exotic sphere Σ^7 of Milnor [Mi4].

Let us apply the cavity construction, based on the collection of 8-manifolds

$$\{Z_i =_{\mathsf{def}} S_i^4 \odot \tilde{D}_i^4 \subset Y^8)\}_{1 \leq i \leq 8},$$

where $S_i^4 \odot \tilde{D}_i^4 \subset S_i^4 \odot D_i^4$ denotes a concentric disk bundle in $S_i^4 \odot D_i^4$. The result of the cavity construction is the 9-manifold $X^9 \subset Y^8 \times [0,9]$, obtained by rounding the conners $\Sigma^7 \times \partial[0,9] \subset Y^8 \times [0,9]$ and deleting from it 8 solid twisted handlebodies

$$\{(S_i^4 \odot \tilde{D}^4) \times [i-0.1,\, i+0.1]\}_{1 \le i \le 8}$$

whose corners $\partial(S_i^4 \odot \tilde{D}^4) \times \partial[i-0.1,\, i+0.1]$ are also rounded.

The resulting manifold X^9 admits a traversally generic and globally 3-convex field v, induced by the constant field, tangent to the fibers of the obvious projection $\pi : Y^8 \times [0,9] \to Y^8$. Its locus $\partial_2^- X(v)$, the union of v-trajectories of the combinatorial type (2), is the exotic sphere $\Sigma^7 \approx \partial Y^8$, or rather $\partial Y^8 \times \{4\} \subset Y^8 \times [0,9]$. The locus

$$\partial_2^+ X(v) = \coprod_{i=1}^{8} \partial(S_i^4 \odot \tilde{D}^4) \times \{i\} \subset Y^8 \times [0,9].$$

The typical v-trajectory through the points of this locus is of the combinatorial type (121). Finally, the locus $X(v,(1221))$, the union of double-tangent trajectories, projects on the disjoint union of seven copies of $S^3 \times S^3$. These copies correspond to seven plumbing regions (topologically $D^4 \times D^4$), where S_i^4 intersects S_j^4 $(i \neq j)$ at a singleton in Y^8. The intersections are in 1-to-1 correspondence with the entries "1" in the upper triangular portion of the symmetric matrix of the bilinear form E_8.

Thus $(11), (2), (121)$ and (1221) are the only combinatorial types of v-trajectories on X.

So the inclusions of connected components of $X(v,(1221))$ into connected components of $X(v,(121))$ "remembers" the intersection form E_8.

We notice that $\mathcal{T}(v)$, the v-trajectory space of X^9, comes equipped with the obvious finitely ramified map $p : \mathcal{T}(v) \to Y^8$. The cardinalities of the p-fibers are: $1, 2$, and 3. ◇

Examples like Example 5.6 lead to few interesting and already familiar questions:

Question 5.3. *Which compact manifolds admit* boundary concave *non-vanishing vector fields? Which compact manifolds admit* boundary concave *non-vanishing* traversing *vector fields?* ◇

Despite the "natural" flavor of these questions, we have a limited understanding of the general answers. It seems clear that the trajectory spaces

$\mathcal{T}(v)$ of such boundary concave traversing vector fields are not collapsible cell complexes.

It is easy to exhibit a vast gallery of manifolds that admit traversing *concave* vector fields. Such manifolds X are manufactured from closed $(n+1)$-manifolds Y by removing a number of $(n+1)$-balls (alternatively by applying the cavity construction to a closed n-manifold). We start with any Morse function $f : Y \to \mathbb{R}$ and delete small convex (in the local Morse coordinates) balls that surround the singularities of f. The gradient fields of f possess the desired concavity and are traversing.

The next theorem connects the *optimal* Morse functions on a smooth $(n+1)$-manifold Y with the traversing boundary concave vector fields on a manifold X, obtained by deleting an appropriate number of $(n+1)$-balls from Y. It relies on the ground-breaking results of Smale [Sm], [Sm1] about the existence of *optimal* Morse functions.

Theorem 5.10. *Let Y be a smooth compact connected $(n+1)$-manifold, $n \geq 5$, with (possibly empty) boundary $\partial Y = Z_0 \sqcup Z_1$. Assume that*

$$\pi_1(Y, x) \approx \pi_1(Z_0, z) \approx \pi_1(Z_1, z')$$

for any choice of the base points x, z, z'. Let

$$m_i =_{\mathsf{def}} \mu\big(H_i(Y, Z_0; \mathbb{Z})\big) + \mu\big(\mathsf{Tor}\, H_{i-1}(Y, Z_0; \mathbb{Z})\big),$$

where $\mu(H)$ denotes the minimal number of generators in a group H.

Put $m =_{\mathsf{def}} \sum_{0 \leq i \leq n+1} m_i$. Consider the cardinality

$$k =_{\mathsf{def}} \#\{i \in [1, n] \big|\; m_i \neq 0\}.$$

Then the manifold $X =_{\mathsf{def}} Y \setminus (\coprod_{j=1}^{m} D^{n+1})$ admits a traversing boundary concave vector flow, whose trajectories are only of the combinatorial types (11) and $\{(1, 2, 2, \dots, 2, 1)\}$, where the number of "2"s does not exceed k. In other words, such manifold X is globally $(k+1)$-convex.

Proof. Here is the familiar key observation which makes the connection between the Morse Theory on Y and traversing vector fields on X quite transparent. Let $x \in \Sigma(f)$ be a critical point of a Morse function $f : Y \to \mathbb{R}$, and B^{n+1}_x a sufficiently small open Euclidean ball in the Morse coordinates, centered on x. Let $D^{n+1}_x \subset B^{n+1}_x$ be a smaller concentric open ball. Consider the shell $Sh_x =_{\mathsf{def}} B^{n+1}_x \setminus D^{n+1}_x$. Then, due to the quadratic nature of f in B^{n+1}_x, the gradient field v of $f : B^{n+1}_x \setminus D^{n+1}_x \to \mathbb{R}$ is concave with respect to the inner boundary ∂D^{n+1}_x of the shell Sh_x. In particular, $\partial_3 Sh_x(v) \cap \partial D^{n+1}_x = \emptyset$.

By Smale's theorem [Sm], Y admits an *optimal* Morse function $f: Y \to \mathbb{R}$ with exactly m_i critical points of each index i and such that $f^{-1}(0) = Z_0, f^{-1}(1) = Z_1$. The optimal Morse function f is self-indexing: the orderings of f-critical points $x \in \Sigma(f)$ by their Morse indices $\{i(x)\}$ and by their critical values $\{f(x)\}$ agree.

By a *broken trajectory* of a vector field v with isolated singularities we mean a finite union of v-trajectories that may have the singular points as their limits. Moreover, the closure of the union of such trajectories must be homeomorphic to a closed interval, and the v-induced orientations of the trajectories are consistent across this interval.

For an optimal self-indexing f, the cardinality of the set of f-critical values of the critical points with Morse indexes different from 0 and $n+1$ equals k. Therefore, any broken v-trajectory visits at most k critical points that are not the extrema of f.

Let us fix a Riemannian metric g on Y which is euclidean in the Morse coordinates for each ball B_x^{n+1}. By a general position argument, we can assume that the gradient field $v = \nabla_g f$ satisfies the Morse-Smale transversality property [Sm]. (The property states that the stable and the unstable manifolds of the critical points are mutually transversal.)

Then no broken v-trajectory connects two critical points x, x' of the same Morse index i. In particular, no broken loop connects a critical point x to itself. Moreover, there is $\epsilon > 0$ so that no "honest" v-trajectory visits the ϵ-balls B_x^{n+1} and $B_{x'}^{n+1}$, provided $i(x) = i(x')$.

By the choice of metric, v is boundary generic in

$$X =_{\mathsf{def}} Y \setminus \Big(\coprod_{x \in \Sigma(f)} D_x^{n+1} \Big)$$

in the sense of Definition 5.1. Since v is the gradient field which is transversal to ∂Y, we conclude that v is traversing in X. (In fact, by Theorem 6.7 and Corollary 6.4, we can perturb v within the space of f-gradient-like fields to obtain a traversally generic gradient field v' which coincides with v in the shells $\{Sh_x\}_{x \in \Sigma(f)}$.)

In the vicinity of each extremal $x \in \Sigma(f)$, any v-trajectory γ is transversal to the sphere ∂D_x^{n+1} and thus contributes 1 to the combinatorial type ω_γ. Since any broken trajectory visits k critical points of Morse indices $i \in [1, n]$ at most, by choosing the balls $\{\partial B_x^{n+1}\}_{x \in \Sigma(f)}$ small enough, each trajectory is tangent to k inner spheres $\{\partial D_x^{n+1}\}_{x \in \Sigma(f)}$ of the shells at most. We leave to the reader the rigorous validation of this claim.

Each tangency has the multiplicity 2, and the flow is transversal to Z_0 and Z_1. Therefore, the combinatorial types of v-trajectories all are of the form (11) and $(1, 2, 2, \dots, 2, 1)$, where the number of "2"s does not exceed k, the number of non-extremal values of f. □

Remark 5.7. Note that any simply-connected closed manifold Y satisfies the hypothesis of the Smale Theorem (stated in Theorem 5.10), and thus admits an optimal Morse function.

Some subtle generalizations of the Smale Theorem are available for certain manifolds Y which violate the hypothesis $\pi_1(Y, x) \approx \pi_1(Z_0, z) \approx \pi_1(Z_1, z')$, but still admit optimal Morse functions (see, for example, the monograph [Sh] of Sharko). For such Y's, the obvious direct analogues of Theorem 5.10 are valid. ◇

Corollary 5.12. *Let Y be a closed smooth* simply-connected *manifold of dimension $n+1 \geq 6$. Let*

$$m = \sum_{i=1}^{n} \Big(\mu(H_i(Z; \mathbb{Z})) + \mu(\mathsf{Tor}\, H_{i-1}(Z; \mathbb{Z}))\Big).$$

Then

$$X =_{\mathsf{def}} Y \setminus (\coprod_{i=1}^{m+2} D^{n+1})$$

admits a traversing concave boundary generic flow whose trajectories are only of the combinatorial types (11) *and* $\{(1, 2, 2, \dots, 2, 1)\}$, *where the number of "2" 's does not exceed k from Theorem 5.10.*

Proof. The proposition instantly follows from Theorem 5.10 by applying it to $Y \setminus (D_0^{n+1} \coprod D_1^{n+1})$, where the two disks correspond to the minimum and maximum of the optimal Morse function f on Y. The roles of Z_0 and Z_1 are played by ∂D_0^{n+1} and ∂D_1^{n+1}, respectively. □

We have seen that the boundary 2-convexity of traversing vector fields on X has strong implications for the topology of X (see Lemmas 5.7–5.8, and Corollaries 5.9–5.10).

Recall that, by itself, the boundary 3-convexity and 3-concavity of traversing fields has no topological significance for the topology of 3-folds: in Theorem 4.6 (see also Theorem 9.5 from [K]), we have proved that, for every 3-fold X, any boundary generic v of the gradient type can be deformed into new such field $\tilde{v}$ with $\partial_3 X(\tilde{v}) = \emptyset$. However, in conjunction with certain topological constraints on $\partial_1^+ X$, the boundary 3-convexity has topological implications (see Corollary 2.5 and [K], Corollary 4.2 and Corollary 4.4).

These observations suggest two general questions:

Question 5.4.

- *Given a compact connected smooth manifold X with boundary, which patterns of the stratifications $\{\partial_j^+ X(v) \subset \partial_j X(v)\}_j$ are realizable by boundary generic* traversing *vector fields v on X?* [13]
- *Given two such vector fields, v_0 and v_1, can we find a continuous path $\{v_t\}_{t\in[0,1]}$ in the space $\mathcal{V}_{\mathsf{trav}}(X)$ that connects v_0 to v_1 and avoids certain types of singularities?* [14] *Specifically, if for some $j > 0$, $\partial_j X(v_0) = \emptyset = \partial_j X(v_1)$, is there a linking path so that $\partial_j X(v_t) = \emptyset$ for all $t \in [0,1]$? So that $\partial_{j+1} X(v_t) = \emptyset$ for all $t \in [0,1]$? What are the obstructions to the existence of such paths? (See Figure 4.2 for an insight to these questions.)* ◇

Remark 5.8. The property of the field v in Question 5.4 being *traversing* (equivalently, non-vanishing and of the gradient type) is the essence of the question. For just boundary generic vector fields, there are no known restrictions on the patterns of $\{\partial_j^+ X(v) \subset \partial_j X(v)\}_j$.

Let us illustrate this remark for the vector fields v such that $\partial_3 X(v) = \emptyset$. We divide the n-dimensional boundary $\partial_1 X$ into two complementary domains, Y^+ and Y^-, which share a common boundary $\partial Y^+ = \partial Y^-$—a closed manifold of dimension $n-1$. It may have several connected components. Next, we divide the $(n-1)$-manifold ∂Y^+ into two complementary closed manifolds, Z^+ and Z^-.

We claim that it is possible to find a boundary generic vector field v with the properties: $\partial_1^\pm X(v) = Y^\pm$, $\partial_2^\pm X(v) = Z^\pm$, and $\partial_3 X(v) = \emptyset$. The construction of such a v is quite familiar (see the arguments in Theorem 5.6). We start with a vector field ν_1 which is normal to ∂Y^+ and points outside of Y^+ along Z^- and inside of Y^+ along Z^+. We extend ν_1 to a vector field v_1, tangent to the boundary $\partial_1 X$, so that v_1 has only isolated

[13] Theorem 5.1 and Corollary 5.1 below gives just a taste of a possible answer.
[14] When $\dim(X) = 3$, Theorem 9.5 in [K] addresses some of these questions.

zeros. Let ν be the outward normal vector field of $\partial_1 X$ in X and $h : \partial_1 X \to \mathbb{R}$ a smooth function such that 0 is its regular value and

$$h^{-1}((-\infty, 0]) = Y^-, \quad h^{-1}([0, +\infty)) = Y^+.$$

Along $\partial_1 X$, form the vector field $v' = v_1 + h \cdot \nu$ and extend it to a field v on X with isolated singularities in $\text{int}(X)$. By its construction, v has all the desired properties. Note that here we do not insist on the property $v \neq 0$ in X. $\diamondsuit$

In our inquiry, we are inspired by the Eliashberg *surgery theory of folding maps* [E1], [E2]. Applying this powerful theory to traversing flows requires to overcome two basic obstacles: (1) not any folding map $q : \partial X \to M$, where the target manifold M is of the dimension $\dim(\partial X)$, extends to a map (even better, a surjection) $Q : X \to M$ with fibers homeomorphic to closed intervals or singletons; (2) the v-trajectory space $\mathcal{T}(v)$, the target space of the v-induced quotient map $\partial X \to \mathcal{T}(v)$, is not a manifold. However, in many cases, Eliashberg's results give criteria for realizing given patterns of the strata $\partial_2^\pm X \subset \partial_1^\pm X$ such that $\partial_3 X = \emptyset$, thus answering Question 5.1 for the 3-convex and 3-concave traversing fields.

Let us state one such strong result, Theorem 5.3 from [E2].

Theorem 5.11 (Eliashberg). *Let $X \subset \mathbb{R}^{n+1}$, $n \geq 2$, be a compact connected smooth submanifold of dimension $(n+1)$. Consider two disjoint closed and nonempty $(n-1)$-submanifolds, Z^+ and Z^-, of ∂X such that their union separates ∂X into two complementary n-manifolds, Y^+ and Y^-. Let ν be the outward normal field of ∂X in X. Denote by* $\deg(\nu)$ *the degree of the Gauss map $G_\nu : \partial X \to S^n$, produced by ν. Let $\pi : \mathbb{R}^{n+1} \to \mathbb{R}^n$ be a linear surjection.*

Then the two constraints

- $\chi(Z^+) - \chi(Z^-) = 0,$ *when* $n \equiv 0 \mod 2$,
- $\chi(Z^+) - \chi(Z^-) = 2 \cdot \deg(\nu),$ *when* $n \equiv 1 \mod 2$,

imposed on the Euler numbers, are necessary and sufficient for the existence of a diffeotopy[15] *$\{h_t : \mathbb{R}^{n+1} \to \mathbb{R}^{n+1}\}_{t.\in[0,1]}$ whose final map $h = h_1$ is an orientation-preserving diffeomorphism $h : \mathbb{R}^{n+1} \to \mathbb{R}^{n+1}$ with the following properties:*

- *$Z^+ \cup Z^-$ is the fold locus of the map $(\pi \circ h) : \partial X \to \mathbb{R}^n$,*

[15] a continuous path in the space of smooth diffeomorphisms that starts with the identity map

- *$\pi \circ h$, being restricted to $Z^+ \cup Z^-$, is a immersion, and the image $(\pi \circ h)(Z^+ \cup Z^-)$ has only transversal self-intersections in $\mathbb{R}^n$,*
- *the differential $D(\pi \circ h)$ takes the normal field $\nu|_{Z^\pm}$ to the field inward/outward normal to $(\pi \circ h)(Z^\pm)$ in $(\pi \circ h)(\partial X)$.* ♢

Considering a traversing non-vanishing vector field v that is tangent to the fibers of the map $\pi \circ h$ from Theorem 5.11, leads instantly to the following implication.

Corollary 5.13. *Under the hypotheses and notations from Theorem 5.11, there exists a boundary generic traversing vector field v on X so that:*

- $\partial_1^\pm X(v) = Y^\pm$,
- $\partial_2^\pm X(v) = Z^\pm$,
- $\partial_3 X(v) = \emptyset$. ♢

Thus, at least for smooth domains $X \subset \mathbb{R}^{n+1}$ and for boundary generic traversing vector fields v, which are both 3-convex and 3-concave, the patterns for the strata

$$\partial_2^+ X(v) \coprod \partial_2^- X(v) \subset \partial_1 X$$

are very flexible indeed. However, it seems that the requirements $Z^+ \neq \emptyset$ and $Z^- \neq \emptyset$ put breaks on application of Corollary 5.13 to boundary generic traversing vector fields v on X, subject to the constraint $\partial_3 X(v) = \emptyset$. In particular, Theorem 5.11 does not validate Conjecture 5.1!

Example 5.7. Let us illustrate how nontrivial the conclusions of Theorem 5.11 and Corollary 5.13 actually are.

Let $X = D^{n+1}$, the standard ball, where $n \geq 2$. When n is odd, take any codimension one submanifold $Z^+ \coprod Z^- \subset S^n$ such that $Z^+ \neq \emptyset$, $Z^- \neq \emptyset$, and $\chi(Z^+) - \chi(Z^-) = 2$.

Then D^{n+1} admits a boundary generic traversing vector field v such that $\partial_2^+ D^{n+1}(v) = Z^+$, $\partial_2^- D^{n+1}(v) = Z^-$, and $\partial_3 D^{n+1}(v) = \emptyset$.

In particular, D^4 admits a boundary generic traversing vector field v such that $\partial_2^- D^4(v) = M^2$, the orientable surface of genus 2, $\partial_2^+ D^4(v) = T^2$, the 2-torus, and $\partial_3 D^4(v) = \emptyset$.

When n is even, we may take any codimension one submanifold $Z^+ \coprod Z^- \subset S^n$ such that $Z^+ \neq \emptyset$, $Z^- \neq \emptyset$, and $\chi(Z^+) = \chi(Z^-)$.

Then D^{n+1} admits a boundary generic traversing vector field v such that $\partial_2^+ D^{n+1}(v) = Z^+$, $\partial_2^- D^{n+1}(v) = Z^-$ and $\partial_3 D^{n+1}(v) = \emptyset$.

For example, for any collection of loops $Z^+ \coprod Z^- \subset S^2$, $Z^+ \neq \emptyset$, $Z^- \neq \emptyset$, the disk D^3 admits a boundary generic traversing vector field v such that $\partial_2^+ D^3(v) = Z^+$, $\partial_2^- D^3(v) = Z^-$, and $\partial_3 D^3(v) = \emptyset$.

A "manual" construction of such traversing fields on disks seems to be equally interesting and challenging. ◇

We suspect that an important for our program generalization of Theorem 5.1 is valid and can be established by the methods as in [E1], [E2]. We formulate this anticipation in the form of a conjecture.

Conjecture 5.2. *Let X be a compact connected smooth manifold of dimension $n+1 \geq 3$, equipped with a traversing vector field v. Let Z^+ and Z^- be two disjoint closed and nonempty $(n-1)$-submanifolds of ∂X, whose union separates ∂X into two n-manifolds, Y^+ and Y^-.*

Then the topological constraints

$$\chi(Y^+) = \chi(X) \qquad \textit{when } n \equiv 0 \mod 2 \tag{5.9}$$

$$\chi(Z^+) - \chi(Z^-) = 2 \cdot \chi(X) \qquad \textit{when } n \equiv 1 \mod 2 \tag{5.10}$$

are necessary and sufficient for the existence of an orientation-preserving diffeomorphism $h : X \to \mathrm{int}(X)$ with the following properties:

- *the restriction of v to the image $h(X)$ is traversing and boundary generic*[16] *(relative to $h(\partial X)$),*
- $\partial_1^{\pm}\big(h(X)\big)(v) = h(Y^{\pm})$,
- $\partial_2^{\pm}\big(h(X)\big)(v) = h(Z^{\pm})$,
- $\partial_3\big(h(X)\big)(v) = \emptyset$.

Moreover, in a given collar U of ∂X in X, there is a U-supported diffeotopy $\{h_t : X \to h_t(X) \subset X\}_{t \in [0,1]}$ with $h = h_1$ as above. ◇

To prove the necessity of the topological constraints (5.9) and (5.10) is straightforward. By the Morse formula (5.2) (see also Corollary 5.1), a necessary condition for the existence of a diffeomorphism h with the desired properties, described in the bullets, is the constraint

$$\chi(h(X)) - \chi(h(Y^+)) + \chi(h(Z^+)) = i(v|_{h(X)}) = 0.$$

[16] *and even traversally generic in the sense of Definition* 6.3

Since h is a homeomorphism, this equation is equivalent to

$$\chi(X) - \chi(Y^+) + \chi(Z^+) = 0. \tag{5.11}$$

If $n \equiv 1 \mod 2$, then

$$\chi(Y^+) = \frac{1}{2}\chi(\partial Y^+) = \frac{1}{2}(\chi(Z^+) + \chi(Z^-)).$$

Therefore, using formula (5.11), the constraint becomes

$$2\chi(X) = \chi(Z^-) - \chi(Z^+)$$

—formula (5.10).

When $n \equiv 0 \mod 2$, since Z^+, Z^- are closed odd-dimensional manifolds, $\chi(Z^+) = 0 = \chi(Z^-)$, and formula (5.11) reduces to $\chi(X) = \chi(Y^+)$—formula (5.9).

Therefore the topological constraints (5.9) and (5.10), imposed on the "candidates" Z^+, Z^- and Y^+ and Y^-, are necessary for the existence of the desired diffeomorphism h.

To prove the sufficiency of these conditions may require a clever application of the h-principle (see [EM]) in the spirit of section 6.3 from [E2].

Corollary 5.14. *Assuming the validity of Conjecture 5.2, any compact smooth connected manifold X with boundary admits a boundary generic traversing field v with the property $\partial_3 X(v) = \emptyset$.*

Proof. By Corollary 5.8, $\mathcal{V}_{\mathsf{trav}}(X) \neq \emptyset$. So we can start with a traversing field v and apply Conjecture 5.2 to it to get the pull-back field $h^*(v)$ with the desired properties. □

The following conjecture reflects our thinking behind Question 5.4. It also influenced the notion of truly generic family of $3D$-flows, introduced in Definition 4.1.

Conjecture 5.3. *Let v_0 and v_1 be two boundary generic traversing vector fields on a smooth compact connected manifold X with boundary, such that $\partial_3 X(v_0) = \emptyset = \partial_3 X(v_1)$. Assume that v_0 is connected by a continuous path to v_1 in the space of traversing vector fields on X.*

Then there is a 1-parameter family of boundary generic traversing vector fields $\{v_t\}_{t\in[0,1]}$ which connects v_0 to v_1 and such that only for finitely many instances $t \in [0,1]$, $\partial_3 X(v_t) \neq \emptyset$. For those exceptional t's, $\partial_4 X(v_t) = \emptyset$.

◊

5.6 On the concave traversing flows whose trajectories are of the combinatorial types (11) and (121) only

We have seen on many occasions that the existence of a boundary 2-convex vector field v on a manifold X puts several restrictions on its topology. Stated differently, if X admits a traversing boundary generic vector field v with trajectories of the combinatorial types (11) and (2) only, the manifold X is diffeomorphic to a compact manifold times an interval, the corners of the product being rounded. The next level of complexity is represented by manifolds X that admit traversing vector fields with trajectories of the combinatorial types (11) and (121) only; such fields are boundary concave! These manifolds X also admit a relatively simple and constructive description, which will preoccupy us for the rest of this chapter.

Lemma 5.9. *If a compact connected smooth $(n+1)$-manifold X with boundary admits a traversing vector field v with trajectories of the combinatorial types* (11) *and* (121) *only*[17]*, then there exists a compact orientable smooth n-manifold Z and a regular embedding $\phi : \partial Z \to \mathrm{int}(Z)$ with trivial normal bundle, so that the canonical thickening of the n-dimensional CW-complex*

$$Z_\phi =_{\mathsf{def}} \big((\partial Z \times [0,1]) \coprod Z\big)/\{z \times \{0\} \sim z;\ z \times \{1\} \sim \phi(z)\}, \quad (5.12)$$

where $z \in \partial Z$, is diffeomorphic to X.

Proof. Take $Z = \partial_1^+ X(v)$ and consider the union W of the downward trajectories streaming from $\partial_2^+ X(v) = \partial(Z)$. Since the only trajectories through $\partial_2^+ X(v)$ are of the type (121), W is diffeomorphic to $\partial Z \times [0,1]$. Then, as in the proof of Theorem 2.5 (see Figure 2.12 and 2.13), X is diffeomorphic to a unique thickening of the gradient spine

$$K(v) =_{\mathsf{def}} \partial_1^+ X(v) \cup W$$

that is consistent with its TN-markers (see Section 2.6).

Now put $\phi(x) = x' \in \gamma_x \cap Z$, where $x \in \partial Z$ and x' is the unique point in $\gamma_x \cap Z$ of multiplicity 1. Then, evidently, $K(v) = Z_\phi$. □

Let $H_*(\sim)$ stands for the homology with an arbitrary abelian group of coefficients.

Lemma 5.9 leads to a few corollaries.

Corollary 5.15. *Let a connected compact smooth $(n+1)$-manifold X with boundary admit a boundary generic traversing vector field v with trajectories*

[17]Such fields are automatically traversally generic in the sense of Definition 6.3.

of the combinatorial types (11) and (121) only. Assume that the natural homomorphisms $H_k(\partial_1 X) \to H_k(X)$ are trivial for all $k \in [1, n-1]$.

Then, for each $k \in [2, n-1]$, there exists a short exact sequence

$$0 \to H_k(X) \to H_{k-1}(\partial_2^+ X(v)) \to H_{k-1}(\partial_1^+ X(v)) \to 0.$$

For $k = 1$, we get the exact sequence

$$0 \to H_1(X) \to H_0(\partial_2^+ X(v)) \to H_0(\partial_1^+ X(v)) \to \mathbb{Z} \to 0,$$

and for $k = n$, the exact sequence

$$0 \to H_n(\partial_1^+ X(v)) \to H_n(X) \to H_{n-1}(\partial_2^+ X(v)) \to H_{n-1}(\partial_1^+ X(v)) \to 0.$$

Thus the difference

$$\mathrm{rk}\big(H_{k-1}(\partial_2^+ X(v))\big) - \mathrm{rk}\big(H_{k-1}(\partial_1^+ X(v))\big) = \mathrm{rk}\big(H_k(X)\big)$$

is a topological invariant for all $k \in [2, n-1]$.

For $k = 1$, we get

$$\mathrm{rk}\big(H_0(\partial_2^+ X(v))\big) - \mathrm{rk}\big(H_0(\partial_1^+ X(v))\big) = \mathrm{rk}\big(H_1(X)\big) - 1.$$

In particular, all these results hold when $\partial_1 X$ is a disjoint union of several H_*-homology n-spheres.

Proof. Let W be the waterfall from $\partial_2^+ X(v)$ and let T be its basis, the locus where W hits $\partial_1^+ X(v)$ transversally[18]. Let $K =_{\mathsf{def}} \partial_1^+ X(v) \cup W$, and let U be a small open regular neighborhood in W of its basis T. Put $U^\circ =_{\mathsf{def}} U \setminus T$. Thus, U° is homeomorphic to $\partial_2^+ X(v) \times (0,1)$. Let $V =_{\mathsf{def}} K \setminus U^\circ$. Evidently, the "resolution" V of K (see Figure 2.19) is homeomorphic to $\partial_1^+ X(v)$. These conclusions rely on the property of v to have trajectories of the combinatorial types (11) and (121) only.

Consider the long exact homology sequence of the pair (K, V):

$$\cdots \to H_k(V) \to H_k(K) \to H_k(K, V) \to H_{k-1}(V) \to \ldots.$$

Using the excision property for the pairs $(K, V) \supset (K \setminus V^\circ, V \setminus V^\circ)$ and the isomorphisms $H_*(K) \approx H_*(X)$, $H_*(V) \approx H_*(\partial_1^+ X(v))$, we rewrite the long exact sequence as

$$\cdots \to H_k(\partial_1^+ X(v)) \to H_k(X) \to H_k(\partial_2^+ X(v) \times I,\ \partial_2^+ X(v) \times \partial I) \\ \to H_{k-1}(\partial_1^+ X(v)) \to \ldots,$$

where $I = [0, 1]$.

[18]In Chapter 2, we have used the notation $s_\dagger(K)$ for the set T.

Employing the suspension isomorphism, we get

$$\cdots \to H_k(\partial_1^+ X(v)) \to H_k(X) \to H_{k-1}(\partial_2^+ X(v)) \to H_{k-1}(\partial_1^+ X(v)) \to \dots. \tag{5.13}$$

Since the natural homomorphism $H_l(\partial_1^+ X(v)) \to H_l(X)$ factors through the natural homomorphism $H_l(\partial_1 X) \to H_l(X)$ and the later one is assumed to be trivial for all $l \in [1, n-1]$, we conclude that $H_l(\partial_1^+ X(v)) \to H_l(X)$ must be trivial as well. Therefore (5.13) splits into short exact sequences

$$0 \to H_k(X) \to H_{k-1}(\partial_2^+ X(v)) \to H_{k-1}(\partial_1^+ X(v)) \to 0$$

for all $k \in [2, n-1]$.

For $k = 1$ or for $k = n$, the exact sequence (5.13) reduces to the two exact sequences that are described in the formulation of Corollary 5.15. □

The next corollary claims that "X"s that admit traversing flows with trajectories of the combinatorial types $(11), (121)$ only tend to have large fundamental groups.

Corollary 5.16. *If a connected compact smooth $(n+1)$-manifold X with boundary admits a boundary generic traversing vector field v with trajectories of the combinatorial types* (11) *and* (121) *only*[19], *then the fundamental group $\pi_1(X)$ admits an epimorphism onto the* free *group $\mathsf{F}_{\chi(v)}$ of rank*

$$\chi(v) =_{\mathsf{def}} 1 + |\pi_0(\partial_2^+ X(v))| - |\pi_0(\partial_1^+ X(v))|.$$

If the natural homomorphism $H_1(\partial_1 X; \mathbb{Z}) \to H_1(X; \mathbb{Z})$ is trivial, then $\chi(v) = \beta_1(X)$, the first Betti number. In such a case, there exists an epimorphism $h_* : \pi_1(X) \to \mathsf{F}_{\beta_1(X)}$.

Proof. Let $K =_{\mathsf{def}} \partial_1^+ X(v) \cup W$ be the gradient spine of X (see the proof of Lemma 5.9 for the description of W). Consider a connected graph G whose vertexes correspond to the connected components $\{\partial_1^+ X(v)_\alpha\}$ of $\partial_1^+ X(v)$ and whose edges correspond to the connected components $\{\partial_2^+ X(v)_\beta\}$ of $\partial_2^+ X(v)$. In forming G, we use that each connected component $\partial_2^+ X(v)_\beta$ belongs to a unique component $\partial_1^+ X(v)_{i(\beta)}$ and is mapped by a map ϕ (as in Lemma 5.9) to a unique component $\partial_1^+ X(v)_{\phi(\beta)}$ of $\partial_1^+ X(v)$, the component that is hit by the waterfall streaming from $\partial_2^+ X(v)_\beta$. So, in G, we connect the vertexes $\alpha = i(\beta)$ and $\alpha' = \phi(\beta)$ by an edge, labeled with β. Since K is

[19] Such v is automatically concave.

connected, so is the graph G. Therefore the number of independent cycles in G is

$$\chi(v) = 1 + |\pi_0(\partial_2^+ X(v))| - |\pi_0(\partial_1^+ X(v))|.$$

As a result, $\pi_1(G)$ is a free group in $\chi(v)$ generators.

Let $h : K \to G$ be the continuous map which takes each component $\partial_1^+ X(v)_\alpha$ to the appropriate vertex α of G and each waterfall $W_\beta \approx \partial_2^+ X(v)_\beta \times [0,1]$ to the edge that connects the vertexes $i(\beta)$ and $\phi(\beta)$. By the construction of h, each cycle γ in G can be lifted against h to a loop $\tilde{\gamma} \subset K$. Therefore the fundamental group $\pi_1(X) \approx \pi_1(K)$ admits an epimorphism h_* onto the free group $\pi_1(G) \approx \mathsf{F}_{\chi(v)}$.

Applying Corollary 5.15 for the case $k = 1$, we see that $\chi(v) = \mathrm{rk}(H_1(X;\mathbb{Z}))$ when the homomorphism $H_1(\partial_1^+ X(v);\mathbb{Z}) \to H_1(X;\mathbb{Z})$ is trivial. The later property holds when $H_1(\partial_1 X;\mathbb{Z}) \to H_1(X;\mathbb{Z})$ is trivial. □

Corollary 5.17. *If the natural homomorphism $H_1(\partial_1 X;\mathbb{Z}) \to H_1(X;\mathbb{Z})$ is trivial and the fundamental group $\pi_1(X)$ does not admit an epimorphism $h_* : \pi_1(X) \to \mathsf{F}_{\beta_1(X)}$, then any traversing boundary generic and* concave *field v on X must have trajectories of the combinatorial type $(1\underbrace{2\dots2}_{k}1)$, where $k \geq 2$.*

Proof. If v is such that $\partial_2 X(v) = \emptyset$, then X is diffeomorphic to a product of a closed manifold $\partial_1^+ X(v)$ with an interval. This outcome contradicts the assumption that $H_1(\partial_1 X;\mathbb{Z}) \to H_1(X;\mathbb{Z})$ is trivial. So, if $\partial_2^- X(v) = \emptyset$ (i.e., v is concave), then $\partial_2^+ X(v) \neq \emptyset$. In other words, v must have a trajectory of the type (121).

Employing the hypotheses, by Corollary 5.16, any traversing boundary generic and *concave* field v on X must have trajectories of combinatorial types different from (11), and (121).

Recall that $\partial_2^- X(v) \cap \partial_2^+ X(v) = \partial_3 X(v)$. Therefore, if $\partial_2^- X(v) = \emptyset$ (the field is concave), then $\partial_3 X(v) = \emptyset$. As a result, $\partial_j X(v) = \emptyset$ for all $j \geq 3$. Therefore, the combinatorial patterns $\omega = (12\dots21)$, where the number of 2's is at least two, must be realized by the v-flow. □

The next Corollary 5.18 should be compared with Corollary 5.10. The latter claims that the homology $H_n(X;\mathbb{Z}) = 0$ if a $(n+1)$-manifold X admits a boundary concave traversing field, unless $X = Y \times [0,1]$, where Y is a closed n-manifold. As you compare the two corollaries, keep in mind that the boundary generic vector fields with trajectories of the types (11) and (121) only *are* boundary concave, but not any boundary concave vector field has such a restricted list of combinatorial types.

Corollary 5.18. *If a connected compact and orientable smooth $(n+1)$-manifold X with boundary is such that $H_n(X) = 0$ and the natural homomorphisms $H_k(\partial_1 X) \to H_k(X)$ are trivial for all $k \in [1, n-1]$, then any boundary generic traversing and* concave *vector field v on X must have trajectories of the combinatorial type* (1221)[20].

Proof. Consider the fragment of the long exact sequence from (5.13):

$$0 \to H_n(\partial_1^+ X(v)) \to H_n(X) \to H_{n-1}(\partial_2^+ X(v)) \to H_{n-1}(\partial_1^+ X(v)) \to 0.$$

Here the group of coefficients is $\mathbb{Z}$. In the exact sequence above, the first term $H_n(\partial_1^+ X(v)) \approx \mathbb{Z}^q$, where q is the number of connected components of $\partial_1^+ X(v)$ that are closed. These components do not contain any points from $\partial_2^+ X(v)$ and define some nontrivial cycles-generators in $H_n(\partial_1 X)$. However, by the hypotheses, $H_n(X) = 0$. So, by the exact sequence, $H_n(\partial_1^+ X(v))$ vanish as well ($q = 0$), and $\mathcal{I} : H_{n-1}(\partial_2^+ X(v)) \to H_{n-1}(\partial_1^+ X(v))$ must be an isomorphism.

Since, for a concave v, $\partial_3 X(v) = \emptyset$, we conclude that $\partial_2^+ X(v)$ is a closed orientable manifold. So the fundamental cycle $\alpha =_{\mathsf{def}} [\partial_2^+ X(v)] \in H_{n-1}(\partial_2^+ X(v))$ must be nontrivial. On the other hand, since α is a boundary of the n-chain $[\partial_1^+ X(v)]$, we conclude that $\mathcal{I}(\alpha) = 0$. This conclusion contradicts to the fact that $\mathcal{I}$ is an isomorphism. Therefore X does not admit a concave vector field with the combinatorial types (11), (121) *only.*

If $\partial_2 X(v) = \emptyset$, then X is diffeomorphic to the product of a *closed* n-manifold M with an interval. If X is orientable, so is M. In such a case, $H_n(X) \approx H_n(M) \neq 0$. Therefore, the hypothesis $H_n(X) = 0$ implies that $\partial_2 X(v) \neq \emptyset$. Using the hypotheses $\partial_2^- X(v) = \emptyset$, we conclude that v has trajectories of the combinatorial type (121).

On the other hand, by an argument as in the proof of Corollary 5.17, which rules out the trajectories of the combinatorial types $\omega = (j_1, j_2, \dots, j_q)$ in which $j_i \geq 3$ are present, v must have multi-tangent trajectories of the combinatorial type $(1\underbrace{2\dots2}_{k}1)$, where $k \geq 2$. In turn, this implies that v must have trajectories of the combinatorial type (1221). □

[20] For a traversally generic v, $\dim\big(\mathcal{T}(v, (1221))\big) = n - 1$.

Corollary 5.19. *Consider the spherical fibrations $\phi : S^{2k+1} \to \mathbb{CP}^k$ and $\psi : S^{4k+3} \to \mathbb{HP}^k$, generated by the classical S^1- and S^3-actions on the corresponding spheres. Let X be the space of the smooth disk bundle, associated with the spherical bundles ϕ or ψ.*

Then X does not admit a traversing boundary generic vector field v with trajectories of the combinatorial types (11) *and* (121) *only. In particular, any concave vector field on such X must have a multi-tangent to the sphere ∂X trajectory.*

Proof. Evidently, the natural homomorphisms $H_*(S^n) \to H_*(X)$ are trivial in all positive dimensions. Also $H_n(X) = 0$, where $n = 2k+1$ or $n = 4k+3$, respectively. By Corollary 5.18, the claim follows. □

Example 5.8. In particular, the manifold $X = \mathbb{CP}^2 \setminus D^4$ does not admit a boundary generic and traversing vector field v with trajectories of the combinatorial types (11) and (121) only. In fact, any such concave v on X must have a double-tangent trajectory of the type (1221).

In contrast, $X = \mathbb{CP}^2 \setminus (D^4 \sqcup D^4 \sqcup D^4)$ does admit a traversing vector field with trajectories of the combinatorial types (11) and (121) only. Such a vector field is the gradient field of a minimal Morse function $f : \mathbb{CP}^2 \to \mathbb{R}$, being restricted to the complement of three small convex balls that contain the critical points of f. From this perspective,

$\mathbb{CP}^2 \setminus D^4$ is *more complex* than $\mathbb{CP}^2 \setminus (D^4 \sqcup D^4 \sqcup D^4)$!

◇

Chapter 6

Traversally Generic and Versal Flows: Semi-algebraic Models of Tangency to Boundary

6.1 In this chapter ...

This chapter serves as an analytical foundation for the investigation of *boundary generic* (see Definition 5.1) and *traversally generic* (see Definition 6.3) vector fields v on compact smooth manifolds X with boundary. The later category of vector fields have a number of pleasing properties. We will show that the trajectory spaces of traversally generic vector fields are compact CW-complexes; in contrast, for vector fields v that vanish on X, the v-trajectory spaces are pathological.

The analytical tools that we are going to develop in this chapter will lead us to semi-algebraic local models of the patterns in which generic nonsingular vector flows on X interact with the boundary ∂X. Here the word "local" refers to the vicinity of any given trajectory γ of the v-flow. The $3D$- and $4D$-versions of these models are familiar to the reader from the first part of the book.

As in low dimensional cases, for smooth boundary generic fields, each intersection point $a \in \gamma \cap \partial X$ comes with a positive integral multiplicity $j(a)$ attached to it. This multiplicity $j(a)$ can be given a number of competing but equivalent definitions, one of which uses the *Morse stratification* (see Definition 5.1 and formula (5.1)), the focus of Chapter 5. Naively, one can think of $j(a)$ as a *multiplicity of tangency* between γ and ∂X. So, surprisingly, the smooth topology of the flow can distinguish between, say, degree 2 and degree 4 tangency patterns!

Lemma 6.1 and Lemma 6.4 describe the models for the boundary ∂X and the field v in the vicinity of a point $a \in \gamma \cap \partial X$ and in the vicinity of a trajectory γ, respectively. It turns out that, for traversally generic fields, in special flow-adjusted coordinates $(u, \vec{x})$, the boundary is given by a real polynomial equation $P(u, \vec{x}) = 0$ of degree that depends on γ and does not exceed $2 \cdot \dim(X)$. The manifold X is given by the polynomial inequality $P(u, \vec{x}) \leq 0$. The polynomial $P(u, \vec{x})$ depends only on the ordered sequence of multiplicities $\{j(a)\}_{a \in \gamma \cap \partial X}$. So, in each dimension m, there are only finitely many semi-algebraic models for the vicinity of v-trajectories γ in X, where $\dim(X) = m$.

Along the way, we introduce accurately a variety of already familiar spaces that correspond to different types of vector fields on X, the space $\mathcal{B}^\dagger(X)$ of boundary generic vector fields, the space $\mathcal{V}^\dagger(X)$ of boundary generic traversing vector fields, and the space $\mathcal{V}^\ddagger(X)$ of traversally generic vector fields are among them.

Two theorems describe our main results: Theorem 6.6 claims that $\mathcal{B}^\dagger(X)$ is an open and dense in the space of all smooth fields $\mathcal{V}(X)$, and Theorem 6.7 claims that $\mathcal{V}^\ddagger(X)$ is open and dense in the space $\mathcal{V}_{\mathsf{trav}}(X)$ of all *traversing* (equivalently, all gradient-like non-vanishing) fields. Recall that traversing fields have only trajectories that are homeomorphic to closed intervals or singletons.

Theorem 6.10, a technical result, claims that the trajectory space of a traversally generic vector flow is a Whitney stratified space, and therefore admits a finite triangulation. This theorem leads to Theorem 6.8 which claims that, for a traversally generic vector field v on X, the trajectory space $\mathcal{T}(v)$ is naturally homotopy equivalent to X.

The Origami Theorem 6.11 shows that any smooth compact connected manifold X with boundary has a traversally generic vector flow, so that its trajectory space $\mathcal{T}(v)$ can be viewed as an origami, folded from a n-dimensional ball, where $n = \dim(\partial X)$. Moreover, X admits an embedding $\alpha : X \to \mathcal{T}(v) \times \mathbb{R}$ which commutes with the obvious projections on $\mathcal{T}(v)$.

Theorem 6.12 describes a glimpse of the phenomenon of *topological holography* for boundary generic traversing flows.

6.2 Morin's local models: how nonsingular flows interact with boundary

Let v be a boundary generic vector field on a smooth compact connected $(n+1)$-manifold X with boundary ∂X. Recall that a boundary generic vector field v on X gives rise to two stratifications as in (5.1), the first stratification by closed submanifolds, the second one by compact submanifolds. Here $\dim(\partial_j X) = \dim(\partial_j^+ X) = n+1-j$. For simplicity, the notations "$\partial_j^\pm X$" do not reflect the dependence of these strata on the vector field v.

When the field varies, we use a more accurate notation "$\partial_j^{\pm} X(v)$".

As before, we can add an external collar to X and smoothly extend the vector field v to a vector field $\hat{v}$ on a larger manifold $\hat{X}$ (which contains X properly) without introducing new singularities.

At each point $x \in \partial_1 X$, the $(-\hat{v})$-flow defines the germ of the projection $p_x : \hat{X} \to S_x$, where S_x is a local transversal section of the $\hat{v}$-flow. The projection p_x is considered in the vicinity of each point $x \in \partial_1 X \subset \hat{X}$. When $\hat{v}$ is a gradient-like field for a submersion $\hat{f} : \hat{X} \to \mathbb{R}$, we can choose the germ of the hypersurface $\hat{f}^{-1}(f(x))$ for the role of S_x.

Let $\mathcal{V}_{\neq 0}(X)$ be the space of smooth *non-vanishing* vector fields, equipped with the C^∞-topology.

A theorem of Morin [Morin] describes all local models of $p_x : \partial_1 X \to S_x$ for a G_δ, or *residual*[1], set of fields in $\mathcal{V}_{\neq 0}(X)$.

Let us introduce and describe these models. For any integer $s \in [1, n+1]$, we form the polynomial

$$Q_s(u_1, u_2, \ldots, u_{n-1}, \mathsf{u}_n) =_{\mathsf{def}} \mathsf{u}_n^s + \sum_{i=0}^{s-2} u_i\, \mathsf{u}_n^i, \tag{6.1}$$

in n real variables, one of which, u_n, will play a distinct role. It is a polynomial of degree s in u_n and linear in the rest of the variables $u_1, u_2, \ldots, u_{n-1}$.

We consider the map $\mu_s : \mathbb{R}^n \to \mathbb{R}^{n+1}$, given by

$$\mu_s : (u_1, u_2, \ldots, u_{n-1}, \mathsf{u}_n) \longrightarrow (u_1, u_2, \ldots, u_{n-1}, Q_s, \mathsf{u}_n). \tag{6.2}$$

Let $(y_1, \ldots, y_n, y_{n+1})$ be coordinates in $\mathbb{R}^{n+1}$. The constant field $e_{n+1} =_{\mathsf{def}} \partial_{y_{n+1}}$, will model the nonsingular field $\hat{v}$ on $\hat{X}$.

Consider the projection $\pi : \mathbb{R}^{n+1} \to \mathbb{R}^n$, defined by the formula

$$\pi : (y_1, y_2, \ldots, y_n, y_{n+1}) \to (y_1, y_2, \ldots, y_n). \tag{6.3}$$

Then the composition $\pi \circ \mu_s$ is given by the formula

$$(u_1, u_2, \ldots, u_{n-1}, \mathsf{u}_n) \longrightarrow (u_1, u_2, \ldots, u_{n-1}, \mathsf{u}_n^s + \sum_{i=0}^{s-2} u_i\, \mathsf{u}_n^i). \tag{6.4}$$

Let us denote by λ the line distribution $\ker(D\pi)$, tangent to the fibers of the projection $\pi : \mathbb{R}^{n+1} \to \mathbb{R}^n$.

The following result [Morin] is of the key importance for us.

[1]that is, a countable intersection of open and dense subsets in $\mathcal{V}_{\neq 0}(X)$

Theorem 6.1 (Morin). *For a G_δ-set of 1-dimensional distributions ℓ on $\hat{X}$ and any point $x \in \partial_1 X$, there is a neighborhood U of x in $\hat{X}$, a diffeomorphism $h : U \to \mathbb{R}^{n+1}$, and an integer $s \in [1, n+1]$ such that*

- *$h(x) = 0 \in \mathbb{R}^{n+1}$,*
- *$h(\partial_1 X \cap U) = \mu_s(\mathbb{R}^n)$, where μ_s is defined by formula (6.2),*
- *the distribution $\ell|_U$ is mapped by the differential $Dh : TU_x \to T\mathbb{R}^{n+1}$ to the distribution λ.* ◇

Note that Morin's theorem allows a priori for the four-fold ambiguity: (1) the distribution ℓ can be given two possible orientations (i.e. the model field can be $\pm e_{n+1}$), and (2) the manifold X can occupy each of the two chambers, in which $\mathbb{R}^{n+1}$ is locally divided by $\mu_s(\mathbb{R}^n)$.

We can describe these local models by replacing their $\vec{u}$-parametric form in formula (6.4) with equations in the new coordinates

$$(u, x_0, x_1 \dots, x_{n-1}) =_{\text{def}} (y_{n+1}, -y_n, y_1, \dots, y_{n-1}).$$

Let

$$P_s(u, x) =_{\text{def}} Q_s(y_1, \dots, y_{n-1}, y_{n+1}) - y_n = u^s + \sum_{i=0}^{s-2} x_i u^i, \qquad (6.5)$$

so that $h(\partial_1 X)$ is given by the equation $\{P_s(u, x_0, x_1, \dots, x_{s-2}) = 0\}$.

For a fixed vector $x =_{\text{def}} (x_0, \dots, x_{n-1})$, P_s is a depressed polynomial in u.

Therefore, for each point $x \in \mathbb{R}^n$, the points of the boundary $\{P(u, x) = 0\}$, residing in the fiber $\pi^{-1}(x)$ of the projection $\pi : \mathbb{R}^{n+1} \to \mathbb{R}^n$, are the real-valued zeros of the u-polynomial $P_s(u, x)$ (which employs only the variables $x_0, \dots, x_{s-2}$).

With each $x \in \mathbb{R}^n$, we associate *the real zero divisor* $D_s(x)$ of the u-polynomial $P_s(u, x)$. Its support resides in $\mathbb{R}$. We will be particularly interested in the *ordered* sequence $\omega(x) = (\omega_1, \dots, \omega_i, \dots, \omega_q)$ of multiplicities that is generated by the divisor $D_s(x)$.

Put $e_{n+1} =_{\text{def}} \frac{\partial}{\partial u}$. We abbreviate $\left(\frac{\partial}{\partial u}\right)^j P_s$ as $P_s^{(j)}$. Let $w =_{\text{def}} (u, x) \in \mathbb{R} \times \mathbb{R}^n \approx \mathbb{R}^{n+1}$.

For the local models as in formula (6.5), the Morse stratification $\{\partial_j^+ X =_{\text{def}} \partial_j^+ X(\partial_u)\}_j$ (see formula (5.1)) can be expressed in terms of the divisor $D_s(x)$ and labeled by the ordered list of multiplicities $\omega(x)$.

Theorem 6.2. *Let X be a compact connected smooth $(n+1)$-dimensional manifold with boundary. For a G_δ-set of smooth vector fields v on X,*

$v|_{\partial_1 X} \neq 0$, and for each point $a \in \partial_1 X$, there exists an integer $s \in [1, n+1]$ such that the models for v and the v-induced strata $\{\partial_j^+ X\}_{1 \leq j \leq n+1}$, in the vicinity of a, are given by one of the four real semi-algebraic sets:

(1) $X = \{w \in \mathbb{R}^{n+1} | \; P_s(w) \geq 0\}$ *and* $v = e_{n+1}$;

- $\partial_j X = \{w \in \mathbb{R}^{n+1} | \; P_s^{(i)}(w) = 0 \text{ for all } i < j\}$
- $\partial_j^+ X = \{w \in \mathbb{R}^{n+1} | \; P_s^{(i)}(w) = 0 \text{ for all } i < j, \text{ and } P_s^{(j)}(w) \geq 0\}$

(2) $X = \{w \in \mathbb{R}^{n+1} | \; P_s(w) \leq 0\}$ *and* $v = e_{n+1}$;

- $\partial_j X = \{w \in \mathbb{R}^{n+1} | \; P_s^{(i)}(w) = 0 \text{ for all } i < j\}$
- $\partial_j^+ X = \{w \in \mathbb{R}^{n+1} | \; P_s^{(i)}(w) = 0 \text{ for all } i < j, \text{ and } P_s^{(j)}(w) \leq 0\}$

(3) $X = \{w \in \mathbb{R}^{n+1} | \; P_s(w) \geq 0\}$ *and* $v = -e_{n+1}$;

- $\partial_j X = \{w \in \mathbb{R}^{n+1} | \; P_s^{(i)}(w) = 0 \text{ for all } i < j\}$
- $\partial_j^+ X = \{w \in \mathbb{R}^{n+1} | \; P_s^{(i)}(w) = 0 \text{ for all } i < j, \text{ and } (-1)^j P_s^{(j)}(w) \geq 0\}$

(4) $X = \{w \in \mathbb{R}^{n+1} | \; P_s(w) \leq 0\}$ *and* $v = -e_{n+1}$;

- $\partial_j X = \{w \in \mathbb{R}^{n+1} | \; P_s^{(i)}(w) = 0 \text{ for all } i < j\}$
- $\partial_j^+ X = \{w \in \mathbb{R}^{n+1} | \; P_s^{(i)}(w) = 0 \text{ for all } i < j, \text{ and } (-1)^j P_s^{(j)}(w) \leq 0\}$.

Proof. The following argument is based on Theorem 6.1; however, in Lemma 6.1, we will present a proof that validated Theorem 6.1. By the second bullet in that theorem, the boundary $\partial_1 X$ is given locally by the equation $\{P_s(w) = 0\}$. By formula (6.5), the gradient

$$\nabla P_s = \big(P_s^{(1)}(u, x), 1, u, u^2, \ldots, u^{s-2}, 0, \ldots 0\big). \tag{6.6}$$

The orthogonality of e_{n+1} and $\nu_1 =_{\mathsf{def}} \nabla P_s$ defines the locus $\partial_2 X$. Hence, the orthogonality is equivalent to the constraint $P_s^{(1)}(x, u) = 0$, and $\partial_2 X$ is determined by two equations: $\{P_s = 0\}$ and $\{P_s^{(1)} = 0\}$. Formulas (6.5) and (6.6) imply that the vector $\nu_2 =_{\mathsf{def}} \nu_1^{(1)} = \nabla[P_s^{(1)}]$ has $P_s^{(2)}$ for its first coordinate. Since $\partial_3 X$ is the locus where e_{n+1} is tangent to $\partial_2 X$ and ν_2 is orthogonal the hypersurface $\{P_s^{(1)} = 0\} \supset \partial_2 X$, the vectors ν_2 and e_{n+1} must be orthogonal along $\partial_3 X$. This leads to the equation $\{P_s^{(2)} = 0\}$.

Using this type of argument repeatedly for the linear independent vector fields $\{\nu_j =_{\mathsf{def}} \nu_1^{(j)}\}_j$, proves the first bullets in claims (1) and (2) of the theorem.

When X is defined by the inequality $\{P_s \geq 0\}$, e_{n+1} points inside X at $w = (u, x) \in \partial_1 X$ if and only if the dot product $e_{n+1} \cdot \nu_1(w) \geq 0$. That is, $\partial_1^+ X$ is defined by $\{P_s^{(1)} \geq 0\}$ together with $\{P_s = 0\}$. On the other hand, when X is defined by the inequality $\{P_s \leq 0\}$, e_{n+1} points inside X at $w \in \partial_1 X$ if and only if $P_s^{(1)}(w) \leq 0$ and $P_s(w) = 0$.

Note that

$$\partial_u(\nu_{j-1} \cdot e_{n+1}) = \partial_u(\nu_{j-1}) \cdot e_{n+1} \; + \; \nu_{j-1} \cdot (\partial_u e_{n+1}) = \nu_j \cdot e_{n+1}. \quad (6.7)$$

Along $\partial_2 X$, in view of formula (6.7), the property $\nu_2 \cdot e_{n+1} \geq 0$ is equivalent to the inequality $\partial_u(\nu_1 \cdot e_{n+1}) \geq 0$. When, along $\partial_2 X$, the Lie derivative

$$\mathcal{L}_{e_{n+1}}(\nu_1 \cdot e_{n+1}) = \partial_u(\nu_1 \cdot e_{n+1}) = \nu_2 \cdot e_{n+1}$$

is nonnegative, the field e_{n+1} points inside $\partial_1^+ X$; otherwise, it points inside $\partial_1^- X$. Therefore, $\partial_2^+ X$ is depicted by the inequality $\{P_s^{(2)} = \nu_2 \cdot e_{n+1} \geq 0\}$, coupled with the pair of equalities $\{P_s = 0, P_s^{(1)} = 0\}$. The general case of

$$\partial_j^+ X = \{P_s^{(i)} = 0\}_{i<j} \cap \{P_s^{(j)} \geq 0\}$$

is analogous.

The argument in case (2) of the theorem is very similar to case (1).

We notice that flipping the direction of the field $v = e_{n+1}$ does not change the polarity of the strata with even j's and reverses the polarity of the strata with the odd ones.

Finally, we combine this description of stratifications $\{\partial_j^+\{P_s \geq 0\}\}_j$, $\{\partial_j^+\{P_s \leq 0\}\}_j$ with Morin's Theorem 6.1 to get the desired local models for a G_δ-set of vector fields in the space $\mathcal{V}(X)$. □

Remark 6.1. Note that Theorem 6.2 does not describe local models for generic Morse data (f, v) $(df(v) > 0)$ in the vicinity of a typical point $a \in \partial_1 X$, just finitely many local models for boundary generic nonsingular fields v; in other words, $\pm e_{n+1}$ mimics v, but the coordinate function $\pm u$, in general, does not represent $f : X \to \mathbb{R}$! ◇

Corollary 6.1. *For the vector fields v that have Morin normal forms as in formula* (6.5) *and in Theorem 6.1, all the strata $\partial_j X$ are manifolds, and the embeddings $\partial_j X \subset \partial_{j-1} X$ are regular. Moreover, these fields are boundary generic in the sense of Definition 5.1.*

In the vicinity of each point $a \in \partial_1 X$ of the s-type, we get $\partial_{s+1} X = \emptyset$, so that the s-type fields are both boundary $(s+1)$-convex and $(s+1)$-concave.

Proof. According to Theorem 6.2, $\partial_j X$ is given by the system of equations

$$\{P_s = 0,\, P_s^{(1)} = 0,\, \dots,\, P_s^{(j-1)} = 0\}.$$

It follows from formulas (6.6) and (6.7) that the gradient vector fields $\nu_1, \dots, \nu_j$, where $\nu_k = \nabla(P_s^{(k-1)})$, are linearly independent. Therefore, $\partial_j X$ is a submanifold of $\partial_1 X$. Moreover, since ν_j is a section of the normal line bundle $T(X_{j-1})/T(X_j)$, transversal to its zero section, and since $e_{n+1} = \sum_k (e_{n+1} \cdot \nu_k)\nu_k$, the constant field e_{n+1} is boundary generic in the sense of Definition 5.1.

To validate the last claim, we notice that $P_s^{(s)} \neq 0$, which implies that $\partial_{s+1} X = \emptyset$. □

Remark 6.2. For local models of vector fields as in Theorem 6.2, the germs of intersections of the strata $\partial_j X$, $j \in [1, n+1]$, with the hypersurfaces of $\{u = c\}$ are *affine subspaces* of $\mathbb{R}^{n+1}$, while the germs of $\partial_j^+ X \cap \{u = c\}$ are *affine half-spaces.* Indeed, for each fixed value of u and any j, the equations $\{P_s^{(i)} = 0\}_{i<j}$ impose *linear* constraints on the rest of the variables; similarly, $\pm P_s^{(j)} \geq 0$ is a linear inequality in $x_0, \dots, x_{n-1}$. Therefore, the equations define a *ruled* real variety, and the inequality picks a semi-algebraic subvariety, ruled by the half-spaces. ◊

Example 6.1. For a 4-dimensional X, the eight local models for the boundary generic flows are:

$$X = \big\{ \pm P_s(u, x_0, x_1, x_2) \geq 0 \big\},\ v = \partial_u = e_4,$$

where $s \in [1, 4]$[2]. They are given by the four polynomials:

- $P_1 = u$,
- $P_2 = u^2 + x_0$,
- $P_3 = u^3 + x_1 u + x_0$,
- $P_4 = u^4 + x_2 u^2 + x_1 u + x_0$.

Let us consider the fibers $\pi^{-1}(x) \cap X$ of the projection $\pi : \mathbb{R}^4 \to \mathbb{R}^3$ in the vicinity of the origin for the P_4-model. Since the coefficients of $P_4(\sim, u)$ are real, the real divisor $D_{\mathbb{R}}(P_4)$ is: (i) either real of degree four, or (ii) real of degree two, or (iii) real of degree zero. Note that, in case (i), the sum of all real roots, taken with their multiplicities, is zero. We call such divisors *balanced.*

[2] To save space, we do not list the other eight cases with $v = -e_4$.

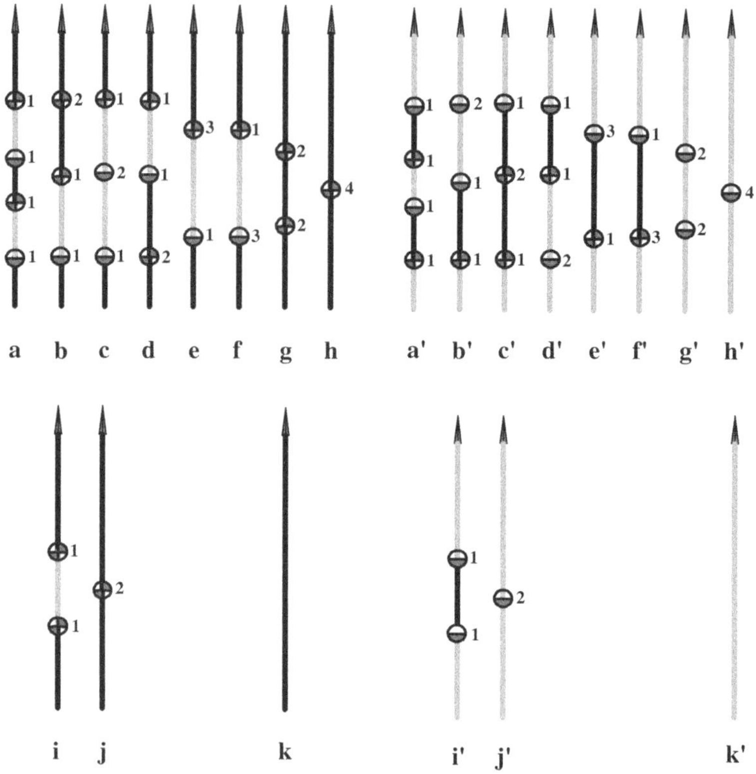

Fig. 6.1 The patterns of solutions for the inequality $\{P_4 \geq 0\}$ (on the left) and for the inequality $\{P_4 \leq 0\}$ (on the right). The numbers 1, 2, 3, 4 indicate the multiplicities of the P_4-roots. Diagrams i, j, i', j' correspond to the case of one pair, and diagrams k, k' to the case of two pairs of complex conjugate roots.

The P_4-model is described by the diagrams a–k and a'–k' in Figure 6.1. In the figure, we do not stress the balanced nature of the divisors $\{P_4 = 0\}$. The shaded portions of the number lines in the figure belong to the 4-fold X; in fact, one can think of X as being disjoint union of these shaded portions. Note the polarities $\{+, -\}$, attached to each root of P_4: they reflect the polarities in the Morse stratification $\partial_j^{\pm} X$. By the Viète Formula, each divisor in diagrams a–h and a'–h' determines a unique point $x = (x_0, x_1, x_2)$ over which it resides. $\diamondsuit$

6.3 Traversally generic and versal vector fields

Guided by the geometry of the local models from Theorem 6.2, we embark on an investigation of the "semi-local" dynamics of generic gradient-like *nonsingular* v-flows.

Here and on, we assume that each v-trajectory does not "end" at a point where it is tangent to the boundary: when possible, it "extends further" in the interior of X. Also, for technical reasons, we do regard a singleton x, the flow "curve" through $x \in \partial_2^- X^\circ$, as a trajectory.

With any boundary generic (see Definition 5.1) and traversing (see Definition 5.10) vector fields field $v \in \mathcal{V}^\dagger(X)$, we associate an ordered sequence of multiplicities with each v-trajectory γ. For any $v \in \mathcal{V}^\dagger(X)$, using the compactness of γ and the transversality, imposed by the property $v \in \mathcal{V}^\dagger(X)$, we conclude that the intersection $\gamma \cap \partial_1 X$ is automatically a finite set $\{a_i\}_i$. For traversing generic fields, the points $\{a_i\}$ are ordered by the field-oriented trajectory γ, and the index i reflects this ordering.

Definition 6.1. Let $v \in \mathcal{V}^\dagger(X)$ be a boundary generic and traversing vector field. Then every v-trajectory γ intersects the boundary $\partial_1 X$ at a finite number of points $\{a_i\}$. Each point a_i belongs to a unique pure stratum $\partial_{j_i} X^\circ$.

Then the *multiplicity* $m(\gamma)$ of γ is defined by the formula

$$m(\gamma) =_{\mathsf{def}} \sum_i j_i. \tag{6.8}$$

The *reduced multiplicity* $m'(\gamma)$ of γ is defined by the formula

$$m'(\gamma) =_{\mathsf{def}} \sum_i (j_i - 1), \tag{6.9}$$

and the *virtual multiplicity* $\mu(\gamma)$ of γ is defined by the formula

$$\mu(\gamma) =_{\mathsf{def}} \sum_i \left\lceil \frac{j_i}{2} \right\rceil, \tag{6.10}$$

where $\lceil \sim \rceil$ denotes the integral part function. ◇

For an open and dense subspace $\mathcal{V}^\ddagger(X)$ of $\mathcal{V}^\dagger(X)$, we may interpret $\mu(\gamma)$ as *the maximal number of tangency points* that any trajectory γ' in the vicinity of γ has (see Theorem 6.6 and Theorem 6.7).

Consider a traversing boundary generic vector field v on a compact connected and smooth manifold X with boundary. Let γ be a typical v-trajectory. It hits the boundary $\partial_1 X$ at a finite number of points

$a_1, a_2, \dots, a_q$, the points being ordered by v. Consider the unique pure stratum $\partial_{j_i} X^\circ(v)$ such that $a_i \in \partial_{j_i} X^\circ(v)$. We associate the integral sequence $\omega_\gamma =_{\mathsf{def}} (j_1, j_2, \dots, j_q)$ with each γ and call it the *combinatorial type* of γ. We will devote Chapter 7 to the somewhat tedious study of the combinatorics of such sequences for a massive class of fields which we call "*traversally generic*" (see Definition 6.3).

Given a combinatorial pattern ω, we introduce two quantities:

$$|\omega| =_{\mathsf{def}} \sum_i j_i \quad \text{and} \quad |\omega|' =_{\mathsf{def}} \sum_i (j_i - 1),$$

the *multiplicity* and the *reduced multiplicity* of ω.

Definition 6.2. We say that a boundary generic traversing vector field v on a compact connected smooth manifold X is *globally k-convex* if, for every v-trajectory γ, the reduced multiplicity $|\omega_\gamma|' < k$. ◇

Note that a globally k-convex vector field is automatically boundary $(k+1)$-convex and concave. Indeed, since for any $x \in \partial_j X(v)$, $|\omega_{\gamma_x}|' \geq j-1$, we get that the inequality $|\omega_{\gamma_x}|' < k$ implies that $j \leq k$.

When $v \neq 0$ along $\partial_1 X$, we can extend it into a larger manifold $\hat{X}$ so that $\hat{X}$ properly contains X and the extension $\hat{v}$ remains nonsingular in in the vicinity of $\partial_1 X \subset \hat{X}$. Throughout this text, we treat the pair $(\hat{X}, \hat{v})$ as a *germ* which extends (X, v).

By Theorem 6.2, for any $v \in \mathcal{V}^\dagger(X)$, each set $\partial_j X$, $j \in [1, n+1]$, is a closed manifold.

Consider the collection of tangent spaces $\{T_{a_i}(\partial_{j_i} X^\circ)\}_i$ to the pure strata $\{\partial_{j_i} X^\circ =_{\mathsf{def}} (\partial_{j_i} X)^\circ\}_i$ that have a nonempty intersection with a given trajectory γ.

By Theorem 6.2, each space $T_{a_i}(\partial_{j_i} X^\circ)$ is transversal to the curve γ.

Let S be a local transversal section of the $\hat{v}$-flow at some point $a_\star \in \gamma$ and let $\mathsf{T}_\star$ be the space, tangent to S at $a_\star$. Each space $T_{a_i}(\partial_j X^\circ)$, with the help of the $\hat{v}$-flow, determines a vector subspace $\mathsf{T}_i = \mathsf{T}_i(\gamma)$ in $\mathsf{T}_\star$. It is the image of the tangent space $T_{a_i}(\partial_j X^\circ)$ under the composition of two maps: (1) the differential of the flow-generated diffeomorphism that takes a_i to $a_\star$, and (2) the linear projection $T_{a_\star}(X) \to \mathsf{T}_\star$ whose kernel is generated by $v(a_\star)$.

Recall that a configuration $\{\mathsf{T}_i\}$ of *affine* subspaces $\mathsf{T}_i \subset \mathsf{T}_\star$ is called *generic* (or *stable*) when all the multiple intersections of spaces from the configuration have the least possible dimensions, consistent with the dimensions of $\{\mathsf{T}_i\}$. In other words,

$$\operatorname{codim}\Big(\bigcap_s \mathsf{T}_{i_s}, \mathsf{T}_\star\Big) = \sum_s \operatorname{codim}(\mathsf{T}_{i_s}, \mathsf{T}_\star)$$

for any subcollection $\{\mathsf{T}_{i_s}\}$ of spaces from the list $\{\mathsf{T}_i\}$.

Consider the case when $\{\mathsf{T}_i\}$ are *vector* subspaces of $\mathsf{T}_\star$. If we interpret each T_i as the kernel of a linear epimorphism $\Phi_i : \mathsf{T}_\star \to \mathbb{R}^{n_i}$, then the property of $\{\mathsf{T}_i\}$ being generic can be reformulated as the property of the direct product map

$$\prod_i \Phi_i : \mathsf{T}_\star \to \prod_i \mathbb{R}^{n_i}$$

being an epimorphism. In particular, for a generic configuration of affine subspaces, if a point belongs to several T_i's, then the sum of their codimensions n_i does not exceed the dimension of the ambient space $\mathsf{T}_\star$.

Example 6.2. The configuration of a line and a plane in $\mathbb{R}^3$, the line being transversal to the plane, is generic; and so is the configuration of two planes in $\mathbb{R}^3$ which share only a line. On the other hand, two lines in $\mathbb{R}^3$, which share a point, do not form a generic configuration. Also, any three lines in $\mathbb{R}^2$, which share a point, do not form a generic configuration. ◇

The definition below resembles and is inspired by the "Condition NC" imposed on, so called, *Boardman maps* between smooth manifolds (see [GG], page 157, for the relevant definitions). In fact, for traversing generic fields v, the v-flow delivers germs of Boardman maps $p(v, \gamma) : \partial_1 X \to \mathbb{R}^n$, available in the vicinity of every trajectory γ.

Definition 6.3. We say that a traversing field v on X is *traversally generic* if:

- the field is boundary generic in the sense of Definition 5.1,
- for each v-trajectory $\gamma \subset X$, the collection of subspaces $\{\mathsf{T}_i(\gamma)\}_i$ is generic in $\mathsf{T}_\star$: that is, the obvious quotient map $\mathsf{T}_\star \to \prod_i \big(\mathsf{T}_\star/\mathsf{T}_i(\gamma)\big)$ is surjective.

We denote by $\mathcal{V}^\ddagger(X)$ the space of all traversally generic fields on X. ◇

Remark 6.3. In particular, the second bullet of the definition implies the inequality

$$\sum_i \operatorname{codim}(\mathsf{T}_i(\gamma), \mathsf{T}_\star) \leq \dim(\mathsf{T}_\star) = n.$$

In other words, for traversally generic fields, the reduced multiplicity of each trajectory γ satisfies the inequality

$$m'(\gamma) = \sum_i (j_i - 1) \leq n. \tag{6.11}$$

Evidently the property of the configuration $\{\mathsf{T}_i(\gamma)\}_i$ being generic in $\mathsf{T}_\star$ does not depend on the choice of the local smooth section S, traversal to the flow. ♢

Remark 6.4. Note that, if a smooth submersion $F : (\tilde{X}, \partial\tilde{X}) \to (X, \partial X)$ is a finite covering map of $(n+1)$-manifolds, then the pull-back $\tilde{v}$ of a traversally generic field v on X is a traversally generic field on $\tilde{X}$.

Indeed, each v-trajectory γ is a segment or a singleton. Therefore $F^{-1}(\gamma)$ is a disjoint union of arcs or points. These are exactly the trajectories of the pull-back field $\tilde{v}$ on $\tilde{X}$. Each of the lifted trajectories has a $\tilde{v}$-adjusted neighborhood $U_{\tilde{\gamma}}$ that projects by the local diffeomorphism F onto a v-adjusted neighborhood U_γ. That field-respecting diffeomorphism $F : U_{\tilde{\gamma}} \to U_\gamma$ maps $\partial\tilde{X} \cap U_{\tilde{\gamma}}$ to $\partial X \cap U_\gamma$, so that all the structures, participating in Definition 6.3, are respected by F. ♢

Let us review the list of notations for various kinds of vector fields on X. Recall that we have introduced the following nested collection of spaces:

$$\mathcal{V}^\ddagger(X) \subset \mathcal{V}^\dagger(X) \subset \mathcal{B}^\dagger(X) \subset \mathcal{V}(X)$$

based on traversally generic, traversing and boundary generic, boundary generic, and arbitrary smooth vector fields on X, respectively. We denote by $\mathcal{V}_{\neq 0}(X)$ the space of non-vanishing fields on X. We are also considering the space $\mathcal{V}_{\mathsf{grad}}(X)$ of gradient-like fields. Finally, $\mathcal{V}_{\mathsf{trav}}(X)$ denotes the space of traversing fields (see Definition 5.10).

In view of Corollary 5.8, the following inclusions follow from the definitions:

$$\mathcal{V}_{\mathsf{grad}}(X) \cap \mathcal{V}_{\neq 0}(X) = \mathcal{V}_{\mathsf{trav}}(X) \subset \mathcal{V}_{\neq 0}(X), \tag{6.12}$$

$$\mathcal{V}^\ddagger(X) \subset \mathcal{V}^\dagger(X) = \mathcal{B}^\dagger(X) \cap \mathcal{V}_{\mathsf{trav}}(X) \subset \mathcal{V}_{\mathsf{trav}}(X). \tag{6.13}$$

Let $\mathsf{Sub}(X, \mathbb{R})$ denote the space of submersions $f : X \to \mathbb{R}$. Recall that, by Phillips' Theorem B [Ph] (see Theorem 4.3), for a fixed Riemannian metric g on X, the gradient map

$$\nabla_g : \mathsf{Sub}(X, \mathbb{R}) \to \mathcal{V}_{\neq 0}(X)$$

is a *weak homotopy equivalence.*

In fact, the image of ∇_g is contained automatically in $\mathcal{V}_{\mathsf{trav}}(X) \subset \mathcal{V}_{\neq 0}(X)$. One might speculate that $\nabla_g : \mathsf{Sub}(X, \mathbb{R}) \to \mathcal{V}_{\mathsf{trav}}(X)$ is a weak homotopy equivalence as well. By Corollary 4.3, if two gradient-like vector fields can be connected by a path in the space of all non-vanishing vector fields, then they can be connected by a path in the space of all traversing vector fields (equivalently, non-vanishing gradient-like vector fields) as well. Therefore, at least the map

$$(\nabla_g)_* : \pi_0(\mathsf{Sub}(X, \mathbb{R})) \to \pi_0(\mathcal{V}_{\mathsf{trav}}(X))$$

is bijective.

For any vector field $v \in \mathcal{V}^\dagger(X)$, the intersection patterns of a v-trajectory γ with the pure Morse strata $\{\partial_k X^\circ\}_k$ reflects "*the order of tangency*" between γ and $\partial_1 X$. The fundamental lemma below establishes this fact. The lemma is a "pedestrian" proof of Morin's Theorem 6.1.

We embed the pair (X, v) into a pair $(\hat{X}, \hat{v})$ in a way that has been described on many occasions previously.

Lemma 6.1. *Let X be a compact smooth $(n+1)$-manifold with boundary. Assume that $v \in \mathcal{V}^\dagger(X)$. Let $z : \hat{X} \to \mathbb{R}$ be a smooth function in the vicinity of $\partial_1 X \subset \hat{X}$ such that:*

(1) 0 is a regular value of z,
(2) $z^{-1}(0) = \partial_1 X$, and
(3) $z^{-1}((-\infty, 0]) = X$.

Then the following properties hold:

- *For each point $a \in \partial_k X(v)^\circ$, the restricted function $z|_{\gamma_a}$ on the $\hat{v}$-trajectory γ_a through a has zero of multiplicity k at a.*
- *In the vicinity of a in $\hat{X}$, there exists a coordinate system (u, x), where $u \in \mathbb{R}$ and $x \in \mathbb{R}^n$, so that:*
  1) each v-trajectory γ is defined by an equation $\{x = \vec{const}\}$,
  2) the boundary $\partial_1 X$ is defined by the equation

$$u^k + \sum_{j=0}^{k-2} x_j\, u^j = 0. \tag{6.14}$$

- *Each point $a \in \partial_k X^\circ$ has a $\hat{v}$-flow adjusted neighborhood $U(a)$ with the following property: there exists a set of indexes $J(a) \subset \{0, 1, \dots, n+1\}$*

such that each $\hat{v}$-trajectory $\gamma \subset U(a)$ hits only some strata $\{\partial_j X^\circ \cap U(a)\}_{j \in J(a)}$ so that

$$\sum_{j \in J(a)} j \leq k, \quad \textit{and} \quad \sum_{j \in J(a)} j \equiv k \mod 2.$$

- *For each $r > 0$, the intersection of the hypersurface $\partial_1 X$ with the boundary $\partial \Pi_r$ of the r-cylinder $\Pi_r =_{\mathsf{def}} \{(u,x) \big| \, \|x\| \leq r\}$ is a smooth $(n-1)$-manifold, regularly embedded in $\partial_1 X$ and in $\partial \Pi_r$. Thus $\partial_1 X \cap \Pi_r$ is a smooth manifold with boundary $\partial_1 X \cap \partial \Pi_r$.*

Proof. Consider the space $\mathbb{R}^1 \times \mathbb{R}^n$ with coordinates $(z, x) =_{\mathsf{def}} (z, x_1, \dots x_n)$ and the standard Euclidean scalar product $\langle \sim, \sim \rangle$. Let the sets $M_0 =_{\mathsf{def}} \{z \geq 0\}$ and $M_1 =_{\mathsf{def}} \{z = 0\}$ represent the germ of the pair $(X, \partial_1 X)$ in the vicinity of a typical point from $\partial_1 X$.

Consider a smooth vector field $v(z,x) = \big(u(z,x),\ w(z,x)\big)$ on $\mathbb{R}^1 \times \mathbb{R}^n$, where the component $u(z,x)$ is parallel to $\mathbb{R}^1$, and $w(z,x)$ to $\mathbb{R}^n$.

In the argument to follow, we view the coordinates (z,x) as "more permanent feature", while analyzing the restrictions on $u(z,x), w(z,x)$, imposed by the desired property of the field $v(x,z)$ being boundary generic. In fact, we may allow for some changes in the coordinates (z,x) as well, as long as the locus $\{z = 0\}$ (which models the boundary $\partial_1 X$) remains fixed.

The field v is tangent to $M_1 \subset \mathbb{R}^1 \times \mathbb{R}^n$ along the locus

$$M_2 =_{\mathsf{def}} \{z = 0,\ u(z,x) = 0\}$$

which models $\partial_2 X = \partial_2 X(v)$. To reflect the generic nature of v, we require that $u(0,x)$, viewed as a section of the obvious projection $\mathbb{R}^1 \times \mathbb{R}^n \to \mathbb{R}^n$, will be transversal to M_1 along M_2. In other words, the gradient $\nabla_x u$ of the function $u(0,x)$ with respect to the coordinates x does not vanish along the manifold M_2. The locus M_3, where v is tangent to M_2, is given by two equations $\{z = 0\}$ and $\{\langle \nabla_x u,\ w\rangle = 0\}$, while the transversality of v to M_3 can be expressed as the linear independence of the two fields, $\nabla_x u$ and $\nabla_x \langle \nabla_x u,\ w\rangle$. The set

$$M_1^+ =_{\mathsf{def}} \{x |\ z = 0,\ u \geq 0\} \subset M_1$$

mimics the locus $\partial_1^+ X$, while the set

$$M_2^+ =_{\mathsf{def}} \{x |\ z = 0,\ u = 0,\ \langle \nabla_x u, w\rangle \geq 0\} \subset M_2$$

mimics $\partial_2^+ X$.

In order to capture the emerging pattern, for each pair of smooth maps $u : \mathbb{R}^1 \times \mathbb{R}^n \to \mathbb{R}$ and $w : \mathbb{R}^1 \times \mathbb{R}^n \to \mathbb{R}^n$, we introduce a sequence of new functions $\psi_k : \mathbb{R}^1 \times \mathbb{R}^n \to \mathbb{R}^1$ in the variables z, x, defined by the recursive formula:

$$\begin{aligned} \psi_1(z,x) &=_{\mathsf{def}} u(z,x), \\ \psi_k(z,x) &=_{\mathsf{def}} \langle \nabla_x \psi_{k-1}(z,x),\, w(z,x) \rangle. \end{aligned} \tag{6.15}$$

For $k > 0$, the locus M_k, a model of $\partial_k X$, is given by the equations

$$\{z = 0,\ \psi_1(z,x) = 0,\ \dots,\ \psi_{k-1}(z,x) = 0\}, \tag{6.16}$$

together with the requirement that, on the solution set $\{(0,x)\}$ of equation (6.16), the vector fields

$$\Big\{\nabla_x \psi_1(0,x),\ \dots,\ \nabla_x \psi_{k-1}(0,x)\Big\} \tag{6.17}$$

are linearly independent (hence, M_k is a manifold). Thus the linear independence of the the gradient fields in (6.17), where $2 \leq k \leq n+1$, is equivalent to v *locally* belonging to the space $\mathcal{B}^\dagger(X)$.

The submanifold $M_k^+ \subset M_k$ is defined by the additional inequality $\{\psi_k(z,x) \geq 0\}$: there the field v points inside of M_{k-1}^+.

Consider the following system of ordinary differential equations, determined by the field $v(z,x)$:

$$\begin{aligned} \frac{dz}{dt} &= u(z,x) \\ \frac{dx}{dt} &= w(z,x). \end{aligned} \tag{6.18}$$

Let us choose a point $x \in M_k^\circ =_{\mathsf{def}} M_k \setminus M_{k+1}$. Let $\gamma(t) = (z(t), x(t))$ be the solution of (6.18) such that $\gamma(0) = (x, 0)$. In order to prove the lemma, we need to verify that the function z, being restricted to $\gamma(t)$, has x as its zero of multiplicity k.

For any j, with the help of equations (6.15) and (6.18), we get

$$\frac{d\psi_j}{dt} = \frac{\partial \psi_j}{\partial z} \cdot \frac{dz}{dt} + \left\langle \nabla_x \psi_j, \frac{dx}{dt} \right\rangle = \frac{\partial \psi_j}{\partial z} \cdot \psi_1 + \psi_{j+1}.$$

Put $\xi_j =_{\mathsf{def}} \dfrac{\partial \psi_j}{\partial z}$. Then the formula above can be rewritten as

$$\frac{d\psi_j}{dt} = \xi_j \cdot \psi_1 + \psi_{j+1}. \tag{6.19}$$

Next, we claim that the function $\dfrac{d^j z}{(dt)^j}$, being restricted to the integral curve γ, can be represented as $\psi_j + \sum_{i=1}^{j-1} \beta_i \cdot \psi_i$ for an appropriate choice of smooth functions $\beta_i = \beta_i(z, x)$. In other words, the functions $\frac{d^j z}{(dt)^j}\Big|_\gamma$ and $\psi_j|_\gamma$ are congruent modulo the ideal of $C^\infty(\gamma, \mathbb{R})$, generated by $\{\psi_i|_\gamma\}_{i<j}$. In view of equations (6.16), this implies that $\frac{d^j z}{(dt)^j}(0) = 0$ for all $j < k$. Since $x \notin M_{k+1}$ translates as $\psi_k(x) \neq 0$, it follows that $\frac{d^k z}{(dt)^k}(0) \neq 0$. So $z|_\gamma$ indeed has a zero of multiplicity k at $\gamma(0) \in M_k^\circ$.

We proceed to prove the formula

$$\frac{d^j z}{(dt)^j}\Big|_\gamma = \Big(\psi_j + \sum_{i=1}^{j-1} \beta_i \cdot \psi_i\Big)\Big|_\gamma$$

by induction in j. The inductive step "$j \Rightarrow j+1$" is carried out by differentiating with respect to t the identity conjectured for j, and then by using formulas (6.19). The basis of induction is represented by the claim $\frac{dz}{(dt)}\Big|_\gamma = \psi_1|_\gamma =_{\mathsf{def}} u|_\gamma$, the first equation in (6.18).

Finally, we notice that all these arguments do not depend on the choice of the auxiliary function z, subject to the properties (1)-(3), described in the lemma. This proves the claim in the first bullet of the lemma.

Next, we switch to *new* local coordinates (u, y), amenable to the v-flow (in contrast with the previous coordinates (z, x) which were adjusted to the boundary $\partial_1 X$). Consider a point $a \in M_k^\circ$ and a germ of a v-trajectory γ passing through a. Let S be a germ of a smooth transversal flow section at a, equipped with coordinates $y = (y_1, \dots, y_n)$, so that a is the origin. Let u be the coordinate, obtained by integrating v so that S is the locus $\{u = 0\}$. The procedure for producing the coordinate u (under the name "$\int_S^x v$") has been described in the proof of Lemma 5.6.

Now we view $z : \hat{X} \to \mathbb{R}$ as a smooth function of (u, y). We have shown that $z|_\gamma = z(u, 0)$ has a zero at $u = 0$ of multiplicity k. Using The Malgrange Preparation Theorem [Mal], there exist an invertible germ of a smooth function $Q(u, y)$ and a germ of the form

$$P(u, y) = u^k + \sum_{j=0}^{k-1} \phi_j(y) u^j,$$

where $\phi_j(0) = 0$, so that $z = P \cdot Q$.

Due to this factorization, in the coordinates (u, y), the locus $\{z = 0\}$

(the boundary $\partial_1 X$) is given by an equation

$$u^k + \sum_{j=0}^{k-1} \phi_j(y)u^j = 0.$$

We can improve the latter equation by modifying the coordinates (u, y) (by re-parameterizing the trajectories): just put $\tilde{u} = u + \frac{1}{k} \cdot \phi_{k-1}(y)$ and $\tilde{y} = y$. Then, in the new coordinates, the equation $z = 0$ is transformed into the equation

$$\tilde{u}^k + \sum_{j=0}^{k-2} \tilde{\phi}_j(y)\tilde{u}^j = 0,$$

a "surrogate" of the Morin canonical form from (6.1) and (6.4).

We still need to bridge the gap between the surrogate and true Morin's canonical forms. With this goal in mind, consider the Jacobi matrix $D\Phi(y)$, formed by the partial derivatives $\left\{\frac{\partial}{\partial y_i}\tilde{\phi}_j(y)\right\}_{i,j}$. If $D\Phi(0)$ has the maximal possible rank $k-1$, then we can choose the functions $\{\tilde{\phi}_j(y)\}_{0 \leq j < k-2}$ for the role of the first new coordinates in the vicinity of the origin. They can be completed to a coordinate system comprising the function $\tilde{u}$ together with some functions $\{\tilde{y}_i\}_{k \leq i \leq n}$, drawn from the list of $\{y_i\}_{1 \leq i \leq n}$.

As before, each trajectory γ is produced by freezing all the coordinates, but $\tilde{u}$. In the new coordinates $(\tilde{\phi}_0, \dots, \tilde{\phi}_{k-2}, \tilde{y}_k, \dots, \tilde{y}_n, \tilde{u})$, the germ of $\partial_1 X$ at a is given by the depressed polynomial equation

$$\tilde{u}^k + \sum_{j=0}^{k-2} \tilde{\phi}_j \tilde{u}^j = 0,$$

the true Morin canonical form. Therefore $\mathrm{rk}(D\Phi(0)) = k - 1$ implies the existence of the Morin canonical coordinates in the vicinity of a.

In fact (see Lemma 6.3), the requirement $\mathrm{rk}(D\Phi(0)) = s$ has an intrinsic meaning. In particular, it is independent on the factorization $z = P \cdot Q$.

Now consider a trajectory $\gamma = \{y = c\}$ (where $c = (c_1, \dots, c_n)$ is a fixed vector) in the vicinity of a point $a \in \partial_1 X \cap \gamma$ together with the zeros of the function z, being restricted to γ. We claim that the two loci

$$\left\{z(u,c) = 0,\ \frac{\partial}{\partial u}z(u,c) = 0,\ \dots,\ \frac{\partial^j}{\partial u^j}z(u,c) = 0\right\},$$

$$\left\{P(u,c) = 0,\ \frac{\partial}{\partial u}P(u,c) = 0,\ \dots,\ \frac{\partial^j}{\partial u^j}P(u,c) = 0\right\} \qquad (6.20)$$

coincide. The argument, inductive in j, employes the fact that $Q(u,c) \neq 0$.

Since we have established the validity of the first bullet in Lemma 6.1, we get that both systems of equations in (6.20) locally define the stratum $\partial_{j+1}X$. Moreover, as in the proof of Theorem 6.2 (in particular, see formula (6.6)), the fact that $v \in \mathcal{B}^\dagger(X)$, in the vicinity of $a \in \partial_k X^\circ$, can be translated as the property of linear independence for the gradients

$$\nabla z,\ \nabla \circ \partial_u(z),\ \ldots,\ \nabla \circ \partial_u^{k-2}(z) \tag{6.21}$$

in the coordinates (u,z) at the origin $(0,0)$[3]. (Here "$\circ$" denotes the composition of differential operators.) Note that these are the gradients of functions in (6.20) that determine the stratum $\partial_k X$.

We claim that this linear independence is equivalent to the linear independence of the gradients

$$\nabla P,\ \nabla \circ \partial_u(P),\ \ldots,\ \nabla \circ \partial_u^{k-2}(P). \tag{6.22}$$

In order to validate this claim, we introduce the u-parameter curve $\alpha(u) =_{\mathsf{def}} (1, u, \ldots, u^{k-2})$. Note that the gradients in (6.22) admit the following representations:

$$\nabla P = \big(\partial_u P,\ D\Phi * \alpha\big),\ \ \nabla \circ \partial_u P = \big(\partial_u^2 P,\ D\Phi * \partial_u \alpha\big),\ \ \ldots\ ,$$

$$\nabla \circ \partial_u^{k-2} P = \big(\partial_u^{k-1} P,\ D\Phi * \partial_u^{k-2}\alpha\big),$$

where $\big(\partial_u^j P,\ D\Phi * \partial_u^{j-1}\alpha\big)$ is the vector whose first component is $\partial_u^j P$, and the second component $D\Phi * \partial_u^{j-1}\alpha$ is the multiplication of the matrix $D\Phi$ by the vector $\partial_u^{j-1}\alpha$.

Since $\partial_u^j P(0,0) = 0$ for all $1 \leq j \leq k-1$, we conclude that the dimension of the space spanned by vectors

$$\nabla P(0,0),\ \nabla \circ \partial_u P(0,0),\ \ldots,\ \nabla \circ \partial_u^{k-2} P(0,0)$$

equals the dimension of the space spanned by vectors

$$D\Phi * \alpha,\ D\Phi * \partial_u \alpha,\ \ldots\ ,\ D\Phi * \partial_u^{k-2}\alpha,$$

being evaluated at $(0,0)$. Since α and its u-derivatives are independent vectors, the dimension of the latter space is equal to the rank of $D\Phi(0)$. Now we employ Lemma 6.3 below to conclude that the dimension of the space, spanned by vectors in formula (6.22), is the same as the dimension of the space spanned by vectors in (6.21), both dimensions being equal to $\mathrm{rk}(D\Phi(0))$.

[3]In the coordinates (u,y), this is an analogue of (6.18).

If $v \in \mathcal{B}^\dagger(X)$, the vectors in formula (6.21) must be independent at the origin 0. Therefore, $\text{rk}(D\Phi(0)) = k - 1$, which implies the existence of the Morin coordinates in the vicinity of $a \in \partial_k X^\circ$. This proves the claim in the second bullet of the lemma.

Since two sets of constraints in formula (6.20) have the same set of solutions, there exists $\epsilon > 0$, so that the sum of zero multiplicities of the smooth u-function $z(u, c)$ in the interval $-\epsilon < u < \epsilon$, is the same as the sum of zero multiplicities of the function $P(u, c)$ in the same interval; the later function is a polynomial of degree k in u. Therefore, in the vicinity of a, the sum of zero multiplicities of $z|_\gamma$ does not exceed k. In other words, if $a \in M_k^\circ$, then each trajectory γ, in the vicinity of a, hits only some strata $\{M_j^\circ\}$ so that $\sum j \leq k$ and $\sum j \equiv k \mod 2$.

Consider the polynomial family $P(u, x)$ in the formula (6.14). There is $\delta > 0$ such that $\|c\| < \delta$ implies that $|x_j(c)| < (\epsilon/\rho)^j$, where the universal constant ρ is introduced in the proof of Lemma 6.2 below. By that lemma, for such a $\delta > 0$, all the real roots u of $P(u, x)$ are confined to the interval $(-\epsilon, \epsilon)$. Therefore, each trajectory $\gamma = \{x = c\}$, passing through the section $S =_{\text{def}} \{u = 0,\ \|c\| < \delta\}$, has *all* the roots of $P(u, c)$ concentrated in the interval $(-\epsilon, \epsilon)$.

Let D_γ be the zero divisor of $P(u, c)$. Thus, not only $\deg(D_\gamma) \leq k$, but also $\deg(D_\gamma) \equiv k$ (2). The third bullet of Lemma 6.1 has been validated.

The validation of the last bullet is based on a direct computation. Consider the standard Euclidean metric in the (u, x)-coordinates. The gradient of P is the vector $\nabla P = (P_u, u^{k-2}, \ldots, u, 1)$, and the gradient of the function $\frac{1}{2}\|x\|^2$ is $(0, x_{k-2}, \ldots, x_1, x_0)$. We will show that ∇P and $\frac{1}{2}\nabla\|x\|^2$ are linearly independent along $\partial\Pi_r \cap \partial_1 X$. Indeed, if the two gradients are proportional, we get

$$\lambda(P_u, u^{k-2}, \ldots, u, 1) = (0, x_{k-2}, \ldots, x_1, x_0)$$

for some $\lambda > 0$. This implies that $P_u = 0$ and

$$(x_{k-2}, \ldots, x_1, x_0) = (\lambda u^{k-2}, \ldots, \lambda u, \lambda).$$

As we substitute these expressions in the pair of equations $\{P = 0, P_u = 0\}$, we get

$$u^k + \lambda u^{2(k-2)} + \cdots + \lambda u^2 + \lambda = 0,$$

$$ku^{k-1} + (k-2)\lambda u^{(k-3)+(k-2)} + \cdots + 2\lambda u = 0.$$

We multiply the second equation by u/k and subtract from it the first equation. The resulting equation is of the form $\lambda \cdot Q(u^2) = 0$, where $Q(T)$

a T-polynomial with all negative coefficients and the free term -1. Thus, $\lambda \cdot Q(u^2) < 0$, a contradiction with the equation $Q(u^2) = 0$. So ∇P and $\frac{1}{2}\nabla\|x\|^2$ are linearly independent along $\partial\Pi_r \cap \partial_1 X$. $\square$

Lemma 6.2. *Put $x = (x_0, \dots, x_{k-1})$. Let*

$$P(u, x) = u^k + \sum_{j=0}^{k-1} x_j u^j$$

be a family of real monic u-polynomials, parametrized by x.

Then, there exists a universal constant $\rho = \rho(k) > 0$ with the following property: for any $\epsilon > 0$ and each polynomial $P(u, x)$ with the coefficients $\{|x_j| < (\epsilon/\rho)^j\}$, all the real roots u of $P(u, x)$ are located in the interval $(-\epsilon, \epsilon)$.

Proof. Consider the complex Viète map $V : \mathbb{C}^k \to \mathbb{C}^k$, defined by elementary symmetric polynomials $\sigma_1, \dots, \sigma_k$ in the variables-roots $\alpha_1, \dots \alpha_k$.

For each $\beta > 0$, consider the set $K(\beta) =_{\mathsf{def}} \{|\sigma_j| < \beta^j\}_{1 \le j \le k}$ in the complex space $\mathbb{C}^k$ with the coordinates $\alpha_1, \dots \alpha_k$. Because each polynomial σ_j is homogeneous of degree j, every *real* ray from the origin intersects with $K(\beta) \subset \mathbb{C}^k$ along a segment. Indeed, any ray that does not belong to the coordinate hyperplanes, hits the hypersurface $\{\sigma_k = \beta^k\}$ at a singleton, any ray that belongs to the coordinate hyperplanes hits the hypersurfaces $\{\sigma_{k-1} = \beta^{k-1}\}$ at a singleton, and so on ... Note that $K(\beta')$ can be obtained from $K(\beta)$ by a conformal scaling with the scaling factor β'/β.

Denote by B the unit polydisk $\{|\alpha_j| \le 1\}_j$ in the complex space $\mathbb{C}^k$ with the coordinates $\alpha_1, \dots \alpha_k$. For each point $\alpha \in \partial B$, consider the real ray $r(\alpha)$ through α, emanating from the origin, and the proportion $\rho(\alpha)$ between the lengths of segments $[0, r(\alpha) \cap \partial K(1)]$ and $[0, r(\alpha) \cap \partial B]$.

Using compactness of ∂B, put $\rho =_{\mathsf{def}} \max_{\alpha \in \partial B}\ \rho(\alpha)$. Now, if $\alpha \in \partial K(\beta)$, then each $|\alpha_j| \le \beta/\rho$.

For a given $\epsilon > 0$, take the coefficients $x_j = \sigma_j$ so that $|x_j| < (\epsilon/\rho)^j$. For such choice of coefficients, all the roots $\{\alpha_j\}$ of the polynomial

$$P(u, x) = u^k + \sum_{j=0}^{k-1} x_j u^j$$

are confined to the interval $(-\epsilon, \epsilon)$. $\square$

Lemma 6.1, as well as some future arguments, relies on the technical lemma below.

Lemma 6.3. *Let $\{\alpha_i\}$ be a finite set of distinct real numbers. Let $z(u,y)$ be a smooth function, where $u \in \mathbb{R}$ and $y = (y_1, \dots y_n) \in \mathbb{R}^n$. Assume that $z(u,y)$ is a product of two smooth functions,*

$$P(u,y) = \prod_i \Big[(u-\alpha_i)^{k_i} + \sum_{l=0}^{k_i-2} \varphi_{il}(y)(u-\alpha_i)^l\Big]$$

and $Q(u,y)$, where $Q(\alpha_i, 0) \neq 0$ and $\varphi_{il}(0) = 0$. Assume that

$$\mu =_{\mathsf{def}} \sum_i (k_i - 1) \leq n.$$

For each i, consider the $((k_i-1)\times n)$-matrix

$$D\Phi_i =_{\mathsf{def}} \Big(\frac{\partial \varphi_{il}}{\partial y_m}(0)\Big).$$

Let $D\Phi$ be the $(m \times n)$-matrix, obtained by stacking all the matrices $\{D\Phi_i\}_i$ one on top of the other. Next, consider the matrix

$$M_i(z) =_{\mathsf{def}} \Big(\frac{\partial^{l+1} z}{\partial u^l\, \partial y_m}(\alpha_i, 0)\Big),$$

where $0 \leq l \leq k_i - 2$ and $1 \leq m \leq n$. Denote by $M(z)$ the matrix, formed by stacking all the matrices $\{M_i(z)\}_i$ one on top of the other.

Then the ranks of the matrices $D\Phi$ and $M(z)$ are equal.

As a result, the restriction $\mathrm{rk}(D\Phi) \leq s$, where $s \leq n$, is given by algebraic constraints, imposed on the coefficients of the degree k_i Taylor polynomials of the function $z(u,y)$ at $(\alpha_i, 0)$. In other words, the constraint $\mathrm{rk}(D\Phi) \leq s$ defines a closed affine subvariety in the space

$$\prod_i \mathsf{Jet}^{k_i}\big((\mathbb{R}\times\mathbb{R}^n, \alpha_i \times 0),\ (\mathbb{R}, 0)\big),$$

the product of k_i-jet spaces.

Proof. To save the pain of complex multi-indexing, we will prove one special case of the lemma, when the polynomial $P(u,y)$ consists of a single multiplier

$$(u-\alpha_1)^{k_1} + \sum_{l=0}^{k_1-2} \varphi_{1l}(y)(u-\alpha_1)^l.$$

By shifting all the relevant functions by α_1, we can replace the variable $u - \alpha_1$ with the variable u. Also put $k =_{\mathsf{def}} k_1$.

The argument is an induction "$k \Rightarrow k+1$" by the degree k of the polynomial

$$P(u,y) = u^k + \sum_{l=0}^{k-2} \phi_{k,l}(y)u^l,$$

with $k = 2$ being the base of induction[4].

Consider the polynomial

$$P_{k+1}(u,y) =_{\mathsf{def}} u^{k+1} + \sum_{l=0}^{k-1} \phi_{k+1,l}(y)u^l$$

with some smooth functional coefficients, where $\phi_{k+1,l}(0) = 0$.

Let $z_{k+1} =_{\mathsf{def}} P_{k+1} \cdot Q_{k+1}$ with $Q_{k+1}(0,0) \neq 0$. Then

$$z_{k+1} = \Big[u\Big(u^k + \sum_{l=1}^{k-1} \phi_{k+1,l}(y)u^{l-1}\Big) + \phi_{k+1,0}(y)\Big] Q_{k+1}(u,y).$$

Define

$$P_k =_{\mathsf{def}} u^k + \sum_{l=1}^{k-1} \phi_{k+1,l}(y)u^{l-1}, \quad \phi =_{\mathsf{def}} \phi_{k+1,0}, \quad Q =_{\mathsf{def}} Q_{k+1},$$

and $z_k =_{\mathsf{def}} P_k \cdot Q$. In the new notations,

$$z_{k+1} = u \cdot z_k + \phi \cdot Q.$$

Then

$$\partial_y z_{k+1}(u,0) = u \cdot \partial_y z_k(u,0) + \partial_y \phi(0) \cdot Q(u,0),$$

where $\partial_y =_{\mathsf{def}} \left(\frac{\partial}{\partial y_1}, \dots, \frac{\partial}{\partial y_n}\right)$.

Applying the operators $\{\partial_u^j\}_j$ to the previous identity, for $j \leq k$, we get

$$\frac{\partial^{j+1} z_{k+1}}{\partial u^j \partial y}(u,0) = j \cdot \frac{\partial^j z_k}{\partial u^{j-1} \partial y}(u,0) + u \cdot \frac{\partial^{j+1} z_k}{\partial u^j \partial y}(u,0) + \frac{\partial \phi}{\partial y}(0) \cdot \frac{\partial^j Q}{(\partial u)^j}(u,0).$$

The substitution $u = 0$ leads to the equations

$$\frac{\partial z_{k+1}}{\partial y}(0,0) = \frac{\partial \phi}{\partial y}(0) \cdot Q(0,0),$$
$$\frac{\partial^{j+1} z_{k+1}}{\partial u^j \partial y}(0,0) = j \cdot \frac{\partial^j z_k}{\partial u^{j-1} \partial y}(0,0) + \frac{\partial \phi}{\partial y}(0) \cdot \frac{\partial^j Q}{\partial u^j}(0,0). \qquad (6.23)$$

[4]Note that here the coefficient $\phi_{k,l}$ is distinct from the coefficient φ_{kl}, present in the formulation of Lemma 6.3!

By inductive assumption, the rank of the $(k-1) \times n$-matrix $M(z_k)$, whose rows are the vectors

$$\Big\{\frac{\partial z_k}{\partial y}(0,0),\ \frac{\partial^2 z_k}{\partial u \partial y}(0,0),\ \dots,\ \frac{\partial^{k-2} z_k}{\partial u^{k-2}\partial y}(0,0)\Big\},$$

is equal to the rank of the Jacobi matrix $J(P_k)$, whose rows are the vectors

$$\Big\{\frac{\partial \phi_{k+1,1}}{\partial y}(0),\ \frac{\partial \phi_{k+1,2}}{\partial y}(0),\ \dots,\ \frac{\partial \phi_{k+1,k-1}}{\partial y}(0)\Big\}.$$

Let $\hat{M}(z_k)$ be the $(k \times n)$-matrix, obtained from $M(z_k)$ by adding the first row of zeros. Consider the column-vector

$$q =_{\mathsf{def}} \Big\{Q(0,0),\ \partial_u Q(0,0),\ \dots,\ \partial_u^{k-1} Q(0,0)\Big\}.$$

Then, using (6.23),

$$M(z_{k+1}) = \hat{M}(z_k) + q \times \frac{\partial \phi}{\partial y}(0),$$

where "×" stands for the matrix multiplication.

Remembering that $Q(0,0) \neq 0$ and that $\phi =_{\mathsf{def}} \phi_{k+1,0}$, in view of the equations above (or equations (6.23)), we get that the space spanned by the vectors

$$\frac{\partial z_{k+1}}{\partial y}(0,0), \text{ and } \Big\{\frac{\partial^{j+1} z_{k+1}}{\partial u^j \partial y}(0,0)\Big\}_{1 \leq j \leq k-1}$$

coincides with the space spanned by the vectors

$$\frac{\partial \phi}{\partial y}(0),\ \frac{\partial z_k}{\partial y}(0,0), \text{ and } \Big\{\frac{\partial^{j+1} z_k}{\partial u^j \partial y}(0,0)\Big\}_{1 \leq j \leq k-2}.$$

Therefore, $\mathrm{rk}(M(z_{k+1})) = \mathrm{rk}(J(P_{k+1}))$.

We let the reader to verify the validity of the lemma for $k = 2$. □

Definition 6.4. Let $\hat{v}$ be a nonvanishing vector field in $\hat{X}$. We say that a set $U \subset \hat{X}$ is *$\hat{v}$-adjusted*, if its intersection with every $\hat{v}$-trajectory $\hat{\gamma}$ is an interval. ◊

Definition 6.5. Let v be a traversing vector field in X, and γ its trajectory. Consider the intersection $\gamma \cap \partial_1 X = \coprod_{1 \leq i \leq p} a_i$, where each point $a_i \in \partial_{j_i} X(v)^\circ$. Let us pick a smooth transversal section S of the $\hat{v}$-flow at a point $b \in \gamma$.

We say that v is *traversally generic* **at** γ (cf. Definition 6.3), if the images of the tangent spaces $\{T_{a_i}(\partial_{j_i} X(v))\}_i$, being projected to the tangent space $T_b(S)$ with the help of the $\hat{v}$-flow, are in general position. ◊

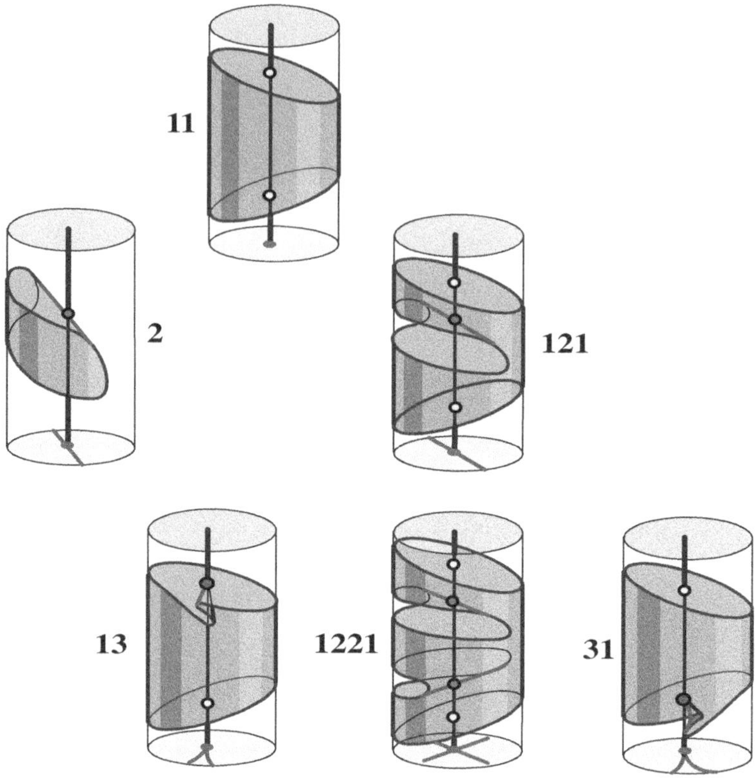

Fig. 6.2 The six geometric patters of traversally generic vector fields on 3-folds, localized to the vicinity of the core trajectory. The patterns are labeled by the six combinatorial tangency types of the core trajectories.

Figure 6.2 shows all the semi-local patterns of traversally generic flows on 3-folds.

The following proposition is a "semi-global" generalization of Morin's Theorem 6.1.

Lemma 6.4. *Let $\hat{v}$ be a traversing vector field in a $\hat{v}$-adjusted neighborhood $U \subset \hat{X}$ of a given v-trajectory $\gamma \subset X$. Assume that v is traversally generic at γ (see Definition 6.5). Let $\gamma \cap \partial_1 X = \cup_i a_i$.*

Then γ has a $\hat{v}$-adjusted neighborhood $V \subset U$ with a special system of coordinates

$$\big(u,\ \underbrace{x_{10}, \dots, x_{1j_1-2}},\ \dots,\ \underbrace{x_{i0}, \dots, x_{ij_i-2}},\ \dots,\ \underbrace{x_{p0}, \dots\ x_{pj_p-2}},\ \underbrace{y_1, \dots, y_{n-m'(\gamma)}}\big)$$

such that:

- $\{u = const\}$ *defines a transversal section of the* $\hat{v}$*-flow,*
- *each* $\hat{v}$*-trajectory in* V *is produced by fixing all the coordinates* $\{x_{il}\}$ *and* $\{y_k\}$,
- *there is an* $\epsilon > 0$ *such that* $V \cap \partial_1 X \subset \coprod_i V_i$*, where*

$$V_i =_{\mathsf{def}} u^{-1}((\alpha_i - \epsilon, \alpha_i + \epsilon)) \cap V,$$

and $\alpha_i = u(a_i)$,
- *for each* i*, the intersection* $V_i \cap \partial_1 X$ *is given by the equation*

$$(u - \alpha_i)^{j_i} + \sum_{l=0}^{j_i - 2} x_{i,l}(u - \alpha_i)^l = 0. \tag{6.24}$$

Proof. Let $\gamma \subset X$ be a given v-trajectory, and $U \subset \hat{X}$ its $\hat{v}$-adjusted neighborhood. With the help of the flow, we introduce a system of coordinates $(u, y) =_{\mathsf{def}} (u, y_1, \dots y_n)$ in U so that:

- $S_i = \{u = \alpha_i\}$ is a transversal section of $\hat{v}$-flow, which contains a_i and is diffeomorphic to a closed disk D^n,
- the locus $\{y = \vec{0}\} \cap X$ is the trajectory γ, and
- each $\hat{v}$-trajectory in U is given by freezing some point y.

Consider a smooth function $z : \hat{X} \to \mathbb{R}$, or rather its germ, in the vicinity of X, such that:

- 0 is a regular value for z,
- $z^{-1}(0) = \partial_1 X$,
- $z^{-1}((-\infty, 0]) = X$.

(6.25)

In the vicinity of each point $a_i \in \gamma \cap \partial_1 X$, the coordinates $(u - \alpha_i, y) =_{\mathsf{def}} (u - \alpha_i, y_1, \dots y_n)$ are available. Since $a_i \in \partial_{j_i} X(v)^\circ$, by Lemma 6.1, the $(j_i - 1)$-jet of $z|_\gamma$ vanishes, but the j_i-jet does not. Thus, in some $\hat{v}$-adjusted neighborhood $W_i \subset U$ of each point a_i, the globally-defined function z in (6.25) can be written as the product $P_i(u - \alpha_i, y) \cdot Q_i(u - \alpha_i, y)$ of two smooth functions, where $Q_i(u - \alpha_i, y) \neq 0$, and the polynomial

$$P_i(u - \alpha_i, y) = (u - \alpha_i)^{j_i} + \sum_{l=0}^{j_i - 2} \phi_{i,l}(y)(u - \alpha_i)^l.$$

Its smooth functional coefficients $\{\phi_{i,l}(y)\}_l$ are such that $\phi_{i,l}(0) = 0$.

We can adjust the size of the W_i's so that, for each pair W_i and W_k, the set of trajectories passing through W_i and the set of trajectories passing through W_k coincide. In other words, we can find a $\hat{v}$-adjusted cylindrical neighborhood $W \subset U$ of γ, such that $W \cap W_i = W_i$ for all i.

Let $\Phi_i : \mathbb{R}^n \to \mathbb{R}^{j_i - 1}$ be given by the functions $\{\phi_{i,l}(y)\}_l$. So we can assume that, in such $\hat{v}$-adjusted neighborhood W of γ, the locus $\{z = 0\}$ is given by *one* of the polynomial in $(u - \alpha_i)$ equations $\{P_i(u - \alpha_i, y) = 0\}_i$.

By Lemma 6.3, $\mathrm{rk}(D\Phi_i(0)) = j_i - 1$, since γ is j_i-tangent to $\partial_1 X$ at a_i.

Since the $\hat{v}$-flow is assumed to be traversally generic at γ, the flow-generated images $\{\mathsf{T}_i\}_i$ of all tangent spaces $\{T_{a_i}(\partial_{j_i} X(v))\}_i$ of the minimal strata $\{\partial_{j_i} X(v) \ni a_i\}_i$ must be in general position in the tangent space T of the section S at the point $a = \gamma \cap S$ (a resides below the lowest point $a_1 \in \gamma \cap \partial_1 X$). The space T_i is of the codimension $j_i - 1$ in T.

Let $\Phi : \mathbb{R}^n \to \mathbb{R}^{m'(\gamma)}$ be the direct product of the maps Φ_i, where $m'(\gamma) = \sum_i (j_i - 1)$. Then, by Definition 6.5 of traversal genericity at γ and by Lemma 6.3, $\mathrm{rk}(D\Phi(0)) = m'(\gamma) \leq n$ (see Remark 6.3).

If $\mathrm{rk}(D\Phi(0)) = m'(\gamma)$, then the desired system of coordinates in a new $\hat{v}$-adjusted neighborhood $V \subset W$ of γ is available: indeed, just put $x_{il} = \phi_{i,l}(y)$ and keep certain $(n - m'(\gamma))$-tuple of the original coordinates $\{y_k\}$ unchanged. These coordinates $\{y_k\}$ are chosen so that, together with Φ, they define a map $\mathbb{R}^n \to \mathbb{R}^{m'(\gamma)} \times \mathbb{R}^{n - m'(\gamma)}$ of the maximal rank n at the origin. □

Remark 6.5. Lemma 6.4 implies that, in such special coordinates, the intersection $V \cap \partial_1 X$ is given by the single polynomial equation

$$P(u, x) =_{\mathsf{def}} \prod_i \Big[(u - \alpha_i)^{j_i} + \sum_{l=0}^{j_i - 2} x_{i,l}(u - \alpha_i)^l\Big] = 0 \tag{6.26}$$

of degree $m(\gamma) = \sum_i j_i$, and $V \cap X$ — by the inequality $P(u, x) \leq 0$. ◇

Lemma 6.5. *Let the smooth manifold $X \subset \mathbb{R} \times \mathbb{R}^d$ be given by the polynomial inequality*

$$\Big\{P(u, x) =_{\mathsf{def}} u^d + \sum_{k=0}^{d-1} x_k u^k \ \leq\ 0\Big\}.$$

Then the field ∂_u is traversally generic with respect to the boundary $\partial X =_{\mathsf{def}} \{P(u, x) = 0\}$.

Proof. Let $\pi : \mathbb{R} \times \mathbb{R}^d \to \mathbb{R}^d$ denote the obvious projection.

For every point $w = (u, x) \in \partial X \subset \mathbb{R} \times \mathbb{R}^d$, there exists a unique integer $j(w) \geq 0$ such that

$$P^{(k)}(w) = 0 \text{ for all } 0 \leq k < j(w), \text{ and } P^{(j(w))}(w) \neq 0,$$

where $P^{(j)}$ denotes the j-th partial derivative of P with respect to u. Evidently, $j(w) \leq d$.

We denote by $\Sigma_{[j]} \subset \mathbb{R} \times \mathbb{R}^d$ the locus $\{w | \, j(w) \geq j\}$. It is given by the j equations

$$\{P(u, x) = 0, \, P^{(1)}(u, x) = 0, \, \ldots, \, P^{(j-1)}(u, x) = 0\}. \tag{6.27}$$

Consider any t-parameter curve $\gamma(t) = (u(t), x(t)) \subset \Sigma_{[j]}$ and let us compute the restrictions imposed by (6.27) on its tangent vector $\dot{\gamma}(0)$ at the point $\gamma(0)$. Differentiating with respect to t the equations (6.27) that define $\Sigma_{[j]}$, we get:

$$\begin{aligned} P^{(1)} \dot{u} + P_x * \dot{x} &= 0, \\ P^{(2)} \dot{u} + P_x^{(1)} * \dot{x} &= 0, \\ &\cdots \\ P^{(j)} \dot{u} + P_x^{(j-1)} * \dot{x} &= 0, \end{aligned} \tag{6.28}$$

where P_x denotes the row vector $(u^{d-1}, \ldots, u, 1)$, $\dot{x}$ denotes the column vector $(\dot{x}_{d-1}, \ldots, \dot{x}_1, \dot{x}_0)$, and $P_x * \dot{x}$ stands for the matrix multiplication. By combining (6.28) with the equations (6.27) of $\Sigma_{[j]}$, we get

$$\begin{aligned} P_x * \dot{x} &= 0, \\ P_x^{(1)} * \dot{x} &= 0, \\ &\cdots \\ P_x^{(j-2)} * \dot{x} &= 0, \\ P^{(j)} \dot{u} + P_x^{(j-1)} * \dot{x} &= 0. \end{aligned} \tag{6.29}$$

These equations imply that for any $\dot{x}$, subject to the first $j - 1$ equations in (6.29), there exists a unique value $\dot{u} = -(P_x^{(j-1)} * \dot{x})/P^{(j)}$, unless $P^{(j)}(u, x) = 0$. The latter vanishing does not happen in the proper stratum $\Sigma_{[j]}^\circ =_{\mathsf{def}} \Sigma_{[j]} \setminus \Sigma_{[j+1]}$.

Therefore, for each vector $(\dot{u}, \dot{x}) \in T_{(u,x)} \Sigma_{[j]}^\circ$, tangent to $\Sigma_{[j]}^\circ$ at a point (u, x), the projection $D\pi((\dot{u}, \dot{x})) = \dot{x}$ belongs to a vector subspace $V_{[j]}(u, x) \subset T_x(\mathbb{R}^d) \approx \mathbb{R}^d$, defined by the first $j - 1$ equations in (6.29). Moreover, every vector $\dot{x}$ at x that belongs to $V_{[j]}(u, x)$ has a unique preimage in the tangent space $T_{(u,x)} \Sigma_{[j]}^\circ$, so that

$$D\pi : T_{(u,x)} \Sigma_{[j]}^\circ \to V_{[j]}(u, x)$$

is a linear isomorphism.

Let us pick now a generic point $x \in \mathbb{R}^d$. Let ℓ_x denote the line $\pi^{-1}(x) \subset \mathbb{R} \times \mathbb{R}^n$. Put $Z_x =_{\mathsf{def}} \ell_x \cap \{P = 0\}$. The u-coordinate maps Z_x to the set of zeros $\{\alpha_i = \alpha_i(x)\}_i$ of the u-polynomial $P(u, x)$. We denote by $j_i = j(\alpha_i, x)$ the multiplicity of the root α_i.

Our goal is to show that π maps the tangent spaces $\{T_{(\alpha_i,x)}\Sigma_{[j_i]}\}_i$ to a general position configuration in $\mathbb{R}^d$. We have shown that the $D\pi$ maps bijectively each tangent space $T_{(\alpha_i,x)}\Sigma^\circ_{[j_i]}$ to the vector space $V(\alpha_i, x) =_{\mathsf{def}} V_{[j_i]}(\alpha_i, x)$—the solution set of the homogeneous linear system

$$\mathcal{L}(\alpha_i, j_i) =_{\mathsf{def}}$$
$$\left\{P_x(\alpha_i) * \dot{x} = 0,\ P_x^{(1)}(\alpha_i) * \dot{x} = 0,\ \dots,\ P^{(j_i-2)}(\alpha_i) * \dot{x} = 0\right\} \quad (6.30)$$

in the variables $\dot{x} = (\dot{x}_0, \dots, \dot{x}_{d-1})$. This space $V(\alpha_i, x)$ is viewed as the subspace of $T_x\mathbb{R}^d$, where (α_i, x) satisfies (6.27) (with u being replaced with α_i).

We need to verify that the system $\mathcal{LL}(\{\alpha_i, j_i\}_i)$, formed by collecting all the systems $\mathcal{L}(\alpha_i, j_i)$ for individual $V(\alpha_i, x)$'s, is of the maximal possible rank, so that the spaces $\{V(\alpha_i, x)\}_i$ are indeed in general position.

Let $m =_{\mathsf{def}} \sum_i (j_i - 1)$. The matrix of the system $\mathcal{LL}(\{\alpha_i, j_i\}_i)$ is a generalized Vandermonde $(m \times d)$-matrix. Its rank is m—the maximal possible since $m < d$.

Let us validate this fact. Consider the auxiliary u-polynomial

$$T(u) =_{\mathsf{def}} P_x * \dot{x} = \sum_{k=0}^{d-1} \dot{x}_k u^k.$$

By (6.30), the solutions $\dot{x}$ of $\mathcal{LL}(\{\alpha_i, j_i\}_i)$ correspond exactly to the u-polynomials $T(u)$ with the coefficients $\dot{x}_0, \dots, \dot{x}_{d-1}$ and the roots $\{\alpha_i\}$ of multiplicities $\{j_i - 1\}$. In other words, $T(u)$ are the u-polynomials that are divisible by the real u-polynomial

$$S(u) =_{\mathsf{def}} \prod_i (u - \alpha_i)^{j_i - 1}$$

of degree m.

The equations (6.30) represent the requirement that $R(u)$, the remainder of this division, vanishes (a priori, $R(u)$ is a polynomial of degree $m-1$ at most). The quotient $T(u)/S(u)$ consists of polynomials of the form $Q(u, y) = \sum_{l=0}^{q} y_l\, u^l$, where $q = d - 1 - m$ and $\{y_l\}_l$ are some real coefficients. We can view them as free variables, since any choice of y_l's produces the polynomial $T(u, y) =_{\mathsf{def}} S(u) \cdot Q(u, y)$, which gives rise to a solution $\dot{x}(y)$

of the system $\mathcal{LL}(\{\alpha_i, j_i\}_i)$. Moreover, different parameters y produce different polynomials $S(u) \cdot Q(u, y)$. Thus y gives a $(d-m)$-parametric representation of the solution space of the system $\mathcal{LL}(\{\alpha_i, j_i\}_i)$.

So (6.30) is solvable, and the rank of its matrix equals m. Therefore, the spaces $\{V(\alpha^\star, x^\star)\}$ form a general position configuration in $\mathbb{R}^d$. As a result, the field ∂_u is traversally generic with respect to the boundary $\partial X = \{P(u, x) = 0\}$. □

Lemma 6.6. *Let $P(u, x)$ be the polynomial as in* (6.26) *of degree $d = \sum_i j_i$. Put $m = \sum_i (j_i - 1)$. Consider the smooth manifold $X \subset \mathbb{R} \times \mathbb{R}^m$, given by the polynomial inequality $\{P(u, x) \leq 0\}$.*

Then, in the vicinity of the line $\ell_0 =_{\mathsf{def}} \{x = \vec{0}\} \subset \mathbb{R} \times \mathbb{R}^m$, the field ∂_u is traversally generic with respect to the boundary $\partial X = \{P(u, x) = 0\}$.

Proof. We follow the outline of the arguments from the previous lemma.

We denote by Z the hypersurface $\{P(u, x) = 0\} \subset \mathbb{R} \times \mathbb{R}^m$.

Let $x^{\{i\}} =_{\mathsf{def}} (x_{i,0}, x_{i,1}, \dots, x_{i,j_i-2})$. By definition, the polynomial $P(u, x)$ from (6.26) of degree $d = \sum_i j_i$ is the product of the factors

$$P_i(u, x^{\{i\}}) =_{\mathsf{def}} (u - \alpha_i)^{j_i} + \sum_{l=0}^{j_i-2} x_{i,l}(u - \alpha_i)^l.$$

Consider the hypersurfaces $Z_i =_{\mathsf{def}} \{P_i(u, x^{\{i\}}) = 0\}$. There exists an $\epsilon > 0$ such that the hypersurfaces $\{Z_i\}_i$ are disjoint in the cylinder

$$\Pi_\epsilon =_{\mathsf{def}} \{\{\|x^{\{i\}}\| < \epsilon\}_i\} \subset \mathbb{R} \times \mathbb{R}^m$$

with the polydisk base $B_\epsilon \subset \mathbb{R}^m$. In other words, for such a choice of $\epsilon > 0$,

$$Z \cap \Pi_\epsilon = \coprod_i (Z_i \cap \Pi_\epsilon).$$

Indeed, the claim follows from Lemma 6.2, since all the roots $\{\alpha_i\}$ of $P(u, 0)$ are distinct (note that the intersection $\ell_0 \cap Z = \coprod_i (\alpha_i, 0)$) and the roots of nearby polynomials $P(u, x)$, $x \in B_\epsilon$, are grouped around the α_i's.

Therefore, for any $x^\star \in B_\epsilon$, the intersection of the line

$$\ell_{x^\star} =_{\mathsf{def}} \{x = x^\star\} \subset \mathbb{R} \times \mathbb{R}^m$$

with Z is of the form $\coprod_i \left(\coprod_{k \in A_i} (\beta_{ik}, x^\star)\right)$, where $\{\beta_{ik}\}_k$ are the real roots of the u-polynomial $P_i(u, x^\star)$ and A_i denotes the appropriate index set.

Let us denote by $j^\star_{ik}$ the multiplicity of the root $\beta^\star_{ik} =_{\mathsf{def}} \beta_{ik}(x^\star)$.

Since, in the vicinity of the point $(\alpha_i, 0)$, the equations $\{P(u,x) = 0\}$ and $\{P_i(u,x) = 0\}$ determine the same solution sets, we may reduce the study of the locus $\Sigma_{[j^\star_{ik}]} \subset Z$ to the study of a similar locus $\Sigma_{[j^\star_{ik}]} \subset Z_i$. As in (6.30), both loci are given by the equations:

$$\{P_i(u,x) = 0,\ P_i^{(1)}(u,x) = 0,\ \dots,\ P_i^{(j^\star_{ik}-1)}(u,x) = 0\}.$$

Thus we have reduced our settings to the ones studied in Lemma 6.5. Let us adapt equations (6.27)-(6.29) to our present environment[5]. We denote by $\tilde{P}_i$ the α_i-shifted polynomial, so that $\tilde{P}_i(u - \alpha_i, x) =_{\mathsf{def}} P_i(u,x)$. In particular, the vector $P_x =_{\mathsf{def}} (u^{d-1}, \dots, u, 1)$ in (6.28) and (6.29) must be replaced by the vector of partial, with respect to the variables from the list $x^{\{i\}}$, derivatives

$$\xi_i(u) =_{\mathsf{def}} (\tilde{P}_i)_{x_{\{i\}}} = \big((u-\alpha_i)^{j_i-2},\ \dots,\ (u-\alpha_i),\ 1\big).$$

We denote by $\{\xi_i^{(l)}(u)\}_l$ its multiple derivatives with respect to the variable u.

As in (6.29), we conclude that each vector $(\dot{u}, \dot{x})$, tangent to the locus $\Sigma_{[j^\star_{ik}]}$ at the point $(\beta_{ik}, x^\star)$ is determined by its $D\pi$-projection $\dot{x} \in T_{x^\star}(\mathbb{R}^m) \approx \mathbb{R}^m$.

Therefore, $D\pi$ maps each tangent space $T_{(\beta^\star_{ik}, x^\star)}(\Sigma^\circ_{[j^\star_{ik}]})$ bijectively to the linear subspace of $T_{x^\star}\mathbb{R}^m$, given by the $j^\star_{ik} - 1$ homogeneous equations

$$V(\beta^\star_{ik}) =_{\mathsf{def}} \big\{\xi_i(\beta^\star_{ik}) * \dot{x}_{\{i\}} = 0,\ \ \xi_i^{(1)}(\beta^\star_{ik}) * \dot{x}_{\{i\}} = 0,\\ \dots,\ \xi_i^{(j^\star_{ik}-2)}(\beta^\star_{ik}) * \dot{x}_{\{i\}} = 0\big\} \qquad (6.31)$$

with respect to the $j_i - 1$ variables $\dot{x}_{\{i\}}$. Note that if some $j^\star_{ik} = 1$, then the simple root $\beta^\star_{ik}$ does not contribute to the equations.

To conclude the proof, we need to show that these spaces $\{V(\beta^\star_{ik})\}_{i,k}$ are in general position in $\mathbb{R}^d$.

Again, the argument is a modification of a similar proof from Lemma 6.5. Consider the entire collection of spaces $\{V(\beta^\star_{ik})\}_{i,k}$ and the collection $\mathcal{LL}(x^\star) =_{\mathsf{def}} \{\mathcal{L}_{ik}(x^\star)\}_{i,k}$ of equations as in (6.31) with respect to the $m =_{\mathsf{def}} \sum_i (j_i - 1)$ variables $\{\dot{x}^{\{i\}}\}_i$. The number of such equations is $m^\star =_{\mathsf{def}} \sum_{i,k}(j^\star_{ik} - 1)$.

So we have to show that the rank of the matrix of the system $\mathcal{LL}(x^\star)$ is $m^\star$, the maximal possible. Note that m is the reduced multiplicity of the ∂_u-trajectory through the point $(0,0)$, while $m^\star \leq m$ is the reduced multiplicity of the ∂_u-trajectory through the nearby point $(0, x^\star)$.

[5]Note the difference: in the present setting, the polynomial $\tilde{P}_i$ is depressed, i.e., the coefficient x_{i,j_i-1} of $(u-\alpha_i)^{j_i-1}$ is zero.

The equations (6.31) are equivalent to the requirement that, for each $x^\star \in B_\epsilon$, the u-polynomial $T_i(u, \dot{x}) =_{\mathsf{def}} \xi_i(u) * \dot{x}^{\{i\}}$ of degree $j_i - 2$ is divisible by the polynomial

$$S_i(u) =_{\mathsf{def}} \prod_k (u - \beta^\star_{ik})^{j^\star_{ik}-1}$$

of degree $m^\star_i =_{\mathsf{def}} \sum_k (j^\star_{ik} - 1)$. So the quotient $Q_i =_{\mathsf{def}} T_i(u, \dot{x})/S_i(u)$ is a u-polynomial of the form

$$Q_i(u, y^{\{i\}}) = \sum_{l=0}^{j_i-2-m^\star_i} y_{il}\, u^l,$$

which depends on $(j_i - 1 - m^\star_i)$ parameters $\{y_{il}\}_l$.

On the other hand, for any choice of these parameters $\{y_{il}\}_l$, we can form the polynomial $S_i(u) \cdot Q_i(u, y_{\{i\}})$ and compute its Taylor polynomial $\sum_{l=0}^{j_i-2} a_{il}(u - \alpha_i)^l$ at the point α_i. Then $\{\dot{x}_{il} = a_{il}\}_l$ must satisfy $\mathcal{L}_i(x^\star)$.

Thus any solution of the system $\mathcal{LL}(x^\star)$ is generated via this linear mechanism by choosing the $m - m^\star$ parameters-variables $\{y^{\{i\}}\}_i$. Different choices of $\{y^{\{i\}}\}_i$ lead to different ordered sets of u-polynomials $\{S_i(u) \cdot Q_i(u, y_{\{i\}})\}_i$ and therefore to different solutions $\{\dot{x}_{il}\}$ of $\mathcal{LL}(x^\star)$. Therefore, the rank of the matrix of $\mathcal{LL}(x^\star)$ is $m - m^\star$, the maximal possible. In turn, this implies that the $\{V(\beta^\star_{ik})\}_{i,k}$ are in general position in $\mathbb{R}^m$. □

Definition 6.6. We say that a vector field $v \in \mathcal{V}_{\mathsf{trav}}(X)$ is *versal*, if each v-trajectory γ has a $\hat{v}$-adjusted neighborhood $U \subset \hat{X}$, equipped with special coordinates

$$\big(u,\ \underbrace{x_{10}, \dots, x_{1j_1-2}},\ \dots\ \underbrace{x_{i0}, \dots, x_{ij_i-2}},\ \dots\ \underbrace{x_{p0}, \dots\ x_{pj_p-2}},\ \underbrace{y_1, \dots, y_{n-m'(\gamma)}}\big)$$

as in Lemma 6.4, in which $\partial_1 X$ is given by the equation (6.26)[6].

We denote by $\mathcal{V}_{\mathsf{vers}}(X) \subset \mathcal{V}_{\mathsf{trav}}(X)$ the subspace of all versal fields on the manifold X. ◇

Corollary 6.2. *If a vector field v is versal in a v-adjusted neighborhood of its trajectory γ, then there is another v-adjusted neighborhood $U \subset \hat{X}$ of γ so that v is traversally generic in U (with respect to $\partial_1 X \cap U$).*

[6]In particular, for any γ, we require that the reduced multiplicity $m'(\gamma) \leq \dim(X) - 1$.

Proof. By the definition of a versal field, in special coordinates with the core γ, the boundary $\partial_1 X$ is given by a polynomial equation $P = 0$ as in (6.26). By applying Lemma 6.6 to P, we conclude that v is traversally generic in a smaller v-adjusted neighborhood $U \subset \hat{X}$ of γ. □

The next corollary claims that if a vector field is traversally generic *at* a trajectory γ, then it is traversally generic in its vicinity.

Corollary 6.3. *Let v be a traversing field on X which is traversally generic* at *a trajectory γ (see Definition 6.5).*

Then there exists a $\hat{v}$-adjusted neighborhood U of γ, such that the field v is traversally generic in U with respect to $\partial_1 X \cap U$.

Proof. If v is traversally generic at γ, then by Lemma 6.4, v is versal in the vicinity of γ. By Corollary 6.2, the field is traversally generic in the vicinity of γ. □

Theorem 6.3. *A vector field on X is versal if and only if it is traversally generic. In other words, $\mathcal{V}_{\mathsf{vers}}(X) = \mathcal{V}^{\ddagger}(X)$.*

Proof. By Lemma 6.4, any traversally generic field is versal. By Corollary 6.2, any versal field is traversally generic in the vicinity of each trajectory γ, and therefore, is traversally generic globally. □

Our ultimate goal in this chapter is to improve upon the Morin Theorem 6.1 by showing that, for a compact connected smooth manifold X with boundary, the space $\mathcal{V}_{\mathsf{vers}}(X) = \mathcal{V}^{\ddagger}(X)$ of versal/traversally generic vector fields is actually *open* and *dense* in the space $\mathcal{V}_{\mathsf{trav}}(X)$ of all traversing fields (see Theorem 6.7 below). We would like to provide the reader with an argument that does not rely heavily on the jet magic of the singularity theory in general, and on the Thom-Boardman Stratification Theory [Bo] in particular. However, reluctantly, we need to use few implications of Boardman's theory.

Let M, N be two smooth manifolds. For a given smooth map $\Phi : M \to N$, consider the locus

$$\Sigma^{(j)}(\Phi) =_{\mathsf{def}} \big\{x \in M|\ \dim\big(\ker(D_x\Phi)\big) \geq j\big\}$$

where the rank of the differential $D\Phi$ drops by j at least.

If $\Sigma^{(j)}(\Phi)$ is a smooth submanifold, then we may consider the smooth map $\Phi| : \Sigma^{(j)}(\Phi) \to N$ and define the locus

$$\Sigma^{(j,k)}(\Phi) =_{\mathsf{def}} \Sigma^{(k)}\big(\Phi| : \Sigma^{(j)}(\Phi) \to N\big).$$

By the definition of $\Sigma^{(j,k)}(\Phi)$, we get $j \leq k$.

In a similar way, we may introduce the locus

$$\Sigma^{(j,k,l)}(\Phi) =_{\mathsf{def}} \Sigma^{(l)}\big(\Phi| : \Sigma^{(j,k)}(\Phi) \to N\big),$$

provided that $\Sigma^{(j,k)}(\Phi)$ is a smooth submanifold. Again, $j \geq k \geq l$.

This recipe can be repeated to generate the *Thom-Boardman stratum* $\Sigma^{\omega}(\Phi)$, where $\omega =_{\mathsf{def}} (j_1 \geq j_2 \geq \cdots \geq j_m)$.

In [Bo], for every such sequence ω, Bordman defines a subbundle $\mathcal{S}_\omega$ of the m-jet bundle $J^m(M,N)$ (relative to the fibration $J^m(M,N) \to M \times N$), which has the following property: if the ℓ-jet extension map $j^\ell\Phi$ is *transversal* to all the manifolds $\mathcal{S}_{t_1,t_2,\dots,t_\ell}$, where $\ell < m$, then $\Sigma^{\omega}(\Phi)$ is well-defined and

$$x \in \Sigma^{\omega}(\Phi) \Leftrightarrow (j^m\Phi)(x) \in \mathcal{S}_\omega.$$

The construction of the subbundle $\mathcal{S}_\omega \subset J^m(M,N)$ is far from being trivial. It depends on the notion of "Jacobean extensions" (see [Bo], [GG], [AGV]).

Definition 6.7. We call $\Phi \in C^\infty(M,N)$ a *Boardman map* if the jet map extension $j^m\Phi$ is transversal to the subbundle $\mathcal{S}_\omega \subset J^m(M,N)$ for all

$$\omega = (j_1 \geq j_2 \geq \cdots \geq j_m).$$

◇

The main theorem in [Bo] claims that Boardman maps form a *residual set* in $C^\infty(M,N)$. When M is compact, this set is open and dense in $C^\infty(M,N)$. In particular, the maps $\Phi \in C^\infty(M,N)$, for which all the loci $\{\Sigma^{\omega}(\Phi)\}_\omega$ are smooth manifolds, form a residual set in $C^\infty(M,N)$.

We denote by $\Sigma_{[j]}(\Phi)$ the locus $\Sigma^{\omega}(\Phi)$ in M, where $\omega = (1,1,1,\dots)$ is of the length j.

Note that although $\Sigma_{[j]}(\Phi) \subset \Sigma^{(j)}(\Phi)$, the two loci are different in general!

Given a smooth map $\Phi : M \to N$, for each $x \in M$, we denote by $j_\Phi(x)$ the maximal integer j such that $x \in \Sigma_{[j]}(\Phi)$.

Definition 6.8. Let M, N be smooth n-dimensional manifolds, and let $\Phi : M \to N$ be a smooth map. We say that Φ is *traversally generic* if

- for each $1 \leq j \leq n$, the singular set $\Sigma_{[j]}(\Phi)$ is a regularly embedded submanifold of M of codimension j,
- the map $\Phi : \Sigma_{[j]}(\Phi) \setminus \Sigma_{[j+1]}(\Phi) \to N$ is an immersion,
- for each $y \in N$, the images of the maps
$$\big\{\Phi| : \Sigma_{[j_\Phi(x)]}(\Phi) \to N\big\}_{\{x \in \Phi^{-1}(y)|\ j_\Phi(x) \geq 1\}}$$
are in general position in the target manifold N.

Equivalently, for each $y \in N$, the images of the tangent spaces
$$\big\{T_x(\Sigma_{[j_\Phi(x)]}(\Phi))\big\}_{\{x \in \Phi^{-1}(y)|\ j_\Phi(x) \geq 1\}}$$
under the linear monomorphisms
$$\big\{D_x\Phi : T_x(\Sigma_{[j_\Phi(x)]}(\Phi)) \to T_y(N)\big\}_{\{x \in \Phi^{-1}(y)|\ j_\Phi(x) \geq 1\}}$$
are in general position in the tangent space $T_y(N)$.

Let us denote the space of traversally generic maps by the symbol $\mathcal{G}^\ddagger(M, N)$. $\diamondsuit$

The next theorem is a well-known but nontrivial implication of the *Theory of Boardman's Maps* (see [Bo], [GG]). In fact, it is the only claim of this theory that we will need for our study of traversally generic vector fields.

Theorem 6.4. *Let M be a n-dimensional compact smooth manifold, and $K \subset M$ a closed set. Assume that $\Phi_0 : M \to \mathbb{R}^n$ is a smooth map which is traversally generic*[7] *in the vicinity of K.*

Then the traversally generic maps $\Phi : M \to \mathbb{R}^n$ that coincide with the given map Φ_0 in the vicinity of K form an open and dense set $\mathcal{G}^\ddagger_{\Phi_0}(M, \mathbb{R}^n)$ in the space $C^\infty_{\Phi_0}(M, \mathbb{R}^n)$ of all smooth maps that coincide with Φ_0 in the vicinity of K.

Proof. When $K = \emptyset$, the theorem is a special case of Theorem 5.2 from [GG]. That theorem claims that the set of all Boardman's maps, satisfying the *normal crossing condition* (the "NC condition" on page 157 in [GG]), is residual (open and dense when M is compact) in $C^\infty(M, N)$. By their very definition, the Boardman maps satisfy the first two bullets of Definition 6.8, while the third bullet is exactly the formulation of the normal crossing property.

[7] in the sense of Definition 6.8

When $K \neq \emptyset$, the openness of the traversally generic maps in the space $C^\infty(M, \mathbb{R}^n)$ implies that the traversally generic maps that coincide with Φ_0 in the vicinity of K are open in the subspace $C^\infty_{\Phi_0}(M, \mathbb{R}^n) \subset C^\infty(M, \mathbb{R}^n)$.

Tracing the proof of Theorem 5.2 from [GG], the density of $\mathcal{G}^\ddagger_{\Phi_0}(M, \mathbb{R}^n)$ in $C^\infty_{\Phi_0}(M, \mathbb{R}^n)$ is validated by the following general observations.

If, for a map $\Phi : M \to \mathbb{R}^n$, its N-jet $j^N(\Phi)$ is transversal to some real algebraic variety $W \subset Jet^N(M, \mathbb{R}^n)$ (the transversality of $j^N(\Phi)$ to this variety is equivalent to the property of f being a Bordman map) at a compact set $K \subset M$, then Φ can be perturbed to a map Φ' such that $\Phi' = \Phi$ in the vicinity of K and $j^N(\Phi')$ is transversal to $W \subset Jet^N(M, \mathbb{R}^n)$ everywhere on M. Using Thom's Multijet Transversality Theorem (see [GG], Theorem 4.13), a similar extension principle applies to maps that satisfy the normal crossing condition.

Let us clarify these claims. If $j^N(\Phi) : M \to Jet^N(M, \mathbb{R}^n)$ is transversal to the variety W on the compact K, then there is a compact neighborhood U of K, $K \subset \mathrm{int}(U)$ on which $j^N(\Phi)$ is transversal to W. We can choose U with a smooth boundary ∂U.

We form a compact smooth collar $C \subset U \setminus K$ of ∂U and consider a smooth partition of unity $\{a, b \in C^\infty(M, \mathbb{R})\}$ on M, subordinate to the cover $\{U,\ (M \setminus U) \cup C\}$ and such that $a = 1$ on $U \setminus C$. Then $b = 1$ on $M \setminus U$.

The maps $\Psi : M \to \mathbb{R}^n$, such that of $j^N(\Psi)$ is transversal to W on a given compact T in M, form an open and dense subset $\mathcal{U}^T \subset C^\infty(M, \mathbb{R}^n)$. Therefore $\mathcal{U}^U_\Phi =_{\mathsf{def}} \mathcal{U}^U$ is an open neighborhood of the map Φ.

Next, we consider the open and dense subsets $\mathcal{U}^M$, $\mathcal{U}^U$, $\mathcal{U}^C$, and $\mathcal{U}^{M \setminus U^\circ}$ of $C^\infty(M, \mathbb{R}^n)$. Note that $\Phi \in \overline{\mathcal{U}^M} \cap \mathcal{U}^U_\Phi$, where $\overline{\mathcal{U}^M}$ denotes the closure of $\mathcal{U}^M$ in $C^\infty(M, \mathbb{R}^n)$.

Our goal is to construct a map $\Theta \in \mathcal{U}^M$ such that $\Theta = \Phi$ on $U \setminus C$.

Let us introduce the space

$$\mathcal{B}_\Phi =_{\mathsf{def}} \{\Psi \in C^\infty(M, \mathbb{R}^n) \big|\ a \cdot \Phi + b \cdot \Psi \in \mathcal{U}^C\}.$$

Since the operator $\Psi \to a \cdot \Phi + b \cdot \Psi$ is continuous in the C^∞-topology and $\mathcal{U}^C$ is an open set, $\mathcal{B}_\Phi$ is an open set. Since $a + b \equiv 1$, $\Phi \in \mathcal{B}_\Phi$; so $\mathcal{B}_\Phi \neq \emptyset$.

Using that $\mathcal{B}_\Phi$ is nonempty and open and $\mathcal{U}^M$ is open and dense, we get $\mathcal{B}_\Phi \cap \mathcal{U}^M \neq \emptyset$. Therefore there exists a map $\Psi_\star \in \mathcal{B}_\Phi \cap \mathcal{U}^M$. Let us form the map

$$\Theta =_{\mathsf{def}} a \cdot \Phi + b \cdot \Psi_\star.$$

By its definition, $\Theta = \Phi$ on $U \setminus C$. Since, by the construction, $\Phi \in \mathcal{U}^{U \setminus C^\circ}$, $\Theta \in \mathcal{U}^C$, $\Psi_\star \in \mathcal{U}^{M \setminus U^\circ}$, and a, b represent a unity partition, we conclude that $\Theta \in \mathcal{U}^M$. □

Let A, B be topological spaces. Recall that a map $\Psi : A \to B$ is called *quasi-open*, if the interior $\mathrm{int}(\Psi(U)) \neq \emptyset$ for any open set $U \subset A$.

The next three simple lemmas will prepare us for the key Lemma 6.10 below.

Lemma 6.7. *Let A and B be two topological spaces. Assume that there exists a pair of continuous maps, $\Psi : A \to B$ and $\Theta : B \to A$, such that $\Psi \circ \Theta = Id_B$. Then Ψ is a quasi-open map.*

Proof. By the continuity of Θ, for any open set $U \subset A$, the set $\Theta^{-1}(U)$ is open in B. The identity $\Psi \circ \Theta = Id_B$ implies that $\Theta^{-1}(U) \subset \Psi(U)$. The same identity implies that if $U \neq \emptyset$ then $\Theta^{-1}(U) \neq \emptyset$. So the Ψ-image of any nonempty open set contains a nonempty open set. □

Lemma 6.8. *Let $\Psi : A \to B$ be a continuous quasi-open map. Let a subset $E \subset B$ be open and dense in B. Then $\Psi^{-1}(E)$ is open and dense in A.*

Proof. By the continuity of Ψ, the set $\Psi^{-1}(E)$ is open. Assume, to the contrary, that $\Psi^{-1}(E)$ is not dense in A. Then there exists an open set $U \subset A$ such that $U \cap \Psi^{-1}(E) = \emptyset$. Consider the set $\Psi(U)$. Since the map Ψ is quasi-open, $\Psi(U)$ must contain a nonempty open subset V. Since E is dense, the open set $V \cap E \neq \emptyset$. Therefore there exists a point $a \in U$ such that $\Psi(a) \in V \cap E$. This implies that there is $a \in \Psi^{-1}(E) \cap U$, a contradiction with the assumption that $\Psi^{-1}(E) \cap U = \emptyset$. □

By combining the previous two lemmas, we get the following corollary.

Lemma 6.9. *Let A and B be two topological spaces. Assume that there exists a pair of continuous maps, $\Psi : A \to B$ and $\Theta : B \to A$, such that $\Psi \circ \Theta = Id_B$. Let a subset $E \subset B$ be open and dense in B. Then $\Psi^{-1}(E)$ is open and dense in A.* ◇

The next Lemma 6.10 claims that each trajectory γ of a traversing and boundary generic vector field v has a pair of special neighborhoods $V \subset U$ such that v admits an arbitrary C^∞-small perturbation which is supported in U and is traversally generic with respect to $\partial_1 X \cap V$. It is important to clarify the nature of the claim: *first* one chooses the right neighborhoods, and *then*, in these neighborhoods, *arbitrary small* perturbations of v with the desired properties are available.

As usually, we extend a given traversing field v on X to a germ of manifold $\hat{X}$ that properly contains X so that the extended field $\hat{v}$ is traversing in $\hat{X}$.

Lemma 6.10. *Let v be a traversing and boundary generic vector field on a compact connected smooth $(n+1)$-manifold X with boundary. Let γ be a v-trajectory. Assume that v is traversally generic in the vicinity of a closed $\hat{v}$-adjusted set $F \subset \hat{X}$.*

Then there is a triple $W \subset V \subset U$ of compact $\hat{v}$-adjusted neighborhoods of γ in $\hat{X}$ with the following properties:

- *there exists an arbitrary C^∞-small and U-supported perturbation $\hat{v}'$ of $\hat{v}$, such that $v' =_{\mathsf{def}} \hat{v}'|_X$ still is traversing and boundary generic,*
- *$\hat{v}'$ is* traversally generic *with respect to $\partial_1 X \cap V$,*
- *for any $\hat{v}'$-trajectory γ' which intersects W, $\gamma' \cap \partial_1 X \subset \partial_1 X \cap V$,*
- *$\hat{v}' = \hat{v}$ in the vicinity of $F \cup (\hat{X} \setminus U)$ in $\hat{X}$.*

Proof. We denote by D^n_r the closed Euclidean n-ball of radius r.

For any traversing vector field $v \in \mathcal{V}^\dagger(X)$, each trajectory γ has regular $\hat{v}$-adjusted neighborhoods $W \subset V \subset U \subset \hat{X}$ such that $W \subset \mathrm{int}(V)$ and $V \subset \mathrm{int}(U)$. With the help of the $\hat{v}$-flow, U, V, and W are diffeomorphic to the co-axial cylinders $[0,1] \times D^n_1$, $[0,1] \times D^n_{0.5}$ and $[0,1] \times D^n_{0.25}$, respectively. Let G denote this diffeomorphism. It maps γ to $[0,1] \times \{\vec{0}\}$.

Let $f : U \to [0,1]$ denote the height function, produced by G, such that $df(\hat{v}) > 0$. Let $S =_{\mathsf{def}} f^{-1}(0)$ be a transversal section of the $\hat{v}$-flow in U.

Put $\partial U =_{\mathsf{def}} G^{-1}(\partial D^n_1 \times [0,1])$ and $\delta U =_{\mathsf{def}} G^{-1}(D^n_1 \times \partial[0,1])$.

Evidently, one can pick the tube U with the core $\hat{\gamma} \supset \gamma$ so narrow that $\delta U \subset \hat{X} \setminus X$. Consider two sets:

$$\partial^F U =_{\mathsf{def}} \partial U \cup (F \cap U) \text{ and } \partial^F S =_{\mathsf{def}} \partial S \cup (F \cap S).$$

Let $\mathcal{V}_f(U, \partial^F U)$ be the space of all smooth vector fields w in U such that:

- the field $w|_{\partial^F U}$ is non-negatively proportional to the given field $\hat{v}|_{\partial^F U}$[8],
- $df(w) > 0$ in $U \setminus \delta U$,

[8]and thus tangent to $\partial^F U$.

- $w|_{\delta U} = 0.$

$$(6.32)$$

Such fields w generate 1-parameter flows $\Phi_w : U \times \mathbb{R} \to U$ for which $F \cap U$ is invariant.

We may assume that $\hat{v} \in \mathcal{V}_f(U, \partial^F U)$ without modifying v in X. The $\hat{v}$-flow $\Phi_{\hat{v}}$ defines the obvious submersion $p_{\hat{v}} : U \to S$.

We denote by $\mathsf{Sub}\big((U, \partial^F U),\, (S, \partial^F S)\big)$ the space of submersions $p : U \to S$ whose restrictions to $\partial^F U$ is equal to the given map $p_{\hat{v}}$. Since each $w \in \mathcal{V}_f(U, \partial^F U)$, the the w-flow defines a submersion p_w, whose restriction to $\partial^F U$ is prescribed and equals to $p_{\hat{v}}$, we get a continuous map

$$J : \mathcal{V}_f(U, \partial^F U) \to \mathsf{Sub}\big((U, \partial^F U),\, (S, \partial^F S)\big). \tag{6.33}$$

This map J is surjective. Indeed, the fibers of any submersion $p \in \mathsf{Sub}\big((U, \partial^F U), (S, \partial^F S)\big)$ form an oriented 1-dimensional foliation $\mathcal{F}(p)$. Using a product metric g on U, we form a unit vector field $w(p)$, tangent to the fibers of $\mathcal{F}(p)$. By multiplying $w(p)$ by an appropriate universal smooth function $\phi : U \to \mathbb{R}_+$, such that $\phi > 0$ in $U \setminus \delta U$ and $\phi|_{\delta U} = 0$, we produce a vector field $\tilde{w}(p) \in \mathcal{V}_f(U, \partial^F U)$. Evidently, $p_{\tilde{w}(p)} = p$.

This construction "$p \Rightarrow \mathcal{F}(p) \Rightarrow \phi \cdot w(p)$" defines a continuous map

$$K : \mathsf{Sub}\big((U, \partial^F U),\, (S, \partial^F S)\big) \to \mathcal{V}_f(U, \partial^F U),$$

such that $J \circ K = Id$, where J is as in (6.33). Therefore, by Lemma 6.7, J is a quasi-open map.

For any $p \in \mathsf{Sub}\big((U, \partial^F U), (S, \partial^F S)\big)$, the restriction of p to $\partial_1 X \cap U$ produces a continuous restriction operator

$$\Psi : \mathsf{Sub}\big((U, \partial^F U),\, (S, \partial^F S)\big) \to C^\infty\big((\partial_1 X \cap U,\, \partial_1 X \cap \partial^F U),\, (S, \partial^F S)\big). \tag{6.34}$$

Let us show that Ψ is a *quasi-open* map. To validate this claim, guided by Lemma 6.9, we will construct a continuous extension operator

$$\Theta : C^\infty\big((\partial_1 X \cap U,\, \partial_1 X \cap \partial^F U),\, (S, \partial^F S)\big) \to C^\infty\big((U, \partial^F U),\, (S, \partial^F S)\big) \tag{6.35}$$

whose composition with the restriction operator Ψ from (6.34) is the identity. The construction of operator Θ will depend on the choice of a particular map $p : (U, \partial^F U) \to (S, \partial^F S)$.

We consider a regular neighborhood $\mathcal{N}(\partial_1 X)$ of the submanifold $\partial_1 X \cap U$ in U. The neighborhood fibers over its core $\partial_1 X \cap U$ with the fiber being a segment. The last bullet in Lemma 6.1 implies that $\partial U \cap \partial_1 X$ is a regularly embedded manifold in ∂U. Hence we can choose a product structure

$$\pi : \mathcal{N}(\partial_1 X) \approx (\partial_1 X \cap U) \times [-1, 1]$$

so that the intersection $\partial_1 X \cap U = \pi^{-1}\big((\partial_1 X \cap U) \times \{0\}\big)$ and $\mathcal{N}(\partial_1 X) \cap \partial U$ is entirely built of π-fibers.

For any smooth function $h : \partial_1 X \cap U \to \mathbb{R}$, we aim to construct its canonical smooth extension $\hat{H} : \mathcal{N}(\partial_1 X) \to \mathbb{R}$. With this goal in mind, consider a 1-parameter family $\{\phi_a : [-1, 1] \to \mathbb{R}\}_a$ of smooth bell-shaped functions such that $\phi_a(-1) = 0 = \phi_a(1)$ and $\phi_a =_{\mathsf{def}} a \cdot \phi_0$ (the parameter a being the height $\phi_a(0)$ of the bell). Thus $\phi_{a+b} = \phi_a + \phi_b$. Note that, for $a > 0$, the function $\phi_a > 0$ in $(0, 1)$; for $a < 0$, the function $\phi_a < 0$ in $(0, 1)$; for $a = 0$, $\phi_0 = 0$.

For a given $h : \partial_1 X \cap U \to \mathbb{R}$, we define a function

$$\tilde{H} : (\partial_1 X \cap U) \times [-1, 1] \to \mathbb{R}$$

by the formula $\tilde{H}(x, t) =_{\mathsf{def}} \phi_{h(x)}(t)$, where $x \in \partial_1 X \cap U$ and $t \in [-1, 1]$. Then we put $\hat{H} =_{\mathsf{def}} \pi \circ \tilde{H}$. This function $\hat{H} : \mathcal{N}(\partial_1 X) \to \mathbb{R}$ extends to a smooth function $H : U \to \mathbb{R}$ by letting H vanish on the complementary set $U \setminus \mathcal{N}(\partial_1 X)$.

Let

$$\mathcal{E} : C^\infty(\partial_1 X \cap U, \mathbb{R}) \to C^\infty(U, \mathbb{R})$$

be the continuous extension operator, defined by the formula $\mathcal{E}(h) =_{\mathsf{def}} H$ for all $h \in C^\infty(\partial_1 X \cap U, \mathbb{R})$. Since $\phi_a =_{\mathsf{def}} a \cdot \phi_0$, the operator $\mathcal{E}$ is linear.

Evidently, the operator $\mathcal{E}$ gives rise to a continuous linear operator

$$\mathcal{E}_n : C^\infty(\partial_1 X \cap U, \mathbb{R}^n) \to C^\infty(U, \mathbb{R}^n).$$

This operator $\mathcal{E}_n$ will be instrumental in the construction of the extension operator Θ from (6.35). In a sense, $\mathcal{E}_n$ can be viewed as "the variation" for the operator Θ.

For a given submersion $p \in \mathsf{Sub}\big((U, \partial^F U),\, (S, \partial^F S)\big)$, we consider its restriction

$$\Psi(p) \in C^\infty\big((\partial_1 X \cap U,\ \partial_1 X \cap \partial^F U),\ (S, \partial^F S)\big).$$

Note that, for a given space Y, the difference $g_1 - g_0$ between two maps, $g_0 : Y \to \mathbb{R}^n$ and $g_1 : Y \to \mathbb{R}^n$, makes sense as a map from Y to $\mathbb{R}^n$.

Since the flow section S is identified with $\mathbb{R}^n$, for any two maps

$$h, h' \in C^\infty\big((\partial_1 X \cap U,\ \partial_1 X \cap \partial^F U),\ (S, \partial^F S)\big),$$

the difference $h' - h : (\partial_1 X \cap U,\ \partial_1 X \cap \partial^F U) \to (\mathbb{R}^n, 0)$ is well-defined.

We take $h =_{\mathsf{def}} \Psi(p)$ in $h' - h$ and consider the map

$$p + \mathcal{E}_n\big(h' - \Psi(p)\big) : (U, \partial^F U) \to (\mathbb{R}^n, \partial^F S).$$

We notice that the restriction of this map to $\partial^F U$ indeed takes $\partial^F U$ to $\partial^F S$: by definition, the restrictions of $\Psi(p)$ and h' to $\partial_1 X \cap \partial^F U$ coincide, and p takes $\partial_1 X \cap \partial^F U$ to $\partial^F S$.

Finally, for any h' as above, we define the operator $\Theta =_{\mathsf{def}} \Theta(p)$ in (6.35) by the formula

$$\Theta(h') =_{\mathsf{def}} p + \mathcal{E}_n\big(h' - \Psi(p)\big).$$

Due to the properties of $\{\phi_a\}_a$, $\mathcal{E}_n(0) = 0$. Thus, $\Theta(\Psi(p)) = p$. By the linearity of the operators $\mathcal{E}_n$ and Ψ and using that $\Psi \circ \mathcal{E}_n = Id$, we get

$$\Psi(\Theta(h')) =_{\mathsf{def}} \Psi\big(p + \mathcal{E}_n(h' - \Psi(p))\big) = \Psi\big(\mathcal{E}_n(h') + p - \mathcal{E}_n(\Psi(p))\big)$$

$$= \Psi(\mathcal{E}_n(h')) + \Psi(p) - (\Psi \circ \mathcal{E}_n)(\Psi(p)) = h' + \Psi(p) - \Psi(p) = h'.$$

Since $\Psi \circ \Theta = Id$, by Lemma 6.7, the restriction operator Ψ is a quasi-open map.

We consider the subspace

$$\mathcal{G}^\ddagger_V \subset C^\infty\big((\partial_1 X \cap U,\ \partial_1 X \cap \partial^F U),\ (S, \partial^F S)\big)$$

such that, for any $h \in \mathcal{G}^\ddagger_V$, the restriction $h| : V \cap \partial_1 X \to S$ is a traversally generic map in the sense of Definition 6.8. By Theorem 6.4, $\mathcal{G}^\ddagger_V$ is an open and dense subset. Since Ψ is a quasi-open map, by Lemmas 6.7-6.9, $\Psi^{-1}(\mathcal{G}^\ddagger_V)$ is open and dense in the space $C^\infty\big((U, \partial^F U),\ (S, \partial^F S)\big)$. Since the space of submersions $\mathsf{Sub}\big((U, \partial^F U),\ (S, \partial^F S)\big)$ is open in the space $C^\infty\big((U, \partial^F U),\ (S, \partial^F S)\big)$, we conclude that the set

$$\mathcal{E}^\ddagger_V =_{\mathsf{def}} \Psi^{-1}(\mathcal{G}^\ddagger_V) \cap \mathsf{Sub}\big((U, \partial^F U),\ (S, \partial^F S)\big)$$

is open and dense in the space $\mathsf{Sub}\big((U, \partial^F U),\ (S, \partial^F S)\big)$.

Let us revisit the map J from (6.33). Recall that J is a quasi-open map. Therefore, $J^{-1}(\mathcal{E}^\ddagger_V)$ is open and dense in the space $\mathcal{V}_f(U, \partial^F U)$ (recall that $f : U \to \mathbb{R}$ is a fixed submersion, an ingredient of definition in (6.32)).

Thus we have shown that the vector fields that are traversally generic with respect $\partial_1 X \cap V$ form an open and dense set in the space $\mathcal{V}_f(U, \partial^F U)$.

Let us spell out what this claim means: for a field $\hat{v}' \in J^{-1}(\mathcal{E}_V^\ddagger)$, each $\hat{v}'$-trajectory γ intersects with $\partial_1 X \cap V$ along a finite set of points $\{a_i \in \partial_{j_i} X(\hat{v}')^\circ\}_i$ in such a way that the differential $Dp_{\hat{v}'}$ places the tangent spaces $\{T_{a_i}(\partial_{j_i} X(\hat{v}')^\circ)\}_i$ in general position in $T(S)$. This behavior still leaves out the following option: some $\hat{v}'$-trajectory, passing through a point of $\partial_1 X \cap V$, may hit $\partial_1 X \cap U$ at a point $b \in \partial_k X(\hat{v}')^\circ$ such that $b \notin V$. So the $Dp_{\hat{v}'}$-image of the tangent space $T_b(\partial_k X(\hat{v}')^\circ)$ may not be in general position with respect to $Dp_{\hat{v}'}$-images of the spaces $\{T_{a_i}(\partial_{j_i} X(\hat{v}')^\circ)\}_i$.

To control better the flows in the vicinity of W, consider the vector fields $w \in \mathcal{V}_f(U, \partial^F U)$, having the following additional "property K": if a w-trajectory γ' has a nonempty intersection with the set W (recall that $W \subset V \subset U$), then the intersection $\gamma' \cap \partial_1 X$ is contained in $\partial_1 X \cap \mathrm{int}(V)$. Let us denote by $\mathcal{K}_{V,W}$ the set of vector fields $w \in \mathcal{V}_f(U, \partial^F U)$ that possess this property K.

Since W is compact and properly contained in the interior of V, the set $\mathcal{K}_{V,W} \subset \mathcal{V}_f(U, \partial^F U)$ is open due to the smooth dependence of the solutions of ODEs on the initial data and coefficients (the relevant vector fields are non-vanishing).

Therefore $J^{-1}(\mathcal{E}_V^\ddagger) \cap \mathcal{K}_{V,W}$ is open in $\mathcal{V}_f(U, \partial^F U)$ and dense in $\mathcal{K}_{V,W}$. Evidently, the original vector field $\hat{v} \in \mathcal{K}_{V,W}$. So $\hat{v}$ admits an arbitrary C^∞-small perturbation $\hat{v}'$ which belongs to $J^{-1}(\mathcal{E}_V^\ddagger) \cap \mathcal{K}_{V,W}$. By the definitions of all relevant spaces, $\hat{v}'$ possesses the properties listed in the lemma. □

Let $v \in \mathcal{V}^\dagger(X)$ and let $z : \hat{X} \to \mathbb{R}$ be as in (6.25). Then along each trajectory γ, in the appropriate coordinates (u, y), the locus $\partial_1 X$ can be described as the zero set of the function

$$z(u,y) = \prod_{a_i \in \gamma \cap \partial_1 X} \Big[(u - u(a_i))^{j_i} + \sum_{l=0}^{j_i - 2} \phi_{i,l}(y)(u - u(a_i))^l\Big], \quad (6.36)$$

where j_i denotes the multiplicity of $z|_\gamma$ at $a_i \in \gamma \cap \partial_1 X$, and $\phi_{i,l}(0) = 0$.

We have seen the crucial role played in the previous arguments by the Jacobi $m'(\gamma) \times n$ matrix $D\Phi_\gamma$ whose rows are the vectors $\{\nabla_y \phi_{i,l}(0)\}_{i,l}$. Based on Lemma 6.3, we can give still another useful interpretation to the condition $\mathrm{rk}(D\Phi_\gamma) = s$.

Let $\gamma \cap \partial_1 X = \{a_i\}$, where $\{a_i\}$ are ordered by v. For each pair $(a_i, a_{i'})$, in the vicinity of γ, consider the germ of the $\hat{v}$-flow, generated diffeomorphism $\Psi_{i,i'}(\gamma) : \hat{X} \to \hat{X}$ that takes $a_{i'}$ to a_i. In fact, the transversal flow sections $S_i, S_{i'}$ at a_i and $a_{i'}$ can be chosen so that $\Psi_{i,i'}(\gamma)(S_{i'}) = S_i$.

With the function $z : \hat{X} \to \mathbb{R}$ as in (6.25) in place, at each $a_i \in \partial_{j_i} X^\circ$, we consider the 1-form dz and its iterated v-directed Lie derivatives

$$\mathcal{L}_v(dz) = d(\mathcal{L}_v z),\ \mathcal{L}_v^2(dz) = d(\mathcal{L}_v^2 z),\ \dots,\ \mathcal{L}_v^{j_i-2}(dz) = d(\mathcal{L}_v^{j_i-2} z),$$

viewed as sections of the cotangent space $T^*_{a_i}(X)$. Note that points $a_i \in \partial_1 X^\circ$ do not contribute to this list (there are two such points at most). By Theorem 6.5 below, $\mathrm{rk}(D\Phi_\gamma)$ is the dimension of the space, spanned by the 1-forms

$$\Big\{\Psi^*_{i,1}(\gamma)\big(dz|_{T_{a_i}S_i}\big),\ \Psi^*_{i,1}(\gamma)\big(\mathcal{L}_v(dz)|_{T_{a_i}S_i}\big),\ \dots$$
$$\dots,\ \Psi^*_{i,1}(\gamma)\big(\mathcal{L}_v^{j_i-2}(dz)|_{T_{a_i}S_i}\big)\Big\}_i \qquad (6.37)$$

in the cotangent space $T^*_{a_1}(S_1)$ of the lowest section S_1.

The next Theorem 6.5 expands the scope of this observation and incorporates the main claim from Theorem 6.3.

Theorem 6.5. *Let X be a compact connected smooth manifold with boundary. For a boundary generic traversing vector field $v \in \mathcal{B}^\dagger(X) \cap \mathcal{V}_{\mathsf{trav}}(X)$ the following properties are equivalent:*

- *v is traversally generic in the sense of Definition 6.3,*
- *v is versal in the sense of Definition 6.6,*
- *for each v-trajectory γ, $\mathrm{rk}(D\Phi_\gamma) = m'(\gamma)$,*[9] *where the Jacobi matrix $D\Phi_\gamma$ is produced as in Lemma 6.3 from the representation* (6.26),
- *for each v-trajectory γ, the dimension of the space, spanned by the* 1-*forms in* (6.37), *is $m'(\gamma)$.*

Proof. By Theorem 6.1, if v is versal, then it is traversally generic and vice versa. So the first two bullets in the theorem are equivalent.

By the arguments from Lemma 6.4, the property of a traversing $v \in \mathcal{V}^\dagger(X)$ being traversally generic implies that the boundary $\partial_1 X$, in the vicinity of each v-trajectory γ and in $\hat{v}$-adjusted coordinates $(u, y) \in \mathbb{R}\times\mathbb{R}^n$, is given by an equation

$$\prod_{a_i \in \gamma\cap\partial_1 X} \Big[(u - u(a_i))^{j_i} + \sum_{l=0}^{j_i-2} \phi_{i,l}(y)(u-u(a_i))^l\Big] = 0.$$

Moreover, the map $\Phi_\gamma : \mathbb{R}^n \to \mathbb{R}^{m'(\gamma)}$, produced by the functions $\{\phi_{i,l}(y)\}_{i,l}$, has the Jacobi matrix $D\Phi_\gamma(0)$ of the maximal rank $m'(\gamma)$ at the

[9]See formula (6.9).

origin $0 \in \mathbb{R}^n$. Furthermore, the reasoning in Lemma 6.4 has established that the property $\mathrm{rk}(D\Phi_\gamma(0)) = m'(\gamma)$ implies the existence of better coordinates $(u, x, \tilde{y})$ in the vicinity of γ, the coordinates in which the boundary $\partial_1 X$ is given by a simpler $\tilde{y}$-independent equation

$$\prod_{a_i \in \gamma \cap \partial_1 X} \left[(u - u(a_i))^{j_i} + \sum_{l=0}^{j_i - 2} x_{i,l}(u - u(a_i))^l\right] = 0.$$

This is exactly the property of v being versal. On the other hand, by the argument as in Lemma 6.6, the versality of a vector field implies that $\mathrm{rk}(D\Phi_\gamma(0)) = m'(\gamma)$.

Thus the equivalence of the properties in the first three bullets has been proven.

Finally, the equivalence of the last bullet with the rest is basically implied by Lemma 6.3. Here is a computation that establishes the last equivalence.

Let $v \in \mathcal{V}^\dagger(X)$. Let us choose special coordinates $(u, \vec{x}) \in \mathbb{R} \times \mathbb{R}^n$ in the vicinity of a typical γ so that the "versal" boundary equation is given by (6.26). Consider a typical multiplier P_i in (6.26). Then the vector

$$\frac{\partial P_i}{\partial \vec{x}} = \sum_{l=0}^{j_i - 2} \big(u - u(a_i)\big)^l \, \frac{\partial x_{i,l}}{\partial \vec{x}}.$$

Thus

$$dP_i = \frac{\partial P_i}{\partial u} du + \Big\langle \frac{\partial P_i}{\partial \vec{x}},\, d\vec{x} \Big\rangle,$$

and its restriction to the section $S_i =_{\mathsf{def}} \{u = u(a_i)\}$ is given by $\langle \frac{\partial P_i}{\partial \vec{x}},\, d\vec{x} \rangle$. Therefore

$$\mathcal{L}^s_{\partial_u}(dP_i)\big|_{S_i} = \sum_{k=1}^{n} \frac{\partial^s}{\partial u^s} \frac{\partial P_i}{\partial x_k}\, dx_k.$$

As a result, $\mathrm{rk}(D\Phi_i(\gamma))$ is equal to the dimension of the space, spanned by

$$\big\{\mathcal{L}^s_{\partial_u}(dP_i)\big|_{S_i} \in T^*_{a_i}(S_i)\big\}_{0 \le s \le j_i - 2}.$$

Similarly, $\mathrm{rk}(D\Phi(\gamma))$ equals the dimension of the space, spanned by

$$\big\{\Psi^*_{a_i, a_1}\big(\mathcal{L}^s_{\partial_u}(dP_i)|_{a_i}\big) \in T^*_{a_1}(X)\big\}_{i,\, 0 \le s \le j_i - 2}.$$

By Lemma 6.3, the latter space coincides with the space, spanned by

$$\big\{\Psi^*_{a_i, a_1}\big(\mathcal{L}^s_{\partial_u}(dz)|_{a_i}\big) \in T^*_{a_1}(X)\big\}_{i,\, 0 \le s \le j_i - 2},$$

where

$$z(u, \vec{x}) = Q(u, \vec{x}) \cdot \prod_i P_i(u, \vec{x}))$$

with $Q(u, 0) \neq 0$.

For any vector field $v \in \mathcal{V}^\dagger(X)$, this proves that the properties, described in the last two bullets, are equivalent. □

It is possible to globalize the local construction, defined by formulas (6.15).

Let X be a $(n+1)$-dimensional smooth compact manifold with boundary. Let $z : \hat{X} \to \mathbb{R}$ be a smooth function as in Lemma 6.1. Consider the sequence of already familiar functions:

$$\psi_0 =_{\mathsf{def}} z,\ \psi_1 =_{\mathsf{def}} \mathcal{L}_v(\psi_0),\ \psi_2 =_{\mathsf{def}} \mathcal{L}_v(\psi_1), \dots,\ \psi_n =_{\mathsf{def}} \mathcal{L}_v(\psi_{n-1}). \quad (6.38)$$

They gives rise to the smooth maps

$$\Psi(v, z) =_{\mathsf{def}} (\psi_0, \dots, \psi_n) : X \to \mathbb{R}^{n+1},$$
$$\Psi^\partial(v, z) =_{\mathsf{def}} (\psi_1, \dots, \psi_n) : \partial_1 X \to \mathbb{R}^n. \quad (6.39)$$

As in (6.15) - (6.17), the locus $\partial_1 X$ is defined by the equation $\{\psi_0 = 0\}$, the locus $\partial_2 X$ by the equations $\{\psi_0 = 0, \psi_1 = 0\}$, and so on. The locus $\partial_j X$ is defined by the equations $\{\psi_0 = 0, \psi_1 = 0, \dots, \psi_{j-1} = 0\}$. We notice that $\partial_j^+ X$ is characterized by the additional inequality $\psi_j \geq 0$. Recall that, unlike the maps in (6.39), these loci do not depend on the choice of the auxiliary function $z : \hat{X} \to \mathbb{R}$.

For a $(n+1)$-dimensional X, the formulas (6.39) give rise to the continuous operators

$$\mathbf{\Psi}_z : \mathcal{V}(X) \to C^\infty(X, \mathbb{R}^{n+1}),$$

$$\mathbf{\Psi}_z^\partial : \mathcal{V}(X) \to C^\infty(\partial_1 X, \mathbb{R}^n).$$

Let $\psi_0, \psi_1, \dots \psi_{l-1}$ be the standard coordinates in $\mathbb{R}^l$. Consider the complete flag

$$\mathsf{F}^l =_{\mathsf{def}} \{\mathbb{R}^l = F_0 \supset F_1 \supset F_2 \cdots \supset F_l = \{0\}\},$$

where $F_j \subset \mathbb{R}^l$ is defined by the equations

$$\{\psi_0 = 0,\ \dots,\ \psi_{j-1} = 0\}.$$

Each space F_j is divided by F_{j+1} into two halves: F_j^+ and F_j^-; the half-space F_j^+ is characterized by the inequality $\psi_j \geq 0$.

Let $\mathsf{Diff}^{\mathsf{F}}_+(\mathbb{R}^l)$ denote the group of smooth diffeomorphisms of $\mathbb{R}^l$ that preserve all the half-spaces $\{F_j^\pm\}$.

Definition 6.9. Let M^k be a smooth compact k-manifold. We say that a map $\Psi : M^k \to \mathbb{R}^l$ is F-*stable* if, for an open neighborhood $\mathcal{O}$ of Ψ in $C^\infty(M^k, \mathbb{R}^l)$ and each $\Psi' \in \mathcal{O}$, there exists a smooth diffeomorphism $\chi : M^k \to M^k$ and a diffeomorphism $\phi \in \mathsf{Diff}^{\mathsf{F}}_+(\mathbb{R}^l)$ such that

$$\phi \circ \Psi' = \Psi \circ \chi.$$

$\diamondsuit$

Remark 6.6. Let $\boldsymbol{\Psi}_z : \mathcal{B}^\dagger(X) \to C^\infty(X, \mathbb{R}^{n+1})$ be the operator which takes each boundary generic vector field v to the map $\Psi(v, z)$ from (6.39).

If the map $\Psi(v, z) : \partial_1 X \to \mathbb{R}^n$ is F^{n+1}-stable, then evidently for each $v' \in \boldsymbol{\Psi}_z^{-1}(\mathcal{O})$, the appropriate χ from Definition 6.9 will map each stratum $\partial_j^\pm X(v')$ to the stratum $\partial_j^\pm X(v)$.

By definition, the F^{n+1}-stable maps form an open set in $C^\infty(X, \mathbb{R}^{n+1})$. However, in general, they do not form a dense subset. Recall that even the common stable maps $f : Y \to Z$, where $\dim(Y) = \dim(Z) = n + 1$, are dense in the space of all smooth maps $C^\infty(Y, Z)$ only for $n < 8$ (see [GG], page 163.)!

Nevertheless, in Theorem 6.6 below, we will establish *the local stability of the Morse stratifications* $\{\partial_j^\pm X(v)\}_j$ in the vicinity of *any* $v \in \mathcal{B}^\dagger(X)$—a much weaker property than the F^{n+1}-stability of the map $\Psi(v, z)$. $\diamondsuit$

By the arguments from Lemma 6.1, for any $v \in \mathcal{B}^\dagger(X)$, the map $\Psi(v, z) : X \to \mathbb{R}^{n+1}$ is transversal to each space F_j from the flag

$$\mathsf{F}^{n+1} =_{\mathsf{def}} \{\mathbb{R}^{n+1} \supset F_1 \supset F_2 \supset \cdots \supset F_{n+1} = \{0\}\},$$

and $\Psi(v, z)^{-1}(F_1) = \partial_1 X$. We describe this transversality by saying that "Ψ is transversal to the flag F^{n+1}".

Question 6.1. *Which maps $\Theta : X \to \mathbb{R}^{n+1}$, transversal to the flag F^{n+1}, have the form $\Psi(v, z)$ for some function z as in Lemma 6.1 and $v \in \mathcal{B}^\dagger(X)$? For some $v \in \mathcal{V}^\dagger(X) = \mathcal{B}^\dagger(X) \cap \mathcal{V}_{\mathsf{trav}}(X)$?* $\diamondsuit$

The answer to this question eludes us. However, if we extend the function list in (6.38) by introducing an additional function $\psi_{n+1} =_{\mathsf{def}} \mathcal{L}_v(\psi_n)$, we will get an extension $\hat{\Psi}(z, v) : X \to \mathbb{R}^{n+2}$ of the map $\Psi(z, v) : X \to \mathbb{R}^{n+1}$.

According to the lemma below, for a given "extended" map $\hat{\Theta} : X \to \mathbb{R}^{n+2}$, the field v such that $\hat{\Theta} = \hat{\Psi}(z, v)$ is often unique.

Lemma 6.11. *Let X be a compact connected smooth $(n+1)$-dimensional Riemannian manifold with boundary. For a given smooth map $\hat{\Theta} : X \to \mathbb{R}^{n+2}$, consider the composition $\Theta =_{\mathsf{def}} \pi \circ \hat{\Theta}$, where $\pi : \mathbb{R}^{n+2} \to \mathbb{R}^{n+1}$ is the projection $(\psi_0, \dots, \psi_{n+1}) \to (\psi_0, \dots, \psi_n)$. Assume that:*

- *$z = \psi_0$ has the properties as in (6.25),*
- *Θ is transversal to the flag F^{n+1},*
- $\mathrm{rk}(D\Theta) = n + 1$ *on a* dense *subset A of X.*

Then there is at most one boundary generic vector field v on X such that $\Theta = \Psi(z, v)$ and $\hat{\Theta} = \hat{\Psi}(z, v)$.

Proof. We denote by θ_j the j-th component of the given map $\hat{\Theta}$. Let g be a Riemannian metric on X, such that $\nabla_g(\theta_0) = \nu$, the unitary inward normal to $\partial_1 X$. Such metric g exists since Θ is transversal to F_1.

Recall that for any smooth function $f : \hat{X} \to \mathbb{R}$ and a Riemannian metric g on $\hat{X}$, we have $\mathcal{L}_v(f) = \langle \nabla_g f, v \rangle$, where $\nabla_g f$ is the gradient of f.

If $\Theta = \Psi(z, v)$, then $\theta_0 =_{\mathsf{def}} z$, and v must satisfy the equations:

$$\theta_1 = \langle \nabla_g \theta_0, v \rangle,\ \theta_2 = \langle \nabla_g \theta_1, v \rangle,\ \dots,\ \ \theta_n = \langle \nabla_g \theta_{n-1}, v \rangle.$$

We impose an additional relation $\theta_{n+1} = \langle \nabla_g \theta_n, v \rangle$, which reflects the assumption that Θ extends to $\hat{\Theta} =_{\mathsf{def}} \hat{\Psi}(z, v)$.

Since $\dim(X) = n+1$ and the vector fields $\nabla_g \theta_0, \nabla_g \theta_2, \dots, \nabla_g \theta_n$ are assumed to be independent on the set $A \subset X$, the vector field v is uniquely determined there by its scalar products with the gradient fields $\{\nabla_g \theta_j\}_{0 \leq j \leq n}$:

$$v = \sum_{i=0}^{n} \langle \nabla_g \theta_i, v \rangle \, \nabla_g \theta_i = \sum_{i=0}^{n} \theta_{i+1} \nabla_g \theta_i.$$

For a given metric g, the gradient fields $\{\nabla_g \theta_i\}_i$ are determined by the functions $\{\theta_i\}_i$. Since A is dense, by continuity, there is a single vector field v on X at most, such that $\hat{\Theta} = \hat{\Psi}(z, v)$. If $A \supset \partial X$, then this $v \in \mathcal{B}^\dagger(X)$.

Note that transversality of the map Θ to the flag (assumed by the second bullet) is equivalent the property $v \in \mathcal{B}^\dagger(X)$. So $A \supset \partial_1 X$. $\square$

Recall that, for a vector field $v \in \mathcal{V}_{\mathsf{trav}}(X) \cap \mathcal{B}^\dagger(X)$ and its trajectory γ, each point $a \in \gamma \cap \partial_1 X$ acquires some multiplicity $j(a)$, the set $\gamma \cap \partial_1 X$ is finite, and the points of $\gamma \cap \partial_1 X$ are ordered. So we get an ordered sequence of points $a_i \in \gamma \cap \partial_1 X$, together with their multiplicities $j(a_i)$. We call such weighted sequence D_γ of points on γ the *divisor of* γ. Its degree is the multiplicity $m(\gamma) =_{\mathsf{def}} \sum_{a \in \gamma \cap \partial_1 X} j(a)$ of γ (see Definition 6.1).

Theorem 6.6. *Let X be a compact connected smooth $(n+1)$-manifold with boundary.*

- *The space $\mathcal{B}^\dagger(X)$ of boundary generic vector fields*[10] *is open and dense in the space $\mathcal{V}(X)$ of all vector fields on X.*
- *Up to the natural action by diffeomorphisms of X, the Morse stratification $\{\partial_j X(v)\}_j$ is locally constant within each path-connected component of the space $\mathcal{B}^\dagger(X)$.*
- *For any vector field $v \in \mathcal{B}^\dagger(X)$, there is a neighborhood E of $\partial_1 X$ in $\hat{X}$, such that the portion $\gamma \cap E$ of every $\hat{v}$-trajectory γ has the E-localized multiplicity $m(\gamma \cap E) \leq n+1$. Moreover, for each point $a \in \partial_k X^\circ(v)$, $m(\gamma \cap E) \leq k$ for all γ's in the vicinity of the point a.*
- *Each point $a \in \partial_k X^\circ(v)$ has a $\hat{v}$-adjusted neighborhood $U \subset \hat{X}$ such that any divisor D in $\mathbb{R}$ with the properties* $\deg(D) \leq k$ *and* $\deg(D) \equiv k \mod 2$ *is realized, up to a diffeomorphism of $\mathbb{R}$, as the divisor $D_{\gamma \cap U}$ of the portion $\gamma \cap U$ for some $\hat{v}$-trajectory γ.*

Proof. We fix a Riemannian metric g on $\hat{X}$. Here we assume that ψ_0 is defined only in the vicinity of $\partial_1 X$ in $\hat{X}$. Consider the functions $\{\psi_i\}$ from (6.38) and the map $\Psi = \Psi(v, z)$ from (6.39) that they generate. As in (6.15)-(6.17), the property of v being boundary-generic is equivalent to the requirement that the gradient vectors $\nabla\psi_0, \nabla\psi_1, \dots, \nabla\psi_{j-1}$ are linearly independent on the solution set of $\{\psi_0 = 0, \psi_1 = 0, \dots, \psi_{j-1} = 0\}$ for all $j \in [1, n+1]$.

In terms of the complete flags F^{n+1} (with $F_0 =_{\mathsf{def}} \mathbb{R}^{n+1}$), a field $v \in \mathcal{B}^\dagger(X)$ if and only if, for each subspace $F_j \subset \mathbb{R}^{n+1}$, the rank of the differential $D\Psi$, being restricted to the bundle, normal to the set $\partial_j X(v) = \Psi^{-1}(F_j)$ in X, is equal to j.

Let us denote by $\mathcal{V}_{\{|_\partial \neq 0\}}(X)$ the space of vector fields that do not vanish on the boundary $\partial_1 X$. Evidently, any smooth section of the tangent bundle $T(X)$ can be approximated by a smooth section that does not vanish on

[10] See Definition 5.1.

$\partial_1 X$. Therefore it suffices to show that $\mathcal{B}^\dagger(X)$ is a dense (in the C^∞-topology) subset of $\mathcal{V}_{\{|\partial\neq 0\}}(X)$ in order to conclude that $\mathcal{B}^\dagger(X)$ is a dense subset of $\mathcal{V}(X)$.

Given any vector field v which does not vanish on $\partial_1 X$, we decompose it along $\partial_1 X$ into the normal component n_1 and the tangent component v_1. Then we extend the decomposition $v = v_1 \oplus n_1$ to a collar of $\partial_1 X$ in $\hat{X}$. It is possible to perturb n_1 to make sure that it defines a section of the normal bundle $\nu(\partial_1 X, \hat{X})$ that is transversal to its zero section. The perturbation can be smoothly extended into a collar of $\partial_1 X$ in $\hat{X}$, where it will be supported. Let us use the same notations for the perturbed vector field.[11]

The normal component n_1 will be fixed in the further perturbations. Its zero locus is the manifold $\partial_2 X$. Next, we consider the orthogonal decomposition $v_1 = v_2 \oplus n_2$, where v_2 is tangent to $\partial_2 X$ and n_2 is a section of the normal bundle $\nu(\partial_2 X, \partial_1 X)$. Again, it is possible to extend this decomposition into a collar of $\partial_2 X$ in $\partial_1 X$. Then we perturb n_2 to make it transversal to $\partial_2 X$ and extend this perturbation first into a collar of $\partial_2 X$ in $\partial_1 X$ and then further into a collar of $\partial_1 X$ in $\hat{X}$. The zero locus of n_2 defines a submanifold $\partial_3 X \subset \partial_2 X$. Continuing this sequence of perturbations, we will produce a vector field from $\mathcal{B}^\dagger(X)$. Therefore such vector fields form a dense set in the space $\mathcal{V}_{\{|\partial\neq 0\}}(X)$, and thus, in the space of all vector fields $\mathcal{V}(X)$ (this fact can be derived from the Morin Theorem 6.1 as well).

To show that $\mathcal{B}^\dagger(X)$ is open, we use the local model, employed in the proof of Lemma 6.1. Evidently, the linear independence of the gradient fields in (6.17) in the vicinity of the solution set of (6.16) is an open property, imposed on the pair of functions $u(z,x), w(z,x)$ from that model. By Lemma 6.1, locally, this independence of fields is exactly the property of v to define transversal sections of the quotient 1-bundles $\{T(\partial_{j-1}X)/T(\partial_j X)\}_j$.

By covering a compact collar E of $\partial_1 X$ in $\hat{X}$ with a finite system of compact coordinate charts (z,x) as in the proof of Lemma 6.1, we conclude that if (6.16) and (6.17) are satisfied by a vector field v in each of the charts, then all sufficiently C^∞-small perturbations of v will satisfy similar conditions. Thus, $\mathcal{B}^\dagger(X)$ is open and dense in the space $\mathcal{V}_{\{|\partial\neq 0\}}(X)$ which, in turn, is open and dense in the space $\mathcal{V}(X)$.

Now we will validate the claim in the second bullet of the theorem. Recall that this claim was stated as Theorem 5.1.

Let us fix a Riemannian metric g on X. Recall that the normal line

[11]In fact, the transversality of n_1 to $\partial_1 X$ defines an open set in the space $\mathcal{V}(\hat{X})$.

bundle ν_j to $\partial_j X(v)$ in $\partial_{j-1}X(v)$ is trivial for all j. We assume that the fibers of ν_j are orthogonal to $\partial_j X(v)$ in the metric $g|_{\partial_{j-1}X(v)}$. For a sufficiently small $\epsilon > 0$, with the help of the exponential map, the bundle ν_j gives rise to a 1-disk bundle structure in a regular ϵ-neighborhood $V_j(\epsilon) \subset \partial_{j-1}X(v)$ of $\partial_j X(v)$. These bundle structures in $\{V_k(\epsilon_k)\}_{k\leq j}$ are useful for picking a particular j-disk bundle structure in a regular neighborhood $W_{j,\epsilon}$ of $\partial_j X(v)$ in X, so that the j-bundle $\xi_j : W_{j,\epsilon} \to \partial_j X(v)$ comes equipped with a framing.

By definition, a typical fiber of the bundle ξ_j over a point $a_j \in \partial_j X(v)$ consists of all points in X that can be reached by moving from a_j first in the direction of ν_j along the geodesic $\gamma_j \subset \partial_{j-1}X(v)$ for a distance l_{j-1}, then by moving from a typical point $a_{j-1} \in \gamma_j$ in the direction of ν_{j-1} along the geodesic $\gamma_{j-1} \subset \partial_{j-2}X(v)$ for a distance l_{j-2}, then by moving from a typical point $a_{j-2} \in \gamma_{j-1}$ in the direction of ν_{j-2} along the geodesic $\gamma_{j-2} \subset \partial_{j-3}X(v)$ for a distance l_{j-3}, and so on. Here $\sum_k l^2_{j-k} \leq \epsilon^2$, where $\epsilon > 0$ being sufficiently small, and the geodesic curves $\gamma_k \subset \partial_{k-1}X(v)$ are taken in the metric $g|_{\partial_{k-1}X(v)}$.

Thanks to this structure of the bundle $\xi_j : W_{j,\epsilon} \to \partial_j X(v)$, the map ξ_j decomposes into a sequence of j surjections:

$$\xi_{j,1} : W_{j,\epsilon} \to W_{j,\epsilon} \cap \partial_1 X,$$

$$\xi_{j,2} : W_{j,\epsilon} \cap \partial_1 X \to W_{j,\epsilon} \cap \partial_2 X(v),$$

$$\xi_{j,3} : W_{j,\epsilon} \cap \partial_2 X(v) \to W_{j,\epsilon} \cap \partial_3 X(v),$$

and so on.

The fibers of the line bundle $\xi_{j,k}$ foliate the space $W_{j,\epsilon} \cap \partial_{j-1}X(v)$.

Consider the complete coordinate flag

$$\mathsf{F}^n = \big\{\mathbb{R}^n = F_1 \supset F_2 \supset \cdots \supset F_{n+1} = \{0\}\big\}$$

in $\mathbb{R}^n$. We denote by $F_j^\perp(b)$ be the affine j-space that is orthogonal to F_j in $\mathbb{R}^n$ at a point $b \in F_j$. Let $\pi_j : \mathbb{R}^n \to F_j^\perp(0)$ be the orthogonal projection whose fiber is F_j.

Recall that, for $v \in \mathcal{B}^\dagger(X)$, the map $\Psi^\partial(v, z)$ in (6.39) is transversal to each $F_j \subset \mathbb{R}^n$, $j \in [2, n+1]$. Therefore, for any boundary generic v, there exists an open j-ball $B^j_\epsilon =_{\mathsf{def}} B^j_\epsilon(v) \subset F_j^\perp$ of radius $\epsilon =_{\mathsf{def}} \epsilon(j)$, centered on the origin, such that the map

$$\pi_j \circ \Psi^\partial : U_{j,\epsilon} \to B^j_\epsilon$$

is a surjection and thus a fibration. Here $U_{j,\epsilon} =_{\mathsf{def}} (\pi_j \circ \Psi^\partial)^{-1}(B^j_\epsilon)$. We choose $\epsilon > 0$ appropriate for all j's.

Next we pick the ϵ'-neighborhoods $\{W_{j,\epsilon'}\}_j$ as above and with $\epsilon' > 0$ so small that $\{W_{j,\epsilon'} \subset U_{j,\epsilon}\}_j$ for all j. The fibers of the j-bundle $\xi_j : W_{j,\epsilon'} \to \partial_j X(v)$ are transversal to its core $\partial_j X(v) = (\pi_j \circ \Psi^\partial)^{-1}(0)$, and the the map $\pi_j \circ \Psi$ is transversal to $0 \in B^j_\epsilon$. Therefore there exists $\epsilon' > 0$ such that, for each j, all the fibers of the surjection $\pi_j \circ \Psi^\partial$ are transversal to the fibers of the disk bundle $\xi_j : W_{j,\epsilon} \to \partial_j X(v)$.

For any map $\Psi^\partial : \partial_1 X \to \mathbb{R}^n$, transversal to the complete flag F^n in $\mathbb{R}^n$, we fix the special neighborhoods $W_{j,\epsilon'} \subset U_{j,\epsilon}$ of $\partial_j X(v)$ as above. Then there exists an open neighborhood $\mathcal{O}_j(\Psi^\partial) \subset C^\infty(\partial_1 X, \mathbb{R}^n)$ of the map Ψ^∂, such that, for each map $\Psi' \in \mathcal{O}_j(\Psi^\partial)$, the following properties are valid:

- $(\Psi')^{-1}(F_j) \subset W_{j,\epsilon'}$,
- $(\Psi')^{-1}(F_j)$ is a smooth section of the trivial normal j-disk bundle $\xi_j : W_{j,\epsilon'} \to \partial_j X(v)$, a section transversal to the fibers of ξ_j.

(6.40)

The existence of the neighborhood $\mathcal{O}_j(\Psi^\partial)$ follows routinely from the openness of transversal families of smooth maps on compact sets. For example, to insure that $(\Psi')^{-1}(F_j)$ is transversal to the fibers $\{D_a\}_{a \in (\Psi^\partial)^{-1}(F_j)}$ of the bundle ξ_j, we need to take Ψ' so close to Ψ^∂ that the space $D\Psi'(TD_x)$ is still complementary in $\mathbb{R}^n$ to the space F_j at the point $\Psi'(x)$ for all $x \in W_{j,\epsilon'} \supset (\Psi')^{-1}(F_j)$.

Next, we form the open set $\mathcal{O}(\Psi^\partial) =_{\mathsf{def}} \bigcap_{j=1}^n \mathcal{O}_j(\Psi^\partial)$ in the space $C^\infty(\partial_1 X, \mathbb{R}^n)$, equipped with the Whitney topology.

Let

$$\boldsymbol{\Psi} : \mathcal{B}^\dagger(X) \to C^\infty(\partial_1 X, \mathbb{R}^n)$$

be the continuous map that takes each vector field $v' \in \mathcal{B}^\dagger(X)$ to the map $\Psi^\partial(v', z) \in C^\infty(\partial_1 X, \mathbb{R}^n)$.

Consider the open neighborhood of v:

$$\mathcal{U}(v) =_{\mathsf{def}} \boldsymbol{\Psi}^{-1}\big(\mathcal{O}(\Psi^\partial(v))\big) \subset \mathcal{B}^\dagger(X).$$

We claim that, for any $v' \in \mathcal{U}(v)$, the Morse stratifications $\{\partial_j X(v)\}_j$ and $\{\partial_j X(v')\}_j$ can be transformed one into another by a diffeomorphism of X (actually, by an isotopy). Let us explain how to construct the matching diffeomorphism.

Put $\Psi' =_{\mathsf{def}} \Psi(v', z)$, $\Psi'^{\partial} =_{\mathsf{def}} \Psi^{\partial}(v', z)$, and $\Psi =_{\mathsf{def}} \Psi(v, z)$, $\Psi^{\partial} =_{\mathsf{def}} \Psi^{\partial}(v, z)$. For any point

$$a \in \partial_j X(v) = (\Psi)^{-1}(F_j),$$

consider the unique point

$$b_j(a) =_{\mathsf{def}} (\Psi')^{-1}(F_j) \cap \xi_j^{-1}(a),$$

where $\xi_j^{-1}(a)$ is the fiber over a of the disk bundle $\xi_j : W_{j,\epsilon'} \to \partial_j X(v)$. The two properties in (6.40) imply that the correspondence $\phi_j : b_j(a) \to a$ is a diffeomorphism which maps $(\Psi')^{-1}(F_j)$ to $\Psi^{-1}(F_j)$.

Consider the linear (in the fibers of the bundle ξ_j) diffeotopy

$$\{\phi_j(a,t) : \partial_j X(v') \to \partial_1 X\}_{t \in [0,1]}$$

that links each point $a \in \partial_j X(v)$ to the point $b_j(a) \in (\Psi')^{-1}(F_j) = \partial_j X(v')$, so that $\phi_j(\sim, 1)$ maps $\partial_j X(v')$ to $\partial_j X(v)$. It can be induced by an ambient isotopy (see [Thom])

$$\{\tilde{\Phi}_j(v, v') : \ (X, \partial_1 X) \times [0,1] \to (X, \partial_1 X)\}, \tag{6.41}$$

supported in the neighborhood $U_{j,\epsilon}$. That isotopy matches $\partial_j X(v)$ with $\partial_j X(v')$.

Now we will improve the previous construction by building a *single* ambient isotopy $\tilde{\Phi} =_{\mathsf{def}} \tilde{\Phi}(v, v')$ that matches at once all the strata $\{\partial_j X(v)\}_j$ with the corresponding strata $\{\partial_j X(v')\}_j$. This will be achieved in stages, indexed by $j = 1, 2, 3, \dots$. In what follows, we assume that $v' \in \mathcal{U}(v)$ and take advantage of the special "foliated nature" of the j-disk bundles $\{\xi_j : W_{j,\epsilon'} \to \partial_j X(v)\}_j$ (recall that each ξ_j is foliated by the fibers of the line bundles $\{\xi_{j,k}\}_k$).

We start with a linear in the fibers of the bundle $\xi_2 : W_{2,\epsilon'} \to \partial_2 X(v)$ diffeotopy $\phi_1 : \partial_2 X(v') \times [0,1] \to \partial_1 X$ that takes $\partial_2 X(v')$ to $\partial_2 X(v)$. It extends to an ambient isotopy $\Phi_1 : \partial_1 X \times [0,1] \to \partial_1 X$. In turn, Φ_1 extends to an isotopy $\tilde{\Phi}_1 : X \times [0,1] \to X$, which preserves $\partial_1 X$ invariant.

Then we construct a linear in the fibers of $\xi_3 : W_{3,\epsilon'} \to \partial_3 X(v)$ diffeotopy $\phi_2 : \Phi_1(\partial_3 X(v')) \times [0,1] \to \partial_2 X(v)$ that takes $\Phi_1(\partial_3 X(v'))$ to $\partial_3 X(v)$. It extends to an ambient isotopy $\Phi_2 : \partial_2 X(v) \times [0,1] \to \partial_2 X(v)$. In turn, by an argument as in [Thom], Φ_2 extends to an isotopy $\tilde{\Phi}_2 : X \times [0,1] \to X$, which preserves $\partial_1 X$ and $\partial_2 X(v)$ invariant.

Next, we construct a linear in the fibers of $\xi_4 : W_{4,\epsilon'} \to \partial_4 X(v)$ diffeotopy of $(\Phi_2 \circ \Phi_1)(\partial_4 X(v'))$ in $\partial_3 X(v)$ that takes $(\Phi_2 \circ \Phi_1)(\partial_4 X(v'))$ to $\partial_4 X(v)$. It extends to an ambient isotopy $\Phi_3 : \partial_3 X(v) \times [0,1] \to \partial_3 X(v)$.

In turn, Φ_3 extends to an isotopy $\tilde{\Phi}_3 : X \times [0,1] \to X$, which preserves $\partial_1 X$, $\partial_2 X(v)$, and $\partial_3 X(v)$ invariant.

Continuing this way, we will construct eventually an ambient isotopy $\tilde{\Phi}_n : X \times [0,1] \to X$, which preserves the strata $\partial_1 X, \partial_2 X(v), \dots, \partial_n X(v)$ invariant and matches $\partial_{n+1} X(v')$ with $\partial_{n+1} X(v)$.

Now, for any $v' \in \mathcal{U}(v)$, the composite diffeomorphism

$$\tilde{\Phi}_n(\sim, 1) \circ \cdots \circ \tilde{\Phi}_2(\sim, 1) \circ \tilde{\Phi}_1(\sim, 1) : X \to X$$

matches all the strata $\{\partial_j X(v')\}_j$ with the corresponding strata $\{\partial_j X(v)\}_j$.

Therefore, the smooth type of the stratification $\{\partial_j X(v)\}_j$ is *locally* stable as the function of $v \in \mathcal{B}^\dagger(X)$. Note that we do not claim that $\Psi^\partial(v, z)$ is a F^n-stable map in the sense of Definition 6.9, a much stronger assertion!

In fact, for any path-connected component of $\mathcal{B}^\dagger(X)$, the smooth topological type of the Morse stratification remains constant. Indeed, if the points-fields v_0 and v_1 are connected by a continuous path $\gamma : [0,1] \to \mathcal{B}^\dagger(X)$, then each point $\gamma(t)$ produces an open neighborhood $\mathcal{U}(\gamma(t)) \subset \mathcal{B}^\dagger(X)$ as above. Using the compactness of image $\gamma([0,1])$, we can cover it by a finite number of open sets $\{\mathcal{U}(\gamma(t_i))\}_i$. By the previous arguments, any pair of Morse stratifications $\{\partial_j X(v)\}_j$ and $\{\partial_j X(v')\}_j$, where $v, v' \in \mathcal{U}(\gamma(t_i))$, can be transformed one into the other by a diffeomorphism of X. Therefore $\{\partial_j X(v_0)\}_j$ and $\{\partial_j X(v_1)\}_j$ can be transformed one into the other by a finite composition of locally available diffeomorphisms.

Now, let us validate the last two bullets of the theorem. Formula (6.14) implies that $m(\gamma \cap E) \leq n+1$ for all segments of trajectories γ that are contained in E. Thus $m'(\gamma \cap E) \leq n$. By the same token, if $a \cap \partial_k X(v)^\circ$, then $m(\gamma \cap U_a) \leq k$ for all γ in a $\hat{v}$-adjusted tubular neighborhood U_a of a. Of course, this implies that $m'(\gamma \cap U_a) \leq k-1$.

The last bullet of the theorem follows from Lemma 6.1, in particular from the existence of special coordinates (u, x) in which formula (6.14) is valid. □

Finally, we have reached the summit of this chapter.

Theorem 6.7. *Let X be a smooth compact connected $(n+1)$-dimensional manifold with boundary.*

- *The subspace $\mathcal{V}^\ddagger(X)$ of traversally generic vector fields is open and dense in the space $\mathcal{V}_{\mathsf{trav}}(X)$ of all traversing vector fields*[12].
- *If $v \in \mathcal{V}^\ddagger(X)$, then for every v-trajectory γ, its multiplicity $m(\gamma) \leq 2(n+1)$ and its reduced multiplicity $m'(\gamma) \leq n$.*

Proof. First we would like to show that the space $\mathcal{V}^\ddagger(X)$ of traversally generic fields is open in the space $\mathcal{V}^\dagger(X) = \mathcal{B}^\dagger(X) \cap \mathcal{V}_{\mathsf{trav}}(X)$ of boundary generic traversing vector fields. By Theorem 6.6, this would imply that $\mathcal{V}^\ddagger(X)$ is open in the space $\mathcal{V}_{\mathsf{trav}}(X)$ of traversing fields.

Let us start with a traversally generic field $v \in \mathcal{V}^\ddagger(X)$. Since, for a given function f, the property $df(v) > 0$ is an open one, the traversing vector fields form an open set in the space of all vector fields. By Theorem 6.6, the first bullet, there exists an open neighborhood $\mathcal{O}^\dagger(v) \subset \mathcal{V}(X)$ of v such that $\mathcal{O}^\dagger(v) \subset \mathcal{V}^\dagger(X)$.

We pick a finite set $\{S_\alpha\}$ of transversal sections of the $\hat{v}$-flow in the vicinity of $X \subset \hat{X}$, each section S_α being diffeomorphic to a closed n-disk. We denote by T_α a closed n-disk which is properly contained in S_α. We pick the sections $\{S_\alpha \supset T_\alpha\}_\alpha$ so that each $\hat{v}$-trajectory that intersects with X hits at least one of the flow sections T_α in its interior. Let U_α be the union of $\hat{v}$-trajectories through S_α, and let V_α be the union of $\hat{v}$-trajectories through T_α (so that $V_a \subset U_\alpha$). Thus, $\{V_\alpha \cap X\}_\alpha$ form a cover of X.

We denote by $p_\alpha(\hat{v}) : U_\alpha \to S_\alpha$ the $\hat{v}$-directed projection, defined by the formula $x \to \hat{\gamma}_x \cap S_\alpha$, where $\hat{\gamma}_x$ is the $\hat{v}$-trajectory through x.

Since $\hat{v}$ is transversal to each S_α, there is an open neighborhood $\mathcal{O}^\star(v) \subset \mathcal{V}^\dagger(X)$ of v such that, for every field $v' \in \mathcal{O}^\star(v)$, each $\hat{v}'$-trajectory hits every section S_α transversally or misses it. Moreover, by the C^∞-continuous dependence of ODEs solutions on the initial values and on the non-vanishing vector field, we may assume that each $\hat{v}'$-trajectory through T_α is contained in the set U_α, and that each $\hat{v}'$-trajectory through the hypersurface $\delta V_\alpha =_{\mathsf{def}} \partial_1 X \cap V_\alpha$ hits S_α transversally at a singleton.

We form the set $\mathcal{O}^{\dagger\star}(v) =_{\mathsf{def}} \mathcal{O}^\star(v) \cap \mathcal{O}^\dagger(v)$, an open neighborhood of v in $\mathcal{V}^\dagger(X)$.

Consider the $\hat{v}$-directed maps $\{p_\alpha(\hat{v}) : \delta V_\alpha \to S_\alpha\}_\alpha$. Since $v \in \mathcal{V}^\ddagger(X)$, each map $p_\alpha(\hat{v})$ is traversally generic in the sense of Definition 6.8 (where $M = \delta V_\alpha$ and $N = S_\alpha$). Examining Definition 6.8 and Definition 6.3,

[12] By definition, traversing vector fields do not vanish on X.

we see that the converse is true as well: if all $\{p_\alpha(\hat{v}) : \delta V_\alpha \to S_\alpha\}_a$ are traversally generic maps, then v is a traversally generic field.

By Theorem 6.4, there is an open neighborhood $\mathcal{U}_\alpha$ of the map $p_\alpha(\hat{v})$ in $C^\infty(\delta V_\alpha, S_\alpha)$ such that each map $\Phi \in \mathcal{U}_\alpha$ is traversally generic.

Consider the map

$$\Xi : \mathcal{O}^{\dagger\star}(v) \longrightarrow \prod_\alpha C^\infty(\delta V_\alpha, S_\alpha)$$

that takes each vector field $v' \in \mathcal{O}^{\dagger\star}(v)$ to the collection of maps $\{p_\alpha(\hat{v}')\}_\alpha$, defined by the $\hat{v}'$-flow. By the continuity of Ξ, the set

$$\mathcal{O}^{\ddagger}(v) =_{\mathsf{def}} \Xi^{-1}\big(\prod_\alpha \mathcal{U}_\alpha\big)$$

is open in $\mathcal{O}^{\dagger\star}(v)$ and thus in $\mathcal{V}^\dagger(X)$.

Note that $\{\mathrm{int}(\delta V_\alpha)\}_\alpha$ form an open cover of $\partial_1 X$, so that each $\hat{v}'$-trajectory through X hits one set $\mathrm{int}(\delta V_\alpha)$ at least. Since the property of a vector field v' being traversally generic can be faithfully expressed in "semi-local" terms of the vicinities of its $\hat{v}'$-trajectories, we conclude that any $v' \in \mathcal{O}^\ddagger(v)$ is traversally generic. In other words, if $\mathcal{O}^\ddagger(v)$ is open in $\mathcal{V}^\dagger(X)$, is also open in $\mathcal{V}^\ddagger(X)$.

Now we would like to show that $\mathcal{V}^\ddagger(X)$ is *dense* in the space $\mathcal{V}^\dagger(X)$. So we start with a boundary generic and traversing field v. By Lemma 5.6, for such a field, there exists a smooth function $f : X \to \mathbb{R}$ so that $df(v) > 0$ in X. Let us denote by $\mathcal{C}_f$ the open neighborhood of v in $\mathcal{V}^\dagger(X)$, defined by the inequality $\{df(v') > 0 |\ v' \in \mathcal{V}^\dagger(X)\}$.

Each $\hat{v}$-trajectory γ has a nested triple $W \subset V \subset U$ of $\hat{v}$-adjusted neighborhoods in $\hat{X}$ with the properties described in Lemma 6.10.

Since X is compact, we can choose a finite collection $\{W_i \subset V_i \subset U_i\}_i$ of such triples so that $\{\mathrm{int}(W_i)\}_i$ form a finite cover of $\partial_1 X$. Let us order the triples.

We denote by S_i a transversal section of the $\hat{v}$-flow in U_i. Let $T_i =_{\mathsf{def}} S_i \cap V_i$, and $Q_i =_{\mathsf{def}} S_i \cap W_i$.

Let us recall some old notations: for a given X-traversing field $\hat{w}$ in $\hat{X}$ and a subset $A \subset \hat{X}$, we denote by $\hat{X}(\hat{w}, A)$ the union of $\hat{w}$-trajectories that pass through A. Let $X(w, A) =_{\mathsf{def}} \hat{X}(\hat{w}, A) \cap X$.

As we perturb the given boundary generic and traversing field $\hat{v}$, we will insist on all the perturbations $\hat{v}'$ being so small that:

(1) $df(v') > 0$ (that is, $v' \in \mathcal{C}_f$),

(2) $v' \in \mathcal{V}^\dagger(X)$,
(3) $\hat{v}'$ is transversal to all the sections S_i,
(4) the $\hat{v}'$-adjusted sets $\{\hat{X}(v', Q_i)\}_i$ cover X,
(5) $\hat{X}(v', Q_i) \cap \partial_1 X \subset \hat{X}(v, T_i) \cap \partial_1 X$ for all i.

(6.42)

We denote by $\hat{\mathcal{U}}^\bullet$ the set of such vector fields $\hat{v}'$ on $\hat{X}$. It depends on the choice of sections $\{S_i \supset T_i \supset Q_i\}_i$ and, via these sections, on the original vector field $\hat{v}$. Evidently, $\hat{v} \in \hat{\mathcal{U}}^\bullet$. In fact, the space $\mathcal{U}^\bullet$, formed by the restrictions to X of the vector fields from $\hat{\mathcal{U}}^\bullet$, is open in the space $\mathcal{V}^\dagger(X)$. Indeed, the openness of sets of fields satisfying (1) and (3) is obvious, satisfying (2) follows from Theorem 6.6, and (4) and (5) follows from the smooth dependence of solutions of ODEs on initial data and on non-vanishing vector fields (on the "coefficients").

Let us pick an arbitrary open neighborhood $\mathcal{W}_{\hat{v}} \subset \mathcal{V}(\hat{X})$ of $\hat{v}$. Put

$$\mathcal{W}^\bullet_{\hat{v}} =_{\mathsf{def}} \mathcal{W}_{\hat{v}} \cap \hat{\mathcal{U}}^\bullet \;\text{ and }\; \mathcal{W}^\bullet_{v} =_{\mathsf{def}} \mathcal{W}_{\hat{v}} \cap \mathcal{U}^\bullet.$$

We intend to find a field $\hat{v}' \in \mathcal{W}^\bullet_{\hat{v}}$ that is traversally generic with respect to $\partial_1 X$. This will prove that traversally generic fields form a dense set in $\mathcal{V}^\dagger(X)$.

Let us order the $\hat{v}$-sections by forming a finite list: $(S_1, S_2, \dots, S_N)$. By an inductive argument in the number i of sections from the list, we will systematically "enlarge" a set of trajectories which are traversally generic with respect to a growing portion of $\partial_1 X$.

Here is how the induction step "$i-1 \Rightarrow i$" works. Assume that we have managed to find a field $\hat{v}' \in \mathcal{W}^\bullet_{\hat{v}}$ such that it is traversally generic, when restricted to the closed $\hat{v}'$-adjusted set

$$F_{i-1}(\hat{v}') =_{\mathsf{def}} \hat{X}(\hat{v}', \coprod_{1 \leq k < i} Q_k).$$

By Lemma 6.10, there is a $\hat{X}(\hat{v}', S_i)$-supported perturbation $\hat{v}'' \in \mathcal{W}^\bullet_{\hat{v}}$ of $\hat{v}'$ such that: (1) $\hat{v}''$ is traversally generic in $\hat{X}(\hat{v}'', Q_i)$ and (2) $\hat{v}'' = \hat{v}'$, when restricted to

$$[\hat{X} \setminus \hat{X}(\hat{v}', S_i)] \;\cup\; F_{i-1}(\hat{v}').$$

Such a field $\hat{v}''$ is traversally generic in $F_i(\hat{v}'')$.

Since the property (4) from (6.42) is enforced through the induction arguments, eventually (for $i = N$), we will construct a field $\hat{w} \in \mathcal{W}^\bullet_{\hat{v}}$ which is traversally generic everywhere in X.

Note that the case $i = 1$, the base of induction, is exactly the claim of Lemma 6.10.

It remains to prove the last bullet of the theorem. For any vector field $v \in \mathcal{V}^{\ddagger}(X)$, we have shown (see (6.11)) that $m'(\gamma) \leq n$ for all trajectories γ.

Unless γ is a singleton, there are only two points of odd multiplicity in the set $\gamma \cap \partial_1 X$—the two ends of γ. Thus $m(\gamma)$ can be written in the form

$$(2t_0 + 1) + \sum_{i=1}^{q} 2s_i + (2t_1 + 1).$$

As a result,

$$m'(\gamma) = 2t_0 + \sum_{i=1}^{q} (2s_i - 1) + 2t_1 \leq n.$$

The latter inequality implies that $q \leq n$. Therefore,

$$m(\gamma) = m'(\gamma) + q + 2 \leq n + q + 2 \leq 2n + 2,$$

twice the dimension of X. □

Corollary 6.4. *Let X be a compact connected smooth manifold with boundary. For a given smooth nonsingular function $f : X \to \mathbb{R}$, the traversally generic f-gradient-like fields form an open and dense set in the space of all f-gradient-like fields.*

Proof. For a fixed nonsingular $f : X \to \mathbb{R}$, the set of f-gradient-like fields is defined by the inequality $df(v) > 0$, and therefore is open in the space of all fields $\mathcal{V}(X)$. With this remark in place, the corollary follows from Theorem 6.7. □

The following lemma was used in Chapter 4.

Lemma 6.12. *Let X be a compact connected smooth n-manifold with boundary, and $\tau : [0, 1] \to \mathcal{V}_{\mathsf{trav}}(X)$ a smooth path. By definition, it generates the smooth vector field $v(x, t) =_{\mathsf{def}} \tau(t)(x)$ on $Y =_{\mathsf{def}} X \times [0, 1]$. Put $\delta Y =_{\mathsf{def}} \partial X \times [0, 1]$.*

- *If, for some $T \in (0, 1)$, the vector field v_T is boundary generic on $X_T =_{\mathsf{def}} X \times \{T\}$ relative to $\partial_1 X_T$, then v is boundary generic on Y relative to δY in the vicinity of X_T.*

- *If, for some $T \in (0,1)$, the vector field v_T is traversally generic on X_T, then the vector field v on Y is* traversally generic *with respect to δY in the vicinity of X_T in Y.*

Proof. Let $z : \hat{X} \to \mathbb{R}$ be a smooth function as in Lemma 6.1. Composing z with the obvious map $\hat{Y} = \hat{X} \times [0,1] \to \hat{X}$, we get a function $\tilde{z}$ on $\hat{Y}$, whose restriction to each slice $\hat{X} \times \{t\}$ is z. Let $\pi : Y \to [0,1]$ be the obvious projection.

Then, by Lemma 6.1, a point $a \in \delta Y$ belongs to the stratum $\partial_j Y(v)$ if $\{(\mathcal{L}_v^{(k)}\tilde{z})(a) = 0\}_{k<j}$ and to the pure stratum $\partial_j Y(v)^\circ$ if, in addition, $(\mathcal{L}_v^{(j)}\tilde{z})(a) \neq 0$. The field v in the vicinity of $a \in \partial_j Y(v)^\circ$ is boundary generic if and only if the differential j-form

$$\Theta =_{\mathsf{def}} d(\tilde{z}) \wedge d(\mathcal{L}_v^{(1)}\tilde{z}) \wedge \cdots \wedge d(\mathcal{L}_v^{(j-1)}\tilde{z}) \in \bigwedge^j T^*(Y)$$

does not vanish along the locus $\partial_j Y(v)$.

Similarly, the vector field v_t on X_t, in the vicinity of $a \in \partial_j X(v_t)^\circ$, is boundary generic if and only if the differential j-form

$$\Theta_t =_{\mathsf{def}} d_X(z) \wedge d_X(\mathcal{L}_{v_t}^{(1)}z) \wedge \cdots \wedge d_X(\mathcal{L}_{v_t}^{(j-1)}z) \in \bigwedge^j T^*(X_t),$$

where d_X stands for the differential of functions on $\hat{X}$, does not vanish along the locus $\partial_j X(v_t)$.

By the construction of v and $\tilde{z}$, we have $\mathcal{L}_v^{(k)}\tilde{z} = \mathcal{L}_{v_t}^{(k)}z$ for all k and t.

Let $T_*(\pi) \to \hat{Y}$ be the bundle, tangent to the fibers of $\pi : \hat{Y} \to [0,1]$. Using the product structure in Y, we may decompose the differential d (taking smooth functions to 1-forms) on Y as the sum $d_t + d_\pi$, where d_t stands for the differential in the variable $t \in [0,1]$, and d_π for the differential in the fibers of $\pi : \hat{Y} \to [0,1]$.

Note that

$$d(\mathcal{L}_v^{(k)}\tilde{z}) = d(\mathcal{L}_{v_t}^{(k)}z) = d_t(\mathcal{L}_{v_t}^{(k)}z) + d_\pi(\mathcal{L}_{v_t}^{(k)}z).$$

Let u be a vector field in $T_*(\pi)$. Then, the evaluation

$$d(\mathcal{L}_v^{(k)}\tilde{z})(u) = d_\pi(\mathcal{L}_{v_t}^{(k)}z)(u),$$

since u is a vertical field and thus $d_t(\mathcal{L}_{v_t}^{(k)}z)(u) = \frac{\partial}{\partial t}(\mathcal{L}_{v_t}^{(k)}z)\,dt(u) = 0$.

Let $w \in \bigwedge^j T_*(\pi)|_a$ be a nonzero polyvector. If at $a \in \partial_j X(v_T)^\circ$ the value $\Theta_T(w) \neq 0$, then, by the argument above, this value is equal to $\Theta(w)$, and thus the later must be different from zero at a as well. Therefore, if for some T, the vector field v_T is boundary generic on X_T, it follows that v is boundary generic in the vicinity of X_T in Y.

To prove the second bullet, we start with a traversally generic vector field v_T on X_T. Since v_T it is boundary generic, by the first bullet, v is boundary generic with respect to δY in the vicinity of $X_T \subset Y$. In particular, by the argument above, $\partial_j Y(v) \cap X_t = \partial_j X(v_t)$ for some $\epsilon > 0$ and all $t \in (T - \epsilon, T + \epsilon)$.

Consider now a v_T-trajectory $\gamma \subset X_T$. Each point $x \in \gamma \cap \delta Y$ belongs to a unique pure stratum $\partial_{j(x)} Y(v) \subset \delta Y$ of a maximum codimension $j(x) \leq n$. Let S be a transversal local section of the v-flow in the vicinity of γ. Put $S_T = S \cap X_T$.

With the help of the v_T-flow, we project the tangent spaces

$$\{T_x(\partial_{j(x)} X_T(v_T))\}_{x \in \gamma \cap \delta Y}$$

on the tangent space $\mathsf{T}^\star_{\gamma,T} =_{\mathsf{def}} T_{\gamma \cap S}(S_T)$ to the section S_T at the point $\gamma \cap S$. Let us denote by $\{\mathsf{T}^\star_T(x)\}_{x \in \gamma \cap \delta Y}$ the images of these spaces under the v_T-flow-generated projection $p_{T,\gamma}$.

Similarly, for $x \in \gamma \cap \delta Y$, let $\mathsf{T}^\star(x)$ denote the image of the tangent space $T_x(\partial_{j(x)} Y(v))$ in $\mathsf{T}^\star_\gamma =_{\mathsf{def}} T_{\gamma \cap S}(S)$ under the projection p_γ, defined by the v-flow.

Since v_T is traversally generic, the spaces $\{\mathsf{T}^\star_T(x)\}_{x \in \gamma \cap \delta Y}$ are in general position in $\mathsf{T}^\star_{\gamma,T}$. We aim to show that the tangent spaces $\{\mathsf{T}^\star(x)\}_{x \in \gamma \cap \delta Y}$ are in general position in $\mathsf{T}^\star_\gamma$ as well. Indeed, since v is boundary generic in the vicinity of X_T, employing Lemma 6.1, we conclude that the differential

$$Dp_\gamma| : T_x(\partial_{j(x)} Y(v)) \longrightarrow T_{\gamma \cap S} S =_{\mathsf{def}} \mathsf{T}^\star_\gamma$$

is a monomorphism. Since $\partial_j Y(v) \cap X_T = \partial_j X(v_T)$ and the intersection is transversal thanks to the first bullet of the lemma, we get that the space $\mathsf{T}^\star_T(x) = \mathsf{T}^\star(x) \cap \mathsf{T}^\star_{\gamma,T}$ is a hyperplane in the space $\mathsf{T}^\star(x)$ for each $x \in \gamma \cap \delta Y$. Also $\mathsf{T}^\star_{\gamma,T}$ is a hyperplane in $\mathsf{T}^\star_\gamma$, so that $\mathsf{T}^\star_T(x) = \mathsf{T}^\star(x) \cap \mathsf{T}^\star_{\gamma,T}$ for all $x \in \gamma \cap \delta Y$. Therefore, counting the codimensions of the relevant intersections, $\{\mathsf{T}^\star(x)\}_{x \in \gamma \cap \delta Y}$ are in general position in $\mathsf{T}^\star_\gamma$.

If a vector field is traversally generic *at* some trajectory γ, by Lemma 6.4, it is traversally generic in its vicinity. So v is traversally generic to δY in the vicinity of X_T. □

To summarize, Theorems 6.6 and 6.7 insure that the traversally generic flows are typical among the traversing flows, thus justifying our future endeavors.

6.4 On the trajectory spaces of traversing and traversally generic flows

The semi-local models of traversally generic vector flows, described in Lemma 6.4, will enable us to study the topology of the trajectory spaces, generated by such flows, an interesting class of CW-complexes that behave as "surrogate manifolds".

For a traversing vector field v on X, let $\Gamma : X \to \mathcal{T}(v)$ be the map that takes each point $x \in X$ to the v-trajectory $\gamma_x \in \mathcal{T}(v)$ that contains x.

Theorem 6.8. *Let X be a compact connected smooth manifold with boundary.*

- *For a traversing vector field v on X, the map $\Gamma : X \to \mathcal{T}(v)$ is a weak homotopy equivalence, provided that $\mathcal{T}(v)$, as a stratified space, is weakly homotopy equivalent to a CW-complex. In particular, by Corollary 7.1, this is the case for any traversally generic v.*
- *For any traversing field v and any local coefficient system $\mathcal{A}$ on X, the map $\Gamma : X \to \mathcal{T}(v)$ is a cohomology equivalence:*

$$\Gamma^* : H^*\big(\mathcal{T}(v);\ \Gamma_*(\mathcal{A})\big) \approx H^*(X;\mathcal{A})$$

for all $* \geq 0$.

Proof. Recall that a continuous map $p : E \to B$ is called a *quasifibration* if, for each point $b \in B$, the map $p : (E, p^{-1}(b)) \to (B, p)$ is a weak homotopy equivalence of pairs.

Our argument is based on Theorem 2.6 from [May] which describes conditions under which a filtration-preserving map $p : E \to B$ of two filtered spaces is a quasifibration.

Consider the decreasing filtaration

$$\{F_k\mathcal{T}(v) =_{\mathsf{def}} \bigcup_{\{\gamma \in \mathcal{T}(v) |\ |\omega_\gamma|' \geq k\}} \gamma\ \}_{k \in [0,n]}$$

of $\mathcal{T}(v)$ and its preimage $\{F_k X(v)\}_k$ under the map $\Gamma : X \to \mathcal{T}(v)$. We assume that the space $\mathcal{T}(v)$ admits a triangulation T, consistent with the stratification of $\mathcal{T}(v)$ by the combinatorial types $\omega_\gamma \in \mathbf{\Omega}^\bullet$ of v-trajectories γ. In particular, T is consistent with the filtration $\{F_k\mathcal{T}(v)\}_k$. By the forthcoming Corollary 6.8, this is the case for any traversally generic v!

With the help of the second barycentric subdivision $\beta^2 T$ of T, for each k, we construct a regular neighborhood U_k of $F_{k+1}\mathcal{T}(v)$ in $F_k\mathcal{T}(v)$ and its canonical piecewise linear deformation retraction $\{r_{k,t} : U_k \to F_{k+1}\mathcal{T}(v)\}_{t\in[0,1]}$.

For each k, we form the neighborhood $V_k = \Gamma^{-1}(U_k) \subset F_kX(v)$ of $F_{k+1}X(v)$. Since the Γ-fibers are closed segments or singletons, for each simplex $\Delta \subset \beta^2 T$, its preimage $\Gamma^{-1}(\Delta)$ is a prism $\Pi = \Delta \times I$ or a portion of Π, bounded by the base Δ of Π and a hyperplane section of Π which projects on the base along the I-factor by a 1-to-1-map. Therefore, the piecewise linear retraction $r_{k,t}$ is uniquely covered by a piecewise linear retraction $\{R_{k,t} : V_k \to F_{k+1}X(v)\}_{t\in[0,1]}$ which takes each Γ-fiber to a Γ-fiber.

Thus the following three properties hold:

(1) $\Gamma : F_kX(v) \setminus F_{k+1}X(v) \to F_k\mathcal{T}(v) \setminus F_{k+1}\mathcal{T}(v)$ is a fibration over each component of $F_k\mathcal{T}(v) \setminus F_{k+1}\mathcal{T}(v)$ (its fiber is a closed segment of a singleton).

(2) $\Gamma \circ R_{k,t} = r_{k,t}$ for all $t \in [0,1]$.

(3) $R_{k,1} : \Gamma^{-1}(\gamma) \to \Gamma^{-1}(r_{k,1}(\gamma))$ is a homotopy equivalence for any point $\gamma \in U_k$.

By Theorem 2.6 from [May], these three properties imply that $\Gamma : X \to \mathcal{T}(v)$ is a quasifibration. In particular, Γ is a weak homotopy equivalence, which constitutes the first claim of the theorem.

Now we will validate the second bullet of the theorem.

Consider any local coefficient system $\mathcal{A}$ (a sheaf) on X with an abelian group A for the stock. We denote by $\Gamma_*(\mathcal{A})$ its push-forward, residing on the trajectory space $\mathcal{T}(v)$.

Let U be an open neighborhood of a typical trajectory γ in X. Using that X is compact and a typical Γ-fiber—a trajectory γ—is closed, the canonical homomorphism

$$\lim_{\{U\supset\gamma\}} H^*(U; \mathcal{A}|_U) \to H^*(\gamma; \mathcal{A}|_\gamma)$$

is an isomorphism (see Theorem 4.11.1 in [God]). Since all γ's are either segments or singletons and $\mathcal{A}$ is a local system, we conclude that $H^*(\gamma; \mathcal{A}|_\gamma) = 0$ for all $* \neq 0$ and $H^0(\gamma; \mathcal{A}|_\gamma) = \mathsf{A}$. Thus, the Leray spectral sequence

$$\big\{E_2^{pq} = H^p\big(\mathcal{T}(v);\, \mathcal{H}^q(\gamma; \mathcal{A})\big)\big\}_{p,q}$$

of the map $\Gamma : X \to \mathcal{T}(v)$ collapses (see Theorem 4.17.1 in [God]). As a result, we get that the map Γ establishes an isomorphism

$$\Gamma^* : H^*(\mathcal{T}(v); \Gamma_*(\mathcal{A})) \approx H^*(X; \mathcal{A}).$$

In particular, for a trivial local system A, we get $H^*(X; \mathsf{A}) \approx H^*(\mathcal{T}(v); \mathsf{A})$. □

As usual, in what follows we assume that the field v on X extends to a non-vanishing field $\hat{v}$ on some open manifold $\hat{X}$ which properly contains X. We treat the extension $(\hat{X}, \hat{v})$ as a germ.

We denote by $X(v, \omega)$ the union of v-trajectories of a given combinatorial type $\omega \in \mathbf{\Omega}^\bullet$. Its closure is denoted by $X(v, \omega_{\succeq_\bullet})$.

For traversing fields v, the trajectory space $\mathcal{T}(v)$ has an interesting feature: it comes equipped with a vector n-bundle $\tau(\mathcal{T}(v))$ which plays the role of "surrogate tangent bundle".

Each pure stratum $\mathcal{T}(v, \omega) \subset \mathcal{T}(v)$ is an open smooth manifold and, as such, has a "conventional" tangent bundle. In particular, the pure strata of maximal dimension n have tangent bundles. It turns out that these "honest" tangent n-bundles extend across the singularities of the space $\mathcal{T}(v)$ to form a n-bundle $\tau(\mathcal{T}(v))$ over $\mathcal{T}(v)$! However, at the singularities, no exponential map that takes a vector from $\tau(\mathcal{T}(v))$ to a point in $\mathcal{T}(v)$ is available—the bundle $\tau(\mathcal{T}(v))$ does not reflect faithfully the local geometry of the trajectory space $\mathcal{T}(v)$.

In order to define the dual of the bundle $\tau(\mathcal{T}(v))$ intrinsically, we need to consider a surrogate of smooth structure on the singular space $\mathcal{T}(v)$.

Definition 6.10. Let v be a traversing vector field on X and $\Gamma : X \to \mathcal{T}(v)$ be the obvious projection.

We say that a function $h : \mathcal{T}(v) \to \mathbb{R}$ is *smooth*, if the composition $h \circ \Gamma$ is smooth on X.

We denote by $C^\infty(\mathcal{T}(v))$ the algebra of all smooth functions on the space $\mathcal{T}(v)$. ◊

Definition 6.11. Let v_1, v_2 be two traversing vector fields on compact connected smooth manifolds X_1, X_2 with boundary, respectively.

- A map $\Phi : \mathcal{T}(v_1) \to \mathcal{T}(v_2)$ is called *smooth*, if for any function $h \in C^\infty(\mathcal{T}(v_2))$, its pull-back $\Phi^*(h) \in C^\infty(\mathcal{T}(v_1))$.
- A homeomorphism $\Phi : \mathcal{T}(v_1) \to \mathcal{T}(v_2)$ is called a *diffeomorphism*, is both Φ and Φ^{-1} are smooth. ◊

For any traversing vector field v, the algebra $C^\infty(\mathcal{T}(v))$ of smooth functions on the trajectory space $\mathcal{T}(v)$ can be identified with the subalgebra of $C^\infty(X)$, formed by functions $f : X \to \mathbb{R}$ with the property $\mathcal{L}_v(f) = df(v) = 0$, where $\mathcal{L}_v$ stands for the v-directional derivative[13]. Such functions are constant along each trajectory $\gamma \subset X$.

For a traversing v, consider the algebra $\mathsf{Ker}(\mathcal{L}_v) \approx C^\infty(\mathcal{T}(v))$ of smooth functions on X that are constants along each v-trajectory.

Question 6.2. *For a boundary generic traversing field v on X, how to characterize the image (trace) of the algebra $\mathsf{Ker}(\mathcal{L}_v)$ in the algebra $C^\infty(\partial X)$?* $\diamondsuit$

Remark 6.7. Let v be a traversing and boundary generic vector field on X. If a given smooth function $h : \partial X \to \mathbb{R}$ is a restriction of a smooth function $f : X \to \mathbb{R}$ such that $\mathcal{L}_v f = 0$, then there are few obvious constraints on h. Evidently, the restriction of h to every set $\gamma \cap \partial X$ must be constant.

In addition, since along the locus $\partial_2 X(v)$, the vector field v is tangent to ∂X, we get $\mathcal{L}_v(h) = 0$ along $\partial_2 X(v)$. Are these constraints on h sufficient for the existence of the appropriate smooth f? Perhaps, not ...

Of course, any smooth h, constant on the traces $\{\gamma \cap \partial X\}_\gamma$ of trajectories, defines a unique *continuous* function $f : X \to \mathbb{R}$ that is constant on each trajectory γ. The issue is whether this f is *smooth*! $\diamondsuit$

We denote by $C^\infty_\gamma(\mathcal{T}(v))$ the algebra of germs of smooth functions from $C^\infty(\mathcal{T}(v))$ at a given point $\gamma \in \mathcal{T}(v)$. Let $\mathsf{m}_\gamma(\mathcal{T}(v)) \lhd C^\infty_\gamma(\mathcal{T}(v))$ denote the maximal ideal, formed by the the germs of functions that vanish at γ, and let $\mathsf{m}^2_\gamma(\mathcal{T}(v))$ denote the square of the ideal $\mathsf{m}_\gamma(\mathcal{T}(v))$.

The quotients $\mathsf{m}_\gamma(\mathcal{T}(v))/\mathsf{m}^2_\gamma(\mathcal{T}(v))$ are real n-dimensional vector spaces. Indeed, since the pull-back of smooth functions on $\mathcal{T}(v)$ are the smooth functions on X that are constants along each trajectory γ, the quotient $\mathsf{m}_\gamma(\mathcal{T}(v))/\mathsf{m}^2_\gamma(\mathcal{T}(v))$ can be canonically identified with the quotient $\mathsf{m}_x(S)/\mathsf{m}^2_x(S)$, where S is a germ of a smooth transversal section of the $\hat{v}$-flow at the point $x = \gamma \cap S$, and $\mathsf{m}_x(S)$ denotes the maximal ideal in the algebra $C^\infty(S)$, an ideal comprised of functions that vanish at x.

It is well-known that $\mathsf{m}_x(S)/\mathsf{m}^2_x(S)$ can be canonically identified with the cotangent space $T^*_x(S)$ via the correspondence $f \Rightarrow df$, where the germ of $f : S \to \mathbb{R}$ at x belongs to the ideal $\mathsf{m}_x(S)$. Therefore the spaces

$$\tau^*_\gamma(\mathcal{T}(v)) =_{\mathsf{def}} \mathsf{m}_\gamma(\mathcal{T}(v))/\mathsf{m}^2_\gamma(\mathcal{T}(v))$$

[13]This property is independent of an extension $(\hat{X}, \hat{v})$ of (X, v).

form a vector n-bundle $\tau^*(\mathcal{T}(v))$ over $\mathcal{T}(v)$[14]. Its pull-back $\Gamma^*\big(\tau^*(\mathcal{T}(v))\big)$ can be identified with the subbundle $\tau^*(v)$ of the cotangent bundle $T^*(X)$, formed by the "horizontal" 1-forms α that vanish on v and have the property $\mathcal{L}_v(\alpha) = 0$. The identification is via the correspondence $\Gamma^*(f) \Rightarrow d(\Gamma^*(f))$, where $f \in \mathsf{m}_\gamma(\mathcal{T}(v))$.

Now we define $\tau(\mathcal{T}(v))$ as the dual bundle of $\tau^*(\mathcal{T}(v))$.

Let $(11) \in \mathbf{\Omega}^\bullet_{'\langle n]}$ denote the unique maximal element of the poset; it labels the trajectories that intersect the boundary $\partial_1 X$ only at a pair of distinct points, where they are transversal to the boundary.

Lemma 6.13. *Let X be a compact connected smooth $(n+1)$-dimensional manifold with boundary.*

For any traversing vector field v, the tangent bundles to the components of the maximal stratum $\mathcal{T}(v,(11))$ extend to a n-dimensional vector bundle $\tau(\mathcal{T}(v))$ over the trajectory space $\mathcal{T}(v)$.

Moreover, for a traversally generic vector field v and each element $\omega \in \mathbf{\Omega}^\bullet_{'\langle n]}$, the tangent bundle of the pure stratum $\mathcal{T}(v,\omega)$ embeds in $\tau(\mathcal{T}(v))|_{\mathcal{T}(v,\omega)}$ as a subbundle with a canonically trivialized complement.

Proof. We already have observed that the pull-back $\Gamma^*\big(\tau^*(\mathcal{T}(v))\big)$ of the cotangent bundle $\tau^*(\mathcal{T}(v))$ can be identified with the bundle $\tau^*(v)$ of the flow-invariant 1-forms on X that vanish on v.

The map $\Gamma : X(v,(11)) \to \mathcal{T}(v,(11))$ is a fibration with a closed segment for the fiber. Therefore Γ admits a smooth section $S_{(11)} \subset X(v,(11))$ which is transversal to the v-trajectories. Consider a decomposition of the $(n+1)$-bundle $T(X)|_{S_{(11)}}$ into the tangent n-bundle $T(S_{(11)})$ and a line bundle L, tangent to the v-trajectories. With the help of this decomposition, the cotangent bundle $T^*(S_{(11)})$ can be identified with the restriction $\tau^*(v)|_{S_{(11)}}$ of $\tau^*(v)$ to $S_{(11)}$. Using the isomorphism

$$\tau^*(v)\big|_{S_{(11)}} \approx \Gamma^*\big(\tau^*(\mathcal{T}(v))\big)\big|_{S_{(11)}},$$

we identify the cotangent bundle $T^*(S_{(11)})$ with the bundle $\tau^*(\mathcal{T}(v))|_{S_{(11)}}$, a bundle that evidently is defined on the whole space $\mathcal{T}(v)$.

A similar conclusion holds for any traversally generic vector field v and each $\omega \in \mathbf{\Omega}^\bullet_{'\langle n]}$: by Lemma 6.4, the map $\Gamma : X(v,\omega) \to \mathcal{T}(v,\omega)$ is a fibration with its base being an open smooth $(n - |\omega|')$-manifold and with a closed segment or a singleton for the fiber, the fiber being consistently oriented by

[14]It is dual to the bundle $\tau(\mathcal{T}(v))$ under construction.

v. Therefore $\Gamma|_{X(v,\omega)}$ admits a smooth section S_ω, transversal to the v-flow. The cotangent bundle $\tau^*(S_\omega)$ can be identified with the cotangent bundle $\tau^*(\mathcal{T}(v,\omega))|_{\mathcal{T}(v,\omega)}$. The later bundle embeds into the bundle $\tau^*(\mathcal{T}(v))$.

The only nontrivial statement of the lemma is the existence of a preferred trivialization in the quotient bundle $\tau(\mathcal{T}(v))|_{\mathcal{T}(v,\omega)}/\tau(\mathcal{T}(v,\omega))$. It follows from the last claim of Theorem 6.9 below. Thus, for a traversally generic v, we get the bundle isomorphism

$$\Psi : \tau(\mathcal{T}(v,\omega)) \oplus \underline{\mathbb{R}}^{|\omega|'} \approx \tau(\mathcal{T}(v))|_{\mathcal{T}(v,\omega)},$$

canonically defined by v. □

Corollary 6.5. *For a traversally generic vector field v on X, the stable characteristic classes of the tangent bundles $\tau(\mathcal{T}(v))$ and $\tau(X)$ coincide via the cohomological isomorphism, induced by the projection $\Gamma : X \to \mathcal{T}(v)$.*

Proof. By Lemma 6.13, we get $T(X) \approx \Gamma^*(\tau(\mathcal{T}(v))) \oplus \underline{\mathbb{R}}$. Therefore, the cohomological isomorphism induced by Γ (see Theorem 6.8) helps to identify the stable characteristic classes of $\tau(\mathcal{T}(v))$ and $T(X)$. □

For a traversally generic v, the space $\mathcal{T}(v)$ comes equipped with two distinct *intrinsically-defined orientations* of its pure strata $\{\mathcal{T}(v,\omega)\}_\omega$. These orientations (already familiar from our descriptions of the $3D$-flows) depend only on v and the preferred orientation of X.

Theorem 6.9. *Let X be an oriented compact connected and smooth $(n+1)$-manifold with boundary, and v a traversally generic vector field. Then*

- *each connected component of any pure stratum $\mathcal{T}(v,\omega)$, where $\omega \in \mathbf{\Omega}^\bullet_{'\langle n]}$ and $|\omega|' > 0$, acquires two distinct orientations, called* preferred *and* versal. *Switching the orientation of X affects both orientations of $\mathcal{T}(v,\omega)$ by the same factor $(-1)^{|sup(\omega)|}$.*
- *With the help of these two orientations, each component of $\mathcal{T}(v,\omega)$ acquires one of the two polarities "$\oplus$" and "$\ominus$". They do not depend on the orientation of X.*
- *Each manifold $X(v,\omega)$ comes equipped with a v-induced normal framing in X. Similarly, the "normal" $|\omega|'$-dimensional bundle*

$$\nu(\mathcal{T}(v,\omega)) =_{\mathsf{def}} \tau(\mathcal{T}(v))|_{\mathcal{T}(v,\omega)} / \tau(\mathcal{T}(v,\omega))$$

acquires a v-induced preferred framing.

Proof. Let us extend the field v on X to a non-vanishing field $\hat{v}$ on $\hat{X} \supset X$.

Local transversal sections S of the $\hat{v}$-flow have a well-defined orientation due to the global orientation of X and the preferred orientation of the v-trajectories.

For a traversally generic v on a $(n+1)$-dimensional X, the vicinity $U \subset \hat{X}$ of each v-trajectory γ of the combinatorial type ω has a special coordinate system

$$(u, x, y) : U \to \mathbb{R} \times \mathbb{R}^{|\omega|'} \times \mathbb{R}^{n-|\omega|'}.$$

By Lemma 6.4, in these coordinates, the boundary $\partial_1 X$ is given by the polynomial equation $P(u, x) = 0$ as in (6.26). At the same time, X is given by the polynomial inequality $\{P(u, x) \leq 0\}$. Each v-trajectory in U is produced by freezing all the coordinates x, y, while letting u to be free.

We order the coordinates $\{x_{i,l}\}_{i,l}$ in (6.26) lexicographically: first we order them by the increasing "i"s; then, for a fixed i, the ordering among $\{x_{i,l}\}_l$ is defined by the increasing powers l of the binomial $(u - \alpha_i)$ in the formula (6.26). This ordering of $\{x_{i,l}\}_{i,l}$, together with the orientation in S (induced with the help of v by the orientation of X) gives rise to an orientation of the y-coordinates. The y-coordinates correspond to the space, tangent to the pure stratum $\mathcal{T}(v, \omega)$ at γ.

We still have to check that this ordering of $\{x_{i,l}\}_l$ is determined by the *geometry of tangency* and does not depend on a particular choice of the special coordinates $\{x_{i,l}\}_l$.

Consider a v-trajectory γ of the combinatorial type ω. Let $\gamma \cap \partial_1 X = \coprod_i a_i$, a finite set of points. For each $a_i \in \partial_{j_i} X^\circ$ (by definition, $\omega_i = j_i$), we divide the local coordinates (u, x, y), participating in the formula (6.26), into new groups $(u, x^{\{i\}}, y^{\{i\}})$, where by definition $x^{\{i\}} = (x_{i,0}, x_{i,1}, \dots, x_{i,j_i-2})$ and $y^{\{i\}}$ denotes the rest of $\{x_{i',l'}\}$-coordinates ($i' \neq i$), together with all the y-coordinates.

In the vicinity of each $a_i \in \partial_{j_i} X^\circ$, we write down the auxiliary function z from (6.25) in two ways:

(1) as $u^{j_i} + \sum_{l=0}^{j_i-2} \phi_l(x^{\{i\}}, y^{\{i\}})\, u^l$ (here $\phi_l(0,0) = 0$ for all l),

(2) and as $\left(u^{j_i} + \sum_{l=0}^{j_i-2} x_{i,l}\, u^l\right) \cdot Q(u, x^{\{i\}}, y^{\{i\}})$, where $Q(u, 0, 0) \neq 0$.

To simplify these notations, we put $j =_{\mathsf{def}} j_i$, $\tilde{x} =_{\mathsf{def}} x^{\{i\}}$, $\tilde{y} =_{\mathsf{def}} y^{\{i\}}$.

We aim to show that, at the origin $(\tilde{x}, \tilde{y}) = (0, 0)$, the following two $(j-1)$-forms are equal:

$$d\phi_0 \wedge d\phi_1 \wedge \dots \wedge d\phi_{j-2} \,|_{(0,0)} = dx_{i,0} \wedge dx_{i,1} \wedge \dots \wedge dx_{i,j-2} \,|_{(0,0)}. \quad (6.43)$$

We regard these two forms as $\pm$ volume forms in the normal bundle $\nu(\partial_j X(v), X)$ at the point a_i. We identify $\nu_{a_i}(\partial_j X(v), X)$ with the $\tilde{x}$-space $\mathbb{R}^{j-1}$. Hence the two orientations of the $\tilde{y}$-space $\mathbb{R}^{n-j+1} \approx T_{a_i}(\partial_j X(v))$, induced by the two coordinate systems, $\{\phi_l\}_l$ and $\{x_{i,l}\}_l$, in the vicinity of a_i, do also agree.

The argument validating (6.43) is similar to the one we have used in Lemma 6.3.

First note that $q =_{\mathsf{def}} Q(u, 0, 0) \neq 0$ must be 1: just plug $\tilde{x} = 0, \tilde{y} = 0$ in the identity

$$u^j + \sum_{l=0}^{j-2} \phi_l(\tilde{x}, \tilde{y}) u^l = \big(u^j + \sum_{l=0}^{j-2} x_{i,l} u^l\big)\, Q(u, \tilde{x}, \tilde{y}). \tag{6.44}$$

We form the row-vector $a(u) =_{\mathsf{def}} (u^{j-2}, \dots, u, 1)$ and the column-vector

$$d\Phi =_{\mathsf{def}} (d\phi_{j-2}, \dots, d\phi_1, d\phi_0)^t,$$

whose entries are 1-forms. Then the functional identity (6.44) can be written as

$$u^j + a * \Phi = (u^j + a * \tilde{x})\, Q,$$

where "$*$" stands for the matrix multiplication. Note that its RHS of this equation belongs to the ideal $\langle u^j, \tilde{x}\rangle$ of $C^\infty(\mathbb{R}^{n+1})$, generated by the functions u^j and $x_{i,0}, \dots, x_{i,j-2}$.

Consider the differential of this identity and take its portion that does not contain the du terms. Then, modulo the ideal $\langle u^{j-1}, \tilde{x}\rangle$, generated by the functions u^{j-1} and $x_{i,0}, \dots, x_{i,j-2}$, that portion produces the identity

$$a * d\Phi = Q \cdot a * d\tilde{x} \quad \mod \langle u^{j-1}, \tilde{x}\rangle.$$

Here "mod $\langle u^{j-1}, \tilde{x}\rangle$" means that we ignore the differential 1-forms whose coefficients belong to the functional ideal $\langle u^{j-1}, \tilde{x}\rangle$.

We apply partial derivatives $\frac{\partial}{\partial u}, \dots, \frac{\partial^{j-2}}{\partial u^{j-2}}$ to the identity above to get a new system of identities:

$$\partial_u^{(k)}(a) * d\Phi = \partial_u^{(k)}(Q \cdot a) * d\tilde{x} \quad \mod \langle u^{j-1-k}, \tilde{x}\rangle,$$

where $k = 0, 1, \dots, j-2$.

Now we substitute $u = 0$ and use that $q = 1$ to get the following triangular system of identities, modulo the ideal $\langle \tilde{x}\rangle$ generated by $x_{i,0}, \dots, x_{i,j-2}$:

$$\begin{aligned}
d\phi_0 &= dx_{i,0} \quad \mod \langle \tilde{x}\rangle\\
d\phi_1 &= dx_{i,1} + b_{1,0}\, dx_{i,0} \quad \mod \langle \tilde{x}\rangle\\
d\phi_2 &= dx_{i,2} + b_{2,0}\, dx_{i,0} + b_{2,1}\, dx_{i,1} \quad \mod \langle \tilde{x}\rangle\\
&\dots\dots\\
d\phi_{j-2} &= dx_{i,j-2} + b_{j-2,0}\, dx_{i,0} + b_{j-2,1}\, dx_{i,1} + \dots\\
&\quad + b_{j-2,j-3}\, dx_{i,j-3} \quad \mod \langle \tilde{x}\rangle
\end{aligned}$$

Here $\{h_{\partial,t}\}$ are some functional coefficients whose computation we leave to the reader. Now (6.43) follows by taking exterior products of the 1-forms on the RHS and LHS of the system above and letting $\tilde{x} = 0$.

Let

$$\theta_i =_{\mathsf{def}} dx_{i,0} \wedge \cdots \wedge dx_{i,j_i-2} \quad \text{and} \quad \theta =_{\mathsf{def}} \wedge_i \theta_i.$$

Then $du \wedge \theta$, together with the volume form in X, define the volume form in the y-coordinates. Therefore the orientation of the space $\tau_\gamma(\mathcal{T}(v,\omega))$, tangent to the pure stratum $\mathcal{T}(v,\omega)$ at its typical point γ (this space can be identified with the space spanned by the vectors $\partial_{y_1}, \dots, \partial_{y_{n-|\omega|'}}$), is determined intrinsically by the local geometry of the v-flow in the vicinity of $\gamma \subset \hat{X}$. For time being, let us call this orientation of $\tau_\gamma(\mathcal{T}(v,\omega))$ *versal.*

On the other hand, each manifold $\partial_j X$, $j > 0$, comes equipped with its own *preferred* orientation, which depends only on the stratification $\{\partial_k^+ X(v)\}_k$ (and not on properties of the field v inside X) and on the preferred orientation of X. Here is the familiar recipe for its construction: the orientation of X, with the help of the inward normals, induces a preferred orientation of $\partial_1 X$, and thus of $\partial_1^\pm X$. In turn, the inward normals to $\partial_2 X = \partial(\partial_1^+ X)$ in $\partial_1^+ X$ produce a preferred orientation of $\partial_2 X$, and thus of $\partial_2^\pm X$. And the process goes on: the preferred orientation of $\partial_{j-1} X$, with the help of the inward normal to $\partial_j X$ in $\partial_{j-1}^+ X$, determines a preferred orientation of $\partial_j X$, and hence of $\partial_j^\pm X$.

So, along each trajectory γ, every space T_i, tangent to $\partial_{j_i} X^\circ$ and transversal to γ at the points $a_i \in \gamma \cap \partial_1 X$, is preferably oriented. For a traversally generic v, the $\hat{v}$-flow propagates these spaces T_i's along γ in such a way that they form complementary vector bundles over γ (see Definition 6.3). We order them by the increasing values of i. This ordering, together with the preferred orientations of the T_i's (based on the orientations of $\partial_{j_i}^+ X$), generates a new preferred orientation of the tangent space $\tau_\gamma(\mathcal{T}(v,\omega))$. This preferred orientation may agree or disagree with the versal orientation of the same space, produced with the help of special coordinates in the vicinity of γ (the versal orientation is based on the increasing powers of $(u - \alpha_i)$'s, a feature of the special coordinates). In the first case, we attach the polarity "$\oplus$" to γ, in the second case, the polarity of γ is defined to be "$\ominus$".

Therefore not only the components of pure strata $\mathcal{T}(v,\omega)$ are canonically oriented open manifolds, but they also come in two flavors: "$\oplus$" and "$\ominus$"!

We fix an auxiliary function $z : \hat{X} \to \mathbb{R}$ as in (6.25). With the help of the v-flow, it produces an ordered collection of $|\omega|'$ linearly independent and globally defined 1-forms, a framing of the quotient bundle

$$\nu^*(\mathcal{T}(v,\omega)) =_{\mathsf{def}} \tau^*(\mathcal{T}(v))|_{\mathcal{T}(v,\omega)}/\, \tau^*(\mathcal{T}(v,\omega)),$$

the "normal cotangent bundle" of $\mathcal{T}(v,\omega)$ in $\mathcal{T}(v)$. Let us explain this observation.

For any $\gamma \in \mathcal{T}(v,\omega)$ and any two points $a, x \in \gamma$, we denote by $\phi_{a,x}$ the germ (taken in the vicinity of $\gamma \subset \hat{X}$) of the unique $\hat{v}$-flow-generated and γ-localized diffeomorphism that maps x to a.

For each point $a_i \in \gamma \cap \partial_1 X$ of multiplicity $j_i > 1$, let us consider the 1-forms

$$\{dz, \mathcal{L}_v(dz), \mathcal{L}_v^2(dz), \dots, \mathcal{L}_v^{j_i-2}(dz)\},$$

taken at the point a_i (that is, view them as elements of $T^*_{a_i}(X)$). Then, with the help of one-parameter family of diffeomorphisms $\{\phi_{a_i,x}\}_{x\in\gamma}$, we spread the forms

$$\{dz|_{a_i}, \mathcal{L}_v(dz)|_{a_i}, \mathcal{L}_v^2(dz)|_{a_i}, \dots, \mathcal{L}_v^{j_i-2}(dz)|_{a_i}\}$$

along γ to get $j_i - 1$ independent sections $\eta_{i,0}$, $\eta_{i,1}$, $\dots$, η_{j_i-2} of $T^*(X)|_\gamma$. By their very construction, these sections are flow-invariant. Moreover, since at points of $\partial_2 X$ the field v is tangent to $\partial_1 X = \{z = 0\}$, we get $dz(v)|_{\partial_2 X} = \mathcal{L}_v(z)|_{\partial_2 X} = 0$. Thus $\eta_{i,0}(v)|_\gamma = 0$ for all i.

Similarly, for each $a_i \in \partial_3 X$ (i.e., $j_i > 2$), the field v is tangent to

$$\partial_2 X = \{z = 0,\ \mathcal{L}_v(z) = 0\}.$$

So, using the identity

$$\mathcal{L}_v(dz) = v \rfloor d(dz) + d(v \rfloor dz) = d(v \rfloor dz),$$

we get $\mathcal{L}_v(dz)(v)|_{\partial_3 X} = 0$. As a result, $\eta_{i,1}(v)|_\gamma = 0$ for all i with $j_i > 2$. Similar considerations show that, for each i, all the sections $\{\eta_{i,k}\}_{k<j_i-1}$, have the property $\eta_{i,k}(v)|_\gamma = 0$—they are *horizontal* 1-forms. Therefore they can be viewed as independent sections of the subbundle $\tau^*(v) \subset T^*(X)$. With the help of $(\Gamma^*)^{-1}$, these sections produce independent sections of the quotient bundle $\nu^*(\mathcal{T}(v,\omega))$.

Now we take all $|\omega|'$ sections $\{\eta_{i,0}, \eta_{i,1}, \dots \eta_{j_i-2}\}_i$ of $T^*(X)|_\gamma$, ordered in groups by the increasing values of i. For a traversally generic v, by Theorem 6.5, these sections of $\tau^*(v) \subset T^*(X)|_\gamma$ are linearly independent.

As long as the combinatorial type ω of γ is fixed, these sections depend smoothly on γ. Since their construction relies only on ω, z, and v, they are globally well-defined linearly independent sections of the conormal bundle

$\nu^*(\mathcal{T}(v,\omega))$, an intrinsically defined trivialization of this bundle. Their duals define independent sections of the normal bundle $\nu(\mathcal{T}(v,\omega))$.

The preferred orientation of each $\partial_j X$, $j \geq 1$, depends only on $v|_{\partial_1 X}$ and the orientation of X. In particular, the preferred orientation of $\partial_1 X$ depends on the orientation of X only. As we flip the orientation of X, the preferred orientation of each $\partial_j X$ flips as well. Therefore the preferred orientation of the tangent bundle $\tau(\mathcal{T}(v,\omega))$ changes, as a result of flipping the orientation of X, only when the cardinality of the intersection $\gamma \cap \partial_1 X$—the integer $|\sup(\omega)|$—is odd.

The versal orientation of $\mathcal{T}(v,\omega)$ behaves similarly under the change of an orientation of X. As a result, the polarity "$\oplus$" or "$\ominus$" of each component of $\mathcal{T}(v,\omega)$ is independent of the orientation of X. □

The statement of the next corollary is familiar from our dealings with the 3D-flows.

Corollary 6.6. *For a traversally generic vector field v, the points of minimal 0-dimensional strata $\mathcal{T}(v,\omega)$ come equipped with two sets of polarities: "$+,-$" and "$\oplus,\ominus$".*

Proof. When ω has the maximal possible reduced multiplicity $|\omega|' = n$, we can compare the versal and preferred orientations at each point γ of the zero-dimensional set $\mathcal{T}(v,\omega)$. When the two agree, we attach the polarity "$\oplus$" to γ; otherwise, its polarity is defined to be "$\ominus$". Of course, the preferred orientation of $\nu(\gamma, X)$ can be compared with the preferred orientation of $\partial_1 X$ at the lowest point in $\gamma \cap \partial_1 X$. This comparison allows for another pair $(+,-)$ of polarities to be attached to γ. □

In the Morse theory, flipping "$v \Rightarrow -v$" of the gradient vector field $v = \nabla f$ leads to geometric manifestations of the Poincaré duality. These manifestations take the form of switching the roles of stable and unstable manifolds of the f-critical points.

Keeping this fact in mind, let us investigate how the flipping $v \Rightarrow -v$ affects the preferred and versal orientations of strata in the trajectory space $\mathcal{T}(v)$. Recall that replacing v with $-v$ affects the Morse stratification according to the formula: $\partial_j^+ X(-v) = \partial_j^\epsilon X(v)$, where $\epsilon = +$, when $(n+1) - j \equiv 0 \mod 2$, and $\epsilon = -$ otherwise.

This reversal of the vector field direction also changes the order in which the points of $\gamma \cap \partial_1 X$ occur along a typical v-trajectory γ. As a result, the combinatorial type ω of γ changes to $\bar{\omega}$, the mirror image of the histogram ω. In other words, $\bar{\omega}(i) =_{\mathsf{def}} \omega(q-i)$, where $q =_{\mathsf{def}} |\sup(\omega)|$.

Note that changing v to $-v$ transforms $\{dz, \mathcal{L}_v(dz), \mathcal{L}_v^2(dz), \mathcal{L}_v^3(dz), \dots\}$ into

$$\{dz, \mathcal{L}_{-v}(dz), \mathcal{L}_{-v}^2(dz), \mathcal{L}_{-v}^3(dz), \dots\} = \{dz, -\mathcal{L}_v(dz), \mathcal{L}_v^2(dz), -\mathcal{L}_v^3(dz), \dots\}.$$

Thus the flip $v \Rightarrow -v$ results in the change of the sign of the differential form

$$\theta_j =_{\mathsf{def}} dz \wedge \mathcal{L}_v(dz) \wedge \mathcal{L}_v^2(dz) \wedge \cdots \wedge \mathcal{L}_v^{j-2}(dz)$$

by the factor $(-1)^{\lceil (j-2)/2 \rceil}$. In addition, the flip results in changing the order of the points from $\gamma \cap \partial_1 X$ to the opposite.

Let $\omega = (j_1, j_2, j_3, \dots)$. Consider a division of the ordered set

$$1, \dots, j_1 - 1;\ j_1, \dots, j_1 + j_2 - 2;\ j_1 + j_2 - 1, \dots, j_1 + j_2 + j_3 - 3;\ \dots$$

into successive groups of the lengths $j_1 - 1, j_2 - 1, j_3 - 1, \dots$ etc. Let σ_ω denote the permutation that changes the order of groups to the opposite, but keeps the relative positions of elements within each group. We denote by $sign(\sigma)$ the parity of permutation σ. So the flip $v \Rightarrow -v$ affects the form $\wedge_i \theta_{\omega(i)}$ by multiplying it with the factor $(-1)^{\epsilon(\omega)}$, where

$$\epsilon(\omega) = sign(\sigma_\omega) + \sum_i \left\lceil \frac{\omega(i) - 2}{2} \right\rceil \tag{6.45}$$

Therefore this formula describes the connection between the versal orientation of $\mathcal{T}(v, \omega)$ and the versal orientation of $\mathcal{T}(-v, \bar{\omega})$. Of course, the trajectory spaces $\mathcal{T}(v)$ and $\mathcal{T}(-v)$ are canonically homeomorphic.

Our next goal is to prove that the trajectory space $\mathcal{T}(v)$ of a traversally generic field v is a *Whitney stratified space* (see Definition 6.12). Prior to establishing this fact in Theorem 6.10 below, we need to prove a few lemmas.

Recall that a function f on a closed subset Y of a smooth manifold X is called *smooth* if it is the restriction of a smooth function, defined in an open neighborhood of Y.

Lemma 6.14. *Let v be a traversing and boundary generic vector field on a compact connected smooth manifold X with boundary, and $\Gamma : X \to \mathcal{T}(v)$ the obvious map. Let $F \subset \mathcal{T}(v)$ be a closed subset and $\psi : F \to \mathbb{R}$ a* *function such that $\Gamma^*(\psi)$ is a smooth function on $\Gamma^{-1}(F) \subset X$ with the property $\mathcal{L}_v(\Gamma^*(\psi)) = 0$ on $\Gamma^{-1}(F)$.*

Then $\psi : F \to \mathbb{R}$ admits an extension $\Psi : \mathcal{T}(v) \to \mathbb{R}$ such that $\Gamma^*(\Psi)$ is a smooth function on X with the property $\mathcal{L}_v(\Gamma^*(\Psi)) = 0$.

Proof. Let $h : X \to \mathbb{R}$ be a smooth function with the property $dh(v) > 0$. By Lemma 5.6, such h exists for any traversing v. Using h, we can find a finite set $\mathcal{S}$ of closed smooth transversal sections $\{S_\alpha \subset h^{-1}(c_\alpha)\}_\alpha$ of the v-flow, such that each trajectory hits some section from the collection $\mathcal{S}$. Moreover, we can assume that all the heights $\{c_\alpha\}$ are distinct and separated by some $\epsilon > 0$. The set $\mathcal{S}$ can be given a poset structure: $\beta \succ \alpha$ if there exists an ascending v-trajectory γ that first pierces S_α and then S_β. Evidently, this implies that $c_\alpha < c_\beta$.

For a given α, consider the set $\mathcal{S}_{\succ\alpha} =_{\mathsf{def}} \{\beta \succ \alpha\}$ and put $c_\alpha^{\Uparrow} =_{\mathsf{def}} \min_{\beta\succ\alpha}\{c_\beta\}$.

Now the proof of the lemma proceeds by an induction on the heights $\{c_\alpha\}$, guided by the partial order in $\mathcal{S}$. Assume that the desired extension

$$\tilde{\Psi}_{\succ\alpha} : h^{-1}\big([c_\alpha^{\Uparrow}, +\infty)\big) \to \mathbb{R}$$

of the function $\Gamma|_{\Gamma^{-1}(F)} \circ \psi$, subject to the property $\mathcal{L}_v(\tilde{\Psi}_{\succ\alpha}) = 0$, already has been constructed. The inductive step calls for an extension of $\tilde{\Psi}_{\succ\alpha}$ to a function on $h^{-1}([c_\alpha, +\infty))$, while keeping it constant on the v-trajectories.

We denote by $X(v, A)$ the union of v trajectories through a closed subset $A \subset X$. Consider two sets:

$$F_\alpha =_{\mathsf{def}} \Gamma^{-1}(F) \cap S_\alpha \quad \text{and} \quad Q_\alpha =_{\mathsf{def}} X\big(v, \coprod_{\beta\succ\alpha} S_\beta\big) \cap S_\alpha.$$

Since $\tilde{\Psi}_{\succ\alpha}$ is constant along each trajectory and S_α is smooth and transversal to the flow, $\tilde{\Psi}_{\succ\alpha}$ produces a well-defined smooth function $\hat{\Psi}_{\succ\alpha} : Q_\alpha \to \mathbb{R}$. On the other hand, the function

$$\tilde{\psi} =_{\mathsf{def}} \psi \circ \Gamma : \Gamma^{-1}(F) \to \mathbb{R}$$

is smooth and constant along the v-trajectories by the lemma hypothesis. In particular, it is a smooth function on the closed set F_α. Moreover, since $\tilde{\Psi}_{\succ\alpha}$ is an extension of $\tilde{\psi}$ to $h^{-1}([c_\alpha^{\Uparrow}, +\infty)) \subset X$, both functions $\hat{\Psi}_{\succ\alpha}$ and $\tilde{\psi}$ agree on $F_\alpha \cap Q_\alpha$. Therefore we have produced a function $\Psi_\alpha^\sharp : F_\alpha \cup Q_\alpha \to \mathbb{R}$ which extends to a smooth function $\tilde{\Psi}_\alpha$ on S_α. In turn, $\tilde{\Psi}_\alpha : S_\alpha \to \mathbb{R}$ defines a smooth function $\hat{\Psi}_\alpha : X(v, S_\alpha) \to \mathbb{R}$ which is constant on each trajectory through S_α. By their construction, $\hat{\Psi}_\alpha$ and $\tilde{\Psi}_{\succ\alpha}$ agree on the set

$$X(v, S_\alpha) \cap h^{-1}([c_\alpha^{\Uparrow}, +\infty)).$$

Together, they produce a smooth function on $h^{-1}([c_\alpha, +\infty))$ which is constant along the trajectories through $\coprod_{\beta\succeq\alpha} S_\beta$ and extends $\tilde{\psi}$. This completes the induction step. □

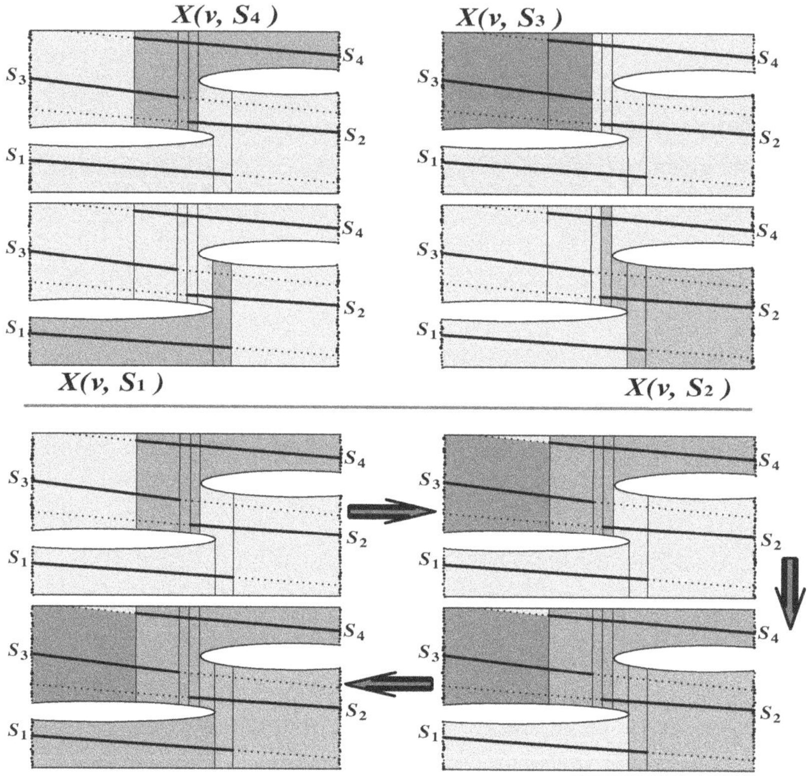

Fig. 6.3 The upper four diagrams show the flow sections $\{S_i\}_{1\le i\le 4}$ and the corresponding sets $\{X(v, S_i)\}_{1\le i\le 4}$. The lower four diagrams show the growth of the domains of ψ-extensions, as they appear in the proof (to simplify the picture, the original set $\Gamma^{-1}(F)$ is not shown).

Next, we introduce the notion of a *Whitney stratified space*.

Definition 6.12. Let Z be a closed subset of an Euclidean space E. Assume that Z decomposes as $\coprod_{\alpha \in \mathcal{S}} Z_\alpha$, where $\mathcal{S}$ a finite poset. Assume that each set Z_α locally is a smooth submanifold of E.

For any pair $Z_\alpha \subset \bar{Z}_\beta$ of strata, where $\bar{Z}_\beta$ stands for the closure of the stratum Z_β, we consider any two sequences $\{x_i \in Z_\beta\}_i$, $\{y_i \in Z_\alpha\}_i$ that converge to the same point $y \in Z_\alpha$. We form the secant lines $\{\ell_i =_{\mathsf{def}} [x_i, y_i]\}_i$ and assume that $\{\ell_i\}_i$ converge to a limiting line ℓ. We also consider a sequence of tangent spaces $\{T_{x_i}(Z_\beta) \subset \mathsf{E}\}_i$ that converges to a limit subspace $\tau \subset \mathsf{E}$ that contains y.

We say that Z is a *Whitney space* if, for any such data, the limit line $\ell \subset \tau$. $\diamondsuit$

In fact, if $Z \subset M$ is a Whitney space, then one can prove that $T_y(Z_\alpha) \subset \tau$ (see [GM2]).

Now we are going to prove that the standard models of traversally generic flows lead to spaces of trajectories which are Whitney spaces. The trajectory space of a traversally generic flow is stratified by strata which are labeled by the elements of the universal poset $\mathbf{\Omega}^\bullet$. The set $\mathbf{\Omega}^\bullet$ will be introduced in Definition 7.3 and the partial order "$\succ_\bullet$" in it in Definition 7.5. However, the exact nature of these definitions is not essential for understanding the next few lemmas; all what one needs to know is that the partial order reflects the inclusion of a stratum in the closure of another stratum.

Lemma 6.15. *Let $\omega \in \mathbf{\Omega}^\bullet$, and let $\omega_{\bullet\preceq} =_{\mathsf{def}} \{\tilde{\omega} \in \mathbf{\Omega}^\bullet \,|\, \tilde{\omega} \succeq_\bullet \omega\}$. Consider the semi-algebraic set*

$$Z_\omega =_{\mathsf{def}} \{P_\omega(u, x) \leq 0,\ \|x\| \leq \epsilon\},$$

where $\epsilon > 0$ is sufficiently small, and P_ω is the polynomial of the even degree $|\omega|$ is as in (6.26)[15]*. Let $\mathcal{T}_\omega$ denote the $(\omega_{\bullet\preceq})$-stratified trajectory space of the constant vector field $v =_{\mathsf{def}} \partial_u$ in Z_ω.*

Then there exists an embedding $K_\omega : \mathcal{T}_\omega \to \mathbb{R}^{2|\omega|'}$, produced by smooth functions on Z_ω, which are constant along each ∂_u-trajectory that resides in Z_ω.

Proof. We view x as a point in the space of coefficients $\mathbb{R}^{|\omega|'}$ for the family of u-polynomials $\{P(u, x)\}_{x \in \mathbb{R}^{|\omega|'}}$ given by (6.26).

[15] Its real divisor has the combinatorial type ω.

Evidently, the x-coordinates $x : Z_\omega \to \mathbb{R}^{|\omega|'}$ provide us with a map $\chi : \mathcal{T}_\omega \to \mathbb{R}^{|\omega|'}$. The map χ is given by the algebraic functions which are constant on the ∂_u-trajectories in Z_ω. However, χ does not separate some trajectories; that is, χ is not an embedding (just a finitely ramified map). We will complement x with another smooth map $\tilde{x} : Z_\omega \to \mathbb{R}^{|\omega|'}$, also constant on the trajectories in Z_ω and such that the pair of maps $(x, \tilde{x})$ will separate the points of $\mathcal{T}_\omega$.

To construct $\tilde{x}$, we will use some facts from Chapter 7, Section 7.4. According to that section, the ball $B_\epsilon = \{\|x\| \leq \epsilon\}$ has a particular *cone structure*. It is given, with the help of the Vieté map, by the the local linear contractions, in the complex plane $\mathbb{C}$, of each "nearby" complex divisor $D_{\mathbb{C}}(P(\sim, x_\star))$ on the complex divisor $D_{\mathbb{C}}(P(\sim, 0))$. This contraction produces a smooth real semi-algebraic curve $A_{x_\star} : [0,1] \to B_\epsilon$ in the coefficient x-space (in fact, $A_{x_\star}([0,1])$ is a generator of the cone), which connects the given point $x_\star \in \mathbb{R}^{|\omega|'}$ to the origin 0. In particular, the combinatorial type of the real divisor $D_{\mathbb{R}}\big(P(\sim, A_{x_\star}(t))\big)$ is constant for all $t \in (0,1]$ (this type belongs to the poset $\mathbf{\Omega}_{\langle|\omega|]}$).

Let $S_{x_\star} =_{\mathsf{def}} \mathbb{R} \times A_{x_\star}$ be the ruled (u,t)-parametric surface that projects along the u-direction onto the curve $A_{x_\star}$. Consider the intersection $\Sigma_{x_\star}$ of the surface $S_{x_\star}$ with the set Z_ω (see Figure 6.4, the left diagram). As $x_\star \in \partial B_\epsilon$ varies, the surfaces $\{\Sigma_{x_\star}\}$ span Z_ω (the trajectory $\{x = 0\}$ serves as the binder of an open book, whose pages are these surfaces).

We will define a new projection $\tilde{x} : \Sigma_{x_\star} \to A_{x_\star}$ as follows. Consider the u-directed line ℓ_x through x. For a typical point $x \in A_{x_\star}$, let $\Pi_x(x_\star) =_{\mathsf{def}} \ell_x \cap \Sigma_{x_\star}$. The set $\Pi_x =_{\mathsf{def}} \Pi_x(x_\star)$ is a disjointed union of closed intervals $\{I_i(x) = [\underline{\alpha}_i(x),\, \bar{\alpha}_i(x)]\}_i$ (where $\underline{\alpha}_i(x) \leq \bar{\alpha}_i(x)$), residing in the line ℓ_x. We order them so that $I_1(x) < I_2(x) < \cdots < I_s(x)$ as sets.

Put

$$\Pi_x^\vee =_{\mathsf{def}} (\ell_x \setminus \Pi_x) \cap [\alpha_{min}(x),\, \alpha_{max}(x)],$$

where $\alpha_{min}(x)$, $\alpha_{max}(x)$ denote the minimal and the maximal real roots of the u-polynomial $P_\omega(u,x)$. Thus $\Pi_x^\vee$ is a finite disjoint union of closed intervals

$$\{I_i^\vee(x) = [\bar{\alpha}_i(x),\, \underline{\alpha}_{i+1}(x)]\}_i,$$

residing in the line ℓ_x. Note that $P_\omega \geq 0$ in each interval $I_i^\vee(x)$. We also order the intervals so that

$$I_1^\vee(x) < I_2^\vee(x) < \cdots < I_{s-1}^\vee(x).$$

Let $\tau_i(x)$ denote the length of the interval $I_i^{\vee}(x)$.

We fix a smooth monotone function $\chi : [0, +\infty) \to [0, 1)$ such that $\chi(0) = 0$ and $\lim_{\tau \to +\infty} \chi(\tau) = 1$ (say, $\chi = \frac{2}{\pi} \tan^{-1}$).

Consider a smooth τ-parametric family ($\tau \in [0, +\infty)$) of smooth monotonically increasing functions $\phi_\tau : [0, 1] \to \mathbb{R}_+$ such that:

(1) $0 < \phi_\tau(t) < t$ for all $t \in (0, 1]$,
(2) the infinite order jet of ϕ_τ of at $t = 0$ coincides with the jet of the identity function $t : [0, 1] \to [0, 1]$,
(3) $\phi_\tau(1) = \chi(\tau)$, and
(4) $\phi_\tau(t)$ is a smooth function in t and τ.

For each i, we map the point

$$\big(\bar{\alpha}_i(A_{x_\star}(t)),\, t\big) \in \partial_1^+ \Sigma_{x_\star}(\partial_u)$$

to the point

$$\big(\underline{\alpha}_i(A_{x_\star}(\phi_{\tau_i(x_\star)}(t)),\, \phi_{\tau_i(x_\star)}(t)\big) \in \partial_1^- \Sigma_{x_\star}(\partial_u).$$

We denote by $\theta_{x_\star, i}$ this correspondence (shown in Figure 6.4). As a function in (u, t), the map $\theta_{x_\star, i}$ is smooth. We notice that, $\phi_{\tau_i(x_\star)}(t) \neq t$ for all $t \in (0, 1]$ and $x_\star \neq \vec{0}$. We also observe that, if the interval $I_i^{\vee}(x_\star)$ shrinks to a singleton as we vary $x_\star$, then the map $\theta_{x_\star, i}$ approaches the identity.

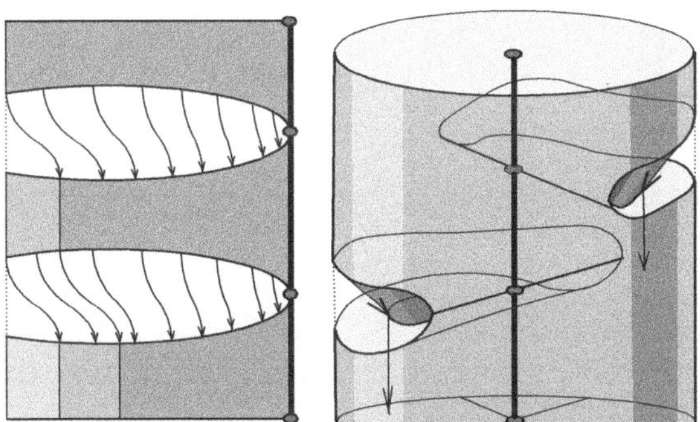

Fig. 6.4 The map $\tilde{x} : \Sigma_{x_\star} \to A_{x_\star}$ over an arc $A_{x_\star}$ (on the left) and the map $\tilde{x} : Z_\omega \to \mathbb{R}^{|\omega|'}$ (on the right).

Now we define $\tilde{x} : \Sigma_{x_\star} \to A_{x_\star}$ by the following formulas:

$$=_{\mathsf{def}}\ x \text{ for all points in } I_1(x),$$
$$=_{\mathsf{def}}\ \theta_{x_\star,1} \text{ for all points in } I_2(x),$$
$$=_{\mathsf{def}}\ \theta_{x_\star,2} \circ \theta_{x_\star,1} \text{ for all points in } I_3(x),$$
$$=_{\mathsf{def}}\ \theta_{x_\star,3} \circ \theta_{x_\star,2} \circ \theta_{x_\star,1} \text{ for all points in } I_4(x),$$
$$=_{\mathsf{def}}\ \cdots \tag{6.46}$$

Since $0 < \phi_\mathcal{T}(t) < t$ for all $t \in (0,1]$, the map $\tilde{x} : Z_\omega \to \mathbb{R}^n$ separates the trajectories that are not distinguished by the map $x : Z_\omega \to \mathbb{R}^n$. Therefore the smooth map

$$J_\omega =_{\mathsf{def}} (x, \tilde{x}(u,x)) : Z_\omega \to \mathbb{R}^{2|\omega|'},$$

being constant on each ∂_u-trajectory *contained in* Z_ω[16], gives rise to a smooth (in the sense of Definition 6.10) embedding $K_\omega : \mathcal{T}_\omega \to \mathbb{R}^{2|\omega|'}$. □

Remark 6.8. It seems that the desired embedding $K_\omega : \mathcal{T}_\omega \to \mathbb{R}^{2|\omega|'}$ cannot be delivered by real analytic functions on Z_ω that are constant on each ∂_u-trajectory in Z_ω. ◇

Corollary 6.7. *The image $K_\omega(\mathcal{T}_\omega) \subset \mathbb{R}^{2|\omega|'}$ is a Whitney $(\omega_{\bullet \preceq})$-stratified space.*

Proof. It is useful to consult continuously with Figure 6.5 that illustrates some key elements in the proof.

Let $\pi : \mathbb{R} \times \mathbb{R}^{2|\omega|'} \to \mathbb{R}^{2|\omega|'}$ denote the obvious projection. Consider the embedding $K =_{\mathsf{def}} K_\omega : \mathcal{T}_\omega \to \mathbb{R}^{2|\omega|'}$, constructed in Lemma 6.15, and the map

$$E =_{\mathsf{def}} E_\omega : Z_\omega \to \mathbb{R} \times \mathbb{R}^{2|\omega|'},$$

given by the formula $E(u,x) =_{\mathsf{def}} (u,\, J(u,x))$. Here $J =_{\mathsf{def}} J_\omega = (x,\, \tilde{x}(u,x))$, and the function $\tilde{x} : Z_\omega \to \mathbb{R}^{|\omega|'}$ is as in the proof of Lemma 6.15. The map E is a regular embedding, given by smooth functions on Z_ω. Let

$$Q : \mathbb{R} \times \mathbb{R}^{2|\omega|'} \to \mathbb{R} \times \mathbb{R}^{|\omega|'},$$

be the projection, given by the formula $Q(u, x, \tilde{x}) =_{\mathsf{def}} (u, x)$. By the definition, $Q(E(Z_\omega)) = Z_\omega$.

[16]But not on any ∂_u-trajectory!

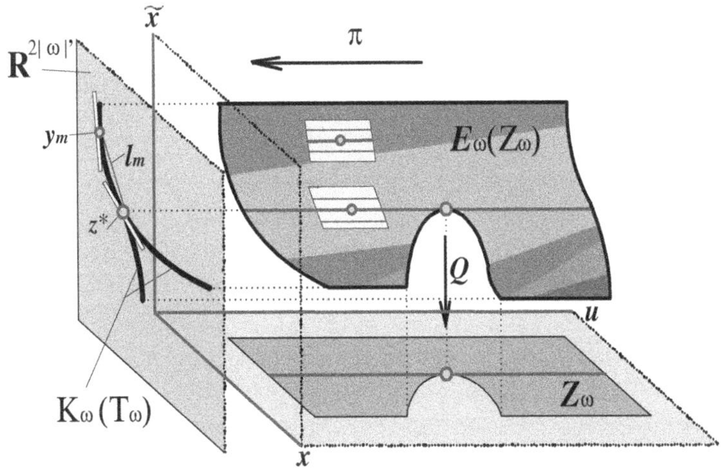

Fig. 6.5 The space $E_\omega(Z_\omega) \subset \mathbb{R} \times \mathbb{R}^{2|\omega|'}$ and its projections π and Q on $K_\omega(\mathcal{T}_\omega) \subset \mathbb{R}^{2|\omega|'}$ and on $Z_\omega \subset \mathbb{R} \times \mathbb{R}^{|\omega|'}$, respectively.

Consider the space $K(\mathcal{T}_\omega) \subset \mathbb{R}^{2|\omega|'}$. Let $\mu \prec_\bullet \nu$ be two elements in the poset $\omega_{\bullet \preceq}$ that labels the combinatorial types of ∂_u-trajectories, residing in Z_ω. Let $\mathcal{K}_\mu, \mathcal{K}_\nu$ be the two pure strata of the space $K(\mathcal{T}_\omega)$, indexed by $\mu \prec_\bullet \nu$ (thus $\mathcal{K}_\mu \subset \bar{\mathcal{K}}_\nu$). Consider an infinite sequence of points $\{y_m \in \mathcal{K}_\nu\}_m$ and an infinite sequence of points $\{z_m \in \mathcal{K}_\mu\}_m$, both converging to a point $z_\star \in \mathcal{K}_\mu$.

We need to verify that, if the tangent spaces $\{T_{y_m}\mathcal{K}_\nu\}_m$ converge in $\mathbb{R}^{2|\omega|'}$ to an affine space $T_\star$ containing $z_\star$, and the sequence of lines $\{\ell_m \supset [z_m, y_m]\}_m$ converges to a line $\ell_\star \subset \mathbb{R}^{2|\omega|'}$, then $\ell_\star \subset T_\star$.

Equivalently, we need to verify that if the spaces

$$\{\mathsf{T}_m =_{\mathsf{def}} \pi^{-1}(T_{y_m}\mathcal{K}_\nu)\}_m$$

converge in $\mathbb{R} \times \mathbb{R}^{2|\omega|'}$ to an affine space

$$\mathsf{T}_\star =_{\mathsf{def}} \pi^{-1}(T_\star) \supset \pi^{-1}(z_\star),$$

and the sequence of 2-planes $\{\mathsf{L}_m =_{\mathsf{def}} \pi^{-1}(\ell_m)\}_m$ converges to a plane

$$\mathsf{L}_\star =_{\mathsf{def}} \pi^{-1}(\ell_\star) \subset \mathbb{R} \times \mathbb{R}^{2|\omega|'},$$

then $\mathsf{L}_\star \subset \mathsf{T}_\star$. Let us call this conjectured property "$\tilde{\mathsf{B}}$".

We notice that all the affine spaces $\mathsf{T}_m, \mathsf{T}_\star, \mathsf{L}_m$, and $\mathsf{L}_\star$, are fibrations with the line fibers, parallel to the direction of $\mathbb{R}$ in the product $\mathbb{R} \times \mathbb{R}^{2|\omega|'}$.

We may think of $E(Z_\omega)$ as a *graph* of a smooth map $\tilde{x}$ from $Z_\omega \subset \mathbb{R} \times \mathbb{R}^{|\omega|'}$ to $\mathbb{R}^{|\omega|'}$ (see Figure 6.5).

Since $Q : E(Z_\omega) \to Z_\omega$ is a $(\omega_{\bullet\preceq})$-stratification-preserving diffeomorphism which respects the ∂_u-induced 1-foliations $\mathcal{F}$ on $E(Z_\omega)$ and $\mathcal{G}$ on Z_ω, the tangent spaces to the ν-indexed pure stratum in $E(Z_\omega)$ are mapped by Q *isomorphically* onto the tangent space to the ν-indexed pure stratum in Z_ω. So, with the help of the graph-manifold $E(Z_\omega)$, any tangent space to the ν-indexed pure stratum in Z_ω determines the corresponding tangent space to the ν-indexed pure stratum in $E(Z_\omega)$.

Let $\tilde{\tau}_\star$ denote the tangent space to $E(Z_\omega)$ at a point $\tilde{z}_\star \in \pi^{-1}(z_\star)$, and let $\tau_\star$ denote the tangent space to Z_ω at the point $Q(z_\star)$. By the very definitions of $\mathsf{T}_\star$ and $\mathsf{L}_\star$ as limit objects, and using that $E(Z_\omega)$ is a smooth manifold[17] carrying the foliation $\mathcal{F}$ (whose leaves are u-parallel lines in $\mathbb{R} \times \mathbb{R}^{2|\omega|'}$), we get that $\mathsf{T}_\star \subset \tilde{\tau}_\star$ and $\mathsf{L}_\star \subset \tilde{\tau}_\star$.

Since $Q : E(Z_\omega) \to Z_\omega$ is a diffeomorphism, $Q : \tilde{\tau}_\star \to \tau_\star$ is an isomorphism of vector spaces. Therefore there exist unique subspaces of $\tilde{\tau}_\star$ that are mapped by Q onto $Q(\mathsf{T}_\star)$ or onto $Q(\mathsf{L}_\star)$; these are exactly the spaces $\mathsf{T}_\star$ and $\mathsf{L}_\star$, respectively. Thus, $Q(\mathsf{L}_\star) \subset Q(\mathsf{T}_\star)$ if and only if $\mathsf{L}_\star \subset \mathsf{T}_\star$.

Therefore the property $\tilde{\mathsf{B}}$ is equivalent to the following property B: "if the spaces $\{Q(\mathsf{T}_m)\}_m$ converge in $\mathbb{R}\times\mathbb{R}^{|\omega|'}$ to the affine space $Q(\mathsf{T}_\star)$, and the sequence of planes $\{Q(\mathsf{L}_m)\}_m$ converges to a plane $Q(\mathsf{L}_\star) \subset \mathbb{R} \times \mathbb{R}^{|\omega|'}$, then $Q(\mathsf{L}_\star) \subset Q(\mathsf{T}_\star)$".

Let $Q' : \mathbb{R}^{2|\omega|'} \to \mathbb{R}^{|\omega|'}$ denote the map given by $(x, \tilde{x}) \to x$. The image $Q'(K(\mathcal{T}_\omega)) = \mathbb{R}^{|\omega|'}$ is stratified by the collection of real discriminant varieties, indexed by various elements $\rho \in \omega_{\preceq}$. In particular, these strata are semi-algebraic sets whose interiors are smooth cells (see Theorem 7.1 and Theorem 7.2). By the fundamental results of [Har1], [Har2], and [Hi], the semianalitic sets are Whitney stratified spaces. As a result, the $(\omega_{\preceq})$-stratified space $Q'(K(\mathcal{T}_\omega))$ is a Whiney space. Thus property B is valid, since all the affine spaces, relevant to B, are fibrations with the line π-fibers over the corresponding spaces in $\mathbb{R}^{|\omega|'} = Q'(K(\mathcal{T}_\omega))$.

On the other hand, each element $\rho \in \omega_{\preceq}$ decomposes canonically into an ordered sequence $\Xi(\rho) =_{\mathsf{def}} \{\rho_\kappa\}_\kappa$ of elements from the set $\omega_{\preceq\bullet}$ (see the combinatorial construction that is depicted in Figure 7.1 and discussed in Example 7.4)[18]. This decomposition $\Xi(\rho)$ is such that, if the ∂_u-trajectory through a point $(u, x) \in Z_\omega$ has a combinatorial type ρ, then u determines

[17] with a piecewise smooth boundary

[18] For example, in Figure 6.5, $\rho = (1111) \in (121)_{\preceq}$ decomposes into $\rho_1 = (11) \in (121)_{\preceq\bullet}$ and $\rho_2 = (11) \in (121)_{\preceq\bullet}$.

the unique type $\rho_\kappa \in \omega_{\preceq_\bullet}$ to which the point $Q^{-1}((u,x)) \cap E(Z_\omega)$ belongs. In that sense the $\omega_{\preceq}$-stratification in Z_ω is finer than the $\omega_{\preceq_\bullet}$-stratification in $E(Z_\omega)$.

Therefore, the validated B-property for the $(\omega_{\preceq})$-stratified space $Q(K(\mathcal{T}_\omega))$ implies the $\tilde{\mathsf{B}}$-property for the $(\omega_{\bullet\preceq})$-stratified space $E(Z_\omega)$. As a result, $K(\mathcal{T}_\omega)$ is a Whitney stratified space in $\mathbb{R}^{2|\omega|'}$. □

Theorem 6.10. *For a traversally generic vector field v on a compact smooth connected $(n+1)$-dimensional manifold X with boundary, the $\mathbf{\Omega}^\bullet_{'\langle n]}$-stratified trajectory space $\mathcal{T}(v)$ can be given the structure of a Whitney space, residing in an Euclidean space $\mathbb{R}^N$.*

Proof. By Lemma 6.4, there exists a finite v-adjusted closed cover $\mathcal{U} = \{U_r\}_r$ of X, such that each $U_r \subset \hat{X}$ admits special coordinates $(u,x,y) =_{\mathsf{def}} (u^{(r)}, x^{(r)}, y^{(r)})$ in which ∂X is given by the polynomial equation $\{P_r(u,x) = 0\}$ as in (6.26). Recall that the equation is determined by the combinatorial type ω_r of the core trajectory $\gamma_r \subset U_r$.

Let us denote by $\mathcal{T}_r$ the space of trajectories of the ∂_u-flow in the domain

$$U_r =_{\mathsf{def}} \{P_r(u,x) \leq 0,\ \|x\| \leq \epsilon,\ \|y\| \leq \epsilon'\}.$$

It is a compact subset of $\mathcal{T}(v)$.

Consider the embeddings

$$K_r : \mathcal{T}_r \to \mathbb{R}^{2|\omega_r|'} \times \mathbb{R}^{n-|\omega_r|'},$$

$$E_r : U_r \to \mathbb{R} \times \mathbb{R}^{2|\omega_r|'} \times \mathbb{R}^{n-|\omega_r|'},$$

given by the formulas

$$K_r\big(\gamma_{\{u^{(r)},\, x^{(r)},\, y^{(r)}\}}\big) =_{\mathsf{def}} \big(x^{(r)},\ \tilde{x}^{(r)}(u^{(r)},\ x^{(r)}),\ y^{(r)}\big),$$

$$E_r\big(u^{(r)},\ x^{(r)},\ y^{(r)}\big) =_{\mathsf{def}} \big(u^{(r)},\ x^{(r)},\ \tilde{x}^{(r)}(u^{(r)},\ x^{(r)}),\ y^{(r)}\big).$$

Here $\gamma_{\{u^{(r)},\, x^{(r)},\, y^{(r)}\}}$ denotes the ∂_u-trajectory in U_r, passing through the point $(u^{(r)}, x^{(r)},\ y^{(r)})$, and $\tilde{x}^{(r)}(u^{(r)}, x^{(r)})$ is a function as in the proof of Lemma 6.15 (see Figure 6.5).

Smooth functions on $\psi : \mathcal{T}_r \to \mathbb{R}$ are exactly the smooth functions on $U_r \cap X$ that are constant along the trajectories. By Lemma 6.14, each such ψ extends to a smooth function on X which is constant on each trajectory. We denote this extension $\hat{\psi}$.

Therefore, employing the local embeddings

$$\{K_r : \mathcal{T}_r \to \mathbb{R}^{2|\omega_r|'} \times \mathbb{R}^{n-|\omega_r|'}\}_r,$$

we extend them to some smooth maps

$$\{\hat{K}_r : \mathcal{T}(v) \to \mathbb{R}^{2|\omega_r|'} \times \mathbb{R}^{n-|\omega_r|'}\}_r.$$

Together they produce a smooth embedding $K : \mathcal{T}(v) \to \mathbb{R}^N$, where

$$K =_{\mathsf{def}} \prod_r \hat{K}_r \quad \text{and} \quad \mathbb{R}^N =_{\mathsf{def}} \prod_r \left(\mathbb{R}^{2|\omega_r|'} \times \mathbb{R}^{n-|\omega_r|'}\right).$$

Let $G : X \to \mathbb{R}^N$ be the composition $\Gamma \circ K$, where $\Gamma : X \to \mathcal{T}(v)$ is the obvious map.

By Lemma 5.6, we can choose a smooth function $h : \hat{X} \to \mathbb{R}$ such that $dh(v) > 0$ in $\hat{X}$. With the help of h, we get a map $E : X \to \mathbb{R} \times \mathbb{R}^N$, given by the formula $E(z) =_{\mathsf{def}} (h(z), G(z))$, where $z \in X$. Since $dh(v) > 0$ and the Jacobian of each map $J_r =_{\mathsf{def}} (x^{(r)}, \tilde{x}^{(r)}, y^{(r)})$ is of the maximal rank n in U_r, the map E is a regular smooth embedding.

Composing E with the obvious projection $\pi : \mathbb{R} \times \mathbb{R}^N \to \mathbb{R}^N$, we get a smooth (see Definition 6.10) embedding $K : \mathcal{T}(v) \to \mathbb{R}^N$.

Our goal is to show that $K(\mathcal{T}(v))$ is a Whitney stratified space in $\mathbb{R}^N$. Since Definition 6.12 of a Whitney space is local, it suffices to check its validity in each local chart $\mathcal{T}_r \subset \mathcal{T}(v)$, that is, to verify that $K(\mathcal{T}_r) \subset \mathbb{R}^N$ is a Whitney space. Next arguments are very similar to the ones used in the proof of Corollary 6.7.

Consider the projection $p_r : \mathbb{R}^N \to \mathbb{R}^{2|\omega_r|'} \times \mathbb{R}^{n-|\omega_r|'}$, produced by omitting the product $\prod_{s\neq r}(\mathbb{R}^{2|\omega_s|'} \times \mathbb{R}^{n-|\omega_s|'})$ from $\prod_s(\mathbb{R}^{2|\omega_s|'} \times \mathbb{R}^{n-|\omega_s|'})$.

Let

$$Q_r : \mathbb{R} \times \mathbb{R}^N \to \mathbb{R} \times \mathbb{R}^{2|\omega_r|'} \times \mathbb{R}^{n-|\omega_r|'}$$

denote the projection $\mathsf{id}_{\mathbb{R}} \times p_r$.

Note that the projection Q_r generates a diffeomorphism between the manifold $E(U_r) \subset \mathbb{R}\times\mathbb{R}^N$ and the manifold $E_r(U_r) \subset \mathbb{R}\times\mathbb{R}^{2|\omega_r|'}\times\mathbb{R}^{n-|\omega_r|'}$, a diffeomorphism which respects the oriented 1-foliations, induced by the v-flow on X, as well as the $(\omega_r)_{\preceq}$-stratifications of $E(U_r)$ and $E_r(U_r)$ by combinatorial types of v-trajectories (or rather, of the π-fibers).

We denote these foliations by $\mathcal{F}_r$ and $\mathcal{G}_r$, respectively.

Let $\mu \prec_\bullet \nu$ be two elements in the poset $(\omega_r)_{\bullet\preceq}$, and $\mathcal{K}_\mu, \mathcal{K}_\nu$ the two pure strata of $K(\mathcal{T}_r) \subset \mathbb{R}^N$, indexed by μ, ν (thus $\mathcal{K}_\mu \subset \bar{\mathcal{K}}_\nu$). Consider a sequence of points $\{y_m \in \mathcal{K}_\nu\}_m$ and a sequence of points $\{z_m \in \mathcal{K}_\mu\}_m$, both converging to a point $z_\star \in \mathcal{K}_\mu$.

We need to verify that, if the tangent spaces $\{T_{y_m}\mathcal{K}_\nu\}_m$ converge in $\mathbb{R}^N$ to an affine space $T_\star$ containing $z_\star$, and the sequence of lines $\{\ell_m \supset [z_m, y_m]\}_m$ converges to a line $\ell_\star \subset \mathbb{R}^N$, then $\ell_\star \subset T_\star$.

Equivalently, we need to verify that if the spaces $\{\mathsf{T}_m =_{\mathsf{def}} \pi^{-1}(T_{y_m}\mathcal{K}_\nu)\}_m$ converge in $\mathbb{R} \times \mathbb{R}^N$ to an affine space $\mathsf{T}_\star =_{\mathsf{def}} \pi^{-1}(T_\star) \supset \pi^{-1}(z_\star)$, and the sequence of 2-planes $\{\mathsf{L}_m =_{\mathsf{def}} \pi^{-1}(\ell_m)\}_m$ converges to a plane $\mathsf{L}_\star =_{\mathsf{def}} \pi^{-1}(\ell_\star) \subset \mathbb{R} \times \mathbb{R}^N$, then $\mathsf{L}_\star \subset \mathsf{T}_\star$. Let us call this conjectured property "$\tilde{\mathsf{A}}$". Note that all the affine spaces $\mathsf{T}_m, \mathsf{T}_\star, \mathsf{L}_m$, and $\mathsf{L}_\star$, are fibrations with the line fibers, parallel to the direction of $\mathbb{R}$ in the product $\mathbb{R} \times \mathbb{R}^N$.

We can think of $E(U_r)$ as a graph of a smooth map from $E_r(U_r)$ to the product $\prod_{s \neq r} \mathbb{R}^{2|\omega_s|'} \times \mathbb{R}^{n-|\omega_s|'}$. Since $Q_r : E(U_r) \to E_r(U_r)$ is a stratification-preserving diffeomorphism, which respects the v-induced 1-foliations $\mathcal{F}_r$ and $\mathcal{G}_r$, the tangent spaces to the ν-indexed pure stratum in $E(U_r)$ are mapped *isomorphically* by Q_r onto the tangent space to the ν-indexed pure stratum in $E_r(U_r)$. So, with the help of the graph-manifold $E(U_r)$, any tangent space to the ν-indexed pure stratum in $E_r(U_r)$ determines the corresponding tangent space to the ν-indexed pure stratum in $E(U_r)$.

Let $\tilde{\tau}_\star$ denote the tangent space to $E(U_r)$ at a generic point $\tilde{z}_\star \in \pi^{-1}(z_\star)$, and let $\tau_\star$ denote the tangent space to $E_r(U_r)$ at the point $Q_r(z_\star)$. By the very definitions of $\mathsf{T}_\star$ and $\mathsf{L}_\star$ as limit objects and using that $E(U_r)$ is a smooth manifold, carrying the foliation $\mathcal{F}_r$ (whose leaves are parallel lines in $\mathbb{R} \times \mathbb{R}^N$), we get that $\mathsf{T}_\star \subset \tilde{\tau}_\star$ and $\mathsf{L}_\star \subset \tilde{\tau}_\star$.

Since $Q_r : E(U_r) \to E_r(U_r)$ is a diffeomorphism, $Q_r : \tilde{\tau}_\star \to \tau_\star$ is an isomorphism of vector spaces. Therefore there exist unique subspaces of $\tilde{\tau}_\star$ that are mapped by Q_r onto $Q_r(\mathsf{T}_\star)$ or onto $Q_r(\mathsf{L}_\star)$; these are exactly the spaces $\mathsf{T}_\star$ and $\mathsf{L}_\star$, respectively. Thus, $Q_r(\mathsf{L}_\star) \subset Q_r(\mathsf{T}_\star)$ if and only if $\mathsf{L}_\star \subset \mathsf{T}_\star$.

Hence the property $\tilde{\mathsf{A}}$ is equivalent to the following property A: if the spaces $\{Q_r(\mathsf{T}_m)\}_m$ converge in $\mathbb{R} \times \mathbb{R}^{2|\omega_r|'} \times \mathbb{R}^{n-|\omega_r|'}$ to the affine space $Q_r(\mathsf{T}_\star)$, and the sequence of planes $\{Q_r(\mathsf{L}_m)\}_m$ converges to a plane $Q_r(\mathsf{L}_\star) \subset \mathbb{R} \times \mathbb{R}^{2|\omega_r|'}$, then $Q_r(\mathsf{L}_\star) \subset Q_r(\mathsf{T}_\star)$.

By Corollary 6.7, $K_r(\mathcal{T}_r) \subset \mathbb{R}^{2|\omega_r|'} \times \mathbb{R}^{n-|\omega_r|'}$ is a Whitney space. Therefore, property A is valid. So the property $\tilde{\mathsf{A}}$ has been validated as well: $K(\mathcal{T}(v))$ is a Whitney stratified space in $\mathbb{R}^N$. □

Remark 6.9. It is desirable to find a more direct proof of Theorem 6.10, the proof that will validate Whitney's property $\tilde{\mathsf{B}}$ geometrically, without relying on the heavy general theorems claiming: "the semianalytic sets are Whitney spaces".

In fact, the discriminant varieties in $\mathbb{R}^d_{\mathsf{coef}}$ that correspond to various combinatorial patterns ω for real divisors of real d-polynomials, do have remarkable intersection patterns for their tangent spaces and cones (see [K11]). Perhaps, these properties of discriminant varieties should be in the basis of any "more geometrical" proof.

It seems that the statement of Theorem 6.10 may be valid for any traversing and boundary generic vector field v. The challenge is to find and validate an analogue of Lemma 6.15 for vector fields from the space $\mathcal{V}^\dagger(X) \supset \mathcal{V}^\ddagger(X)$. The semi-local models of such vector fields depend on some continuous parameters and not only on the combinatorial type of the core trajectory. $\diamondsuit$

Corollary 6.8. *Let X be an $(n+1)$-dimensional compact smooth manifold, carrying a traversally generic vector field v. Then the following claims are valid:*

- *The space of trajectories $\mathcal{T}(v)$ admits the structure of finite cell/simplicial complex.*
- *For each $\omega \in \Omega_{\bullet'\langle n]}$, the locus $\mathcal{T}(v, \omega_{\succeq_\bullet})$ is a codimension $|\omega|'$ subcomplex of $\mathcal{T}(v)$.*
- *With respect to an appropriate cellular/simplicial structure in X, the obvious map $\Gamma : X \to \mathcal{T}(v)$ is cellular/simplicial.*
- *Moreover, $\Gamma : X \to \mathcal{T}(v)$ is a homotopy equivalence.*

Proof. By Theorem 6.10, the trajectory space $\mathcal{T}(v)$ of a traversally generic flow admits a structure of a Whitney space, being embedded in some ambient Euclidean space.

The fundamental results of [Gor], [Jo], and [Ve] claim that the Whiney spaces Y admit smooth triangulations $\tau : T \to Y$, amenable to their stratifications. The adjective "smooth" here refers to the homeomorphism τ being smooth on the interior of each simplex Δ (remember, the pure strata $\mathcal{T}(v, \omega)$ are smooth manifolds!). With respect to such triangulations, the strata are subcomplexes.

Therefore $\mathcal{T}(v)$ admits a finite triangulation so that each locus $\mathcal{T}(v, \omega_{\succeq_\bullet})$ is a subcomplex.

For traversing fields v, over each open simplex $\Delta^\circ \subset \mathcal{T}(v)$, the map $\Gamma : X \to \mathcal{T}(v)$ is a trivial fibration whose fibers are either closed segments, or singletons. Thus each set $\Gamma^{-1}(\Delta^\circ)$ is homeomorphic either to the cylinder $\Delta^\circ \times [0,1]$, or to Δ°. This introduces a cellular structure on X so that Γ becomes a cellular map. With a bit more work, one can refine the cellular structures in X and $\mathcal{T}(v)$, so that Γ becomes a simplicial map.

Since, by Theorem 6.8, $\Gamma : X \to \mathcal{T}(v)$ is a weak homotopy equivalence and both spaces are CW-complexes, we conclude that Γ is a homotopy equivalence [White]. □

Remark 6.10. Although $\Gamma^\partial : \partial_1 X \to \mathcal{T}(v)$ is a finitely ramified covering of $\mathcal{T}(v)$, and $\partial_1 X$ is a smooth manifold (and thus admits a triangulation), it does not follow automatically that the trajectory space $\mathcal{T}(v)$ admits a triangulation! In fact, there are examples of smooth free involutions $\eta : S^4 \to S^4$ such that the quotient space S^4/η (homeomorphic to the standard $\mathbb{R}P^2$) is not triangulable [CS]! ◇

Remark 6.11. Most likely, $\mathcal{T}(v)$ is a compact CW-complex for any traversing and boundary generic (and not necessary *traversally* generic!) field v. However, for such fields, we do not have the "open book" algebraic models (as in Figure 6.4) for their interactions with boundary in the vicinity of a typical trajectory. So we do not know how to extend the previous arguments to vector fields $v \in \mathcal{V}^\dagger(X)$. ◇

6.5 The ball-based Origami theorem for traversally generic flows

Let X be an $(n+1)$-dimensional compact connected smooth manifold, carrying a traversally generic vector field v. Abusing notations, we use the same symbol "γ" for the v-trajectory in X and for the point in the trajectory space $\mathcal{T}(v)$ it represents.

We introduce a new filtration $\{\mathcal{T}^+_{\{max \geq k\}}(v)\}_{k\in[1,n+1]}$ of the trajectory space $\mathcal{T}(v)$ by closed subspaces (actually, by cellular subcomplexes). By definition, a trajectory $\gamma \in \mathcal{T}^+_{\{max\geq k\}}(v)$ if $\gamma \cap \partial_1^+ X$ contains at least one point x of multiplicity greater than or equal to k. Moreover, we insist that this $x \in \partial_k^+ X$ (and not in $\partial_k^- X^\circ$). In other words, $\mathcal{T}^+_{\{max\geq k\}}(v)$ is exactly the image of $\partial_k^+ X(v)$ in the trajectory space under the obvious map $\Gamma : X \to \mathcal{T}(v)$. In particular, $\mathcal{T}^+_{\{max \geq 1\}}(v) = \mathcal{T}(v)$. This stratification $\{\mathcal{T}^+_{\{max \geq k\}}(v)\}_{k \in [1, n+1]}$ is cruder than the stratification $\{\mathcal{T}(v, \omega_{\succeq \bullet})\}_{\omega \in \mathbf{\Omega}^{\bullet}_{'\langle n]}}$.

We notice that Corollary 5.7, Theorems 6.6 and 6.7 imply that there is an nonempty open subset $\mathcal{D}(X) \subset \mathcal{V}^{\ddagger}(X)$ such that, for each field $v \in \mathcal{D}(X)$, all the strata $\{\partial^+_j X\}_j$ are diffeomorphic to closed balls, except for $\partial^+_n X$ (which is a finite union of 1-balls) and for the finite set $\partial^+_{n+1} X$.

So let us consider a model filtration

$$Z^0 \subset Z^1 \subset D^2 \subset D^3 \subset \dots D^{n-1} \subset D^n \tag{6.47}$$

of a closed ball D^n, such that:

(1) each ball $D^j \subset \partial D^{j+1}$,

(2) Z^1 is a disjoint union of finitely many arcs in ∂D^2,

(3) $Z^0 \subset \partial Z^1$ is a finite set.

It turns out that, for $v \in \mathcal{D}(X)$, the trajectory space $\mathcal{T}(v)$ can be produced by an origami-like folding of the ball D^n (see Figure 2.15 and Figure 2.16 for an example of an origami map on a 2-ball).

The result below should be compared with Theorem 1 from [FR], claiming that a closed 3-manifold X has a spine which is the image of an immersed 2-sphere in general position in X. Theorem 6.11 should be compared also with Theorem 2.6, the latter dealing with the flow-generated spines, not trajectory spaces.

Theorem 6.11 (Trajectory spaces as the ball-based origami). *Any compact connected smooth $(n+1)$-manifold X with boundary admits a traversally generic vector field v such that:*

- *its trajectory space $\mathcal{T}(v)$ is the image of a closed ball D^n under a continuous cellular map $\Gamma : D^n \to \mathcal{T}(v)$ which is $(n+1)$-to-1 at most. A generic fiber of Γ is a singleton.*
- *The Γ-image in $\mathcal{T}(v)$ of each ball D^k from the filtration (6.47), is the space $\mathcal{T}^+_{\{max \geq n+1-k\}}(v)$, and the restriction $\Gamma|_{D^k}$ is a $\lceil \frac{n}{n-k} \rceil$-to-1 map at most.*

  For $n > 2$, the maps

  $$\Gamma : Z^1 \to \mathcal{T}^+_{\{max \geq n\}}(v) \;\text{ and }\; \Gamma : Z^0 \to \mathcal{T}^+_{\{max \geq n+1\}}(v)$$

  are both bijective.
- *The restrictions of Γ to $\partial D^{k+1} \setminus D^k$ are 1-to-1 maps for all $1 < k < n$, and so are the restrictions of Γ to $\partial D^2 \setminus Z^1$ and $\partial Z^1 \setminus Z^0$.*

- *The vector fields v, for which the above properties hold, form an open nonempty set in the space $\mathcal{V}^{\ddagger}(X)$ of traversally generic fields on X, and thus in the space $\mathcal{V}_{\mathsf{trav}}(X)$ of all traversing fields.*

Proof. If $\partial_1 X$ has several connected components, we pick one of them, say, $\partial_1^{\star} X$. The union of the remaining boundary components is denoted by $\partial_1^{\star\star} X$.

We can construct a Morse function $f : X \to \mathbb{R}$ so that it is locally constant on $\partial_1^{\star\star} X$, and these constants are the local maxima of f in a collar U of $\partial_1^{\star\star} X$ in X. Then by finger moves (as in the proof of Lemma 5.4) we eliminate all critical points of f without changing f in U. We pick a Riemmanian metric on X and let v be the gradient field of f. Evidently, $\partial_1^{\star\star} X \subset \partial_1^- X(v)$ and $\partial_1^+ X(v) \subset \partial_1^{\star} X$.

By an argument as in Corollary 5.7, in the vicinity of $\partial_1^{\star} X$, we can deform the field v to a new f-gradient-like field so that all the manifolds $\partial_1^+ X$, $\partial_2^+ X, \dots, \partial_{n-1}^+ X$, residing in the component $\partial_1^{\star} X$, will be diffeomorphic to balls, and $Z^1 =_{\mathsf{def}} \partial_n^+ X$ will consist of a number of arcs. The argument in Corollary 5.7 constructs such a v to be boundary generic in the sense of Definition 5.1. Moreover, by Theorems 6.6 and 6.7, we can further perturb v, without changing the smooth topological type of the Morse stratification $\{\partial_j^+ X(v)\}_j$, so that the new perturbation will be a traversally generic field. Abusing notations, we continue to denote the new field by v.

For $k \geq 2$, the locus $\partial_{n+1-k}^+ X(v)$ is diffeomorphic to the disk D^k. We notice that, for each point $x \in D^k$, the v-trajectory γ_x has at least one tangency point of multiplicity $n + 1 - k$, residing in $\partial_1^+ X$, namely x itself. Thus, for $k \geq 2$, Γ maps D_k onto $\mathcal{T}^+_{\{max \geq n+1-k\}}(v)$. Similarly, $\Gamma : Z^1 \to \mathcal{T}^+_{\{max \geq n\}}(v)$ and $\Gamma : Z^0 \to \mathcal{T}^+_{\{max \geq n+1\}}(v)$ are surjective maps.

We notice that, due to the convexity of the flow in their neighborhoods, the points of $\partial_j^- X$ are "protected" in the following sense: no v-trajectory can reach $\partial_2^- X(v)$, unless the trajectory is a singleton which belongs to $\partial_2^- X(v)$ in the first place. Similarly, no v-trajectory can reach $\partial_3^- X(v)$, unless the trajectory is a singleton $\partial_3^- X(v)$, and so on ... (see Figure 5.1). In particular, no v-trajectory through a point of $\partial_1^+ X(v)$ can reach $\partial_2^- X(v)$, unless the trajectory is a singleton which belongs to $\partial_2^- X(v)$ in the first place, no v-trajectory through a point of $\partial_2^+ X(v)$ can reach $\partial_3^- X(v)$, unless the trajectory is a singleton $\partial_3^- X(v)$, and so on ... The claim also follows from Theorem 6.2.

Therefore all the maps $\{\Gamma : \partial_j^- X(v) \setminus \partial_{j+1} X(v) \to \mathcal{T}(v)\}_j$ are 1-to-1. Thus the claim in the third bullet has been validated.

For a traversally generic v, by Corollary 7.2, the map $\Gamma : \partial_1 X \to \mathcal{T}(v)$ is $(n+2)$-to-1 at most. Since each trajectory, distinct from a singleton, must exit through $\partial_1^- X$ at a point of an odd multiplicity, the same argument shows that $\Gamma : \partial_1^+ X \to \mathcal{T}(v)$ is $(n+1)$-to-1 at most. Its generic fiber is a singleton.

For $v \in \mathcal{V}^\ddagger(X)$, the tangent spaces to $\partial_j^+ X^\circ$ along each trajectory γ must form, with the help of the flow, a stable configuration in the germ of a n-section S, transversal to γ (see Definition 6.3). Since $\dim(T_x(\partial_j^+ X)) = n+1-j$ for every point $x \in \gamma \cap \partial_j^+ X^\circ$, and the flow-generated images of the spaces $\{T_x(\partial_j^+ X)\}_{x \in \gamma \cap \partial_j^+ X^\circ}$ must be in general position in a n-dimensional space S, the cardinality of the set $\gamma \cap \partial_j^+ X^\circ$ cannot exceed $\lceil \frac{n}{j-1} \rceil = \lceil \frac{n}{n-k} \rceil$, provided $k < n$. The statement in second bullet has been established.

By the second bullet of Theorem 6.6, the smooth topological type of the stratification $\{\partial_j X(v)\}_j$ is stable under perturbations of v within the space $\mathcal{B}^\dagger(X)$ of boundary generic fields. The same argument shows that $\{\partial_j^+ X(v)\}_j$ is stable as well. Thus, for all fields v' sufficiently close to v, the stratification $\{\partial_j^+ X(v')\}_j$ will remain as in (6.47). By Theorem 6.7, all vector fields, sufficiently close to a traversally generic vector field, will remain traversally generic. Therefore this fact gives the desired control of the cardinality for the fibers of the maps $\Gamma : \partial_j^+ X(v') \to \mathcal{T}(v')$ and of the smooth topology of the stratification $\{\partial_j^+ X(v')\}_j$ within an open neighborhood of v in $\mathcal{V}^\ddagger(X)$. $\square$

Remark 6.12. Recall that the trajectory space $\mathcal{T}(v)$ in the Origami Theorem 6.11 is not only weakly homotopy equivalent to the manifold X (see Theorem 6.8), but also shares all the *stable characteristic classes* with it (Lemma 6.13). So all this information about X is hidden in a subtle way in *the geometry of the origami map* $\Gamma : D^n \to \mathcal{T}(v)$. $\diamondsuit$

The Origami Theorem 6.11 oddly resembles *The Noether Normalization Lemma* in the Commutative Algebra [No], however, with the direction of the ramified morphism being reversed. Recall that, in its algebra-geometric formulation, the Normalization Lemma states that any affine variety is a branched covering over an affine space. In contrast, in our setting, a massive class of trajectory spaces $\mathcal{T}(v)$—rather intricate objects—have a simple and *universal* resolution—the ball.

To explain this analogy, for a traversally generic field v, consider the Lie derivation $\mathcal{L}_v$ of the algebra $C^\infty(X)$ of smooth functions on X. Its kernel $C^\infty(\mathcal{T}(v))$[19] is a part of the long exact sequence of vector spaces:

$$0 \to C^\infty(\mathcal{T}(v)) \to C^\infty(X) \xrightarrow{\mathcal{L}_v} C^\infty(X) \to \dots,$$

where $C^\infty(\mathcal{T}(v))$, the algebra of smooth functions on the space of trajectories, can be identified with the algebra of all smooth functions on X that are constant along each v-trajectory.

Perhaps, $\mathcal{L}_v$ in the exact sequence is always a surjective map...

When a traversally generic v is such that $\partial_1^+ X(v)$ is diffeomorphic to D^n, then employing Theorem 6.11 and with the help of the finitely ramified surjective map

$$\Gamma_\partial : D^n = \partial_1^+ X(v) \subset X \xrightarrow{\Gamma} \mathcal{T}(v),$$

we get the induced monomorphism $\Gamma_\partial^* : C^\infty(\mathcal{T}(v)) \to C^\infty(D^n)$ of algebras. Here the target algebra $C^\infty(D^n)$ of smooth functions on the n-ball is *universal* for a given dimension n.

Any point-trajectory $\gamma \in \mathcal{T}(v)$ gives rise to the maximal ideal $\mathsf{m}_\gamma \lhd C^\infty(\mathcal{T}(v))$, comprising smooth functions on $\mathcal{T}(v)$ that vanish at γ. On the other hand, if $\mathsf{m} \lhd C^\infty(\mathcal{T}(v))$ is a maximal ideal and a function $h \in \mathsf{m}$ does not vanish on the compact $\mathcal{T}(v)$, then the function $1 = (\frac{1}{h}) \cdot h \in \mathsf{m}$, so that $\mathsf{m} = C^\infty(\mathcal{T}(v))$. Thus every maximal ideal $\mathsf{m} \lhd C^\infty(\mathcal{T}(v))$, distinct from the algebra itself, is of the form m_γ.

The map Γ^∂ is finitely ramified with fibers of cardinality $(n+1)$ at most (see Corollary 7.2). Therefore, for any maximal ideal $\mathsf{m} \lhd C^\infty(\mathcal{T}(v))$, its Γ^∂-induced image $(\Gamma^\partial)^*(\mathsf{m})$ is the intersection $\bigcap_i \mathsf{n}_i$ of $(n+1)$ maximal ideals $\mathsf{n}_i \lhd C^\infty(D^n)$ at most.

One can think of smooth vector fields on X as *derivatives* of the algebra $C^\infty(X)$. We denote the space of such operators by the symbol $\mathcal{D}(X)$.

Let $\mathcal{C}^+(X) \subset C^\infty(X)$ denote the open cone, formed by all strictly positive functions. The gradient-like fields v correspond to derivatives $\mathcal{L}_v \in \mathcal{D}(X)$ such that $\mathcal{L}_v(f) \in \mathcal{C}^+(X)$ for some $f \in C^\infty(X)$.

By Theorem 6.7, the traversally generic fields form a nonempty open set $\mathcal{V}^\ddagger(X)$ in the space of all vector fields and an open and dense set in the space of all traversing vector fields. Therefore, the previous considerations lead to the following reformulation of Theorem 6.11.

---

[19] $C^\infty(\mathcal{T}(v))$ is just a subalgebra of $C^\infty(X)$, not an ideal.

Corollary 6.9 (The origami resolutions $\mathbf{C^\infty(D^n)}$ for the kernels of special derivatives of the algebra $\mathbf{C^\infty(X)}$).

Let $C^\infty(D^n)$ denote the algebra of smooth functions on the n-ball.

For any $(n+1)$-dimensional smooth connected and compact manifold X with boundary, let us denote by $\mathcal{D}er(X)$ the space of derivatives $\mathcal{L} : C^\infty(X) \to C^\infty(X)$.

Then there exists an open nonempty subset $\mathcal{D}er^{\odot}(X) \subset \mathcal{D}er(X)$ that possess the following properties:

- *for each $\mathcal{L} \in \mathcal{D}er^{\odot}(X)$, there exists a function $f \in C^\infty(X)$ such that $\mathcal{L}(f) \in \mathcal{C}^+(X)$, the positive cone,*
- *for each $\mathcal{L} \in \mathcal{D}er^{\odot}(X)$, there exists a monomorphism of algebras*[20]
$$(\Gamma^\partial)^* : \ker(\mathcal{L}) \hookrightarrow C^\infty(D^n)$$
such that, for any maximal ideal $\mathsf{m} \lhd \ker(\mathcal{L})$, the image $(\Gamma^\partial)^*(\mathsf{m})$ is contained in an intersection $\bigcap_i \mathsf{n}_i$ of $n+1$ maximal ideals $\mathsf{n}_i \lhd C^\infty(D^n)$ at most. For each m, the maximal set $\{\mathsf{n}_i\}$ of such ideals is unique. $\diamondsuit$

We would like very much to learn the answer to the following question:

Question 6.3. *Let X be a compact connected smooth manifold with boundary. Let v be a traversing and boundary generic vector field on X.*

Describe the image under the restriction map, induced by the inclusion $\partial X \subset X$, of the algebra $\ker(\mathcal{L}_v)$ in the algebra $C^\infty(\partial X)$. $\diamondsuit$

6.6 A glimpse of holography

We devote this section to the fundamental phenomenon of the *holography* for the traversing flows. Crudely, we are concerned with the ability to reconstruct the manifold X and the traversing flow v (or rather, the 1-dimensional oriented foliation $\mathcal{F}(v)$, generated by v) on it in terms of some data, generated by the flow on the boundary ∂X.

The main result of this section, Theorem 6.12, describes some boundary data sufficient for a reconstruction of the pair $(X, \mathcal{F}(v))$, up to a homeomorphism. The reader interested in further developments of these ideas may glance at the papers [K4], [K7], and the forthcoming book [K13].

First, we introduce one basic construction (see Figure 6.6) which will be very useful throughout our investigations.

[20]that is induced by the origami map $\Gamma : D^n \to \mathcal{T}(v)$

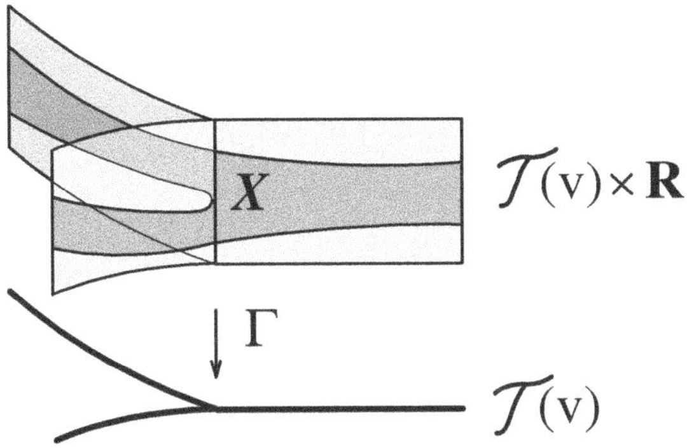

Fig. 6.6 The embedding $\alpha(f, v)$ of X into the product $\mathcal{T}(v) \times \mathbb{R}$.

Lemma 6.16. *Consider a traversing vector field v on a compact smooth connected manifold X with boundary and a Lyapunov function $f : X \to \mathbb{R}$, such that $df(v) > 0$.*

Any such pair (v, f) generates an embedding $\alpha(v, f) : X \subset \mathcal{T}(v) \times \mathbb{R}$, where $\mathcal{T}(v)$ denotes the trajectory space.

For any smooth map $\beta : \mathcal{T}(v) \to \mathbb{R}^N$, the composite map

$$A(v, f) : X \xrightarrow{\alpha} \mathcal{T}(v) \times \mathbb{R} \overset{\beta \times id}{\longrightarrow} R^N \times \mathbb{R}$$

is smooth.

Any two embeddings, $\alpha(f_1, v)$ and $\alpha(f_2, v)$, are isotopic through homeomorphisms, provided that $df_1(v) > 0, df_2(v) > 0$.

Proof. Since f is strictly increasing along the v-trajectories, any point $x \in X$ is determined by the v-trajectory γ_x through x and the value $f(x)$. Therefore, x is determined by the point $\gamma_x \times f(x) \in \mathcal{T}(v) \times \mathbb{R}$. By the definition of topology in $\mathcal{T}(v)$, the correspondence $\alpha(f, v) : x \to \gamma_x \times f(x)$ is a continuous map.

In fact, $\alpha(f, v)$ is a smooth map in the spirit of Definition 6.10: more accurately, for any map $\beta : \mathcal{T}(v) \to \mathbb{R}^N$, given by N smooth functions on $\mathcal{T}(v)$, the composite map $A(v, f) : X \to \mathbb{R}^N \times \mathbb{R}$ is smooth. The verification of this fact is on the level of definitions.

For a fixed v, the condition $df(v) > 0$ defines an open convex cone $\mathcal{C}^+(v)$ in the space $C^\infty(X)$. Thus, f_1 and f_2 can be linked by a path in $\mathcal{C}^+(v)$, which results in $\alpha(f_1, v)$ and $\alpha(f_2, v)$ being homotopic through homeomorphisms. □

Remark 6.13. By examining Figure 6.6 and Figure 1.7, A, we observe an interesting phenomenon: the embedding $\alpha : X \subset \mathcal{T}(v) \times \mathbb{R}$ does not extend to an embedding of a larger manifold $\hat{X} \supset X$, where $\hat{X} \setminus X \approx \partial_1 X \times [0, \epsilon)$. In other words, $\alpha(\partial_1 X)$ has no outward "normal field" in the ambient $\mathcal{T}(v) \times \mathbb{R}$. In that sense, $\alpha(\partial_1 X)$ is *rigid* in $\mathcal{T}(v) \times \mathbb{R}$! ◇

Corollary 6.10. *Any compact connected smooth* $(n+1)$*-manifold* X *with boundary admits an embedding* $\alpha : X \to \mathcal{T} \times \mathbb{R}$*, were* $\mathcal{T}$ *is a CW-complex that is the image of the* n*-ball* D^n *under a simplicial map, whose fibers are of the cardinality* $n+1$ *at most. Moreover,* α *is a homotopy equivalence.*

Proof. We combine the Origami Theorem 6.11 with Lemma 6.16 to validate the first claim of the lemma.

Since $\Gamma = p \circ \alpha$, where $p : \mathcal{T} \times \mathbb{R} \to \mathcal{T}$ is the obvious projection, and Γ is a homotopy equivalence by Theorem 6.8, so is the map α. □

Corollary 6.11. *Let* v *be a traversing vector field on a compact smooth connected manifold* X *with boundary and* $f : X \to \mathbb{R}$ *its Lyapunov function. Let* X° *denote the interior of* X.

Then the embedding

$$\alpha(f, v) : \partial_1 X \longrightarrow \big(\mathcal{T}(v) \times [0,1]\big) \setminus \alpha(f,v)(X^\circ)$$

is a homology equivalence. As a result, the space

$$\big(\mathcal{T}(v) \times [0,1]\big) \setminus \alpha(f,v)(X^\circ)$$

is a Poincaré complex of the formal dimension $\dim(X) - 1$.

Proof. Put $\alpha =_{\mathsf{def}} \alpha(f, v)$. Let us compare the homology long exact sequences of the two pairs:

$$X \supset \partial_1 X \text{ and } \mathcal{T}(v) \times [0,1] \supset \big(\mathcal{T}(v) \times [0,1]\big) \setminus \alpha(X^\circ).$$

They are connected by the vertical homomorphisms that are induced by α. Using the excision property,

$$\alpha_* : H_*(X, \partial_1 X) \to H_*\big(\mathcal{T}(v) \times [0,1],\ (\mathcal{T}(v) \times [0,1]) \setminus \alpha(X^\circ)\big)$$

are isomorphisms. On the other hand, since by Theorem 6.8, $\Gamma : X \to \mathcal{T}(v)$ is a homology equivalence, $\alpha_* : H_*(X) \to H_*(\mathcal{T}(v) \times [0,1])$ are isomorphisms. Therefore by the Five Lemma,

$$\alpha_* : H_*(\partial_1 X) \to H_*\big((\mathcal{T}(v) \times [0,1]) \setminus \alpha(X^\circ)\big)$$

must be isomorphisms as well. Since $\partial_1 X$ is a closed n-manifold, it is a Poincaré complex of formal dimension n, and thus so is the space

$$(\mathcal{T}(v) \times [0,1]) \setminus \alpha(f,v)(X^\circ).$$

□

We denote by $\mathcal{F}(v)$ the oriented 1-dimensional foliation on X, produced by the $\hat{v}$-trajectories.

Definition 6.13. Let v be a traversing vector field on X. Given two points $x, y \in \partial_1 X$, we write $y \succ_v x$ if both points belong to the same v-trajectory $\gamma \subset X$ and, moving from x in the v-direction along γ, we can reach y.

The relation $y \succ_v x$ introduces a *partial order* $\succ_v$ in the set $\partial_1 X$. ◇

Adding an extra ingredient to the partial order $\succ_v$ (equivalently, to the origami construction), allows for a reconstruction of the topological type of the pair $(X, \mathcal{F}(v))$ from the flow-generated information, *residing on the boundary* ∂X. The new ingredient is the restriction of the Lyapunov function $f : X \to \mathbb{R}$ to the boundary.

In a more qualitative formulation of the holography principle, for a typical ODE on a given compact manifold X, which admits a Lyapunov function, the solutions of the appropriately staged boundary value problems are topologically rigid.

Theorem 6.12 (Topological Holography of Traversing Flows). *Let a v be a traversing vector field v on a compact connected smooth manifold X with boundary, and let $f : X \to \mathbb{R}$ be its Lyapunov function.*

Then the partial order $\succ_v$ on $\partial_1 X$, together with the restriction $f^\partial : \partial X \to \mathbb{R}$ of a Lyapunov function $f : X \to \mathbb{R}$, allows for a reconstruction of the topological type of the pair $(X, \mathcal{F}(v))$[21].

Proof. The validation of the theorem is based on Lemma 6.16.

First, we observe that the partial order $\succ_v$ allows for a reconstruction of the trajectory space $\mathcal{T}(v)$ and the quotient map $\Gamma^\partial : \partial_1 X \to \mathcal{T}(v)$. Indeed, we declare two points $x, y \in \partial_1 X$ *equivalent* if $y \succ_v x$ or $x \succ_v y$. This equivalence relation $\sim_v$ produces the quotient map $\Gamma^\partial : \partial_1 X \to (\partial_1 X)/\sim_v$, whose target may be identified with the space $\mathcal{T}(v)$. This claim is valid

[21] With more effort, we can prove [K4] that the partial order $\succ_v$ *alone* is sufficient for the reconstruction of the topological type of the pair $(X, \mathcal{F}(v))$.

since, for a traversing v, every trajectory γ is determined by its intersection $\gamma \cap \partial_1 X$.

As in Lemma 6.16, using f^∂, we construct an embedding

$$\alpha^\partial = \alpha(v, f^\partial)| : \partial X \subset \mathcal{T}(v) \times \mathbb{R}.$$

Then $\alpha^\partial(\partial X)$ divides $\mathcal{T}(v) \times \mathbb{R}$ into two domains, one of which is compact. That compact domain $\mathcal{X}$ is $\alpha(v, f)(X)$. Since $\alpha(v, f) : X \to \mathcal{X}$ is a homeomorphism, we managed to reconstruct the topological type of X from the boundary data $(\succ_v, f^\partial)$ (in the end, from $(\Gamma^\partial, f^\partial)$).

Evidently, $\mathcal{X}$ is equipped with a 1-dimensional foliation $\mathcal{G}$, generated by the product structure in the ambient $\mathcal{T}(v) \times \mathbb{R}$.

By its construction, the homeomorphism $\alpha(v, f)$ maps each leaf of $\mathcal{F}(v)$ to a leaf of $\mathcal{G}$. Thanks to $\alpha(v, f)$, the pair $(\mathcal{X}, \mathcal{G})$, which we have recovered from the boundary data $(\succ_v, f^\partial)$, has the same topological type as the original pair $(X, \mathcal{F}(v))$.

Note that, for a given pair $(\succ_v, f^\partial)$, the homeomorphism $\alpha(v, f)$ is far from being unique. Even, for a fixed pair (X, v), we may vary the Lyapunov function f, while keeping f^∂ fixed. However, the space $\mathsf{Lyap}(v, f^\partial)$ of such Lyapunov functions f is convex, and thus contractible. Therefore, for any two $f_1, f_2 \in \mathsf{Lyap}(v, f^\partial)$, the embeddings $\alpha(v, f_1)$ and $\alpha(v, f_2)$ are homotopic through homeomorphisms that map $\mathcal{F}(v)$ to $\mathcal{G}$. □

Example 6.3. Let Y be a closed surface of genus g. Let X be obtained from Y by deleting an open 2-ball. Consider a traversally generic vector field v on the surface X together with its Lyapunov function $f : X \to \mathbb{R}$ (such a pair (v, f) does exist). Then Theorem 6.12 claims that the order relation $\succ_v$ on the circle ∂X, together with the function $f^\partial : \partial X \to \mathbb{R}$, is sufficient for a reconstruction of the pair $(X, \mathcal{F}(v))$, up to a homeomorphism that is the identity on ∂X. In particular, the genus g may be recovered from the boundary data $(\succ_v, f^\partial)$. ◊

Example 6.4. Let Y be a closed surface, equipped with a smooth vector field w. Consider the product $\hat{X} =_{\mathsf{def}} Y \times \mathbb{R}$. Let $\hat{v}$ be a vector field on $\hat{X}$, whose Y-component is the given w and whose $\mathbb{R}$-component is 1. Evidently, $\hat{v}$ admits a Lyapunov function $T : \hat{X} \to \mathbb{R}$, defined by the obvious projection.

Let $X \subset \hat{X}$ be a compact connected domain with a smooth boundary. Thanks to T, the $\hat{v}$-flow is traversing on the 3-fold X and thus defines the partial order $\succ_{\hat{v}}$ on the surface ∂X. We denote by v the restriction of $\hat{v}$ on X.

Then Theorem 6.12 claims that the order relation $\succ_v$ on the surface ∂X, together with the function $T^\partial : \partial X \to \mathbb{R}$, allows for a reconstruction of the pair $(X, \mathcal{F}(v))$, up to a homeomorphism that is the identity on ∂X.

If we treat T as time and Y as space, it is natural to interpret $\succ_v$ as *the causality relation* between the events-points on the boundary ∂X. $\diamondsuit$

Remark 6.14. The question whether the data $(\succ_v, f^\partial)$ (or even the data $\succ_v$ alone) are sufficient for a reconstruction of the *differentiable* or even *smooth* topological type of the pair $(X, \mathcal{F}(v))$ seems to be much more delicate. We suspect that the positive answer to it will depend on our ability to answer Question 6.3.

Under certain assumptions about v (such as some restrictions on the combinatorial types of v-trajectories), the answer is positive [K4]. $\diamondsuit$

Corollary 6.12. *Let a traversally generic vector field v on X be such that $\partial_1^+ X(v) \approx D^n$* [22]. *Then the origami map $\Gamma^\partial : D^n \to \mathcal{T}(v)$, together with the restriction $f_1^\partial : D^n \to \mathbb{R}$ of the Lyapunov function $f : X \to \mathbb{R}$, allow for a reconstruction of the topological type of the pair $(X, \mathcal{F}(v))$.*

Proof. To validate the claim of the theorem, we combine Theorems 6.11 and 6.12.

Given a traversing v and a function $f_+^\partial : \partial_1^+ X(v) \to \mathbb{R}$, let $\mathsf{Lyap}(v, f_+^\partial)$ be the space of Lyapunov functions $f : X \to \mathbb{R}$ such that $f|_{\partial_1^+ X(v)} = f_+^\partial$. Again, $\mathsf{Lyap}(v, f_+^\partial)$ is a convex contractible space.

We assume that the function f_+^∂ is known and is generated by *some* (unknown) $f \in \mathsf{Lyap}(v)$. By the properties of the Lyapunov function f, we may assume that f_+^∂ extends to a function $f^\partial : \partial_1 X \to \mathbb{R}$ so that, for any $x, y \in \partial_1 X$, $y \succ_v x$, the inequality $f(x) < f(y)$ is valid.

By Theorem 6.11, the image $\Gamma^\partial(D^n)$ of the Origami map Γ^∂ is the trajectory space $\mathcal{T}(v)$, and the fibers of Γ^∂ may be identified with the $(\sim_v)$-equivalence classes of points in D^n. Moreover, since v is traversally generic, by Theorem 6.10, $\mathcal{T}(v)$ is a compact CW-complex.

[22]By Theorem 6.12, such a vector field exists.

Now, as in Lemma 6.16, using f^∂, we construct an embedding $\alpha(v, f^\partial) : \partial X \subset \mathcal{T}(v) \times \mathbb{R}$. The rest of the argument is similar to the argument we used to prove Theorem 6.12. $\square$

The semi-local models of traversally generic flows, developed in this chapter, will form a foundation of our investigations in the next chapter of rich and universal combinatorics that governs such flows.

Chapter 7

Combinatorics of Tangency: the Stratified Spaces of Real Polynomials as Local Models for the Spaces of Trajectories

7.1 In this chapter ...

Employing the local analytic models from Chapter 6 for traversally generic flows on smooth $(n+1)$-manifolds X with boundary, we embark on a detailed and somewhat tedious study of *the universal combinatorics* of their tangency patterns with respect to the boundary ∂X. This combinatorics is captured by a universal poset $\mathbf{\Omega}^\bullet_{\prime\langle n]}$ which depends only on the dimension of X. It is intimately linked with the combinatorial patterns of real divisors of real univariate polynomials of degrees which do not exceed $2(n+1)$. Such patterns are elements of another interesting poset $\mathbf{\Omega}_{\langle 2n+2]}$ that describes the ways in which the real roots merge, divide, appear, and disappear under deformations of real polynomials. The space of real degree d polynomials $\mathcal{P}^d$ is stratified so that its pure strata are cells, labelled by the elements of the poset $\mathbf{\Omega}_{\langle d]}$. This cellular structure in $\mathcal{P}^d$ is interesting on its own right (see Theorem 7.1 and Theorem 7.2). Moreover, it helps to understand the *localized* structure of the trajectory spaces $\mathcal{T}(v)$ for traversally generic fields v, the main subject of Theorem 7.3 and Theorem 7.4.

7.2 Bifurcations of the real polynomial divisors: the universal combinatorics of tangency for traversing flows

In Chapter 6, we have seen an ample evidence that, for smooth traversally generic (or even traversing boundary generic) vector fields v on a compact smooth manifold X with boundary, the combinatorial patterns of tangency along v-trajectories resemble the combinatorics of real divisors, produced by real univariate polynomials of an even degree $d \leq 2\dim(X)$ (see Lemma 6.4). Now we are going to devote time to somewhat tedious investigations of this combinatorics.

For any polynomial $P(z)$ with real coefficients, we denote by $D_{\mathbb{R}}(P)$ its divisor in $\mathbb{R}$, and by $D_{\mathbb{C}}(P)$ its complex conjugation-invariant divisor in $\mathbb{C}$. Points in $\sup(D_{\mathbb{R}}(P))$, the support of $D_{\mathbb{R}}(P)$, inherit the natural order from $\mathbb{R}$. The set $\sup(D_{\mathbb{C}}(P)) \subset \mathbb{C}$ is invariant under the complex conjugation.

Let $\mathcal{D}_d$ denote the space of all divisors D in $\mathbb{R}$ of degree $|D| = d$. The space $\mathcal{D}_d$ can be identified with the d-th symmetric power $\mathsf{Sym}^d(\mathbb{R})$ of the real number line $\mathbb{R}$. Alternatively, $\mathcal{D}_d$ can be introduced as the domain Π_d in $\mathbb{R}^d$, given by the inequalities $x_1 \leq x_2 \leq \cdots \leq x_d$, imposed on the coordinates $(x_1, x_2, \dots, x_d)$.

Let $\mathbb{N}$ be the set of all natural numbers, and $\mathbb{Z}_+$ the set of all non-negative integers. Consider the set $\mathbf{\Omega}$ of all maps $\omega : \mathbb{N} \to \mathbb{Z}_+$ with finite support and such that $\omega(i) \neq 0$ implies $\omega(j) \neq 0$ for all $j < i$.

The real divisors $D \in \mathcal{D}_d$ have *combinatorial models* from $\mathbf{\Omega}$, represented by maps

$$\omega_D : \{1, 2, 3, \dots\} \to \{0, 1, 2, 3, \dots\},$$

where $\omega_D(i) \geq 1$ is the multiplicity of the i-th point in $\sup(D)$, so that $\sum_i \omega_D(i) = d$.

Consider a real polynomial P of degree d and its complex divisor $D_{\mathbb{C}}(P)$. There is $\epsilon > 0$ such that the ϵ-neighborhood U_ϵ of the support $\sup(D_{\mathbb{C}}(P)) \subset \mathbb{C}$ is a union of *disjoint* open disks. Then any real d-polynomial Q, sufficiently close to P, will have all its roots residing in U_ϵ.

The radial contraction of each disk from U_ϵ to its center commutes with the complex conjugation in $\mathbb{C}$ and defines a conjugation-equivariant *deformation* of the divisor $D_{\mathbb{C}}(Q)$ into the divisor $D_{\mathbb{C}}(P)$. This produces a deformation of Q to P, a curve Q_s, $s \in [0, 1]$, in the space of real polynomials of degree d. At all stages of this deformation, but the last one ($s = 1$), the divisor $D_{\mathbb{C}}(Q_s)$ has the same multiplicity pattern as the one of $D_{\mathbb{C}}(Q)$; moreover, the divisor $D_{\mathbb{R}}(Q_s)$ has the same $\mathbb{R}$-ordered multiplicity pattern as the one of $D_{\mathbb{R}}(Q)$.

Now imagine this deformation process from the viewpoint of an observer residing in $\mathbb{R}$, so that morphings of all the roots, residing in $\mathbb{C} \setminus \mathbb{R}$, are "invisible".

Deformations of P within the space of real polynomials of degree $\deg(P)$ change its real divisor $D_{\mathbb{R}}(P)$ by sequences of two *elementary operations* and their inverses:

(1) merging of two adjacent points from the support $\sup(D_{\mathbb{R}}(P))$ (their multiplicities add up);
(2) inserting a point of multiplicity 2 into the set $\mathbb{R} \setminus \sup(D_{\mathbb{R}}(P))$.

The second elementary operation corresponds to a pair of simple complex-conjugate roots merging at a point of $\mathbb{R} \subset \mathbb{C}$.

These operations have combinatorial analogues. We define the *elementary merge operation* $\mathsf{M}_j : \mathbf{\Omega} \to \mathbf{\Omega}$ by the formula

$$\begin{aligned} \mathsf{M}_j(\omega)(i) &= \omega(i) \ \text{ for } \ \text{all } i < j, \\ \mathsf{M}_j(\omega)(j) &= \omega(j) + \omega(j+1), \\ \mathsf{M}_j(\omega)(i) &= \omega(i+1) \ \text{ for } \ \text{all } i > j+1, \end{aligned} \tag{7.1}$$

where $\omega \in \mathbf{\Omega}$ and $1 \leq j \leq |\sup(\omega)|$.

We define the *elementary insert operation* $\mathsf{I}_j : \mathbf{\Omega} \to \mathbf{\Omega}$ by the formula

$$\begin{aligned} \mathsf{I}_j(\omega)(i) &= \omega(i) \ \text{ for } \ \text{all } i < j, \\ \mathsf{I}_j(\omega)(j) &= 2, \\ \mathsf{I}_j(\omega)(i) &= \omega(i-1) \ \text{ for } \ \text{all } i > j \geq 1, \end{aligned} \tag{7.2}$$

and

$$\begin{aligned} \mathsf{I}_0(\omega)(1) &= 2, \\ \mathsf{I}_0(\omega)(i) &= \omega(i-1), \text{for } \ \text{all } i > 1. \end{aligned} \tag{7.3}$$

We also allow for the merge of several adjacent points in $\sup(D_{\mathbb{R}}(P))$ and for the merge of groups of conjugate roots (of any multiplicity) at a point(s) of $\mathbb{R}$. However, we tend to view these morphings as compositions of the elementary operations $\{\mathsf{M}_j\}$ and $\{\mathsf{I}_j\}$.

Note that the *parity* of the degree $\deg(D_{\mathbb{R}}(P))$ is preserved under the merge and insert operations.

Guided by Definition 6.1, we introduce the combinatorial analogues of the quantities from that definition:

- the l_1-*norm* $|\omega|$ of ω by the formula $\sum_i \omega(i)$,[1]
- the *reduced norm* $|\omega|'$ of ω, by the formula $\sum_i(\omega(i) - 1)$,
- the *virtual multiplicity* $\mu(\omega)$ of ω, by the formula $\sum_i \lceil \omega(i)/2 \rceil$, where $\lceil \sim \rceil$ denotes the integral part of a real number.

Now we are in position to define, via the elementary merge and insert operations, a *partial order* "$\succ$" in the set $\mathbf{\Omega}$:

[1]This number represents $\deg(D_{\mathbb{R}}(P))$.

Definition 7.1. For $\omega_1, \omega_2 \in \mathbf{\Omega}$, we write $\omega_1 \succ \omega_2$ if ω_2 can be obtained from ω_1 by a sequence of merge operations $\{\mathsf{M}_j\}$ and insert operations $\{\mathsf{I}_j\}$ as in (7.1)-(7.3). ◇

Let $\mathbf{\Omega}_d \subset \mathbf{\Omega}$ denote the set of all ω's such that $|\omega| = d$. Let $\mathbf{\Omega}_{\langle d]} \subset \mathbf{\Omega}$ denote the set of all ω's such that $|\omega| \leq d$ and $|\omega| \equiv d$ (2). The set $\mathbf{\Omega}_{\langle d]}$ inherits its partial order from the ambient poset $(\mathbf{\Omega}, \succ)$.

For all our applications we will need only the case of an even d.

We may give an interpretation to the poset $(\mathbf{\Omega}, \succ)$ in the spirit of category theory. In this interpretation, the elements of $\mathbf{\Omega}$ become objects of some category $\underline{\mathbf{\Omega}}$ so that the relation $\omega_1 \succ \omega_2$ is transformed into the property $\mathrm{Mor}(\omega_1, \omega_2) \neq \emptyset$.

The following definition mimics the map of complex conjugation-invariant divisors $D_{\mathbb{C}}(Q) \to D_{\mathbb{C}}(P)$, induced by the radial retraction of the ϵ-neighborhood $U_\epsilon(\sup(D_{\mathbb{C}}(P)))$ to its core, $\sup(D_{\mathbb{C}}(P))$, as observed from within $\mathbb{R} \subset \mathbb{C}$.

Definition 7.2. For any two elements $\omega_1, \omega_2 \in \underline{\mathbf{\Omega}}$ we define the set $\mathrm{Mor}(\omega_1, \omega_2)$ as the set of maps $\alpha : \sup(\omega_1) \to \sup(\omega_2)$ such that:

(1) for each pair $i < i'$ in $\sup(\omega_1)$, $\alpha(i) \leq \alpha(i')$,
(2) $\sum_{j \in \alpha^{-1}(i)} \omega_1(j) \leq \omega_2(i)$ for all $i \in \sup(\omega_2)$[2],
(3) $\sum_{j \in \alpha^{-1}(i)} \omega_1(j) \equiv \omega_2(i) \mod 2$ for all $i \in \sup(\omega_2)$[3]. ◇

For example, $\mathrm{Mor}((121), (11)) = \emptyset$, while $\mathrm{Mor}((11), (121))$ consists of three injective maps.

Examining properties (1)-(3) above, we see that there is a natural pairing

$$\mathrm{Mor}(\omega_1, \omega_2) \times \mathrm{Mor}(\omega_2, \omega_3) \to \mathrm{Mor}(\omega_1, \omega_3),$$

defined by the composition of maps, provided $\mathrm{Mor}(\omega_1, \omega_2) \neq \emptyset$ and $\mathrm{Mor}(\omega_2, \omega_3) \neq \emptyset$.

Lemma 7.1. *In the category* $\underline{\mathbf{\Omega}}$, *the set* $\mathrm{Mor}(\omega_1, \omega_2) \neq \emptyset$, *if and only if,* $\omega_1 \succeq \omega_2$ *in the poset* $(\mathbf{\Omega}, \succ)$.

---

[2]This restriction is vacuous when $\alpha^{-1}(i) = \emptyset$.
[3]In particular, if $\alpha^{-1}(i) = \emptyset$, then $\omega_2(i) \equiv 0 \mod 2$.

Proof. Any elementary operation M_j in (7.1) gives rise to an element $\mu_j \in \mathrm{Mor}(\omega, \mathsf{M}_j(\omega))$ which maps the pair $j, j+1 \in \sup(\omega)$ to the single element $j \in \sup(\mathsf{M}_j(\omega))$; the rest of the elements in $\sup(\omega)$ are mapped bijectively by μ_j. Evidently, the properties (1)-(3) in Definition 7.2 are satisfied. Similarly, any elementary operation I_j in (7.2) gives rise to an element $\nu_j \in \mathrm{Mor}(\omega, \mathsf{I}_j(\omega))$ which maps $\sup(\omega)$ bijectively and in a monotone fashion to $\sup(\mathsf{I}_j(\omega))$ so that $j \in \sup(\mathsf{I}_j(\omega))$ is the only element that is not in the image of ν_j. Again, the properties (1)-(3) in Definition 7.2 are satisfied.

Therefore if two elements, ω_1 and ω_2, are linked by a sequence of elementary operations of the types M_j and I_k, then there exists an element $\alpha \in \mathrm{Mor}(\omega_1, \omega_2)$ which is obtained by composing the chain of corresponding maps μ_j and ν_k. Hence, $\mathrm{Mor}(\omega_1, \omega_2) \neq \emptyset$.

On the other hand, if $\mathrm{Mor}(\omega_1, \omega_2) \neq \emptyset$, then any $\alpha \in \mathrm{Mor}(\omega_1, \omega_2)$ can be obtained in such a way via elementary operations. Indeed, for each $i \in \sup(\omega_2)$, consider the set $\alpha^{-1}(i)$. Put

$$\alpha_i =_{\mathsf{def}} \omega_2(i) - \sum_{j \in \alpha^{-1}(i)} \omega_1(j).$$

By Definition 7.2, $\alpha_i \geq 0$ and $\alpha_i \equiv 0 \mod 2$. We apply a sequence of $\alpha_i/2$ insert operations I_k to ω_1, localized to the set $\alpha^{-1}(i)$. They will add $\alpha_i/2$ copies of "2"s to the sequence ω_1. The location of these "2"s relative to the elements of $\alpha^{-1}(i)$ is unimportant. The resulting ω_1' now has the property $\mathrm{Mor}(\omega_1, \omega_1') \neq \emptyset$. These insertions $\{\mathsf{I}_k\}$ define a map $\gamma \in \mathrm{Mor}(\omega_1, \omega_1')$. Next we merge all the original elements of the set $\alpha^{-1}(i)$ and the locations of newly inserted "2"s together into a singleton $i_\star$ by a sequence of elementary merges $\{\mathsf{M}_j\}$. Again, the order in which the elementary merges are performed is unimportant. The resulting ω_1'' is such that there exists $\delta \in \mathrm{Mor}(\omega_1, \omega_1'')$ with the property

$$\sum_{j \in \delta^{-1}(i_\star)} \omega_1(j) = \omega_1''(i_\star) = \omega_2(i).$$

Finally, we apply this procedure to every element $i \in \sup(\omega_2)$. The result of these operations transforms ω_1 into ω_2 by a sequence of elementary operations. Therefore, $\omega_1 \succ \omega_2$ in the poset $\mathbf{\Omega}$. □

The polynomial inequality $P(z) \leq 0$ splits the support $\sup(D_{\mathbb{R}}(P))$ into a number of disjoint sets: each set is formed by the maximal string of consecutive roots so that, in the closed interval, bounded by the maximal and the minimal root from the string, the inequality $P(z) \leq 0$ is valid. For a polynomial of an even degree, each maximal string of roots either is a singleton whose multiplicity is even, or a sequence whose maximal and minimal elements have odd multiplicities, while the rest of roots have even multiplicities. These observations motivate the following combinatorial models.

Definition 7.3. Let $\mathbf{\Omega}^\bullet$ denote the set of maps $\omega \in \mathbf{\Omega}$ that satisfy the following properties:

- either
  (1) $\omega(1)$, $\omega(q)$ are odd numbers, where $q = |\sup(\omega)|$, and
  (2) $\omega(i)$ is even for $1 < i < q$;
- or $\sup(\omega) = \{1\}$, and $\omega(1) \equiv 0 \mod 2$. ◊

A priori q, the cardinality of the support of $\omega \in \mathbf{\Omega}^\bullet$, is not fixed.

Definition 7.4. Let $\mathbf{\Omega}^\bullet_{\prime d}$ denote the set of maps $\omega \in \mathbf{\Omega}^\bullet$ such that

$$|\omega|' =_{\mathsf{def}} \sum_i (\omega(i) - 1) = d.$$

Let

$$\mathbf{\Omega}^\bullet_{\prime [k,d]} =_{\mathsf{def}} \coprod_{k \leq j \leq d} \mathbf{\Omega}^\bullet_{\prime j}.$$

We also will use the shorter notation "$\Omega^\bullet_{\prime\langle d]}$" for the set $\mathbf{\Omega}^\bullet_{\prime [0,d]}$. ◊

Recall that on a $(n+1)$-manifold X, any trajectory γ of a boundary generic field $v \in \mathcal{V}^\dagger(X)$ gives rise to an element $\omega_\gamma \in \mathbf{\Omega}^\bullet$ according to the rule: $\omega_\gamma(i) = m(a_i)$, the multiplicity of the i-th point a_i in the v-ordered set $\gamma \cap \partial_1 X$[4]. By Theorem 6.7, for any traversally generic field $v \in \mathcal{V}^\ddagger(X)$, we get $|\omega_\gamma|' \leq n = \dim(\partial_1 X)$, so that $\omega_\gamma \in \mathbf{\Omega}^\bullet_{\prime\langle n]}$.

In the end of the proof of Theorem 6.7 we have established the following proposition:

Lemma 7.2. *For* $\omega \in \mathbf{\Omega}^\bullet_{\prime\langle n]}$, *we have* $|\omega| =_{\mathsf{def}} \sum_i \omega_\gamma(i) \leq 2n + 2$. ◊

We are going to introduce a partial order among the elements of the set $\mathbf{\Omega}^\bullet_{\prime\langle d]}$, which will match the changing geometry of trajectories for traversally generic fields. Crudely, the order is induced from the ambient poset $\mathbf{\Omega} \supset \mathbf{\Omega}^\bullet_{\prime\langle d]}$, but then enhanced. For a more accurate description, we turn to few auxiliary combinatorial constructions.

[4]That is, $a_i \in \partial_{m(a_i)} X(v)^\circ$.

Given any $\omega \in \mathbf{\Omega}$, we would like to "chop it" into a number of "strings" and "atoms": each string belongs to some $\mathbf{\Omega}'^{\bullet}_{d}$ (as in the first bullet of the Definition 7.3), while each atom has a singleton for its support and takes there an even value (see Figure 7.1). Prior to defining this canonical deconstruction $\Xi(\omega)$ of ω, we need to introduce some notations.

Let $\omega \in \mathbf{\Omega}$ be such that $|\omega| \equiv 0 \mod 2$. Denote by $\sup_{odd}(\omega)$ the points $l \in \mathbb{N}$ in the support of ω such that $\omega(l) \equiv 1 \mod 2$, and by $\sup_{ev}(\omega)$ the points l in the support of ω such that $\omega(l) \equiv 0 \mod 2$. (In the geometric context of traversing flows, the number $|\sup_{odd}(\omega)|$ is even.) We count the elements of $\sup_{odd}(\omega)$ as they appear in the list; some of them acquire odd numerals, others acquire even ones. A *string* in $A \subset \sup(\omega)$ is formed by all the elements that are bounded on the left by an element of $i \in \sup_{odd}(\omega)$ with an odd numeral, and on the right by the next element $j \in \sup_{odd}(\omega)$ (it has an even numeral attached to it). In other words, a string $A =_{\mathsf{def}} [i, j]$ is formed by such i, j and all the elements from $\sup_{ev}(\omega)$ that lie in-between. The points from $\sup_{ev}(\omega)$ that do not belong to any string are called "*atoms*". We can compute the values of ω at the elements of each string A as well as at each atomic support. This gives rise to a unique ordered sequence $\Xi(\omega)$ of strings, interrupted by a number of atoms.

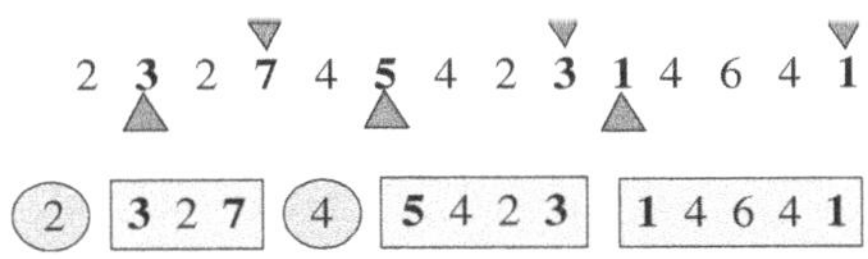

Fig. 7.1 An example of the deconstruction $\omega \Rightarrow \Xi(\omega)$. The strings are boxed, the atoms are circled.

Definition 7.5. We define the partial order "$\prec_{\bullet}$" in $\mathbf{\Omega}^{\bullet}$ as follows: for $\omega_1, \omega_2 \in \Omega^{\bullet}$, we write "$\omega_1 \prec_{\bullet} \omega_2$" if ω_2 occurs as a string or an atom from the list $\Xi(\omega)$ for some $\omega \succ \omega_1$, where $\omega \in \mathbf{\Omega}$ and the last ordering is considered in the poset $(\mathbf{\Omega}, \succ)$. $\diamondsuit$

Example 7.1. Consider the poset $(\mathbf{\Omega}'^{\bullet}_{[0,3]}, \prec_{\bullet})$, shown in Figure 7.2. Then $(1, 4, 1) \prec_{\bullet} (3, 1)$ since $(3, 1)$ is present as a string in $(1, 1, 3, 1) \succ (1, 4, 1)$, the order "$\succ$" being the one from the poset $\mathbf{\Omega}_{[0,6]}$. $\diamondsuit$

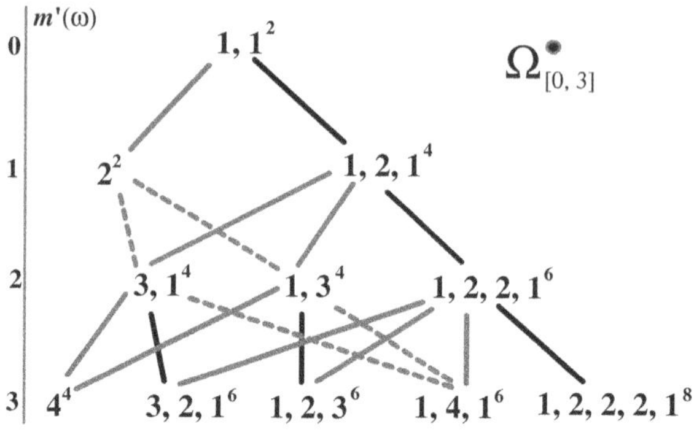

Fig. 7.2 The poset $(\mathbf{\Omega}^\bullet_{'[0,3]}, \prec_\bullet)$ with the "height" function $m'(\omega) = |\omega|'$. The bold dark lines indicate the insert operations, the lighter bold lines the merge operations. The dotted lines represent the order relations, as in Definition 7.5, that are not directly induced from the poset $\mathbf{\Omega}_{[0,8]} \supset \mathbf{\Omega}^\bullet_{'[0,3]}$. The upper index next to each ω shows the value of its norm $|\omega|$.

Lemma 7.3. *For any d, the sub-poset of $\mathbf{\Omega}^\bullet_{'\langle d]}$, defined by the constraint $|\omega|' < d$, is canonically isomorphic to the poset $\mathbf{\Omega}^\bullet_{'\langle d-1]}$. As a result, the direct limit*

$$\lim_{d \to +\infty} \mathbf{\Omega}^\bullet_{'\langle d]} = \mathbf{\Omega}^\bullet$$

as posets.

Proof. Let $\omega_1 \succ_\bullet \omega_2$ in $\mathbf{\Omega}^\bullet_{'\langle d]}$. This implies that ω_1 is string or an atom in some $\omega \succ \omega_2$ in $(\mathbf{\Omega}, \succ)$. If $|\omega_2|' < d$, then $|\omega|' < d$ as well. In turn, $|\omega|' < d$ implies that $|\omega_1|' < d$. Therefore $\omega_1 \succ_\bullet \omega_2$ in $\mathbf{\Omega}^\bullet_{'\langle d-1]}$ as well. □

Similar to the poset $(\mathbf{\Omega}, \succ)$, the poset $(\mathbf{\Omega}^\bullet, \succ_\bullet)$ allows for an interpretation in terms of a category $\underline{\mathbf{\Omega}}^\bullet$ whose objects are elements of $\mathbf{\Omega}^\bullet$. Compare the next definition with Definition 7.2; in particular, note the absence in Definition 7.6 of constraint (3) from Definition 7.2.

Definition 7.6. For any two elements $\omega_1, \omega_2 \in \underline{\mathbf{\Omega}}^\bullet$ we define $\mathrm{Mor}^\bullet(\omega_1, \omega_2)$ as the set of maps $\alpha : \sup(\omega_1) \to \sup(\omega_2)$ such that:

(1) for each pair $i < i'$ in $\sup(\omega_1)$, $\alpha(i) \leq \alpha(i')$,

(2) $\sum_{j \in \alpha^{-1}(i)} \omega_1(j) \leq \omega_2(i)$ for all $i \in \sup(\omega_2)$[5]. ◊

[5]This restriction is vacuous when $\alpha^{-1}(i) = \emptyset$.

Lemma 7.4. *In the category* $\underline{\mathbf{\Omega}}^\bullet$, *the set* $\mathrm{Mor}^\bullet(\omega_1, \omega_2) \neq \emptyset$, *if and only if,* $\omega_1 \succeq_\bullet \omega_2$ *in the poset* $(\mathbf{\Omega}^\bullet, \succ_\bullet)$.

Proof. If $\omega_1 \succeq_\bullet \omega_2$, then by the definition of the order $\succ_\bullet$, there exists an element $\tilde{\omega}_1 \in \mathbf{\Omega}$ such that $\tilde{\omega}_1 \succ \omega_2$, and ω_1 is a string or an atom in $\tilde{\omega}_1$. By Lemma 7.1, $\mathrm{Mor}(\tilde{\omega}_1, \omega_2) \neq \emptyset$. When we restrict a map $\tilde{\alpha} \in \mathrm{Mor}(\tilde{\omega}_1, \omega_2)$ to the subset $\sup(\omega_1) \subset \sup(\tilde{\omega}_1)$ we produce a map $\alpha : \sup(\omega_1) \to \sup(\omega_2)$. Under this restriction, the properties (1) and (2) from Definition 7.2, valid for $\tilde{\alpha}$, are evidently preserved for its restriction α: the support of a string consists of consecutive indices, and $\tilde{\alpha}^{-1}(i) \supset \alpha^{-1}(i)$ for all $i \in \sup(\omega_2)$, while the property (3) may be violated. Thus $\alpha \in \mathrm{Mor}^\bullet(\omega_1, \omega_2)$, a nonempty set.

On the other hand, if $\mathrm{Mor}^\bullet(\omega_1, \omega_2) \neq \emptyset$, then it is possible to construct $\tilde{\omega}_1 \in \mathbf{\Omega}$ such that $\mathrm{Mor}(\tilde{\omega}_1, \omega_2) \neq \emptyset$. Indeed, consider some $\alpha : \sup(\omega_1) \to \sup(\omega_2)$ having properties (1) and (2) from Definition 7.6. If, in addition, property (3) from Definition 7.2 holds, we are done. Note that this parity property can only be violated if one or both ends $\{1\}, \{q\}$ in the support of the string ω_1 are mapped by α to the location $\alpha(1)$ or $\alpha(q)$ such that $\omega_2(\alpha(1)) \equiv 0 \mod 2$, or $\omega_2(\alpha(q)) \equiv 0 \mod 2$.

Let us consider the case $\omega_2(\alpha(1)) \equiv 0 \mod 2$. The monotonicity of α implies that the minimum element $\{1_\star\}$ in the support of ω_2 is not in the image of α. We can append a string $\omega' = (11)$ below (to the left) of the string ω_1 to form a new element $\tilde{\omega}_1 \in \mathbf{\Omega}$ (not a string!) so that $\sup(\tilde{\omega}_1) = \sup((11)) \coprod \sup(\omega_1)$. Then we extend the map α to a map $\tilde{\alpha} : \sup((11)) \coprod \sup(\omega_1) \to \sup(\omega_2)$ by sending the first element of the extended support to the minimal element $\{1_*\} \in \sup(\omega_2)$, and the second and the third elements both to $\alpha(1)$ (note that the third element comes from the element $\{1\} \in \sup(\omega_1)$). This will repair the parity defect of the original α. A similar treatment applies when $\omega_2(\alpha(q)) \equiv 0 \mod 2$. Thus, $\mathrm{Mor}^\bullet(\omega_1, \omega_2) \neq \emptyset$ implies that $\mathrm{Mor}(\tilde{\omega}_1, \omega_2) \neq \emptyset$.

By Lemma 7.1, we get that $\tilde{\omega}_1 \succ \omega_2$ in $(\mathbf{\Omega}, \succ)$. Therefore, $\mathrm{Mor}^\bullet(\omega_1, \omega_2) \neq \emptyset$ implies that $\omega_1 \succ_\bullet \omega_2$. □

The following lemma can be viewed as an a posteriori justification for introducing the poset $(\mathbf{\Omega}^\bullet, \succ_\bullet)$. For a boundary generic vector field $v \in \mathcal{V}^\dagger(X)$ and its trajectory γ, in what follows, we view the set $\gamma \cap \partial_1 X$, together with the multiplicities attached to its points, as a divisor D_γ on the trajectory γ.

Lemma 7.5. *Let $v \in \mathcal{V}^\dagger(X)$. If a sequence $\{x_k \in X\}_k$ converges to a point x, then the reduced norms of the tangency divisors (the reduced multiplicities of the corresponding trajectories) have the property*

$$|D_{\gamma(x)}|' \geq \overline{\lim}_{k\to+\infty} |D_{\gamma(x_k)}|'.$$

In other words, the reduced norm $|D_{\gamma(\sim)}|'$ is an upper semi-continuous function on X. Moreover, if the divisors $\{D_{\gamma(x_k)}\}_k$ all share a combinatorial type ω, and $\omega_1 \neq \omega$ is the combinatorial type of $D_{\gamma(x)}$, then we get:

$$\omega \succ_\bullet \omega_1, \quad |\omega_1| \geq |\omega|, \textit{ and } |\omega_1|' > |\omega|'.$$

Proof. The trust of the argument is that any merge or insert operations (see (7.1)-(7.3)) that change the combinatorics of the tangency divisors D_γ cannot *decrease* the multiplicity $m(\gamma)$, as well as the reduced multiplicity $m'(\gamma)$ (defined in (6.8)-(6.10)). In fact, any such operation does increase $m'(\gamma)$.

In the argument below, we use our understanding of local models (6.14) (see Lemma 6.1) of vector fields $v \in \mathcal{V}^\dagger(X)$ in the vicinity of a given point from $\partial_1 X$. They imply that, as we vary γ, its divisor D_γ (with the support $\gamma \cap \partial_1 X$) can only change locally as the real divisors of the family of u-polynomials in (6.14) do: that is, via resolutions that either preserve their degree, or drop it by an even number. By the same token, as $\{x_k\}$ converge to x, in the vicinity of x, the divisor $D_{\gamma(x)}$ is obtained from $\{D_{\gamma(x_k)}\}$ only via: (1) merging adjacent distinct points from $\sup(D_{\gamma(x_k)})$, or (2) by inserting new points of even multiplicity, or (3) by increasing the odd multiplicity of the end point from $\sup(D_{\gamma(x_k)})$ by an odd number (this happens when the ends of two distinct trajectories merge at x).

Let us take a closer look at these mechanisms. By Lemma 6.1, for a boundary generic v and all its trajectories γ, the multiplicity of each point from $\gamma \cap \partial_1 X$ does not exceed $n+1$, where $n = \dim(\partial X)$. Moreover, by the same lemma and a compactness argument, no trajectory γ has an infinite set $\gamma \cap \partial_1 X$. Thus the multiplicity $m(\gamma) < \infty$ for each γ.

Consider any infinite set of trajectories $\{\gamma_k\}_k$ such that the set of multiplicities $\{m(\gamma_k)\}_k$ is unbounded. Since the multiplicity of each point of tangency is uniformly bounded, this implies that the set

$$A =_{\mathsf{def}} \bigcup_{1\leq k<\infty} (\gamma_k \cap \partial_1 X)$$

is infinite and $\lim_{k\to\infty} |\gamma_k \cap \partial_1 X| = \infty$. Let y be a limit point for A. Consider the trajectory γ_y through y. Using the compactness of γ_y and the local models from Lemma 6.1, γ_y has a neighborhood in which any trajectory $\tilde{\gamma}$ has a uniformly bounded cardinality of the set $\tilde{\gamma} \cap \partial_1 X$. This contradicts to the assumption that $\lim_{k\to\infty} |\gamma_k \cap \partial_1 X| = \infty$. Therefore there is a number d so that, for every trajectory γ in X, its combinatorial type belongs to the set $\mathbf{\Omega}_{\langle d]}$.

Consider the combinatorial types of $\{D_{\gamma(x_k)}\}_k$, a finite set. At least one of them, say ω, must occur infinitely many times in a subsequence $\{D_{\gamma(x_{k_i})}\}_i$ of $\{D_{\gamma(x_k)}\}_k$. For each $j \in \sup(\omega)$, consider the sequence of points $\{y_{j,i} \in \gamma(x_{k_i})\}_i$ which occupy the j-th position in $\gamma(x_{k_i}) \cap \partial_1 X$. By its definition, $y_{j,i} \in \partial_{\omega(j)} X^\circ$. Since $\partial_{\omega(j)} X$ is compact, there exists a subsequence $\{y_{j,i_l}\}_l$ that converges to a point $y_j \in \partial_{\omega(j)} X$. Its multiplicity, $m(y_j)$, is $\omega(j)$ at least.

Consider the segments $\gamma_{j,j+1;i}$ of $\gamma(x_{k_i})$ bounded by the adjacent pairs $(y_{j,i}, y_{j+1,i})$. The interiors of the segments do not intersect the boundary $\partial_1 X$. By the continuous dependence of v-integral curves on their end points, $\{\gamma_{j,j+1;i_l}\}_l$ converge to a trajectory segment $\gamma_{j,j+1}$ that connects y_j with y_{j+1} (that segment may contain "new" points from $\partial_1 X$ in its interior). As a result, all the y_j's belong to the same trajectory γ. Since each $x_{k_{i_l}}$ belongs to some segment $\gamma_{j,j+1;i_l}$ and $\lim_{l\to+\infty} x_{k_{i_l}} = x$, we get $\gamma(x) = \gamma$.

By this argument, $\sup(D_{\gamma(x)})$ may contain new points that are not in the limit of $\bigcup_l \sup(D_{\gamma(x_{k_{i_l}})})$. This can only boost the reduced multiplicity $|D_{\gamma(x)}|'$ in comparison with $|D_{\gamma(x_{k_{i_l}})}|'$. On the other hand, some points in $\sup(D_{\gamma(x)})$ are the result of merging of two or several consecutive points $\{y_{j,i_l}\}_j$ from $\gamma(x_{k_{i_l}})$. In the process, thanks to the local models (6.14), their multiplicities add; so again, the reduced multiplicity of the limiting point exceeds the sum of the reduced multiplicities of the corresponding merging points. Therefore,

$$m'(\gamma(x)) \geq \overline{\lim}_{k\to+\infty}\, m'(\gamma(x_k)).$$

Moreover, if the combinatorial type ω of $\gamma(x_{k_i}) \cap \partial_1 X$ differs from the combinatorial type ω' of $\gamma(x) \cap \partial_1 X$, then

$$m'(\gamma(x)) > \overline{\lim}_{k\to+\infty}\, m'(\gamma(x_k)).$$

Similarly, $m(\gamma(x)) \geq \overline{\lim}_{k\to+\infty}\, m(\gamma(x_k))$.

The same argument shows that the relation between the types ω and ω' is exactly the order relation $\omega \succ_\bullet \omega'$ in $\mathbf{\Omega}^\bullet$, as introduced in Definition 7.5. □

7.3 On stratified spaces

Let $(\mathcal{S}, \succ)$ be a *poset* (a partially ordered set). Given a subset $\Theta \subset \mathcal{S}$ in the poset $(\mathcal{S}, \succ)$, we denote by $\Theta_{\succeq}$ the set

$$\{a \in \mathcal{S} |\ b \succeq a \text{ for some } b \in \Theta\}.$$

Let $\Theta_{\succ} =_{\mathsf{def}} \Theta_{\succeq} \setminus \Theta$. Similarly, we introduce

$$\Theta_{\preceq} =_{\mathsf{def}} \{a \in \mathcal{S} |\ a \succeq b \text{ for some } b \in \Theta\} \text{ and } \Theta_{\prec} =_{\mathsf{def}} \Theta_{\preceq} \setminus \Theta.$$

In particular, for any $\omega \in \mathbf{\Omega}^\bullet$, we will consider routinely the finite posets $\omega_{\succeq_\bullet}$, $\omega_{\succ_\bullet}$, and, for any $\omega \in \mathbf{\Omega}$, the finite posets $\omega_{\succeq}$, $\omega_{\succ}$.

Definition 7.7. A $\mathcal{S}$-*stratification* of a topological space X is a collection of closed topological subspaces $\{X_{\preceq\omega}\}_{\omega\in\mathcal{S}}$ such that $\omega \prec \omega'$ implies the inclusion $X_{\preceq\omega} \subset X_{\preceq\omega'}$. Moreover, we require that each point $x \in X$ which belongs to some stratum $X_{\preceq\omega}$, $\omega \in \mathcal{S}$, belongs to a *unique* smallest stratum $X_{\preceq\omega(x)}$ labelled by a *minimal element* $\omega(x)$ from the poset $\{\preceq \omega\} \subset \mathcal{S}$. ♢

When X itself is a member of the collection $\{X_{\preceq\omega}\}_{\omega\in\mathcal{S}}$, we assume that $\mathcal{S}$ has a unique maximal element $\omega_\star$ and that $X_{\preceq\omega_\star} = X$.

For an $\mathcal{S}$-stratified space X, we define

$$X_{\prec\omega} =_{\mathsf{def}} \bigcup_{\omega' \prec \omega} X_{\preceq\omega'} \text{ and } X_\omega =_{\mathsf{def}} X_{\preceq\omega} \setminus X_{\prec\omega}.$$

In general, for any subset $\Theta \subset \mathcal{S}$, put $X_{\preceq\Theta} =_{\mathsf{def}} \cup_{\omega\in\Theta} X_{\preceq\omega}$.

Definition 7.8. An $\mathcal{S}$-*stratified map* $f : X \to Y$ between $\mathcal{S}$-stratified spaces X and Y is a continuous map such that $f(X_{\preceq\omega}) \subset Y_{\preceq\omega}$ for each $\omega \in \mathcal{S}$.

A $\mathcal{S}$-*stratified homotopy* between $\mathcal{S}$-stratified maps $f_0 : X \to Y$ and $f_1 : X \to Y$ is an $\mathcal{S}$-stratified map $F : X \times [0,1] \to Y$ such that $F(x,0) = f_0(x)$ and $F(x,1) = f_1(x)$ for all $x \in X$. ♢

7.4 Spaces of real univariate polynomials, stratified by the combinatorial types of their real divisors

Next, we would like to investigate one natural stratification in the coefficient space $\mathbb{R}^d_{\mathsf{coef}}$ of real monic polynomials $P(z)$ of a given degree d. This stratification is generated by different combinatorial patterns of zero divisors $D_{\mathbb{R}}(P)$ in $\mathbb{R}$. Its importance for our program is justified by the local models for boundary generic and traversally generic fields that have been described in Lemmas 6.1 and 6.4, respectively.

Let $\tau : \mathbb{C} \to \mathbb{C}$ denote the complex conjugation. Let $\mathsf{Sym}^d(Y)$ stand for the d-th symmetric power $\underbrace{Y \times \cdots \times Y}_{d} / \mathsf{S_d}$ of a given space Y, where $\mathsf{S_d}$ denotes the symmetric group of order d.

The complex Viète map

$$V : \mathbb{C}^d_{\mathsf{root}} =_{\mathsf{def}} \mathsf{Sym}^d\mathbb{C} \to \mathbb{C}^d_{\mathsf{coef}}$$

is defined by the elementary symmetric polynomials $\sigma_1, \dots, \sigma_d$ in the coordinates of the *root space* $\mathbb{C}^d_{\mathsf{root}}$. These elementary symmetric polynomials serve as coordinates in the *coefficient space* $\mathbb{C}^d_{\mathsf{coef}}$.

Via the Viète Map V, the coefficient space $\mathbb{R}^d_{\mathsf{coef}}$ may be identified with the space $(\mathsf{Sym}^d\mathbb{C})^\tau$ of τ-invariant divisors in $\mathbb{C}$ of degree d.

Each function $\omega : \mathbb{N} \to \mathbb{Z}_+$ of the l_1-norm $|\omega| \leq d$ and such that $|\omega| \equiv d$ mod 2 defines a pure stratum $(\mathsf{Sym}^d\mathbb{C})^\tau_\omega$ in $(\mathsf{Sym}^d\mathbb{C})^\tau$ and therefore, in $\mathbb{R}^d_{\mathsf{coef}}$. We denote this stratum $V((\mathsf{Sym}^d\mathbb{C})^\tau_\omega)$ by R_ω. In $\mathbb{R}^d_{\mathsf{coef}}$, it represents all monic real polynomials P such that the combinatorics of the real divisor $D_\mathbb{R}(P)$ is prescribed by ω. We denote by $\mathsf{R}_{\omega_\succeq}$ the closure of the stratum R_ω.

The depressed form of the polynomial in (6.14) and (6.26) calls for the introduction of similar spaces built out of, so called, *balanced divisors*.

Let $\alpha \in \mathbb{R}$. A divisor $D = \oplus_i\, m(i) z_i$, with $z_i \in \mathbb{C}$ and $m(i) \in \mathbb{N}$, is called *α-balanced*, if

$$\sum_i m(i) z_i = \alpha \cdot \sum_i m(i)$$

in $\mathbb{C}$. In other words, α is the center of gravity of the configuration of weighted points-particles in $\mathbb{C}$ representing D. In the root space $\mathsf{Sym}^d\mathbb{C} = \{z_1, \dots, z_d\}$ (where $d = |\omega|$), such divisors are characterized by the equation

$$\sum_{k=1}^{d} z_k = \alpha \cdot d.$$

The conjugation-invariant α-balanced divisors from $(\mathsf{Sym}^d(\mathbb{C}))^\tau$ are described by the equation

$$\sum_{k=1}^{d} \mathsf{Re}(z_k) = d \cdot \alpha.$$

They form a real hypersurface $(\mathsf{Sym}^d_\alpha \mathbb{C})^\tau$ in $(\mathsf{Sym}^d \mathbb{C})^\tau$. Zero-balanced divisors (i.e., $\alpha = 0$) are simply called *balanced.*

The image of $(\mathsf{Sym}^d_\alpha \mathbb{C})^\tau$ under the Viète map V consists of real monic polynomials whose z^{d-1}-coefficient is α. In a similar way, we introduce the stratification

$$\big\{\mathsf{R}_{\alpha,\omega} =_{\mathsf{def}} V\big((\mathsf{Sym}^d_\alpha \mathbb{C})^\tau_\omega\big)\big\}_\omega$$

in the $(d-1)$-dimensional affine space $\mathbb{R}^d_{\mathsf{coef},\alpha}$ of such polynomials.

Let us stress that here we use ω's that attach multiplicities only to *real* roots (their norms $|\omega|$ do not exceed d and $|\omega| \equiv d\ (2)$)!

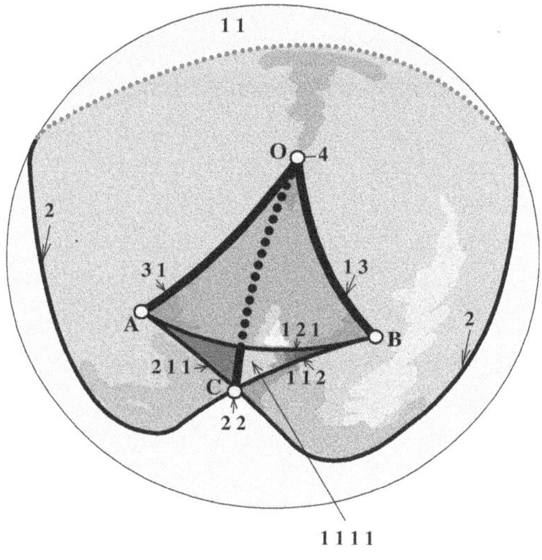

Fig. 7.3 The Swallow Tail Singularity is linked to the Whitney projection of the hypersuface $\{z^4 + x_2 z^2 + x_1 z + x_0 = 0\} \subset \mathbb{R}^4$ onto the space $\mathbb{R}^3_{\mathsf{coef},0}$ with the coordinates (x_0, x_1, x_2). The strata in $\mathbb{R}^3_{\mathsf{coef},0}$ are indexed by elements $\omega \in \mathbf{\Omega}_{\langle 4]}$. They divide the target space into three 3-cells, four 2-cells, three 1-cells, and one 0-cell.

Consider the closed upper half plane

$$\mathbb{H} =_{\mathsf{def}} \{z \in \mathbb{C} |\ \mathsf{Re}(z) \geq 0\}$$

and the open upper half plane

$$\mathbb{H}^\circ =_{\mathsf{def}} \{z \in \mathbb{C} |\ \mathsf{Re}(z) > 0\}.$$

Let $B^2 \subset \mathbb{C}$ be the unit ball, centered on the origin. Put $B^2_+ =_{\mathsf{def}} B^2 \cap \mathbb{H}$. Similarly, for any $\alpha \in \mathbb{R}$, let $B^2(\alpha) \subset \mathbb{C}$ be the unit ball centered on α, and $B^2_+(\alpha) =_{\mathsf{def}} B^2(\alpha) \cap \mathbb{H}$.

Lemma 7.6. *The symmetric product* $\mathsf{Sym}^m\mathbb{H}$ *is homeomorphic to the half-space* $\mathbb{R}^{2m}_+$*, bounded by a hyperplane in* $\mathbb{R}^{2m}$.

Let B^2 *denotes a closed 2-ball. The symmetric product* $\mathsf{Sym}^m B^2$ *is homeomorphic to a closed* $2m$*-ball* B^{2m}.

Proof. First, we validate the second claim of the lemma.

The origin-centered dilation $t : \mathbb{C} \to \mathbb{C}$ induces an action ψ_t on the root space $\mathsf{Sym}^m\mathbb{C}$: under this action, the support of each divisor is scaled by the factor $t > 0$. With the help of the Viète homeomorphism $V : \mathsf{Sym}^m\mathbb{C} \to \mathbb{C}^m_{\mathsf{coef}}$, this ψ_t-action induces a Ψ_t-action on the space $\mathbb{C}^m_{\mathsf{coef}}$.

The boundary of $\mathsf{Sym}^m B^2$ consists of divisors D, whose support is contained in B^2 and has a nonempty intersection with the circle ∂B^2.

We notice that each ψ_t-trajectory through a point $D \in \mathsf{Sym}^m B^2$, distinct from the divisor $m\{0\}$, has a unique point of intersection with the boundary $\partial(\mathsf{Sym}^m B^2)$. Indeed, for each configuration $\Delta \neq \{0\}$ of points in B^2, there is a unique origin-centered dilation $t : \mathbb{C} \to \mathbb{C}$ such that $t(\Delta) \cap \partial B^2 \neq \emptyset$ and $t(\Delta) \subset B^2$.

Since the Viète map $V : \mathsf{Sym}^m\mathbb{C} \to \mathbb{C}^m_{\mathsf{coef}}$ is a smooth homeomorphism, $V(\mathsf{Sym}^m B^2) \subset \mathbb{C}^m_{\mathsf{coef}}$ is homeomorphic to $\mathsf{Sym}^m B^2$. Therefore each Ψ_t-trajectory, except for the trivial trajectory through $m\{0\}$, hits the boundary $\partial V(\mathsf{Sym}^m B^2) = V\big(\partial(\mathsf{Sym}^m B^2)\big)$ transversally at a singleton. The same property is shared by any $(2m-1)$-sphere in $\mathbb{C}^m_{\mathsf{coef}}$ that is centered on the origin $\mathbf{0}$. Therefore, with the help of Ψ_t, $\partial V(\mathsf{Sym}^m B^2)$ is homeomorphic to a $(2m-1)$-sphere, and $V(\mathsf{Sym}^m B^2)$ to a closed ball B^{2m}. Hence $\mathsf{Sym}^m B^2 \approx B^{2m}$ topologically.

A similar argument helps to analyze the topology of $\mathsf{Sym}^m(\mathbb{H})$.

The boundary of the closed unit half-ball $B^2_+ =_{\mathsf{def}} B^2 \cap \mathbb{H}$ consists of the segment $I = [-1, 1] \subset \mathbb{R}$ and the arc $A =_{\mathsf{def}} \{z \in \mathbb{C} |\ |z| = 1,\ \mathsf{Im}(z) \geq 0\}$.

Consider the set $\mathcal{S}$ of divisors $D \in \mathsf{Sym}^m\mathbb{H}$ such that their support $\Delta \subset B^2_+$ and $\Delta \cap A \neq \emptyset$. Again, there is a unique origin-centered dilation $t : \mathbb{C} \to \mathbb{C}$, $t > 0$, such that $t(\Delta) \cap A \neq \emptyset$ and $t(\Delta) \subset B^2_+$, provided that $D \neq m\{0\}$. Therefore each ψ_t-trajectory through a point $D \in \mathsf{Sym}^m\mathbb{H}$, except for the trivial trajectory through $m\{0\}$, has a unique point of intersection with $\mathcal{S}$.

The boundary $\partial(\mathsf{Sym}^m\mathbb{H})$ in $\mathsf{Sym}^m\mathbb{C}$ consists of divisors D such that their support Δ has a nonempty intersection with the real line $\mathbb{R} \subset \mathbb{H}$. Evidently, $\partial(\mathsf{Sym}^m\mathbb{H})$ is invariant under ψ_t-flow. Thus, $\partial(\mathsf{Sym}^m\mathbb{H})$ topologically is a cone over $\partial(\mathsf{Sym}^m\mathbb{H}) \cap \mathcal{S}$, the set of divisors whose support $\Delta \subset B^2_+$ and such that $\Delta \cap I \neq \emptyset$.

Now, we take a point $p \in \partial B^2$ and a small *open* ϵ-ball $B_\epsilon(p) \subset \mathbb{C}$ with the center at p. Let $H =_{\mathsf{def}} B_\epsilon(p) \cap B^2$. Since H is homeomorphic to the half-plane $\mathbb{H}$, we get a homeomorphism $\mathsf{Sym}^m\mathbb{H} \approx \mathsf{Sym}^m H$.

Consider $m \cdot p$ as a point in $\partial(\mathsf{Sym}^m B^2) \approx \partial B^{2m}$ (by the already validated second claim of the lemma). Then $\mathsf{Sym}^m H$ can be viewed as an open ϵ-neighborhood of the point $m \cdot p$ in the space $\mathsf{Sym}^m B^2$. Here the distance in the closed ball $\mathsf{Sym}^m B^2$ being the Hausdorff distance between S_m-orbits in $(B^2)^m \subset \mathbb{C}^m$. At least for a sufficiently small $\epsilon > 0$, that neighborhood $\mathsf{Sym}^m H$, being an ϵ-neighborhood of a point $m \cdot p \in \partial(B^{2m})$, is homeomorphic to $\mathbb{R}^{2m}_+$, and so is $\mathsf{Sym}^m\mathbb{H}$. □

Definition 7.9.

- For $\omega \in \mathbf{\Omega}_{\langle d]}$, consider the space

$$e_\omega =_{\mathsf{def}} \mathsf{Sym}^{|\,\sup(\omega)|}\mathbb{R} \times \mathsf{Sym}^{\frac{d-|\omega|}{2}}\mathbb{H}. \tag{7.4}$$

- We denote by σ_ω the subset of e_ω that consists of *pairs* $D' \times D''$ of divisors from (7.4) such that $\sup(D') \cup \sup(D'') \subset B^2 \subset \mathbb{C}$, but not in the interior of B^2.
- Let the subset $e_{\alpha,\omega} \subset e_\omega$ be defined by the constraint: the divisor $D' + D'' + \tau(D'')$ is α-balanced. Similarly, let $\sigma_{\alpha,\omega} \subset \sigma_\omega$ be defined by the property of $D' + D'' + \tau(D'')$ being α-balanced. ♢

In what follows, by a "*CW-complex*" we mean a Hausdorff topological space Z which is a disjoint union of its subsets $\{e^\circ_\alpha\}_\alpha$, where each e°_α is homeomorphic to an open ball B°_α of some dimension d_α. Let B_α be either a closed ball, or an open ball, together with a northern hemisphere in its spherical boundary[6]. The homeomorphism $B^\circ_\alpha \to e^\circ_\alpha$ must extend to a continuous map $\phi_\alpha : B_\alpha \to Z$, so that the image $\phi_\alpha(\partial B_\alpha)$ is contained in a union of finitely many cells e_β of dimensions lower than d_α.

[6]This deviation from the standard definition of CW-complex is due to our need to consider *germs* of classical CW-complexes.

By definition, $Y \subset Z$ is closed if, for each α, $\phi_\alpha^{-1}(Y \cap \phi_\alpha(B_\alpha))$ is closed in B_α. Note that e_α, the closure of e_α° in Z, coincides with $\phi_\alpha(B_\alpha)$. We do not require that e_α will be homeomorphic to a closed ball!

Let $\bar{\mathcal{P}}^d$ denote the one point compactification of the space $\mathcal{P}^d = \mathbb{R}^d_{\mathsf{coef}}$. It may be identified with a d-sphere $S^d_{\mathsf{coef}} =_{\mathsf{def}} \bar{\mathbb{R}}^d_{\mathsf{coef}}$.

The next proposition describes one particularly important CW-structure on the space $\mathbb{R}^d_{\mathsf{coef}}$ (or rater on its one-point compactification $\bar{\mathbb{R}}^d_{\mathsf{coef}}$), the structure induced by the combinatorial types ω of the real divisors of polynomials from $\mathbb{R}^d_{\mathsf{coef}}$. In particular, for any cell $\mathsf{R}_\omega \subset \mathbb{R}^d_{\mathsf{coef}}$, we need to describe explicitly the attaching maps that take the boundaries of closed/half-closed k-balls, where $k = d - |\omega|'$, to the $(k-1)$-skeleton of the CW-complex structure on $\mathbb{R}^d_{\mathsf{coef}}$. This will require to resolve appropriately each affine variety $\mathsf{R}_{\omega_\succeq} \subset \mathbb{R}^d_{\mathsf{coef}}$ into a cone over a closed/half-closed ball.

Theorem 7.1. *Let $\omega \in \mathbf{\Omega}_{\langle d]}$, and let $\alpha \in \mathbb{R}$. Then the following $\mathbf{\Omega}_{\langle d]}$-stratifications of $\mathbb{R}^d_{\mathsf{coef}}$ are available:*

- *Each pure stratum $\mathsf{R}_\omega \subset \mathbb{R}^d_{\mathsf{coef}}$ is homeomorphic to an open ball*[7] *of codimension $|\omega|'$.*
- *The space σ_ω from Definition 7.9 is homeomorphic to a closed $(d - |\omega|' - 1)$-ball, and the space $\sigma_{\alpha,\omega}$ to a closed $(d - |\omega|' - 2)$-ball. The space e_ω from (7.4) is a positive cone over σ_ω, and $e_{\alpha,\omega}$ is a positive cone over $\sigma_{\alpha,\omega}$*[8].
- *For each $\hat{\omega} \in \mathbf{\Omega}_{\langle d]}$, the strata $\{\mathsf{R}_\omega\}_{\omega \preceq \hat{\omega}}$, define the structure of a cell complex on the real affine variety $\mathsf{R}_{\hat{\omega}_\succeq} \subset \mathbb{R}^d_{\mathsf{coef}}$. The attaching maps*

$$\Phi^\partial_\omega : \partial e_\omega \to \mathsf{R}_{\omega_\succeq} \setminus \mathsf{R}_\omega$$

  for the cells e_ω are described in formulas (7.7) below.
- *The space $\mathbb{R}^d_{\mathsf{coef}}$ admits a 1-parameter flow Ψ_t which has a single stationary point $\mathbf{0}$ (a source), is transversal to each sphere $S^{d-1}_{\mathsf{coef}} \subset \mathbb{R}^d_{\mathsf{coef}}$, centered on $\mathbf{0}$, and preserves each stratum R_ω. Thus, the cellular structure on $\mathbb{R}^d_{\mathsf{coef}}$ (described in the third bullet) is a cone over a similar cellular structure on the sphere $S^{d-1}_{\mathsf{coef}} \subset \mathbb{R}^d_{\mathsf{coef}}$.*

[7]In general, the intersection of $\bar{\mathsf{R}}_\omega$ with a ball $B^d \subset \mathbb{R}^d$, centered at the origin, topologically may not be a ball.

[8]See Definition 7.9, introducing these spaces.

Proof. Any real divisor comes with a particular linear order among the points in its support. Let P be a real monic polynomial of degree d. Its real divisor $D_{\mathbb{R}}(P)$ of the combinatorial type ω, with the help of ω, is determined by the support $\sup(D_{\mathbb{R}}(P)) \subset \mathbb{R}$. Put $p =_{\mathsf{def}} |\sup(\omega)|$,

We consider the chamber $\Pi^p_\circ$ of $\mathbb{R}^p = \{(y_1, \dots, y_p)\}$, defined by the inequalities $y_1 < y_2 < \dots < y_p$. In fact, $\Pi^p_\circ$ is one of the 2^{p-1} chambers-cells in which the hyperplanes $\{y_i = y_{i+1}\}_i$ divide $\mathbb{R}^p$.

The set $\Pi^p_\circ$, admits an obvious embedding into the space $\mathbb{R}^{|\omega|}$ with the coordinates $(x_1, \dots, x_{|\omega|})$. It is defined there by the system of ω-dependent equations and inequalities:

$$\underbrace{x_1 = x_2 = \dots = x_{\omega(1)}} < \underbrace{x_{\omega(1)+1} = x_{\omega(1)+2} = \dots = x_{\omega(1)+\omega(2)}} < \underbrace{\dots}$$

We denote by Π°_ω the solution set of this system. Note that Π°_ω is homeomorphic to the open chamber $\Pi^p_\circ \subset \mathbb{R}^p$. Let Π_ω be the closure of Π°_ω in $\mathbb{R}^{|\omega|}$. That closure is homeomorphic to $\Pi^p \approx \mathsf{Sym}^p\mathbb{R}$. Indeed, each orbit of the natural action of the symmetric group S_p on $\mathbb{R}^p$ intersects the closed chamber Π^p at a singleton.

With its real roots being fixed, a polynomial P is determined by the unordered pairs of non-real complex-conjugate roots (possibly with multiplicities). Then the conjugate non-real pairs can be identified with points of the space $\mathsf{Sym}^m\mathbb{H}^\circ$, where $m = \frac{1}{2}(d - |\omega|)$. Therefore, $(\mathsf{Sym}^d\mathbb{C})^\tau_\omega$ is homeomorphic to the space $\Pi^p_\circ \times \mathsf{Sym}^m\mathbb{H}^\circ$, where $p =_{\mathsf{def}} |\sup(\omega)|$. Since $\mathbb{H}^\circ$ is homeomorphic to $\mathbb{C}$ and $\mathsf{Sym}^m\mathbb{C}$ is homeomorphic to $\mathbb{C}^m$ via the Viète Map V, we conclude that $\mathsf{Sym}^m\mathbb{H}^\circ$ is homeomorphic to $\mathbb{C}^m$. Thus,

$$(\mathsf{Sym}^d\mathbb{C})^\tau_\omega \approx \Pi^p_\circ \times \mathsf{Sym}^m\mathbb{H}^\circ \approx \Pi^p_\circ \times \mathbb{C}^m,$$

an open cell of dimension

$$p + 2m = |\sup(\omega)| + (d - |\omega|) = d - |\omega|'.$$

Note that, the Viète map V generates a smooth homeomorphism between the spaces $(\mathsf{Sym}^d\mathbb{C})^\tau_\omega \approx \mathbb{R}^{d-|\omega|'}$ and $\mathsf{R}_\omega = (\mathbb{R}^d_{\mathsf{coef}})_\omega$. Thus the claim of the first bullet is validated.

The flow $\Psi_t : \mathbb{R}^d_{\mathsf{coef}} \to \mathbb{R}^d_{\mathsf{coef}}$ (see the fourth bullet) is induced by the flow ψ_t in the root space $(\mathsf{Sym}^d\mathbb{C})^\tau$ that applies dilatations (centered on the zero) by real factors $t > 0$ to each τ-symmetric root configuration in $\mathbb{C}$. The transplantation of ψ_t to $\mathbb{R}^d_{\mathsf{coef}}$ is done via the Viète Map V, a smooth homeomorphism $(\mathsf{Sym}^d\mathbb{C})^\tau \to \mathbb{R}^d_{\mathsf{coef}}$. Evidently, each stratum $(\mathsf{Sym}^d\mathbb{C})^\tau_\omega$ is ψ_t-invariant; therefore each stratum R_ω must be invariant under Ψ_t.

We leave to the reader to verify that, for any polynomial $P \neq z^d$, its Ψ_t-trajectory is transversal to the spheres $S^{d-1}_{\mathsf{coef}} \subset \mathbb{R}^d_{\mathsf{coef}}$, centered on $\mathbf{0}$ (the verification is a straightforward computation). This proves the claim in the forth bullet.

Next, we turn our attention to the second bullet of the theorem. Each divisor $D \in \mathsf{Sym}^m\mathbb{H}$ produces a unique divisor $D_{\mathbb{R}}$, the part of D that is supported in $\mathbb{R} \subset \mathbb{H}$. We view $D_{\mathbb{R}}$ as an element of $\mathsf{Sym}^l\mathbb{R}$, $l \leq m$. For a typical D, $D_{\mathbb{R}}$ will have an empty support, so we interpret $\mathsf{Sym}^0\mathbb{R}$ as an empty set.

Recall that $e_\omega =_{\mathsf{def}} \Pi_\omega \times \mathsf{Sym}^m\mathbb{H}$, where $m = \frac{1}{2}(d - |\omega|)$ (see (7.4)). In the proof of Lemma 7.6, we have seen that e_ω is homeomorphic to $\Pi_\omega \times \mathbb{R}^{d-|\omega|}_+$—a $|\sup(\omega)|$-dimensional pyramid times a half-space in $\mathbb{R}^{d-|\omega|}$.

The boundary of the unit half-disk $B^2_+ =_{\mathsf{def}} \{z \in \mathbb{C} : \; |z| \leq 1, \, \mathsf{Im}(z) \geq 0\}$ consists of the segment $I = [-1, 1] \subset \mathbb{R}$ and the arc

$$A =_{\mathsf{def}} \{z \in \mathbb{C} : \; |z| = 1, \, \mathsf{Im}(z) \geq 0\}.$$

The multiplicative group $\mathbb{R}^*_+$ of positive real numbers acts semi-freely on the space $\Pi_\omega \times \mathsf{Sym}^m\mathbb{H}$ by the diagonal ψ_t-action which is induced by the origin-centered dilations in $\mathbb{C}$. Its only fixed point is the origin $\mathbf{0} = \{0\} \times \{0\}$. This $\mathbb{R}^*_+$-action on $e_\omega \setminus \mathbf{0}$ admits a compact section σ_ω that consists of points $D' \times D'' \in e_\omega$ such that $\sup(D') \cup \sup(D'')$ is contained in the unit half-disk $B^2_+ \subset \mathbb{H}$ (centered on the origin) and has a nonempty intersection with the arc A.

Thus σ_ω is the set of pairs $D' \in \mathsf{Sym}^p(I), D'' \in \mathsf{Sym}^m(B^2_+)$ such that either $\sup(D') \cap \partial I \neq \emptyset$, or $\sup(D'') \cap A \neq \emptyset$, or both. Here $p = |\sup(\omega)|$. Note that if $\sup(D') \cap \partial I \neq \emptyset$, then $\sup(D') \cap A \neq \emptyset$ since $\partial A = \partial I$. Therefore, σ_ω can be also described as a set of pairs (D', D'') such that $\sup(D' + D'') \subset B^2_+$ and $\sup(D' + D'') \cap A \neq \emptyset$.

Recall that $\mathsf{Sym}^p I \approx \Delta^p$, a p-simplex. By Lemma 7.6, $\mathsf{Sym}^m B^2_+ \approx B^{2m}_+$, a half-ball. Let $\delta B^{2m}_+ \subset \partial(B^{2m}_+)$ denote the northern hemisphere in the boundary of the ball $B^{2m} \supset B^{2m}_+$. It corresponds to the divisors D'' with the property $\sup(D'') \cap A \neq \emptyset$.

In the new notations, the section σ_ω is the set of pairs $D' \in \Delta^p, D'' \in B^{2m}_+$ such that either $D' \in \partial\Delta^p$, or $D'' \in \delta B^{2m}_+$, or both. As a result, we get a homeomorphism

$$\sigma_\omega \approx (\partial\Delta^p \times B^{2m}_+) \bigcup_{\partial\Delta^p \times \delta B^{2m}_+} (\Delta^p \times \delta B^{2m}_+),$$

whose target topologically is a $(d - |\omega|' - 1)$-ball. Indeed, as the formula above testifies, σ_ω is obtained from the sold torus $T = S^{p-1} \times D^{2m}$ of dimension $d - |\omega|' - 1$ by attaching a p-handle along $S^{p-1} \times D^{2m-1} \subset \partial T$.

Therefore e_ω, an infinite positive cone over the ball σ_ω, is homeomorphic to a positive cone in $\mathbb{R}^{d-|\omega|'}$ with a closed $(d - |\omega|' - 1)$-ball base.

Now we proceed to describe the attaching maps for the cells e_ω (see (7.6) - (7.8) below). With this goal in mind, we will "partially compactify" the pure stratum R_ω in order to form an "honest" $(d - |\omega|')$-cell e_ω and will show how to attach its boundary ∂e_ω to the strata $\{\mathsf{R}_{\tilde\omega}\}_{\tilde\omega \in \omega_\succeq}$ of dimensions smaller than $\dim(\mathsf{R}_\omega)$. This cell e_ω can be regarded as a *resolution* of the real variety $\mathsf{R}_{\omega_\succeq}$.

Let us consider the maps

$$\Theta_\omega : e_\omega =_{\mathsf{def}} \Pi_\omega \times \mathsf{Sym}^{\frac{d-|\omega|}{2}} \mathbb{H} \longrightarrow (\mathsf{Sym}^d \mathbb{C})^\tau, \tag{7.5}$$

defined by the formula

$$\Theta_\omega(D' \times D'') = D' + D'' + \tau(D''),$$

where $D' \in \Pi_\omega$, $D'' \in \mathsf{Sym}^{\frac{d-|\omega|}{2}} \mathbb{H}$, $\tau(D'')$ stands for the complex-conjugate of the divisor D'', and "+" denotes the sum of divisors in $\mathbb{C}$.

As formula (7.5) testifies, each map Θ_ω doubles the real part $D''_\mathbb{R}$ of each divisor $D'' \in \mathsf{Sym}^m \mathbb{H}$ and thus mimics the merging of conjugate pairs of complex roots into the appropriate real roots of even multiplicity.

We notice that the restriction of Θ_ω to $e^\circ_\omega =_{\mathsf{def}} \Pi^\circ_\omega \times \mathsf{Sym}^{\frac{d-|\omega|}{2}} \mathbb{H}^\circ$, the interior of e_ω, is a homeomorphism onto the pure stratum $(\mathsf{Sym}^d \mathbb{C})^\tau_\omega$.

The restriction Θ^∂_ω of Θ_ω to the boundary ∂e_ω provides us with the attaching maps

$$\left\{\Theta^\partial_\omega : \partial e_\omega \to (\mathsf{Sym}^d \mathbb{C})^\tau\right\}_{\omega \in \mathbf{\Omega}_{\langle d]}}. \tag{7.6}$$

By the very construction of e_ω, the Θ_ω-image of ∂e_ω belongs to the union of strata $\left\{(\mathsf{Sym}^d \mathbb{C})^\tau_{\tilde\omega}\right\}_{\tilde\omega}$, where $\tilde\omega \in \omega_\succeq$. Indeed, if $D' \times D'' \in \partial e_\omega$, then either $D' \in \partial \Pi_\omega$, or $D'' \in \partial(\mathsf{Sym}^{\frac{d-|\omega|}{2}} \mathbb{H})$. In the first case, D' is obtained from some $D \in \Pi^\circ_\omega$ via merge operations; thus $(D' + D'' + \tau(D''))_\mathbb{R}$ is obtained from $(D + D'' + \tau(D''))_\mathbb{R}$ by the same merges. In the second case, $(D' + D'' + \tau(D''))_\mathbb{R}$ can be obtained from D' by inserting $(D'' + \tau(D''))_\mathbb{R}$.

Therefore the maps $\{\Theta^\partial_\omega\}_\omega$ from (7.6) define cellular structures on the real affine variety $(\mathsf{Sym}^d \mathbb{C})^\tau$ and its subvarieties $\{(\mathsf{Sym}^d \mathbb{C})^\tau_{\hat\omega_\succeq}\}_{\hat\omega}$.

We notice that all the maps Θ_ω are equivariant under the $\mathbb{R}^*_+$-actions ψ_t and Ψ_t in the source and target spaces, respectively. So the attaching maps are consistent with the cone structures of e_ω and of the strata $\{(\mathsf{Sym}^d\mathbb{C})^{\tau}_{\hat{\omega}}\}$. Moreover, the sections σ_ω are mapped by Θ_ω to some sections $\mathsf{S}_\omega \subset \mathsf{R}_\omega$ of the flow Ψ_t. The space S_ω is defined as the set of real polynomials with all their roots residing in the ball $B^2 \subset \mathbb{C}$, but not in its interior, and with the combinatorics of the real roots being prescribed by the poset $\omega_{\succeq} \subset \mathbf{\Omega}_{\langle d]}$.

With the help of the Viète homeomorphism V, the maps

$$\{\Phi_\omega : e_\omega \stackrel{V\circ\Theta_\omega}{\longrightarrow} \mathsf{R}_{\omega_{\succeq}} \subset \mathbb{R}^d_{\mathsf{coef}}\}_{\omega\in\mathbf{\Omega}_{\langle d]}}, \tag{7.7}$$

$$\{\Phi^\partial_\omega : \partial e_\omega \stackrel{V\circ\Theta^\partial_\omega}{\longrightarrow} \mathsf{R}_{\omega_{\succ}} \subset \mathbb{R}^d_{\mathsf{coef}}\}_{\omega\in\mathbf{\Omega}_{\langle d]}}$$

define cellular structures in $\mathbb{R}^d_{\mathsf{coef}}$ and its subvarieties $\{\mathsf{R}_{\hat{\omega}_{\succeq}}\}_{\hat{\omega}}$. Again, the attaching maps $\{\Phi_\omega, \Phi^\partial_\omega\}_\omega$ are $\mathbb{R}^*_+$-equivariant. They are consistent with the cone structures in the strata $e_{\hat{\omega}}$ and $\mathsf{R}_{\hat{\omega}}$. In particular, we get the maps:

$$\{\Phi_\omega : \sigma_\omega \stackrel{V\circ\Theta_\omega}{\longrightarrow} \mathsf{S}_{\omega_{\succeq}} \subset S^{d-1}_{\mathsf{coef}}\}_{\omega\in\Omega_{\langle d]}}, \tag{7.8}$$

$$\{\Phi^\partial_\omega : \partial\sigma_\omega \stackrel{V\circ\Theta^\partial_\omega}{\longrightarrow} \mathsf{S}_{\omega_{\succ}} \subset S^{d-1}_{\mathsf{coef}}\}_{\omega\in\Omega_{\langle d]}}.$$

Here S^{d-1}_{coef} denotes the space of degree d real monic polynomials whose roots are in B^2, but not in its interior. With the help of Ψ_t, the space S^{d-1}_{coef} is diffeomorphic to the standard sphere S^{d-1}. This completes the proof of the third bullet. □

Let us fix $\alpha \in \mathbb{R}$. Results, similar to the ones described in Theorem 7.1, are valid for the space of real degree d monic polynomials with a fixed coefficient $d \cdot \alpha$ of z^{d-1}.

Theorem 7.2. *Let $\omega \in \mathbf{\Omega}_{\langle d]}$ and $\alpha \in \mathbb{R}$. We denote by $\omega_\star : 1 \to d$ the minimal element of the poset $\mathbf{\Omega}_{\langle d]}$. The following properties hold:*

- *the stratum $\mathsf{R}_{\alpha,\omega} \subset \mathbb{R}^{d-1}_{\mathsf{coef},\alpha}$ is an open ball of codimension $|\omega|'$,*
- *for each $\hat{\omega} \in \mathbf{\Omega}_{\langle d]}$, the strata $\{\mathsf{R}_{\alpha,\omega}\}_{\omega\preceq\hat{\omega}}$ give rise to a cellular structure on the affine variety $\mathsf{R}_{\alpha,\hat{\omega}_{\succeq}} \subset \mathbb{R}^{d-1}_{\mathsf{coef},\alpha}$. The attaching maps*

$$\Phi^\partial_{\alpha,\omega} : \partial e_{\alpha,\omega} \to \mathsf{R}_{\alpha,\omega_{\succeq}} \setminus \mathsf{R}_{\alpha,\omega}$$

  for the cells $e_{\alpha,\omega}$ are described by formulas similar to formulas (7.7)-(7.8),
- *the space $\mathbb{R}^{d-1}_{\mathsf{coef},\alpha}$ admits a 1-parameter flow Ψ^α_t which has a single stationary point O_α (a source), is transversal to each sphere $S^{d-2}_{\mathsf{coef},\alpha}$, centered on O_α, and preserves each stratum $\mathsf{R}_{\alpha,\omega}$. Thus, the $\mathbf{\Omega}_{\langle d]}$-labeled cellular structure on $\mathbb{R}^{d-1}_{\mathsf{coef},\alpha}$ is a cone over a similar $(\mathbf{\Omega}_{\langle d]} \setminus \omega_\star)$-labelled cellular structure on the sphere $S^{d-2}_{\mathsf{coef},\alpha}$.*

Proof. We turn to the α-balanced divisors, which tell a similar story.

Consider the flow

$$\phi_t : \{z_1, \dots, z_d\} \to \{z_1 - \frac{\alpha}{d}t, \dots, z_d - \frac{\alpha}{d}t\}$$

in $(\mathsf{Sym}^d\mathbb{C})^\tau$. For $t = 1$, it maps $\{z_1, \dots, z_d\}$ to $\{z_1 - \frac{\alpha}{d}, \dots, z_d - \frac{\alpha}{d}\}$, a α-balanced configuration. Note that ϕ_t preserves the ω-stratification in $(\mathsf{Sym}^d\mathbb{C})^\tau$. Hence, with the help of the Viète map V, ϕ_t gives rise to a retraction of the stratum R_ω onto the stratum $\mathsf{R}_{\alpha,\omega}$. Therefore, $\mathsf{R}_{\alpha,\omega}$ is an open ball of dimension $d - |\omega|' - 1$ and codimension $|\omega|'$ in $\mathbb{R}^d_{\mathsf{coef},\alpha}$.

Next, consider the flow $\Psi^\alpha_t : \mathbb{R}^d_{\mathsf{coef},\alpha} \to \mathbb{R}^d_{\mathsf{coef},\alpha}$, induced with the help of the Viète map V by a flow $\{\psi^\alpha_t\}_{t>0}$ in the root space $(\mathsf{Sym}^d_\alpha\mathbb{C})^\tau$. The flow ψ^α_t applies t-dilatations that are centered on the point $o_\alpha =_{\mathsf{def}} \alpha/d \in \mathbb{R}$ to each τ-symmetric and α-balanced divisor D in $\mathbb{C}$. The dilatation ψ^α_t keeps the center of gravity of the weighted configuration $\psi^\alpha_t(D)$ at the point $o_\alpha \in \mathbb{C}$, so that $\psi^\alpha_t(D)$ is α-balanced for all t. Evidently, these dilatations preserve the ω-type of each α-balanced configuration in $\mathbb{C}$. As a result, each $\mathsf{R}_{\alpha,\omega}$ is Ψ^α_t-invariant.

By a direct computation, any Ψ^α_t-trajectory is transversal to the spheres $S^{d-2}_{\mathsf{coef},\alpha} \subset \mathbb{R}^d_{\mathsf{coef},\alpha}$ centered on the point $O_\alpha =_{\mathsf{def}} V(o_\alpha, \dots, o_\alpha)$. The exception is the trivial trajectory through O_α.

We denote by $e_{\alpha,\omega}$ the subset of e_ω (see (7.4)) that consists of pairs (D', D''), where $D' \in \Pi_\omega$, $D'' \in \mathsf{Sym}^m\mathbb{H}$, and such that the divisor

$$D' + D'' + \tau(D'')$$

is α-balanced. In other words, if $D' = \sum_i m_i x_i$ and $D'' = \sum_k \tilde{m}_k z_k$, then $e_{\alpha,\omega}$ is the space of a fibration over the base $\mathsf{Sym}^m\mathbb{H} \approx \mathbb{R}^{2m}_+$ with the cell-like fiber $F_{D''} \subset \Pi_\omega$ (over the point D'') that is defined by the constraint

$$\sum_i m_i x_i = d \cdot \alpha - \sum_k 2\tilde{m}_k\, \mathsf{Re}(z_k).$$

We denote by $\sigma_{\alpha,\omega}$ the subset of $e_{\alpha,\omega}$ that consists of α-balanced pairs (D', D'') such that $\sup(D') \cup \sup(D'')$ is contained in the α-centered unit ball $B^2_+(\alpha)$, but not in its interior. As in the "unbalanced" case, $\sigma_{\alpha,\omega}$ is a section of the ψ^α_t-flow in $e_{\alpha,\omega}$. This results in $e_{\alpha,\omega}$ being a positive cone over the base $\sigma_{\alpha,\omega}$. By an argument as in the proof of Theorem 7.1, $\sigma_{\alpha,\omega}$ is homeomorphic to a closed $(d - 2 - |\omega|')$-ball.

For any real α, the maps Θ_ω in (7.5), being restricted to $e_{\alpha,\omega} \subset e_\omega$, produce the maps $\Theta_{\alpha,\omega} : e_{\alpha,\omega} \to (\mathsf{Sym}^d_\alpha\mathbb{C})^\tau$ which, with the help of V, give rise to the maps

$$\Phi_{\alpha,\omega} : e_{\alpha,\omega} \longrightarrow \mathsf{R}_{\alpha,\omega_\succeq} \subset \mathbb{R}^{d-1}_{\mathsf{coef},\alpha},$$

already familiar from the "unbalanced" formulas (7.7). They define a cellular structure on the real variety $\mathbb{R}^{d-1}_{\mathsf{coef},\alpha}$ and its subvarieties $\mathsf{R}_{\alpha,\tilde{\omega}_{\succeq}}$.

Analogously, the maps

$$\Phi_{\alpha,\omega} : \sigma_{\alpha,\omega} \longrightarrow \mathsf{S}_{\alpha,\omega_{\succeq}} \subset S^{d-2}_{\mathsf{coef},\alpha}$$

define a cellular structure on the sphere $S^{d-2}_{\mathsf{coef},\alpha}$ and its strata $\mathsf{S}_{\alpha,\tilde{\omega}_{\succeq}}$. Here $S^{d-2}_{\mathsf{coef},\alpha}$ is the space of real monic polynomials $P(z)$ of the form

$$z^d + \alpha z^{d-1} + \dots$$

such that the P-roots are contained in $B^2(\alpha)$, but not in its interior. Note that $P(z) \neq (z - \frac{\alpha}{d})^d$—the apex of the cone. Similarly, $\mathsf{S}_{\alpha,\omega_{\succeq}} \subset S^{d-2}_{\mathsf{coef},\alpha}$, is the set of such polynomials $P(z)$ whose real roots have the combinatorics that is prescribed by the poset $\omega_{\succeq}$. □

Example 7.2. Let $\omega_\star = (d)$ be the minimal element of the poset $\mathbf{\Omega}_{\langle d]}$. Theorem 7.1 claims that the sphere S^{d-1} admits a cellular structure whose cells $\{\mathsf{S}_\omega\}$ of codimension $|\omega|'$ are indexed by the elements of the poset $(\mathbf{\Omega}_{\langle d]} \setminus \omega_\star, \succ)$. Moreover, $\mathsf{S}_{\omega'_{\succeq}} \subset \mathsf{S}_{\omega''_{\succeq}}$ if and only if $\omega'' \succeq \omega'$. In other words, the poset $(\mathbf{\Omega}_{\langle d]} \setminus \omega_\star, \succ)$ provides a complete set of instructions for assembling S^{d-1}! Note that this cellular structure is not a regular one: the closures $\mathsf{S}_{\omega_{\succeq}}$ of cells S_ω are not necessarily closed balls.

You may glance at Figure 7.3 to examine how the assembly works for of $d = 4$ and the balanced case of $S^2 = S^2_{\mathsf{coef},0}$. ◇

The next lemma describes the attaching maps $\Phi^\partial_\omega : \partial e_\omega \to \mathbb{R}^d_{\mathsf{coef}}$ for the cells e_ω as being finitely ramified over their images. The degrees of ramification are described by some combinatorics-flavored numbers $\{o(\omega,\tilde{\omega})\}$, whose exact Definition 7.10 will be provided below.

Lemma 7.7. *Each map $\Phi_\omega : e_\omega \to \mathbb{R}^d_{\mathsf{coef}}$ from (7.7) has finite fibers, and is a bijection on the interior of the cell e_ω.*

For any point-polynomial $Q \in \Phi_\omega(\partial e_\omega)$, the cardinality of the fiber $\Phi_\omega^{-1}(Q)$ is equal to the natural number $o(\omega,\tilde{\omega})$, introduced in Definition 7.10. Here $\tilde{\omega}$ denotes the combinatorial pattern of the real divisor $D_{\mathbb{R}}(Q)$.

Proof. Let us fix a degree d conjugation-invariant divisor

$$D = \Theta_\omega((D', D'')) =_{\mathsf{def}} D' + D'' + \tau(D''),$$

where $D' \in \Pi_\omega$, $D'' \in \mathsf{Sym}^m \mathbb{H}$, and $m =_{\mathsf{def}} (d - |\omega|)/2$. We denote by $\tilde{\omega}$ the combinatorial type of $D_{\mathbb{R}}$. There exist only finitely many pairs (D', D'')

that deliver D. Indeed, since $\sup(D') \subset \mathbb{R}$, the divisor $D'' - D''_{\mathbb{R}}$ with the support in $\mathbb{H}^\circ$ is uniquely determined by D. Note that $D_{\mathbb{R}}$ and D' (with the support in $\mathbb{R}$) differ by $2D''_{\mathbb{R}}$. This leaves only a finite set of choices for $D' = D_{\mathbb{R}} - 2D''_{\mathbb{R}}$, whose support must be contained in $\sup(D_{\mathbb{R}})$ and whose degree is bounded by $\deg(D_{\mathbb{R}})$.

If $(D', D'') \in \partial e_\omega$, then either $D' \in \partial \Pi_\omega$, or $D'' \in \partial(\mathsf{Sym}^m \mathbb{H})$ (i.e. $\sup(D'') \cap \mathbb{R} \neq \emptyset$), or both. In the first case, $\omega(D')$, the combinatorial type of D', is obtained from ω by a sequence of merge operations $\{\mathsf{M}_j\}$. In the second case, $\tilde{\omega} = \tilde{\omega}(D_{\mathbb{R}})$, the combinatorial type of the Θ_ω-image of (D', D''), is obtained from ω by a sequence of insert operations $\{\mathsf{I}_k\}$. In the third mixed case, one applies both types of operations.

As a result, the combinatorics of reconstructing (D', D'') from $D = D' + D'' + \tau(D'')$ can be described as follows. If $\tilde{\omega}$ is the combinatorial pattern of $D_{\mathbb{R}}$, then ω', the combinatorial pattern of $D' = D'_{\mathbb{R}}$ can be obtained by: (1) subtracting from $\tilde{\omega}$ a non-negative function $2\omega''$, such that $\tilde{\omega} - 2\omega''$ is again a non-negative function (ω'' represents the divisor $D''_{\mathbb{R}}$), and (2) deleting all the positions i, where $(\tilde{\omega} - 2\omega'')(i) = 0$.

Let us denote by K this "repackaging" operator from (2); it takes any histogram and deletes from it all the columns of zero height.

Since $D' \in \partial \Pi_\omega$, we conclude that $\omega' = K(\tilde{\omega} - 2\omega'')$ should be obtainable from ω via a sequence of merge operations alone.

For each $\omega \in \mathbf{\Omega}$, let us denote by $\omega_{\succeq \mathsf{M}}$ the subset of $\mathbf{\Omega}$ that consists of elements that can be obtained from ω by the merge operations $\{\mathsf{M}_j\}$ alone.

In these notations, we get that $K(\tilde{\omega} - 2\omega'') \in \omega_{\succeq \mathsf{M}}$.

To complete the proof, we need the following definition.

Definition 7.10. For any pair $\omega \succeq \tilde{\omega}$ in $\mathbf{\Omega}$, consider the subset $O(\omega, \tilde{\omega}) \subset \mathbf{\Omega}$ such that, for any $\omega'' \in O(\omega, \tilde{\omega})$, the following properties hold:

- the function $\tilde{\omega} - 2\omega'' \geq 0$
- $K(\tilde{\omega} - 2\omega'') \in \omega_{\succeq \mathsf{M}}$.

Let $o(\omega, \tilde{\omega})$ denote the cardinality of $O(\omega, \tilde{\omega})$. $\diamondsuit$

Recall that the Viète map $V : (\mathsf{Sym}^d \mathbb{C})^\tau \to \mathbb{R}^d_{\mathsf{coef}}$ is a stratification-preserving homeomorphism of the $\mathbf{\Omega}_{\langle d]}$-stratified spaces; in particular, it is bijective. Therefore the cardinality of each fiber $\Phi_\omega^{-1}(Q)$, where $Q \in \mathbb{R}^d_{\mathsf{coef}}$, is equal to the cardinality of the fiber $\Theta_\omega^{-1}(V^{-1}(Q))$.

Pick $Q \in \Phi_\omega(\partial e_\omega)$ and let $\tilde{\omega}$ denote the combinatorial type of the divisor $D_{\mathbb{R}}(Q)$. Then, by the considerations above, we have proved that $|\Phi_\omega^{-1}(Q)| =$

$|\Theta_{\omega}^{-1}(V^{-1}(Q))| = o(\omega, \tilde{\omega})$. □

For each $\omega \in \mathbf{\Omega}_{\langle d]}$, let $\omega_{\prec\rightsquigarrow k}$ denote the set of elements $\tilde{\omega} \in \mathbf{\Omega}_{\langle d]}$ that can produce ω as a result of a sequence of k elementary operations $\{\mathsf{M}_i, \mathsf{I}_j\}$, being applied to $\tilde{\omega}$ (each such $\tilde{\omega}$ is the maximal element in a chain of ω-predecessors in $\mathbf{\Omega}_{\langle d]}$ of length k). Similar, let $\omega_{\preceq\rightsquigarrow k}$ be the set of elements in $\mathbf{\Omega}_{\langle d]}$ that can produce ω by a sequence of k elementary operations at most.

Let $\mathcal{S}_\alpha$ be a given pure connected stratum in a stratified space $\mathcal{X}$. Consider the union of all connected strata $\{\mathcal{S}_\beta\}_\beta$ whose closures contain $\mathcal{S}_\alpha$. Put $St(\mathcal{S}_\alpha) =_{\mathsf{def}} \bigcup_{\beta \succeq \alpha} \mathcal{S}_\beta$ and call this union the *star* of $\mathcal{S}_\alpha$. When $\mathcal{S}_\alpha$ has a normal bundle ν in $\mathcal{X}$, we define the *normal star* $St_\nu(\alpha)$ of $\mathcal{S}_\alpha$ by intersecting $St(\mathcal{S}_\alpha)$ with a small ball in a typical fiber of ν. Let the stratified space $\mathcal{X}$ be such that the stratified topological type of the normal star does not depend on the fiber of ν and the size of the normal ball in it.

We denote by $St_\nu(\omega)$ the star, normal in $\mathbb{R}^d_{\mathsf{coef}}$ to the cell R_ω at a generic point $P_\omega \in \mathsf{R}_\omega$.

The next lemma is an extension of Theorem 7.1 and Theorem 7.2. It spells out the local arrangement of cells in the direction, normal to a given pure stratum R_ω in $\mathbb{R}^d_{\mathsf{coef}}$ or to $\mathsf{R}_{\alpha,\omega}$ in $\mathbb{R}^{d-1}_{\mathsf{coef},\alpha}$.

Lemma 7.8.

- *For any $\omega \in \mathbf{\Omega}_{\langle d]}$, the normal star $St_\nu(\omega)$ is homeomorphic to the space $\mathbb{R}^{|\omega|'}$ and inherits the cellular structure from the $\mathbf{\Omega}_{\langle d]}$-stratified space $\mathbb{R}^d_{\mathsf{coef}}$. The strata $\{\mathsf{f}_{\tilde{\omega}} =_{\mathsf{def}} St_\nu(\omega) \cap \mathsf{R}_{\tilde{\omega}}\}_{\tilde{\omega}}$ of this stratification of the dimension $|\omega|' - |\tilde{\omega}|'$ are labeled by the elements*
$$\tilde{\omega} \in \omega_{\preceq} =_{\mathsf{def}} \{\tilde{\omega} \in \mathbf{\Omega}_{\langle d]} \,|\, \tilde{\omega} \succeq \omega\}.$$
- *The incidence of the strata $\{\mathsf{f}_{\tilde{\omega}}\}$ in $St_\nu(\omega)$ is prescribed by the partial order in the poset $\omega_{\preceq}$. Specifically, any element $\tilde{\omega} \in \omega_{\prec\rightsquigarrow k}$ gives rise to a stratum $\mathsf{f}_{\tilde{\omega}} \subset St_\nu(P)$ of the dimension k. So the total number of strata $\mathsf{f}_{\tilde{\omega}} \subset St_\nu(P)$ of the dimension k is the cardinality of the set $\omega_{\prec\rightsquigarrow k}$.*
- *The connected components $\{\mathsf{f}_{\tilde{\omega},\beta}\}_\beta$ of $\mathsf{f}_{\tilde{\omega}}$ in the vicinity of the apex P_ω are in 1-to-1 correspondence with the elements of the set* $\mathrm{Mor}(\tilde{\omega}, \omega)$ *from Definition 7.2. As a result, each k-dimensional cell $\mathsf{f}_{\tilde{\omega},\beta} \subset St_\nu(\omega)$ is contained in exactly $|\tilde{\omega}|' + \#(\tilde{\omega}^{-1}(2))$ cells $\{\mathsf{f}_{\tilde{\omega}',\beta'} \subset St_\nu(\omega)\}_{\beta'}$ of the dimension $k+1$, localized to the vicinity of a point $P_{\tilde{\omega},\beta} \in \mathsf{f}_{\tilde{\omega},\beta}$.*

- *The cellular structure of* $\mathbb{R}^d_{\mathsf{coef}}$ *at* P_ω *as a product of this cellular structure in* $St_\nu(\omega)$, *localized to the vicinity of the apex* P_ω, *times the open cell* R_ω.
- *Similar properties hold for the strata* $\{St_\nu(\omega) \cap \mathsf{R}_{\alpha,\tilde\omega}\}_{\tilde\omega}$ *in the* α-*balanced polynomial space* $\mathbb{R}^d_{\mathsf{coef},\alpha}$.

Proof. Let $St(\mathsf{R}_\omega)$ be the union of all cells $\{\mathsf{R}_{\tilde\omega}\}$ in $\mathbb{R}^d_{\mathsf{coef}}$ such that $\mathsf{R}_{\tilde\omega_{\succeq}} \supset \mathsf{R}_\omega$[9].

By Theorem 7.1, $St_\nu(\omega)$, the germ of the vector space, normal to the cell R_ω at a generic point $P_\omega \in \mathsf{R}_\omega$, has the dimension $|\omega|'$. It is transversal at P_ω to each cell $\mathsf{R}_{\tilde\omega}$ whose closure contains R_ω. Thus, the stratum $\mathsf{f}_{\tilde\omega} =_{\mathsf{def}} St_\nu(\omega) \cap \mathsf{R}_{\tilde\omega}$ is of dimension $|\omega|' - |\tilde\omega|'$.

By the proof of Lemma 7.5, the connected components of the stratum $\mathsf{R}_{\tilde\omega} \supset \mathsf{R}_\omega$ in the vicinity of R_ω are in 1-to-1 correspondence with the elements of the set $\mathrm{Mor}(\tilde\omega, \omega)$, introduced in Definition 7.2.

This reveals the cellular structure of $\mathbb{R}^d_{\mathsf{coef}}$ and of $\mathbb{R}^{d-1}_{\mathsf{coef},\alpha}$ at P_ω as a product of the cellular structure of $St_\nu(\omega)$ in the vicinity of the apex P_ω times the open cell R_ω. Therefore, the category $\underline{\mathbf{\Omega}}_{\langle d]}$, generated by the poset $(\mathbf{\Omega}_{\langle d]}, \succ)$ governs both the localized cellular structures of $St_\nu(\omega)$ and of $St(\mathsf{R}_\omega)$. This combinatorics may be also expressed in terms of the elementary operations from (7.1)-(7.3).

Recall that any elementary operation M_j or I_j, being applied to an element of $\mathbf{\Omega}$, lowers its reduced norm by 1. Thus each stratum $\mathsf{R}_{\tilde\omega}$ from $St(\mathsf{R}_\omega)$ has $\mathrm{codim}(\mathsf{R}_\omega, \mathsf{R}_{\tilde\omega_{\succeq}}) = k$, if and only if, $\tilde\omega \in \omega_{\prec \rightsquigarrow k}$; that is, if ω can be obtained from $\tilde\omega$ by a sequence of k elementary operations $\{\mathsf{M}_j\}$ and $\{\mathsf{I}_j\}$. In particular, there are exactly as many R_ω-localized connected components of the collection $\{\mathsf{R}_{\tilde\omega} \supset \mathsf{R}_\omega\}_{\tilde\omega}$, where $\dim(\mathsf{R}_{\tilde\omega}) = \dim(\mathsf{R}_\omega) + 1$, as there are elementary resolutions and 2-reductions of ω. Each value $\omega(i)$ can be "resolved" in $\omega(i) - 1$ distinct ways:

$$(\omega(i) - 1, 1),\ (\omega(i) - 2, 2),\ \dots,\ (1, \omega(i) - 1)$$

(the order does matter!), and each $\omega(i) = 2$ can be "deleted" or "reduced". Therefore, the total number of elementary resolutions is $\sum_i (\omega(i) - 1) = |\omega|'$. We notice that distinct resolutions of ω produce distinct $\tilde\omega$'s. The total number of reductions is $\#(\omega^{-1}(2))$, the cardinality of $\omega^{-1}(2)$. All together, there are $|\omega|' + \#(\omega^{-1}(2))$ inverses of the elementary operations, applicable

[9] In contrast with the standard definition of the star, here we ignore the cells $\mathsf{R}_{\tilde\omega}$ such that $\mathsf{R}_{\tilde\omega_{\succeq}} \cap \mathsf{R}_\omega \neq \emptyset$, but $\mathsf{R}_{\tilde\omega_{\succeq}}$ does not contain R_ω.

to ω. We notice also that deleting 2 from a string of several *consecutive* 2's will lead to the same $\tilde{\omega}$. however, different 2-deletions from the same string of 2's will produce different elements of the set $\mathrm{Mor}(\tilde{\omega}, \omega)$ and thus different connected components of the corresponding $\mathsf{R}_{\tilde{\omega}}$ in the vicinity of R_ω. Using the product structure in the normal bundle to R_ω, similar conclusion is valid for the connected components of the strata in $St_\nu(\omega)$.

Stated differently, the *multiplicity* of each cell R_ω—the number of connected components of dimension $(\dim(\mathsf{R}_\omega)+1)$ that contain it—is the *codimension* of R_ω in $\mathbb{R}^d_{\mathsf{coef}}$ (or in $\mathbb{R}^{d-1}_{\mathsf{coef},\alpha}$) plus $\#(\omega^{-1}(2))$.

By a similar argument, each cell $\mathsf{f}_{\tilde{\omega},\beta} \subset St_\nu(\omega)$, a connected component of the stratum $\mathsf{f}_{\tilde{\omega}} \subset St_\nu(\omega)$, is contained in exactly $|\tilde{\omega}|' + \#(\tilde{\omega}^{-1}(2))$ cells of the next dimension.

Finally, using the natural deformation retraction of $\mathbb{R}^d_{\mathsf{coef}}$ onto the α-balanced polynomial space $\mathbb{R}^d_{\mathsf{coef},\alpha}$, the retraction which preserves the $\mathbf{\Omega}_{\langle d]}$-stratifications of the two spaces, we see that the same arguments hold for the α-balanced polynomials. $\square$

Remark 7.1. Note that the combinatorics of these cellular structures is different from the combinatorics of the standard simplex Δ^d: the multiplicity of each subsimplex $\Delta' \subset \Delta^d$ (the number of $(\dim(\Delta')+1)$-simplexes that contain Δ') is equal to its codimension of Δ' (to $|\tilde{\omega}|'$ in our notations); thus the "defect" $\#(\tilde{\omega}^{-1}(2))$ measures the deviation of the $\mathbf{\Omega}_{\langle d]}$-labeled cellular structures in $\mathbb{R}^d_{\mathsf{coef}}$ from the standard simplicial one in the simplex Δ^d. $\diamondsuit$

7.5 Combinatorics of multi-tangent trajectories

We continue to investigate of the topology and combinatorics of traversing flows that we have started in the end of Chapter 6.

As before, let X be a compact connected smooth $(n+1)$-manifold with boundary. For a boundary generic vector field v, each v-trajectory γ intersects the boundary $\partial_1 X$ at a finite number of points a of multiplicities $m(a) \leq n+1$. Recall that formulas (6.8)-(6.10) attach the multiplicity $m(\gamma)$, the reduced multiplicity $m'(\gamma)$, and the virtual multiplicity $\mu(\gamma)$ to each trajectory γ. For traversally generic fields, in addition, we have $m(\gamma) \leq 2(n+1)$ and $m'(\gamma) \leq n$ for every γ (see Theorem 6.7).

The space of v-trajectories $\mathcal{T}(v)$ is given the quotient topology, so that the obvious quotient map $\Gamma : X \to \mathcal{T}(v)$ is continuous. Let us stress again that when v has singularities, the trajectory space $\mathcal{T}(v)$ is quite pathological (non-separable). In contrast, by Corollary 7.1, for (nonsingular) traversally generic vector fields, $\mathcal{T}(v)$ is a decent space, a compact CW-complex.

For any sub-poset $\mathbf{\Theta} \subset \mathbf{\Omega}^\bullet_{'\langle n]}$ and a traversally generic field v on X, let us consider the subsets $X(v, \mathbf{\Theta}) \subset X$ and $\mathcal{T}(v, \mathbf{\Theta}) \subset \mathcal{T}(v)$, formed by points $x \in X$ or by trajectories $\gamma_x \in \mathcal{T}(v)$ whose divisors D_{γ_x} have the combinatorial models, prescribed by the elements of $\mathbf{\Theta}$. In particular, the webs of subspaces

$$\{X(v, \omega_{\succeq})\}_{\omega \in \mathbf{\Omega}^\bullet_{'\langle n]}} \quad \text{and} \quad \{\mathcal{T}(v, \omega_{\succeq})\}_{\omega \in \mathbf{\Omega}^\bullet_{'\langle n]}}$$

form a remarkable geometric structure. It will preoccupy us for the rest of this book.

A cruder stratification (filtration) of X and $\mathcal{T}(v)$ is provided by the spaces

$$\{X(v, \mathbf{\Omega}^\bullet_{'[k,n]})\}_{0 \leq k \leq n} \quad \text{and} \quad \{\mathcal{T}(v, \mathbf{\Omega}^\bullet_{'[k,n]})\}_{0 \leq k \leq n},$$

respectively.

Combining Theorem 6.10 and Theorem 6.8 leads to an useful implication.

Corollary 7.1. *Let X be an $(n+1)$-dimensional compact connected smooth manifold, carrying a traversally generic vector field v. Then the following claims are valid:*

- *The space of trajectories $\mathcal{T}(v)$ admits the structure of finite cell/simplicial complex.*
- *For each $\omega \in \mathbf{\Omega}^\bullet_{'\langle n]}$, the stratum $\mathcal{T}(v, \omega_{\succeq_\bullet})$ is a codimension $|\omega|'$ subcomplex of $\mathcal{T}(v)$.*
- *With respect to an appropriate cellular/simplicial structure in X, the obvious map $\Gamma : X \to \mathcal{T}(v)$ is cellular/simplicial. As a result, Γ is a homotopy equivalence.*

Proof. By Theorem 6.10, the trajectory space $\mathcal{T}(v)$ of a traversally generic flow admits a structure of a Whitney space, being embedded in some ambient Euclidean space.

The fundamental results of [Gor], [Jo], and [Ve] claim that Whiney spaces Y admit smooth triangulations $\tau : T \to Y$, amenable to their stratifications. The adjective "smooth" here refers to the homeomorphism τ

being smooth on the interior of each simplex $\Delta \subset T$ (remember, the pure strata $\mathcal{T}(v, \omega)$ are smooth manifolds!). With respect to such triangulations, the strata are subcomplexes.

Therefore $\mathcal{T}(v)$ admits a finite triangulation so that each stratum $\mathcal{T}(v, \omega_{\succeq\bullet})$ is a subcomplex.

For traversing fields v, over each adjusted to the stratification open simplex $\Delta^\circ \subset \mathcal{T}(v)$, the map $\Gamma : X \to \mathcal{T}(v)$ is a trivial fibration whose fibers are either closed segments, or singletons. Thus each set $\Gamma^{-1}(\Delta^\circ)$ is homeomorphic either to the cylinder $\Delta^\circ \times [0, 1]$, or to Δ°. Since the faces of Δ belong to the $(\mathbf{\Omega}^\bullet_{'\langle n]})$-strata of $\mathcal{T}(v)$ and the Γ-fibers are segments or singletons, $\Gamma^{-1}(\Delta)$ is homeomorphic to a portion of the prism $\Delta \times [0, 1]$, truncated by a hyperplane Π that passes through the verticies of $\Delta \times \{0\}$ or/and of $\Delta \times \{1\}$. The section of $\Delta \times [0, 1]$ by Π projects *onto* the base $\Delta \times \{0\}$. This introduces a cellular structure on X so that Γ becomes a cellular map. With a bit more work, one can refine the cellular structures in X and $\mathcal{T}(v)$, so that Γ becomes a simplicial map.

Therefore, by the first bullet from Theorem 6.8, $\Gamma : X \to \mathcal{T}(v)$ is a weak homotopy equivalence. Since both spaces are CW-complexes, we conclude that Γ is a homotopy equivalence. □

Lemma 6.4 and Theorem 6.7 lead to the following corollary.

Corollary 7.2. *Let X be a compact connected smooth $(n+1)$-manifold with boundary. For any traversally generic vector field v on X and for each $\omega \in \mathbf{\Omega}^\bullet_{'\langle n]}$, the restriction of $\Gamma^\partial : \partial_1 X \to \mathcal{T}(v)$ to the subspace $(\Gamma^\partial)^{-1}(\mathcal{T}(v, \omega))$ is a $|\sup(\omega)|$-to-1 map.*

Moreover, $\Gamma^\partial : (\Gamma^\partial)^{-1}(\mathcal{T}(v, \omega)) \cap \partial_1 X \to \mathcal{T}(v, \omega)$ is a covering map with the trivial monodromy and a fiber of cardinality $|\sup(\omega)|$.

As a result, the map $\Gamma^\partial : \partial_1 X \to \mathcal{T}(v)$ is $(n+2)$-to-1 at most, and the map $\Gamma^\partial : \partial_1^+ X \to \mathcal{T}(v)$ is $(n+1)$-to-1 at most. At the same time, $\Gamma^\partial : \partial_2 X \to \mathcal{T}(v)$ is n-to-1 at most.

Proof. For a traversally generic field, by Theorem 6.7, $m(\gamma) \leq 2(n+1)$. Each trajectory, but a singleton, has exactly two end points of odd multiplicities, the rest of the points are tangency points of even multiplicities. Their number does not exceed n. Thus $\Gamma^\partial : \partial_1 X \to \mathcal{T}(v)$ is $(n+2)$-to-1 at most, $\Gamma^\partial : \partial_1^+ X \to \mathcal{T}(v)$ is $(n+1)$-to-1 at most, and $\Gamma|_{\partial_2 X}$ is n-to-1 at most.

The statement dealing with the cardinalities of the fibers of the map

$$\Gamma^{\partial} : \Gamma^{-1}(\mathcal{T}(v,\omega)) \cap \partial_1 X \to \mathcal{T}(v,\omega)$$

follows instantly from the definitions of the relevant spaces.

When $\omega = (2k)$, where $k \in [1, d/2]$, then $|\omega| = 1$, and the monodromy is trivial. So let us consider any ω whose support is not a singleton. Let β be a loop in $\mathcal{T}(v,\omega)$, and let $E_\beta =_{\mathsf{def}} \Gamma^{-1}(\beta) \subset X$. Note that $\Gamma : E_\beta \to \beta$, thanks to the orientation by v, is a cylinder with the base β (and not a Möbius band). Consider the intersection $E_\beta \cap \partial_1 X$. Since β is contained in the pure stratum $\mathcal{T}(v,\omega)$, $\Gamma : E_\beta \cap \partial_1 X \to \beta$ is a covering map with a finite fiber. Because its space $E_\beta \cap \partial_1 X$ is contained in the cylinder E_β, we conclude that $\Gamma : E_\beta \cap \partial_1 X \to \beta$ must be a trivial covering. □

Now, for a traversing v, we would like to understand better the stratified topology of the trajectory space $\mathcal{T}(v)$, localized to the vicinity of given point-trajectory γ. Since $\mathcal{T}(v)$ is a quotient space of X, it is harder to propose a natural local algebraic model for $\mathcal{T}(v)$. We will attempt, at least partially, to overcame this difficulty.

Let X be a compact smooth $(n+1)$-dimensional manifold. As usual, we operate within an extension germ $(\hat{X}, \hat{v})$ of (X, v). For a boundary generic traversing v and any trajectory γ, the intersection $\gamma \cap \partial_1 X = \coprod_i a_i$ is a v-ordered finite set of points. Assume that each $a_i \in \partial_{j_i} X^{\circ}$.

Then, in the vicinity of γ, there exists a local system of coordinates $(u, x) : \hat{X} \to \mathbb{R}^{n+1}$ so that X is given by the polynomial inequality

$$P(u,x) =_{\mathsf{def}} \prod_i \left[(u-\alpha_i)^{j_i} + \sum_{l=0}^{j_i-2} \phi_{i,l}(x)(u-\alpha_i)^l\right] \leq 0, \qquad (7.9)$$

where $\alpha_i = u(a_i)$ and $\phi_{i,l} : \mathbb{R}^n \to \mathbb{R}$ are some smooth functions which vanish at the origin; moreover, for each i, the rank of the differential of the map Φ_i, defined by $\{\phi_{i,0}, \dots, \phi_{i,j_i-2}\}$, is $j_i - 1$, the maximal possible.

Unfortunately, for a boundary generic traversing v, the combinatorial type $\omega = (j_1, \dots, j_q)$ alone is not sufficient to determine the γ-localized topology of the trajectory space $\mathcal{T}(v)$: it may depend on some continuous parameters.

Assuming now that v is traversally generic, our immediate goal is to describe one particular *localized* cellular structure of the trajectory space $\mathcal{T}(v)$. It is governed by the combinatorics of the divisors in $\mathbb{R}$ of real degree $\leq 2(n+1)$ polynomials.

For a traversally generic v, by Theorem 6.7 (see also Lemma 6.4), this tangency pattern $\omega = (j_1, j_2, \dots)$ is an element of the poset $\mathbf{\Omega}^{\bullet}_{'\langle n]} \cap \mathbf{\Omega}_{\langle 2n+2]}$.

By Lemma 6.4 and formula (6.26), in special coordinates (u, x, y) in some $\hat{v}$-adjusted tube surrounding γ, the manifold X is given by the inequality

$$P(u, x) =_{\mathsf{def}} \prod_i \; \Big[(u - \alpha_i)^{j_i} + \sum_{l=0}^{j_i - 2} x_{i,l}(u - \alpha_i)^l\Big] \;\leq\; 0, \qquad (7.10)$$

where $\alpha_i = u(a_i)$ and $x = \{x_{i,l}\}$. In the local analysis of traversally generic flows, we may pick the canonical model of X in the vicinity of γ by assuming that each $\alpha_i = i$.

If $|\omega| \equiv 0 \mod 2$, for each fixed value of the coordinates (x, y), the solution set of (7.10) is a disjoint union of several closed intervals and singletons, residing in the u-line $\hat{\gamma}_x$ (see Figure 6.1), the union depending on x alone. Each of these intervals and singletons represent a v-trajectory, suspended over (x, y) (for some x, $\hat{\gamma}_x$ may be empty!). So to get the space of trajectories $\mathcal{T}(v)$ in the vicinity of γ, we need to collapse each interval to an appropriate point-marker that resides in it. Let us formalize the collapsing procedure.

Consider the solution set $\mathcal{E}_\omega$ of the inequality (7.10). We say that two points $(u, x, y), (u', x', y') \in \mathcal{E}_\omega$ are equivalent ("$\sim$"), if $x = x', y = y'$, and the interval $([u, u'], x, y) \subset \mathcal{E}_\omega$. Now we define the space $\mathcal{T}_\omega$ as the quotient space $\mathcal{E}_\omega / \sim$. Actually, this construction of the model trajectory space makes sense using the solution of the inequality (7.9). However, the resulting trajectory space will depend on the u-polynomial $P(u, x)$ in the ways that is difficult to control.

The space $\mathcal{T}_\omega$ comes equipped with the map

$$p : \mathcal{T}_\omega \to \mathbb{R}^{|\omega|'} \times \mathbb{R}^{n - |\omega|'},$$

induced by the projection $(u, x, y) \to (x, y)$. Since X is compact, for each x, the polynomial $P(u, x)$ in (7.10) has finitely many closed intervals, where it is non-negative. Thus p is a ramified map with finite fibers.

For any fixed x, the u-polynomial in $P(u, x)$ in (7.10) can be viewed also as an element of the space $\mathcal{P}^{|\omega|} =_{\mathsf{def}} \mathbb{R}^{|\omega|}_{\mathsf{coef}}$. As such, it belongs to a unique pure stratum $\mathsf{R}_{\omega'} =_{\mathsf{def}} \mathsf{R}_{\omega'}(x) \subset \mathcal{P}^{|\omega|}$, where $\omega' \succeq \omega$ in $\mathbf{\Omega}_{\langle |\omega|]}$. Therefore, with the help of (7.10), each $x \in \mathbb{R}^{|\omega|'}$ has a well-defined combinatorial type $\omega(x) = \omega' \in \mathbf{\Omega}_{\langle|\omega|]}$, associated to it. As a result, $\mathbb{R}^{|\omega|'}$ is an $\mathbf{\Omega}_{\langle|\omega|]}$-stratified space, and so is, via the obvious projection onto $\mathbb{R}^{|\omega|'}$, the space $\mathcal{E}_\omega \subset \mathbb{R} \times \mathbb{R}^{|\omega|'} \times \mathbb{R}^{n-|\omega|'}$.

Recall that the ∂_u-trajectories, residing in $\mathcal{E}_\omega$, must have the combinatorial patterns $\omega' \in \mathbf{\Omega}^\bullet$, where $\omega' \succeq_\bullet \omega$. So, the space $\mathcal{E}_\omega$ admit a "more intrinsic" stratification, which is labeled by the elements of the poset $\mathbf{\Omega}^\bullet_{'\langle k]}$, where $k = |\omega|'$. In a sense, this stratification is cruder than the $\mathbf{\Omega}_{\langle|\omega|]}$-stratification of $\mathbb{R}^{|\omega|'}$.

Here is the description of this $\mathbf{\Omega}^\bullet_{'\langle k]}$-stratification. For each point $(u_\star, x, y) \in \mathcal{E}_\omega$, there is a unique closed interval $I_{u_\star, x} =_{\mathsf{def}} [a, b]$ such that: (1) $u_\star \in [a, b]$, (2) $P(u, x) \leq 0$ for all $u \in [a, b]$, and $I_{u_\star, x}$ is the maximal closed interval possessing these two properties.

Consider the real zero divisor $D_{(u_\star, x)}$ of the u-polynomial $P(u, x)$, being *restricted* to the interval $I_{u_\star, x}$. Its combinatorial type $\omega(u_\star, x) \in \mathbf{\Omega}^\bullet$ and is independent on the choice of $u_\star$ within the interval $I_{u_\star, x}$. Thus, $\omega(u_\star, x)$ depends only on the equivalence class of $(u_\star, x, y) \in \mathcal{E}_\omega$, a point in $\mathcal{T}_\omega$.

On the other hand, if $\omega(x) = \omega(x')$ for some $x, x' \in \mathbb{R}^{|\omega|'}$, then there exist $(u_\star, x, y), (u'_\star, x', y) \in \mathcal{E}_\omega$ such that $\omega(u_\star, x) = \omega(u'_\star, x')$. Moreover, if we order the connected components of the sets $P^{-1}((-\infty, 0], x)$ and $P^{-1}((-\infty, 0], x')$ by the u-values, then the intervals $I_{u_\star, x}$ and $I_{u'_\star, x'}$ share the same ordinal. Stated differently, the combinatorial type $\omega(x) \in \mathbf{\Omega}_{\langle|\omega|]}$ determines *the unique ordered sequence* $\Xi(\omega(x))$ *of types* $\{\omega(u_\star, x) \in \mathbf{\Omega}^\bullet\}_{u_\star}$ for points in the fiber $p^{-1}(x)$ (cf. the discussion, preceding Figure 7.1).

Theorem 7.3. *Let X be a compact connected smooth $(n+1)$-manifold X with boundary, and v a traversally generic vector field v on X. Pick $\omega \in \mathbf{\Omega}^\bullet_{'\langle n]}$ and put $k = |\omega|'$.*

Then, for any v-trajectory γ with the tangency pattern ω, the $\mathbf{\Omega}^\bullet_{'\langle k]}$-stratified topological type of the germ $\mathcal{T}_\omega$ of the trajectory space $\mathcal{T}(v)$ at the point-trajectory γ is determined by ω alone.

Proof. By Theorem 6.5, any $\hat{v}$-trajectory γ has a flow-adjusted neighborhood in which X is given by the inequality (7.10). By the arguments that precede Theorem 7.3, the construction (in terms of the space $\mathcal{E}_\omega$ and its projection on $\mathbb{R}^k \times \mathbb{R}^{n-k}$) of the space $\mathcal{T}_\omega$ and of its $\mathbf{\Omega}^\bullet_{'\langle|\omega|']}$-stratification depend (with the help of Ξ) only on the combinatorial pattern $\omega \in \mathbf{\Omega}^\bullet$. □

Example 7.3. For the traversally generic vector fields v on smooth compact connected 4-folds X with boundary, there are 11 distinct local topological types for $\mathcal{T}(v)$. They are labeled by the *marked* elements of the poset from Figure 7.5 (each small triangle in the figure is a marker). $\diamondsuit$

Next, we will employ similar considerations to describe in some detail the germ at γ of a cellular structure in $\mathcal{T}_\omega$, the structure subordinate to the $\Omega^\bullet_{'\langle k]}$-stratification of $\mathcal{T}_\omega$, where $k = |\omega|'$. Eventually, these investigations will culminate in Theorem 7.4 below.

By choosing an appropriately narrow $\hat{v}$-adjusted neighborhood $U \subset \hat{X}$ of γ, thanks to Lemma 6.2, for any x, sufficiently close to the origin, the real roots of $P(u, x)$ from (7.10) can be separated into disjoint groups. These groups correspond to the real roots $\{\alpha_i\}_i$ of the polynomial $P(u, 0)$[10] and reside in their vicinity.

Thus, for any $\hat{v}$-trajectory $\hat{\gamma} \subset U$, the real zero divisor $D_{\hat{\gamma}}$ splits canonically into several real divisors $\{D_{\hat{\gamma},i}\}_i$. Their combinatorial types are described by some elements $\hat{\omega}_i \in \mathbf{\Omega}_{\langle\omega(i)]}$ such that $|\hat{\omega}_i| \leq \omega(i)$ and $|\hat{\omega}_i| \equiv \omega(i)$ mod 2. In other words, $\omega_{\hat{\gamma}}$, the combinatorial type of $D_{\hat{\gamma}}$, is determined by an element of the product $\prod_{i \in \mathsf{sup}(\omega)} \mathbf{\Omega}_{\langle\omega(i)]}$. Evidently, there is a canonical assembly map

$$\kappa : \prod_{i \in \mathsf{sup}(\omega)} \mathbf{\Omega}_{\langle\omega(i)]} \longrightarrow \mathbf{\Omega}_{\langle|\omega|]}$$

that places the elements $\{\hat{\omega}_i\}_i$ in a single array $\omega_{\hat{\gamma}} =_{\mathsf{def}} (\hat{\omega}_1, \hat{\omega}_2, \dots)$. It is the inverse map to the decomposition Ξ (shown in Figure 7.1).

For each i, we denote by $x_{\{i\}}$ the ordered subset $\{x_{i,1}, x_{i,2}, \dots\}$ of the x-coordinates, amenable to the formula (7.10).

The decomposition of $P(u, x) = \prod_i P_i(u - \alpha_i, x_{\{i\}})$ of the left-hand side of (7.10) into a product of monic depressed polynomials $P_i(u - \alpha_i, x_{\{i\}})$ in $u - \alpha_i$ of degrees $j_i = \omega(i)$ gives rise a continuous map

$$A_\omega : \mathbb{R}^{|\omega|'} \longrightarrow \prod_{i \in \mathsf{sup}(\omega)} \mathbb{R}^{\omega(i)-1}_{\mathsf{coef},\alpha_i} \tag{7.11}$$

from the space of x-coordinates $\mathbb{R}^{|\omega|'}$, to the product of polynomial spaces $\{\mathbb{R}^{\omega(i)-1}_{\mathsf{coef},\alpha_i}\}_i$. The dimension of the product is also $|\omega|'$. If $x \neq x'$, then at least for one i, the vectors $x_{\{i\}}$, $x'_{\{i\}}$ are distinct, since the groups of coordinates $\{x_{\{i\}}\}_i$ are distinct. Thus the monic $(u - \alpha_i)$-polynomials $P_i(u - \alpha_i, x_{\{i\}})$ and $P_i(u - \alpha_i, x'_{\{i\}})$ must have distinct coefficients. As a result, A_ω is a

[10] which is a product of linear polynomials over $\mathbb{R}$

1-to-1 map. Evidently, by (7.10), A_ω is surjective. So A_ω establishes a smooth homeomorphism between the two spaces from (7.11).

With the help of Theorems 7.1 and 7.2, we get the cellular structures

$$\{\Phi_{\alpha_i,\hat{\omega}_i} : e_{\alpha_i,\hat{\omega}_i} \longrightarrow \mathsf{R}_{\alpha_i,\hat{\omega}_i}\}_{\hat{\omega}_i \in \mathbf{\Omega}_{\langle \omega(i)]}}$$

for each space $\mathbb{R}^{\omega(i)-1}_{\mathsf{coef},\alpha_i}$. By the homeomorphism A_ω^{-1} from (7.11), this product of cellular structures in the target space of (7.11) gives rise to a cellular structure in the space $\mathbb{R}^{|\omega|'}$ of x-coordinates.

Let $\hat{\omega} =_{\mathsf{def}} \prod_{i \in \mathsf{sup}(\omega)} \hat{\omega}_i$. Employing the assembly map κ, the cellular maps $\{\Phi_{\alpha_i,\hat{\omega}_i}\}_i$, and the homeomorphism A_ω, we get the cellular maps

$$\Psi_{\hat{\omega}} =_{\mathsf{def}} A_\omega^{-1} \circ \Big(\prod_{i \in \mathsf{sup}(\omega)} \Phi_{\alpha_i,\hat{\omega}_i} \Big) : \quad \prod_{i \in \mathsf{sup}(\omega)} e_{\alpha_i,\hat{\omega}_i} \longrightarrow \mathbb{R}^{|\omega|'} \qquad (7.12)$$

that are consistent with the stratification of $\mathbb{R}^{|\omega|'}$, labeled by the elements $\{\hat{\omega}_i\}_i$ of the poset $\prod_{i \in \mathsf{sup}(\omega)} \mathbf{\Omega}_{\langle \omega(i)]}$. Using κ, we may identify this product with the poset $\mathbf{\Omega}_{\langle |\omega|]}$.

In what follows, the cellular structure in the product $\mathbb{R}^{|\omega|'} \times \mathbb{R}^{n-|\omega|'}$ is chosen to be this cellular structure in $\mathbb{R}^{|\omega|'}$ (given by (7.12)), being multiplied by a single open cell $e^{n-|\omega|'} \approx \mathbb{R}^{n-|\omega|'}$.

So far, the cells in $\mathbb{R}^{|\omega|'}$ are labelled by the elements of the poset $\omega_{\preceq} \subset \mathbf{\Omega}_{\langle |\omega|]}$, and not by elements of the poset $\omega_{\bullet \preceq} \subset \mathbf{\Omega}^{\bullet}_{'\langle |\omega|']}$, appropriate for the trajectories in $\mathcal{E}_\omega$ (see the discussion preceding Theorem 7.3)!

Next, using the ramified map $p : \mathcal{T}_\omega \to \mathbb{R}^{|\omega|'} \times \mathbb{R}^{n-|\omega|'}$ with finite fibers, we will employ the cellular structure in the target space $\mathbb{R}^{|\omega|'}$, given by (7.12), to produce a cellular structure in the source space $\mathcal{T}_\omega$, so that p will become a cellular map. That task will preoccupy us for a while...

With this goal in mind, we introduce *markers*, a new combinatorial device (see Figure 7.1 and Figure 7.4).

For each $\omega \in \mathbf{\Omega}_{\langle d]}$, consider the auxiliary u-polynomial

$$\wp_\omega(u) = \prod_{i=1}^{|\sup(\omega)|} (u - i)^{\omega(i)}$$

of the degree $|\omega|$ that does not exceed d and shares the same parity with it.

We interpret the support $\sup(\omega)$ as the ordered set $\{1, \dots, |\sup(\omega)|\} \subset \mathbb{N}$. Evidently, it coincides with the zero set of $\wp_\omega(u)$.

For a given ω, we consider a pair (ω, k), where the *marker* $k \in \sup(\omega)$ is chosen so that:

- either $\omega(k) \equiv 0 \mod 2$, $\wp_\omega(k - 0.5) > 0$, and $\wp_\omega(k + 0.5) > 0$,
- or $\omega(k) \equiv 1 \mod 2$, $\wp_\omega(k - 0.5) > 0$, and $\wp_\omega(k + 0.5) < 0$.

We denote by $\mathbf{\Omega}^\mu_{\langle d]}$ the set of all *marked pairs* (ω, k) as above. Let $\Upsilon(\omega) \subset \{1, \dots, |\sup(\omega)|\}$ be the set of all markers k, associated with ω.

Each marker $k \in \Upsilon(\omega)$, as an element of linearly ordered set $\Upsilon(\omega)$, acquires its ordinal $p = p(k)$. We will denote the p-th marker in $\Upsilon(\omega)$ by k_p. Let $\Upsilon_p(\omega) \subset \mathbb{N}$ be the maximal set of consecutive natural numbers $j \geq k_p$ such that $\wp_\omega(u) \leq 0$ for all $u \in [k_p, j]$.

With the markers in place, it is possible to extend the elementary operations M_j and I_j (merge and insert), introduced in (7.1)-(7.3) for the poset $\mathbf{\Omega}_{\langle d]}$, to the elements of the new set $\mathbf{\Omega}^\mu_{\langle d]}$.

The new merge operation $\mathsf{M}^\mu_j : \mathbf{\Omega}^\mu_{\langle d]} \to \mathbf{\Omega}^\mu_{\langle d]}$ is shown in Figure 7.4 that illustrate the behavior of markers under merges. Let

$$\mathsf{M}^\mu_j(\omega, k) =_{\mathsf{def}} (\mathsf{M}_j(\omega), \mu_j(k)), \tag{7.13}$$

where the new marker $\mu_j(k) \in \Upsilon(\mathsf{M}_j(\omega))$ is defined formally as follows.

If both j and $j+1$ belong to the same set $\Upsilon_p(\omega)$, and the marker $k \in \Upsilon_q(\omega)$, where $q \neq p$, then $\mu_j(k) = k$ as elements of the set $\Upsilon_q(\mathsf{M}_j(\omega)) = \Upsilon_q(\omega)$.

If j and $j+1$ belong to $\Upsilon_p(\omega)$, and the marker $k \in \Upsilon_p(\omega)$, then again $\mu_j(k) = k$ as elements of the set $\Upsilon_p(\mathsf{M}_j(\omega)) \subset \Upsilon_p(\omega)$.

At the same time, if $j \in \Upsilon_p(\omega)$, $j+1 \in \Upsilon_{p+1}(\omega)$, and $k \in \Upsilon_{p+1}(\omega)$, then $\mu_j(k)$ is defined to be the unique minimal element in $\Upsilon_p(\mathsf{M}_j(\omega))$. When $k \in \Upsilon_q(\omega)$ and $q \neq p+1$, the marker keeps its minimal position within the subset $\Upsilon_q(\omega)$.

Similarly, the new insert operation

$$\mathsf{I}^\mu_j(\omega, k) =_{\mathsf{def}} (\mathsf{I}_j(\omega), \mu_j(k)) \tag{7.14}$$

is described formally as follows. Under I^μ_j, ω is subjected to the old insert operation I_j, while the marker $k \in \Upsilon_p(\omega)$ keeps its minimal position within the set $\Upsilon_p(\omega) \subseteq \Upsilon_q(\mathsf{I}_j(\omega))$ (where q is uniquely determined by p and j), that is, $\mu_j(k) = k$ as the minimal elements of the appropriate sets.

Here is a slightly more geometrical look at the markers and their behavior. It is motivated by interactions of the traversing vector flows with the boundary $\partial_1 X$. Let $\hat{\gamma} \subset \hat{X}$ be a typical $\hat{v}$-trajectory in the vicinity of a given trajectory $\gamma \subset X$. The intersection $\hat{\gamma} \cap \partial_1 X = \{a_l\}_{1 \leq l \leq q}$, allows us to *shade* some of the intervals $[a_l, a_m]$ in which $\partial_1 X$ divides $\hat{\gamma}$: an interval is shaded if it belongs to X. Using any auxiliary function $z : \hat{X} \to \mathbb{R}$ (as in

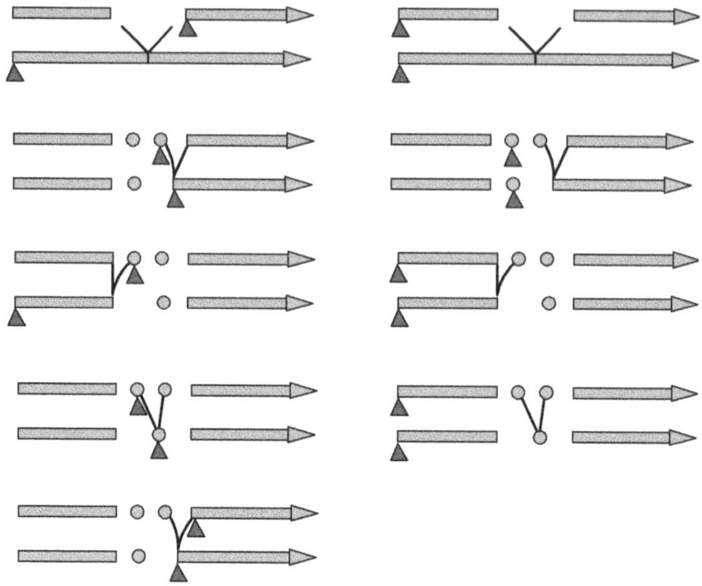

Fig. 7.4 The rules by which the markers (the bold triangles) evolve under the merge operations (depicted as the "V-shaped passages" from the top to the bottom bars in each of the seven diagrams).

Lemma 6.1), the shading is defined as the locus where $z|_\gamma \leq 0$.[11] Therefore, the ordered sequence of the multiplicities $\omega_{\hat{\gamma}} = (m(a_1), m(a_2), \dots, m(a_q))$ uniquely determines which intervals $[a_l, a_m]$ along $\hat{\gamma}$ are shaded (cf. Figure 6.1, where the shading corresponds to "strings" and "atoms"): each shaded interval can be marked with a unique lowest *odd*-multiplicity point in it; also each atom can be marked.

As in Section 7.2 (see the discussion preceding Figure 7.1), we denote by $\sup_{odd}(\omega)$ the points $l \in \mathbb{N}$ in the support of ω such that $\omega(l) \equiv 1 \mod 2$, and by $\sup_{ev}(\omega)$ the points l in the support of ω such that $\omega(l) \equiv 0 \mod 2$. We can count the elements of $\sup_{odd}(\omega)$ and pick only the ones that acquire odd numerals in that count (in this way, we pick half of the elements in $\sup_{odd}(\omega)$). We denote this set by $\sup^+_{odd}(\omega)$ and its complement by $\sup^-_{odd}(\omega)$. We divide the points of $\sup_{ev}(\omega)$ also into two complementary sets: the first one, $\sup^-_{ev}(\omega)$, contains points that are bounded from below

[11]Recall that z is chosen to possess the following properties: 1) 0 is a regular z-value and $z^{-1}(0) = \partial_1 X$; 2) $z^{-1}((-\infty, 0]) = X$.

by a point from $\sup^+_{odd}(\omega)$ and from above by a point from $\sup^-_{odd}(\omega)$; the second one, $\sup^+_{ev}(\omega)$, contains points that are bounded from below by a point from $\sup^-_{odd}(\omega)$ and from above by a point from $\sup^+_{odd}(\omega)$. In fact, $\Upsilon(\omega) = \sup^+_{odd}(\omega) \cup \sup^+_{ev}(\omega)$.

Consider $\omega' \succeq \omega$, where $\omega', \omega \in \mathbf{\Omega}_{\langle d]}$, and two elements-markers $k' \in \Upsilon(\omega')$ and $k \in \Upsilon(\omega)$. Our next task is to define a relation "$k' \rightsquigarrow k$" between k' and k. Recall that ω can be obtained from ω' by a sequence of elementary merges and multiplicity 2 inserts as described in (7.1)-(7.3). Given a marker $k' \in \Upsilon(\omega')$, we will describe its evolution under these elementary transformations of ω'.

An insertion of an even multiplicity point does not affect any marker below the insertion and shifts the location of the marker above the insertion by one. If two points from $\sup^+_{ev}(\omega')$ merge and one of them is marked, the maker is placed at the location where the merge took place. If a marked point from $\sup^+_{ev}(\omega')$ is merging with a point from $\sup^+_{odd}(\omega')$, the location of the merge is marked. On the other hand, if a marked point from $\sup^+_{ev}(\omega')$ is merging with a point from $\sup^-_{odd}(\omega')$, then the marker is placed at the first point of $\sup^+_{odd}(\omega')$, located below the merge—the marker "travels down until it reaches a point of $\sup^+_{odd}(\omega')$". Finally, if a marked point from $\sup^+_{odd}(\omega')$ merges with the adjacent point from $\sup^-_{odd}(\omega')$, the marker is placed at the first point of $\sup^+_{odd}(\omega')$, located below the merge. If one inserts a point of an even multiplicity, the marker does not change its location.

Therefore, when $\omega' \succeq \omega$, the marker k' defines a unique marker $k \in \Upsilon(\omega)$. In such a case, we say that k' *collapses* to k and write "$k' \rightsquigarrow k$".

There is a natural map $\Xi : \mathbf{\Omega}^\mu \to \mathbf{\Omega}^\bullet$ which acts from the set of all marked pairs (ω, k) to the set $\mathbf{\Omega}^\bullet$ of strings and atoms (see Figure 7.1). By definition, Ξ takes each pair (ω, k), $k \in \Upsilon(\omega)$, to the restriction of ω to the unique maximal interval $[k, j(k)]$ of indices $j \in \sup(\omega)$ with the property $\wp_\omega(j) \leq 0$ for all $j \in [k, j(k)]$. Then, by a shift of indices, we reinterpret $\omega : [k, j(k)] \to \mathbb{N}$ as a map $\omega : [1, j(k) - k - 1] \to \mathbb{N}$, an element of $\mathbf{\Omega}^\bullet$. In fact, $\Xi : \mathbf{\Omega}^\mu_{\langle 2n+2]} \to \mathbf{\Omega}^\bullet_{'\langle n]}$ for each n.

In the following theorem, we combine the acquired knowledge about one particular cellular structure of the model spaces $\{\mathcal{T}_\omega\}$ with the local models for traversally generic vector fields (described in Lemma 6.4). This leads to a *local* purely combinatorial description of trajectory spaces for traversally generic fields. Regretfully, the formulation of the theorem is lengthy.

Theorem 7.4. *Let X be a compact connected smooth $(n+1)$-manifold with boundary. Let v be a traversally generic vector field on X and γ its trajectory of the combinatorial type $\omega \in \mathbf{\Omega}^{\bullet}_{'\langle n]}$.*

Then the following statements hold:

- *In the vicinity of γ, the trajectory space $\mathcal{T}(v)$ has a structure of the model $|\omega|'$-dimensional finite CW-complex $\mathcal{T}_\omega$, being multiplied by the vector space $\mathbb{R}^{n-|\omega|'}$. The first factor in the product models the normal star of $\mathcal{T}(v,\omega)$ in $\mathcal{T}(v)$, the second factor models the stratum $\mathcal{T}(v,\omega)$.*
- *Each open cell $\mathsf{E}_{\hat\omega,\,k} \subset \mathcal{T}_\omega$ (of the dimension $|\omega|' - |\hat\omega|'$) is indexed by an element*
$$\hat\omega = \prod_{i\in \mathsf{sup}(\omega)} \hat\omega_i \in \prod_{i\in \mathsf{sup}(\omega)} \mathbf{\Omega}_{\langle \omega(i)]}{}^{12},$$
together with a marker $k \in \Upsilon(\kappa(\hat\omega))$. Here $\kappa(\prod_i \hat\omega_i) \succeq \omega$.

  The cell $\mathsf{E}_{\hat\omega,\,k}$ models the pure stratum $\mathcal{T}\big(v,\, \Xi(\kappa(\hat\omega),\, k)\big)$ of $\mathcal{T}(v)$, labeled by the element $\Xi(\kappa(\hat\omega),\, k)\big) \in \mathbf{\Omega}^{\bullet}_{'\langle n]}$.
- *$\mathsf{E}_{\hat\omega',\,k'} \subset \mathsf{E}_{\hat\omega_{\succeq},\,k}$ if and only if the following two conditions are satisfied:* **(1)** *$\hat\omega \succeq \hat\omega'$ in the product poset $\prod_{i\in \mathsf{sup}(\omega)} \mathbf{\Omega}_{\langle\omega(i)]}$,* **(2)** *the markers k and k' satisfy the collapsing relation "$k \rightsquigarrow k'$"[13].*
- *Employing the attaching maps*
$$\Psi_{\hat\omega} : \, e_{\hat\omega} =_{\mathsf{def}} \prod_{i\in \mathsf{sup}(\omega)} e_{\hat\omega_i} \longrightarrow \mathbb{R}^{|\omega|'}$$
from (7.12), the space $\mathcal{T}_\omega$ can be assembled from the union of marked cells $\{(e_{\hat\omega},\, k)\}_{\hat\omega,\,k}$ according to the rules described in Figure 7.4. Each closed stratum $\mathsf{E}_{\hat\omega_{\succeq},\,k}$ is the image of the cell $(e_{\hat\omega},\, k)$ under the attaching map $\Psi_{\hat\omega,\,k} : \coprod_{\hat\omega,k}(e_{\hat\omega},\, k) \to \mathcal{T}_\omega$, based on (7.12).
- *Each closed stratum $\mathsf{E}_{\hat\omega_{\succeq},\,k}$, where $\kappa(\hat\omega) \succ \omega$, topologically is a positive cone over a compact stratum $\mathsf{S}_{\hat\omega_{\succeq},\,k}$. These strata $\{\mathsf{S}_{\hat\omega_{\succeq},\,k}\}$ form a link of the base point $\gamma \in \mathcal{T}_\omega$. The rules that describe their incidences are similar to the incidence rules for $\{\mathsf{E}_{\hat\omega_{\succeq},\,k}\}$'s in $\mathcal{T}_\omega$ (see Lemma 7.8).*
- *$\mathcal{T}_\omega$ is equipped with a ramified cellular map $p : \mathcal{T}_\omega \to \prod_{i\in \mathsf{sup}(\omega)} \mathbb{R}^{\omega(i)-1}_{\mathsf{coef},\,i}$ whose fibers are of the cardinality $|\omega|/2$ at most.*

  The image $\mathsf{E}_{\hat\omega} =_{\mathsf{def}} p(\mathsf{E}_{\hat\omega,\,k})$ is the product of the cells $\{\mathsf{E}_{\hat\omega_i} \subset \mathbb{R}^{\omega(i)-1}_{\mathsf{coef},\,i}\}_i$. The cellular structure $\{\mathsf{E}_{\hat\omega_i}\}_{\hat\omega_i \in \mathbf{\Omega}_{\langle\omega(i)]}}$ in each space $\mathbb{R}^{\omega(i)-1}_{\mathsf{coef},\,i}$ is described in Theorem 7.1 and Lemma 7.8 in terms of the poset $(\mathbf{\Omega}_{\langle\omega(i)]}, \succ)$.

---

[12]where the partial order in $\prod_i \mathbf{\Omega}_{\langle\omega(i)]}$ is the one of the posets product.
[13]see Figure 7.4.

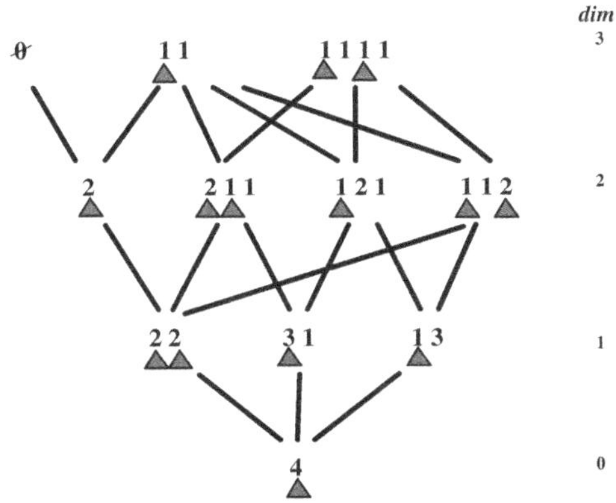

Fig. 7.5 The poset $\mathbf{\Omega}_{\langle 4]}$ with all possible markers k, denoted here by the symbol "$\triangle$". The space $\mathcal{T}_{(4)}$ is the trajectory space of the ∂_u-flow in the domain $\mathcal{E}_{(4)} = \{u^4 + x_2u^2 + x_1u + x_0 \leq 0\} \subset \mathbb{R}^4$. The diagram indicates that the 3-complex $\mathcal{T}_{(4)}$ is assembled from three 3-cells, six 2-cells, four 1-cells and one 0-cell.

Proof. As usual, we extend (X, v) to an external pair $(\hat{X}, \hat{v})$.

For $v \in \mathcal{V}^\dagger(X)$ and each trajectory $\hat{\gamma} \in \hat{X}$, the intersection $\hat{\gamma} \cap X$ consists of a number of closed ("shaded") intervals and few isolated points of even multiplicities. Each of these intervals or isolated points defines a singleton in the space $\mathcal{T}(v)$. We place a single marker "$\triangle$" at one of the lower ends of the shaded intervals or at one of the isolated points, so that a trajectory $\hat{\gamma} \in \hat{X}$ and the marker $\triangle \in \hat{\gamma} \cap X$ will determine a point in $\mathcal{T}(v)$. Thus a pair $(\hat{\gamma}, \triangle)$, where $\hat{\gamma} \in \mathcal{T}(\hat{v})$, may be viewed a point of $\mathcal{T}(v)$. Here we distinguish between the combinatorial marker k and its geometrical realization $\triangle \in \partial_1 X \cap \hat{\gamma}$.

Let S denotes a local section of the $\hat{v}$-flow through a point on γ, which resides slightly below the first intersection $a_1 \in \gamma \cap \partial_1 X$. In the vicinity of a given trajectory $\gamma \subset X$, each trajectory $\hat{\gamma} \subset \hat{X}$ is determined by the intersection point $\hat{\gamma} \cap S$. By Theorems 7.1 and 7.2, the combinatorial types $\hat{\omega} \in \prod_i \mathbf{\Omega}_{\langle \omega(i)]}$, or rather their κ-images, label strata $\mathsf{E}_{\hat{\omega}}$ of codimension $|\kappa(\hat{\omega})|'$ in S.

To define the corresponding cell decomposition of $\mathcal{T}(v)$ in the vicinity of γ, we need to take multiple copies $\{\mathsf{E}_{\hat{\omega},k}\}_k$ of each cell $\mathsf{E}_{\hat{\omega}}$, the copies that are indexed by all admissible markers k. Their number is the cardinality

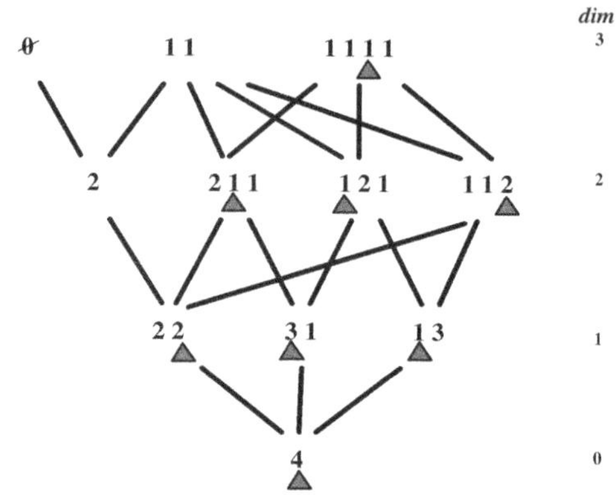

Fig. 7.6 The poset $\mathbf{\Omega}_{\langle 4]}$ with the downwards evolution "$\rightsquigarrow$" of a particular marker $\triangle$, attached to the third entry of $\omega = (1111)$. The diagram indicates that the marked 3-cell $e_{(1,1,1_\triangle,1)}$ in $\mathcal{T}_{(4)}$ shares its boundary with three marked 2-cells: $e_{(2,1_\triangle,1)}$, $e_{(1,2_\triangle,1)}$, $e_{(1,1,2_\triangle)}$, and that each of these is attached to two marked 1-cells, and so on...

of the set $\Upsilon(\kappa(\hat{\omega}))$. For each $\hat{\gamma}$, there is a canonic correspondence between the markers $k \in \Upsilon(\kappa(\hat{\omega}))$ and the connected components $\hat{\tau} \in \pi_0(\hat{\gamma} \cap X)$. So we will use elements $k \in \Upsilon(\kappa(\hat{\omega}))$ and $\hat{\omega} \in \prod_i \mathbf{\Omega}_{\langle \omega(i)]}$ to label all the cells in $\mathcal{T}(v)$.

We are going to show that, for any pair $\hat{\omega} \succ \hat{\omega}'$ from $\prod_i \mathbf{\Omega}_{\langle \omega(i)]}$, a k-marked cell $\mathsf{E}_{\hat{\omega},k}$ is incident to a k'-marked cell $\mathsf{E}_{\hat{\omega}',k'}$, if and only if $\mathsf{E}_{\hat{\omega}}$ is incident to $\mathsf{E}_{\hat{\omega}'}$ in S and $k \rightsquigarrow k'$.

In the vicinity of γ, we employ the special coordinates (u, x, y) as in Lemma 6.4. The next argument is similar to the proof of Lemma 7.5. Consider a smooth path $\{w(t)\}_{0 \leq t < 1}$ in the pure stratum $S(\kappa(\hat{\omega})) \subset S$ so that $\lim_{t \to 1} w(t) =_{\mathsf{def}} w_1 \in S(\kappa(\hat{\omega}'))$. As $t \to 1$, the corresponding trajectories $\gamma_{w(t)} \subset \hat{X}$ converge to the trajectory γ_{w_1} in such a way that the sets $\gamma_{w(t)} \cap \partial_1 X$ go through merges of adjacent points or/and insertions of even multiplicity points. Examining the evolution of each element $\hat{\tau}_t \in \pi_0(\gamma_{w(t)} \cap X)$ into an element $\hat{\tau}' \in \pi_0(\gamma_{w_1} \cap X)$, we see that it is captured by the marker collapse $\triangle_t \rightsquigarrow \triangle'$, as depicted in Figure 7.4. Also glance at Figure 7.5 and Figure 7.6, which give examples of such an evolution.

With the rigid rules for the collapses $\tau \rightsquigarrow \tau'$ from Figure 7.4 in place, we can reconstruct the germ $\mathcal{T}_\gamma$ of $\mathcal{T}(v)$ at γ from the cell complex S (whose cells are indexed by the elements of the poset $(\prod_i \mathbf{\Omega}_{\langle \omega(i)]}, \succ)$), and ultimately, from $\omega = \omega(\gamma)$ itself. Speaking informally, in order to attach a cell $\mathsf{E}_{\omega_\succeq, k}$ to a cell $\mathsf{E}_{\omega', k'}$, first we verify whether $k \rightsquigarrow k'$, and if the verification leads to a positive answer, we use the same map that attaches $\mathsf{E}_{\omega_\succeq}$ to $\mathsf{E}_{\omega'}$ in S.

Let us describe a more formal construction of $\mathcal{T}_\gamma \approx \mathcal{T}_\omega \times \mathbb{R}^{n-|\omega|'}$.

For $\hat\omega = \{\hat\omega_i \in \mathbf{\Omega}_{\langle\omega(i)]}\}_i$, let $e_{\hat\omega} = \prod_i e_{\hat\omega_i}$, where the hypersurface

$$e_{\hat\omega_i} \subset \mathsf{Sym}^{|\,\sup(\hat\omega_i)|}(\mathbb{R}) \times \mathsf{Sym}^{\frac{\omega(i)-|\hat\omega_i|}{2}}(\mathbb{H}) \tag{7.15}$$

consists of divisors (D'_i, D''_i) such that $D'_i + D''_i + \tau(D''_i)$ is i-balanced.

Then $\mathcal{T}_\omega$ is the quotient of the space

$$\mathcal{Z}_\omega =_{\mathsf{def}} \coprod_{\hat\omega \in \prod_i \mathbf{\Omega}_{\langle\omega(i)]},\ k \in \Upsilon(\kappa(\hat\omega))} (e_{\hat\omega}, k) \tag{7.16}$$

by the equivalence relation "$(z,k) \sim (z',k')$" that is defined as follows:

- $\hat\omega \succ \hat\omega'$ in the poset $\prod_i(\mathbf{\Omega}_{\langle\omega(i)]}, \succ)$,
- $z \in \partial e_{\hat\omega}$, $z' \in e^\circ_{\hat\omega'}$ are such that $\Psi_{\hat\omega}(z) = \Psi_{\hat\omega'}(z')$ in the space $\prod_i \mathbb{R}^{\omega(i)-1}_{\mathsf{coef},\, i}$, where the Ψ-maps are defined by (7.12),
- $k \rightsquigarrow k'$.

Each closed cell $\mathsf{E}_{\hat\omega_\succeq, k} \subset \mathcal{T}_\omega$ is defined as the equivalence class of the set $(e_{\hat\omega}, k)$ in the quotient space $\mathcal{Z}_\omega/\sim$. Its interior is homeomorphic to an open ball $e^\circ_{\hat\omega}$ of dimension $|\omega|' - |\hat\omega|'$.

As a result, for a traversally generic v-flow, we have precise instructions for assembling the germ of a cell complex $\mathcal{T}_\gamma$, associated with each trajectory $\gamma \in \mathcal{T}(v)$, or rather, with its combinatorial type ω.

In the vicinity of γ, the projection $p : \mathcal{T}(v) \to S$ is evidently a finitely-ramified cellular map with respect to the cellular structures $\{\mathsf{E}_{\hat\omega,k}\}_{\hat\omega,k}$ and $\{\mathsf{E}_{\hat\omega}\}_\omega$.

Note that not every cell in S is the p-image of some cell from $\mathcal{T}(v)$: the cells in S that are pierced by the trajectories $\hat\gamma \subset \hat X$ with the property $\hat\gamma \cap X = \emptyset$ are not in $p(\mathcal{T}(v))$. However, each $\mathsf{E}_{\hat\omega}$ with $\sup(\hat\omega) \neq \emptyset$ belongs to $p(\mathcal{T}(v))$.

For each $\hat\omega$, the p-fiber over the pure stratum $S(\kappa(\hat\omega))^\circ \subset S$ is of the cardinality $\#[\Upsilon(\kappa(\hat\omega))]$. We would like to get an upper bound for $\#[\Upsilon(\kappa(\hat\omega))]$ in terms of the ω. In fact, the half of multiplicity $m(\gamma)/2 = |\omega|/2$ is such an upper bound. Indeed, the divisor D_γ can be resolved in a divisor with $m(\gamma)$ *simple* roots. Recall that $m(\gamma)$ must be even, and, by Theorem

6.5 and Theorem 6.7, any potential resolution of D_γ (with the combinatorics described by elements of $\kappa(\prod_i \mathbf{\Omega}_{\langle\omega(i)]}))$ is realized in the vicinity of γ. Therefore $m(\gamma)/2$ shaded intervals do occur in the vicinity of γ. In fact, for a traversally generic v, the number $m(\gamma)/2 \leq \dim(X)$ is the maximal possible cardinality of the p-fibers in the vicinity of γ.

Next we turn to describing the positive cone structure of $\mathcal{T}(v)$ at γ, which is consistent with its cellular structure $\{\mathsf{E}_{\hat\omega,k}\}_{\hat\omega,k}$. In order to construct a base $\sigma_{\hat\omega}$ for the cone structure in each cell $e_{\hat\omega} = \prod_i e_{\hat\omega_i}$ (see (7.8) and (7.12)), consider a set of disjoint closed 2-balls $\{B(i) \subset \mathbb{C}\}_i$ of radii $1/3$ and with the centers at $i \in \mathbb{R}$ [14].

We denote by $\sigma_{\hat\omega} \subset e_{\hat\omega}$ the set of complex conjugation-invariant divisor pairs $(D', D'') = \oplus_i(D'_i, D''_i)$, where D'_i is a real divisor, the divisor $D'_i + D''_i + \tau(D''_i)$ of degree $\omega(i)$ is i-balanced, and the $\sup(D'_i + D''_i) \subset B(i)$; furthermore, at least for one i, $\sup(D'_i + D''_i)$ is not contained in the interior of $B(i)$. The group $\mathbb{R}^*_+$ acts semi-freely on $e_{\hat\omega} = \prod_i e_{\hat\omega_i}$ by the diagonal action that applies a t-dilatation, centered on the point $i \in \mathbb{R} \subset \mathbb{C}$, to each pair $(D'_i, D''_i) \in e_{\hat\omega_i}$. Evidently, for each pair (D', D'') there is a single $t \in \mathbb{R}^*_+$ so that $t(D', D'') \in \sigma_{\hat\omega}$. This $\mathbb{R}^*_+$-action extends to the space $\mathcal{Z}_\omega$ in (7.16), the action on the markers being trivial.

By Theorem 7.1 and with the help of the Viète map V, a similar semi-free action is available on the space $\prod_i \mathbb{R}^{\omega(i)-1}_{\mathsf{coef},i}$. Since the maps $\prod_i \Phi_{i,\hat\omega_i}$ are $\mathbb{R}^*_+$-equivariant, the quotient space $\mathcal{T}_\omega = \mathcal{Z}_\omega/\sim$ inherits a semi-free $\mathbb{R}^*_+$-action for which γ is the only fixed point.

Let $\mathsf{S}_{\hat\omega,k}$ be the image of $\sigma_{\hat\omega} \times k$ under the obvious map $\mathcal{Z}_\omega \to \mathcal{T}_\omega$. Any nontrivial $\mathbb{R}^*_+$-trajectory meets $\mathsf{S}_{\hat\omega,k}$ at a singleton. Therefore $\mathsf{E}_{\hat\omega,k}$ is a positive cone over $\mathsf{S}_{\hat\omega,k}$. By their constructions, these cone structures in the individual cells $\mathsf{E}_{\hat\omega,k}$ are well-correlated and produce a positive cone structure in $\mathcal{T}_\omega$. As a result, $\{\mathsf{S}_{\hat\omega,k}\}_{\hat\omega,k}$ define a cellular structure in the link of γ in $\mathcal{T}_\gamma = \mathcal{T}_\omega \times \mathbb{R}^{n-|\omega|'}$. □

Example 7.4. Let $\omega = (122\ldots21)$, where the number of "2"s is $m \leq n$. We are going to describe explicitly the model space $\mathcal{T}_\omega$. In view of (7.10), for traversally generic flow, the vicinity E_ω of a trajectory γ of the combinatorial type ω is given by the polynomial inequality

$$\wp_\omega(u) = u\Big\{\prod_{i\in[1,m]}\big[(u-i)^2 + x_i\big]\Big\}(u-m-1) \leq 0.$$

[14] Here i is an index, not $\sqrt{-1}$.

The inequalities $\{x_i \leq 0\}_i$ divide the coordinate space $\mathbb{R}^m = \{(x_1, \dots x_m)\}$ into m half-spaces $\{V_i^-\}$, bounded by the hyperplanes

$$\{H_i =_{\mathsf{def}} \{x_i = 0\}\}_{1 \leq i \leq m}.$$

Crossing H_i from V_i^- into its complement V_i^+ decreases by one the number of connected components of the solution set $\{\wp_\omega(u, x) \leq 0\}$. Over a points $x = (x_1, \dots x_m)$ in the interior of the quadrant $\cap_i V_i^-$ the inequality $\wp_\omega(u) \leq 0$ has $m+1$ connected components, while over $x \in \cap_i V_i^+$ —a single component.

Recall that we use markers k to identify a particular component of the set $\{\wp_\omega(u, x) \leq 0\} \subset \mathbb{R}$, where $x \in \mathbb{R}^m$. Thus, for a given $x \in \mathbb{R}^m$, a marker k_0 may be placed at the point $0 \in \mathbb{R}$—the smallest zero of $\wp_\omega(u, x)$, and at any point $k_i =_{\mathsf{def}} i + \sqrt{-x_i}$, where i is such that $x_i \leq 0$.

Let $B_\epsilon \subset \mathbb{R}^m$ denote the m-dimensional ϵ-ball, centered on the origin and marked with k_0. Evidently, the marker k_i "spreads" over the half-ball $B^-_{\epsilon,i} =_{\mathsf{def}} V_i^- \cap B_\epsilon$ and "disappears" over the interior of $B^+_{\epsilon,i} =_{\mathsf{def}} V_i^+ \cap B_\epsilon$. Therefore, by Theorem 7.4, the local model for $\mathcal{T}_\omega$ is obtained by attaching to B_ϵ the half-balls $\{B^-_{\epsilon,i}\}_{1 \leq i \leq m}$ of dimension m along the coordinate $(m-1)$-balls $B_\epsilon \cap H_i$.

Thus $\mathcal{T}_\omega$ is a cone over a cell complex $\mathcal{S}_\omega$ with $2m$ vertices (the vertices are in 1-to-1 correspondence with the points where the coordinate axes pierce the boundary of B_ϵ), and $2^m + m$ cells of the top dimension $m-1$.

By Corollary 6.7, $\mathcal{T}_\omega$ embeds in $\mathbb{R}^{2m}$ as a Whiney subspace. ♢

Here is a short summary of what we have established in this chapter: a traversally generic v-flow generates a stratification $\{\mathcal{T}(v, \tilde{\omega})\}_{\tilde{\omega} \in \mathbf{\Omega}^\bullet_{\prime\langle n]}}$ of the trajectory space $\mathcal{T}(v)$, which is consistent (in a subtle way!) with, but cruder than, the γ-local cellular structure

$$\left\{\mathsf{E}_{\hat{\omega},k} \times \mathbb{R}^{n-|\omega|'}\right\}_{\hat{\omega} \in \prod_i \mathbf{\Omega}_{\langle \omega(i)]},\ k \in \Upsilon(\kappa(\omega))}$$

in the model space $\mathcal{T}_\gamma \approx \mathcal{T}_\omega \times \mathbb{R}^{n-|\omega|'}$ that we have described in Theorem 7.4. Each cell $\mathsf{E}_{\hat{\omega},k} \times \mathbb{R}^{n-|\omega|'}$ belongs to the stratum $\mathcal{T}(v, \tilde{\omega})$, where $\tilde{\omega} = \Xi(\kappa(\hat{\omega})) \in \mathbf{\Omega}_{\bullet'\langle n]}$. In the process, distinct cells could acquire the same label $\tilde{\omega} \in \mathbf{\Omega}_{\bullet'\langle n]}$.

This understanding of the cellular structure of the spaces of real polynomials of the degree $2n+2$—the structure which reflect the universal posets $\mathbf{\Omega}_{\langle 2n+2]}$ and $\mathbf{\Omega}_{\bullet'\langle n]}$—will enable us, in the forthcoming book [K13], to define a rich variety of new characteristic classes of traversally generic smooth flows on $(n+1)$-manifolds with boundary.

Chapter 8

Complexity of Shadows and Traversing Flows in Terms of the Simplicial Volume

8.1 In this chapter ...

This chapter is based on results from [AK], [K6], [A]. These papers were strongly influenced by our conversations with Larry Guth. Further developments are contained in [CaSa]. As these articles, the chapter draws its inspiration from the paper of Gromov [Gr1] where, among other things, the machinery of amenable localization has been developed.

In Section 8.4, we combine Gromov's amenable localization with the Poincaré duality to study the *shadows* and *traversally generic* vector flows on smooth compact manifolds X with boundary.

In Section 8.2, we study maps from compact connected *piecewise linear* ("PL" for short) manifolds X with boundary onto special CW-complexes K, where $\dim(K) = \dim(X)-1$. The local topology (the types of singularities) of K is prescribed a priori; it is independent of X. By definition, the fibers of $F : X \to K$ are PL-homeomorphic to closed segments or to singletons. We call such maps F the *shadows* of X. This setting is mimicking the familiar maps $\Gamma : X \to \mathcal{T}(v)$, generated by traversally generic fields v on smooth manifolds X with boundary.

The target spaces K of shadows F come equipped with a natural stratification, defined by the local topology of their singular loci. With the help of F, that stratification induces stratifications of X and of its double $DX =_{\mathsf{def}} X \cup_{\partial X} X$.

Expending some ideas about complexity of $3D$-flows from Chapter 3, we introduce the j-th *complexities* of F, X, and DX as the number of connected components of the strata of a fixed dimension j in each of the stratified spaces $F(X), X$, and DX (thus the word "complexity" in the title of this chapter).

The main results of this section are Theorems 8.3 and 8.4, named "*the amenable localization of the Poincaré duality*". They deal with the localizations of the Poincaré duality homomorphism to the F-induced strata of a given codimension in X or DX.

Theorem 8.5 links the singularities of shadows $F : X \to K$ with the "reduced homology" $H^{\Delta}_*(X;\mathbb{R})$ and $H^{\Delta}_*(DX;\mathbb{R})$ of X and of its double DX. The spaces $H^{\Delta}_*(X;\mathbb{R}) \subset H^{\Delta}_*(\pi_1(X);\mathbb{R})$ and $H^{\Delta}_*(DX;\mathbb{R}) \subset H^{\Delta}_*(\pi_1(DX);\mathbb{R})$ are close relatives of the *normed* homology spaces $H_*(X;\mathbb{R})$ and $H_*(DX;\mathbb{R})$, respectively. The norm here is the Gromov simplicial semi-norm $\| \sim \|_\Delta$ in homology [Gr]. For example, in Theorem 8.5, the F-induced j-th complexity of DX is estimated from below by the ranks of the groups $H^{\Delta}_{n+1-j}(DX;\mathbb{R})$, where $n+1 = \dim(X)$.

We stress that our results are *vacuous* when the fundamental groups $\pi_1(X)$ or $\pi_1(DX)$ are amenable (see Definition 8.9).

In Section 8.4, we apply the results from Section 8.3 to the shadows, produced by the traversally generic vector fields. These applications result in Theorem 8.8 and Theorem 8.11. The v-flow canonically generates some well-understood stratifications of the spaces $\mathcal{T}(v)$, X, and DX. As in the category of shadows, these stratifications lead to few competing notions of *complexity* for traversally generic flows. In Theorem 8.11, we get universal lower estimates of the numbers of connected components of these flow-generated strata of any given dimension. Again, these universal bounds are basically expressed in terms of the *normed* homology $H^{\Delta}_*(DX;\mathbb{R})$ and $H^{\Delta}_*(X;\mathbb{R})$. This phenomenon looks very much as an amenably localized version of the Morse inequalities.

It turns out that some of these homology spaces form *universal obstructions* to the existence of the *globally k-convex* traversally generic vector flows on a given X!

Then we apply these results to the traversally generic flows on punctured hyperbolic manifolds (see Theorem 8.10 and Theorem 8.13). In fact, this application is the source of motivation for our investigations in this chapter.

Finally, Section 8.5 connects the complexity of the fundamental group $\pi_1(X/\partial X)$ to the number of minimal connected components of the X-stratification by the combinatorial types of v-trajectories.

8.2 On spines and cospines

This section provides a motivation for introducing in Section 8.3 the notions of *cospines* and *shadows*. We start with reviewing few classical connections between spines of manifolds with boundary and their Riemannian geometry.

The reader is familiar already from Section 2.5 with the notions of a spine of a $(n+1)$-dimensional PL-manifold X (see Definition 2.3).

According to [Ma1], piecewise linear compact $(n+1)$-manifolds X with boundary admit n-spines K which are naturally stratified cell complexes with a quite particular local structure. Let us describe this structure.

Let Δ_k denote the standard k-simplex. We denote by Δ_{l+1}^{l-1} the $(l-1)$-skeleton of the $(l+1)$-dimensional simplex Δ_{l+1}. We consider the model spaces $\{C(\Delta_{l+1}^{l-1})\}_l$, where $C(Y)$ stands for the *cone* over a space Y.

Definition 8.1. Let X be a compact connected $(n+1)$-dimensional PL-manifold with boundary. Its spine $K \subset X$ is called *simple*, if each point $x \in K$ has a compact neighborhood $U \subset X$ such that $V =_{\mathsf{def}} U \cap K$ is PL-homeomorphic to the product $\Delta_k \times C(\Delta_{l+1}^{l-1})$, where $l+k=n$. Moreover, this homeomorphism extends to a PL-homeomorphism of the pairs

$$(U, V) \to (\Delta_k \times \Delta_{l+1}, \Delta_k \times C(\Delta_{l+1}^{l-1})).$$

◇

The closure of the set of points in K, whose neighborhoods are modeled after $\Delta_{n-k} \times C(\Delta_{k+1}^{k-1})$, is denoted by K_{-k}. So, K is stratified by the subcomplexes

$$K = K_0 \supset K_{-1} \supset \cdots \supset K_{-n}.$$

The pure stratum $K_{-k}^{\circ} =_{\mathsf{def}} K_{-k} \setminus K_{-(k+1)}$ topologically is an open manifold of dimension $n-k$.

Definition 8.2. A simple spine $K \subset X$ is called *special* if the closure of each connected component of $K_0 \setminus K_{-1}$ is a n-cell[1]. ◇

The following proposition plays the foundational role in the theory of spines ([Ma1], Theorems 2-4).

Theorem 8.1 (Matveev). *For $n \geq 2$, any $(n+1)$-dimensional compact connected* PL*-manifold X with boundary has a* special *simple spine K.*

[1]Some authors use the stricter notion of special spines by insisting that connected components of all strata $\{K_{-k}^{\circ}\}_k$ are cells.

If X_1 is another PL*-manifold of dimension $n+1$ with a* special *simple spine K_1 which is* PL*-homeomorphic to K, then X and X_1 are* PL*-homeomorphic manifolds.*

Moreover, for $n \geq 2$, any simple polyhedron K^n with the local topology as in Definition 8.1 and such that, for each k, all the pure connected strata $\{K_k^\circ\}_k$ are cells, *is a spine for some* PL*-manifold X.* $\diamondsuit$

As we have seen in Chapters 2 and 3, the category of simple spines is too restrictive when we deal with spines that are generated by traversing flows. The the notion of a *quasi-simple spine* bellow seems more appropriate.

Definition 8.3. A *quasi-simple* spine $K \subset X$ is a spine. Each point $x \in K$ has a closed neighborhood $U \subset X$ such that the pair $(U, U \cap K)$ is PL-homeomorphic to the pair $(\Delta_k \times \Delta_{l+1}, \Delta_k \times C(L))$, where $k + l = \dim(K)$ and $L \subset \Delta_{l+1}^{l-1}$ is a $(l-1)$-dimensional subcomplex of the skeleton Δ_{l+1}^{l-1}. $\diamondsuit$

There is a beautiful and subtle interplay between the geometry of a compact Riemannian manifold X with boundary and its spines. It turns out that a rather wide class of Riemannian metrics on X naturally give rise to simple spines!

Let X be a compact $(n+1)$-manifold with a smooth complete Riemannian metric g. Let K_X denotes the *sectional curvature* of X, and $\kappa_{\partial X}$ the *normal curvature* of the boundary ∂X.

The *inradius* (denoted by "$inr(X, \partial X)$") is defined as

$$inr(X, \partial X) =_{\mathsf{def}} \sup_{x \in X} \{dist(x, \partial X)\},$$

where "$dist$" stands for the g-induced geodesic distance in X. Alternatively, the inradius is the supremum of radii of metric balls in X that do not intersect with ∂X.

The *curvature normalized inradius*, a scale-free invariant, is defined by the formula

$$Inr(X, \partial X) =_{\mathsf{def}} inr(X, \partial X) \cdot \max\{\sup \sqrt{|K_X|},\ \sup |\kappa_{\partial X}|)\}, \quad (8.1)$$

where K_X is the sectional curvature tensor on X and $\kappa_{\partial X}$ the normal Gaussian curvature of the boundary ∂X[2].

For each point $p \in \partial X$, there exists a maximal geodesic segment $\bar{\gamma}$ that starts at p, is perpendicular there to ∂X, and such that $dist(x, \partial X) = dist(x, p)$ for all $x \in \bar{\gamma}$. Its other end q is called the *cut point* of ∂X with

[2]defined in terms of the second fundamental form of ∂X

the base at p. The geodesic segment $\bar{\gamma} = [p, q]$ is called the *minimizer* for the cut point q.

The *cut locus* K of ∂X is defined to be the union of such cut points q, produced for each $p \in \partial X$. A point q from the cut locus K either has a finite number of distinct minimizers $\{\gamma_i = [p_i, q]\}_{1 \leq i \leq l(q)}$, or is *focal*. In the first case, the number $l = l(q)$ is called the *branching number* of q.

Let K_l° denote the set of cut points with the branching number l. Thus, for a compact X, we get

$$K = F \cup \big(\cup_{l>0} K_l^\circ\big),$$

where F stands for the set of focal points (see [AB]). It follows from [AB] that K_l° is an open manifold, locally closed in X.

We say that K is *k-branched* if $K_k^\circ \neq \emptyset$, $K_l^\circ = \emptyset$ for all $l > k$, and $F = \emptyset$.

The remarkable theorems of Alexander and Bishop [AB], [AB1] link the curvature normalized inradius $Inr(X, \partial X)$ to the cut locus K of ∂X and the branching numbers of points in K.

Theorem 8.2. (Alexander, Bishop) *There exist a sequence of universal positive constants*[3] $a_2 < a_3 < a_4 < \cdots < 0.195...$ *such that, if a connected complete Riemannian manifold X with boundary ∂X has the curvature-normalized inradius $Inr(X, \partial X) < a_k$, then the cut locus K of ∂X has no focal points and is a k-branched simple spine of X (numerically, $a_2 \approx 0.075$, and $a_3 \approx 0.108$).*

Moreover, if $Inr(X, \partial X) < a_k$, then, for every $q \in K$, the angles between distinct minimizers $\{[p_i, q]\}_i$ are greater than $\cos^{-1}(-1/k)$. ♢

The next definition is motivated by the objects studied in [Gr1].

Definition 8.4. Let X be a $(n+1)$-dimensional connected and compact PL-manifold with boundary, and K a compact n-dimensional polyhedron K with the local structure as in Definition 8.3. For each subcomplex $L \subset \Delta_{k+1}^{k-1}$, we denote by K_L° the set of points in K whose normal star is PL-homeomorphic to the cone over L.

Then K is called a *cospine* of X if there exists a simplicial surjective map $F : X \to K$ such that:

- the fibers of F are PL-homeomorphic to *finite trees* with a single root in $X \setminus \partial X$ and finitely many leaves that reside in ∂X,

[3] These constants do not depend even on dim X!

- for each subcomplex $L \subset \Delta^{k-1}_{k+1}$, the map $F| : F^{-1}(K^\circ_L) \to K^\circ_L$ is a locally trivial PL-fibration. ♢

Due to the second bullet, the topological type of the F-fiber over each connected component of K°_L is constant; so the number l of branches, forming the fiber, is a function of that component.

By the same token, the structure group $\mathsf{Homeo}_{\mathsf{PL}}(F^{-1}(\star))$ of the fibration $F| : F^{-1}(K^\circ_L) \to K^\circ_L$, being restricted to each connected component of K°_L, has as many connected components, as the number of permutations of the finite set of the leaves in the fiber $F^{-1}(\star)$.

In fact, $F : X \to K$ from Definition 8.4 is a homotopy equivalence, since the fibers is are contractible graphs.

Example 8.1. Let Δ_n be the standard n-simplex, and Δ^k_n its k-skeleton. We denote by $\beta\Delta_n$ the barycentric subdivision of Δ_n.

Note that the cone $C(\Delta^k_n)$ over the skeleton Δ^k_n embeds naturally as a subcomplex in the $(k+1)$-skeleton $(\beta\Delta_{n+1})^{k+1}$. In particular, we have the imbedding

$$\rho_n : C(\Delta^{n-2}_n) \subset (\beta\Delta_{n+1})^{n-1}$$

and the simplicial map

$$F : \beta\Delta_{n+1} \longrightarrow \rho_n(C(\Delta^{n-2}_n)) =_{\mathsf{def}} K$$

which projects each n-face Δ' of Δ_{n+1} homeomorphically onto $\partial(C(\Delta')) \setminus \mathrm{int}(\Delta')$. The map F takes the barycenter of each n-face to the barycenter of Δ_{n+1} and is the identity on the rest of the vertices from $\beta\Delta_{n+1}$. By linearity, this rule determines the F-image of each simplex from $\beta\Delta_{n+1}$.

The fiber of F over the barycenter $b_\star$ of Δ_{n+1} is a tree with the root at $b_\star$ and $(n+2)$ branches. The fiber of F over the points from the skeleton Δ^{n-1}_{n+1} is a singleton. ♢

The main observation that flows out of Example 8.1 is that spines in the sense of [Ma1] are also cospines in the sense of Definition 8.4.

Applying Theorem 8.2, we get the following claim.

Corollary 8.1. *Let X be a connected $(n+1)$-dimensional complete smooth Riemannian manifold with boundary, such that its curvature-normalized inradius $Inr(X, \partial X) < a_{n+2}$, the universal constant from Theorem 8.2.*

Then the construction of boundary cut locus K produces a simplicial map $F : X \to K$ whose fibers are trees with a single root and $n+2$ leaves *at most. Moreover, this map $F : X \to K$ delivers a simple cospine in the sense of Definition 8.4.* $\diamondsuit$

In the next two sections, we will use traversing flows (instead of Riemannian metrics as in the Alexander-Bishop Theorem 8.2) to manufacture geometrically the *shadows* $F : X \to K$ of X, some close relatives of the cospines. The F-fibers of shadows are modeled after different types of contractible graphs: they are just segments of singletons.

8.3 Shadows of manifolds with boundary

The notion and properties of *shadows* (see Definition 8.5), the main subject of this section, are inspired by the maps $\Gamma : X \to \mathcal{T}(v)$, where the field v is traversally generic.

We are going to pick a fixed and carefully chosen class of compact n-dimensional CW-complexes K, whose *local* topological structure is prescribed a priori.

Let X be a compact connected PL-manifold of dimension $n+1$ with boundary.

We will consider a variety of surjective maps $\{F : X \to K\}$ with the F-fibers being *contractible* 1-dimensional complexes of a particular type (in this chapter, segments and singletons).

We think of such CW-complexes $K = F(X)$ as "shadows" of a given manifold X on a variety of $(\dim(X)-1)$-dimensional "screens". We consider singularities in K of particular types, labeled by a finite set of indexes $\{\omega\}$. We denote by $K(\omega)$ the locus of such singularities. Our goal is to count $\{\#\pi_0(K(\omega))\}_\omega$, the numbers of connected components of the strata $\{K(\omega)\}_\omega$. We view this count as measuring the *complexity* of the surjection F. Then we minimize these numbers over all F's to get various notions of complexity for the given manifold X.

Let us start with a quite general setting, motivated by our experience with the maps $\Gamma : X \to \mathcal{T}(v)$ for traversally generic vector fields v.

Let $\mathcal{S}$ be a poset, equipped with two maps: a map $\mu : \mathcal{S} \to \mathbb{Z}_+$ and an order-preserving map $\mu' : \mathcal{S} \to \mathbb{Z}_+$. By definition, for each $\omega \in \mathcal{S}$, $\mu'(\omega) < \mu(\omega)$. For each $n \in \mathbb{Z}_+$, we assume that the poset $\mathcal{S}_n =_{\mathsf{def}} (\mu')^{-1}([0, n])$ is finite.

With each element $\omega \in \mathcal{S}_n$ we associate a model compact CW-complex T_ω of dimension $\mu'(\omega)$ and a model compact PL-ball E_ω of dimension $\mu'(\omega)+$

1. They are linked by a surjective PL-map $p_\omega : \mathsf{E}_\omega \to \mathsf{T}_\omega$, whose fibers are *closed intervals* or *singletons*.

In what follows, we will list additional properties of the two collections, $\mathsf{E} =_{\mathsf{def}} \{\mathsf{E}_\omega\}_{\omega\in\mathcal{S}}$ and $\mathsf{T} =_{\mathsf{def}} \{\mathsf{T}_\omega\}_{\omega\in\mathcal{S}}$ (exhibiting topologically distinct T_ω's). We will do it in a recursive fashion.

We consider a set $\mathsf{Cosp}(\mathsf{T}, n)$[4] of n-dimensional compact simplicial complexes K, considered up to PL-homeomorphisms. We assume that each point $y \in K$ has a neighborhood which is PL-homeomorphic to the product $\mathsf{T}_\omega \times D^{n-\mu'(\omega)}$, where $\omega \in \mathcal{S}$, $\mu'(\omega) \leq n$, and $D^{n-\mu'(\omega)}$ denotes the standard ball.

We require that each model space T_ω topologically is a *cone* over a space S_ω, which belongs to the set $\mathsf{Cosp}(\mathsf{T}, n-1)$.

By definition, $\mathsf{Cosp}(\mathsf{T}, 1)$ consists of finite graphs whose vertices are of valencies 1 and 3 only.

We denote by $K(\omega)$ the set of points in K whose regular neighborhoods are modeled after the space $\mathsf{T}_\omega \times D^{n-\mu(\omega)}$. It follows that each $K(\omega) \subset K$ is a locally closed PL-manifold.

Let us also consider a filtration

$$K =_{\mathsf{def}} K_{-0} \supset K_{-1} \supset \cdots \supset K_{-n}$$

of K by the closed subcomplexes

$$K_{-j} =_{\mathsf{def}} \bigcup_{\{\omega\in\mathcal{S}_n \mid\ \mu'(\omega)\geq j\}} K(\omega) \tag{8.2}$$

of dimensions $n-j$. Note that $K_{-j} = \emptyset$ implies $K_{-(j+1)} = \emptyset$.

Definition 8.5. Let X be a compact connected PL-manifold of dimension $n+1$. We assume that $\partial X \neq \emptyset$. Consider the set $\mathsf{Shad}(X, \mathsf{E} \Rightarrow \mathsf{T})$ of surjective PL-maps $F : X \to K$ such that:

- K is a compact simplicial complex of the type from $\mathsf{Cosp}(\mathsf{T}, n)$,
- each fiber of F is PL-homeomorphic to a closed interval I or to a singleton pt, where ∂I and pt reside in ∂X,
- $F^\partial =_{\mathsf{def}} F| : \partial X \to K$ is a surjective map with finite fibers,
- for each $\omega \in \mathcal{S}$, the map
$$F| : F^{-1}(K(\omega)) \to K(\omega)$$
is a trivial fibration with an orientable manifold base, and the map
$$F^\partial| : (F^\partial)^{-1}(K(\omega)) \to K(\omega)$$
is a trivial covering with the fiber of cardinality $\mu(\omega) - \mu'(\omega)$,

[4] "Cosp" is an abbreviation of "cospine".

- each point $y \in K(\omega)$ has a regular neighborhood $V_y \subset K$ such that the map

$$F : (F^{-1}(V_y),\, F^{-1}(\partial V_y)) \to (V_y,\, \partial V_y)$$

is conjugate to the model map

$$p_\omega \times id : (\mathsf{E}_\omega, \delta\mathsf{E}_\omega) \times D^{n-\mu(\omega)} \to (\mathsf{T}_\omega, \mathsf{S}_\omega) \times D^{n-\mu(\omega)}$$

via a PL- homeomorphism which preserves the $\mathcal{S}_n$-stratifications in both spaces. Here $\delta\mathsf{E}_\omega$ denotes the portion of the boundary $\partial\mathsf{E}_\omega$ that is mapped to S_ω.

For a given map $F \in \mathsf{Shad}(X,\, \mathsf{E} \Rightarrow \mathsf{T})$, we call the CW-complex $F(X)$ a $\mathcal{S}$-*shadow* of X. ♢

Definition 8.6. We say that a compact connected PL-manifold X with boundary is *globally k-convex* if it has a shadow $F \in \mathsf{Shad}(X,\, \mathsf{E} \Rightarrow \mathsf{T})$ with the property $F(X)_{-k} = \emptyset$.

Note that the global k-convexity implies global $(k+1)$-convexity. ♢

In part, the next section is devoted to finding obstructions to the global k-convexity, selected from the tool set of algebraic topology.

8.4 Simplicial semi-norms, amenable localization, and complexity of shadows

We will employ now *Gromov's simplicial semi-norms* [Gr] to give lower estimates of the complexities of shadows and of traversing flows on a given $(n+1)$-manifold X. The number of v-trajectories of the (maximal) reduced multiplicity $\dim(X) - 1$ can serve as an example of such complexity. First, let us recall a few relevant definitions (see [Gr], [Gr1]).

Definition 8.7. Let $X \supset Y$ be topological spaces. Given a relative homology class $h \in H_k(X, Y; \mathbb{R})$, consider all relative singular cycles $c = \sum_i r_i\sigma_i$ that represent h. Here $r_i \in \mathbb{R}$ and $\sigma_i : \Delta^k \to X$ are singular simplicies, where each $(k-1)$-simplex from the algebraic boundary ∂c is mapped to the subspace Y.

We assume that, for any compact $K \subset X$, only finitely many images $\{\sigma_i(\Delta)\}$ intersect K. Put

$$\|c\|_{l_1} =_{\mathsf{def}} \sum_i |r_i|.$$

Following Gromov [Gr], we define the *simplicial semi-norm* of a homology class h by the formula:

$$\|h\|_\Delta =_{\mathsf{def}} \inf_c \{\|c\|_{l_1}\}.$$

$\diamondsuit$

This construction introduces a *semi-norm* $\| \sim \|_\Delta$ on the vector space $H_k(X, Y;\, \mathbb{R})$. The semi-norm $\| \sim \|_\Delta$ is monotone decreasing under continuous maps of pairs of topological spaces. In other words, for any continuous map $f : (X_1, Y_1) \to (X_2, Y_2)$ and any $h \in H_k(X_1, Y_1\ \mathbb{Z})$, we get

$$\|h\|_\Delta \geq \|f_*(h)\|_\Delta.$$

Moreover, if $f : X_1 \to X_2$ is a continuous map such that $f_* : \pi_1(X_1) \to \pi_1(X_2)$ is an *isomorphism* of the fundamental groups, then $\|f_*(h)\|_\Delta = \|h\|_\Delta$ for any $h \in H_k(X_1; \mathbb{R})$ ([Gr], The Mapping Theorem and its Corollary on page 248).

If M is a closed oriented *hyperbolic* manifold, then

$$\mathrm{Vol}(M) = c(n) \cdot \|[M]\|_\Delta, \tag{8.3}$$

where $[M]$ denotes the fundamental class of M and $c(n)$ is an universal positive constant (see *The Proportionality Theorem*, page 11 of [Gr]). For this reason, the simplicial norm of the fundamental class $[X]$ is often called the *simplicial volume.*

Gromov's Localization Lemma 8.1 below relies on the notion of the *stratified simplicial semi-norm*, available for stratified spaces X and pairs $X \supset Y$ of stratified spaces. Here we deal with stratifications $\mathcal{S}$ whose strata partition X into *disjoint* subsets.

We consider stratified spaces such that, if a stratum S intersects the closure $\overline{S'}$ of another stratum S', then $S \subseteq \overline{S'}$. In this case, we write $S \preceq S'$. If neither $S \preceq S'$ nor $S' \preceq S$, then we say the two strata are *incomparable.*

Recall that the *corank* of a stratum Y_ω in a $\mathcal{S}$-stratification of a given space Y is the maximal length k of a filtration $Y_\omega \subset \bar{Y}_{\omega_1} \subset \dots \subset \bar{Y}_{\omega_k}$ by the distinct strata whose closures contain Y_ω.

Definition 8.8. Let X be a $\mathcal{S}$-stratified topological space. Given a real homology class $h \in H_k(X; \mathbb{R})$, consider all singular cycles $c = \sum_i r_i\sigma_i$ that represent h, where $r_i \in \mathbb{R}$ and $\sigma_i : \Delta^k \to X$ are singular simplicies which are consistent with the stratification $\mathcal{S}$, in the following sense[5]:

[5]Gromov gives two conditions: *ord*(er) and *int*(ernality) ([Gr1], p. 27). We use these two conditions plus two more!

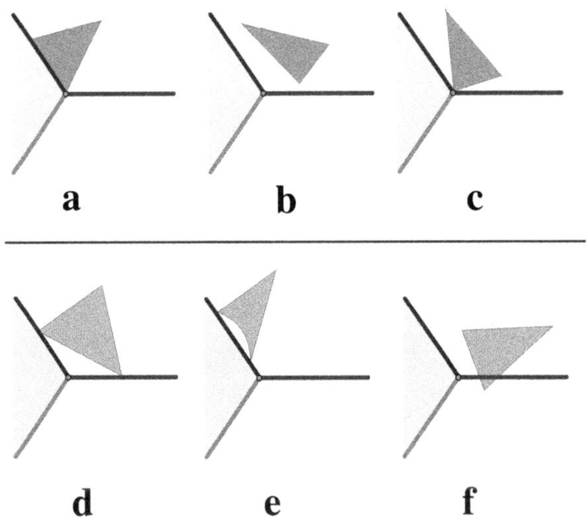

Fig. 8.1 Examples of a singular 2-simplex in relation to a stratification of the plane by a single stratum of codimension 2, three strata of codimension 1 (the rays), and two strata of codimension 0 (the two non-shaded sectors). Diagrams (a), (b), (c) are consistent with the four bullet list from Definition 8.8; diagram (d) violates the second bullet, diagram (e) violates the third bullet, diagram (f) violates the first bullet.

- We require that for each simplex σ_i of c, the image of the interior of each face (of any dimension) must be contained in one pure stratum. We call this the *cellular* condition.
- The (*ord*) condition states that the image of each simplex of c must be contained in a *totally ordered chain of strata*; that is, the simplex does not intersect any two incomparable strata.
- The (*int*) condition states that for each simplex of c, if the boundary of a face (of any dimension) maps into a stratum S, then the whole face maps into S.
- For technical reasons (involving the Amenable Reduction Lemma in [AK]), we require that if two vertices of a simplex σ_i map to the same point $v \in X$, then the edge between them must be constant at v. We call this the *loop* condition.

We define the $\mathcal{S}$-*stratified* simplicial semi-norm of h by the formula:

$$\|h\|^{\mathcal{S}}_{\Delta} =_{\mathsf{def}} \inf_{c} \{\|c\|_{l_1}\},$$

where c runs over all the cycles $c = \sum_i r_i \sigma_i$ that represent h, subject to the four properties above.

Similar definition of $\|h\|_\Delta^{\mathcal{S}}$ is available for relative homology classes $h \in H_k(X, Y; \mathbb{R})$, where the inclusion $Y \subset X$ is a $\mathcal{S}$-stratified map. ◇

The next lemma relies on the notion of amenable group, a central concept for the investigations in this chapter.

Definition 8.9. A discrete group G is called *amenable* if, for every finite subset $S \subset G$ and every $\epsilon > 0$, there exists a finite nonempty set $A \subset G$ such that the proportion of cardinalities

$$\frac{|((g \cdot A) \cup A) \setminus ((g \cdot A) \cap A)|}{|A|} < \epsilon$$

for all $g \in S$. ◇

Finally, we are in position to state the pivotal Gromov's Localization Lemma from [Gr1], page 772. Its proof there is a bit rough; a detailed proof can be found in [AK].

We denote by $H_*^c(\sim)$ the homology with compact support and real coefficients.

Lemma 8.1 (Gromov's Amenable Localization Lemma).
Let X be an open $(n+1)$-manifold with a stratification $\mathcal{S}$ consisting of finitely many connected locally closed submanifolds.

Pick an integer j from the interval $[0, n+1]$. Let $X_{-j} \subseteq X$ denote the union of strata of codimension j at least. Let U be a neighborhood of X_{-j} in X. We denote by ∂U its boundary.

Let Z be a space with contractible universal cover[6], *and let $\alpha : X \to Z$ be a continuous map such that the α-image of the fundamental group of each stratum of codimension* less *than j is an amenable subgroup of $\pi_1(Z)$.*

Then the α-image of every j-dimensional homology class $h \in H_j^c(X)$ satisfies the upper bound

$$\|\alpha_*(h)\|_\Delta \leq \|h_U\|_\Delta^{\mathcal{S}},$$

where $h_U \in H_j^c(U, \partial U)$ denotes the restriction of h to U, obtained via the composite homomorphism

$$H_j^c(X) \to H_j^c(X, X \setminus U) \to H_j^c(U, \partial U),$$

the last map being the excision isomorphism. ◇

[6]that is, a $K(\pi, 1)$-space

Let X be a compact oriented manifold with boundary. For technical reasons, related to the application of Lemma 8.1, many arguments to follow will deal with the *double* $DX =_{\mathsf{def}} X \cup_{\partial X} X$ of X, instead of the pair $(X, \partial X)$. The orientation on X extends to an orientation on its double. Then there is an orientation-reversing involution $\tau : DX \to DX$ whose orbit space is X and whose fixed point set $DX^\tau = \partial X$.

Note that any absolute homology class $\tilde{h} \in H_*(DX)$, via the excision property, gives rise to a relative class $h \in H_*(X, \partial X)$. Conversely, every relative class h, with the help of the involution τ, gives rise to an absolute class $\tilde{h} \in H_*(DX)$ whose restriction to $X \subset DX$ produces h. Therefore, $H_*(X, \partial X)$ can be viewed as a direct summand of $H_*(DX)$.

However, the relation between the simplicial semi-norms of $\tilde{h} \in H_*(DX)$ and of $h \in H_*(X, \partial X)$ is not so transparent. For example, if $X \approx S^1 \times [0, 1]$ is the surface of a cylinder and $h \in H_1(X, \partial X)$ is a generator, then $\|h\|_\Delta = 1$, while $\|\tilde{h}\|_\Delta = 0$. Indeed, since homologically the longitude $\tilde{h}$ of the torus DX can be represented by the singular rational cycle $\frac{1}{N} \cdot \{f : [0, 1] \to DX\}$, where f wraps the segment $[0, 1]$ N times around the longitude. Therefore $\|\tilde{h}\|_\Delta \leq \frac{1}{N}$ for all N.

In this example, $X/\partial X$ is a pinched torus with one of the meridians of T^2 being collapsed to a point. The simplicial semi-norm of the longitude in $X/\partial X$ is zero. Therefore, although $H_1(X, \partial X)$ is canonically isomorphic to the reduced homology $H_1(X/\partial X,\ \partial X/\partial X)$, the isomorphism is not an isometry with respect to the two simplicial semi-norms!

At the same time, similar considerations imply that, for the cylinder X, the simplicial semi-norms on both $H_2(X, \partial X)$ and $H_2(DX)$ are trivial.

So it seems that the excision property of homology is not friendly to the simplicial semi-norms ...

Lemma 8.2. *Let X be a connected compact and oriented $(n+1)$-manifold with boundary. Let $\beta : DX \to X/\partial X$ be the degree 1 map which collapses the second copy $\tau(X) \subset DX$ of X to the point $\star = \partial X/\partial X \in X/\partial X$, and let $\alpha : (X, \partial X) \to (X/\partial X, \star)$ be the obvious map of pairs.*

Then, for any $h \in H_j(X, \partial X)$, we get

$$2\|h\|_\Delta \geq \|h - \tau_*(h)\|_\Delta \geq \|\alpha_*(h)\|_\Delta,$$

where the middle term in the inequalities is the simplicial semi-norm of the class $h - \tau_*(h) \in H_j(DX)$.

In particular, for the fundamental class $h = [X, \partial X] \in H_{n+1}(X, \partial X; \mathbb{R})$, the class $h - \tau_*(h) = [DX]$ is fundamental. So we get

$$2\|[X, \partial X]\|_\Delta \geq \|[DX]\|_\Delta \geq \|[X/\partial X]\|_\Delta.$$

Proof. If $c = \sum_i r_i \sigma_i$ is a relative cycle with the l_1-norm $\|c\|$ being ϵ-close to $\|h\|_\Delta$, then the chain $\sum_i r_i\sigma_i - \sum_i r_i \tau(\sigma_i)$ is an absolute cycle in DX whose l_1-norm is $2\|c\|$ at most. Thus $2\|h\|_\Delta \geq \|h - \tau_*(h)\|_\Delta$.

On the other hand, consider the degree 1 map $\beta : DX \to X/\partial X$. By [Gr], the simplicial volume does not increase under continuous maps. Therefore, observing that $\alpha_*(h) = \beta_*(h - \tau_*(h))$, we get $\|h - \tau_*(h)\|_\Delta \geq \|\alpha_*(h)]\|_\Delta$. □

Let $\mathcal{S}$ be a stratification of X by strata $\{S\}$ such that the sets $\{S \cap \partial X\}_{S \in \mathcal{S}}$ form a stratification $\mathcal{S}^\partial$ of the boundary ∂X. Then $\mathcal{S}$ gives rise to a stratification $\mathcal{DS}$ of the double DX so that any stratum $S \in \mathcal{DS}$ either belongs to ∂X (in which case $S \in \mathcal{S}^\partial$) or to $DX \setminus \partial X$. Conversely, any τ-invariant stratification $\mathcal{DS}$ of the double induces a stratification $\mathcal{S}$ of X.

Next, we are going to associate few differential chain complexes and their homology groups with shadows of manifolds X with boundary. Eventually, they will become useful instruments in our investigations of various notions of complexity of traversing flows.

Let $F \in \mathsf{Shad}(X, \mathsf{E} \Rightarrow \mathsf{T})$ be a shadow (see Definition 8.5) of a compact orientable PL-manifold X of dimension $n+1$. Put $K =_{\mathsf{def}} F(X)$. We consider the finite $\mathcal{S}_n$-stratification $\{K(\omega)\}_{\omega \in \mathcal{S}_n}$ of K. Each stratum $K(\omega)$ is formed by points that share the model normal (to the strata $K(\omega)$) space $\{\mathsf{T}_\omega\}$ of dimension $\{\mu'(\omega)\}$. We denote by $X(F, \omega)$ the F-preimage of the pure stratum $K(\omega)$. These sets form a stratification $\mathcal{S}_F(X)$ of X.

The CW-complex $K \in \mathsf{Cosp}(\mathsf{T}, n)$ comes equipped with a filtration

$$K =_{\mathsf{def}} K_{-0} \supset K_{-1} \supset \dots \supset K_{-n}$$

that has been introduced in (8.2) and employs the $\mathcal{S}_n$-stratification.

Let $\mathcal{A}$ be an abelian coefficient system on K. As a default, $\mathcal{A} = \mathbb{Z}$, or $\mathbb{R}$, the trivial coefficient system with the integers or reals for its stalk.

For each $j \in [0, n]$, consider the relative homology groups[7]

$$\big\{\mathsf{C}^{\mho}_j(K) =_{\mathsf{def}} H_j(K_{-(n-j)}, K_{-(n-j+1)}; \mathcal{A})\big\}_j, \tag{8.4}$$

associated with the filtration. Note that $\dim(K_{-(n-j)}) = j$, so $\mathsf{C}^{\mho}_j(K)$ is the top homology of the quotient $K_{-(n-j)}/K_{-(n-j+1)}$.

These homology groups can be organized into a *differential complex*

$$\mathbf{C}^{\mho}_*(K) =_{\mathsf{def}} \big\{0 \to \mathsf{C}^{\mho}_n(K) \xrightarrow{\partial_n} \mathsf{C}^{\mho}_{n-1}(K) \xrightarrow{\partial_{n-1}} \cdots \xrightarrow{\partial_1} \mathsf{C}^{\mho}_0(K) \to 0\big\}, \tag{8.5}$$

[7] In our notations, we suppress the dependance of these homology groups on the coefficients $\mathcal{A}$.

where the differentials $\{\partial_j\}$ are the boundary homomorphisms from the long exact homology sequences of the triples

$$\{K_{-(n-j)} \supset K_{-(n-j+1)} \supset K_{-(n-j+2)}\}_j.$$

For $K = \mathcal{T}(v)$, the differential complex as in (8.5), produced with the help of a traversally generic vector field $v \in \mathcal{V}^\ddagger(X)$, will become in [K13] the central object of study.

Since the pure strata $K^\circ_{-(n-j)} =_{\mathsf{def}} K_{-(n-j)} \setminus K_{-(n-j+1)}$ are open manifolds of dimension j, the $\mathcal{A}$-modules $\{\mathsf{C}^{\mho}_j(K)\}_j$ are *free* and finitely generated, the number of generators being the number of connected components of the locus $K^\circ_{-(n-j)}$.

We call the homology groups $\mathsf{H}^{\mho}_*(K)$, associated with the differential complex (8.5), the $\mho$-*homology* of K.

In fact, the $\mho$-*homology* is an ingredient of the homology spectral sequence, associated with the filtration $\{K_{-j}\}_j$ and converging to the regular homology $H_*(K) \approx H_*(X)$ for any shadow or cospine $K = F(X)$. Therefore the is a canonical homomorphism

$$A^{\mho}_* : \mathsf{H}^{\mho}_*(K) \to Gr(H_*(K)),$$

where $Gr(H_*(K))$ stands for the graded group, associated with the filtration $\{H_*(K_{-j}) \to H_*(K) \approx H_*(X)\}_j$.

Definition 8.10. With any filtered CW-complex $K \in \mathsf{Shad}(\mathsf{T}, n)$ and its filtration as in (8.2), we associate the ordered collection of ranks

$$\{c^{\mho}_j(K) =_{\mathsf{def}} \mathrm{rk}_{\mathcal{A}}(\mathsf{C}^{\mho}_j(K))\}_{0 \leq j \leq n},$$

where the groups $\{\mathsf{C}^{\mho}_j(K))\}_j$ were introduced in (8.4).

We call the integer $c^{\mho}_j(K)$ *the* j*-th* $\mho$*-complexity* of K.

◇

Definition 8.11. Let X be a compact connected PL-manifold with boundary, $\dim(X) = n + 1$. Consider the variety of maps $F : X \to K$ from the set $\mathsf{Shad}(X, \mathsf{E} \Rightarrow \mathsf{T})$ as in Definition 8.5.

Each F produces the sequence of $\mho$-complexities:

$$\mathbf{c}(F) =_{\mathsf{def}} \{c^{\mho}_0(F(X)), c^{\mho}_1(F(X)), \dots, c^{\mho}_n(F(X))\}.$$

Consider the *lexicographical minimum*

$$\mathbf{c}^{\mathsf{shad}}(X, \mathsf{E} \Rightarrow \mathsf{T})) =_{\mathsf{def}} \text{lex.min}_{\{F \in \mathsf{Shad}(X, \mathsf{E} \Rightarrow \mathsf{T})\}} \mathbf{c}(F(X)). \qquad (8.6)$$

We call this ordered sequence of complexities *the lexicographic shadow complexity* of X.

We denote by $c^{\mathsf{shad}}_j(X,\, \mathsf{E} \Rightarrow \mathsf{T})$ the $(j+1)$-component of the vector $\mathbf{c}^{\mathsf{shad}}(X, \mathsf{E} \Rightarrow \mathsf{T})$. ♢

Remark 8.1. Of course, the previous definitions rely on the set $\mathsf{Shad}(X,\, \mathsf{E} \Rightarrow \mathsf{T})$ being nonempty, a nontrivial fact which requires a carefully designed poset $\mathcal{S}$ and a system of models $\{p_\omega : \mathsf{E}_\omega \to \mathsf{T}_\omega\}_{\omega \in \mathcal{S}}$ (as the ones introduced in Chapters 6 and 7). ♢

Remark 8.2. If a compact connected PL-manifold X of dimension $n+1$ is *globally k-convex* (see Definition 8.6), then for some shadow F and all $j \geq k$, we get $F(X)_{-j} = \emptyset$. Thus $\mathsf{C}^{\mho}_{n-j}(F(X)) = 0$ for all $j \geq k$, implying $c^{\mathsf{shad}}_{n-j}(X,\, \mathsf{E} \Rightarrow \mathsf{T}) = 0$ for all $j \geq k$. In other words, $c^{\mathsf{shad}}_l(X,\, \mathsf{E} \Rightarrow \mathsf{T}) = 0$ for all $l \leq n-k$. ♢

Our next goal is to find some lower estimates of the complexities $\{c^{\mathsf{shad}}_j(X,\, \mathsf{E} \Rightarrow \mathsf{T}))\}_j$ in terms of the algebraic topology of X.

Let us consider a refinement $\mathcal{S}^\bullet_F(X)$ of the stratification $\mathcal{S}_F(X)$ of X, formed by the *connected components* of the sets:

$$\begin{aligned}\big\{X^\circ(F,\omega) &=_{\mathsf{def}} X(F,\omega) \setminus (\partial X \cap X(F,\omega)),\\ &X^\partial(F,\omega) =_{\mathsf{def}} \partial X \cap X(F,\omega)\big\}_{\omega \in \mathcal{S}}.\end{aligned} \tag{8.7}$$

Figure 1.5 shows an example of such a stratification $\mathcal{S}^\bullet_F(X)$ on a surface X. The shadow of X is the graph $\mathcal{T}(v)$, whose vertices have valencies 1 and 3.

The double DX acquires the τ-equivariant stratification $\mathcal{S}_F(DX)$, whose pure strata are:

$$\{X^\circ(F,\omega),\ \tau(X^\circ(F,\omega)),\ X^\partial(F,\omega)\}_\omega,$$

and its refinement $\mathcal{S}^\bullet_F(DX)$, formed by the *connected components* of the strata from $\mathcal{S}_F(DX)$.

The stratifications $\mathcal{S}^\bullet_F(X)$ and $\mathcal{S}^\bullet_F(DX)$ give rise to the filtrations:

$$X =_{\mathsf{def}} X^F_{-0} \supset X^F_{-1} \supset \cdots \supset X^F_{-(n+1)},$$

$$DX =_{\mathsf{def}} DX^F_{-0} \supset DX^F_{-1} \supset \cdots \supset DX^F_{-(n+1)}$$

by the union of strata of a fixed codimension, each pure stratum being an open manifold.

Analogously to (8.4), we consider the homology and cohomology groups with coefficients in $\mathcal{A}$:

$$\begin{aligned}\{\mathsf{C}^{\mho}_j(DX,F) &=_{\mathsf{def}} H_j(DX^F_{-(n+1-j)}, DX^F_{-(n+2-j)}; \mathcal{A})\}_j,\\ \{\mathsf{C}_{\mho}^j(DX,F) &=_{\mathsf{def}} H^j(DX^F_{-(n+1-j)}, DX^F_{-(n+2-j)}; \mathcal{A})\}_j.\end{aligned} \tag{8.8}$$

These homology groups can be organized into a differential complex:

$$\mathbf{C}^{\mho}_*(DX,F) =_{\mathsf{def}} \tag{8.9}$$

$$\{0 \to \mathsf{C}^{\mho}_{n+1}(DX,F) \xrightarrow{\partial_{n+1}} \mathsf{C}^{\mho}_n(DX,F) \xrightarrow{\partial_n} \cdots \xrightarrow{\partial_1} \mathsf{C}^{\mho}_0(DX,F) \to 0\},$$

where the differentials $\{\partial_j\}$ are the boundary homomorphisms from the long exact homology sequences of the triples

$$\{DX^F_{-(j-1)} \supset DX^F_{-j} \supset DX^F_{-(j+1)}\}_j.$$

Similarly, we introduce the dual differential complex

$$\mathbf{C}_{\mho}^*(DX,F) =_{\mathsf{def}} \tag{8.10}$$

$$\{0 \leftarrow \mathsf{C}_{\mho}^{n+1}(DX,F) \xleftarrow{\partial^*_{n+1}} \mathsf{C}_{\mho}^n(DX,F) \xleftarrow{\partial^*_n} \cdots \xleftarrow{\partial^*_1} \mathsf{C}_{\mho}^0(DX,F) \leftarrow 0\}.$$

If $\mathcal{A} = \mathbb{Z}$, since $DX^F_{-(n-j-1)} \setminus DX^F_{-(n-j)}$ is an open orientable j-manifold, by [Hat], Corollary 3.28, the torsion

$$\mathrm{tor}\{H_{j-1}(DX^F_{-(n-j-1)}, DX^F_{-(n-j)}; \mathbb{Z})\} = 0,$$

which results in the natural isomorphism

$$\mathsf{C}_{\mho}^j(DX,F) \approx \mathrm{Hom}_{\mathbb{Z}}(\mathsf{C}^{\mho}_j(DX,F), \mathbb{Z}).$$

In the notations to follow, we drop the dependence of our constructions on the coefficient group $\mathcal{A}$, but presume that $\mathcal{A}$ is $\mathbb{Z}$ or $\mathbb{R}$. So the complexes $\mathbf{C}^{\mho}_*(DX,F)$ and $\mathbf{C}_{\mho}^*(DX,F)$ are comprised of free $\mathcal{A}$-modules whose generators $\{[\sigma]\}$ are in 1-to-1 correspondence with the strata $\sigma \in \mathcal{S}^\bullet_F(DX)$ (see (8.7)).

We denote by $\mathsf{Z}^{\mho}_j(DX,F)$ the kernels of the differentials ∂_j and by $\mathsf{B}^{\mho}_j(DX,F)$ the images of the differentials ∂_{j+1} from (8.9).

We form the *$\mho$-homology*

$$\mathsf{H}^{\mho}_j(DX,F) =_{\mathsf{def}} \mathsf{Z}^{\mho}_j(DX,F)/\mathsf{B}^{\mho}_j(DX,F)$$

of the double DX, associated with its stratification $\mathcal{S}^\bullet_F(DX)$.

Since the differential complex (8.10) is the dual of the differential complex (8.9), the natural paring $\mathsf{C}_{\mho}^{n-k}(DX,F) \otimes \mathsf{C}^{\mho}_{n-k}(DX,F) \to \mathbb{R}$ produces an isomorphism

$$\Phi : \mathsf{C}_{\mho}^{n-k}(DX,F)/\mathsf{B}_{\mho}^{n-k}(DX,F) \approx (Z^{\mho}_{n-k}(DX,F))^*,$$

where $(Z^{\mho}_{n-k}(DX, F))^*$ denotes the dual space of $Z^{\mho}_{n-k}(DX, F)$.

To understand how the differential complexes $\mathbf{C}^{\mho}_*(F(X))$ in (8.5) and $\mathbf{C}^{\mho}_*(DX, F)$ in (8.9) vary, as F runs over the set of all shadows $\mathsf{Shad}(X, \mathsf{E} \Rightarrow \mathsf{T})$, is a crucial challenge. Such understanding would allow to treat the appropriate equivalence class (bordism class) of a particular differential complex $\mathbf{C}^{\mho}_*(F(X))$ as a new invariant of X. We will devote a part of [K13] to such investigations.

Now we are going to introduce a *localization construction*, central to the rest of this section. Using the Poincaré duality $\mathcal{D}$ (over $\mathbb{Z}$ or $\mathbb{R}$) on the closed oriented manifold DX, for each $k \in [-1, n]$, we form the localization transfer map

$$\mathcal{L}^F_{k+1} : H_{k+1}(DX) \xrightarrow{\approx \mathcal{D}} H^{n-k}(DX) \xrightarrow{i^*_{\mathsf{loc}}} H^{n-k}(DX^F_{-(k+1)}),$$

where $i_{\mathsf{loc}} : DX^F_{-(k+1)} \subset DX$ is the embedding.

Because the locus $DX^F_{-(k+1)}$ is a $(n-k)$-dimensional CW-complex, the top cohomology group $H^{n-k}(DX^F_{-(k+1)})$ can be identified with the quotient $\mathsf{C}^{n-k}_{\mho}(DX, F)/\mathsf{B}^{n-k}_{\mho}(DX, F)$ from the complex in (8.10). This fact follows from the diagram chase in the two long exact cohomology sequences of the pairs

$$(DX^F_{-(k+1)}, DX^F_{-(k+2)}),\ (DX^F_{-(k+2)}, DX^F_{-(k+3)}).$$

The first pair gives rise to the fragment of the exact sequence

$$\to H^{n-k-1}(DX^F_{-(k+2)}) \xrightarrow{\delta} \mathsf{C}^{n-k}_{\mho}(DX, F) \xrightarrow{j^*} H^{n-k}(DX^F_{-(k+1)}) \to 0,$$

while the second pair to the fragment

$$\to \mathsf{C}^{n-k-1}_{\mho}(DX, F) \xrightarrow{J^*} H^{n-k-1}(DX^F_{-(k+2)}) \to 0.$$

So the composition $J^* \circ \delta$ coincides with the boundary map

$$\delta^{n-k-1} : \mathsf{C}^{n-k-1}_{\mho}(DX, F) \to \mathsf{C}^{n-k}_{\mho}(DX, F)$$

from the differential complex $\mathsf{C}^*_{\mho}(DX, F)$. Therefore

$$H^{n-k}(DX^F_{-(k+1)}) \approx \mathsf{C}^{n-k}_{\mho}(DX, F)/\mathsf{B}^{n-k}_{\mho}(DX, F).$$

Similar diagram chase in homology leads to an isomorphism

$$H_{n-k}(DX^F_{-(k+1)}) \approx \mathsf{Z}^{\mho}_{n-k}(DX, F).$$

As a result, for each shadow F, we get the *localized* (to the locus $DX^F_{-(k+1)}$) Poincaré homomorphism

$$\mathcal{L}^{F,\mho}_{k+1} : H_{k+1}(DX) \xrightarrow{\approx \mathcal{D}} H^{n-k}(DX) \xrightarrow{i^*_{\mathsf{loc}}} H^{n-k}(DX^F_{-(k+1)}) \\ \xrightarrow{\approx} \mathsf{C}^{n-k}_{\mho}(DX, F)/\mathsf{B}^{n-k}_{\mho}(DX, F). \tag{8.11}$$

Our next goal is to introduce a norm $\|[\sim]\|_{\mho}$ in the target space $\mathsf{C}^{n-k}_{\mho}(DX, F)/\mathsf{B}^{n-k}_{\mho}(DX, F)$ of the homomorphism $\mathcal{L}^{F,\mho}_{k+1}$ and to show that, for any shadow $F \in \mathsf{Shad}(X, \mathsf{E} \Rightarrow \mathsf{T})$ and any homology class $h \in H_{k+1}(DX)$, the inequality

$$\|h\|_{\Delta} \leq const(n) \cdot \|[\mathcal{L}^{F,\mho}_{k+1}(h)]\|_{\mho} \tag{8.12}$$

holds. Here $const(n)$ is some *universal* positive constant, which depends only on the list of model maps $\{\mathsf{E}_\omega \to \mathsf{T}_\omega\}_\omega$ as in the last bullet of Definition 8.5.

Consider the l_1-norm $\| \sim \|_{l_1}$ on the space $\mathsf{C}^{n-k}_{\mho}(DX, F; \mathbb{R})$ in the unitary basis $\{[\sigma^*]\}$, labeled by the strata $\sigma \in \mathcal{S}^\bullet_F(DX)$ of dimension $n-k$, and dual to the preferred basis $\{[\sigma]\}$ of $\mathsf{C}^{\mho}_{n-k}(DX, F; \mathbb{Z})$. Then the norm $\| \sim \|^\bullet_{\mho}$ on

$$\mathsf{C}^{n-k}_{\mho}(DX, F; \mathbb{R})/\mathsf{B}^{n-k}_{\mho}(DX, F; \mathbb{R})$$

is, by definition, the quotient norm, induced by the norm $\| \sim \|_{l_1}$ on the space $\mathsf{C}^{n-k}_{\mho}(DX, F; \mathbb{R})$. The unit ball in the l_1-norm is a regular polyhedron $\diamondsuit$, dual to the hypercube; its vertices are located at the points of the lattice $\mathsf{C}^{n-k}_{\mho}(DX, F; \mathbb{Z})$. The unit ball $\diamondsuit_{\mho}$ in the norm $\| \sim \|^\bullet_{\mho}$ is an image under the quotient map

$$\mathsf{C}^{n-k}_{\mho}(DX, F; \mathbb{R}) \to \mathsf{C}^{n-k}_{\mho}(DX, F; \mathbb{R})/\mathsf{B}^{n-k}_{\mho}(DX, F; \mathbb{R})$$

of the polyhedron $\diamondsuit$. In fact, it is the convex hull of the images, under the quotient map, of the vertices of $\diamondsuit$.

We may regard the maximal proportion

$$\sup_{\{h \in H_{k+1}(DX),\ \|h\|_\Delta \neq 0\}} \left\{ \|[\mathcal{L}^{F,\mho}_{k+1}(h)]\|_{\mho} / \|h\|_\Delta \right\}$$

as the norm of the operator $\mathcal{L}^{F,\mho}_{k+1}$. So the inequality (8.12) (see also (8.13) below) testifies that these norms admit a positive lower bound, universal for all X of a given dimension $n+1$ and for their shadows F.

As the reader examines the hypotheses of the next theorem, it is helpful to keep in mind the following simple example. Let X be a 2-dimensional ball with two holes. Then DX is a closed surface of genus 2. Note that the fundamental group of each component of ∂X is abelian. So its image in $\pi_1(DX)$ is an amenable group.

Theorem 8.3 (The first amenable localization of the Poincaré Duality).
Let X be a compact connected $(n+1)$-dimensional PL-manifold with boundary. Assume that for each connected component of the boundary ∂X, the image of its fundamental group in $\pi_1(DX)$ is an amenable group[8].

Let $\lambda > 0$ be a constant that depends only on n, the given poset $\mathcal{S}$, and the given list $\mathsf{E} \Rightarrow \mathsf{T}$ of model maps $\{\mathsf{E}_\omega \to \mathsf{T}_\omega\}_{\omega \in \mathcal{S}}$, as described in formula (8.18).

Then, for any shadow $F \in \mathsf{Shad}(X, \mathsf{E} \Rightarrow \mathsf{T})$, the space

$$\mathsf{C}^{n-k}_{\mho}(DX, F)/\mathsf{B}^{n-k}_{\mho}(DX, F)$$

admits a norm $\|[\sim]\|_{\mho}$ so that

$$\|h\|_\Delta \leq \lambda \cdot \|[\,\mathcal{L}^{F,\mho}_{k+1}(h)\,]\|_{\mho} \tag{8.13}$$

for all $h \in H_{k+1}(DX)$. Here $\mathcal{L}^{F,\mho}_{k+1}$ denotes the localized Poincaré duality operator, introduced in (8.11).

Proof. Let U be a regular neighborhood of the set $DX^F_{-(k+1)}$ in DX. So the PL-manifold U has a homotopy type of a $(n-k)$-dimensional CW-complex $DX^F_{-(k+1)}$.

Let $\Pi =_{\mathsf{def}} \pi_1(DX)$. In order to apply Localization Lemma 8.1 to the neighborhood U, the homology class $h \in H_{k+1}(DX; \mathbb{Z})$, and the classifying map $\beta : DX \to K(\Pi, 1)$, we need to check that, for each stratum $\sigma \in \mathcal{S}^\bullet_F(DX)$, the subgroup $\beta_*\pi_1(\sigma)$ of Π is amenable. If $\sigma \subseteq \partial X$, this is true by the hypotheses. Indeed, for each connected component of ∂X, the image of its fundamental group in Π is amenable, and every subgroup of an amenable group is amenable.

Otherwise, σ is contained in some locus $X^\circ(F,\omega)$ or in some locus $\tau(X^\circ(F,\omega))$. By the fourth bullet of Definition 8.5, this preimage $X^\circ(F,\omega)$ is an oriented bundle $F : X^\circ(F,\omega) \to K(\omega)$ whose fiber is an open interval.

[8]If, for each connected component of the boundary ∂X, the image of its fundamental group in $\pi_1(X)$ is amenable, then it is automatically amenable in $\pi_1(DX)$.

Therefore $F : \sigma \to F(\sigma) \subset K(\omega)$ is a trivial fibration with an open interval for the fiber. As a result, $F_* : \pi_1(\sigma) \to \pi_1(F(\sigma))$ is an isomorphism.

On the other hand, $F : X^\partial(F, \omega) \to K(\omega)$ is a trivial covering. Therefore it admits a section $\rho : K(\omega) \to X^\partial(F, \omega)$ such that $\rho(K(\omega))$ intersects with the closure $\mathrm{cl}(\sigma)$ of σ in X. Put $\bar{\sigma} =_{\mathsf{def}} \sigma \cup \rho(F(\sigma))$. The map $F : \bar{\sigma} \to F(\sigma)$ is a trivial fibration with a semi-open interval for the fiber. Thus the imbedding $j : \sigma \subset \bar{\sigma}$ is a homotopy equivalence, and so is the obvious imbedding $q : \rho(F(\sigma)) \subset \bar{\sigma}$. Hence $j_* : \pi_1(\sigma) \to \pi_1(\bar{\sigma})$ and $q_* : \pi_1(\rho(F(\sigma))) \to \pi_1(\bar{\sigma})$ are isomorphisms. Since $\rho(F(\sigma)) \subset \partial X$ and, for every choice of the base point $x_\star \in \partial X$, the β_*-image of each fundamental group $\pi_1(\partial X, x_\star)$ in Π is amenable, so is the β_*-image of $\pi_1\big(\rho(F(\sigma))\big)$; indeed, a subgroup of an amenable group is amenable. By the previous arguments, the β_*-image of $\pi_1(\sigma)$ is amenable.

Now, applying localization Lemma 8.1 to the classifying map $\beta : DX \to K(\Pi, 1)$, gives the inequality

$$\|h\|_\Delta \;\leq\; \|h_U\|_\Delta^{\mathcal{S}^\bullet_F(U)}, \tag{8.14}$$

where h_U denotes the restriction of the absolute homology class h to $(U, \partial U)$. Indeed, by [Gr], page 248, $\|\beta_*(h)\|_\Delta = \|h\|_\Delta$ since, by its construction, β induces an isomorphism of the fundamental groups of DX and $K(\Pi, 1)$.

For each $(n-k)$-dimensional stratum $\sigma \in \mathcal{S}^\bullet_F(DX)$, consider an oriented disk

$$(D^{k+1}_\sigma,\, \partial D^{k+1}_\sigma) \subset (U,\, \partial U),$$

normal to the open manifold $\sigma \subset U$ at its typical point. Taking a smaller regular neighborhood U of the subcomplex $DX^F_{-(k+1)} \subset DX$ if necessarily, we can arrange for $\{D^{k+1}_\sigma\}_\sigma$ to be disjoint, so that each disk D^{k+1}_σ hits its stratum σ transversally at a singleton and misses the rest of the strata. Note that the relative integral homology classes $\{[D^{k+1}_\sigma, \partial D^{k+1}_\sigma]\}_\sigma$ may be dependent in $H_{k+1}(U, \partial U)$.

We claim that any element $h_U \in H_{k+1}(U, \partial U)$ can be written as a linear combination of relative disks-cycles $\{[D^{k+1}_\sigma, \partial D^{k+1}_\sigma] \in H_{k+1}(U, \partial U)\}_\sigma$:

$$h_U \;=\; \sum_{\sigma \in \mathcal{S}^\bullet_F(DX),\, \dim\sigma = n-k} r_\sigma \cdot [D^{k+1}_\sigma, \partial D^{k+1}_\sigma], \tag{8.15}$$

where, depending on the choice of coefficients R, $r_\sigma \in \mathbb{Z}$ or $r_\sigma \in \mathbb{R}$.

In fact, $H_{k+1}(U, \partial U)$ can be recovered from the free R-module M, generated by the elements $\{[D^{k+1}_\sigma, \partial D^{k+1}_\sigma]\}_\sigma$, by factoring M by the appropriate relations.

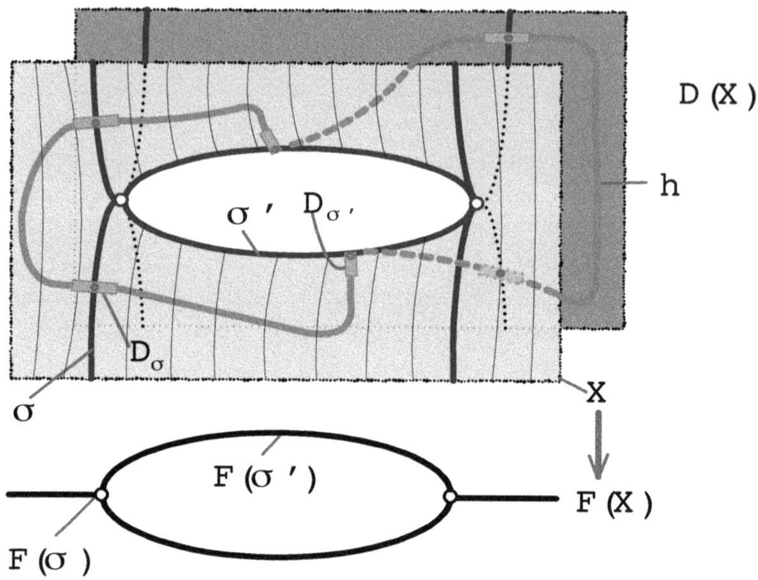

Fig. 8.2 The double DX of a surface X, the plane with a hole, and a graph-shadow $F(X)$ of X. The diagram shows a cycle $h \in H_1(DX)$, its transversal intersections with the strata $\{\sigma \subset DX^F_{-1}\}$, and the normal disks $\{(D^1_\sigma, \partial D^1_\sigma)\}_\sigma$, which represent the localization $h_U \in H_1(U, \partial U)$ of h.

To justify the presentation in (8.15), we notice that, since $DX^F_{-(k+1)}$ is a deformation retract of U, any $(n-k)$-dimensional homology class in U is represented by a cycle z in $DX^F_{-(k+1)}$, a combination $\sum_\sigma n_\sigma \cdot \sigma$ of the top strata $\{\sigma\}$ of the $(n-k)$-dimensional CW-complex $DX^F_{-(k+1)}$. The algebraic intersection $z \circ D^{k+1}_\sigma$ is equal to n_σ, since $\sigma \circ D^{k+1}_\sigma = 1$ and $\sigma' \circ D^{k+1}_\sigma = 0$ for any $\sigma' \neq \sigma$ by the very choice of the normal disks D^{k+1}_σ's.

By the Poincaré duality, any $(k+1)$-dimensional homology class $\tilde{h} \in H_{k+1}(U, \partial U)$ is determined by its algebraic intersections with the cycles z from a free R-module $H_{n-k}(U) \approx H_{n-k}(DX^F_{-(k+1)})$. Therefore any $\tilde{h}$ is a linear combination of $\{[D^{k+1}_\sigma, \partial D^{k+1}_\sigma]\}_\sigma$.

To simplify our notations, put $[D^{k+1}_\sigma] =_{\mathsf{def}} [D^{k+1}_\sigma, \partial D^{k+1}_\sigma]$.

We introduce the norm $\|[\sim]\|$ of the class h_U by the formula

$$\|[\, h_U \,]\| =_{\mathsf{def}} \inf_{\{\text{representations of}\, h_U\}} \Big\{ \sum_{\sigma \in \mathcal{S}^\bullet_F(DX),\, \dim \sigma = n-k} |r_\sigma| \Big\}, \quad (8.16)$$

the minimum being taken over all representations of h_U as in (8.15).

Applying (8.14) to any presentation of h_U as in (8.15), we get

$$\|h\|_\Delta \leq \sum_{\sigma \in \mathcal{S}^\bullet_F(DX),\ \dim\sigma = n-k} |r_\sigma| \cdot \|\,[D^{k+1}_\sigma]\,\|_\Delta^{\mathcal{S}^\bullet_F(U)}. \tag{8.17}$$

Thus,

$$\|h\|_\Delta \leq \lambda \cdot \sum_{\sigma \in \mathcal{S}^\bullet_F(DX),\ \dim\sigma = n-k} |r_\sigma|,$$

where

$$\lambda =_{\mathsf{def}} \max_{\sigma \in \mathcal{S}^\bullet_F(DX),\ \dim\sigma = n-k} \Big\{\|\,[D^{k+1}_\sigma]\,\|_\Delta^{\mathcal{S}^\bullet_F(U)}\Big\}. \tag{8.18}$$

Here we stratify the normal disk D^{k+1}_σ by intersecting it with the $\mathcal{S}^\bullet_F(DX)$-stratification in the ambient space DX.

Employing Definition 8.5, the normal to σ disk D^{k+1}_σ can be viewed as a subspace of the double $D\mathsf{E}_{\omega_\sigma}$ or of the space $\mathsf{E}_{\omega_\sigma}$, both stratified with the help of the model map

$$p_{\omega_\sigma} : (\mathsf{E}_{\omega_\sigma}, \delta\mathsf{E}_{\omega_\sigma}) \to (\mathsf{T}_{\omega_\sigma}, \mathsf{S}_{\omega_\sigma}).$$

The normal disk acquires its stratification from the ambient space

$$D\mathsf{E}_{\omega_\sigma} =_{\mathsf{def}} E_{\omega_\sigma} \cup_{\{\partial\mathsf{E}_{\omega_\sigma} \setminus \delta\mathsf{E}_{\omega_\sigma}\}} \mathsf{E}_{\omega_\sigma}.$$

Hence, as a stratified topological space, D^{k+1}_σ depends only on the position of the stratum σ in the canonical stratification of $D\mathsf{E}_{\omega_\sigma}$.

As a result, the stratified simplicial norms $\big\{\|\,[D^{k+1}_\sigma]\,\|_\Delta^{\mathcal{S}^\bullet_F(U)}\big\}_\sigma$ take a *finite set of values*, which depend only on the list of model maps $\{p_\omega : \mathsf{E}_\omega \to \mathsf{T}_\omega\}_\omega$.

Therefore, in line with formula (8.17) and the definition in (8.18), we get

$$\|h\|_\Delta \leq \lambda \cdot \|[\,h_U\,]\| \tag{8.19}$$

where $\lambda > 0$ is universal for all X, F, and h.

Using (8.19) we notice that, if $\|h\|_\Delta > 0$ for some $h \in H_{k+1}(DX)$, then $\lambda > 0$; moreover, $\|[\,h_U\,]\| > 0$ as well.

So computing the proportion $\|h\|_\Delta / \|[\,h_U\,]\|$ for some X, F, and $h \in H_{k+1}(DX)$ gives us an *upper* estimate of the universal λ. It seems more challenging to get some *lower* estimates of the universal λ.

Next, we are going to reinterpret the norm $\|[\,h_U\,]\|$ in terms of the differential complexes $\mathbf{C}^*_{\mho}(DX, F)$ and $\mathbf{C}^{\mho}_*(DX, F)$ in order to make it "more computable".

Let $\{[\sigma]^*\}_{\sigma\in\mathcal{S}^\bullet(DX_{-(n-k)},F)}$ be the basis of $\mathsf{C}^{n-k}_{\mho}(DX,F)$, dual to the basis $\{[\sigma]\}_{\sigma\in\mathcal{S}^\bullet(DX_{-(n-k)},F)}$ in $\mathsf{C}^{\mho}_{n-k}(DX,F)$.

The $|[\sim]|^{\mho}$-norm of a class

$$\eta^* \in \mathsf{C}^{n-k}_{\mho}(DX,F)/\mathsf{B}^{n-k}_{\mho}(DX,F)$$

is the *quotient norm*, induced by the l_1-norm on $\mathsf{C}^{n-k}_{\mho}(DX,F)$. It is defined by the formula (which resembles the formula in (8.16))

$$|[\eta^*]|^{\mho} =_{\mathsf{def}} \inf_{\{\zeta^*\equiv\eta^* \mod \mathsf{B}^{n-k}_{\mho}(DX,F)\}} \Big\{ \sum_{\sigma\in\mathcal{S}^\bullet_F(DX),\ \dim\sigma=j} |r_\sigma| \Big\},$$

where

$$\zeta^* = \sum_{\sigma\in\mathcal{S}^\bullet_F(DX),\ \dim\sigma=j} r_\sigma\cdot[\sigma]^*.$$

Recall that in (8.15) we have considered an epimorphism $A:\mathsf{C}^\dagger_{k+1}\to H_{k+1}(U,\partial U)$, where $\mathsf{C}^\dagger_{k+1}$ denotes the free module over $\mathbb{R}$ (or over $\mathbb{Z}$), generated by the normal relative disks $\{D^{k+1}_\sigma\}_\sigma$. Put $\mathsf{R}^\dagger_{k+1} =_{\mathsf{def}} \ker(A)$, so that

$$\mathsf{C}^\dagger_{k+1}/\mathsf{R}^\dagger_{k+1} \approx H_{k+1}(U,\partial U).$$

On the other hand, the Poincaré duality $\mathcal{D}_U$ produces an isomorphism

$$B:\ H_{k+1}(U,\partial U) \xrightarrow{\approx\mathcal{D}_U} H^{n-k}(U) \xrightarrow{\approx} \mathsf{C}^{n-k}_{\mho}(DX,F)/\mathsf{B}^{n-k}_{\mho}(DX,F).$$

The composition $B\circ A$ takes each generator $D^{k+1}_\sigma\in\mathsf{C}^\dagger_{k+1}$ to the class of $[\sigma]^*$ and identifies the space of relations $\mathsf{R}^\dagger_{k+1}$ with the space $\mathsf{B}^{n-k}_{\mho}(DX,F)$.

Let $\mathcal{D}_U(h_U)$ denotes the Poincaré dual in U of the relative homology class h_U.

Examining the definitions of the norm in (8.16) and of the quotient norm $|[\sim]|^{\mho}$ and tracing the nature of the Poincaré duality (in terms of the intersections of relative and absolute cycles of complementary dimensions in U), we see that the norm $|[h_U]|$ in (8.16) coincides with the norm $|[\mathcal{D}_U(h_U)]|^{\mho}$, where

$$\mathcal{D}_U(h_U)\in H^{n-k}(DX^F_{-(k+1)}) \approx \mathsf{C}^{n-k}_{\mho}(DX,F)/\mathsf{B}^{n-k}_{\mho}(DX,F).$$

Therefore, combining this observation with (8.19), we get the desired inequality:

$$\|h\|_\Delta \leq \lambda\cdot|[\mathcal{L}^{F,\mho}_{k+1}(h)]|_{\mho}. \tag{8.20}$$

□

Theorem 8.3 has a "more geometric" interpretation (see Figure 8.2).

Corollary 8.2. *Let a homology class $h \in H_{k+1}(DX;\mathbb{Z})$ be realized my a singular pseudo-manifold* [9] $f : M \to DX$, $\dim(M) = k+1$.

Under the hypotheses of Theorem 8.3, the number of intersections of the cycle $f(M)$ with the locus $DX^F_{-(k+1)}$ is greater than or equal to $\lambda^{-1} \cdot \|h\|_\Delta$, provided that f is in general position with the subcomplex $DX^F_{-(k+1)} \subset DX$.

Proof. Recall that any integral homology class $h \in H_{k+1}(DX;\mathbb{Z})$ can be realized by a pseudo-manifold $f : M \to DX$ ([Hat], pages 108-109).

By a small perturbation of f, we can assume that $f(M)$ intersects transversally only with the pure $(n-k)$-dimensional strata in $DX^F_{-(k+1)}$ from the poset $\mathcal{S}^\bullet_F(DX)$.

Consider the transversal intersections from the set $f(M) \cap DX^F_{-(k+1)}$. Then, for a sufficiently narrow regular neighborhood U of $DX^F_{-(k+1)}$, the localized class h_U has a representation

$$h_U = \sum_{x_\sigma \in f(M) \cap DX^F_{-(k+1)}} r_\sigma \cdot [D^{k+1}_{x_\sigma}],$$

where the relative disk $(D^{k+1}_{x_\sigma}, \partial D^{k+1}_{x_\sigma}) \subset (U, \partial U) \cap f(M)$ and $r_\sigma = \pm 1$.

By (8.17), the norm

$$\|[h_U]\| \leq \sum_{x_\sigma \in f(M) \cap DX^F_{-(k+1)}} 1 = \#\Big(f(M) \cap DX^F_{-(k+1)}\Big).$$

Again, by (8.17), we get $\|h\|_\Delta \leq \lambda \cdot \#(f(M) \cap DX^F_{-(k+1)})$. Thus

$$\#(f(M) \cap DX^F_{-(k+1)}) \geq \lambda^{-1} \cdot \|h\|_\Delta, \tag{8.21}$$

where the positive constant λ^{-1} depends only on n, the poset $\mathcal{S}$, and the list of model maps $\mathsf{E} \Rightarrow \mathsf{T}$ in the way that is described in formula (8.18). □

In order to get similar results about *absolute* homology classes $h \in H_{k+1}(X)$, we consider the F-induced filtration

$$X^{F^\circ}_{-0} \supset X^{F^\circ}_{-1} \supset \cdots \supset X^{F^\circ}_{-n}$$

of $X^\circ = \mathsf{int}(X)$ by the *connected components* of the strata $\{X^\circ(F,\omega)\}_{\omega \in \mathcal{S}_n}$ as in (8.7), together with the F-induced filtration

$$X^{F^\bullet}_{-0} \supset X^{F^\bullet}_{-1} \supset \cdots \supset X^{F^\bullet}_{-(n+1)}$$

[9]a compact simplicial complex whose singular set has at least codimension 2

of X by the connected components of the strata

$$\{X^\circ(F,\omega)\}_{\omega\in\mathcal{S}_n} \text{ and } \{\partial X \cap X(F,\omega)\}_{\omega\in\mathcal{S}_n}$$

of a given codimension.

We introduce the groups

$$\begin{aligned}
&\big\{\mathsf{C}^{\mho}_{n+1-j}(X^\circ, F) =_{\mathsf{def}} H_{n+1-j}\big(X^{F^\bullet}_{-j},\, X^{F^\bullet}_{-(j+1)} \cup (X^{F^\bullet}_{-j} \cap \partial X);\, \mathsf{R}\big)\big\}_j,\\
&\big\{\mathsf{C}^{n+1-j}_{\mho}(X^\circ, F) =_{\mathsf{def}} H^{n+1-j}\big(X^{F^\bullet}_{-j},\, X^{F^\bullet}_{-(j+1)} \cup (X^{F^\bullet}_{-j} \cap \partial X);\, \mathsf{R}\big)\big\}_j.
\end{aligned} \tag{8.22}$$

They come with bases, whose elements correspond to the relative fundamental cycles of the codimension j connected components of $\{X^\circ(F,\omega)\}$ from the stratification $\mathcal{S}^\circ_F(X)$ of $\mathsf{int}(X)$.

Using the excision property, the long exact sequences of the triples

$$\big\{X^{F^\bullet}_{-(j-1)} \cup \partial X \supset X^{F^\bullet}_{-j} \cup \partial X \supset X^{F^\bullet}_{-(j+1)} \cup \partial X\big\}_j$$

help to organize the groups in (8.22) into differential complexes $\mathbf{C}^{\mho}_*(X^\circ, F)$ and $\mathbf{C}^*_{\mho}(X^\circ, F)$, similar to the ones in (8.9) and (8.10).

Employing the Poincaré duality $\mathcal{D}$ again, we introduce the localization homomorphism

$$\mathcal{M}^F_{k+1} : H_{k+1}(X) \xrightarrow{\approx\mathcal{D}} H^{n-k}(X, \partial X) \xrightarrow{i^*_{\mathsf{loc}}} H^{n-k}\big(X^{F^\bullet}_{-(k+1)}, X^{F^\bullet}_{-(k+1)} \cap \partial X\big).$$

Its target may be identified with the quotient $\mathsf{C}^{n-k}_{\mho}(X^\circ, F)/\mathsf{B}^{n-k}_{\mho}(X^\circ, F)$ by an isomorphism

$$H^{n-k}\big(X^{F^\bullet}_{-(k+1)},\, X^{F^\bullet}_{-(k+1)} \cap \partial X\big) \approx \mathsf{C}^{n-k}_{\mho}(X^\circ, F)/\mathsf{B}^{n-k}_{\mho}(X^\circ, F).$$

The isomorphism can be validated in a way that is similar to the one that led to (8.11). Indeed, put

$$A =_{\mathsf{def}} X^{F^\bullet}_{-(k+1)},\; B =_{\mathsf{def}} (X^{F^\bullet}_{-(k+1)} \cap \partial X) \cup X^{F^\bullet}_{-(k+2)},\; C =_{\mathsf{def}} (X^{F^\bullet}_{-(k+1)} \cap \partial X).$$

Consider the fragment $J : H^{n-k}(A, B) \to H^{n-k}(A, C)$ of the long exact sequence of the triple $A \supset B \supset C$. By definition, $H^{n-k}(A, C)$ is the target of $\mathcal{M}^F_{k+1}$, and $H^{n-k}(A, B) = \mathsf{C}^{n-k}_{\mho}(X^\circ, F)$. For dimensional reasons, $H^{n-k}(B, C) = 0$; so J is an epimorphism.

Similar arguments, which involve a different triple of spaces, imply that $\ker J$ can be identified with the image

$$\mathrm{im}\big(\delta^{n-k-1} : \mathsf{C}^{n-k-1}_{\mho}(X^\circ, F) \to \mathsf{C}^{n-k}_{\mho}(X^\circ, F)\big).$$

As a result, for each shadow F, we get a localization of the Poincaré duality:

$$\mathcal{M}^{F,\mho}_{k+1} : H_{k+1}(X) \stackrel{\approx\mathcal{D}}{\longrightarrow} H^{n-k}(X, \partial X) \stackrel{I^*}{\longrightarrow} \mathsf{C}^{n-k}_{\mho}(X^\circ, F)/\mathsf{B}^{n-k}_{\mho}(X^\circ, F). \tag{8.23}$$

As before, the target space of $\mathcal{M}^{F,\mho}_{k+1}$ admits the quotient norm $\|[\sim]\|_\mho$. It is induced by the l_1-norm on the based space $\mathsf{C}^{n-k}_{\mho}(X^\circ, F)$, the base being indexed by the connected components of $\{X^\circ(F,\omega)\}_\omega$.

These considerations lead to a twin of Theorem 8.3.

Theorem 8.4 (The second amenable localization of the Poincaré duality).
Let X be a compact connected $(n+1)$-dimensional PL*-manifold with boundary. Assume that for each connected component of the boundary ∂X, the image of its fundamental group in $\pi_1(X)$ is an amenable group.*

Then, there exists an universal *constant $\mu > 0$ such that, for every X and its shadow $F \in \mathsf{Shad}(X, \mathsf{E} \Rightarrow \mathsf{T})$, the space $\mathsf{C}^{n-k}_{\mho}(X^\circ, F)/\mathsf{B}^{n-k}_{\mho}(X^\circ, F)$ admits a norm $\|[\sim]\|_\mho$ so that*

$$\|h\|_\Delta \leq \mu \cdot \|[\mathcal{M}^{F,\mho}_{k+1}(h)]\|_\mho \tag{8.24}$$

for all $h \in H_{k+1}(X)$. Here the Poincaré duality localizing operator $\mathcal{M}^{F,\mho}_{k+1}$ is introduced in (8.23), and the constant μ depends only on n, the poset $\mathcal{S}$, and the list of model maps $\{\mathsf{E}_\omega \to \mathsf{T}_\omega\}_{\omega\in\mathcal{S}}$ in the way that is described in formula (8.25).

Proof. The cycle h is represented by a singular cycle, whose image is contained in $\mathsf{int}(X)$. So it interacts only with the strata from $\mathcal{S}^\circ_F(X)$ (and misses the strata from $\mathcal{S}^\bullet_F(\partial X)$). Therefore, we can pick a regular neighborhood U of $X^{F^\circ}_{-(k+1)}$ in $\text{int}(X)$ so small that the localized class $h_U \in H_{k+1}(U, \partial U)$ is a linear combination of finitely many disjoint disks

$$\{(D^{k+1}_{x_\sigma}, \partial D^{k+1}_{x_\sigma}) \subset (U, \partial U)\}_\sigma,$$

normal to the strata σ in $X^{F\circ}_{-(k+1)}$.

By the argument used in the proof of Theorem 8.3, the images of the fundamental groups of the strata from $\mathcal{S}^\circ_F(X)$ are amenable in $\pi_1(X) \approx \pi_1(\text{int}(X))$. Thus we may apply the Localization Lemma 8.1 to the image of the class h under the classifying map $\alpha : \text{int}(X) \to K(\pi_1(X), 1)$.

Since only the disks $\{(D^{k+1}_{x_\sigma}, \partial D^{k+1}_{x_\sigma})\}_{\sigma \in \mathcal{S}^\circ_F(X)}$ participate in the localization of h to U, the constant λ in the estimate (8.18) must be replaced by the constant

$$\mu =_{\mathsf{def}} \max_{\sigma \in \mathcal{S}^\circ_F(X),\ \dim \sigma = n-k} \{\| \, [D^{k+1}_\sigma] \, \|^{\mathcal{S}^\circ_F(U)}_\Delta \}. \tag{8.25}$$

It is smaller than or equal to λ from (8.18), but still positive. Like λ, μ also depends on $\mathcal{S}$, n, and k only.

The rest of the arguments are similar to the ones from the proof of Theorem 8.3. □

Retracing the proof of Theorem 8.4, we get the following:

Corollary 8.3. *Under the amenability hypotheses of Theorem 8.4, if $h \in H_{k+1}(X; \mathbb{Z})$ is represented by a singular pseudo-manifold $f : M \to X$, then, for any shadow F of X, the number of transversal intersections of the absolute cycle $f(M)$ with the locus $X^{F^\circ}_{-(k+1)} \subset \mathrm{int}(X)$ is greater than or equal to $(\mu)^{-1} \cdot \|h\|_\Delta$, where the positive constant μ is universal for all X and their shadows F.* ◇

For a pair of topological spaces $Z \supset W$ and an integer $j \geq 0$, consider the subspace/subgroup

$$\mathcal{K}^{\Delta=0}_j(Z, W) \subset H_j(Z, W),$$

formed by the homology classes $h \in H_j(Z, W)$ whose simplicial semi-norms $\|h\|_\Delta$ vanish. Then $\| \sim \|_\Delta$ becomes a *norm* on the quotient space

$$H^{\mathbf{\Delta}}_j(Z, W) =_{\mathsf{def}} H_j(Z, W)/\mathcal{K}^{\Delta=0}_j(Z, W).$$

Assume that Z admits a continuous map $\phi : (Z, W) \to (Z, W)$ such that its action $\phi_* : H_j(Z, W) \to H_j(Z, W)$ on the homology has an eigen-element h, so that $\phi_*(h) = \lambda \cdot h$ for a scalar λ, subject to $|\lambda| > 1$. Then

$$\|h\|_\Delta \;\geq\; \|\phi_*(h)\|_\Delta \;=\; |\lambda| \cdot \|h\|_\Delta.$$

In such case, we get $\|h\|_\Delta = 0$; so such an element $h \in H_j(Z, W)$ dies in the quotient $H^{\mathbf{\Delta}}_j(Z, W)$.

In fact, for any Z, we get $H^{\mathbf{\Delta}}_1(Z) = 0$. However, if Z is a closed surface of genus $g > 1$, then $H^{\mathbf{\Delta}}_2(Z) \neq 0$. Moreover, when Z is a product of several closed surfaces of genera $g(M_i) \geq 2$, then the group $H^{\mathbf{\Delta}}_2(Z) \neq 0$ is quite rich.

Recall that, for a continuous map $f : Z \to Y$ of topological spaces, $\|h\|_\Delta \geq \|f_*(h)\|_\Delta$. Therefore, f induces a continuous linear map of normed spaces $f_* : H_j^\Delta(Z) \to H_j^\Delta(Y)$ whose operator norm is ≤ 1.

It is not difficult to verify that when f is a homotopy equivalence, then this map f_* is an *isometry* (in the simplicial norm) between the normed spaces $H_j^\Delta(Z)$ and $H_j^\Delta(Y)$. In particular, *the shape of the unit ball* in $H_j^\Delta(Z)$ is an invariant of the homotopy type of Z.

We can take this observation one step further. The Mapping Theorem from [Gr], section 3.1, implies that the classifying map $f : Z \to K(\pi_1(Z), 1)$ induces an *isometry*

$$f_* : H_j^\Delta(Z) \to H_j^\Delta\big(K(\pi_1(Z), 1)\big),$$

so that f_* is a monomorphism.

The integral classes $H_*(X; \mathbb{Z}) \subset H_*(X; \mathbb{R})$ form a lattice. For a closed 3-fold X, Thurston proved (see [Th1], Theorem 2) that, if $\| \sim \|_\Delta$ is a norm, then the unit ball $B \subset H_2^\Delta(X; \mathbb{R}) = H_2(X; \mathbb{R})$ is a finite polyhedron, given by integral inequalities. This result motivates the following question.

Question 8.1.

- *For any compact CW-complex X, is the unit ball $B_j \subset H_j^\Delta(X; \mathbb{R})$ a finite polyhedron? Are its vertices among the images of the lattice $H_j(X; \mathbb{Z})$ under the quotient map $H_j(X; \mathbb{R}) \to H_j^\Delta(X; \mathbb{R})$?*
- *For a given finitely presented group π, how to describe the shape of the unit ball B_j in the normed space $H_j^\Delta(K(\pi, 1); \mathbb{R})$ in terms of π (say, in terms of its subgroups or in terms of the representation theory)? Is B_j a polyhedron for a finitely-presented π?* ◇

It turns out that formula (8.20) gives a nontrivial lower boundary for the number of strata $\sigma \in \mathcal{S}_F^\bullet(DX)$ of dimension $n - k$ for *any* shadow $F \in \mathsf{Shad}(X, \mathsf{E} \Rightarrow \mathsf{T})$. This universal boundary is expressed in terms of the group $H_{k+1}^\Delta(DX)$.

The next theorem is also a manifestation of the amenable localization, coupled with the Poincaré duality. It may be viewed as a version of *Morse Inequalities*, where a Morse function $f : X \to \mathbb{R}$ is replaced by a shadow F, the f-critical points are replaced by the F-induced strata in the double DX, and the homology of X by the "homology" $H_*^\Delta(DX)$. In particular, it shows that the groups $H_{k+1}^\Delta(DX)$ provide lower bounds of the ranks of the groups $\mathsf{C}_{\mho}^{n-k}(DX, F)$ from the differential complex in (8.10).

Theorem 8.5. *Let X be a compact connected $(n+1)$-dimensional* PL*-manifold with boundary.*

- *Assume that, for each connected component of the boundary ∂X, the image of its fundamental group in $\pi_1(DX)$ be amenable.*

  Then, for each $k \in [-1, n]$ and every shadow $F \in \mathsf{Shad}(X, \mathsf{E} \Rightarrow \mathsf{T})$, the Poincaré duality localizing operator $\mathcal{L}^{F,\mho}_{k+1}$ from (8.11) has the property

$$\mathrm{rk}\big(\mathrm{im}(\mathcal{L}^{F,\mho}_{k+1})\big) \;\geq\; \mathrm{rk}\big(H^{\boldsymbol{\Delta}}_{k+1}(DX)\big). \tag{8.26}$$

  As a result, $\mathrm{rk}\big(\mathsf{C}^{n-k}_{\mho}(DX, F)\big)$—the number of connected components of the strata $\sigma \in \mathcal{S}^{\bullet}_{F}(DX)$ of dimension $(n-k)$—is greater than or equal to $\mathrm{rk}\big(H^{\boldsymbol{\Delta}}_{k+1}(DX)\big)$.

- *Assume that, for each connected component of the boundary ∂X, the image of its fundamental group in $\pi_1(X)$ be amenable.*

  Then, for for each $k \in [-1, n-1]$ and every shadow $F \in \mathsf{Shad}(X, \mathsf{E} \Rightarrow \mathsf{T})$, the Poincaré duality localizing operator $\mathcal{M}^{F,\mho}_{k+1}$ from (8.23) has the property

$$\mathrm{rk}\big(\mathrm{im}(\mathcal{M}^{F,\mho}_{k+1})\big) \;\geq\; \mathrm{rk}\big(H^{\boldsymbol{\Delta}}_{k+1}(X)\big). \tag{8.27}$$

  Thus, $\mathrm{rk}\big(\mathsf{C}^{n-k}_{\mho}(X^{\circ}, F)\big)$—the number of connected components of the strata $\sigma \in \mathcal{S}^{\circ}_{F}(X)$ of dimension $(n-k)$—is greater than or equal to $\mathrm{rk}\big(H^{\boldsymbol{\Delta}}_{k+1}(X)\big)$.

Proof. Put $\mathcal{K}^{\Delta=0}_{k+1} =_{\mathsf{def}} \mathcal{K}^{\Delta=0}_{k+1}(DX)$, the space of homology classes whose simplicial norm vanishes. If $h \in H_{k+1}(DX)$ belongs to the kernel of the homomorphism $\mathcal{L}^{F,\mho}_{k+1}$, then evidently

$$\|[\,\mathcal{L}^{F,\mho}_{k+1}(h)\,]\|_{\mho} = \|[\,0\,]\|_{\mho} = 0.$$

By the inequality (8.20), $\|h\|_{\Delta} = 0$. Therefore, $h \in \mathcal{K}^{\Delta=0}_{k+1}$. In other words,

$$\ker(\mathcal{L}^{F,\mho}_{k+1}) \subset \mathcal{K}^{\Delta=0}_{k+1}.$$

Hence,

$$\mathrm{rk}\big(H_{k+1}(DX)/\mathcal{K}^{\Delta=0}_{k+1}\big) \leq \mathrm{rk}\big(H_{k+1}(DX)/\ker(\mathcal{L}^{F,\mho}_{k+1})\big) = \mathrm{rk}\big(\mathrm{im}(\mathcal{L}^{F,\mho}_{k+1})\big).$$

So we get

$$\begin{aligned}\mathrm{rk}\big(H^{\boldsymbol{\Delta}}_{k+1}(DX)\big) \;&\leq\; \mathrm{rk}\big(\mathsf{C}^{n-k}_{\mho}(DX, F)/\mathsf{B}^{n-k}_{\mho}(DX, F)\big)\\ &\leq\; \mathrm{rk}\big(\mathsf{C}^{n-k}_{\mho}(DX, F)\big),\end{aligned} \tag{8.28}$$

the later rank is the number of strata σ of dimension $n-k$ in the stratification $\mathcal{S}^{\bullet}_F(DX)$.

Similar arguments, based on Theorem 8.4, prove that

$$\begin{aligned}\mathrm{rk}\big(H^{\Delta}_{k+1}(X)\big) &\leq \mathrm{rk}\big(\mathsf{C}^{n-k}_{\mho}(X^\circ, F)/\mathsf{B}^{n-k}_{\mho}(X^\circ, F)\big)\\ &\leq \mathrm{rk}\big(\mathsf{C}^{n-k}_{\mho}(X^\circ, F)\big).\end{aligned} \tag{8.29}$$

□

Remark 8.3. Since $H^{\Delta}_1(DX) = 0$ and $H^{\Delta}_1(X) = 0$ for all X, the statements of Theorem 8.5 are vacuous for $k = 0$. Also, for $k = n$, the statement in the second bullet is vacuous since $H^{\Delta}_{n+1}(X) = 0$.

Since the classifying map $\beta : DX \to K(\Pi, 1)$, where $\Pi =_{\mathsf{def}} \pi_1(DX)$, induces an isomorphism of the fundamental groups, $\beta_* : H_*(DX) \to H_*(K(\Pi, 1))$ is an isometry (see [Gr]). As a result, the induced map

$$\tilde{\beta}_* : H^{\Delta}_*(DX) \to H^{\Delta}_*(K(\Pi, 1))$$

is a monomorphism. Therefore, *the best lower estimate* that formula (8.28) could provide is equal to

$$\mathrm{rk}\big(H^{\Delta}_{k+1}(K(\Pi,\, 1))\big) \;\geq\; \mathrm{rk}\big(H^{\Delta}_{k+1}(DX)\big).$$

Similarly, with $\pi = \pi_1(X)$, *the best lower estimate* that formula (8.29) could provide is equal to

$$\mathrm{rk}\big(H^{\Delta}_{k+1}(K(\pi,\, 1))\big) \;\geq\; \mathrm{rk}\big(H^{\Delta}_{k+1}(X)\big).$$

◇

Question 8.2. *Can one probe faithfully the shape of the unit ball $\tilde{B}_{k+1}$ in the quotient norm $\|\sim\|_\Delta$ on $H^{\Delta}_{k+1}(DX)$ by counting the cardinalities $\#\big(f(M) \cap DX^F_{-(k+1)}\big)$ for various shadows F and singular pseudo-manifolds $f : M \to DX$?*

A similar question may be posed about the shape of the unit ball B_{k+1} in the quotient norm $\|\sim\|_\Delta$ on $H^{\Delta}_{k+1}(X)$. ◇

In one interesting special case that deals with the strata of maximal codimension, Theorems 8.3-8.5 can be made more explicit. The proposition below is a generalization of Theorem 2 from [AK].

Theorem 8.6. *Let X be a compact connected and oriented $(n+1)$-dimensional* PL*-manifold with boundary. Assume that, for each connected* *component of the boundary ∂X, the image of its fundamental group in $\pi_1(DX)$ is amenable.*

Then there is a X-independent universal constant $\theta > 0$ such that, for any shadow $F \in \mathsf{Shad}(X, \mathsf{E} \Rightarrow \mathsf{T})$, the cardinality of the finite set $F(X)_{-n}$ satisfies the inequality

$$\#\big(F(X)_{-n}\big) \;\geq\; \theta \cdot \|\, [DX] \,\|_\Delta. \tag{8.30}$$

Proof. The argument is based on the proof of Theorem 8.3. Note that the fundamental class $[DX] \in H_{n+1}(DX)$, being restricted to the regular neighborhood

$$U =_{\mathsf{def}} \coprod_{\{\sigma \in \mathcal{S}^\bullet_F(DX) |\; \mu'(\omega_\sigma) = n\}} D^{n+1}_\sigma$$

of the locus $DX^F_{-(n+1)}$, equals the sum

$$\sum_{\{\sigma \in \mathcal{S}^\bullet_F(DX) |\; \mu'(\omega_\sigma) = n\}} [D^{n+1}_\sigma, \partial D^{n+1}_\sigma].$$

Therefore, by (8.16), $[\| \, [DX]_U \, \|]_\mho = \#\big(DX^F_{-(n+1)}\big)$. Employing inequality (8.20), we get

$$\|\, [DX] \,\|_\Delta \;\leq\; \lambda \cdot \#\big(DX^F_{-(n+1)}\big) \;\leq\; \lambda \cdot \kappa \cdot \#\big(F(X)_{-n}\big),$$

where

$$\kappa =_{\mathsf{def}} \max_{\{\omega |\; \mu'(\omega) = n\}} \big(\mu(\omega) - \mu'(\omega)\big)$$

is the maximal cardinality of the fibers of the map

$$F:\; DX^F_{-(n+1)} = F^{-1}\big(F(X)_{-n}\big) \cap \partial X \longrightarrow F(X)_{-n}.$$

Choosing $\theta = (\lambda \cdot \kappa)^{-1}$ completes the argument.

Note that X has no absolute fundamental cycle and that $X^{F^\circ}_{-(n+1)} = \emptyset$; therefore applying the second bullet of Theorem 8.5 leads to a tautology.

□

The next implication of Theorem 8.5 reveals the groups $H^{\mathbf{\Delta}}_{k+1}(DX)$ and $H^{\mathbf{\Delta}}_k(X)$ as *obstructions* to the existence of *globally k-convex shadows* of X (see Definition 8.6).

Corollary 8.4. *Let X be a compact connected $(n+1)$-dimensional* PL-*manifold with boundary. Assume that, for each connected component of the* *boundary ∂X, the image of its fundamental group in $\pi_1(X)$ is an amenable group.*

If X is globally k-convex, then the simplicial semi-norm is trivial on $H_{j+1}(DX)$ and on $H_j(X)$ for all $j \geq k$.

In particular, if an oriented X admits a shadow $F \in \mathsf{Shad}(X, \mathsf{E} \Rightarrow \mathsf{T})$ such that $F(X)_{-n} = \emptyset$, then $\| [DX] \|_\Delta = 0$ and $H_n^{\Delta}(X) = 0$.

Proof. By Remark 8.2, $F(X)_{-k} = \emptyset$ implies $F(X)_{-j} = \emptyset$ for all $j \geq k$. Therefore $DX^F_{-(j+1)} = \emptyset$ for all $j \geq k$, since the $(n-j)$-dimensional strata from $\mathcal{S}^\bullet_F(DX)$ are contributed only by the strata of $F(X)$ of dimensions $(n-j)$ and $(n-j-1)$. The exception to this rule is provided by the 0-dimensional strata from $\mathcal{S}^\bullet_F(DX)$: they are contributed by the points from $F(X)_{-n}$ only.

In contrast, only the $(n-j)$-dimensional strata of $F(X)$ contributes to the $(n+1-j)$-dimensional strata from $\mathcal{S}^\circ_F(X)$. So $F(X)_{-j} = \emptyset$ implies $X^{F^\circ}_{-j} = \emptyset$. In turn, this implies $\mathsf{C}^{n+1-j}_{\mho}(X^\circ, F) = 0$. Therefore, by (8.29), $H_j^{\Delta}(X) = 0$ for all $j \geq k$. $\square$

Definition 8.12. Let X be a compact connected $(n+1)$-dimensional PL-manifold with boundary, and $\mathcal{A}$ denotes an abelian group or a field. For any shadow $F \in \mathsf{Shad}(X, \mathsf{E} \Rightarrow \mathsf{T})$ and each $j \in [0, n+1]$, let

$$(\underline{\Sigma c})^j(X, F) =_{\mathsf{def}} \mathrm{rk}_{\mathcal{A}}\big(\mathsf{C}^j_{\mho}(DX, F)\big).$$

We call this integer the j-th *suspension $\mho$-complexity* of the shadow F.

Let

$$(\underline{\Sigma c})^j_{\mathsf{shad}}(X) =_{\mathsf{def}} \min_{F \in \mathsf{Shad}(X,\, \mathsf{E} \Rightarrow \mathsf{T})} (\underline{\Sigma c})^j(X, F).$$

We call $(\underline{\Sigma c})^j_{\mathsf{shad}}(X)$ the j-th *suspension shadow complexity* of X. $\diamondsuit$

These non-negative integers can be organized into the sequence

$$(\underline{\boldsymbol{\Sigma} \mathbf{c}})_{\mathsf{shad}}(X) =_{\mathsf{def}} \big((\underline{\Sigma c})^0_{\mathsf{shad}}(X),\ (\underline{\Sigma c})^1_{\mathsf{shad}}(X),\ \dots,\ (\underline{\Sigma c})^{n+1}_{\mathsf{shad}}(X)\big).$$

Note that this optimal sequence may not be realizable by a single shadow! To avoid this fundamental difficulty, we will need a more manageable version of the previous definition.

Definition 8.13. Let X be a compact connected $(n+1)$-dimensional PL-manifold with boundary. For any shadow $F \in \mathsf{Shad}(X, \mathsf{E} \Rightarrow \mathsf{T})$, we form the sequence of non-negative integers

$$(\underline{\boldsymbol{\Sigma} \mathbf{c}})(X, F) =_{\mathsf{def}} \big((\underline{\Sigma c})^0(X, F),\ (\underline{\Sigma c})^1(X, F),\ \dots,\ (\underline{\Sigma c})^{n+1}(X, F)\big).$$

and take the *lexicographic* minimum

$$(\underline{\mathbf{\Sigma c}})^{\mathsf{lex}}_{\mathsf{shad}}(X) =_{\mathsf{def}} \text{lex.min}_{F \in \mathsf{Shad}(X, \mathsf{E} \Rightarrow \mathsf{T})} \, (\underline{\mathbf{\Sigma c}})(X, F).$$

We denote by $(\underline{\Sigma c})^{\mathsf{lex},\, j}_{\mathsf{shad}}(X)$ the $(j+1)$-th component of the vector $(\underline{\mathbf{\Sigma c}})^{\mathsf{lex}}_{\mathsf{shad}}(X)$ and call it the j-th *suspended lexicographic shadow complexity* of X. ♢

Remark 8.4. By its very definition, the lexicographically optimal sequence of complexities is delivered by some shadow F!

Evidently, for each j, $(\underline{\Sigma c})^{\mathsf{lex},\, j}_{\mathsf{shad}}(X) \geq (\underline{\Sigma c})^{j}_{\mathsf{shad}}(X)$. ♢

Theorem 8.4 has an immediate implication.

Corollary 8.5. *Let X be a compact connected $(n+1)$-dimensional* PL-*manifold with boundary. Assume that, for each connected component of the boundary ∂X, the image of its fundamental group in $\pi_1(DX)$ is an amenable group.*

Then, for any $k \in [-1, n]$, the suspended shadow complexity of X satisfies the inequality

$$(\underline{\Sigma c})^{n-k}_{\mathsf{shad}}(X) \; \geq \; \text{rk}\big(H^{\mathbf{\Delta}}_{k+1}(DX)\big).$$

♢

Let us consider the sequence of non-negative integers

$$\mathbf{rk}\big(H^{\mathbf{\Delta}}_{*}(DX)\big) =_{\mathsf{def}} \Big(\text{rk}\big(H^{\mathbf{\Delta}}_{n+1}(DX)\big),\, \text{rk}\big(H^{\mathbf{\Delta}}_{n}(DX)\big),\, \dots,\, \text{rk}\big(H^{\mathbf{\Delta}}_{0}(DX)\big)\Big).$$

Then Corollary 8.5 can be expressed in its compressed form as

$$(\underline{\mathbf{\Sigma c}})^{\mathsf{lex}}_{\mathsf{shad}}(X) \geq (\underline{\mathbf{\Sigma c}})_{\mathsf{shad}}(X) \geq \mathbf{rk}\big(H^{\mathbf{\Delta}}_{*}(DX)\big), \tag{8.31}$$

where the vectorial inequality is understood as the inequality among all the the corresponding components of the participating vectors.

We notice that the RHS of inequality (8.31) depends only on the homotopy type of DX, while its LHS a priori depends on its PL-topological type.

Let X be a compact connected $(n+1)$-dimensional PL-manifold with boundary, and $F : X \to K$ its shadow. Recall that, by Definition 8.5, the cardinality of the fiber $F : X(F, \omega) \cap \partial X \to K(\omega)$ depends only on $\omega \in \mathcal{S}$: it is the difference $\mu(\omega) - \mu'(\omega)$.

Definition 8.14. Consider a stratified simplicial complex $K \in \mathsf{Shad}(\mathsf{T}, n)$ and its filtration as in (8.2). Put

$$\kappa_j(n) =_{\mathsf{def}} \max_{\{\omega|\, \mu'(\omega) = n-j\}} \big(\mu(\omega) - \mu'(\omega)\big).$$

The *weighted j-th complexity* of K is defined by the formula

$$^\sharp c^j(K) =_{\mathsf{def}} \kappa_j(n) \cdot c^j(K).$$

$\diamondsuit$

The next proposition helps to link the suspension complexities $\{(\underline{\Sigma c})^j(X, F)\}_j$ to the weighted ones $\{^\sharp c^j(F(X))\}_j$ and thus motivates Definition 8.14.

Corollary 8.6. *Let X be a compact connected $(n+1)$-dimensional* PL-*manifold with boundary. Assume that, for each connected component of the boundary ∂X, the image of its fundamental group in $\pi_1(X)$ is an amenable group.*

Then, for any shadow $F \in \mathsf{Shad}(X, \mathsf{E} \Rightarrow \mathsf{T})$ and each $k \in [-1, n]$,

$$^\sharp c^{n-k}(F(X)) + 2 \cdot {^\sharp c^{n-k-1}}(F(X)) \geq \mathrm{rk}\big(H^{\boldsymbol{\Delta}}_{k+1}(DX)\big), \tag{8.32}$$

$$^\sharp c^{n-k-1}(F(X)) \geq \mathrm{rk}\big(H^{\boldsymbol{\Delta}}_{k+1}(X)\big).$$

Proof. Let $K = F(X)$ be a shadow of X. Any connected component of $K(\omega)$ of dimension $n - \mu'(\omega)$ (see Definition 8.5) gives rise to $\mu(\omega) - \mu'(\omega)$ strata $\sigma \in \mathcal{S}^\bullet_F(DX)$ of dimension $n - \mu'(\omega)$ and to $2(\mu(\omega) - \mu'(\omega) - 1)$ strata $\sigma \in \mathcal{S}^\bullet_F(DX)$ of dimension $n + 1 - \mu'(\omega)$. Therefore (see Figure 8.2) the number of strata $\sigma \in \mathcal{S}^\bullet_F(DX)$ of a dimension $j = n - k$ is given by the formula:

$$\sum_{\omega|\, \mu'(\omega) = k} \big(\mu(\omega) - k - 1\big) \cdot \#\big(\pi_0\big(K(\omega)\big)\big)$$
$$+ \sum_{\omega|\, \mu'(\omega) = k+1} 2\big(\mu(\omega) - k\big) \cdot \#\big(\pi_0\big(K(\omega)\big)\big). \tag{8.33}$$

By Definition 8.14 of $\{\kappa_j(n)\}_j$, the number in (8.33) is smaller than or equal to

$$\kappa_{n-k}(n) \cdot \sum_{\omega|\, \mu'(\omega) = k} \#\big(\pi_0\big(K(\omega)\big)\big) + 2\kappa_{n-k-1}(n) \cdot \sum_{\omega|\, \mu'(\omega) = k+1} \#\big(\pi_0\big(K(\omega)\big)\big),$$

where the fist sum is the number of strata in $F(X)$ of dimension $n-k$, and the second sum is the number of strata of dimension $n-k-1$. Thus the previous formula is an upper bound of the number of components in $\mathcal{S}^\bullet_F(DX)$ of dimension $n-k$.

Similar upper estimate of the number of connected components in $X^{F^\circ}_{-k}$ leads to the LHS of the second inequality in (8.32).

Finally, Theorem 8.5 implies formulas (8.32). □

Corollary 8.6 has the following two immediate implications, which reveal the nontriviality of the groups/spaces $H^{\Delta}_{k+1}(DX) \subset H^{\Delta}_{k+1}(\Pi)$ and $H^{\Delta}_k(X) \subset H^{\Delta}_k(\pi)$ as *obstructions* to the existence of shadows F of vanishing $(n-k-1)$-complexity. The first implication is just a repackaging of Corollary 8.4.

Corollary 8.7. *Under the hypotheses of Corollary 8.6, if $H^{\Delta}_{k+1}(DX) \neq 0$, then either $c^{n-k}(F(X)) \neq 0$ or $c^{n-k-1}(F(X)) \neq 0$ for* any *shadow F.*

If $H^{\Delta}_{k+1}(X) \neq 0$, then $c^{n-k-1}(F(X)) \neq 0$ for any *shadow F.* ◊

Corollary 8.8. *Let X be a compact connected $(n+1)$-dimensional* PL*-manifold with boundary. Assume that, for each connected component of the boundary ∂X, the image of its fundamental group in $\pi_1(X)$ is amenable.*

Then, for each $k \in [-1, n]$,

$$^\sharp c^{n-k}_{\mathsf{shad}}(X, \mathsf{E} \Rightarrow \mathsf{T}) + 2 \cdot {}^\sharp c^{n-k-1}_{\mathsf{shad}}(X, \mathsf{E} \Rightarrow \mathsf{T}) \geq \mathrm{rk}\big(H^{\Delta}_{k+1}(DX)\big),$$

$$^\sharp c^{n-k-1}_{\mathsf{shad}}(X, \mathsf{E} \Rightarrow \mathsf{T}) \geq \mathrm{rk}\big(H^{\Delta}_{k+1}(X)\big). \tag{8.34}$$

◊

8.5 Complexity of traversally generic vector fields and the simplicial semi-norms

Now let us examine how these results about the complexities of shadows apply to the traversally generic flows on smooth manifolds with boundary. In fact, the entire general setting in Section 8.4 was designed with these applications in mind. As we will show next, any proposition about shadows of PL-manifolds with boundary and the simplicial semi-norms has a straightforward analogue for the smooth manifolds with boundary that carry *traversally generic* vector fields. In short, the traversally generic fields are a good source of shadows.

We will apply Theorems 8.3-8.5, and their corollaries to traversally generic vector fields v. In this setting, the poset

$$\mathcal{S} =_{\mathsf{def}} \mathbf{\Omega}^\bullet, \ \mu =_{\mathsf{def}} |\sim|, \ \mu' =_{\mathsf{def}} |\sim|',$$

the role of shadows is played by the obvious maps $\Gamma : X \to \mathcal{T}(v)$ which belong to an appropriate set $\mathsf{Shad}(X, \mathsf{E} \Rightarrow \mathsf{T})$.

Theorem 8.7. *Let X be a smooth compact connected and oriented $(n+1)$-manifold with boundary.*

Then any traversally generic vector field v gives rise to a shadow $\Gamma : X \to \mathcal{T}(v)$ in the sense of Definition 8.5. The the model projections $\{p_\omega : \mathsf{E}_\omega \to \mathsf{T}_\omega\}_{\omega \in \mathbf{\Omega}^\bullet_{\langle n]}}$ are described in Theorems 7.3 and 7.4, utilizing special coordinates as in Lemma 6.4, equations (6.24).

Proof. Recall that, for a traversally generic v, the trajectory space $\mathcal{T}(v)$ is a Whiney stratified space (see Theorem 6.10), which implies that Γ is a simplicial map with respect to the appropriate triangulations of the source and target spaces. Moreover, the triangulation of X can be chosen to be smooth.

By the very definition of a traversing flow, each fiber $\Gamma^{-1}(\gamma)$, $\gamma \in \mathcal{T}(v)$ is either a closed segment, or a singleton.

By Corollary 7.2, the map $\Gamma : \partial X \to \mathcal{T}(v)$, being restricted to the preimage of each proper stratum $\mathcal{T}(v, \omega)$, is a cover with the trivial monodromy and fibers of cardinality

$$\#(\sup(\omega)) = |\omega| - |\omega|' = \mu(\omega) - \mu'(\omega).$$

Moreover, by Theorem 6.9, each pure stratum $\mathcal{T}(v, \omega)$ is an open orientable smooth manifold.

By Lemma 6.4, Theorems 7.3 and 7.4, each point $\gamma \in \mathcal{T}(v, \omega)$, has a regular neigborhood $V_\gamma \subset \mathcal{T}(v)$, so that $\Gamma : \Gamma^{-1}(V_\gamma) \to V_\gamma$ is PL-homeomorphic to the model projection $p_\omega : \mathsf{E}_\omega \to \mathsf{T}_\omega$.

This completes the checklist of bullets from Definition 8.5. □

Let X be a smooth compact connected and oriented $(n+1)$-manifold with boundary, and v a traversally generic vector field.

Let us consider the filtration

$$\mathcal{T}(v) = \mathcal{T}(v)_{-0} \supset \mathcal{T}(v)_{-1} \supset \cdots \supset \mathcal{T}(v)_{-n}$$

of the trajectory space $\mathcal{T}(v)$ by the closed subcomplexes

$$\mathcal{T}(v)_{-j} =_{\mathsf{def}} \bigcup_{\{\omega \in \mathbf{\Omega}^\bullet_{'\langle n]} \mid\, |\omega|' \geq j\}} \mathcal{T}(v, \omega) \tag{8.35}$$

of dimensions $n - j$.

Recall that $v \in \mathcal{V}^\ddagger(X)$ is called *globally k-convex* if $\mathcal{T}(v)_{-k} = \emptyset$.

Let $\mathcal{A}$ be an abelian coefficient system on $\mathcal{T}(v)$ (equivalently, on X). As a default, $\mathcal{A} = \mathbb{R}, \mathbb{Z}$. For each $j \in [0, n]$, consider the relative homology groups[10]

$$\big\{\mathsf{C}^{\mho}_{n-j}(\mathcal{T}(v)) =_{\mathsf{def}} H_{n-j}\big(\mathcal{T}(v)_{-j},\, \mathcal{T}(v)_{-(j+1)};\, \mathcal{A}\big)\big\}_j \tag{8.36}$$

associated with the filtration. As in the context of shadows, $\mathsf{C}^{\mho}_j(\mathcal{T}(v))$ is the top homology of the quotient $\mathcal{T}(v)_{-j}/\mathcal{T}(v)_{-(j+1)}$.

These homology groups can be organized into a *differential complex*

$$\mathbf{C}^{\mho}_*(\mathcal{T}(v)) =_{\mathsf{def}} \tag{8.37}$$

$$\big\{0 \to \mathsf{C}^{\mho}_n(\mathcal{T}(v)) \xrightarrow{\partial_n} \mathsf{C}^{\mho}_{n-1}(\mathcal{T}(v)) \xrightarrow{\partial_{n-1}} \dots \xrightarrow{\partial_1} \mathsf{C}^{\mho}_0(\mathcal{T}(v)) \to 0\big\},$$

where the differentials $\{\partial_j\}$ are the boundary homomorphisms from the long exact homology sequences of the triples

$$\{\mathcal{T}(v)_{-(j-1)} \supset \mathcal{T}(v)_{-j} \supset \mathcal{T}(v)_{-(j+1)}\}_j.$$

The modules $\mathsf{C}^{\mho}_k(\mathcal{T}(v))$ are free and finitely generated, the number of generators being the number of connected components of the set

$$\mathcal{T}(v)^\circ_{-(n-k)} =_{\mathsf{def}} \mathcal{T}(v)_{-(n-k)} \setminus \mathcal{T}(v)_{-(n-k+1)}.$$

This observation is valid, since by Theorem 6.9, the components of $\mathcal{T}(v)^\circ_{-(n-k)}$ are orientable smooth manifolds.

Therefore any traversally generic field v on X gives rise, with the help of the differential complex (8.37), to its homology groups. We call them *$\mho$-homology groups* of v and denote them

$$\mathsf{H}^{\mho}_k(v) =_{\mathsf{def}} \mathsf{H}^{\mho}_k(\mathcal{T}(v)).$$

As in the category of shadows, the traversally generic vector fields give rise to several notions of complexity.

Definition 8.15. With any traversally generic vector field v on a smooth compact connected and oriented $(n+1)$-manifold X with boundary, we associate the ordered collection of ranks

$$\big\{\underline{tc}_j(v) =_{\mathsf{def}} \mathrm{rk}_{\mathcal{A}}\big(\mathsf{C}^{\mho}_j(\mathcal{T}(v))\big)\big\}_{0 \leq j \leq n},$$

where the groups $\mathsf{C}^{\mho}_j(\mathcal{T}(v)))$ have been introduced in (8.36).

We call $\underline{tc}_j(v)$ *the j-th traversing complexity*[11] of the field v. ◇

---

[10] In our notations, we suppress the dependance of these homology groups on the coefficients $\mathcal{A}$.

[11] "tc" for short

Remark 8.5. Reviewing (8.36), we notice that $\{(\underline{tc})_{n-k}(v) = 0\}$ implies that $\mathcal{T}(v)^\circ_{-k} = \emptyset$, since $\mathcal{T}(v)^\circ_{-k}$ is an orientable manifold. Examining the local models $\{p_\omega : \mathsf{E}_\omega \to \mathsf{T}_\omega\}_\omega$, we observe that, if a stratum of a particular combinatorial type ω' is missing in $\mathcal{T}(v)$, then all the smaller combinatorial types $\omega'' \prec \omega'$ (of greater codimensions) are missing as well (see Chapter 7). Therefore $\{\mathcal{T}(v)^\circ_{-k} = \emptyset\}$ implies that $\{\mathcal{T}(v)_{-j} = \emptyset\}_{j \geq k}$. As a result, if the complexity $\underline{tc}_{n-k}(v) = 0$, then the complexities $\underline{tc}_{n-j}(v) = 0$ for all $j \geq k$. ♢

Definition 8.16. Let X be a compact connected and oriented smooth $(n+1)$-manifold with boundary. For each traversally generic field $v \in \mathcal{V}^\ddagger(X)$, we form the sequence of traversing complexities:

$$\underline{\mathbf{tc}}(v) =_{\mathsf{def}} \{\underline{tc}_0(v), \underline{tc}_1(v), \dots, \underline{tc}_n(v)\}.$$

Consider the *lexicographical minimum*

$$\underline{\mathbf{tc}}^{\mathsf{lex}}(X) =_{\mathsf{def}} \text{lex.min}_{\{v \in \mathcal{V}^\ddagger(X)\}} \underline{\mathbf{tc}}(v). \tag{8.38}$$

We call this vector *the lexicographic traversing complexity* of X.

We denote by $\underline{tc}_j^{\mathsf{lex}}(X)$ the $(j+1)$-component of the vector $\underline{\mathbf{tc}}^{\mathsf{lex}}(X)$. ♢

Remark 8.6. If a compact connected and oriented smooth manifold X of dimension $n+1$ is globally k-convex in the sense of Definition 8.6, being applied to shadows $\{X \to \mathcal{T}(v)\}_{v \in \mathcal{V}^\ddagger(X)}$, then evidently $\underline{tc}_j^{\mathsf{lex}}(X) = 0$ for all $j \leq n-k$ in accordance with Remark 8.5. ♢

Remark 8.7. Thanks to Theorem 8.7, traversally generic vector fields produce a particular kind of shadows from the appropriate set $\mathsf{Shad}(X, \mathsf{E} \Rightarrow \mathsf{T})$. Therefore,

$$\mathbf{c}^{\mathsf{lex}}_{\mathsf{shad}}(X) \leq \underline{\mathbf{tc}}^{\mathsf{lex}}(X).$$

♢

As with the shadows, we will need "suspensions" of these notions and constructions.

We consider the stratification $\Omega^\bullet(X, v)$ of X by the connected components of the strata

$$\begin{aligned} \{X^\circ(v,\omega) &=_{\mathsf{def}} X(v,\omega) \setminus (\partial X \cap X(v,\omega)), \\ X^\partial(v,\omega) &=_{\mathsf{def}} \partial X \cap X(v,\omega)\}_{\omega \in \mathbf{\Omega}^\bullet_{'\langle n]}} \end{aligned} \tag{8.39}$$

and the τ-invariant stratification $\Omega^\bullet(DX, v)$ of the double DX, which is induced by $\Omega^\bullet(X, v)$.

Let

$$X^v_{-(k+1)} =_{\mathsf{def}} \Big(\bigcup_{\{\omega \mid\, |\omega|' \geq k+1\}} X^\circ(v,\omega)\Big) \bigcup \Big(\bigcup_{\{\omega \mid\, |\omega|' \geq k\}} X^\partial(v,\omega)\Big).$$

The stratifications $\Omega^\bullet(X, v)$ and $\Omega^\bullet(DX, v)$ give rise to the filtrations:

$$X =_{\mathsf{def}} X^v_{-0} \supset X^v_{-1} \supset \dots \supset X^v_{-(n+1)},$$

$$DX =_{\mathsf{def}} DX^v_{-0} \supset DX^v_{-1} \supset \dots \supset DX^v_{-(n+1)}$$

by the union of strata of a fixed codimension, each pure stratum being an open manifold.

Analogously to (8.8), we consider the homology and cohomology groups with coefficients in $\mathcal{A} = \mathbb{R}$ or $\mathcal{A} = \mathbb{Z}$:

$$\begin{aligned} \{\mathsf{C}^{\mho}_j(DX, v) &=_{\mathsf{def}} H_j(DX^v_{j-n-1}, DX^v_{j-n}; \mathcal{A})\}_j, \\ \{\mathsf{C}^j_{\mho}(DX, v) &=_{\mathsf{def}} H^j(DX^v_{j-n-1}, DX^v_{j-n}; \mathcal{A})\}_j. \end{aligned} \tag{8.40}$$

As in (8.9) and (8.10), they can be organized into a differential complex:

$$\mathbf{C}^{\mho}_*(DX, v) =_{\mathsf{def}} \tag{8.41}$$

$$\{0 \to \mathsf{C}^{\mho}_{n+1}(DX, v) \xrightarrow{\partial_{n+1}} \mathsf{C}^{\mho}_n(DX, v) \xrightarrow{\partial_n} \dots \xrightarrow{\partial_1} \mathsf{C}^{\mho}_0(DX, v) \to 0\},$$

where the differentials $\{\partial_j\}$ are the boundary homomorphisms from the long exact homology sequences of triples $\{DX^v_{-j+1} \supset DX^v_{-j} \supset DX^v_{-j-1}\}_j$.

Similarly, we introduce the dual differential complex

$$\mathbf{C}^*_{\mho}(DX, v) =_{\mathsf{def}} \tag{8.42}$$

$$\{0 \leftarrow \mathsf{C}^{n+1}_{\mho}(DX, v) \xleftarrow{\partial^*_{n+1}} \mathsf{C}^n_{\mho}(DX, v) \xleftarrow{\partial^*_n} \dots \xleftarrow{\partial^*_1} \mathsf{C}^0_{\mho}(DX, v) \leftarrow 0\}.$$

These differential complexes produce the *suspension $\mho$-homology* $\mathbf{H}^{\mho}_*(DX, v)$ and the *suspension $\mho$-cohomology* $\mathbf{H}^*_{\mho}(DX, v)$ of traversally generic v-flows on X.

Definition 8.17. Let X be a compact connected oriented and smooth $(n+1)$-dimensional manifold with boundary. For any traversally generic vector field $v \in \mathcal{V}^\ddagger(X)$ and each $j \in [0, n+1]$, let

$$\underline{\Sigma tc}^j(X, v) =_{\mathsf{def}} \mathrm{rk}_{\mathcal{A}}\big(\mathsf{C}^j_{\mho}(DX, v)\big).$$

We call this integer the j-th *suspension $\mho$-complexity* of the field v. ◇

Definition 8.18. Let X be a compact connected oriented and smooth $(n+1)$-dimensional manifold with boundary. For any traversally generic field $v \in \mathcal{V}^\ddagger(X)$, we form the sequence

$$\underline{\mathbf{\Sigma tc}}(X, v) =_{\mathsf{def}} \big(\underline{\Sigma tc}^0(X, v),\, \underline{\Sigma tc}^1(X, v),\, \dots,\, \underline{\Sigma tc}^{n+1}(X, v)\big),$$

and take the *lexicographic* minimum

$$\underline{\mathbf{\Sigma tc}}^{\mathsf{lex}}_{\mathsf{trav}}(X) =_{\mathsf{def}} \text{lex.min}_{v \,\in\, \mathcal{V}^\ddagger(X)}\, \underline{\mathbf{\Sigma tc}}(X, v).$$

We denote by $\underline{\Sigma tc}^{\mathsf{lex},\,j}_{\mathsf{trav}}(X)$ the $(j+1)$ component of the vector $\underline{\mathbf{\Sigma tc}}^{\mathsf{lex}}_{\mathsf{trav}}(X)$ and call it the j-th *lexicographic suspended traversing complexity* of X. $\diamondsuit$

Remark 8.8. By its very definition, the lexicographically optimal sequence of complexities is delivered by some traversally generic vector field!

Evidently,

$$\underline{\mathbf{\Sigma tc}}^{\mathsf{lex}}_{\mathsf{trav}}(X) \geq \mathbf{\Sigma c}^{\mathsf{lex}}_{\mathsf{shad}}(X),$$

where "shad" refers to the shadows, based on the list of model maps $\{\mathsf{E}_\omega \to \mathsf{T}_\omega\}_{\omega \in \mathbf{\Omega}^\bullet_{\langle n]}}$ exemplifying the local structure of traversally generic flows.

A priori, both vectors, $\underline{\mathbf{tc}}^{\mathsf{lex}}_{\mathsf{trav}}(X)$ and $\underline{\mathbf{\Sigma tc}}^{\mathsf{lex}}_{\mathsf{trav}}(X)$, are invariants of the *smooth* topological type of X, while $\mathbf{c}^{\mathsf{lex}}_{\mathsf{shad}}(X)$ and $\mathbf{\Sigma c}^{\mathsf{lex}}_{\mathsf{shad}}(X)$ are invariants of the PL-topological type. $\diamondsuit$

Consider the space $\mathcal{V}^\ddagger(X)$ of traversally generic vector fields on X and its subspace $\mathcal{V}^\ddagger_{\mathsf{fold}}(X)$, formed by vector fields for which the multiplicity $m(a) \leq 2$ for any v-trajectory γ at each point $a \in \gamma \cap \partial X$. Locally, for such vector fields v, $\Gamma : \partial X \to \mathcal{T}(v)$ is a folding map.

Definition 8.19. Let X be a compact connected oriented and smooth $(n+1)$-dimensional manifold with a boundary.

For any traversally generic field $v \in \mathcal{V}^\ddagger_{\mathsf{fold}}(X)$ we define the *lexicographic folding complexity* of X by

$$\underline{\mathbf{tc}}^{\mathsf{lex}}_{\mathsf{fold\,trav}}(X) =_{\mathsf{def}} \text{lex.min}_{v \,\in\, \mathcal{V}^\ddagger_{\mathsf{fold}}(X)}\, \underline{\mathbf{tc}}(v).$$

$\diamondsuit$

Clearly,

$$\underline{\mathbf{tc}}^{\mathsf{lex}}_{\mathsf{fold\,trav}}(X) \geq \underline{\mathbf{tc}}^{\mathsf{lex}}_{\mathsf{trav}}(X),$$

provided $\mathcal{V}^\ddagger_{\mathsf{fold}}(X) \neq \emptyset$.

When X is a 3-fold, by Theorem 4.6,

$$\underline{tc}^0_{\mathsf{fold\,trav}}(X) = \underline{tc}^0_{\mathsf{trav}}(X).$$

Besides this equality and the previous inequality, we know little about the relation between $\underline{\mathbf{tc}}^{\mathsf{lex}}_{\mathsf{fold\,trav}}(X)$ and $\underline{\mathbf{tc}}^{\mathsf{lex}}_{\mathsf{trav}}(X)$.

Now let us examine how the marriage of amenable localization and Poincaré Duality works in the environment of traversing fields.

For each $k \in [-1, n]$, by composing the Poincaré duality map $\mathcal{D}$ with the localization to the subsets $DX^v_{-(k+1)} \subset DX$ and $X^{v\circ}_{-(k+1)} \subset X$, we get two localization maps

$$\mathcal{L}^v_{k+1} : H_{k+1}(DX) \stackrel{\approx\mathcal{D}}{\longrightarrow} H^{n-k}(DX) \stackrel{i^*}{\longrightarrow} H^{n-k}(DX^v_{-(k+1)}),$$

$$\mathcal{M}^v_{k+1} : H_{k+1}(X) \stackrel{\approx\mathcal{D}}{\longrightarrow} H^{n-k}(X, \partial X)$$

$$\stackrel{i^*}{\longrightarrow} H^{n-k}(X^v_{-(k+1)}, X^v_{-(k+2)} \cup (\partial X \cap X^v_{-(k+1)})),$$

whose targets can be identified with the quotients

$$\mathsf{C}^{n-k}_{\mho}(DX, v)/\mathsf{B}^{n-k}_{\mho}(DX, v) \text{ and } \mathsf{C}^{n-k}_{\mho}(X^\circ, v)/\mathsf{B}^{n-k}_{\mho}(X^\circ, v),$$

respectively.

We denote by

$$\begin{aligned} \mathcal{L}^{v,\mho}_{k+1} &: H_{k+1}(DX) \to \mathsf{C}^{n-k}_{\mho}(DX, v)/\mathsf{B}^{n-k}_{\mho}(DX, v), \\ \mathcal{M}^{v,\mho}_{k+1} &: H_{k+1}(X) \to \mathsf{C}^{n-k}_{\mho}(X^\circ, v)/\mathsf{B}^{n-k}_{\mho}(X^\circ, v) \end{aligned} \tag{8.43}$$

the resulting operators.

Now Theorem 8.5, being combined with Theorem 8.7, delivers the following result.

Theorem 8.8 (Amenable localization of the Poincaré duality for traversing flows).
Let X be a compact connected oriented and smooth $(n+1)$-manifold X with boundary. Let v be a traversally generic vector field v on X. Pick an integer $k \in [-1, n]$.

- *Assume that, for each connected component of the boundary ∂X, the image of its fundamental group in $\pi_1(DX)$ is amenable.*

  Then the space $\mathsf{C}^{n-k}_{\mho}(DX, v)/\mathsf{B}^{n-k}_{\mho}(DX, v)$ admits a norm $\| \sim \|^{\bullet}_{\mho}$ so that

$$\|h\|_\Delta \leq \lambda \cdot \| \mathcal{L}^{v,\mho}_{k+1}(h) \|^{\bullet}_{\mho} \tag{8.44}$$

for any $h \in H_{k+1}(DX)$. Here the localized Poincaré duality operator $\mathcal{L}^{v,\mho}_{k+1}$ is introduced in (8.43), and the (X, v)-universal constant $\lambda > 0$ depends only on the list of model maps $\{p_\omega : \mathsf{E}_\omega \to \mathsf{T}_\omega\}_{\omega \in \mathbf{\Omega}^\bullet_{'\langle n]}}$ in the way that is described in formula (8.18).

- *Assume that, for each connected component of the boundary ∂X, the image of its fundamental group in $\pi_1(X)$ is an amenable group.*

  Then the space $\mathsf{C}^{n-k}_{\mho}(X^\circ, v)/\mathsf{B}^{n-k}_{\mho}(X^\circ, v)$ admits a norm $\| \sim \|^\bullet_{\mho}$ so that

$$\|h\|_\Delta \;\leq\; \mu \cdot \big\| \mathcal{M}^{v,\mho}_{k+1}(h) \big\|^\bullet_{\mho} \tag{8.45}$$

  for any $h \in H_{k+1}(X)$. Here the localized Poincaré duality operator $\mathcal{M}^{v,\mho}_{k+1}$ is introduced in (8.43), and the (X, v)-universal constant $\mu > 0$ depends only on the list of model maps $\{p_\omega : \mathsf{E}_\omega \to \mathsf{T}_\omega\}_{\omega \in \mathbf{\Omega}^\bullet_{'\langle n]}}$ in the way that is described in formula (8.25). ♢

Theorem 8.7, together with Corollaries 8.1 and 8.2 leads to:

Corollary 8.9. *Assume the hypotheses from Theorem 8.8.*

- *Let an integral homology class $h \in H_{k+1}(DX)$ be realized by a singular oriented pseudo-manifold $f : M \to DX$.*

  Then the number of intersections of the cycle $f(M)$ with the locus $DX^v_{-(k+1)}$ is greater than or equal to $\lambda^{-1} \cdot \|h\|_\Delta$, provided that f is in general position with the subcomplex $DX^v_{-(k+1)} \subset DX$.
- *Let an integral class $h \in H_{k+1}(X)$ be realized by a singular oriented pseudo-manifold $f : M \to X$. Then $(\mu)^{-1} \cdot \|h\|_\Delta$ gives a lower bound of the number of transversal intersections of the absolute cycle $f(M)$ with the locus $X^{\circ v}_{-(k+1)} =_{\mathsf{def}} \mathrm{int}(X) \cap X^v_{-(k+1)}$.* ♢

The following example is a product of the author's conversations with Larry Guth.

Example 8.2. Consider a fibration $p : E \to M$ whose base M is a closed oriented hyperbolic manifold of dimension $k+1$ and whose fiber F is a closed manifold. Assume that the $(n+1)$-dimensional manifold E is oriented and that p admits a section $s : M \to E$. In the complement to $s(M)$, pick a smooth codimension zero submanifold V such that $\{\pi_1(\partial V, pt)\}_{pt \in \partial V}$ are amenable groups. Let $X =_{\mathsf{def}} E \setminus \mathrm{int}(V)$.

Recall that $\|[M]\|_\Delta$ is proportional to the hyperbolic volume vol(M) with an universal positive proportionality constant [Gr]. Since s is a section, and the simplicial semi-norms do not increase under the continuous maps, it follows that the simplicial norm of $s_*([M]) \in H_{k+1}(X)$ is proportional to vol(M) with the same proportionality constant.

For any traversally generic vector field v on X, consider the locus $X^v_{-(k+1)}$. We can perturb the section s so that the intersections of the cycle $s(M)$ is transversal to $X^v_{-(k+1)}$.

According to Corollary 8.9, there exits a universal constant $\rho > 0$ so that, for any such X and any traversally generic vector field v on X, the transversal intersections of the perturbed cycle $s(M)$ with the locus $X^v_{-(k+1)}$ has $\rho \cdot \mathrm{vol}(M)$ intersections at least.

By the choice of V, $s(M) \cap \partial V = \emptyset$. Therefore, for each $v \in \mathcal{V}^\ddagger(X)$, there exist at least $\rho \cdot \mathrm{vol}(M)$ trajectories γ of the reduced multiplicity $m'(\gamma) = k+1$, each of which has a nonempty intersection with the section $s(M) \subset \mathrm{int}(X)$ (and evidently with $\partial X = \partial V$). ◇

It worth describing the important special case "$h = [DX]$" of formula (8.18), being applied to the shadows that are produced by traversally generic flows.

Theorem 8.9. *Let X be a smooth compact connected and oriented $(n+1)$-manifold with boundary. Assume that, for each connected component of the boundary ∂X, the image of its fundamental group in $\pi_1(DX)$ is an amenable group. Let v be a traversally generic vector field on X.*

Then

$$\sum_{\omega \in \mathbf{\Omega}^\bullet_{'[n]}} \theta(\omega) \cdot \#(\mathcal{T}(v,\omega)) \;\geq\; \|\,[DX]\,\|_\Delta, \tag{8.46}$$

where $\#(\mathcal{T}(v,\omega))$ is the cardinality of the 0-dimensional stratum $\mathcal{T}(v,\omega)$, and the universal constant

$$\theta(\omega) \;=_{\mathsf{def}}\; \|\,[D\mathsf{E}_\omega, D(\delta\mathsf{E}_\omega)]\,\|_\Delta^{\mathbf{\Omega}^\bullet_{'\langle n]}}$$

is the $\mathbf{\Omega}^\bullet_{'\langle n]}$-stratified simplicial volume of the model pair $(D\mathsf{E}_\omega, D(\delta\mathsf{E}_\omega))$.

Proof. The v-induced 0-dimensional strata of the $\mathbf{\Omega}^\bullet_{'\langle n]}$-stratification of DX of are mapped by $\Gamma : X \to \mathcal{T}(v)$ to the 0-dimensional strata of $\mathcal{T}(v)$.

Formula (8.46) is validated by applying the arguments of Theorem 8.7 to the fundamental class $h = [DX]$. In this case, it is simpler to formulate the result of the amenable localization, if we chose a regular neighborhood of the double of the graph $\Gamma^{-1}(\mathcal{T}(v)_{-n})$[12] in DX for the role of the neighborhood U of the zero-dimensional locus $DX^v_{-(n+1)}$ and then localize h to h_U. $\square$

Although desirable, an exact computation of the universal constants $\{\theta(\omega)\}$ in formula (8.46) is somewhat tricky.

The $\mathbf{\Omega}^\bullet_{'\langle n]}$-stratification in each pair $[\mathsf{E}_\omega, \delta\mathsf{E}_\omega]$ is described in Theorems 7.3 and 7.4. Using the first bullet from Definition 8.8, we conclude that $\| [D\mathsf{E}_\omega, D(\delta\mathsf{E}_\omega)] \|_\Delta^{\mathbf{\Omega}^\bullet_{'\langle n]}}$ is greater than or equal to the number of $(n+1)$-dimensional cells in $D\mathsf{E}_\omega$, which is twice the number of $(n+1)$-dimensional cells in E_ω. The latter number is also equal to the number of n-dimensional cells in the model trajectory space $\mathcal{T}_\omega$.

Let us estimate this number from below. Consider the model u-polynomial

$$\wp(u, \vec{x}) =_{\mathsf{def}} \prod_{i=1}^{|\sup(\omega)|} \Big((u-i)^{\omega_i} + \sum_{j=0}^{\omega_i - 2} x_{i,j}(u-i)^j\Big).$$

By Theorem 7.4, the $(n+1)$-dimensional cells in E_ω are labeled by the sequences $\mu = \{\mu_i\}_{i\in[1,q]}$ of integers $\mu_i \geq 0$, together with a marker k, specific to each sequence μ. Here $\mu_i \leq \omega_i$, $\mu_i \equiv \omega_i \mod 2$, and $q =_{\mathsf{def}} |\sup(\omega)|$.

For a generic choice of a sufficiently small $\vec{x} = \{x_{i,j}\}_{i,j}$ that belongs to the image in $\mathbb{R}^{|\omega|'}$ of a particular chamber-cell from E_ω, each μ_i represents the number of *simple* roots of $\wp(u, \vec{x})$ that reside in the vicinity of the root i of the polynomial $\wp(u, \vec{0})$.

Each sequence $\mu = \{\mu_i\}_{i\in[1,q]}$ admits $\frac{1}{2}|\mu|$ distinct markers, where $|\mu| = \sum_i \mu_i$. Thus the total number of $(n+1)$-cells in the double $D\mathsf{E}_\omega$ is

$$\eta(\omega) =_{\mathsf{def}} \sum_{\{\mu \mid 0\leq \mu_i \leq \omega_i,\ \mu_i \equiv \omega_i \mod 2 \text{ for all } i\in[1,q]\}} |\mu|. \tag{8.47}$$

Therefore $\eta(\omega) \leq \| [D\mathsf{E}_\omega, D(\delta\mathsf{E}_\omega)] \|_\Delta^{\mathbf{\Omega}^\bullet_{'\langle n]}}$. In fact, the latter norm must be bigger than $\eta(\omega)$ in order for the $(n+1)$-dimensional simplicial chain, which realizes the norm, to satisfy the second bullet (the order condition) from Definition 8.8. By subdividing each cell "barycentrically", *conjecturally*, one may get the exact value of the norm $\| [D\mathsf{E}_\omega, D(\delta\mathsf{E}_\omega)] \|_\Delta^{\mathbf{\Omega}^\bullet_{'\langle n]}}$.

[12]This graph is a finite collection of necklaces of touching circles "...ooo...".

Consider a prism Π_n whose base is a n-simplex Δ^n. Our aim is to subdivide Π_n economically into $(n+1)$-simplicies in such a way that this subdivision $\beta\Pi_n$ will be invariant under the action of the permutation group S_{n+1} on Π_n (the action is induced by the obvious S_{n+1}-action on Δ^n). We call this triangulation $\beta\Pi_n$ of Π_n *the barycentric subdivision.* It can be introduced inductively. Assume that $\beta\Pi_{n-1}$ has been constructed. Then we use its construction for each of the $n+1$ side faces of Π_n. We add the top and the bottom of the prism Π_n to get a triangulation $T^\partial(n)$ of the boundary $\partial\Pi_n$. We take the barycenter b of Π_n and use b as the tip for building the cone over the triangulation $T^\partial(n)$. This procedure will generate the desired barycentric triangulation $\beta\Pi_n$.

Let us introduce a function $\beta(n)$ that counts the number of $(n+1)$-dimensional simplicies in $\beta\Pi_n$. Mirroring the geometric procedure for constructing $\beta\Pi_n$, we define $\beta(n)$ recursively by the formulas

$$\beta(n) = (n+1)\cdot\beta(n-1)+2, \quad \beta(0)=2. \tag{8.48}$$

In one important special case of ω, we can also crudely estimate *form above* the stratified simplicial norms of the relative fundamental cycles of the model spaces $\mathcal{T}_\omega, \mathsf{E}_\omega$, and $D\mathsf{E}_\omega$.

Lemma 8.3. *For* $\omega_\star =_{\mathsf{def}} (1\underbrace{2\dots2}_{n}1)$, *we get the following estimates:*

$$2^n+n \leq \left\| [\mathcal{T}_{\omega_\star}, \mathcal{S}_{\omega_\star}] \right\|_\Delta^{\boldsymbol{\Omega}^\bullet_{'\langle n]}} \leq 1.5\cdot 2^n\cdot n!,$$

$$2^n+n \leq \left\| [\mathsf{E}_{\omega_\star}, \delta\mathsf{E}_{\omega_\star}] \right\|_\Delta^{\boldsymbol{\Omega}^\bullet_{'\langle n]}} \leq 2^n\cdot(n+1)!\cdot\beta(n),$$

$$2^{n+1}+2n \leq \left\| [D\mathsf{E}_{\omega_\star}, D(\delta\mathsf{E}_{\omega_\star})] \right\|_\Delta^{\boldsymbol{\Omega}^\bullet_{'\langle n]}} \leq 2^{n+1}\cdot(n+1)!\cdot\beta(n),$$

where $\beta(n)$ *is defined by (8.48).*

Proof. We start with the model space $\mathcal{T}_{\omega_\star}$, a cone over the compact space $\mathcal{S}_{\omega_\star}$. As established in Example 7.4, the space $\mathcal{T}_{\omega_\star}$ is formed by attaching n copies of a half-ball B^n_+ to a ball B^n along the coordinate hyperplanes which divide B^n into 2^n quadrants. We denote by $B^n_\dagger$ this stratified ball.

So the number of n-dimensional cells-strata in $\mathcal{T}_{\omega_\star}$ is 2^n+n. Therefore, the pull-back of the $\boldsymbol{\Omega}^\bullet_{'\langle n]}$-stratification in $\mathcal{T}_{\omega_\star}$ divides the model space $\mathsf{E}_{\omega_\star}$ also into 2^n+n chambers-cells of dimension $n+1$. By the first bullet from Definition 8.8, the stratified simplicial norms must satisfy the inequalities

$$\left\| [\mathcal{T}_{\omega_\star}, \mathcal{S}_{\omega_\star}] \right\|_\Delta^{\boldsymbol{\Omega}^\bullet_{'\langle n]}} \geq 2^n+n, \qquad \left\| [\mathsf{E}_{\omega_\star}, \delta\mathsf{E}_{\omega_\star}] \right\|_\Delta^{\boldsymbol{\Omega}^\bullet_{'\langle n]}} \geq 2^n+n,$$

$$\big\|[D\mathsf{E}_{\omega_\star}, D(\delta\mathsf{E}_{\omega_\star})]\big\|_\Delta^{\boldsymbol{\Omega}^\bullet_{\prime\langle n]}} \geq 2^{n+1} + 2n.$$

Next, let us estimate $\big\|\,[\mathcal{T}_{\omega_\star}, \mathcal{S}_{\omega_\star}]\,\big\|_\Delta^{\boldsymbol{\Omega}^\bullet_{\prime\langle n]}}$ from *above*, a more consequential bound!

Each chamber Δ of $B^n_\dagger$ (topologically a n-simplex) is a cone over a $(n-1)$-dimensional simplex Δ', the apex of the cone is the vertex γ of Δ that is not in Δ' and represents the only trajectory of the type $\omega_\star$. However, this n-simplex Δ does not satisfy the second bullet from Definition 8.8 (the order condition). By taking the barycentric subdivision $\beta\Delta'$ of Δ' and applying the cone construction (with apex at γ) over $\beta\Delta'$, we satisfy the order condition. The barycentric subdivision $\beta\Delta'$ contains $n!$ simplicies of dimension $n-1$. So for each chamber Δ, the $n!$ simplices that form $\Delta = \text{cone}(\beta\Delta')$ will deliver a chain that satisfies all the bullets of Definition 8.8. To produce such a chain that will serve the entire $B^n_\dagger$, it will suffice to employ $2^n \cdot (n!)$ singular simplicies of dimension n.

Let $\{H_i\}_{1\leq i\leq n}$ be the coordinate hyperplanes that divide the ball B^n and generate the stratified ball $B^n_\dagger$. The space $\mathcal{T}_{\omega_\star}$ is assembled from the stratified n-dimensional ball $B^n_\dagger$ and n copies of half-balls $(B^n_+)_i$, attached to $B^n \cap H_i$. That stratified intersection may be identified with $B^{n-1}_\dagger$ and thus is a union of $2^{n-1}\cdot(n-1)!$ simplicies of dimension $(n-1)$. The half-ball $(B^n_+)_i$, being a cone over $B^{n-1}_\dagger$, could be subdivided into the same number of n-simplicies. There are n such half-balls $(B^n_+)_i$. Therefore, the stratified simplicial norm of $[\mathcal{T}_{\omega_\star}, \mathcal{S}_{\omega_\star}]$ does not exceed $2^n\cdot n! + 2^{n-1}\cdot n! = 1.5\cdot 2^n\cdot n!$.

In order to estimate $\big\|[\mathsf{E}_{\omega_\star}, \delta\mathsf{E}_{\omega_\star}]\big\|$, let us describe the geometry of the $(\omega_\star)_{\preceq}$-stratified space $\mathsf{E}_{\omega_\star}$ in some detail (see Figure 6.2, the diagram labeled with (1221)).

Recall that the space $\mathsf{E}_{\omega_\star}$, in the vicinity of the u-axis, is defined by the inequalities $\{\wp_{\omega_\star}(u, \vec{x}) \leq 0;\ \|\vec{x}\| \leq \epsilon\}$, where $\wp_{\omega_\star}$ is the polynomial from Example 7.4 (with $m = n$). We denote by $\partial_1\mathsf{E}_{\omega_\star}$ the portion of the boundary, defined by $\{\wp_{\omega_\star}(u, \vec{x}) = 0;\ \|\vec{x}\| \leq \epsilon\}$.

Let $\{\mathsf{f}_i\}_{1\leq i\leq n}$ be the connected components of the folding locus of the obvious projection $p_{\omega_\star} : \partial_1\mathsf{E}_{\omega_\star} \to \mathbb{R}^n$. They are labeled by the zeros $1,\dots,n$ of the polynomial $\wp_{\omega_\star}(u, \vec{0})$. Each f_i resides in the vicinity of the points $(i, \vec{0}) \in \partial_1\mathsf{E}_{\omega_\star}$. In fact, f_i is a $(n-1)$-dimensional ball, which is mapped by the immersion $p_{\omega_\star} : \mathsf{f}_i \to \mathbb{R}^n$ onto the intersection of the coordinate hyperplane $H_i = \{x_i = 0\}$ in $\mathbb{R}^n$ with the ϵ-ball $B_\dagger(\epsilon) \subset \mathbb{R}^n$.

We denote by $B^\pm_i$ the connected component of the locus $\partial^\pm_1\mathsf{E}_{\omega_\star}(\partial_u)$ that resides in the vicinity of the point $(i, \vec{0})$. For $i \neq 0, n+1$, each $B^\pm_i$ is homeomorphic to a n-dimensional half-ball with the equator f_i, while B^+_0

and B^-_{n+1} are balls.

For each $i \in [1, n]$, we choose a new n-dimensional half-ball $B^\#_i \subset \mathsf{E}_{\omega_\star}$ with the equator f_i so that $S^\pm_i =_{\mathsf{def}} B^\#_i \cup_{\mathsf{f}_i} B^\pm_i$ is a section of $p_{\omega_\star}$ and its $p_{\omega_\star}$-image is the ball $B_\dagger(\epsilon) \subset \mathbb{R}^n$.

The sections $\{S^\pm_i\}_{1 \leq i \leq n}$ subdivide the space $\mathsf{E}_{\omega_\star}$ into $n+1$ portions $\{\mathsf{F}_i\}_{0 \leq i \leq n}$, each portion being homeomorphic to a cylinder. The base of F_i is the n-ball S^+_i, its top is the n-ball S^-_{i+1}. The generators of the cylinder F_i are segments of the $p_{\omega_\star}$-fibers. The union $B^+_i \cup B^-_i \cup B^\#_i$ is a three-paged open book with f_i for the binder.

We stratify the ball $B_\dagger(\epsilon)$ by the various multiple intersections of the coordinate hyperplanes and denote this stratification by $\mathcal{S}_n$. We stratify each cylinder F_i by the $p_{\omega_\star}$-preimages of the $\mathcal{S}_n$-strata, together with the $p_{\omega_\star}$-preimages of the $\mathcal{S}_n$-strata, being restricted to the bottom S^+_i and the top S^-_{i+1}. Such a stratification of F_i is *finer* than the $(\omega_\star)_\preceq$-stratification of $\mathsf{E}_{\omega_\star}$, being restricted to F_i.

Consider a triangulation T of $B_\dagger(\epsilon)$ by n-simplicies Δ that satisfy the constraints of Definition 8.8 for the $\mathcal{S}_n$-stratification of $B_\dagger(\epsilon)$. Such a triangulation consists of $2^n \cdot n!$ simplicies Δ at most. Each Δ is a cone with the tip at the origin.

The $p_{\omega_\star}$-preimage of each n-simplex Δ in F_i is a prism $\Pi(\Delta) \approx \Delta \times [0, 1]$. By an argument above, such the prism F_i may be subdivided barycentrically into $\beta(n)$ simplicies of dimension $n + 1$[13]. Because the barycentric subdivision $\beta\Pi$ is invariant under the S_n-permutations, the barycentric triangulations of different prisms $\{\Pi(\Delta)\}_{\Delta \in T}$ will match automatically.

Since $\mathsf{E}_{\omega_\star}$ consists of $(n+1)$ portions $\{\mathsf{F}_i\}_i$, we get that $[\mathsf{E}_{\omega_\star}, \delta\mathsf{E}_{\omega_\star}]$ may be realized by a chain which consists of

$$(n+1) \cdot 2^n \cdot n! \cdot \beta(n) = 2^n \cdot (n+1)! \cdot \beta(n)$$

simplicies that satisfy the constraints of Definition 8.8.

To realize $[D\mathsf{E}_{\omega_\star}, D(\delta\mathsf{E}_{\omega_\star})]$ we just double the chain that delivers the relative fundamental cycle of $\mathsf{E}_{\omega_\star}$. □

Corollary 8.10. *Let X be as in Theorem 8.8. Let v be a traversally generic vector field on X with the property $\partial_3 X(v) = \emptyset$[14].*

Then Theorem 8.8 is valid for $\lambda = 2^{k+2} \cdot (k+2)! \cdot \beta(k+1)$ (see formula (8.44)) and $\mu = 2^{k+1} \cdot (k+2)! \cdot \beta(k+1)$ (see formula (8.45)).

[13]It is possible to triangulate $\Pi(\Delta)$ by $(n+1)$ simplicies only; however, such optimal triangulations $\{\Pi(\Delta)\}_\Delta$ may not agree across the faces of the prisms.

[14]In other words, v is boundary 3-concave and 3-convex.

Moreover, Theorem 8.9 is valid for $\theta(\omega) = 2^{n+1} \cdot (n+1)! \cdot \beta(n)$.

Proof. When $\partial_3 X(v) = \emptyset$, the stratum $DX^v_{-(k+1)}$ is formed by connected components of the locus $X(v, (1\underbrace{2\dots2}_{k+1}1))$, its image under the involution $\tau : DX \to DX$, and the locus $\partial X(v, (1\underbrace{2\dots2}_{k}1)) = X(v, (1\underbrace{2\dots2}_{k}1)) \cap \partial_1 X$. Therefore, in the proof of Theorem 8.8 participate only the local models $\mathsf{DE}_{\omega_\star}$, where $\omega_\star = (1\underbrace{2\dots2}_{k+1}1)$ and $\omega_\star = (1\underbrace{2\dots2}_{k}1)$. By Lemma 8.3, we may pick $\lambda = 2^{k+2} \cdot (k+2)! \cdot \beta(k+1)$ in formula (8.44).

A similar argument justifies the choice $\mu = 2^{k+1} \cdot (k+2)! \cdot \beta(k+1)$ in formula (8.45).

Revisiting Theorem 8.9, by the same token, we get that $\theta(\omega) = 2^{n+1} \cdot (n+1)! \cdot \beta(n)$, provided that v is boundary 3-concave and 3-convex. □

Let us restate Corollary 8.9 using Lemma 8.3.

Corollary 8.11. *Let X be a smooth compact connected and oriented $(n+1)$-manifold with boundary such that the image in $\pi_1(X)$ of the fundamental group of each boundary component is amenable. Let v be a traversally generic vector field with the property $\partial_3 X(v) = \emptyset$. Let $f : M \to X$ be an integral singular k-cycle that realizes a class $h \in H_k(X; \mathbb{Z})$. By a general position argument, we may assume that $f(M)$ is transversal to the locus $X(v, (1\underbrace{2\dots2}_{k}1))$ and misses the closure of the locus $X(v, (1\underbrace{2\dots2}_{k+1}1))$.*

Then $f(M)$ intersects $X(v, (1\underbrace{2\dots2}_{k}1))$ at least at

$$\left(2^k \cdot (k+1)! \cdot \beta(k)\right)^{-1} \cdot \|h\|_\Delta \tag{8.49}$$

points.

In particular, if M is a closed hyperbolic manifold, and there is a continuous map $g : X \to M$ such that $g \circ f = \mathsf{Id}_M$, then $f(M)$ intersects $X(v, (1\underbrace{2\dots2}_{k}1))$ at least at

$$\left(2^k \cdot (k+1)! \cdot \beta(k)\right)^{-1} \cdot \mathrm{Vol}(M)/\mathrm{Vol}(\Delta^k) \tag{8.50}$$

points. Here $\mathrm{Vol}(M)$ *denotes the hyperbolic volume of M and* $\mathrm{Vol}(\Delta^k)$ *the hyperbolic volume of an ideal simplex Δ^k in the hyperbolic space $\mathbf{H}^k$.*

Proof. We repeat the argument in the proof of Corollary 8.10, coupled with the arguments from Corollary 8.9 (see also Example 8.2). □

Remark 8.9. In this chapter, we are trying to estimate *from below* the number of connected components of various flow-induced stratifications; the estimates are given in terms of the simplicial norms $\|h\|_\Delta$ of the appropriate homology classes h. In our estimates, the universal coefficients next to $\|h\|_\Delta$, as functions of $\dim h$, are very small. So very often the property $\|h\|_\Delta \neq 0$ does not imply the multitude of connected components of a particular combinatorial type; all we can conclude is that such a component does exist. We suspect that this weakness of the estimates does not reflect the nature of the relation between the size of simplicial norms and the richness of the flow-induced stratifications; it probably reflects the limitations of our techniques.

However, as $\|h\|_\Delta$ grows (for example, as we replace h with its multiple $\ell \cdot h$), the estimates become stronger.

As an illustration, let us reconsider formula (8.50). For large k, the hyperbolic volume of an ideal k-simplex is approximatively $\frac{\sqrt{k}}{k!} \cdot e$. Thus the M-independent coefficient next to $\mathrm{Vol}(M)$ in (8.50) is, as $k \to \infty$, asymptotically

$$\sim \big(e \cdot (k+1) \cdot \sqrt{k} \cdot 2^k \cdot \beta(k)\big)^{-1}.$$

◊

Now let us introduce the number

$$\rho(n) =_{\mathsf{def}} \big(\max_{\omega \in \mathbf{\Omega}^\bullet_{\prime\langle n]}} \{\theta(\omega)\}\big)^{-1}.$$

Theorem 8.9 implies Theorem 2 from [AK], which we state below.

Theorem 8.10 (Alpert, Katz). *Let X be a smooth compact connected and oriented $(n+1)$-manifold with boundary. Assume that, for each connected component of the boundary, the image of its fundamental group in $\pi_1(DX)$ is an amenable group. Let v be a traversally generic vector field on X.*

Then there is an universal constant $\rho(n) > 0$ such that, for any X and v, the cardinality of the set $\mathcal{T}(v)_{-n}$, formed by the trajectories of the maximal reduced multiplicity n, satisfies the inequality

$$\#(\mathcal{T}(v)_{-n}) \geq \rho(n) \cdot \| [DX] \|_\Delta. \tag{8.51}$$

◊

Example 8.3. Consider a disk X with $g \geq 2$ holes. We may view X as an annulus A with $g-1$ holes. We take the radial vector field $\hat{v}$ on A and restrict it to $X \subset A$. Let us denote by v this restriction. We may assume that the holes in A are convex with respect to $\hat{v}$ and no radius in A is tangent to two holes. In such a case, v is boundary generic on X. The number of v-trajectories of the combinatorial type (121) is $2(g-1)$. Since the genus of the surface DX is exceeds 1, DX admits a hyperbolic metric. So, by [Gr], $\| [DX] \|_\Delta = |\chi(DX)| = 2g-2$. Therefore the proportion

$$\#(\mathcal{T}(v)_{-1})/\| [DX] \|_\Delta = \frac{2(g-1)}{2g-2} = 1.$$

By an indentation of ∂X against the background of $\hat{v}$, we can add a pair of points of $\hat{v}$-tangency that have opposite polarities. The deformation will add 2 to the cardinality $\#(\mathcal{T}(v)_{-1})$. So this cardinality may take any value $2g-2+2s$, and the proportion

$$\#(\mathcal{T}(v)_{-1})/\| [DX] \|_\Delta = \frac{2g-2+2s}{2g-2} \geq 1.$$

Therefore, *in this set of examples,*

$$\min_{(X,v)} \#(\mathcal{T}(v)_{-1})/\| [DX] \|_\Delta = 1.$$

As a result, the *universal* constant $\rho(1)$ must satisfy the inequality $\rho(1) \leq 1$. Perhaps, $\rho(1) = 1$? ♢

Combining Theorem 8.5 with Theorem 8.7, we get the main result of this section.

Theorem 8.11. *Let X be a smooth compact connected and oriented manifold with boundary, $\dim(X) = n+1$. Let v be a traversally generic vector field on X.*

- *Assume that, for each connected component of the boundary ∂X, the image of its fundamental group in $\pi_1(DX)$ is an amenable group. Then, for each $k \in [-1, n]$,*

$$\mathrm{rk}\big(\mathrm{im}(\mathcal{L}^{v,\mho}_{k+1})\big) \geq \mathrm{rk}\big(H^{\mathbf{\Delta}}_{k+1}(DX)\big). \tag{8.52}$$

As a result, $\mathrm{rk}\big(\mathsf{C}^{n-k}_{\mho}(DX,v)\big)$—the number of connected components of the pure $(n-k)$-dimensional strata in the stratification $\Omega^\bullet(DX,v)$[15] of DX —is greater than or equal to $\mathrm{rk}\big(H^{\mathbf{\Delta}}_{k+1}(DX)\big)$.

[15]See (8.39).

- *Assume that, for each connected component of the boundary ∂X, the image of its fundamental group in $\pi_1(X)$ is an amenable group. Then, for each $k \in [-1, n-1]$,*

$$\mathrm{rk}\big(\mathrm{im}(\mathcal{M}^{v,\mho}_{k+1})\big) \geq \mathrm{rk}\big(H^{\mathbf{\Delta}}_{k+1}(X)\big). \tag{8.53}$$

Thus, $\mathrm{rk}\big(\mathsf{C}^{n-k}_{\mho}(X^\circ, v)\big)$—the number of connected components of pure $(n-k)$-dimensional strata in the stratification $\Omega^\bullet(X^\circ, v)$—is greater than or equal to $\mathrm{rk}\big(H^{\mathbf{\Delta}}_{k+1}(X)\big)$. ◇

The first bullet in Theorem 8.11, together with the definitions of suspension complexities, leads instantly to the following result.

Theorem 8.12. *Let X be a smooth compact connected and oriented manifold with boundary, $\dim(X) = n+1$. Assume that, for each connected component of the boundary ∂X, the image of its fundamental group in $\pi_1(DX)$ is an amenable group. Then, for each $k \in [-1, n]$,*

$$\mathbf{\Sigma tc}^{\mathsf{lex}}_{\mathsf{trav}}(X) \geq \mathbf{\Sigma c}^{\mathsf{lex}}_{\mathsf{shad}}(X) \geq \mathbf{rk}\big(H^{\mathbf{\Delta}}_*(DX)\big), \tag{8.54}$$

where the vectorial inequality is understood as the inequality among all the the corresponding components of the participating vectors. The vector $\mathbf{rk}\big(H^{\mathbf{\Delta}}_*(DX)\big)$ has been introduced prior to formula (8.31). ◇

The next lemma rephrases inequalities (8.54) in terms of the traversing complexities $\underline{tc}^*(v)$, related to the trajectory spaces.

Lemma 8.4. *Let v be a traversally generic vector field on a connected compact smooth $(n+1)$-manifold X. Then, for any $k \in [-1, n]$, the following inequalities are valid:*

$$\mathrm{rk}\big(H^{\mathbf{\Delta}}_{k+1}(DX)\big) \leq \underline{\Sigma tc}^{n-k}(X, v) \leq (n+2)\big(\underline{tc}^{n-k}(v) + 2 \cdot \underline{tc}^{n-k-1}(v)\big),$$

$$\mathrm{rk}\big(H^{\mathbf{\Delta}}_{k+1}(X)\big) \leq (n+2)\big(\underline{tc}^{n-k-1}(v)\big). \tag{8.55}$$

Proof. Recall that, for $(n+1)$-dimensional X and a traversally generic v on it, $\#\sup(\omega) \leq n+2$ for all ω's (see Theorem 6.7). By the construction of the stratification $\Omega^\bullet(DX, v)$ of the double DX (see Figure 1.5 and Figure 8.2), we get a formula which is similar to (8.33). It connects the combinatorics of the stratified space $\Omega^\bullet(DX, v)$ to the combinatorics of the stratification $\Omega^\bullet(\mathcal{T}(v))$ of $\mathcal{T}(v)$ by the connected components of the strata $\mathcal{T}(v,\omega)$:

$$\begin{aligned}\#\pi_0(DX^v_{-k}) &= \sum_{\{\omega|\ |\omega|'=k\}} \#\sup(\omega)\cdot\#\pi_0(\mathcal{T}(v,\omega)) \\ &+ 2\cdot \sum_{\{\hat\omega|\ |\hat\omega|'=k+1\}} (\#\sup(\hat\omega)-1)\cdot\#\pi_0(\mathcal{T}(v,\hat\omega)) \\ \le (n+2)\Big(\sum_{\{\omega|\ |\omega|'=k\}} &\#\pi_0(\mathcal{T}(v,\omega)) + 2\cdot \sum_{\{\hat\omega|\ |\hat\omega|'=k+1\}} \#\pi_0(\mathcal{T}(v,\hat\omega))\Big) \\ &= (n+2)\Big(\#\pi_0(\mathcal{T}(v)^\circ_{-k}) + 2\cdot\#\pi_0(\mathcal{T}(v)^\circ_{-(k+1)})\Big).\end{aligned}$$

Therefore, the suspended complexity of v can be estimated in terms of the traversing v-complexities:

$$\underline{\Sigma tc}^{n+1-k}(X,v) \le (n+2)\big(\underline{tc}^{n-k}(v) + 2\cdot\underline{tc}^{n-k-1}(v)\big).$$

With this inequality being validated, by Theorem 8.11 and under its hypotheses, for each $k \in [-1,n]$, we get the inequality (8.55). □

Let $\pi =_{\mathsf{def}} \pi_1(X)$ and $\Pi =_{\mathsf{def}} \pi_1(DX)$. Utilizing formulas (8.54) and (8.56), for each $k \in [-1,n]$, Theorems 8.8 and 8.11 expose the groups $H^{\Delta}_{k+1}(DX) \subset H^{\Delta}_{k+1}(\Pi)$ and $H^{\Delta}_k(X) \subset H^{\Delta}_k(\pi)$ as *obstructions* to the existence of *globally k-convex*[16] traversing flows on X.

Corollary 8.12. *Let X be a smooth compact connected and oriented $(n+1)$-manifold with boundary. Assume that X admits a globally k-convex traversally generic vector field v.*

If, for each connected component of the boundary ∂X, the image of its fundamental group in $\pi_1(DX)$ is amenable, then the simplicial semi-norm on $H_{j+1}(DX)$ is trivial for all $j \ge k$.

If, for each connected component of the boundary ∂X, the image of its fundamental group in $\pi_1(X)$ is amenable, then he simplicial semi-norm is trivial on $H_j(X)$ for all $j \ge k$.

In particular, if X admits a traversally generic vector field v with no trajectories of the maximal codimension n,[17] *then $\|[DX]\|_\Delta = 0$.*

---

[16] That is, by Definition 6.2, $\mathcal{T}(v)_{-k} = \emptyset$.
[17] Equivalently, if $\mathcal{T}(v)_{-n} = \emptyset$.

Proof. We notice that if $\underline{tc}^{n-k}(v) = 0$, then by Remark 8.5, $\underline{tc}^{n-k-1}(v) = 0$. Therefore, by Lemma 8.4, $\underline{tc}^{n-k}(v) = 0$ implies $H^{\mathbf{\Delta}}_{k+1}(DX) = 0$ and $H^{\mathbf{\Delta}}_k(X) = 0$. □

Example 8.4. Let $\{M_i\}_{1 \leq i \leq N}$ be a collection of closed orientable surfaces of genera $g(M_i) \geq 2$. Put $Y_i =_{\mathsf{def}} M_i \times S^1$. We consider the connected sum

$$Y =_{\mathsf{def}} Y_1 \# Y_2 \# \dots \# Y_N,$$

and let $Z =_{\mathsf{def}} Y \setminus B$, the complement to several smooth 3-balls in the 3-fold Y.

We denote by Z_i the 3-fold $Y_i \setminus (Y_i \cap B)$.

Our goal is to show that, for any $v \in \mathcal{V}^{\ddagger}(X)$, the number of 1-dimensional connected components in the $\mathbf{\Omega}^{\bullet}_{\langle 2]}$-stratification of DZ is $2N$ at least.

We may assume that the solid 1-handles $\{\mathbf{H}_i \approx D^3 \times D^1\}_{1 \leq i \leq N-1}$, participating in the connected sum construction of Y, are attached to the complements of the surfaces $M_i \times \star_i$ (where $\star_i \in S^1$ is a base point) in $Y_i = M_i \times S^1$ and to the complement of the balls $B \subset Y$. Thus,

$$Z = Z_1 \# Z_2 \# \dots \# Z_N.$$

Let $H_i \approx S^2 \times D^1$ be the portion of $\partial \mathbf{H}_i$ that is contained in Y. We denote by $J_i = \partial \mathbf{H}_i \setminus \mathrm{int}(H_i)$ the disjoint union of two 3-balls, residing in $Y_i \sqcup Y_{i+1}$, and by $K_i \subset \mathbf{H}_i$ the core of the solid handle. Its boundary ∂K_i is contained in the interior of J_i.

Consider the following two spaces:

$$Z^{\bullet} =_{\mathsf{def}} (\cup_i \mathbf{H}_i) \cup_{\{\cup_i J_i\}} Z, \qquad Z^{\#} =_{\mathsf{def}} (\cup_i K_i) \cup_{\{\cup_i \partial K_i\}} Z.$$

We will construct in stages a map $f_i : Z \to Z_i$ which is a homeomorphism on $M_i \times \star_i$. First, we map $Z^{\bullet}$ to $Z^{\#}$ so that each handle H_i is mapped to the core $K_i \subset \mathbf{H}_i$, and each set J_i to ∂K_i, the two centers of the two balls that form J_i. Then, for $i \in [2, N-1]$, we map $Z^{\#}$ to Z_i by collapsing the component of $Z^{\#} \setminus Z_i$ that contains Z_{i-1} to the point $\partial K_{i-1} \cap Y_i$, and the component that contains Z_{i+1} to the point $\partial K_i \cap Y_i$. The collapse is a homeomorphism on $M_i \times \star_i$. The composition of the maps $Z \subset Z^{\bullet} \to Z^{\#}$ and $Z^{\#} \to Z_i$ is the desired map $f_i : Z \to Z_i$, whose restriction to $M_i \times \star_i$ is also a homeomorphism.

The constructions of f_N is similar.

We observe that, since the embedding $Z_j \subset Y_j$ induces an isomorphism of the fundamental groups, the natural homomorphism $H_2(Z_j) \to H_2(Y_j)$

is an isometry in the simplicial norms [Gr]. This implies that $H_2^{\Delta}(Z_j) \to H_2^{\Delta}(Y_j) \approx \mathbb{R}$ is an isomorphism (the new 2-cycles in Z_j are the spheres that form ∂Z_j, each of which has a vanishing simplicial norm) .

Let $I : \coprod_i M_i \to Z$ be the obvious embedding. We claim that the 2-dimensional classes $\{I_*[M_i] \in H_2^{\Delta}(Z)\}_i$, are linearly independent. Indeed, assume that some combination $h =_{\mathsf{def}} \sum_i r_i I_*[M_i]$ has the property $\|h\|_\Delta = 0$. Then, for each j, $\|(f_j)_*(h)\|_\Delta = 0$. On the other hand, by the construction of f_j,

$$(f_j)_*(h) = r_j \cdot (I_j)_*[M_j] \in H_2^{\Delta}(Z_j) \approx \mathbb{R},$$

where the imbedding $I_j : M_j \subset Z_j$, being composed with f_j, produces the identity map of M_j. Therefore, using that the genus of M_j is greater than one, we get $\|(I_j)_*[M_j]\|_\Delta = \|[M_j]\|_\Delta \neq 0$. However, $r_j \cdot \|(I_j)_*[M_j]\|_\Delta = 0$ since $\|(f_j)_*(h)\|_\Delta = 0$. Thus $r_j = 0$, and the classes $\{I_*[M_i]\}_i$ are independent in $H_2^{\Delta}(Z)$.

Let X be a compact smooth 3-fold which is homotopy equivalent to Z and such that ∂X is a disjoint union of 2-spheres. Let v be a traversally generic vector field on X. Since $\pi_1(\partial X) = 0$, by Theorem 8.10 and following the arguments that lead to (8.32) and (8.54), we count the 1-dimensional connected strata in the stratification $\Omega^\bullet(DX, v)$ to get the inequality

$$6 \cdot \#\mathcal{T}(v, (1221)) + 2 \cdot \#\mathcal{T}(v, (13)) + 2 \cdot \#\mathcal{T}(v, (31))$$

$$+ 3 \cdot \#\pi_0(\mathcal{T}(v, (121))) + \#\pi_0(\mathcal{T}(v, (2))) \geq \mathrm{rk}\big(H_2^{\Delta}(DZ)\big) \geq 2N.$$

The integral coefficients in the formula above are produced as:

$$6 = 2(\#\sup(1221) - 1),\ 2 = 2(\#\sup(13) - 1),\ 2 = 2(\#\sup(31) - 1),$$

$$3 = \#\sup(121),\ 1 = \#\sup(2).$$

Since Y_j admits a self-map of degree $\neq \pm 1$, we get $\|[Y_j]\|_\Delta = 0$ for all j.

The simplicial semi-norm is additive under the connected sum operation (see [Gr], page 10, formula (**)), provided that the dimension of manifolds is 3 at least. Thus $\|[Y]\|_\Delta = 0$. The 3-fold Z is produced from Y by deleting a number balls D^3, whose semi-norms $\|[D^3, S^2]\|_\Delta$ vanish. By [Gr], Theorem on page 55, $\|Z\|_\Delta = \|Y\|_\Delta = 0$. Therefore, by the same Gromov theorem, we conclude that $\|[DX]\|_\Delta = 0$. So, regrettably, for $k + 1 = 3$, Theorem 8.11 provides us with only the trivial estimate

$$4 \cdot \#\mathcal{T}(v, (1221)) + 2 \cdot \#\mathcal{T}(v, (13)) + 2 \cdot \#\mathcal{T}(v, (31)) \geq 0,$$

for the number of 0-dimensional strata in $\Omega^\bullet(DX, v)$. ◇

Example 8.5. Let $f : Y \to \mathbb{R}$ be a Morse function on a closed smooth 4-manifold Y. Let $\hat{v}$ be a gradient-like vector field on Y. Let X be obtained from Y by deleting small convex balls, centered on all the critical points of f. By perturbing $\hat{v}$ if necessary, we may assume that $v = \hat{v}|_X$ is a traversally generic vector field on X such that $df(v) > 0$.

Then, by Theorem 8.12 and the proof of Lemma 8.4, and revisiting the stratification (8.39) of the double DX, we get the following estimates:

$$\bullet\ \mathrm{rk}\Big(H_2^{\mathbf{\Delta}}(DX)\Big) \leq \mathrm{rk}\Big(\mathsf{C}^2_{\mho}(DX, v)\Big)$$

$$= 6 \cdot \#\pi_0(\mathcal{T}(v, (1221))) + 3 \cdot \#\pi_0(\mathcal{T}(v, (121))),$$

$$\bullet\ \mathrm{rk}\Big(H_3^{\mathbf{\Delta}}(DX)\Big) \leq \mathrm{rk}\Big(\mathsf{C}^1_{\mho}(DX, v)\Big)$$

$$= 8 \cdot \#\mathcal{T}(v, (12221)) + 4 \cdot \#\pi_0(\mathcal{T}(v, (1221))),$$

$$\bullet\ \mathrm{rk}\Big(H_4^{\mathbf{\Delta}}(DX)\Big) \leq \mathrm{rk}\Big(\mathsf{C}^0_{\mho}(DX, v)\Big) = 5 \cdot \#\mathcal{T}(v, (12221)),$$

where

$$6 = 2(\#\sup(1221) - 1), \;\; 3 = \#\sup(121);$$

$$8 = 2(\#\sup(12221) - 1), \;\; 4 = \#\sup(1221); \;\; \text{and} \;\; 5 = \#\sup(12221).$$

We notice that $\mathrm{rk}(H_4^{\mathbf{\Delta}}(DX))$ is either 1 or 0. In particular, if $\mathrm{rk}(H_4^{\mathbf{\Delta}}(DX)) = 1$, then any such v-flow must have a triple-tangent trajectory.

Let us take a closer look how these estimates work for a specific family of 4-folds. Let $Z_i \to M_i$ be a fibration over a closed oriented surface M_i of a genus exceeding 1 and with a closed surface fiber F_i. Assume that $Z_i \to M_i$ admits a section s_i. We form the connected sum

$$Z = Z_1 \# Z_2 \# \ldots \# Z_N.$$

Let us consider a smooth closed 4-fold Y which is homeomorphic to Z with the help of a homeomorphism $h : Z \to Y$.

Let $f : Y \to \mathbb{R}$ be a Morse function which satisfies the Morse-Smale transversality property and such that the h-preimages of all its critical points have an empty intersection with the surfaces $\coprod_i s_i(M_i) \subset Z$.

Finally, let X be a compact smooth 4-fold, obtained by deleting from Y several small convex 4-balls, centered on all the f-critical points.

Then, applying the first bulleted inequality to the traversally generic gradient field $v = \nabla f$ on X, we get

$$6 \cdot \#\pi_0(\mathcal{T}(v,(1221))) + 3 \cdot \#\pi_0(\mathcal{T}(v,(121))) \geq N,$$

since the images of the classes $\{h_*[s_i(M_i)]\}_i$ are independent in $H_2^{\mathbf{\Delta}}(X)$.

At least when the fibrations $\{Z_i \to M_i\}_i$ are trivial, we suspect that $H_3^{\mathbf{\Delta}}(DX) = 0$, so the estimates in the second bullet probably are trivial.

To validate the estimate in the third bullet, we use a slightly modified argument from Example 8.4.

At least when all the fibrations $Z_i \to M_i$ are trivial and their fibers F_i are surfaces of genera greater than one, one can prove that $\mathrm{rk}\big(H_4^{\mathbf{\Delta}}(DX)\big) = 1$, which implies that $\#\mathcal{T}(v,(12221)) \neq 0$. In such a case, the v-flow on X must have a triple-tangent trajectory. ♢

The next theorem, proven in [AK], should be compared with Theorem 8.11 (see also Theorem 7.5 from [K], its older $3D$-relative). In a way, that theorem is the source of motivation for developing the machinery of amenable localization in [AK] and in this chapter. For 3-folds M with a simply-connected boundary, Theorem 8.13 below is known with $c(2) = 1/\mathrm{Vol}(\Delta^3)$, the inverse of the volume of the ideal tetrahedron in the hyperbolic space $\mathbf{H}^3$.

Theorem 8.13 (Alpert, Katz). *Let M be a closed, oriented* hyperbolic *manifold of dimension $n+1 \geq 2$. Let X be the space, obtained by deleting a domain U from M, such that U is contained in a $(n+1)$-ball $B \subset M$. Let v be a traversally generic vector field on X.*

Then the cardinality of the set of v-trajectories of the maximal reduced multiplicity n satisfies the inequality

$$\#\big(\mathcal{T}(v)_{-n}\big) \geq c(n) \cdot \mathrm{Vol}(M), \tag{8.56}$$

where $c(n) > 0$ is an universal constant, and $\mathrm{Vol}(M)$ *denotes the hyperbolic volume of M.*

Proof. Recall that a closed hyperbolic manifold M has a contractible universal covering space; in other words, it is a $K(\pi_1(M), 1)$-space.

Until now, we did not use the full flexibility of the Localization Lemma 8.1: in the previous arguments, implicitly the map $\alpha : X \to Z$ in the lemma was always presumed to be the classifying map $\alpha : DX \to K\big(\pi_1(DX), 1\big)$ or the classifying map $\alpha : X \to K\big(\pi_1(X), 1\big)$. Now we construct a degree-one map $\alpha : DX \to M$ that sends the boundary ∂X to a single point. Given such a map, we have $\alpha_* \pi_1(\partial X) = 0$ (and 0 is an amenable group). So Theorem 8.10 gives

$$\#(\mathcal{T}(v)_{-n}) \geq c(n) \cdot \|\alpha_*[DX]\|_\Delta = c(n) \cdot \|[M]\|_\Delta \geq c(n) \cdot \mathrm{Vol}(M),$$

where the value of $c(n)$ is not fixed and may change between the inequalities, but is always positive and M-independent.

To construct α, let B be an open ball containing $\overline{U}$, and let B' be a slightly smaller ball with $\overline{U} \subset B' \subset \overline{B'} \subset B$. There is a degree-one map $\beta : M \to M$, obtained by collapsing $\overline{B'}$ to a single point $*$, and stretching the shell $B \setminus B'$ to fill in $B \setminus *$. We define α on X as the restriction of β to X, and we define α on the second copy of X as the constant map to $*$. Then α on all of DX has degree 1, and we get $\alpha(\partial X) = *$. □

Corollary 8.13. *The inequality (8.56) from Theorem 8.13 is valid for any X, obtained by deleting a number of $(n+1)$-balls from a closed hyperbolic manifold M.*

Proof. When the domain U from Theorem 8.12 is a union of $(n+1)$-balls, we can encapsulate them into a single $(n+1)$-ball B. Therefore, Theorem 8.12 is applicable to $X = M \setminus U$. □

Definition 8.20. A *k-broken trajectory* of a gradient vector field v with isolated singularities is the union of k trajectories $\gamma_1, \dots \gamma_{k+1}$ such that each γ_i asymptotically approaches a pair of singularities (both in positive and negative times); moreover, the union $\cup_{i=1}^{k+1} \bar{\gamma}_i$ of their closures $\{\bar{\gamma}_i\}_i$ topologically is a segment which admits an orientation, consistent with the v-induced orientations of the γ_i's. ◇

Let us apply Corollary 8.13 to the landscape of Morse functions on closed aspherical or hyperbolic manifolds. Moving towards this goal, we formulate the following conjecture.

Conjecture 8.1. *Let $f : M \to \mathbb{R}$ be a Morse function on a closed manifold M of dimension $(n+1)$, and v its gradient-like vector field. Assume that v satisfies the Morse-Smale transversality condition ([Sm], [Sm1]).*

Then, for all sufficiently small $\epsilon > 0$, the field v on the complement

$$X =_{\mathsf{def}} M \setminus \coprod_{x \in \Sigma_f} B_\epsilon(x)$$

to the union of convex (in the local Morse coordinates) balls $B_\epsilon(x)$ is traversally generic. The combinatorial types of the v-trajectories on X are drawn from the list: $(11), (121), (1221), \dots, (1\underbrace{2\dots2}_{n}1)$.

Moreover, there exists a universal constant $a(n) > 0$ such that the number of n-broken v-trajectories on M, and the number of n-tangent v-trajectories in X are related by the formula

$$\#\big(\mathcal{T}(v, \omega_\star)\big) = a(n) \cdot \#\big(\mathsf{broken}_n(v)\big), \tag{8.57}$$

where $\omega_\star = (1\underbrace{2\dots2}_{n}1)$. ◊

Some evidence for this conjecture may be found in the proof of Theorem 3.8 and in Figure 3.2 that illustrates the proof.

Combining Conjecture 8.1 with Theorem 8.10 and Theorem 8.13, one could arrive to an estimate from below of $\#(\mathsf{broken}_{(n+1)}(v))$ for Morse-Smale gradient fields v on closed aspherical (in particular, on hyperbolic) $(n+1)$-manifolds.

Fortunately, regardless whether Conjecture 8.1 is valid or not, the cardinality $\#\big(\mathsf{broken}_{(n+1)}(v)\big)$ has a lower boundary, delivered by the simplicial (normalized hyperbolic) volume! The beautiful proposition below has been recently proven by Hannah Alpert [A]. The proof involves the same circle of amenable localization techniques as in [Gr1], [AK], being applied to intricate stratifications of certain natural compactifications of the unstable manifolds of the gradient v-flow.

Theorem 8.14 (Alpert). *Let $f : M \to \mathbb{R}$ be a Morse function on a closed aspherical manifold M of dimension $(n+1)$, and v its gradient-like field. Assume that v satisfies the Morse-Smale transversality condition. Then*

$$\#\big(\mathsf{broken}_n(v)\big) \geq \|M\|_\Delta.$$

In particular, if M is hyperbolic, then

$$\#\big(\mathsf{broken}_n(v)\big) \geq \mathrm{Vol}(M)/\mathrm{Vol}(\Delta^{n+1}),$$

where $\mathrm{Vol}(\Delta^{n+1})$ denotes the volume of an ideal hyperbolic simplex Δ^{n+1}.

◊

Campagnolo and Sauer, using some ideas from [A], proved recently another nice theorem about counting maximally broken trajectories of gradient flows on aspherical closed manifolds [CaSa]. It involves $\mathsf{tors}(H_*(\sim;\mathbb{Z}))$, the cardinality of the homology torsion group.

Theorem 8.15 (Campagnolo, Sauer). *Let M be a closed orientable aspherical $(n+1)$-dimensional Riemannian manifold, equipped with a Morse-Smale function $f : M \to \mathbb{R}$.*

Then the number of maximally broken trajectories of the gradient vector field $v = \nabla f$, satisfies the inequality

$$\#\big(\mathsf{broken}_n(v)\big) \geq \frac{1}{2^{n+3}(n+3)!} \cdot \sum_{i=0}^{n+1} \mathsf{size}\big(H_i(M;\mathbb{Z})\big),$$

where $\mathsf{size}(A)$ *of a group A is defined as* $\mathsf{rank}(A) + \log(\mathsf{tors}(A))$. ◊

8.6 Complexity of the fundamental group as a lower bound for the complexity of traversing flows

We just have internalized the crucial role that the non-amenable fundamental groups $\pi_1(X)$ and $\pi_1(DX)$ play in delivering some lower bounds of the traversing complexities $\underline{\mathbf{tc}}^{\mathsf{lex}}_{\mathsf{trav}}(X)$ and $\underline{\mathbf{\Sigma tc}}^{\mathsf{lex}}_{\mathsf{trav}}(X)$.

Now we will connect the complexity of the fundamental group $\pi_1(X/\partial X)$ with the number of *minimal* connected components in the $\mathbf{\Omega}^\bullet$-stratification $\{X(v,\omega)\}_\omega$ of X. The following considerations are quite elementary; in particular, the amenability properties do not play any role here.

Each v-trajectory γ of the combinatorial type ω defines a loop or a bouquet of loops in the quotient space $X/\partial X$. Similarly, each γ defines its double $D(\gamma) \subset DX$. The double $D(\gamma)$ is a chain of loops (like "**o**", or "**oo**", or "**ooo**", etc), the number of loops in the chain being equal to $\#(\gamma \cap \partial X) - 1$.

Therefore each trajectory γ produces an element $[\gamma] \in \pi_1(X/\partial X)$, represented by the bouquet $\gamma/(\gamma \cap \partial X)$. We denote by $[[\gamma]]$ the subgroup of $\pi_1(X/\partial X)$, generated by the loops from the bouquet $\gamma/(\gamma \cap \partial X)$. The group $[[\gamma]]$ is equipped with the ordered set of $(|\sup(\omega)| - 1)$ generators—the v-ordered loops of the bouquet $\gamma/(\gamma \cap \partial X)$. Here $|\sup(\omega)|$ stands for the cardinality of the set $\gamma \cap \partial X$. The element $[\gamma]$ and the subgroup $[[\gamma]]$ are constant within each connected component $X(v,\sigma)$ of the pure stratum

$X(v, \omega)$, the union of trajectories of the combinatorial type ω. By Corollary 7.2, this follows from the fact that the finite covering

$$\Gamma : X(v, \sigma) \cap \partial X \to \Gamma\big(X(v, \sigma) \cap \partial X\big)$$

is trivial.

Let us denote by $\mathcal{S}^\bullet(v)$ the poset whose elements are the connected components $\{X(v, \sigma)\}$. Let $\mathcal{S}^\bullet_{\min}(v)$ denote the set of minimal elements in $\mathcal{S}^\bullet(v)$.

So we get a system of groups $\{[[\gamma_\sigma]]\}_{\sigma \in \mathcal{S}^\bullet(v)}$. By Lemma 7.5, they are linked by homomorphisms

$$\psi_{\sigma,\sigma'} : [[\gamma_\sigma]] \to [[\gamma_{\sigma'}]]$$

for any pair $\sigma \succ \sigma'$ in $\mathcal{S}^\bullet(v)$.

This construction leads to the following claim.

Lemma 8.5. *For a traversally generic vector field v on X, the "minimal" subgroups $\big\{[[\gamma_\sigma]]\big\}_{\sigma \in \mathcal{S}^\bullet_{\min}(v)}$ generate $\pi_1(X/\partial X)$.*

Proof. We need to show that any loop $\rho : I/\partial I \to X/\partial X$ through the point $\star = \partial X/\partial X$ is a product of loops form the groups $[[\gamma_\sigma]]$, where σ runs over the minimal elements of the poset $\mathcal{S}^\bullet(v)$. Equivalently, it will suffice to show that any path $(\rho, \partial\rho) : (I, \partial I) \to (X, \partial X)$ can be homotoped, relatively to the boundary ∂X, to an ordered union several *segments* of oriented trajectories γ, labeled with the elements of the minimal set $\mathcal{S}^\bullet_{\min}(v)$. Here the segments of trajectories γ are bounded by two points from $\gamma \cap \partial X$, and the orientations of segments are not necessarily the ones induced by v. The segments are linked by arcs which reside in ∂X, so that the oriented union of the segments and arcs represents the path ρ.

Note that each oriented segment of γ belongs to the subgroup $[[\gamma]]$.

First, with the help of the $(-v)$-flow, we homotop $(\rho, \partial\rho)$ to a path that is realized by several segments of trajectories, intermingled with some paths residing in ∂X.

Here is a sketch of how to construct the desired homotopy. By dimensional arguments, we may assume that the path ρ misses all the loci $X(v, \omega)$ with $|\omega|' \geq 2$ and is transversal to all the loci $X(v, \omega)$ with $|\omega|' = 1$.

If $\rho(0)$ reside in $\partial_1^- X(v)^\circ$, we homotop ρ in the vicinity of $0 \in [0, 1]$ so that the new path $\tilde\rho$ has the properties: $\tilde\rho(0) = \rho(0)$ and $\tilde\rho([0, \epsilon])$ is a segment of the $(-v)$-trajectory γ through $\rho(0)$, and $\tilde\rho(\epsilon) \in \partial_1^+ X(v)^\circ$. If $\rho(1)$ resides in $\partial_1^- X(v)^\circ$, by a similar argument, we may assume that:

$\tilde\rho(1) = \rho(1)$, $\tilde\rho([1-\epsilon, 1])$ is a segment of the $(-v)$-trajectory γ through $\rho(1)$, and $\tilde\rho(1-\epsilon) \in \partial_1^+ X(v)^\circ$.

Thus it will suffice to deal with paths ρ that start and end at points of $\partial_1^+ X(v)^\circ$ and are in general positions with respect to the strata $\{X(v,\omega)\}_\omega$ of codimensions 1 and 2. In particular, we may assume that ρ misses the closure of the locus $X(v, (1221))$ of codimension 2. Therefore, the set of downward trajectories through the curve $\rho([0,1])$ is a "histogram" as in Figure 2.10. As a result, with the help of $-v$, we may homotope ρ to a new path that consists of segments of trajectories through the points of $\partial_2^+ X(v)$ and segments that reside in $\partial_1^+ X(v)$. Thus any ρ is homotopic (modulo its end points) to a new path $\tilde\rho$ which is represented in $\pi_1(X/\partial_1 X)$ by a composition of loops that are generated by segments of v-trajectories.

The rest of argument is an induction, based on the order $\succ$ in $\mathcal{S}^\bullet(v)$. We can replace any segment of $\gamma \subset X(v,\sigma)$ with the corresponding segment of any other trajectory from that stratum. This replacement is possible since $\Gamma : X(v,\sigma) \cap \partial X \to \mathcal{T}(v,\sigma)$ is a trivial covering. Moreover, we can homotope such segment of γ to a segment of some trajectory γ', residing in any stratum $X(v,\sigma')$ that belongs to the closure of $X(v,\sigma)$. In other words, we may replace any segment of $\gamma \subset X(v,\sigma)$ with a segment of some trajectory $\gamma' \subset X(v,\sigma')$, where $\sigma' \prec \sigma$. Following these replacements, eventually we will arrive to an ordered finite collection of segments of $\{\gamma_\sigma \subset X(v,\sigma)\}_{\sigma \in \mathcal{S}^\bullet_{\min}(v)}$; the collection will represent the relative homotopy class of the path ρ (some of the the segments of trajectories γ_σ's in this representation may appear with integral multiplicities). □

Given a finitely presentable group π, we denote by $c_{\mathsf{gen}}(\pi)$ the *minimal number of its generators*, the minimum is taken over all finite presentations of π.

Theorem 8.16. *For a traversally generic vector field v on a connected compact smooth manifold X with boundary,*

$$c_{\mathsf{gen}}\big(\pi_1(X/\partial X)\big) \leq \sum_{\sigma \in \mathcal{S}^\bullet_{\min}(v)} \Big(|\sup(\omega(\sigma))| - 1\Big),$$

where $\mathcal{S}^\bullet_{\min}(v)$ is the set of connected components of the minimal strata $\{X(v,\omega)\}$, and $\omega(\sigma)$ is the combinatorial type of a typical v-trajectory passing through the connected component, labeled by σ .

Proof. By Lemma 8.5, any element $\beta \in \pi_1(X/\partial X)$ produces a word in the alphabet, whose letters are the loops, generated by the segments of special trajectories that correspond to the minimal connected strata of $\mathcal{S}^\bullet_{\min}(v)$. Each minimal stratum $X(v,\sigma)$, $\sigma \in \mathcal{S}^\bullet_{\min}(v)$, contains a unique special trajectory. The number of loops-letters in each trajectory is $|\sup(\omega(\sigma))| - 1$. Therefore the total number of letters in the alphabet is given by the RHS of the formula above. □

Corollary 8.14. *For a traversally generic vector field v on a connected compact smooth $(n+1)$-dimensional manifold X with boundary, the number of minimal connected components in $\mathcal{T}(v)$ satisfies the inequality:*

$$\#\big(\mathcal{S}^\bullet_{\min}(v)\big) \geq c_{\mathsf{gen}}\big(\pi_1(X/\partial X)\big)/(n+1).$$

In particular, if ∂X consists of m connected components, then

$$\#\big(\mathcal{S}^\bullet_{\min}(v)\big) \geq (m-1)/(n+1).$$

Proof. Note that for a traversally generic v, $|\sup(\omega(\sigma))| \leq n+2$, so that the number of segments, in which a typical trajectory γ_σ is divided by ∂X, is $n+1$ at most. Utilizing the arguments from Lemma 8.5, Theorem 8.16 implies the first inequality of the corollary.

To prove the second one, we notice that $X/\partial X$ admits a continuous map F onto a connected graph G with two vertices, a and b, and m edges. Indeed, let U be a collar of ∂X in X. We send $X \setminus \mathrm{int}(U)$ to $a \in G$, ∂X to b, and each component of the collar U to the corresponding edge. Each basic loop in G lifts against F to a loop in $X/\partial X$. Therefore there exists an epimorphism $\pi_1(X/\partial X) \to \mathsf{F}_{m-1}$, where F_{m-1} denotes the free group in $m-1$ generators. Thus $c_{\mathsf{gen}}\big(\pi_1(X/\partial X)\big) \geq m-1$, which implies $\#\big(\mathcal{S}^\bullet_{\min}(v)\big) \geq (m-1)/(n+1)$. □

Remark 8.10. We observe that, by definition, $\#\big(\mathcal{S}^\bullet_{\min}(v)\big) \geq \#\big(\mathcal{T}(v)_{-n}\big)$. Thus, under the hypotheses of Theorem 8.13 and by that theorem, there exists an universal constant $c(n) > 0$ such that $\#\big(\mathcal{S}^\bullet_{\min}(v)\big) \geq c(n)\cdot \mathrm{Vol}(M)$, where the hyperbolic volume $\mathrm{Vol}(M)$ depends only on $\pi_1(M)$.

If X is obtained from a closed hyperbolic M by deleting a single ball, then $\pi_1(M) \approx \pi_1(X/\partial X)$. In such a case, $\#\big(\mathcal{S}^\bullet_{\min}(v)\big) \geq c_{\mathsf{gen}}\big(\pi_1(M)\big)/(n+1)$. So for the hyperbolic $X = M \setminus D^{n+1}$, both lower bounds for $\#\big(\mathcal{S}^\bullet_{\min}(v)\big)$ are expressed essentially in terms of $\pi_1(M)$. ◇

Chapter 9

The Burnside Ring-valued Morse Formula for Equivariant Vector Fields

9.1 In this chapter ...

This chapter suggests that, in the presence of a smooth symmetry, some results from Chapter 5 have their equivariant analogues. In what follows, we will only present the equivariant analogue of the Morse Formula for vector fields. Basically, the chapter describes our investigation in [K7], launched by a valuable conversation which Wolfgang Lück.

It turns out that the Morse stratifications, in the presence of smooth symmetry, are very special: the main orbit-type of the action is present in any Morse stratum.

Let G be a compact Lie group (not necessarily connected). We consider the *left* actions of G on topological spaces. We denote by $A(G)$ the *Burnside Ring* of the group G. Let us recall the definition of $A(G)$, which is due to tom Dieck [D].

Definition 9.1. Let X and Y be two compact G-manifolds. We say that X and Y are *Burnside-equivalent* if, for each subgroup $H \subset G$, the Euler numbers of the H-fixed sets, X^H and Y^H, coincide:

$$\chi(X^H) = \chi(Y^H).$$

By definition, $A(G)$ is generated over $\mathbb{Z}$ by Burnside equivalence classes of compact differentiable G-manifolds. $\diamondsuit$

In fact, $A(G)$ is a ring with the addition, induced by the disjoint unions of G-spaces, and the multiplication by their Cartesian products, the G-action on products being diagonal.

For a compact smooth n-dimensional[1] G-manifold X, equipped with a boundary generic G-invariant smooth vector field v, we prove that

$$\mathsf{Ind}^G(v) = \sum_{k=0}^{n} (-1)^k \chi^G(\partial_k^+ X),$$

an equivariant analog of the Morse formula in Theorem 5.3, taking its values in the ring $A(G)$. Here $\mathsf{Ind}^G(v) \in A(G)$ denotes the equivariant index of the field v, $\{\partial_k^+ X\}$ the v-induced Morse stratification (as in Chapter 5) of the boundary ∂X, and $\chi^G(\partial_k^+ X)$ the Burnside class of the $(n-k)$-dimensional G-manifold $\partial_k^+ X$ [2] in $A(G)$.

In Theorem 9.1, we investigate quite strong restrictions on the nature of the Morse stratification $\{\partial_j X\}$ (and thus of $\{\partial_j^+ X\}$), imposed by a G-symmetry of the vector field v.

We examine applications of this formula to the equivariant real polynomial fields v in compact domains $X \subset \mathbb{R}^n$, defined via a generic polynomial inequality. Next, we link the equivariant Morse formula with the equivariant degrees of certain Gauss maps. This link is an equivariant generalization of Gottlieb's formulas [Got], [Got1] (as in Theorem 5.4).

9.2 G-equivariant vector fields and Morse stratifications

We consider the *left* actions of a given compact Lie group G on topological spaces X. Violating the conventions, we denote the orbit space by X/G (instead of the common notation "$G\backslash X$"). This is done to accommodate other labels that we routinely attach to Morse strata. At the same time, we use the conventional notation "G/H" for the orbit-types of a left action.

From now and on, we assume that G-actions are *smooth*.

When a vector field v on X is both boundary generic and G-invariant, the Morse stratification $\{\partial_k^+ X(v)\}$ is invariant as well. In fact this conclusion is valid when the field v is G-invariant only up to a positive conformal factor, a smooth function on X. As a result, for such a v, the G-symmetry of X gives rise to a G-symmetry of the trajectory space $\mathcal{T}(v)$.

Using the G-symmetry of v, the next lemma describes an interaction of two stratifications in $\partial_1 X$: the Morse stratification, generated by v, and the stratification by the orbit-types of the G-action.

[1] Note the change of our convention from "$\dim(X) = n+1$" to "$\dim(X) = n$" in this chapter.

[2] By definition, $\partial_0^+ X = X$.

Lemma 9.1. *For a G-invariant boundary generic vector field v and each compact subgroup $H \subset G$, v is tangent to $X^H \subset X$ and*

$$\partial_k^+ X^H(v|_{X^H}) = X^H \cap \partial_k^+ X(v).$$

Proof. The key observation here is that, for each compact subgroup $H \subset G$, any G-invariant field v on X is *tangent* to the submanifold $X^H \subset X$. Indeed, in a G-equivariant metric g, the normal to X^H component ν_H of v must be H-invariant, and thus vanishes for $H \neq 1$ (since no nontrivial vector, normal to X^H, is H-fixed).

Let $G_x \subset G$ denote the stationary subgroup, formed by the elements that fix a given point $x \in X$.

Each submanifold X^H meets $\partial_1 X$ transversally: just consider a linearization of the G_x-action, $G_x \supset H$, at a typical point $x \in X^H \cap \partial_1 X$. Since v is tangent to X^H, evidently, $\partial_1^+ X^H(v|_{X^H}) = X^H \cap \partial_1^+ X(v)$.

We proceed by induction in k. Assume that, for all $s < k$, the manifolds X^H and $\partial_s X(v)$ are transversal,

$$\partial_s X^H(v|_{X^H}) = X^H \cap \partial_s X(v), \text{ and } \partial_s^+ X^H(v|_{X^H}) = X^H \cap \partial_s^+ X(v).$$

If $x \in X^H \cap \partial_k^+ X(v)$ and $v(x)$ is tangent to

$$\partial_{k-1} X(v) \cap X^H = \partial_{k-1} X^H(v|_{X^H}),$$

then $v(x)$ points inside of $\partial_{k-1}^+ X(v)$ if and only if it points inside $\partial_{k-1}^+ X^H(v|_{X^H})$. Thus,

$$\partial_k X^H(v|_{X^H}) = X^H \cap \partial_k X(v),$$

which completes the induction step. □

Consider a smooth G-action on a compact connected manifold X with the boundary $\partial_1 X$. In general, the orbit space $Y =_{\mathsf{def}} \partial_1 X/G$ may have several connected components $\{Y_\alpha\}_\alpha$. We denote by $\partial_{1,\alpha} X$ the union of connected components of $\partial_1 X$ that are mapped by the quotient map to Y_α.

By Theorem 3.1 from [B], Chapter IV, the G-action on $\partial_{1,\alpha} X$ has a unique *main* orbit-type $[G/H_\alpha]$. Moreover, the union of the orbits of the type $[G/H_\alpha]$ is open and dense in $\partial_{1,\alpha} X$, and its image in Y_α is connected.

Let v be a G-invariant and boundary generic vector field on X. Since $\partial_k X(v)$ is a G-invariant closed submanifold, the same analysis applies to each connected component $Y_{k,\beta}$ of the orbit space $Y_k =_{\mathsf{def}} \partial_k X(v)/G$. Of course, each component $Y_{k,\beta}$ is contained in a unique set Y_α, where $\alpha = \alpha(k, \beta)$.

Let $\partial_{k,\beta}X(v)$ denote the union of connected components of $\partial_k X(v)$ that are mapped to $Y_{k,\beta}$. Thus for each $k > 0$ and β, there exists a main orbit-type $[G/H_{k,\beta}]$ for the G-action on $\partial_{k,\beta}X(v)$. Again, such orbits form an open and dense set in $\partial_{k,\beta}X(v)$. The image of this set of orbits in $Y_{k,\beta}$ is connected.

The next lemma shows that, for a G-invariant and boundary generic vector field v, the strata $\{\partial_k X(v)\}_{k>1}$ must contain the main orbit-type.

Lemma 9.2. *Let v be a G-invariant and boundary generic vector field v on a compact connected smooth manifold X.*

Then, for each $k \geq 1$ and index β, labeling the connected components of the orbit space $\partial_k X(v)/G$, the main orbit-type $[G/H_{k,\beta}]$ of the G-space $\partial_{k,\beta}X(v) \subset \partial_k X(v)$ coincides with the main orbit-type $[G/H_{\alpha(k,\beta)}]$ of the G-space $\partial_{1,\,\alpha(k,\beta)}X(v) \subset \partial_1 X$, provided $\partial_{k,\beta}X(v) \neq \emptyset$. Here the index $\alpha(k,\beta)$ labels the connected component of $\partial_1 X/G$ that contains the β-component of $\partial_k X(v)/G$.

Proof. We pick $x \in \partial_k X^\circ(v) =_{\mathsf{def}} \partial_k X(v) \setminus \partial_{k+1}X(v)$ and let $H =_{\mathsf{def}} G_x$, the stationary group of x. Because v is H-invariant, a linearization of the H-action at x must leave the flag F_x of tangent half-spaces $\{T_x^+(\partial_l^+ X(v))\}_{l \leq k}$ invariant as well. Let us choose a G-invariant metric on X. Consider the unique "projective frame" ν_x, formed by k mutually orthogonal rays $r_1, \dots, r_k$. The rays are generated by intersecting the flag F_x with the space N_x^k, normal in $T_x(X)$ to $T_x(\partial_k X(v))$. So, for all $l < k$, the intersection $N_x^k \cap T_x^+(\partial_{k-l}^+ X(v))$ is the positive cone $Span^+\{r_1, \dots, r_l\}$. The frame ν_x must be H-invariant. Thus, in a basis consistent with ν_x, the H-action is *diagonal* with the positive eigenvalues. Since H is compact, its action on the space N_x^k (spanned by the frame ν_x) is trivial.

We apply this conclusion to each $x \in \partial_{k,\beta}X(v)$ whose orbit type is $[G/H_{k,\beta}]$, the main orbit-type in $\partial_{k,\beta}X(v)$.

By the Tube Theorem 4.2 from [B], a tubular neighborhood in $\partial_1 X$ of the orbit through $x \in \partial_{k,\beta}X(v)$ is equivariantly homeomorphic to the balanced product $G \times_{H_{k,\beta}} V_{k,\beta}$, where $V_{k,\beta}$ denotes the space of a representation $\psi_{k,\beta} : H_{k,\beta} \to \mathsf{GL}(V_{k,\beta})$. By the same token, a the tubular neighborhood in $\partial_k X(v)$ of the orbit through x is equivariantly homeomorphic to the balanced product $G \times_{H_{k,\beta}} W_{k,\beta}$, where $W_{k,\beta}$ denotes a $\psi_{k,\beta}(H_{k,\beta})$-invariant linear subspace of $V_{k,\beta}$. Let us denote by $W_{k,\beta}^\perp$ the orthogonal complement to $W_{k,\beta}$ in $V_{k,\beta}$, the scalar product on $V_{k,\beta}$ being $\psi_{k,\beta}(H_{k,\beta})$-invariant.

We notice that, if an orbit $\mathcal{O}$ of a smooth G-action on a closed manifold Z is of the *main* orbit-type $[G/H]$, then an equivariant tubular neighborhood of $\mathcal{O}$ in Z is of the form $G \times_H V$, where V is the space of the *trivial* H-representation. Therefore, the $H_{k,\beta}$-action on $W_{k,\beta}$ must be trivial.

By identifying the normal bundle $\nu(\partial_k X(v), \partial_1 X)$ with $G \times_{H_{k,\beta}} W_{k,\beta}^{\perp}$ locally and by the previous argument (that involves the flag F_x), we conclude that the $H_{k,\beta}$-action on $W_{k,\beta}^{\perp}$ must be trivial as well. Therefore, the $H_{k,\beta}$-action on $V_{k,\beta}$ is trivial. As a result, $[G/H_{k,\beta}]$ is the main orbit-type in the open tube $G \times_{H_{k,\beta}} V_{k,\beta}$. Since the G-action on $\partial_{1,\alpha(k,\beta)} X$ has a single main orbit-type $[G/H_{1,\alpha(k,\beta)}]$, we conclude that $H_{k,\beta} = H_{1,\alpha(k,\beta)}$ for all $k > 0$ and β. □

As the following theorem testifies, the Morse stratifications of equivariant vector fields are very special.

Theorem 9.1. *Let a compact Lie group G act on a smooth compact $(n+1)$-manifold X with boundary. Assume that $[G]$ is the main orbit-type of the G-action on the boundary $\partial_1 X$.*

Then, for any boundary generic and G-equivariant vector field v on X and each $k > 0$, the following statements are valid:

- *if $\partial_k X(v) \neq \emptyset$, the main orbit-type of the space $\partial_k X(v)$ is $[G]$,*
- *hence, for all $k > n + 1 - \dim(G)$, $\partial_k X(v) = \emptyset$*[3]*,*
- *the sets $\partial^{\pm}_{n+1-\dim(G)} X(v)$ are disjoint unions of* free *G-orbits,*
- *if G is connected and the G-action on the boundary preserves its orientation, the sets $\partial^{\pm}_{n-\dim(G)} X(v)$ are disjoint unions of the following G-spaces:*
  - *(1) $G \times [0,1]$,*
  - *(2) $G \times S^1$,*
  - *(3) the mapping cones $\mathcal{C}(G \to G/H)$, where $H \approx \mathsf{SO}(2)$ or $\mathsf{SU}(2)$,*
  - *(4) the union of two mapping cones*

$$\mathcal{C}(G \to G/H) \cup_G \mathcal{C}(G \to G/K),$$

where $H, K \approx \mathsf{SO}(2)$ or $\mathsf{SU}(2)$.

Proof. Since $[G]$ is the main orbit-type of the boundary $\partial_1 X$, the first bullet follows from Lemma 9.2, being applied to the case $H_{\alpha(k,\beta)} = 1$.

[3] That is, v is boundary $(n + 2 - \dim(G))$-convex and concave.

By a dimensional argument, the second bullet is an instant implication of the first one.

By the second bullet, $\partial_{n+2-\dim(G)}X(v) = \emptyset$. Therefore the two sets $\partial^{\pm}_{n+1-\dim(G)}X(v)$ are *closed* G-invariant manifolds. By Lemma 9.2, each of them contains an open subset of trajectories of the main orbit-type $[G]$. By a dimensional argument, $\partial^{\pm}_{n+1-\dim(G)}X(v)$ is a disjointed union of free trajectories. Hence the third bullet is valid.

By the third bullet, the boundary $\partial_{n+1-\dim(G)}X(v)$ of $\partial^{\pm}_{n-\dim(G)}X(v)$ is a disjoint union of free orbits. The topology of $\partial^{\pm}_{n-\dim(G)}X(v)$ must be special since the orbit-space $(\partial^{\pm}_{n-\dim(G)}X(v))/G$ is 1-dimensional.

When G is connected, the compact connected G-manifolds of dimension $\dim(G)+1$ and with the main orbit-type $[G]$ are very rigid: they all are either products $G \times S^1$, $G \times [0,1]$, or mapping cylinders $\mathcal{C}(G \to G/H)$, or unions of two mapping cylinders $\mathcal{C}(G \to G/H)$, $\mathcal{C}(G \to G/K)$ with H and K being diffeomorphic to a sphere [B], [K12]. The two mapping cylinders share the same "top" G. This conclusion is based on the fact that the only spheres among the Lie groups are: $\mathsf{O}(1) \approx \mathbb{Z}_2, \mathsf{SO}(2), \mathsf{SU}(2)$. However, the $\mathbb{Z}_2$-option must be excluded: it leads to a non-orientable manifold, while all $\partial_k X(v)$ are oriented for $k > 0$.

This leaves us with the models $G \times S^1$ and

$$\mathcal{C}(G \to G/H) \cup_G \mathcal{C}(G \to G/K),$$

where $H, K = \mathsf{SO}(2), \mathsf{SU}(2)$, for the components of $\partial_{n-1-\dim(G)}X$.

Similarly, the models for $\partial^{\pm}_{n-1-\dim(G)}X(v)$ are:

$$\mathcal{C}(G \to G/H) \cup_G \mathcal{C}(G \to G/K),\ \ \mathcal{C}(G \to G/H),\ G \times S^1, \text{ and } G \times [0,1],$$

where $H, K = \mathsf{SO}(2), \mathsf{SU}(2)$. □

Corollary 9.1. *Let X, v be as in Theorem 9.1.*

- *If $G = \mathsf{SO}(2)$, then $\partial_{n+1}X(v) = \emptyset$, each $\partial^{\pm}_n X(v)$ is a union of circles on which G acts freely, and each $\partial^{\pm}_{n-1}X$ is a union of tori $\mathsf{SO}(2) \times S^1$, cylinders $\mathsf{SO}(2) \times [0,1]$, 2-disks, and 2-spheres. The $\mathsf{SO}(2)$-action on each disk or sphere is semi-free.*
- *If G is finite, then, for some integer $\ell \geq 0$, $|\partial_{n+1}X(v)| = 2\ell \cdot |G|$, provided $|G|$ being odd, and $|\partial_{n+1}X(v)| = \ell \cdot |G|$, provided $|G|$ being even.*

Proof. Both claims are based on the fact that the Morse strata must contain free orbits $[G]$. The first case $G = \mathsf{SO}(2)$ follows instantly from Theorem 9.1.

Consider the case of a finite G. When $|G| = 1 \mod 2$, the parity condition, imposed on $|\partial_{n+1}X(v)|/|G|$ follows from the fact that $\partial_{n+1}X$ is a boundary of an equivariant 1-manifold $\partial_n^+ X$, a union of intervals and loops. Thus, $|\partial_{n+1}X| = 0 \mod 2$. □

Along the lines of Question 5.4, we are compelled to formulate its equivariant analogue:

Question 9.1. *For a given compact smooth G-manifold X of dimension $n+1$, such that the main orbit-type of its boundary is $[G]$, what is the minimal number of free trajectories that form the sets $\partial_{n+1-\mathsf{dim}(G)}X(v)$ and $\partial^{\pm}_{n+1-\mathsf{dim}(G)}X(v)$? The minimum is taken over the space of all boundary generic G-invariant vector fields v as in Theorem 9.1.* ◊

9.3 Equivariant Morse formula for vector fields

For a compact Lie group G and its subgroup H, we denote by (H) the conjugacy class of H in G.

Any compact G-manifold X gives rise to an element $\chi^G(X) \in A(G)$ of the Burnside Ring $A(G)$ (see Definition 9.1).

Let

$$ch : A(G) \longrightarrow \prod_{(H)\,\in\,\mathsf{conj}(G)} \mathbb{Z}$$

be a ring homomorphism, induced by the correspondence

$$\chi^G(X) \longrightarrow \{\chi(X^H)\}_{(H)\,\in\,\mathsf{conj}(G)},$$

where $\mathsf{conj}(G)$ denotes the set of conjugacy classes of compact subgroups in G, and $\chi(\sim)$ denotes the Euler number. It turns out that ch is a *monomorphism* [D]. We denote by ch_H the (H)-indexed component of the map ch.

Let NH denote the normalizer of H in G. We consider only the conjugacy classes (H) with the *finite* quotients $WH =_{\mathsf{def}} NH/H$ and denote by $\Phi(G)$ the set of such classes (H).

Next, we form a free abelian group $A'(G)$, generated by elements of $\Phi(G)$ (equivalently, by the orbit-types $\{G/H\}_{(H)\in\Phi(G)}$, regarded as G-spaces).

By [D], Theorem 1, the natural homomorphism $A'(G) \to A(G)$ is an *isomorphism*. As a result, any element $[\mathcal{X}] \in A(G)$ is detected by the collection $\{ch_H([\mathcal{X}])\}_{(H)\in\Phi(G)}$ alone.

Let U_Z be a G-invariant regular neighborhood[4] of the zero set

$$Z =_{\mathsf{def}} Z(v) = \{x \in X|\ v(x) = 0\}.$$

U_Z is chosen to be a smooth manifold with boundary. Assume that v is boundary generic with respect to $\partial_1 U_Z =_{\mathsf{def}} \partial U_Z$. Note that, for each $H \subset G$, the H-fixed point sets U_Z^H and Z^H are homotopy equivalent.

Inspired by [Mo], we propose the following definition.

Definition 9.2. For a smooth vector field v on a compact smooth n-dimensional manifold X, let

$$\begin{aligned} \mathsf{ind}^G(v) &=_{\mathsf{def}} \sum_{k=0}^{n} (-1)^k \chi^G(\partial_k^+ U_Z) \\ &= \chi^G(Z) + \sum_{k=1}^{n} (-1)^k \chi^G(\partial_k^+ U_Z). \end{aligned} \tag{9.1}$$

◇

Lemma 9.3. *For each* $(H) \in \mathsf{conj}(G)$,

$$ch_H(\mathsf{ind}^G(v)) = \mathsf{ind}(v|_{X^H}),$$

the non-equivariant index of the field v *on* X^H *(equivalently, on* U_Z^H*).*

Proof. By Lemma 9.1, v is tangent to $X^H \supset U_Z^H$ and $(\partial_k^+ U_Z)^H = \partial_k^+(U_Z^H)$. Since v is tangent to X^H, all the zeros of $v|_{X^H}$ are among the zeros of v on X, and no new zeros appear. By the Morse formula [Mo] and Lemma 9.1,

$$\mathsf{ind}(v|_{U_Z^H}) = \sum_{k=0}^{n} (-1)^k \chi(\partial_k^+(U_Z^H)) = \sum_{k=0}^{n} (-1)^k \chi(\partial_k^+(U_Z)^H).$$

Thus,

$$\begin{aligned} ch_H\big(\mathsf{ind}^G(v)\big) &= \sum_{k=0}^{n} (-1)^k ch_H(\partial_k^+ U_Z) \\ &= \mathsf{ind}\big(v|_{U_Z^H}\big) = \mathsf{ind}\big(v|_{X^H}\big). \end{aligned} \tag{9.2}$$

□

[4]In particular, Z is an equivariant deformation retract of U_Z.

For G-invariant vector fields, there is an alternative definition of the equivariant index. It is more involved and mimics the classical definition of index as the sum of local degrees, produced by the field v in the vicinity of its zero set. Let us describe this definition along the lines of [Lu], [LR].

The tangent space $T_{(x,0)}(TX)$ decomposes as a direct sum of the subspaces $T_xX \oplus T_xX$, where the first factor is thought as being tangent to the zero section $\epsilon : X \to TX$ and the second one to the fiber of the bundle $TX \to X$. Let $\alpha : T_xX \oplus T_xX \to T_{(x,0)}(TX)$ denote this G_x-equivariant isomorphism.

Consider a G-equivariant vector field v on X, viewed as a section $v : X \to TX$ of the tangent G-bundle TX. Assume that v is transversal to the zero section ϵ (due to the "doubling" nature[5] of the G_x-space $T_{(x,0)}(TX)$, this equivariant transversality of v and ϵ is available by a general position argument). If $v(x) = 0$ for some $x \in X$, then $T_{(x,0)}(TX)$ decomposes a direct sum of $Dv(T_xX)$ and $D\epsilon(T_xX)$ as well. Here Dv and $D\epsilon$ denote the differentials of the respective maps-sections. Consider the G_x-equivariant bundle isomorphism

$$\mathcal{D}_{x,v} : T_xX \xrightarrow{Dv} T_{(x,0)}(TX) \xrightarrow{\alpha^{-1}} T_xX \oplus T_xX \xrightarrow{pr_2} T_xX, \tag{9.3}$$

where pr_2 stands for the projection on the second factor.

$\mathcal{D}_{x,v}$ induces a G_x-map $\mathcal{D}^c_{x,v} : T_xX^c \to T_xX^c$ on the one point compactification T_xX^c of T_xX. For such a map, its *degree*,

$$Deg^{G_x}(\mathcal{D}^c_{x,v}) \in A'(G_x) \approx A(G_x),$$

is defined in terms of the equivariant *Lefschetz classes* (see (9.6))

$$\Lambda^{G_x}(\mathcal{D}^c_{x,v}), \ \ \Lambda^{G_x}(\mathsf{Id}_{T_xX^c}) \in A'(G_x),$$

by the formula

$$Deg^{G_x}(\mathcal{D}^c_{x,v}) = \left[\Lambda^{G_x}(\mathcal{D}^c_{x,v}) - 1\right]\left[\Lambda^{G_x}(\mathsf{Id}_{T_xX^c}) - 1\right] \tag{9.4}$$

(see [Lu], [LR] for the relevant definitions and facts), where $1 \in A(G_x)$ stands for the class of a point.

In Section 9.5, we will return briefly to the definition of an equivariant degree $Deg^G(f)$ for a general G-map $f : X \to Y$ of two G-manifolds.

Meanwhile, let us recall the definition of the equivariant Lefschetz class $\Lambda^G(f) \in A'(G)$ of a G-map $f : X \to X$, where X is a finite G-CW-complex[6]. In particular, for each $(H) \in \Phi(G)$, the orbit-space ${}_{WH}\backslash X^H$ has the structure of a finite CW-complex.

[5]It satisfies the so called *gap hypotheses* [Br].

[6]For example, see [O] for the definition of a G-CW-complex.

We denote by $X^{>H}$ the locus of points whose stationary groups contain H and differ from it. We consider the finitely-generated free $\mathbb{Z}[WH]$-module $C_p(X^H, X^{>H}; \mathbb{Z})$ of the relative cellular p-chains of the pair $(X^H, X^{>H})$ and the f-induced automorphism $\mathcal{A}_p^H(f)$ of the module $C_p(X^H, X^{>H}; \mathbb{Z})$.

Let

$$\Lambda^G(f) =_{\mathsf{def}} \sum_{(H)\in\Phi(G)} L^{\mathbb{Z}[WH]}(f^H, f^{>H}) \cdot \chi^G(G/H), \tag{9.5}$$

where $\chi^G(G/H)$ is the Burnside class of the homogeneous space G/H in $A'(G)$, and the integer-valued Lefschetz trace is defined by the formula

$$L^{\mathbb{Z}[WH]}(f^H, f^{>H}) =_{\mathsf{def}} \sum_{p\geq 0} (-1)^p \cdot tr_{\mathbb{Z}[WH]}\big(\mathcal{A}_p^H(f)\big). \tag{9.6}$$

In (9.6), we employ the $\mathbb{Z}$-valued trace $tr_{\mathbb{Z}[WH]}$ of the f-induced automorphism $\mathcal{A}_p^H(f)$ (see [LR] for details). This trace is obtained by taking the trace (with the values in $\mathbb{Z}[WH]$) of the $\mathbb{Z}[WH]$-matrix $\mathcal{A}_p^H(f)$ and then applying the augmentation homomorphism $\epsilon : \mathbb{Z}[WH] \to \mathbb{Z}$ to it. Recall that ϵ takes any element $\sum_\kappa n_\kappa \cdot g_\kappa \in \mathbb{Z}[WH]$ to the integer $\sum_\kappa n_\kappa$.

Lemma 9.4. *The equivariant Lefschetz class $\Lambda^G(f)$ from (9.5) is detected by the non-equivariant Lefschetz numbers $\{\Lambda(f^H) \in \mathbb{Z}\}_{(H)\in\Phi(G)}$.*

Proof. By [D], Theorem 2, only the orbit-types G/H with WH being finite contribute to the homomorphism $ch : A(G) \to \prod_{(H)} \mathbb{Z}$. Therefore, we will consider only the orbit-types from $\Phi(G)$.

In (9.5), we took into account only the orbit-types $(H) \in \Phi(G)$ with non-vanishing coefficients $L^{\mathbb{Z}[WH]}(f^H, f^{>H})$ (defined by (9.6)). Among them, pick a minimal[7] orbit-type (K) and apply ch_K to $\Lambda^G(f)$. Note that

[7] i.e. a *maximal* subgroup $K \subset G$

$(G/H)^K = \emptyset$ for all $H \subset K$, $H \neq K$. Therefore,

$$\begin{aligned} ch_K(\Lambda^G(f)) &= \sum_{(H)\in\Phi(G)} \Big\{\sum_{p\geq 0} (-1)^p\, tr_{\mathbb{Z}[WH]}\big(\mathcal{A}_p^H(f)\big)\Big\} \cdot \chi\big((G/H)^K\big) \\ &= \Big\{\sum_{p\geq 0} (-1)^p\, tr_{\mathbb{Z}[WK]}\big(\mathcal{A}_p^K(f)\big)\Big\} \cdot \chi\big((G/K)^K\big) \\ &= \Big\{\sum_{p\geq 0} (-1)^p\, tr_{\mathbb{Z}[WK]}\big(\mathcal{A}_p^K(f)\big)\Big\} \cdot |WK| \\ &= \sum_{p\geq 0} (-1)^p\, tr_{\mathbb{Z}}\big(\mathcal{A}_p^K(f)\big) \\ &=_{\mathsf{def}} \Lambda(f^K). \end{aligned} \tag{9.7}$$

In other words, for such a minimal $(K) \in \Phi(G)$, we get

$$L^{\mathbb{Z}[WK]}(f^K, f^{>K}) = |WK|^{-1}\Lambda(f^K) = |WK|^{-1} ch_K(\Lambda^G(f)).$$

The rest of the argument is performed inductively, the induction step being applied to a minimal orbit-type (K') in $\Phi(G) \setminus (K)$ with a nonzero coefficient $L^{\mathbb{Z}[WK']}(f^{K'}, f^{>K'})$. Indeed, we compute $ch_{K'}$ of

$$\beta =_{\mathsf{def}} \Lambda^G(f) - L^{\mathbb{Z}[WK]}(f^K, f^{>K}) \cdot \chi^G(G/K)$$

to conclude, in a similar way, that $ch_{K'}(\beta)$ equals $\Lambda(f^{K'})$. □

Corollary 9.2. *The equivariant degree $Deg^{G_x}(\mathcal{D}^c_{x,v}) \in A(G_x)$ is determined by the $\mathbb{Z}$-valued degrees $\{ Deg(\mathcal{D}^{H,c}_{x,v}) \}_{(H)\in\Phi(G_x)}$ of the spherical maps*

$$\mathcal{D}^{H,c}_{x,v} =_{\mathsf{def}} \mathcal{D}^c_{x,v}| : (T_xX^c)^H \to (T_xX^c)^H.$$

For all subgroups $H \subset G_x$, where $H \notin \Phi(G_x)$, we have

$$ch_H[Deg^{G_x}(\mathcal{D}^c_{x,v})] = 0.$$

Proof. We apply the arguments in (9.7) of Lemma 9.4 (with G being replaced by G_x) to formula (9.4). □

Assume that the zero set $Z(v)$ is a finite union of G-orbits. Now define

$$\mathsf{Ind}^G(v) =_{\mathsf{def}} \sum_{Gx \in Z(v)/G} G \times_{G_x} [Deg^{G_x}(\mathcal{D}^c_{x,v})], \tag{9.8}$$

the summation being executed over the set of distinct G-orbits through the zeros of the vector field v.

Lemma 9.5. $\mathsf{Ind}^G(v)$, *defined by (9.8), and* $\mathsf{ind}^G(v)$, *defined by (9.1), are equal.*

Proof. Let $Z(v)$ stand for the zero locus of v. For each x from the orbit-space ${}_G\backslash Z(v)$, we consider a small G_x-equivariant n-disk $D_x \subset X$, centered on x. We notice that, when $Z(v)$ is discrete, the quotient G/G_x must be finite.

Evidently, $\partial_k^+ U_Z \approx \coprod_x G \times_{G_x} (\partial_k^+ D_x)$, where the summation is carried over the set of $x \in Z(v)$, representing the distinct orbits Gx. Therefore, in order to prove $\mathsf{Ind}^G(v) = \mathsf{ind}^G(v)$, it will suffice to verify that, for each orbit $Gx \in Z(v)/G$,

$$Deg^{G_x}(\mathcal{D}^c_{x,v}) = \sum_k (-1)^k \chi^{G_x}(\partial_k^+ D_x)$$

in $A'(G_x)$. Using the validity of non-equvariant Morse formulas as in Corollary 5.1, the last equality follows from Corollary 9.2 (with G being replaced by $H \subset G_x$ and X by D_x). □

Theorem 9.2. *Let G be a compact Lie group, and let X[8] be a compact orientable smooth n-dimensional G-manifold with boundary $\partial_1 X$. Let v be a boundary generic G-invariant vector field on X, whose zero locus is a finite union of G-orbits. Then*

- *the equivariant Morse formula*

$$\mathsf{Ind}^G(v) = \mathsf{ind}^G(v) = \sum_{k=0}^{n} (-1)^k \chi^G(\partial_k^+ X(v)) \tag{9.9}$$

  is valid in the ring $A(G)$.
- *If the main orbit-type of the G-action on $\partial_1 X$ is $[G]$, then* $\dim(G) < n$ *and (9.9) reduces to*

$$\mathsf{Ind}^G(v) = \mathsf{ind}^G(v) = \sum_{k=0}^{n-\dim(G)} (-1)^k \chi^G(\partial_k^+ X(v)). \tag{9.10}$$

- *When a connected G of a positive dimension acts on $\partial_1 X$ by orientation-preserving transformations, and the main orbit-type of this action is $[G]$, then the contribution of the minimal stratum $\partial^+_{n-\dim(G)} X(v)$ to $\mathsf{Ind}^G(v)$ (see (9.9)) is trivial, and the contribution of the stratum $\partial^+_{n-1-\dim(G)} X(v)$ is quite special:*

$$\chi^G(\partial^+_{n-1-\dim(G)} X(v)) = \sum_{\{(H) \in \Phi(G) \mid H \approx \mathsf{SO}(2),\ or\ \mathsf{SU}(2)\}} n_H \cdot \chi^G(G/H),$$

[8] also denoted by $\partial_0^+ X$

where $\{n_H\}$ are some non–negative integers. In particular, if a connected Lie group G of a positive dimension is such that, for each $H \subset G$ that is isomorphic to $\mathsf{SO}(2)$ or to $\mathsf{SU}(2)$, the group WH is infinite, then the contribution of the stratum $\partial^+_{n-1-\dim(G)}X(v)$ vanishes as well.

Proof. To simplify the notations, we drop the dependence of Morse stratifications on v.

By Lemma 9.4, $\mathsf{Ind}^G(v) = \mathsf{ind}^G(v)$. Since the elements of $A(G)$ are detected by the character map $ch : A(G) \to \prod_{(H)\in\Phi(G)} \mathbb{Z}$, it will suffice to check the validity of (9.9) by applying ch to both sides of the conjectured equation

$$\mathsf{ind}^G(v) = \sum_{k=0}^{n}(-1)^k\chi^G(\partial_k^+ X).$$

By Lemma 9.1, for any $H \subset G$,

$$ch_H\big(\chi^G(\partial_k^+ X)\big) =_{\mathsf{def}} \chi((\partial_k^+ X)^H) = \chi(\partial_k^+(X^H)).$$

Thus,

$$ch_H\Big(\sum_{k=0}^{n}(-1)^k\chi^G(\partial_k^+ X)\Big) = \sum_{k=0}^{n}(-1)^k\chi(\partial_k^+(X^H)). \tag{9.11}$$

On the other hand, by Lemma 9.1 and formula (9.2), $ch_H(\mathsf{ind}^G(v)) = \mathsf{ind}(v|_{X^H})$. Since the Morse formula claims that

$$\mathsf{ind}(v|_{X^H}) = \sum_{k=0}^{n}(-1)^k\chi(\partial_k^+(X^H)),$$

we get the desired equality

$$ch_H(\mathsf{ind}^G(v)) = ch_H\Big(\sum_{k=0}^{n}(-1)^k\chi^G(\partial_k^+ X)\Big)$$

for all $(H) \in conj(G)$.

In order to derive formula (9.10) from formula (9.9), we employ Theorem 9.1 to conclude that $\partial_k^+ X = \emptyset$ for all $k > n-\dim(G)$. Moreover, $\partial^+_{n-\dim(G)}X$ is the union of free G-orbits, and thus $\chi^G(\partial^+_{n-\dim(G)}X) = 0$, provided $\dim(G) > 0$.

The following observation validates the last claim of the theorem. Models (1) and (2) from Theorem 9.1 have zero classes in $A(G)$, model (3) is

G-homotopy equivalent to G/H, and model (4), by the additivity of Euler characteristics, produces the element $\chi^G(G/H) + \chi^G(G/K)$.

If a connected G is such that, for each $H \subset G$ that is isomorphic to $\mathsf{SO}(2)$ or $\mathsf{SU}(2)$, the group WH is infinite, then the space G/H admits a *free* S^1-action, where $S^1 \subset WH$. Thus, $\chi^G(G/H) = 0$. As a result, for such a G, we get a simplification of (9.10):

$$\mathsf{Ind}^G(v) = \sum_{k=0}^{n-2-\dim(G)} (-1)^k \chi^G(\partial_k^+ X). \tag{9.12}$$

□

Corollary 9.3. *Let X and v be as in Theorem 9.1. We denote by v_1^+ an orthogonal projection (with respect to a G-invariant metric on X) of the vector field $v|_{\partial_1^+ X}$ on the tangent bundle $T(\partial_1^+ X)$. We assume that the zero locus of v_1^+ consists of finitely many G-trajectories. Then the following formula holds in $A(G)$:*

$$\chi^G(X) = \mathsf{ind}^G(v) + \mathsf{ind}^G(v_1^+). \tag{9.13}$$

Proof. In view of the recursive nature of the Morse stratification $\{\partial_k^+ X\}$, (9.9) and (9.13), applied to all consecutive terms in the stratification, are equivalent formulas. □

9.4 Polynomial vector fields in polynomial domains

Now, we turn our attention to polynomial vector fields v in compact domains in $\mathbb{R}^n$ that are defined by polynomial inequalities.

Let us consider a real n-dimensional vector space and a polynomial vector field v in it with polynomial components $\{P_i\}_{1 \leq i \leq n}$. Assume that $\deg(P_i) \leq m_i$. We denote by $\tilde{P}_i$ the homogenized versions of P_i so that $\tilde{P}_i(1, x_1, \ldots, x_n) = P_i(x_1, \ldots, x_n)$.

Definition 9.3. Consider a polynomial $Q : \mathbb{R}^n \to \mathbb{R}$ of degree d and a polynomial vector field v in $\mathbb{R}^n$. We say that the pair (v, Q) is *non-degenerated*, if:

(1) all zeros of v are simple,

(2) the system $\{x_0 = 0, \ \tilde{P}_i(x_0, x_1, \dots, x_n) = 0\}_{1 \le i \le n}$ has only the trivial solution[9],

(3) the hypersurface $\{Q = 0\}$ does not contain the zeros of v. ◊

For any non-increasing sequence $m_1, m_2, \dots, m_n$ of natural numbers, we form the parallelepiped $\Pi(m_1, \dots, m_n)$ in $\mathbb{R}^n$, defined by the inequalities

$$\{0 \le x_i \le m_i - 1\}_{1 \le i \le n}. \tag{9.14}$$

Let $O(d, m_1, \dots, m_n)$ be the number of *integral lattice points* $(x_1, \dots, x_n)$ in $\Pi(m_1, \dots m_n)$, subject to the additional inequalities

$$\begin{aligned} \frac{1}{2}(m_1 + \dots + m_n - d - n) \ &\le \ x_1 + \dots + x_n \\ &\le \ \frac{1}{2}(m_1 + \dots + m_n - n). \end{aligned} \tag{9.15}$$

We consider the domain

$$X_Q =_{\mathsf{def}} \{w \in \mathbb{R}^n | \ Q(w) \ge 0\}.$$

Then, according to Khovanskii (Theorem 1, [Kh]), for a non-degenerated pair (v, Q) as in Definition 9.3, $|\mathsf{ind}(v)|$—the absolute value of the index of the vector field v in X_Q— is bounded from above by the integer $O(d, m_1, \dots, m_n)$. Moreover, this estimate is sharp!

Combining the Khovanskii Theorem and the Morse Formula from Theorem 5.3, we get the following result.

Theorem 9.3. *Consider a polynomial vector field $v = (P_1, \dots, P_n)$ in $\mathbb{R}^n$, where $\deg(P_i) \le m_i$. Assume that the pair (Q, v) is generic in the sense of Definition 9.3 and that the compact domain $X_Q \subset \mathbb{R}^n$ has a smooth boundary. Assume also that v is boundary generic with respect to ∂X_Q* [10].

Then

$$\left| \sum_{k=0}^{n} (-1)^k \chi(\partial_k^+ X_Q) \right| \ \le \ O(d, m_1, \dots, m_n). \tag{9.16}$$

Moreover, there exists a pair (Q, v) for which the inequality (9.16) *can be replaced by the equality.* ◊

Let $V \subset \mathbb{R}^n$ be a vector subspace of dimension l. Let $d_V \le d$ denote the degree of the polynomial Q, being restricted to the subspace V. Consider a subspace $U \subset \mathbb{R}^n$ that is spanned by some set of l basic vectors $\{e_{j_1}, \dots e_{j_l}\}$

[9] I.e. no zero of v escapes to infinity.

[10] We suspect that this requirement can be dropped.

in $\mathbb{R}^n$ and such that the obvious orthogonal projection $p_U : \mathbb{R}^n \to U$, being restricted to V, is onto. For a given V, we denote by $\mathcal{U}(V)$ the finite set of such U's.

Put

$$O(V; d_V, m_1, \dots, m_n) =_{\mathsf{def}} \min_{U \in \mathcal{U}(V)} \{O(d_V, m_{j_1}, \dots, m_{j_l})\}, \quad (9.17)$$

where $U =_{\mathsf{def}} span\{e_{j_1}, \dots, e_{j_l}\}$ and $O(d_V, m_{j_1}, \dots, m_{j_l})$ is defined as in (9.15).

We consider a real n-dimensional vector space W and a G-representation $\Psi : G \to \mathsf{GL}_{\mathbb{R}}(W)$. We denote by $\mathcal{P}_\Psi$ the algebra of Ψ-invariant polynomials on W. Let v be a G-invariant (or rather $\Psi(G)$-invariant) vector field on $W \approx \mathbb{R}^n$. It has polynomial components $\{P_i \in \mathcal{P}_\Psi\}_{1 \leq i \leq n}$. Assume that $\deg(P_i) \leq m_i$.

The theorem below generalizes the estimates (9.16) for the equivariant setting.

Theorem 9.4. *Let G be a compact Lie group. Pick a sequence*

$$m_1 \geq m_2 \geq \dots \geq m_n > 0$$

of integers. Consider a representation $\Psi : G \to \mathsf{GL}(n, \mathbb{R})$, a $\Psi(G)$- invariant polynomial $Q(x_1, \dots, x_n)$ of degree d, and a non-degenerate $\Psi(G)$-invariant polynomial field

$$v = (P_1(x_1, \dots, x_n), \dots, P_n(x_1, \dots, x_n))$$

such that $\deg(P_i) \leq m_i$. *Assume that $X_Q =_{\mathsf{def}} \{\vec{x} \in \mathbb{R}^n |\; Q(\vec{x}) \geq 0\}$ is a compact*[11] *domain with a smooth boundary $\partial_1 X_Q$ and that v is boundary generic vector field with respect to $\partial_1 X_Q$.*

Then the image $\prod_{(H) \in \Phi(G)} z_{(H)}$ of the element

$$\sum_{k=0}^{n} (-1)^k \chi(\partial_k^+ X_Q) \in A(G)$$

under the character monomorphism $ch : A(G) \to \prod_{(H)\in\Phi(G)} \mathbb{Z}$ belongs to the parallelepiped $\mathcal{P}_{\Phi(G)}$, defined by the inequalities:

$$\Big\{ |z_{(H)}| \leq O\big((\mathbb{R}^n)^H;\ d_{(\mathbb{R}^n)^H},\ m_1, \dots, m_n\big) \Big\}_{(H)\in\Phi(G)}, \quad (9.18)$$

where the number $O\big((\mathbb{R}^n)^H;\ d_{(\mathbb{R}^n)^H},\ m_1, \dots, m_n\big)$ was introduced in (9.17).

[11]For example, take Q being equal to $a_1 x_1^d + \dots + a_n x_n^d \in \mathcal{P}_\Psi$, where all $a_i < 0$, plus a lower degree polynomial.

Proof. Let $l = \dim((\mathbb{R}^n)^H)$. For each $U = span\{e_{j_1}, \dots, e_{j_l}\}$ as above with the property $p_U : (\mathbb{R}^n)^H \to U$ being onto, the projection p_U induces an invertible linear transformation of the pairs $(v,\ X_Q^H)$ and $(p_U(v),\ p_U(X_Q^H))$, where $X_Q^H = X_Q \cap (\mathbb{R}^n)^H$.

Let I be the set of indices, complementary to the set $J =_{\mathsf{def}} \{j_1, \dots, j_l\}$. Then the components $\tilde{P}_j$ of $p_U(v)$ are obtained by substituting $\{x_i = 0\}_{i \in I}$ into $\{P_j(x_1, \dots, x_n)\}_{j \in J}$. Note that $\deg(\tilde{P}_j) \leq \deg(P_j) \leq m_j$. Also, $p_U(X_Q^H) \subset Span\{e_j\}_{j \in J}$ is defined by a polynomial inequality $\tilde{Q}(x_{j_1}, \dots, x_{j_l}) \geq 0$ that is obtained from $Q(x_1, \dots, x_n) \geq 0$ by a substitution that expresses each x_i, $i \in I$, as linear combination of the $\{x_j\}_{j \in J}$. Evidently, $\deg(\tilde{Q}) = d_{(\mathbb{R}^n)^H} \leq d$. Since $v|_{(\mathbb{R}^n)^H}$ is parallel to $(\mathbb{R}^n)^H$, p_U also maps $\partial_k^+(X_Q^H)$ onto $\partial_k^+[p_U(X_Q^H)]$.

By [Kh] and Theorem 9.2, $|\mathsf{Ind}(p_U(v))|$ in $p_U(X_Q^H)$ (equivalently,

$$\Big|\sum_{k=0}^{n} (-1)^k \chi(\partial_k^+ p_U(X_Q))\Big|\ )$$

has an upper boundary $O(d_{(\mathbb{R}^n)^H}, m_{j_1}, \dots, m_{j_l})$. Via the linear diffeomorphism p_U, $\mathsf{Ind}(p_U(v))$ in $\{\tilde{Q} \geq 0\}$ and $\mathsf{Ind}(v|_{X_Q^H})$ in X_Q^H are equal. Thus,

$$|\mathsf{Ind}(p_U(v))| \leq O(d_{(\mathbb{R}^n)^H}, m_{j_1}, \dots, m_{j_l}).$$

In view of formula (9.16) and Lemma 9.3,

$$|ch_H(\mathsf{Ind}^G(v))| = |\mathsf{Ind}(v|_{X_Q^H})| \leq O(d_{(\mathbb{R}^n)^H}, m_{j_1}, \dots, m_{j_l}),$$

and thus, using formula-definition (9.17), we get

$$|ch_H(\mathsf{Ind}^G(v))| \leq O\big((\mathbb{R}^n)^H; d_{(\mathbb{R}^n)^H}, m_1, \dots, m_n\big).$$

□

Corollary 9.4. *Let $Q_r =_{\mathsf{def}} r^2 - x_1^2 - \cdots - x_n^2$ and let $B_r = \{Q_r \geq 0\}$. Let $\Phi : G \to \mathsf{O}(n, \mathbb{R})$ be an orthogonal representation. Put $d_H = \dim(\mathbb{R}^n)^H$. Assume that a polynomial vector field v is as in Theorem 9.4 with respect to some ball $B_{r_\star}$. Then, for each class $(H) \in \Phi(G)$,*

$$\big|ch_H(\mathsf{Ind}^G(v|_{B_{r_\star}}))\big| = \Big|\sum_{k=0}^{d_H} (-1)^k \chi(\partial_k^+ B_{r_\star}^H)\Big| \leq O\big((\mathbb{R}^n)^H; 2, m_1, \dots, m_n\big).$$

◊

9.5 Gottlieb's and Gauss-Bonnet's equivariant formulas

Our next goal is to reinterpret, within an equivariant setting, Gottlieb's formulas [Got1] for indices of pull-back vector fields .

Let us recall the notion of a *pull-back* field (see [G1]). Let $F : X \to Y$ be a differentiable map of two n-dimensional Riemannian manifolds, and w a vector field on Y. Let F^*w be a field on X defined by the formula

$$\langle F^*w,\, u\rangle_X \;=\; \langle w,\, DF(u)\rangle_Y,$$

where $u(x) \in T_xX$ is a generic vector, and $\langle\ \sim, \sim\ \rangle$ denotes the scalar product in the appropriate tangent space. In other words, if ω is a 1-form, dual to w in Y, then F^*w is dual to $F^*\omega$ in X.

In the case of a gradient field $w = \nabla f$ on Y, where $f : Y \to \mathbb{R}$ is a smooth function, $F^*w = \nabla(f \circ F)$.

We notice that, when F is an equivariant map, both metrics on X and Y are G-invariant, and w is an invariant field, then F^*w is invariant as well.

We recall the notion of an *equivariant degree* $Deg^G(F) \in A(G)$ of a G-map $F : X \to Y$ between two compact G-manifolds of the same dimension (cf. [Lu]). Crudely, it is an element of $\prod_{(H)\in conj(G)} \mathbb{Z}$ whose (H)-component is the usual $\deg(F^H)$, where $F^H : X^H \to Y^H$. In fact, such an element $Deg^G(F) \in A(G)$. This naive construction runs into some complications because of the ambiguities in choosing orientations of fixed point components, both in the source and the target. One wants to coordinate, in a sense, the orientations of X^H and Y^H (when they are of the same dimension). The issues with the coherent orientations can be resolved by introducing some additional synchronizing structure, called in [Lu] "the $O(G)$-transformation of the fiber transports". This structure assigns to each $x \in X$ a transfer G-map $(T_{F(x)}Y)^c \to (T_xX)^c$, where the upper index c denotes the one point compactification.

Fortunately, we need to employ $Deg^G(F)$ in a particular situation, where the synchronization of the orientations can be achieved by pedestrian means. Consider an equivariant immersion f of a closed oriented $(n-1)$-dimensional G-manifold Z into a real n-dimensional space V of an orthogonal G-representation. We denote by $S(V)$ the unit sphere, centered on the origin. To each $x \in Z$ we assign the unit vector $n(x)$, tangent to V at $f(x)$ and normal to the f-image of a small neighborhood $U_x \subset Z$ of x. The orientations of V and Z help to resolve the two-fold ambiguity in picking $n(x)$.

Since f is a G-immersion, for any $H \subset G$, we get $Z^H = f^{-1}(f(Z) \cap V^H)$. Moreover, because $n(x)$ is orthogonal to $f(U_x)$ at $f(x) \in V^H$, $n(x)$ must be parallel to V^H. Indeed, if $n(x)$ would have a nontrivial component ν_x normal to V^H, ν_x must be moved by elements of $H \setminus \{1\}$; on the other hand, $n(x)$ is H-invariant since $T_{f(x)}(U_x)$ is.

In fact, $f^H : Z^H \to V^H$ is an immersion as well. Therefore an orientation of V^H, with the help of the vector $n(x)$ (we assume that $x \in Z^H \setminus Sing(f^H)$), picks a particular orientation of Z^H. In the following, we assume that the orientations of V^H and Z^H are always synchronized in this way.

Thus the Gaussian map $\gamma^H : Z^H \to S(V^H)$ is well-defined for any conjugacy class (H) that occurs as a stationary group of the G-space Z. Unless $\dim(V^H) \leq 1$, $S(V^H)$ is connected. In such case, the degree $deg(\gamma^H)$ is defined as a sum of degrees of maps $\{\gamma^H_\alpha : Z^H_\alpha \to S(V^H)\}_\alpha$, where Z^H_α denotes a typical connected component of Z^H. When $\dim(V^H) = 1$, $S(V^H) = a \coprod b$, and $deg(\gamma^H)$ is defined as the sum of degrees of the two obvious maps with the singleton targets a and b.

The theorem below is an equivariant version of the "Topological Gauss-Bonnet's Theorem" from [Got], page 466 (cf. Theorem 5.4).

Theorem 9.5. *Let V be a real vector space of dimension n on which a compact Lie group G acts orthogonally. We assume that V admits a G-invariant non-vanishing vector field w*[12]*. Let X a compact smooth n-dimensional G-manifold with an oriented boundary $\partial_1 X$. We consider a smooth G-equivariant map $F : X \to V$ whose Jacobian is nonzero along $\partial_1 X$.*

Let $v =_{\mathsf{def}} F^* w$ be the the pull-back of the vector field w under F. We denote by $\{\partial_k^+ X(v)\}_{0 \leq k \leq n}$ the Morse stratification of X, induced by v.

Then the degree of the Gauss map $\gamma : \partial_1 X \to S(V)$ with values in the Burnside ring $A(G)$ can be computed by the formula

$$Deg^G(\gamma) = \chi^G(X) - \mathsf{Ind}^G(v) = -\sum_{k=1}^{n} (-1)^k \chi^G(\partial_k^+ X(v)). \quad (9.19)$$

Hence, $\mathsf{Ind}^G(v)$ and the RHS of (9.19) *are w-independent*[13]*.*

[12]For example, such $w \neq 0$ exists when $V = U \oplus \mathbb{R}^1$, where U is the space of a G-representation: just put $w = \nabla(f)$, where $f : U \oplus \mathbb{R}^1 \to \mathbb{R}^1$ is the obvious projection.

[13]Note that (9.19) proves that $Deg^G(\gamma)$, a priori just an element of $\prod_{(H) \in \mathsf{conj}(G)} \mathbb{Z}$, actually belongs to $A(G)$.

Proof. Since $DF|_{\partial_1 X}$ is of the maximal rank, we can pull-back the G-invariant Riemannian metric g in V to an equivariant collar of $\partial_1 X \subset X$ and then, by an averaging procedure, extend the pull-back $F^*(\mathsf{g})$ to an invariant metric on X. Let n be the unitary field, outward normal to $\partial_1 X$. Then the Gauss map $\gamma : x \to DF(n(x)) \in S^{n-1}$, where $x \in \partial_1 X$, is well-defined, equivariant, and helps to pick coherent orientations of components in $\partial_1(X^H)$. We have noticed already that $n(x)$ is contained in $T_x(X^H)$ and is normal to $\partial_1(X^H)$, provided $x \in X^H$. Also, for $x \in X^H$, $v(x) \in T_x(X^H)$. We can apply the non-equivariant Gottlieb's formula to each Gaussian map $\gamma^H : \partial_1(X^H) \to S(V^H)$ to conclude that

$$deg(\gamma^H) = \chi(X^H) - \mathsf{Ind}(v|_{X^H}) = - \sum_{k=1}^{\dim(V^H)} (-1)^k \chi(\partial_k^+(X^H)).$$

By Lemma 9.1, the latter sum is the (H)-component ch_H of

$$- \sum_{k=1}^{n} (-1)^k \chi^G(\partial_k^+(X)) \in A(G).$$

Finally, $\{deg(\gamma^H)\}_{(H)\in\Phi(G)}$ detect the element $Deg^G(\gamma)$. Along the way, we have shown that $Deg^G(\gamma) \in A(G)$.

We notice that the degree of the Gauss map γ depends on F, but not on the vector field w on V. Therefore the RHS of 9.19 is w-independent! □

Remark 9.1. Formula (9.19) tells us that we can *define* $Deg^G(\gamma) \in A(G)$ by the formula

$$- \sum_{k=1}^{n} (-1)^k \chi^G(\partial_k^+ X(F^*(w))) \tag{9.20}$$

for any choice of an invariant non-vanishing vector field w in V, thus a priori avoiding all the troubles with the orientations.

On the other hand, this definition of $Deg^G(\gamma)$ makes perfect sense for any equivariant immersion $f : Z \to V$ of a closed oriented G-manifold Z of codimension one: just define Z^+ to be the locus $\{x \in Z |\ \langle n(x), w(x) \rangle \leq 0\}$, and then proceed as in (5.1).

We conjecture that, in general, $\sum_{k=0}^{n-1} (-1)^k \chi^G(\partial_k^+ Z)$ is w-independent. By Theorem 9.5, to prove this conjecture will suffice to construct an equivariant coboundary X for Z and to extend $f : Z \to V$ to a G-map $F : X \to V$. ◇

Theorem 9.5 leads to:

Corollary 9.5. *Let* g *denotes a G-invariant Riemannian metric on V. Under notations and hypotheses of Theorem 9.5, we get*

$$\int_{\partial_1 X^H} K_H \, d\mu_H \; = \; - \sum_{k=1}^{\dim(V^H)} (-1)^k \chi\big(\partial_k^+ X^H(v)\big). \tag{9.21}$$

Here we are employing the pull-back $v =_{\mathsf{def}} F^* w$ of the equivariant vector field $w \neq 0$ on V under the map $F : X \to V$ and the pull-back metric $F^*(\mathsf{g})$ in vicinity of $\partial_1 X \subset X$ to generate the volume form $d\mu_H$ on $\partial_1 X^H$ and K_H, the normal in X^H curvature of $\partial_1 X^H$. ♢

9.6 Speculations about differently flavored Burnside rings

We end this chapter with few speculations.

One can refine the definition Birnside ring $A(G)$ in a number of ways. For example, following Tom Dieck [D], I.10.3, one can introduce the *component category* $\Pi_0(G, X)$ of a G-space X, whose objects are G-maps $x : G/H \to X$. A morphism σ from $x : G/H \to X$ to $y : G/K \to X$ is a G-map $\sigma : G/H \to G/K$ such that x and $y \circ \sigma$ are G-homotopic.

We denote by $\mathsf{Is}\,\Pi_0(G, X)$ the set of isomorphism classes $[x]$ of objects $x : G/H \to X$. Let

$$\mathcal{A}^G(X) =_{\mathsf{def}} \mathbb{Z}\big[\mathsf{Is}\,\Pi_0(G, X)\big],$$

where $\mathbb{Z}[S]$ stands for a free abelian group with the basis S. In fact, there is a bijection

$$\mathsf{Is}\,\Pi_0(G, X) \to \coprod_{(H) \in conj(G)} {}_{WH}\backslash \pi_0(X^H).$$

This construction defines a covariant functor $\mathcal{A}^G(\sim)$ on the category of G-spaces X with values in the category of abelian groups, a functor which is sensitive to the connected component structure of the sets $\{X^H\}$. We conjecture that all previous results can be restated and refined in terms of $\mathcal{A}^G(\sim)$ along the lines of [LR].

However, we would like to propose now a different generalization of the Burnside Ring $A(G)$ and to speculate about the corresponding Morse Formulas for equivariant vector fields. In this generalization, one pays a close attention to the H-representations $\{\psi_\alpha^H\}_\alpha$ arising in the normal bundles $\nu(X^H, X)$.

Let G be a compact Lie group. For each $(H) \in \Phi(G)$, let us fix a set $\Psi(H)$ of distinct isomorphism classes of representations $\{\psi_\alpha : H \to \mathsf{GL}(V_\alpha)\}_\alpha$ with the property $V_\alpha^{\psi_\alpha(H)} = \{0\}$. Moreover, for any $\psi_\alpha \in \Psi(H)$ and $K \subset H$, the representation

$$\overline{Res}_K(\psi_\alpha) : K \to \mathsf{GL}\big(V_\alpha / V_\alpha^{\psi_\alpha(K)}\big),$$

generated by $\psi_\alpha : H \to \mathsf{GL}(V_\alpha)$, is required to be in $\Psi(K)$. We call a collection

$$\mathcal{F} =_{\mathsf{def}} \{\psi_\alpha\}_{(H)\in\Phi(G),\ \alpha\in\Psi(H)}$$

of such representations a *normal family.*

Let $X^H_{\psi_a}$ denote the set of points in X^H with the normal H-representations isomorphic to ψ_α.

Definition 9.4. Let $\mathcal{F} = \mathcal{F}(G)$ be a normal family of representations. Two compact smooth G-manifolds X and Y are $\mathcal{F}$-equivalent if, for each conjugacy class $(H) \in \Phi(G)$ and each representation $\psi_a \in \Psi(H) \subset \mathcal{F}(G)$, we have

$$\chi(X^H_{\psi_a}) = \chi(Y^H_{\psi_a}).$$

We denote by $A(G,\, \mathcal{F})$ the group of such equivalence classes. ◊

Conjecture 9.1. *All the equivariant Morse formulas above are valid in the refined Burnside group* $A(G,\, \mathcal{F})$. ◊

Bibliography

[A] Alpert, H., *Using Simplicial Volume to Count Maximally Broken Morse Trajectories*, Geometry and Topology 20 (2016), 2997-3018.

[AB] Alexander, S.B. and Bishop, R.L., *Spines and homology of thin Riemannian manifolds with boundary*, Advances in Math 155, 23-48 (2000).

[AB1] Alexander, S.B. and Bishop, R.L., *Spines and Topology of Thin Riemannian Manifolds*, Transactions of AMS, vol. 355, no. 12, (2003) 4933-4954.

[AK] Alpert, H., Katz, G., *Using Simplicial Volume to Count Multi-tangent Trajectories of Traversing Vector Fields*, Geometriae Dedicata, 180 (1), 323–338, 2015

[AkKi] Akbulut, S., Kirby, R., *Mazur manifolds*. Michigan Math. J. 26 (3): 259284 (1979).

[Ar] Arnold, V.I., *Topological Invariants of Plane Curves and Caustics*, University Lecture Series 5, Publications AMS, 1994.

[AGV] Arnold, V., Gusein-Zade, S., Varchenko, A., *Singularities of Differentiable Maps, Volume 1*, Bikhäuser, Boston-Basel-Stuttgart, 1985.

[Ba] Barannikov, S. A., *The framed Morse complex and its invariants*, Advances in Soviet Mathematics, Vol. 21, V. I. Arnold Ed., 1994, 93-115.

[BP] Benedetti, R., Petronio, C., *Branched Standard Spines of 3-manifolds*, Lecture Notes in Mathematics 1653, Springer (1997).

[BP1] Benedetti, R., Petronio, C., *Combed 3-Manifolds with Concave Boundary, Framed Links, and Pseudo-Legandrian Links* arXiv:math/0001162 v1[math.GT].

[Bo] Boardman, J. M., *Singularities of Differentiable Maps*, Publ. Math. I.H.E.S. 33 (1967), 21-57.

[BNR] Borodzik, M., Ne'methi, A., Ranicki, A., *Morse Theory for Manifolds with Boundary*, ArXiv:1207.3066v4 [mathGT] 16 Sep 2014.

[Bri] Brieskorn, E. V., *Examples of singular normal complex spaces which are topological manifolds*, Proceedings of the National Academy of Sciences, 55 (6): 13951397 (1966).

[B] Bredon, G., *Introduction to Compact Transformation Groups*, Academic Press, New York - Lndon, 1972.

[Br] Browder, W., *Isovariant vs. equivariant homotopy equivalences*, Prince-

ton preprint.
[CaSa] Campagnolo, C., Sauer, R., *Counting Maximally Broken Trajectories on Aspherical Manifolds*, arXiv: 1808.08737 [math. GT] Aug 2018.
[Ca] Calabi, E., *An Intrinsic Characterization of Harmonic 1-Forms*, Global Analysis, Papers in Honor of K. Kodaira, D.C. Spenser and S. Iyanaga, Eds., 1969, 101-117.
[CS] Cappell, S., Shaneson J., *Some New Four-manifolds*, Annals of Mathematics, 104 (1976), 61-72.
[CF] Conner, P.E., Floyd, E.E., *Differentiable Periodic Maps*, Springer-Verlag, 1964.
[C] Cohen, R.L., *Topics in Morse Theory*, Stanford University, 1991.
[De] Dehn, M., Uber die Topologie des dreidimensionalen Raumes, Math. Ann. 69 , 137-168 (1911).
[D] tom Dieck, T., *The Burnside Ring of a Compact Lie Group. I*, Math. Ann. 215, 235-250 (1975).
[E1] Eliashberg, Y., *Singularities of Folding Type*, Izv. Akad Nauk, 34 (1970), 1110-1126.
[E2] Eliashberg, Y., *Surgery of Singularities of Smooth Mappings*, Izv. Akad Nauk, 36 (1972), 1321-1347.
[EM] Eliashberg, Y., Mishachev, N., *Introduction to the h-Principle*, Graduate Studies in Mathematics, v 48, Publication of AMS, Providence, Rhode Island.
[F] Farber, M., *Topology of Closed One-Forms*, Mathematical Surveys and Monographs 108, AMS, 2004.
[FKL] Farber, M., Katz, G., Levine, J., *Morse Theory of Harmonic Forms*, Topology 37 (1998), 469-483.
[FR] Fenn, R., Rourke, C., *Nice spines of 3-manifolds*, Topology of Low-Dimensional Manifolds, Lecture Notes in Mathematics, no. 722, 31-36.
[GR] Gilman, D., Rolfsen, D., *The Zeeman conjecture for standard spines is equivalent to the Poincaré conjecture*, Topology 22 (1983), 315-323.
[GR1] Gilman, D., Rolfsen, D., *Three-manifolds embed in small 3-complexes*, Int. J. Math. 3 (1992), 179-183.
[GR2] Gilman, D., Rolfsen, D., *Manifolds and their special spines*, Contemporary Mathematics, vol. 20, 1983, 145-151.
[God] Godement, R., *Topologie Algébraique et Théorie des Faisceaux*, Publications de l'Institut de mathématique de Universite de Strasbourg XIII, Hermann, Paris, 1958.
[GG] Golubitsky, M., Guillemin, V., *Stable Mappings and Their Singularities*, Graduate Texts in Mathematics 14, Springer-Verlag, New York Heidelberg Berlin, 1973.
[GM] Goresky, M., MacPherson, R., *Stratified Morse Theory*, Proceedings of Symposia in Pure Mathematics, Vol. 40 (1983), Part 1, 517-533.
[GM1] Goresky, M., MacPherson, R., Morse theory for the intersection homology groups, Analyse et Topologie sur les Espaces Singulieres, Astérisque #101 (1983), 135-192, Société Mathématique de France.
[GM2] Goresky, M., MacPherson, R., Stratified Morse Theory, Springer Ver-

lag, N. Y. (1989), Ergebnisse vol. 14. Also translated into Russian and published by MIR Press, Moscow, 1991.

[Gor] Goresky, M., *Triangulations of stratified objects*, Proc. Amer. Math. Soc., 72 (1978), 193-200.

[Got] Gottlieb, D.H., *All the Way with Gauss-Bonnet and the Sociology of Mathematics*, Math. Monthly, 103 (1996), 457-469.

[Got1] Gottlieb, D.H., *On the Index of Pullback Vector Fields*, Differential Topology Proceedings, Lecture Notes in Mathematics 1350, 167-170 (1988).

[Gr] Gromov, M., *Volume and bounded cohomology* Publ. Math. I.H.E.S., tome 56 (1982), 5-99.

[Gr1] Gromov, M., *Singularities, Expanders and Topology of Maps. Part I: Homology versus Volume in the Spaces of Cycles*, Geom. Funct. Anal. vol 19 (2009), 743-841.

[Gu] Guth, L., *Minimal number of self-intersections of the boundary of an immersed surface in the plane*, arXiv:0903.3112v1 [math.DG] 18 Mar 2009.

[Ha] Hardon, D.G.C.,*The Morse Complex for a Morse Function on a Manifold with Corners*, arXiv:math.GT/0406486 v1, 2004.

[Har1] Hart, R., *Topological properties of subanalytic sets*, Trans. Amer. Math. Soc., 211 (1975), 57-70.

[Har2] Hart, R., *Stratifications of real analytic maps and images*, Invent. Math. 28 (1975), 193-208.

[Hat] Hatcher, A., *Algebraic Topology*, Cambridge University Press, 2002.

[He] Hempel, J., *3-Manifolds*, Annals of Mathematics Studies, Princeton University Press,1976.

[Hi] Hironaka, H., *Subanalytic sets*, Number Theory, Algebraic Geometry and Commutative Algebra, volume in honor of A. Akizuki, Kinokunya Tokyo 1973, 453-493.

[H] Hirsch, M.W., *Differential Topology*, Springer-Verlag, New York - Heidelberg - Berlin, 1976. *

[Hopf] Hopf, H., *Vectorfelder in n-dimensionalen Mannigfaltigkeiten*, Math. Annalen 96 (1937), 225-250.

[Hu] Hu, S.T., *Homotopy Theory*, Academic Press, New York & London, 1959.

[Hud] Hudson, J.F.P., *Piecewise Linear Topology*, W. A. Benjamin, Inc., New York — Amsterdam, 1969.

[I] Ishii, I., *Flows and Spines*, Tokyo J. Math. 9, No. 2, 505-525 (1986).

[I1] Ishii, I., *Moves for Flows-Spines and Topological Invariants of 3-Manifolds*, Tokyo J. Math. 15, No. 2, 297-312 (1992).

[J] Jaco, W.H., *Lectures on 3-Manifold Topology*, AMS Regional Conference Series in Mathematics 43 (1980).

[Jo] Johnson, F., *On the triangulation of stratified sets and singular varieties*, Trans. Amer. Math.Soc. 275 (1983), 333-343.

[K] Katz, G., *Convexity of Morse Stratifications and Gradient Spines of 3-Manifolds*, JP Journal of Geometry and Topology, vol. 9, no 1 (2009),

1-119.

[K1] Katz, G., *Stratified Convexity & Concavity of Gradient Flows on Manifolds with Boundary*, Applied Mathematics, 2014, vol. 5, 2823-2848. http://www.scirp.org/journal/am.

[K2] Katz, G., *Traversally Generic & Versal Flows: Semi-algebraic Models of Tangency to the Boundary*, Asian J. of Math., vol. 21, No. 1 (2017), 127-168 (arXiv:1407.1345v1 [mathGT] 4 July, 2014)).

[K3] Katz, G., *The Stratified Spaces of Real Polynomials & Trajectory Spaces of Traversing Flows*, JP J. of Geometry and Topology, v. 19, No. 2 (2016), 95-160 (arXiv:1407.2984v3 [mathGT] 6 Aug 2014).

[K4] Katz, G., *Causal Holography of Traversing Flows*, arXiv:1409.0588v1 [mathGT] (2 Sep 2014).

[K5] Katz, G., *Flows in Flatland: a Romance of Few Dimensions*, Arnold Math. J., DOI 10.1007/s40598-016-0059-1 (2016).

[K6] Katz, G., *Complexity of Shadows & Traversing Flows in Terms of the Simplicial Volume*, J. of Topology and Analysis, vol. 8, No.3 (2016), 501-543 (arXiv:1503.09131v2 [mathGT] 24 Apr 2015).

[K7] Katz G., *Causal Holography in Application to the Inverse Scattering Problem*, Inverse Problems and Imaging Journal, June 2019, 13(3): 597-633.

[K8] Katz G., *Applying Gromov's amenable localization to geodesic flows*, arXiv:1710.06151v1 [mathGT] 17 Oct 2017.

[K9] Katz, G., *The Burnside Ring-Valued Morse Formula for Vector Fields on Manifolds with Boundary*, J. of Topology and Analysis, vol 1, No 1 (2009), 13-27.

[K10] Katz G., *Harmonic Forms and Near-minimal Singular Foliations* Comment. Math. Helv. 77 (2002) 39-77.

[K11] Katz G., *How Tangents Solve Algebraic Equations or a Remarkable Geometry of Discriminant Varieties*, Expositiones Mathematicae, vol. 21 (2003), 219-261.

[K12] Katz, G., *The Wall Groups of Finite Groups and the G-Signatures of Manifolds*, Math. USSR Sbornik, 27: 2 (1975), 163-181.

[K13] Katz, G., *Holography and Homology of Traversing Flows*, to be published by World Scientific.

[Ke] Kervaire, M. A. (1969). *Smooth homology spheres and their fundamental groups*. Trans. Amer. Math. Soc. 144: 6772 (1969).

[Kh] Khovanskii, A. G., *Index of a Polynomial Vector Field*, Functional Analysis and its Applications (Russ), 13:1 (1979), 49-58.

[Ko] Koda, Y., *Branched spines and Heegard genus of 3-manifolds*, manuscripta math. 123, 285-299 (2007).

[Kos] Kosinski, A., *Differential Manifolds*, Academic Press, Boston, San Diego, New York, London, Sydney, Tokyo, Toronto, 1992.

[KM] Kronheimer, P.B., Mrowka, T.S., *Monopoles and three-manifolds*, Cambridge University Press, 2007.

[La] Laudenbach, F., *Morse Complex on Manifolds with Boundary*, Geom. Dedicata, (2011), 153, 47-57.

[Lo] Lojasiewicz, S., *Triangulations of semi-algebraic sets*, Ann. Scu. Norm. di Pisa (1965), 449-474.

[Lu] Lück, W., *The Equivariant Degree*, Proceedings of Algebraic Topology and Transformation Groups (Göttingen 1987), Lecture Notes in Mathematics 1361, 123-165, Springer-Verlag.

[LR] Lück, W., Rosenberg, J., *The equivariant Lefschetz fixed point theorem for proper cocompact G-manifolds*, High-dimensional manifold topology, 2003 - World Scientific.

[Mal] Malgrange, B., *The preparation theorem for differential functions*, Differential Analysis (Papers presented at the Bombay Colloquium, 1964), 203-208.

[Ma] Matveev, S. M., Algorithmic Topology and Classification of 3-Manifolds, Algorithms and Computation in Mathematics, v. 9, Springer, 2003.

[Ma1] Matveev, S. M., *Special spines of piecewise linear manifolds*, Mat. Sb. (N.S.), vol. 92(134), No 2(10) (1973), 282-293.

[MP] Matveev, S. M., Pervova, E.L., Lower bounds for the complexity of three-dimensional manifolds (Russian), Dokl. Akad. Nauk, 378, no 2, 151-152 (2001).

[Maz] Mazur, B., *A note on some contractible 4-manifolds*. Ann. of Math. 73: 221228 (1961).

[May] P.J. May, *Weak equivalences and quasifibrations* Lecture Notes in Mathematics, Vol. 1425, 1990, pp 91-101.

[Me] Metzler, W., On the Andrews-Curtis-Conjecture and related problems, Contemporary Mathematics, v. 44 (1985) , 35-50.

[Mi] Milnor, J., *Two complexes that are homeomorphic, but combinatorially distinct*, Ann. of Math. 74, 575-590 (1961).

[Mi2] Milnor, J., *Lectures on the h-cobordism Theorem*, Princeton University Press, Princeton, New Jersey, 1965.

[Mi3] Milnor, J., *Morse Theory*, Annals of Mathematics Studies, No 51, Princeton University Press, Princeton, New Jersey, 1963.

[Mi4] Milnor, J., *On manifolds homeomorphic to the 7-sphere*, Annals of Mathematics, 64 (2): 399405 (1956).

[MiS] Milnor, J.H., Stasheff, J.D., *Characteristic Classes*, Annals of Mathematics Studies 76, Princeton University Press, 1974.

[Moi] Moise, E. E., *Geometric Topology in Dimensions 2 and 3*, New York, Springer-Verlag, (1977)

[MT] Morgan, J., Tian, G., *Ricci Flow and the Poincaré Conjecture*, Clay Mathematics Monographs, vol. 3, 2007.

[Mo] Morse, M. *Singular points of vector fields under general boundary conditions*, Amer. J. Math. 51 (1929), 165-178.

[MS] Morse, M., Van Schaack, G. B., *The critical point theory under general boundary conditions*, Ann. of Math. (2) vol. 35 (1934), no. 3, pp. 545-571.

[Morin] Morin, B., *Formes canoniques des singularities d'une application differentiable*, Comptes Rendus Acad. Sci., Paris 260, (1965) 5662-5665, 6503-6506.

[Mu] Munkres, J., *Elementary Differential Topology*, Princeton University

Press, Princeton, 1966.

[No] Noether, E., *Der Endlichkeitsatz der Invarianten endlicher linearer Gruppen der Charakteristik p*, Nachrichten von der Gesellschaft der Wissenschaften zu Gttingen: (1926) 28-35.

[O] Oliver, R., *A Transfer for Compact Lie Group Actions* Proc. of AMS, Vol. 97, No. 3 (Jul., 1986), pp. 546-548.

[Ph] Phillips, A., *Submersions of open manifolds*, Topology 6 (1967), 171-206.

[P1] Perelmann, G., *The Entropy Formula for Ricci Flow and its Geometric Applications*, arXiv:math.DG/0303109, (2002).

[P2] Perelmann, G., *Ricci flow with Surgery on Three-manifolds*, arXiv:math.DG/0303109, (2003).

[Po] Pontriagin, L.S., *Smooth Manifolds and their Applications in Homotopy Theory*, Trudy Mat. Inst. Steklov 45 (1955). (= American Math. Soc. Transl. (2) 11, 1955.)

[Poe] Poenaru, V., *Les decompositions de l'hypercube en produit topologique*, Bull. Soc. Math. France. 88: 113129 (1960).

[Ra] Rapaport, E.S., *Groups of Order 1: Some properties of Presentations*, Acta Math. 121(1968), 127-150.

[Re] Reidemeister, K., *Homotopieringe und LinsenrHaume*, Abh. Math. Sem. Univ. Hamburg 11 (1935), 102-109.

[Re1] Reidemeister, K., *Elementare Begründung der Knotentheorie*, Abh. Math. Sem. Univ. Hamburg 5 (1926), 24-32.

[Rokh] Rokhlin, V. A., *New results in the theory of four-dimensional manifolds*, Doklady Acad. Nauk. SSSR (N.S.) 84 (1952) 221224.

[RS] Rourke, C.P., Sanderson, B.J., *Introduction to Piecewise-Linear Topology*, Springer-Verlag, Berlin Heidelberg New York, 1972.

[S] Sard, A., *The measure of the critical values of differentiable maps*, Bulletin of the American Mathematical Society 48 (12) (1942): 883890.

[Sh] Sharko, V.V., *Functions on Manifolds, Algebraic and Topological Aspects*, Translations of Mathematical Monographs, vol. 13, AMS publication, 1993.

[Sm] Smale, S., *On the structure of manifolds*, Amer. J. Math., 84 (1962), 387-399.

[Sm1] Smale, S., *Generalized Poincare's conjecture in dimensions greater than four*, Ann. of Math. (2) 74 (1961), 39-406.

[Te] Teschi, G., *Ordinary Differential Equations and Dynamical Systems*, Graduate Studies in Math., 140, AMS publication, 2012.

[Thom] Thom, R., *La classification des immersions*, Sémin. Bourbaki 157, 1957-58.

[Thom1] Thom, R., L'équivalence d'une fonction différentiable et d'un polynôme, Topology 3, suppl. 2, (1965), 297-307.

[Th] Thurston, W.P., *Three-dimensional Geometry and Topology, vol. I*, Princeton Univ. Press, Princeton, NJ, 1997.

[Th1] Thurston, W.P., *A norm for the homology of 3-manifolds*, Mem. Amer. Math. Soc., 59 (33): i-vi, 99-130 (1986).

[TV] Turaev, V.G., Viro, O.Ya, *State Sum Invariants of 3-Manifolds and*

Quantum 6j-symbols, Topology, 31, no. 4,865-902 (1992).

[T] Turaev, V., *Torsion Invariants of Spin^c-Structures on 3-Manifolds*, Math. Res. Lett. 4 (1997) 679-662.

[Ve] Verona, A., *Triangulation of stratified fiber bundles*, Manuscripta Math., 30 (1979), 425-445.

[W] Whitney, H., *On Singularities of Mappings of Euclidian Spaces*, Ann. Math., 62 (1955), 374-410.

[W1] Whitney, H., *On regular closed curves in the plane*, Comp. Math. 4 (1937) 276-284.

[Wh] Whitehead, J. H. C., *On C^1-complexes*, Ann. of Math, 41 (4): 809-824.

[White] Whitehead, G.W., *Elements of Homotopy Theory*, Graduate Texts in Mathematics 61, Springer-Verlag, New York Heidelberg Berlin, 1978.